Three decades of research using IGISOL technique at the University of Jyväskylä

Juha Äystö • Tommi Eronen • Ari Jokinen
Anu Kankainen • Iain D. Moore
Heikki Penttilä
Editors

Three decades of research using IGISOL technique at the University of Jyväskylä

A Portrait of the Ion Guide Isotope Separator On-Line Facility in Jyväskylä

Previously published in *Hyperfine Interactions* Volume 223, Issues 1-3, 2014 and *The European Physical Journal A* Volume 48, Issue 4, 2012

Springer

Editors
Juha Äystö
Helsinki Institute of Physics
University of Helsinki
Finland

Tommi Eronen
Max-Planck-Institut für Kernphysik
Heidelberg
Germany

Ari Jokinen
Department of Physics
University of Jyväskylä
Finland

Anu Kankainen
University of Edinburgh
Edinburgh
United Kingdom

Iain D. Moore
Department of Physics
University of Jyväskylä
Finland

Heikki Penttilä
Department of Physics
University of Jyväskylä
Finland

ISBN 978-94-024-0061-8 ISBN 978-94-007-5555-0 (eBook)
DOI 10.1007/978-94-007-5555-0
Springer Dordrecht Heidelberg New York London

Softcover reprint of the hardcover 1st edition 2014

Printed on acid-free paper

Springer is part of Springer Science+Business Media (www.springer.com)

Contents

An IGISOL portrait

Guest editors
J Äystö, T Eronen, A Jokinen, A Kankainen, I D Moore, H Penttilä

This IGISOL Portrait contains 31 articles describing scientific work related to the research at the ion guide isotope separator on-line facility IGISOL at the University of Jyväskylä, Finland. Ten contributions containing a significant amount of original results, earlier unpublished, were presented as a Topical Collection in the European Physical Journal (EPJA 48 issue No 4, 2012). The other 21 articles were published in Hyperfine Interactions (number xxx). The current portrait combines the two special volumes, giving a complete and consistent overview of the progress of methods applied and the physics studied at the IGISOL facility during three decades.

In the early 1980s, the Ion Guide Isotope Separator On-Line (IGISOL) concept was realized at the 20 MeV proton cyclotron of the Physics Department at the University of Jyväskylä (JYFL). For the first time, the technique allowed the production of high quality ISOL beams from nuclear reaction products with millisecond delay times without chemical restrictions. By providing access to short-lived isotopes of refractory elements it opened up new regions in the nuclear chart for nuclear decay- and ground-state spectroscopy.

During the 1980s, the main research activity at IGISOL was decay spectroscopy, and in particular, of fission products produced in proton-induced fission of natural uranium. Approximately 40 new isotopes and isomers were discovered and studied employing beta-, gamma-, and conversion electron spectroscopy. These experiments covered a broad range of neutron-rich nuclei from yttrium (Z=39) to cadmium (Z=48) and revealed structural changes between highly deformed zirconium through triaxial ruthenium isotopes to weakly deformed vibrational cadmium isotopes.

The original IGISOL facility was shut down in October 1991 and was reinstalled as IGISOL-2 at the science campus housing the laboratory for the K=130 MeV heavy ion cyclotron. A new window of opportunity for extending research to neutron-deficient nuclei far from stability was opened and had a strong impact in extending the scope of the IGISOL research program. The first two decades of IGISOL were reviewed in Nuclear Physics A 609 (2001) 477.

At the start of the new millennium novel ion trap installations were introduced at the IGISOL facility. The radiofrequency cooler and buncher trap (RFQ) was installed to improve the beam quality for the collinear laser spectroscopy station. The ability to bunch the ion beam dramatically suppressed the scattered photon background, a technique that has since been adapted at other international facilities, e.g. ISOLDE at CERN. The drive to produce isobarically clean beams was realized with the construction of a double Penning trap system. Utilizing the other trap for precision mass measurements would later lead to a very successful mass measurement program.

In 2003 the front end of the separator was upgraded, implying the birth of the IGISOL-3 facility. With improved yields and the aid of the new instrumentation, precision experiments on both neutron-rich fission products as well as neutron-deficient nuclei produced in light- and heavy-ion fusion reactions could be performed. The experiments covered both decay spectroscopy, often enhanced by Penning trap purification of the reaction products, and accurate measurements of ground state properties such as charge radii and masses.

In particular, the Penning trap mass spectroscopy of the neutron-rich fission products and neutron-deficient fusion-evaporation residues resulted in improved atomic mass values for over 300 isotopes or isomeric states. These measurements revealed large inaccuracies in adopted atomic masses, thus having influence on the future development of theoretical mass models. Precisely measured two-neutron separation energies allowed the investigation of shape transitions around N=60 and evolution of shell gaps at N=50 and N=82. These measurements were often supported by both laser- and decay spectroscopy studies, thus providing a complementary picture of the evolution of nuclear structure of exotic nuclei.

Operation at IGISOL-3 concluded in June 2010. During the subsequent two years the whole facility was moved, upgraded and reconstructed in a new accelerator hall which houses a new, high intensity light-ion cyclotron MCC30. Primary beams can be delivered to the new IGISOL-4 facility from both the new cyclotron as well as the K=130 machine. Since the MCC30 cyclotron is almost solely dedicated for IGISOL, time available for experiments will be significantly increased. The move has also offered an opportunity to dramatically improve the overall layout in order to overcome several shortcomings of the previous facility, as well as make use of a considerable increase in floor space for future developments and sophisticated detector setups.

The increase in available primary beam intensity has led to a complete re-design of the ion guide mounting system as well as the construction of a second ion source station located on the second floor of the facility. This removes constraints imposed at IGISOL-3 on the cooling down time required after on-line experiments. Optical transport to the target area has been significantly improved, with direct access to both in-source and in-jet laser ionization. The gas purification and transport system has been completely renovated in order to reduce impurity levels to below sub-parts-per-billion. Downstream of a new electrostatic switchyard at the mass separator focal plane, the layout of the facility has changed dramatically. The collinear laser beam line is now situated on the ground floor and the beam from the RFQ can be either transported left towards the laser spectroscopy station or right towards JYFLTRAP. Such separation of collinear and trap beam lines enables direct access of laser light into the RFQ for optical manipulation, ion resonance ionization or even polarization of ion beams. Additional electrical isolation between the Penning traps and the RFQ promises more flexible operation, with parallel running being possible during collinear laser spectroscopy experiments. Research and development has been a priority at IGISOL for the last three decades and with the new facility this will continue well into the coming decade and beyond.

As guest editors of the EPJA and Hyperfine Interactions special issues we would like to thank over 100 authors who contributed their time and effort to make this portrait a reality. We are also very grateful to our colleagues for participating in the development of instrumentation and techniques, as well as for proposing, running and analyzing the data of the experiments. We would finally like to invite all our collaborators, existing and new ones, to join us for a successful 4th decade and beyond at IGISOL-4.

DOI 10.1007/978-94-007-5555-0_1
Reprinted from *Hyperfine Interactions* Volume 223, Issues 1-3, 2014
and *The European Physical Journal A* Volume 48, Issue 4, 2012

The layout of the IGISOL 3 facility
(3 December 2003–29 June 2010)

H. Penttilä

Published online: 6 April 2012

Abstract The front end of the IGISOL facility was upgraded in 2003 by increasing its pumping capacity and by improving the radiation shielding. This marked the birth of IGISOL 3. In late 2005, the traditional conical skimmer electrode of the mass separator was permanently replaced by a radiofrequency sextupole ion beam guide, SPIG, which further improved the mass separator efficiency almost an order of magnitude. The IGISOL 3 facility is described here such as it appeared after all the improvements had been installed.

Keywords Mass spectrometer · Ion guide · Laboratory facilities

1 Prehistory of IGISOL 3

The layout of the IGISOL facility has been constantly modified throughout the years, which evolution includes some total reconstructions. The original IGISOL (Ion Guide Isotope Separator On Line) was created by introducing the ion guide technique as the on-line ion source of the already existing mass separator in the MC-20 cyclotron laboratory in early 1980's [1–4]. A major change was the move to a new laboratory at K-130 heavy ion cyclotron in 1991–1992. The mass separator beam distribution system inflated from about two meters of beam tube after the dipole magnet to a full beamyard with four beamlines including a vertical section to collinear laser spectroscopy setup [5]. In the mass separator front end, the differential pumping capacity was increased by replacing the single Roots blower with the Roots blower array. The pumping of the separator extraction region was left unchanged in the move [6].

During the new IGISOL facility era 1993–2003 the expansion of the facility towards a multitude of ion manipulation instruments continued. The ion beam

H. Penttilä (✉)
Department of Physics, University of Jyväskylä, P.O.Box 35, 40014 Jyväskylä, Finland
e-mail: penttila@jyu.fi

quality requirements of laser spectroscopy lead to construction of the radiofrequency quadrupole (RFQ) ion cooler and buncher trap [7, 8], and the quest to improve the mass resolving power of the system to the level of separating isobars was answered by introducing a Penning trap system [9, 10]. The rise of the novel beam handling techniques would have soon made the separator front end the weakest link of the facility. Furthermore, the way the heavy ion guide, HIGISOL, was developed during 1990's [11], was not properly foreseen. More space was needed for the external target inside the target chamber. In addition, the need of selective ionization techniques such as laser ionization [12] was recognized. All this eventually lead to launching a front end upgrade project.

The upgrade took place in 2003 and resulted in a new edition of the facility, IGISOL 3 [13, 14]. The main goals were increasing of the mass separator beam intensity by means of higher pumping efficiency in the extraction region and having a better radiation shielding to fully benefit from the high intensity light ion beams available from the K130 cyclotron. The technical requirements for laser ionisation as well as for utilisation of heavy ion induced reactions were taken into account in the new front end design. The IGISOL 3 facility was completed in 2005 by the permanent replacement of the conventional skimmer electrode system by a double RF sextupole (SPIG) electrode [15, 16], which significantly increased the efficiency of the mass separator.

In the following, the technical realization of the IGISOL 3 facility is described such as it appeared after all the improvements had been installed. The reader is assumed to be familiar with the basic principles of the ion guide technique, which is described in detail in several articles in this Portrait. In accordance with this, the performance of different ion guides will not be discussed in this article.

2 IGISOL 3 layout

The layout of the IGISOL 3 facility is shown in Fig. 1. A majority of the experiments in 2004-10 took place in the central beamline. Towards the end of the decade an increasing fraction of experiments was utilizing the JYFLTRAP facility. The laser beam line going upstairs is described in a more detail by Cheal et al. in this Portrait.

3 Vacuum system

The ion guide based separator vacuum system consists of two regions. In the differential pumping region a relatively high gas load need to be pumped away as efficiently as possible to be able maintain a good vacuum elsewhere. In addition, the harsh radiation conditions in the IGISOL 3 front end area called for some special arrangements. Only diffusion pumps were used in the front end. The control units of the turbomolecular pumps were learned to be too sensitive to the radiation, while the diffusion pumps are robust and practically immune to radiation damage.

The differential pumping section of IGISOL 3 consisted of the 600 mm $\times$ 1000 mm $\times$ 330 mm target chamber, which housed the ion guide, and the adjacent extraction electrode chamber (Figs. 2 and 3). The target chamber was made of aluminium to minimise the activation of material. It was made large enough to be able reserve

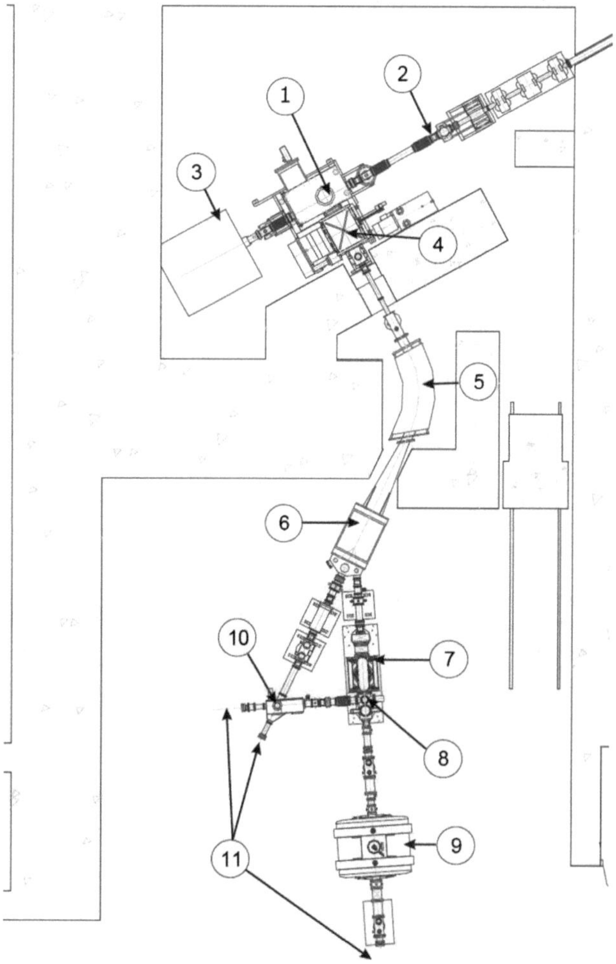

Fig. 1 The IGISOL 3 facility floor level layout after the front end upgade. *1* Target chamber. *2* K-130 cyclotron beam line. *3* Light ion beam dump. *4* Extraction and acceleration chambers. *5* Dipole magnet. *6* Switchyard. *7* RFQ, radiofrequency cooler. *8* Miniquadrupole beam deflector. *9* JYFLTRAP. *10* Electrostatic deflector and beamline upstairs to collinear laser experiments. *11* Beam lines for experimental setups

300 mm space for the HIGISOL target system in front of the ion guide and 200 mm between the primary beam line and the extraction chamber for the SPIG electrode. These space reservations removed the narrow pumping channel in the ion guide exit nozzle region that was presumed to limit the pressure inside the ion guides before 2003 [17].

The target chamber was evacuated through a 300 mm diameter, 2.0 m long stainless steel tube by a Roots array, consisting of two parallel Edwards 4200 Mechanical Booster Pumps which were backed by a third similar pump. These were backed by a 250 m^3/h Edwards E2M275 two-stage rotary vacuum pump. The evacuated helium was either pumped to a recycling reservoir or released through a filtering system. The lowest pressure reached with the Roots array without any gas load from the ion guide was of the order of 10^{-5} mbar (1 mPa). Failure to reach this pressure turned out to be an excellent indicator of a leak or pump failure.

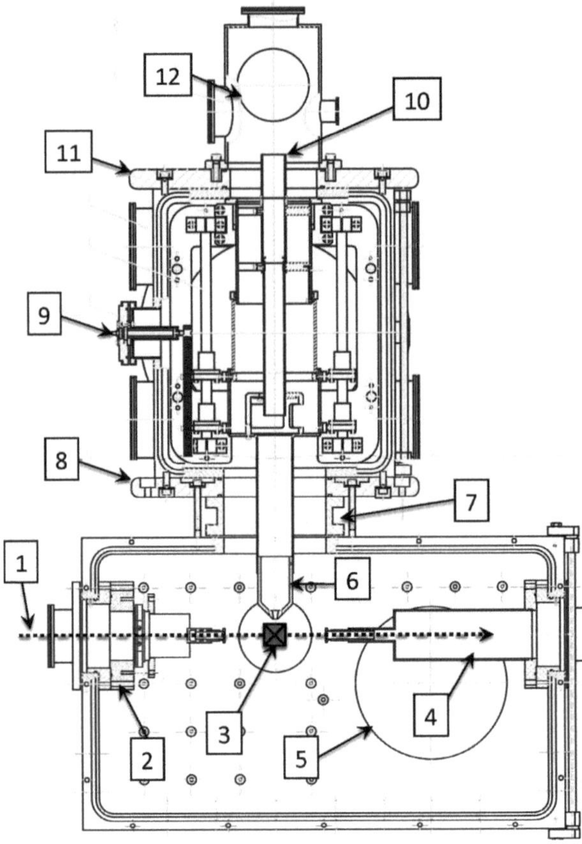

Fig. 2 The technical drawing of the target and the extraction chambers. In this figure, the extraction electrode is shown at its closest position to the ion guide and the cyclotron beam line, which both are marked in the figure. Note that the differential pumping electrode (either a SPIG or a skimmer electrode) is **not** shown. *1* Cycloton beam. *2* Water-cooled beam collimator. *3* Position of the ion guide. *4* Beam enclosure, leading to the beam dump. *5* 300 mm pumping channel to the roots array. *6* Extraction electrode. *7* 200 mm gate valve. *8* Insulator between target chamber and the extraction chamber. *9* Extraction electrode moving mechanism. *10* Ground electrode. *11* Insulator between extraction chamber and ground. *12* Beam diagnostics chamber

When the ion guide was pressurised to ≈ 100 mbar, the pressure in the ion guide exit nozzle region was of the order of a few millibars. In these conditions, the pressure in the half way to the Roots array in the 300 mm pumping channel was typically below 0.1 mbar, and in between of these values elsewhere. The low vacuum of the target chamber was coupled to high vacuum regions by means of differential pumping.

Firstly, the cyclotron beam line high vacuum was connected to the target chamber without beam windows by differential pumping with a water cooled 2000 l/s Alcatel Crystal 200 oil diffusion pump. The windowless operation was important for heavy ion induced reactions, since the intensity of heavy ion beam was not limited by the heating of the beam windows.

In the direction of the mass separator, the extraction chamber was connected to the target chamber by a 6.0 mm diameter aperture, selected large enough to allow the laser beam of the FURIOS laser ion source to enter from the separator side. The 7.0 mm aperture in the extractor electrode determined the final differential pumping stage. The extraction region was evacuated by 8000 l/s Alcatel 6400 oil diffusion pump, backed by 128 m^3/h Pfeiffer Duo 120 two-stage rotary vane pump. The effective pumping speed in the extraction region of IGISOL 3 was limited by the conductance of the pumping channel between the extractor electrode and the

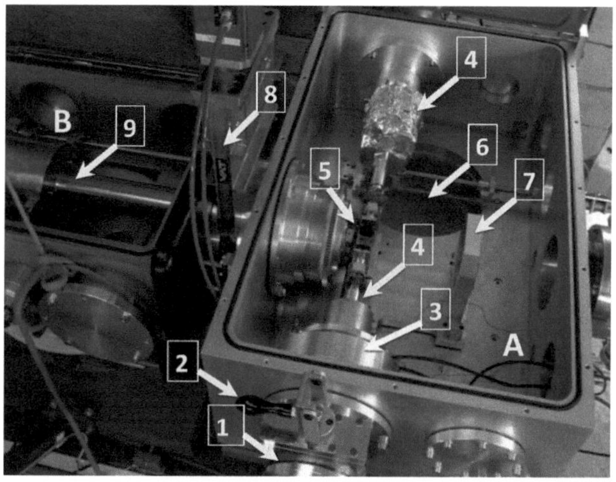

Fig. 3 The target (*A*) and the extraction (*B*) chambers of IGISOL 3 facility. Width of the target chamber is 600 mm and length 1000 mm. *1* Beam line from K-130 cyclotron. *2* Manual ISO-100 gate valve between the beam line and the target chamber. *3* Water-cooled collimator. *4* Beam tubes enclosing the light ion primary beam. *5* Position of the ion guide. *6* 300 mm diameter pumping channel for ion guide buffer gas removal. *7* Ion guide barrel holder. *8* Remote controlled ISO-200 gate valve between the target and the extraction chambers. *9* Remotely movable extractor electrode. Aluminum foil and tapes on the top of the beam tube on the beam dump side hold dosimeters for neutron flux determination

cylinder holding the SPIG. It was calculated to be about ≈ 4500 l/s. The acceleration chamber, which vacuumvise included also the inside of the extractor electrode, was pumped by water cooled Edwards Speedivac EO6 1300 l/s diffusion pump, backed by 20 m^3/h Pfeiffer Duo 20 two-stage rotary vane pump. A similar pump was evacuating also the lens chamber just before the dipole magnet. The typical pressure achieved in the extraction chamber when the ion guide was pressurised was $\approx 10^{-4}$ mbar, in the acceleration chamber $\approx 10^{-5}$ mbar and in the lens chamber $\approx 10^{-6}$ mbar. Since the pressure in the switchyard was also $\approx 10^{-6}$ mbar, the pressure in the dipole magnet vacuum tank where there were no gauges must have also been roughly of the same order.

The target chamber could be isolated from the Roots pumping line by a remotely controlled 300 mm gate valve, from the extraction chamber by a 200 mm gate valve, also remotely controlled, and from the cyclotron beam line by two manually operated 100 mm gate valves. The target box could thus be vented independently of the other vacuum lines.

The vacuum in the switchyard and in the separator beamlines was provided by turbomolecular pumps. The capacity of a typical beamline pump was of the order of 500 l/s. Vacuum in the regular beamlines was typically of the order of $\approx 10^{-7}$ mbar. The ion manipulation devices, the RFQ and Penning trap had their own special vacuum arrangements. The operation of RFQ is based on helium buffer gas in $\approx 10^{-1}$ mbar pressure, and the differential pumping was used to maintain the low pressure in the beam lines. JYFLTRAP operation requires ultra-high vacuum ($\approx 10^{-8}$ mbar) in the precision trap. The RFQ was pumped with two and JYFLTRAP

Fig. 4 The fission ion guide. Buffer gas feeding is in the center tube. The thin metal tubes are for cooling water circulation in the copper cooling blocks on the top and the bottom of the ion guide. The wires are electricity leads to heating elements. Diameter of the beam window is 10 mm

with one 2000 l/s turbomolecular pump. Since the gas load was extremely small, it was sufficient to back them with diaphragm vacuum pumps.

4 Primary beam delivery

The primary beams from the K-130 cyclotron were focused to the target through a 7 mm water-cooled collimator using a magnetic quadrupole lens and an XY-steering magnet. Due to space limitation, the quadrupole lens was slightly closer to the target than its ion optical optimum. It was however possible to focus the beam to a spot of a few mm in diameter.

Ion guides located in the target box were attached to aluminium barrels, which the cooling water and helium feeding tubes run within. This construction allowed changing the light ion fusion ion guide target without breaking the vacuum by rotating the barrel 180 degrees. The ion guide and the barrel formed a compact unit that could be changed as a one block.

In the HIGISOL technique the target is outside of the gas cell and only the heavy ion induced fusion reaction products enter the ion guide through a large metal window. The separation of the reaction products from the primary beam is based on their larger angular distribution [18]. At IGISOL 3, a rotating wheel target prevented the heating of the target becoming an issue. The primary heavy ion beam was stopped in a 7 mm diameter carbon block in front of the ion guide, while a significant fraction of reaction products passed the block to the ion guide. The primary beam current was measured from the carbon block.

In the light ion fusion and in the fission ion guides (Fig. 4) the targets were placed inside the ion guide. The projectile ions penetrated the entrance window, the target and the exit window in the back of the ion guide, leaving only a fraction of their energy in the window and target foils. The intensity of the used proton beam was sometimes as high as 40 μA. The heating of the beam windows or the target was however not a problem, since they were cooled by the flowing helium buffer gas.

Instead, the ionization of the residual gas in the target chamber due to the primary beam had been observed cause separator beam instability and in general reduce the

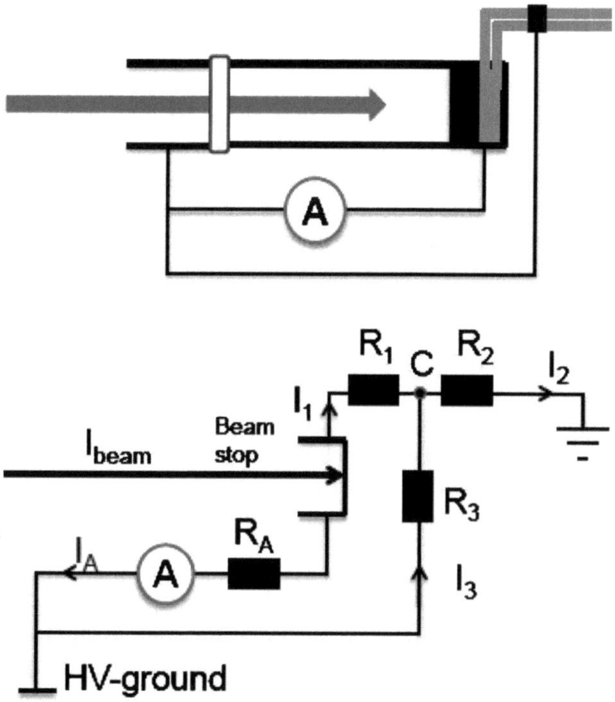

Fig. 5 Beam current reading from the beam stop. The beam stop was electrically connected also to ground via the water cooling water lines. The leak current through the slightly conducting cooling water was compensated by connecting the water lines to HV ground at about 30 cm from the beam stop. The *lower part* of the figure shows the equivalent electrical diagram. Since $R_3 < R_A << R_1 << R_2$, point C is essentially at the same high voltage as the beam stop. The current I_1 through R_1 is negligible, and $I_3 \approx I_2$ and $I_{beam} \approx I_A$

ion transmission to extractor region. The primary light ion beam was thus enclosed to its own volume, a 100 mm in diameter aluminium tube with a 10 mm in diameter Havar foil beam window and a carbon collimator at the end just before the ion guide. A similar tube was installed at the ion guide exit window. The primary light ions that passed through the ion guide, exited the target chamber into the tube through a carbon collimator and a Havar foil and were eventually stopped in a water-cooled aluminum beam stopper at the end of the beam line, 1.5 m downstream behind the target box.

Since the vacuum of the beam dump was separated from the target chamber, it was evacuated by a sliding vane pump, providing a vacuum of the order of 10^{-4} mbar. The beam stop and the last 1.5 m of the beam line were electrically insulated from the target box, and they served as a Faraday cup for beam current reading. The end of the beam line was enclosed in a 150 mm diameter plastic tube to electrically isolate it from the beam dump. It was kept in the same high voltage as the target chamber. It was connected to virtual ground in high voltage through a Keithley 485 Picoammeter. The effect of the leak current from the beam stop to real ground through the slightly conducting cooling water was canceled by connecting a point about 30 cm from the beam stop in the plastic water tubes to the virtual high voltage ground, see Fig. 5. In addition to the end of the beam line, the beam current was read also from the collimator. The beam current could additionally be measured by a shielded Faraday cup before the IGISOL area.

5 Radiation shielding

Radiation shielding improvements were one of the key issues in IGISOL 3. Proton beams with intensities higher than 50 μA were developed for medical isotope production, but before IGISOL 3 such beams could not be utilised since the radiation level in the adjoined experimental area behind the 1.5 m thick concrete wall exceeded 1 μSv/hour due to beam related secondary neutrons that emerged in the direction of the primary beam. In the upgrade the attenuation of neutrons was improved by four orders of magnitude by increasing the shielding thickness to 2.5 m of concrete.

The beam stop was constructed of aluminium and designed to allow 2 kW beam power. The cooling water was circulated so that it is in direct contact with the beam stopping block. The beam stop was surrounded by a radiation shielding consisting of a 400 mm × 500 mm × 500 mm iron core with a 200 mm diameter bore for the beam stop. Iron was surrounded by a 200 mm layer of boron-loaded paraffin and the paraffin by a 200 mm thick layer of concrete blocks. Iron was to moderate the fastest neutrons which were further moderated and thermalized in the paraffin and finally removed via neutron capture in boron, which was mixed with paraffin in the form of boron nitride.

Towards the separator beam line the most intense neutron emission was due to proton-induced fission. In contrary to the beam related neutrons emitted at forward angle with energies up to the same as the primary proton beam, the energy distribution of the fission neutrons does not extend above 10 MeV and reaches a maximum intensity below 1 MeV. Thus, no fast neutron moderator was needed in this direction. The opening for separator beam tube in the 1 m thick concrete wall was filled up with a 500 mm thick layer of boron loaded paraffin and 200 mm of lead to shield gamma rays and brehmstrahlung.

Although the radiation shielding concept of IGISOL 3 was found sufficient to absorb the neutrons, the softening of paraffin at relatively low temperatures often caused a lot of concern. The commercially available boron loaded polyethylene looks as much better material for shielding and absorbing the neutrons.

6 Buffer gas purification system

The purity of the carrier gas is crucial for the ion guide operation, see discussion for example in [19] and articles elsewhere in this Portrait. In the IGISOL 3 system the absolute concentration of impurities was in the sub-ppm range. The facility was equipped with a gas purification system, consisting of two liquid nitrogen cooled, active carbon filled traps, pressurized typically to about 3 bar. Only one of the traps was used at any given time, the other being regenerated by baking. If further purification was needed, it was achieved by a Saes MonoTorr PS4-MT15-R-2 getter purifier. The purified gas was fed to the ion guide via a bakeable stainless steel pipe (Fig. 6).

The gas used in most experiments was recycled helium that had been purified by a liquid nitrogen cooled active carbon trap in 140 bar pressure. After this purification the helium met the purity requirements of commercial 4.6 grade helium. The main source of recycled helium was the liquid helium used for sample and equipment pre-cooling in the nanophysics laboratory. When the supply was not sufficient,

Fig. 6 The gas purification units. Two gas feeding lines (on the *right*) allow changing the buffer gas bottle without interrupting the run. The manual gas flow control panel shows trap and trap bypass lines. The liquid nitrogen cooled purification traps are below the control panel. The right one is shown in a regenerator oven. On the *top left corner* of the figure the insulated gas lines go to Saes MonoTorr getter purifier unit that also can be used for the gas purification

commercial 4.6 grade helium was used, or, when a large amount of helium was needed, it was evaporated from liquid helium purchased entirely for this purpose. Most of the time, the helium from IGISOL could not be recycled, because of the potential danger that all the radioactive isotopes could not be properly filtered from the exhaust helium gas.

7 Skimmer and SPIG electrodes

A conical, 7 mm tall, 45° skimmer electrode with a 2 mm diameter aperture had for several years proved quite versatile device to drag the ions to the mass separator extractor stage. The typical distance between the ion guide nozzle and the skimmer was 10 mm, and typical skimmer voltage 200–300 V. In addition, the ion guides were equipped with a ring electrode [11] between the ion guide exit nozzle and the skimmer to compensate the effect of space charge between skimmer and the exit nozzle.

The use of skimmer was continued after the upgrade, until it was permanently replaced in 2005 by a double radiofrequency sextupole ion beam guide. The SPIG was originally designed to be operated as a laser ion storage trap (LIST) device [20] but it had, in addition to capability to be operated with ionisation lasers, other clear advantages as well. It improved differential pumping by moving the aperture between target chamber and extraction chamber about 20 times farther from the ion guide nozzle. The energy spread of the accelerated beam was lower with the SPIG, which provided better mass resolving after the dipole magnet. Lower energy spread also made it easier to stop the beam in the RFQ, though the difference here was not significant. Most importantly, the SPIG allows much higher ion current from the ion guide before showing any sign of saturation (Fig. 7).

Fig. 7 The RF-SPIG electrode attached to its support ring

8 Separator beam distribution system

In the extraction chamber the ions were accelerated with a conical extraction electrode to the acceleration stage of the mass separator. The aperture of the extraction electrode was exchangeable, but 7.0 mm aperture was used practically in all experiments. The extraction electrode could be remotely moved by a dc motor. The voltage between the extraction electrode and the SPIG backplate depended on the extractor distance; a typical field strength in the extraction region was ≈ 2 kV/cm. The total acceleration voltage used at IGISOL was typically ≈ 30 kV in the experiments applying JYFLTRAP and ≈ 40 kV in the experiments in the central line and in the laser line. The 30 keV voltage in the trap experiments was used due to the discharge problems in the trap region with higher voltages. A slightly poorer MRP achieved with lower acceleration voltage was usually not an issue for Penning trap experiments, where the purification trap provided mass resolving power of tens or even hundreds thousands after the trap.

From the switchyard the ions could be directed either to central beam line, or through a 30° bend to the deceleration stage of the RFQ. The typical MRP after the dipole magnet was of the order of 500. The modest MRP was due to the energy spread of the ions, the magnet itself was known to be capable to provide a MRP of the order of several thousand. The highest ion beam intensity was available in the central beam line, while going through the RFQ provided a bunched, ion optically high quality beam, at the cost of losing about half of the beam intensity. The beam quality and beam bunching were however essential for the experiments in the collinear laser line and with JYFLTRAP. It is worth noticing that the RFQ did not improve the MRP since the mass selection took place before the RFQ. JYFLTRAP, in addition to precision mass measurements, could be used as a mass filter to provide isotopic or even isomeric purification. Examples of experiments employing the Penning trap are found elsewhere in this Portrait.

9 Separator beam tuning and monitoring

Faraday cups and ions of stable isotopes were used for the first tuning of the separator. Some of these beams were present due to natural impurities in the buffer gas. The impurities were ionised by collisions of the primary beam, or, in the case of the fission ion guide, where the primary beam has been isolated from the stopping gas cell, by the collisions of fission fragments. Since a great effort was taken to remove these impurities from the gas, it was usually necessary to intentionally leak noble gases to the buffer gas for mass calibration. Xenon was typically used; since the performance of for example the SPIG and the RFQ is in fact mass dependent, it was sometimes necessary to use lighter ions for the tuning.

One large, remotely controlled Faraday cup was placed right after the acceleration chamber. It was used to measure the current of ions before the mass separation by the dipole magnet. Other Faraday cups were were placed in the switchyard after the mass separation with the dipole magnet, after the 30° electrostatic bend before the RFQ, and at the end of each beam line.

The yield of the typically radioactive nuclear reaction products could be measured by three methods. The direct ion counting could be extended to rates unmeasurable with a Faraday cup using a multichannel plate detector (MCP). A MCP can tolerate a count rate up to several million ions per second, which is about two orders of magnitude less than the lowest rates that were observable with a Faraday cup. A typical rate of nuclear reaction products was below this level. The usability of the MCP detectors was however limited due to ion beams of both atomic and molecular impurities essentially in every mass number that were of the same order or larger than the ion beams from the studied nuclear reaction. These impurity beams could be distinguished from the reaction products only after separation with the JYFLTRAP purification Penning trap, which provided an unambiguous identification of each isotope—in many cases also for the stable and long-lived isotopes.

Detection of the decay of radioactive isotopes usually provided more convenient way of tuning the separator. The isotopes for the tuning preferably decay with a short half-life to a stable or very long-lived isotope. Their decay was normally detected by a silicon detector with a thin foil (aluminum or nickel) in front of the detector. The ion implantation foils were installed in a ladder to be able to move a fresh foil in front of the silicon detector whenever necessary.

There were identical silicon detector setups installed in the IGISOL beam lines. One setup was installed permanently in the switchyard at the entrance of the central beam line and another after the 30° electrostatic bend just before the RFQ. The third silicon detector could be installed in the appropriate beam line. These detectors have been used to determine the transmission of the RFQ cooler as well as the transmission of the purification Penning trap, but they were also providing a very effective and accurate means for determining the yield of the radioactive ions. The efficiency of the silicon detector setup of $(33\pm1)\%$ was determined using beta-gamma coincidences.

The drawback of the silicon detector was that it could not resolve the beta decay of one isotope from the beta decay of another. The half-life information could sometimes be utilised. The third way to determine the yield of radioactive isotopes, gamma ray detection, easily resolves different isotopes. Observing and counting gamma rays was a superior method for the identification of the mass separated isotopes. On the other side, the efficiency of gamma detection (below 0.5%) was considerably lower than that of beta detection, and some isotopes were difficult to observe due to unfavorable gamma ray branching. The efficiency calibration of a germanium detector was also less accurate than that of the silicon beta detector, which all in all lead to higher uncertainty of the determined yields.

Acknowledgements This work has been supported by the Academy of Finland under the Finnish Centre of Excellence Programmme 2006–2011 (Project No. 213503, Nuclear and Accelerator Based Physics Programme at JYFL).

References

1. Ärje, J., Äystö, J., Honkanen, J., Valli, K., Hautojärvi, A.: Nucl. Instrum. Methods **186**, 149 (1981)
2. Ärje, J., Äystö, J., Hyvönen, H., Taskinen, P., Koponen, V., Honkanen, J., Hautojärvi, A., Vierinen, K.: Phys. Rev. Lett. **54**, 99 (1985)
3. Ärje, J., Äystö, J., Hyvönen, H., Taskinen, P., Koponen, V., Honkanen, J., Hautojärvi, A., Valli, K., Vierinen, K.: Nucl. Instrum. Methods Phys. Res., A **247**, 431 (1986)
4. Äystö, J., Taskinen, P., Yoshii, M., Honkanen, J., Jauho, P., Ärje, J., Valli, K.: Nucl. Instrum. Methods Phys. Res., B **26**, 394 (1987)
5. Huhta, M.: PhD Thesis, Department of Physics, University of Jyväskylä Research Report No. 1/1997, Jyväskylä (1997)
6. Penttilä, H., et al.: Nucl. Instrum. Methods Phys. Res., B **126**, 213 (1997)
7. Nieminen, A., et al.: Phys. Rev. Lett. **88**, 094801 (2002)
8. Nieminen, A., et al.: Nucl. Instrum. Methods Phys. Res., B **204**, 563 (2003)
9. Kolhinen, V., Eronen, T., Hager, U., Hakala, J., Jokinen, A., Kopecky, S., Rinta-Antila, S., Szerypo, J., Äystö, J.: Nucl. Instrum. Methods Phys. Res., A **528**, 776 (2004)
10. Rinta-Antila, S., Kopecky, S., Kolhinen, V.S., Hakala, J., Huikari, J., Jokinen, A., Nieminen, A., Äystö, J., Szerypo, J.: Phys. Rev., C **70**, 011301 (2004)
11. Huikari, J., Dendooven, P., Jokinen, A., Nieminen, A., Penttilä, H., Peräjärvi, K., Popov, A., Rinta-Antila, S., Äystö, J.: Nucl. Instrum. Methods Phys. Res., B **222**, 632 (2004)
12. Moore, I.D., et al.: J. Phys. G **31**, S1499 (2005)
13. Penttilä, H., et al.: Eur. Phys. J. A **25**(Supplement 1), 745 (2005)
14. Karvonen, P., et al.: Nucl. Instrum. Methods Phys. Res., B **266**, 4454 (2008)
15. Karvonen, P., et al.: Eur. Phys. J. Special Topics **150**, 283 (2007)
16. Karvonen, P., Moore, I.D., Sonoda, T., Kessler, T., Penttilä, H., Peräjärvi, K., Ronkanen, P., Äystö, J.: Nucl. Instrum. Methods Phys. Res., B **266**, 4794 (2008)
17. Penttilä, H., et al.: In: Proceedings of the Third International Conference of Fission and Properties of Neutron-Rich Nuclei, Sanibel Island, Florida, USA, 3–9 November 2002, p. 611
18. Bèraud, R., et al.: Nucl. Instrum. Methods Phys. Res., A **346**, 196 (1994)
19. Moore, I.D.: Nucl. Instrum. Methods Phys. Res., B **266**, 4794 (2008)
20. Blaum, K., Geppert, C., Kluge, H.-J., Mukherjee, M., Schwarz, S., Wendt, K.: Nucl. Instrum. Methods Phys. Res., B **204**, 331 (2003)

DOI 10.1007/978-94-007-5555-0_2
Reprinted from *Hyperfine Interactions* Volume 223, Issues 1-3, 2014
and *The European Physical Journal A* Volume 48, Issue 4, 2012

The IGISOL technique—three decades of developments

I. D. Moore · P. Dendooven · J. Ärje

Published online: 2 May 2013

Abstract The Ion Guide Isotope Separator On-Line (IGISOL) technique, conceived in the early 1980s as a novel variation to the helium-jet method, has been used to provide radioactive ion beams of short-lived exotic nuclei for fundamental nuclear structure research and applications for three decades. This direct on-line mass separation of primary recoil ions from nuclear reactions has achieved similar extraction efficiencies for both volatile and non-volatile elements throughout the periodic table. The evolution of the ion guide has been driven by the pursuit of physics research on both sides of the valley of beta stability. The gradual improvement in the primary beam intensities of light ions, as well as the ever increasing availability of a range of heavy ions has been matched with the development of novel ion manipulation techniques in order to maximise the output of the IGISOL facility. In this article we describe the continual development of the facility to match the needs of the scientific programme, a relationship which has proceeded hand-in-hand over the years. We will show that this strategy has been and continues to be very important for the huge success of the ion guide method for radioactive ion beam production.

Keywords IGISOL technique · Ion guide · On-line mass separation · Radioactive ion beams

Laboratory portrait. Three decades of research using the IGISOL technique at the University of Jyväskylä, Finland.

I. D. Moore (✉) · J. Ärje
Department of Physics, University of Jyväskylä, P.O. Box 35 (YFL), 40014 Jyväskylä, Finland
e-mail: iain.d.moore@jyu.fi

P. Dendooven
KVI, University of Groningen, Zernikelaan 25, 9747 Groningen, The Netherlands

1 IGISOL–1: Development of the original ion guide technique

1.1 Background

The ion guide method developed for on-line isotope separation is based on a variation of the helium-jet technique. This method enables primary recoil ions from nuclear reactions to be mass separated on-line. When the development of the ion guide originally started it was already known that due to the processes involved in radioactive decays and/or nuclear reactions the emerging recoils are initially highly charged ions. It was also known that in a stopping medium, solid or gas, the charge state of the ions is proportional to the speed in the stopping material [1].

As far as we know, the idea to try a direct mass separation without a conventional ion source originated from Albert Ghiorso's experiments during the development of the helium-jet technique in Lawrence Berkeley National Laboratory (LBL) in the late 1950s and early 1960s. Ghiorso supposed that some of the recoils thermalized in helium may remain in a charged form and thus could be exploited in direct mass separation [2].

P.C. Stevenson from the Lawrence Radiation Laboratory in Livermore, encouraged by Ghiorso, tried to identify fission fragments using a method in which fragments emitted from ^{244}Cm and ^{252}Cf sources were thermalized in atmospheric helium. In a large helium chamber (~25 liters) an electric field was implemented by a set of electrodes. The field was to guide the charged recoils through a very small nozzle ($\varnothing = 0.05$ mm) and skimmer ($\varnothing = 0.3$ mm) into a mass spectrometer. Test results showed, however, that only a very small fraction of the fragments were charged and additionally massive molecular ions were produced. The total efficiency of the system ($\sim 10^{-11}$) turned out to be unsatisfactory [3]. In addition, J.M. Nitschke in LBL, Berkeley, performed experimental studies in a direct mass separation in the late 1960s and concluded, "Because of the uncertainty in charge and the possible formation of heavy molecules, a direct mass analysis by simply skimming off the He and injecting the recoils into a mass spectrometer seems to be somewhat adventurous" [4].

At JYFL, studies had been conducted within the laboratory since 1970 for the purpose of developing helium-jet transport techniques and their integral coupling to an on-line mass separator. During the course of measurements transporting radioactive recoils with pure helium at room temperature [5], at cryogenic temperatures (80 K) [6] and with carrier-loaded helium [7], the infrastructure to connect the target chamber to a mass separator was constructed. Figure 1 illustrates a schematic drawing of the target assembly connected to the beam line of the MC-20 cyclotron. Such a system was coupled to different ion sources via the capillary, with the majority of the helium removed by means of two consecutive skimmers and differential pumping. Overall efficiencies of between 0.01 and 1.0 % were measured for several nuclides under various experimental conditions [8]. Over the next few years, investigations proceeded towards improving the lifetime of the ion sources which would quickly become corroded due to the build up of the heavy clusters used in the helium jet. In parallel, spark plugs were used to create pulsed discharges which resulted in the ability to measure the ion time-of-flight through the mass separator. Finally, an idea was formed to install the target in a gas-filled chamber and to connect this directly to the mass separator, completely removing the use of the capillary (thus the helium-jet transport) and ion source. The ion guide was born.

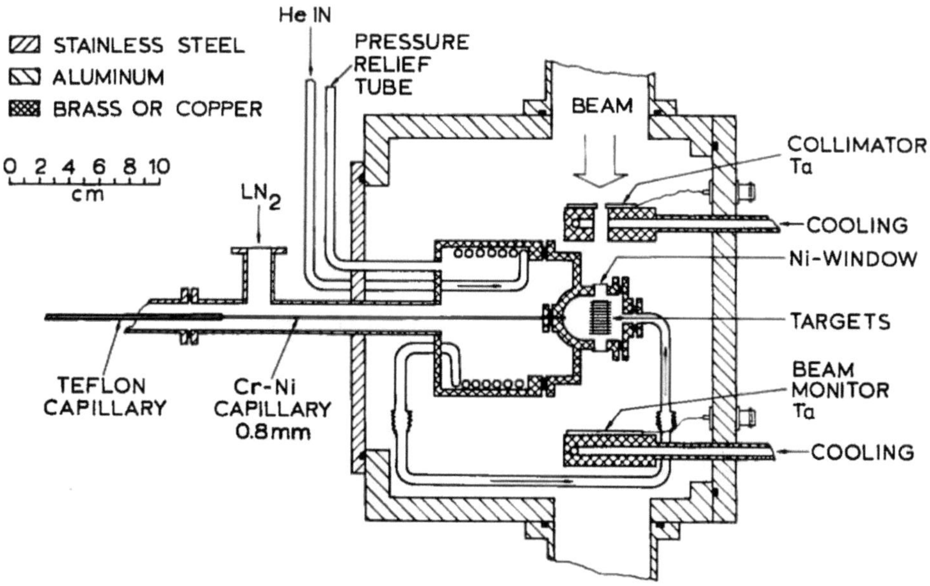

Fig. 1 A schematic drawing of the helium jet target assembly and connection to the MC20 cyclotron [8]

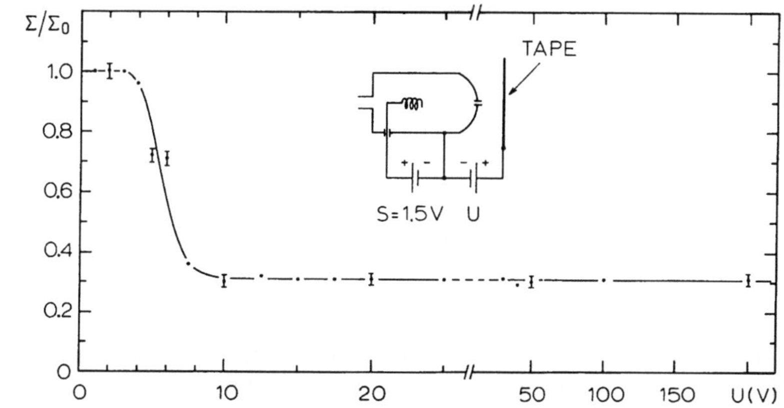

Fig. 2 The relative number of observed α disintegrations as a function of the potential difference between the nozzle and collector tape [9]

1.2 Off-line work with recoil ions from a source of α-active ^{227}Ac

As in many earlier experiments using the helium-jet technique, the development work of the ion guide was initiated with a radioactive source. An α-active ^{227}Ac ($\tau_{1/2} = 21.8$ y) source was installed in a recoil chamber filled with approximately 100 mbar of helium. The decay products were thermalized in the helium and transported with the gas flow through a small nozzle (0.35 mm) to an adjacent vacuum chamber. There, a few millimeters downstream, the activity was collected on

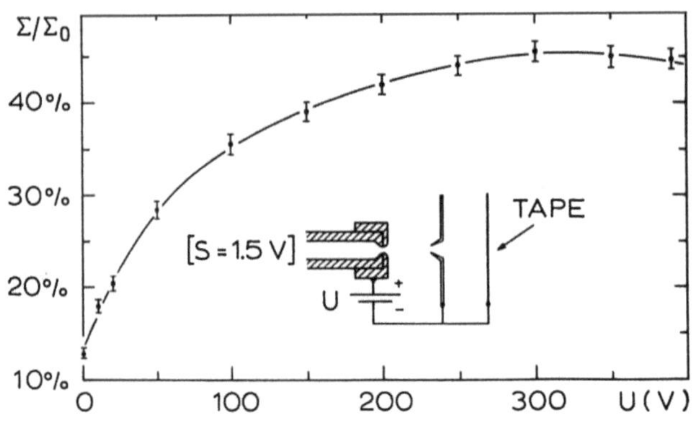

Fig. 3 The relative number of observed α disintegrations as a function of the skimmer voltage [9]

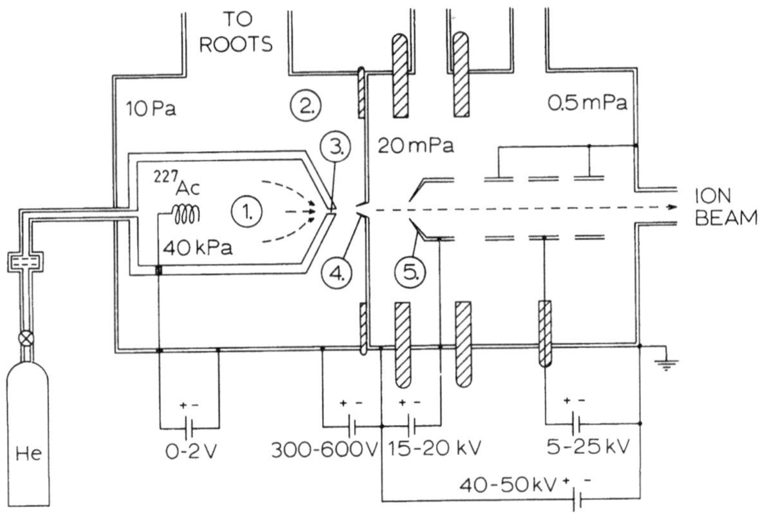

Fig. 4 Principal drawing of the ion guide connected to the accelerating section of the isotope separator [9]. The labels are as follows: *1* Ion guide chamber, *2* Vacuum chamber, *3* Exit hole, *4* Skimmer electrode, and *5* Extraction electrode

a small aluminium plate, later replaced by a movable mylar tape. In the first phase an analysis of the tape was made to see if any fraction of the transported radioactive recoil products appeared as ions. Figure 2 shows the experimental arrangement (inset) and the result in which approximately 70 % of the collected activity was found to be in the form of positive ions.

Following this, a skimmer holder was constructed between the chamber nozzle and the collector tape. A second-hand diopter sight was connected to the recoil chamber support in order to have x-y tuning for accurate alignment of the nozzle and the skimmer. Figure 3 reveals that around 45 % of the total activity collected could be guided through the skimmer hole. Whether the ions were singly or multiply charged, atomic or molecular was still an open question which could be verified in the next phase when the ion guide system was connected to the isotope separator.

Figure 4 shows the principle drawing of how the He-jet ion guide was originally connected to the accelerating stage of the isotope separator at JYFL. The collector tape was installed in a chamber downstream of the mass-to-charge analyzing magnet,

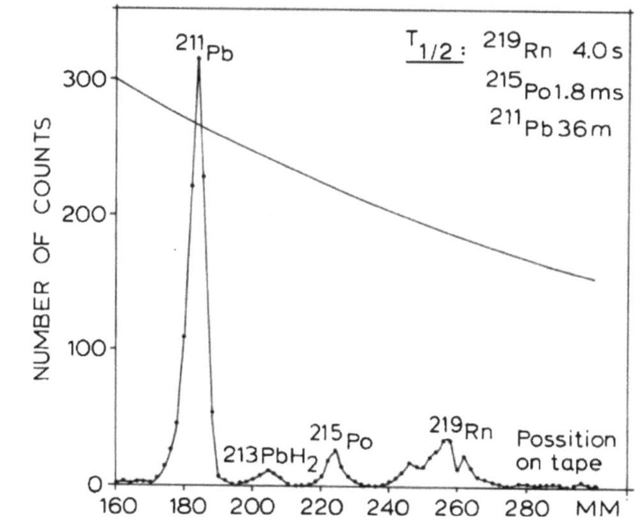

Fig. 5 Mass spectrum of the actinium series separated by the IGIS [9]

Fig. 6 A photograph of the front end of the original off-line Ion Guide Isotope Separator, IGIS (summer 1980). The labels are as follows: **a** Roots blower, **b** Helium pumping pipe with HV reduction network, **c** Vacuum chamber containing recoil chamber, **d** Acceleration stage of separator, and **e** Front end of analyzing magnet

not shown in the figure. During the very first mass separation experiments activity was found on the collector tape, an example of which is illustrated in Fig. 5. This milestone took place in December 1979. The mass spectrum of Fig. 5 showed that the charge states of the recoil ions were +1 and that the main fraction of the particles was in atomic form. The weak peaks at mass positions $A = 213$ and 227 were found to be $^{213}(PbH_2)^+$ and $^{227}(PbO)^+$ molecular ions by an activity analysis. Doubly charged ions were not found in the mass separated samples. The total efficiency was determined to be 13 ± 5 % and the separator resolving power around 340 at the FWHM [9].

During the early days of the development of the Ion Guide Isotope Separator, high voltage (HV) breakdowns caused problems when operating at 40 kV. Figure 6 shows the front end of the IGIS at the time. The vacuum chamber (c) was at high voltage and the Roots blower (a) was sitting at ground potential. The majority of the helium was pumped through a plastic drainpipe (b) which had a step-by-step HV reduction network inside. Breakdowns took place in this pipe, some being rather severe. To overcome the discharging problem, the Roots blower and its pre-pump were later operated at the high voltage of the separator.

1.3 On-line experiments with light-ion induced reactions: the first Ion Guide Isotope Separator On-Line, IGISOL

A beam-line connecting the MC-20 cyclotron to the IGIS recoil chamber was completed in 1982 and thus enabled the start of on-line experiments with a new setup, called the IGISOL [10]. During the construction phase an in-depth literature survey was made in order to understand the processes expected to happen in the recoil chamber which would result in singly-charged, atomic ions from radioactive decays in high-pressure helium [11]. Very intensive development work with the on-line facility took place in the fall 1982–spring 1983. Over 800 h of beam-time was used in less than half a year. It is believed that this was one of the key factors for the success of the development work and such a statement still holds true today.

Over the course of a large number of experiments it turned out that a higher gas flow rate rather than a higher gas pressure led to a better efficiency. A higher flow rate, i.e. a faster evacuation of the target chamber, also enabled the study of radioactive nuclei with shorter half-lives. Additionally, a shorter evacuation time resulted in less atomic physics processes which could lead to neutralization or molecular formation of recoil ions (this is discussed in more detail in connection to IGISOL–3, Section 3). In order to achieve this goal the target chamber volume was decreased from its original size of 75 cm^3 to less than 2 cm^3 and the 0.35 mm nozzle was finally replaced by a thin 1.2 mm exit hole. With the original setup the calculated evacuation time of the chamber was 1600 ms, reducing to about 4 ms with the new chamber. Such numbers can be compared to charge recombination times in high pressure helium which are in the range of milliseconds to tens of milliseconds. The helium pressures used were varied in the range of 100–400 mbar.

The very first on-line mass separation experiment was performed with the reaction $^{20}Ne(p,n)^{20}Na$ in which ^{20}Na ($\tau_{1/2} = 446$ ms) was produced with a proton energy of 20 MeV [12]. This took place in November 1982. In this run neon worked not only as a target, but also as a stopping and transporting medium.

The first "real" IGISOL experiment using helium buffer gas and a solid target took place in January 1983. In this experiment, the target was ^{24}Mg and via a (p,n)-reaction ^{24}Al ($\tau_{1/2} = 129$ ms) was mass separated for the first time (Fig. 7). This result finally convinced other international laboratories of the possibilities of this novel on-line isotope separation technique. Later, the main reaction in the development work was $^{54}Fe(p,n)^{54}Co$, $\tau_{1/2} = 193$ ms. During the years 1983–1984 almost 40 different radioactive isotopes were studied [13].

The experiment which established the two pre-assumed important properties of the ion guide, very fast mass separation and independence of recoil volatility, was the production of the 5.5 ms ^{180m}W isomer in the reaction $^{180}Ta(p,2n)$ (Fig. 8). The melting point of tungsten is 3407 °C, the second highest of all elements after carbon. The shortest half-life ever measured in on-line mass separation, as far as we know, is the 64 μs ^{204}Tl isotope. This experiment took place in February 1987.

The overall efficiency of the IGISOL technique was measured for some heavier elements in which the recoil ranges are short (as in the case of ^{227}Ac α-source recoils),

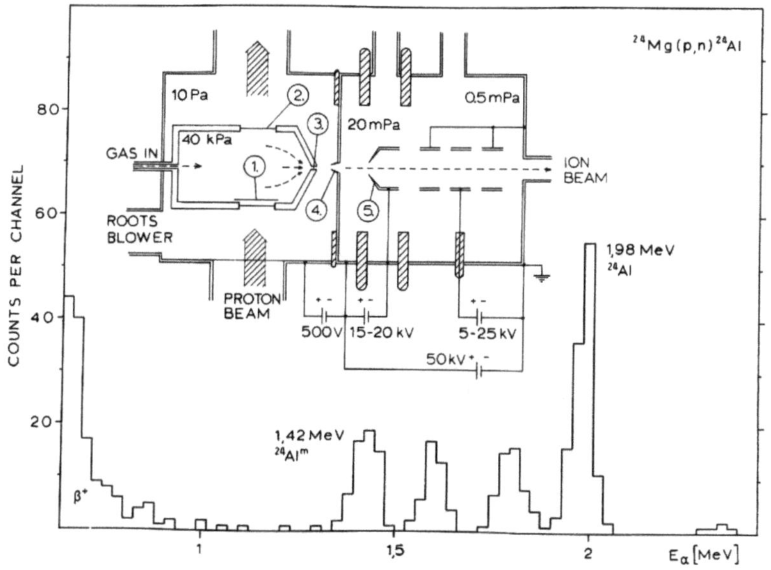

Fig. 7 α-spectrum from the first mass separation using helium buffer gas and a solid target with the IGISOL separator. The labels are as follows: *1* Target, *2* Window, *3* Exit hole of ion guide, *4* Skimmer electrode, and *5* Extraction electrode [12]

Fig. 8 The gamma spectrum from the decay of the mass-separated 5.5 ms 8^--isomer of ^{180}W produced via the 20 MeV proton + ^{nat}Ta reaction [14]

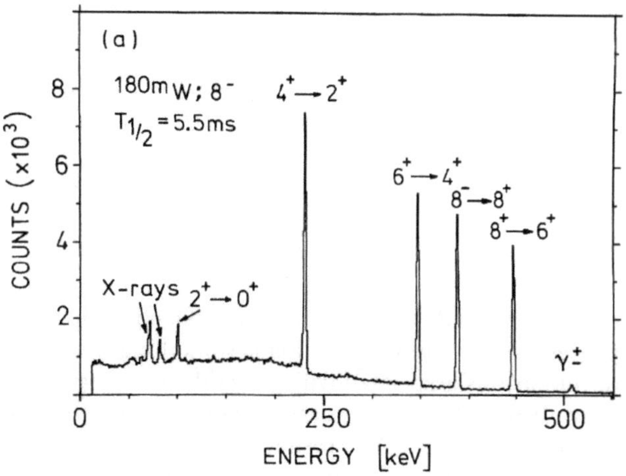

well within the stopping distance, i.e. the diameter of the target chamber (~10 mm). Figure 9 shows that the best values reached almost 10 %. On the other hand, the stopping efficiency for ^{54}Co was around 30 %.

In one important series of measurements, the radioactive ion survival in different buffer gases was studied (Fig. 10). Gases heavier than helium have much higher stopping efficiencies for recoils. It was found that the survival of the ionic charge in the buffer gas rather than the gas nature appeared to be a key parameter. For example, about five times more ions were thermalized in nitrogen than in helium but

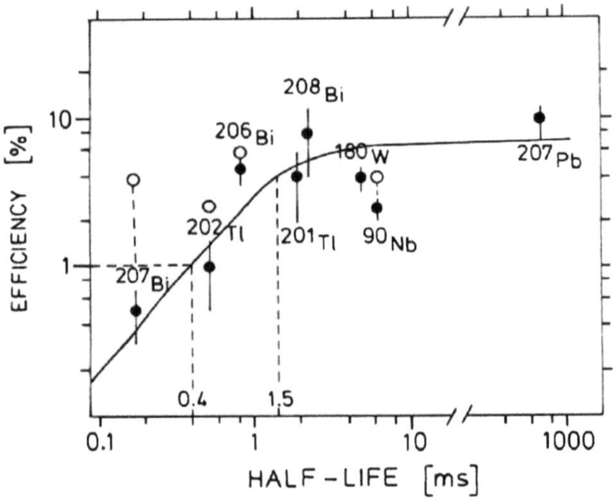

Fig. 9 Overall transmission efficiency for some heavy nuclides in their isomeric states as a function of half-life. The error bars are statistical only [14]

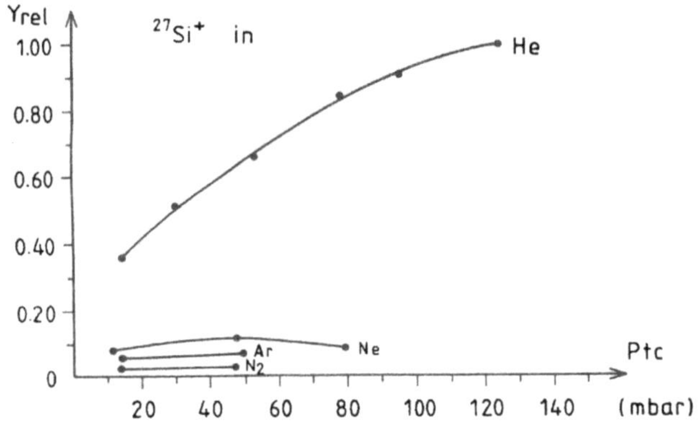

Fig. 10 Relative yields of ^{27}Si recoil ions thermalized in He, Ne, Ar and N_2 buffer gases [15]

only a very small fraction of them survived during the thermalization and transport processes. The survival fraction is something we will return to in connection to later developments associated with IGISOL–3 (Section 3).

In 1986, after three years of studies, the number of targets used in IGISOL experiments was 17. As a result, 45 different chemical elements were mass separated, including both volatile and very short-lived isotopes of highly non-volatile elements such as boron, yttrium, niobium and tungsten. The half-lives of the nuclei varied from 301 ms to 20.2 ms. The IGISOL technique was successfully applied to studies of nuclear properties of the mirror nuclei ^{47}Cr, ^{51}Fe and ^{55}Ni [16]. In addition, the intruder state in ^{203}Bi was found following IGISOL separation [17].

1.4 Development of the fission ion guide

Inspired by the success of the rapid separation of reaction products from light-ion induced reactions, a research programme aimed at applying the same technique to the on-line mass separation of fission products was initiated. Despite the progress in conventional ion source-based separator systems, the availability of short-lived fission isotopes was very limited, in particular for the refractory elements. Neutron-rich isotopes of these elements, the so-called transitional nuclei, are only produced weakly in thermal neutron-induced fission. Since most of the fission-product separators were based at reactors, these elements were not available as directly produced beams at on-line isotope separators. The coupling of the ion guide to the MC-20 cyclotron in Jyväskylä was strongly motivated by a wish to provide a competitive technique for the study of exotic neutron-rich species. By using charged particle-induced fission, a much less asymmetric product mass distribution can be obtained than that of thermal neutron-induced fission.

The very first fission experiments were already performed in 1985. Initially, a similar setup as used for light-ion induced reactions was used for the production of fission fragments, namely a small (1 cm^3) efficiently evacuated, effective volume. The fragments were produced by bombarding three 12 mg/cm^2 thick natural uranium targets with 20 MeV protons. The target thicknesses were selected to be more than the ranges of the fragments so that a continuous energy distribution was obtained. With the operated helium pressure of about 200 mbar (limited by the pumping capacity of the system), only about 1 % of the fragments were stopped. Nevertheless, despite the inadequate stopping, the ion guide technique proved to be an effective tool at producing very neutron-rich nuclides in the $A = 110$–120 mass region for nuclear spectroscopic applications, with several new activities observed and no chemical selectivity present even for the most non-volatile elements [18].

Due to the high density of free electrons in the weakly ionized helium plasma created by the heavy fission fragments, the average lifetime of an ion against recombination can be far shorter than the evacuation time from the gas cell. In order to significantly reduce the recombination losses, the fission ion guide chamber was later modified such that two chambers were housed in the one (larger volume) gas cell. Based on the nearly isotropic distribution and the long range of fission fragments, a 1 mg/cm^2 curved nickel foil was added which divided the ion guide into two parts, the target chamber (housing four 15 mg/cm^2 ^{238}U targets) and the thermalizing recoil chamber (shown in the left panel of Fig. 11). By this method the foil separates the plasma created by the beam from the thermalizing volume, thus dramatically reducing recombination losses.

Two important parameters which determine the performance of an on-line isotope separator are the yield of radioactive nuclei and the mass resolving power of the total system. It turned out that both qualities depend sensitively on the skimmer voltage (the yield also depends on the skimmer distance to the ion guide exit hole as well as the skimmer shape) as shown in the right panel of Fig. 11. The mass resolving power can be determined by adding a trace amount of Xe gas into the helium buffer gas, the atoms of which are ionized via interactions with the fission fragments thus ideally simulating the performance of the system to radioactive ions. The energy loss and dispersion for ions accelerated by the skimmer electric field, in the presence of the expanding buffer gas, are proportional to the skimmer voltage. A low electric field can give a reasonably good mass resolution, but at the cost of transmission efficiency.

With a total production rate of about 10^9 fragments/s with a 1 μA 20 MeV proton beam, a 20 mbarn cross section per mass number for the symmetric mass split, a 10 % intrinsic efficiency and a 1 % stopping efficiency (with limited solid angle for the fragment emission), an estimated yield of 10^4 ions/s per mass number was expected. In general, the yields observed were slightly lower than this estimate. By 1989,

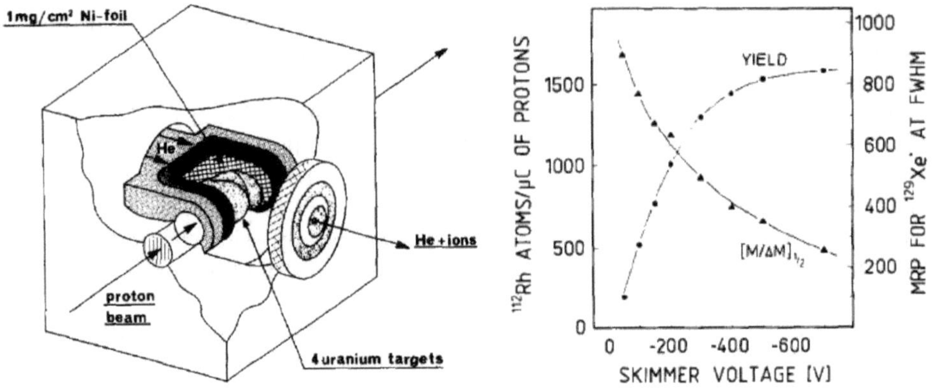

Fig. 11 *Left*: A cut-out view of the fission ion guide chamber. The Ni foil screens the beam-produced plasma from the thermalization volume. *Right*: The yield of $^{112}Rh^{+}$ ions and the mass resolving power for an impurity $^{129}Xe^{+}$ ion beam as a function of skimmer voltage. The helium pressure was 80 mbar [19]

short-lived isotopes of elements such as Y, Zr, Nb, Mo, Tc, Ru, Rh and Pd were available as radioactive beams for the first time [20]. Interestingly, the purification of the gas and the gas handling system itself was becoming of interest due to the molecular oxide formation among some fission products, notably Y, Zr and Nb ions. Gas purity would be an issue which started to be addressed in IGISOL–2 (Section 2) and which was studied in more detail using IGISOL–3 (Section 3).

1.5 IGISOLs built in international laboratories

The first foreign IGISOL was constructed in the Cyclotron and Radioisotope Center (CYRIC, Tohoku University) in Sendai, Japan. The facility was connected on-line to the K = 40 MeV cyclotron which provided beams of p, d, ^{3}He, and also heavy ion beams of ^{14}N and ^{16}O. Isotope separations started in December 1984. The configuration of the CYRIC-IGISOL [21] was similar to the JYFL-IGISOL with a somewhat larger Roots pumping capacity. Test experiments during the spring of 1985 were made using the reactions ^{nat}Zn(p,n)^{64}Ga, $\tau_{1/2} = 2.63$ m and ^{58}Ni(p,n)^{58}Cu, $\tau_{1/2} = 3.2$ s. Figure 12 shows an important off-line result which revealed a lack of pumping capacity on the separator side. In these experiments, positively charged ^{64}Ga ions were collected just after the exit hole while increasing the target chamber pressure. It was found that by using higher pressure than was permitted by the extraction section of the separator, the ion yield increased linearly until around double the pressure was reached. There was then a smooth saturation effect which was not studied further. It was also not clear whether the ions preserved an atomic form at higher pressures. The CYRIC-IGISOL was applied to studies of nuclear properties of the mirror nuclei ^{45}V and ^{57}Cu [22].

The third ion guide-based separator was the LIGISOL [23, 24] connected on-line to the CYCLONE K = 130 MeV cyclotron at the Université Catholique de Louvain in Louvain-la-Neuve, Belgium. The LIGISOL was constructed and operated by the group from the Katholieke Universiteit Leuven, Belgium. It was the first IGISOL facility designed for operation with heavy ions and started test experiments in June

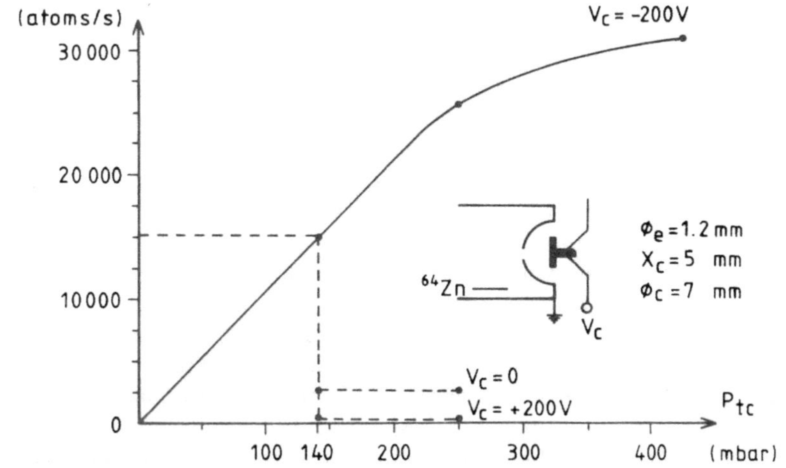

Fig. 12 Absolute yields of ^{64}Ga recoil atoms as a function of helium buffer gas pressure, measured with three different voltages (V_c) between the target chamber exit and collector. The *dashed line* at $P_{tc} = 140$ mbar corresponds to the maximum target chamber pressure permitted by the pumping speed in the extraction chamber [15]

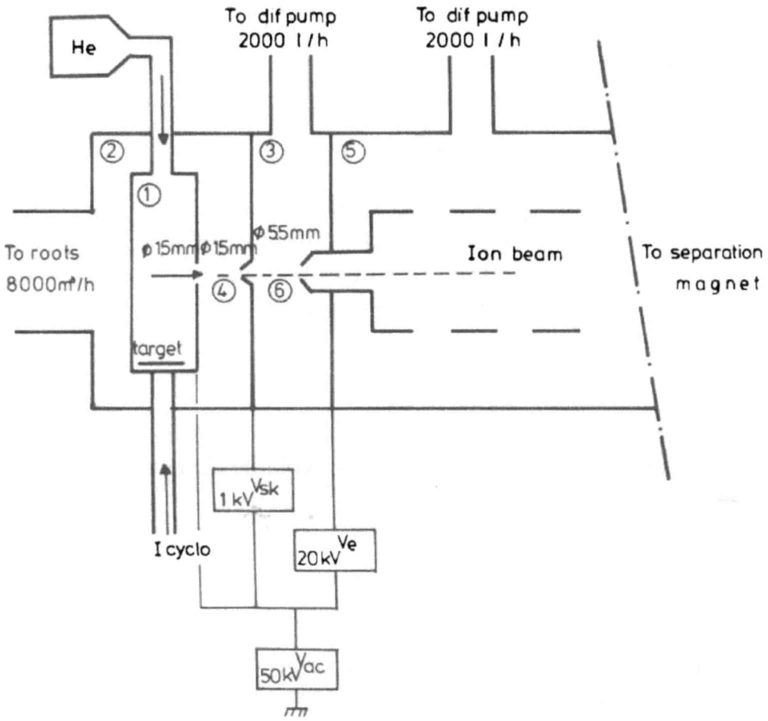

Fig. 13 Schematic overview of the Leuven Ion Guide Isotope Separator [24]

1985. The LIGISOL differed from the original IGISOL in having a four times larger Roots pumping capacity (8000 m^3/h) and a totally different design in the target chamber which was an 8.5 cm long cylinder with a volume of 12.5 cm^3 (Fig. 13).

The purpose of the extra length was to give a longer stopping path for higher energy recoils. In order to evacuate the target chamber quickly, a 2.0 mm exit hole was used. Also here the available evacuation capacity of the Roots blower could not be exploited due to the similar, limited pumping capacity in the extraction section of the separator as in the JYFL and CYRIC IGISOLs.

In one of the first successful heavy ion experiments, 23 ions/s of the isotope ^{100}Ag was mass separated from the reaction $^{93}Nb(^{16}O,5n)$. The energy of the heavy ion beam was 130 MeV and the current 0.3 μA [25]. In one series of experiments, the projectile beam was $^{20}Ne^{7+}$ with an energy of 240 MeV and intensity of 1.1 μA. The yields of 22 radioactive isotopes of 13 elements were determined at the skimmer position using two different target chamber pressures, 400 mbar and 667 mbar, and three different voltages between the collector plate and the target chamber, −200 V, 0 V and +200 V. For the first time it was discovered that a remarkable fraction of negative ions, in addition to positive ions and neutral recoils, were emerging from the target chamber. A possible formation of molecular recoils in these conditions could not be verified. In the case of 400 mbar pressure, the ratio between positive ions and neutrals was 1.56 and for negative ions it was 1.24. These ratios were averages for the 22 isotopes. The increase in target chamber pressure by a factor of 1.67 resulted in relatively higher yields of radioactivity, for positive ions a factor of 1.81, for negative ions 2.81 and for neutral atoms 2.13. Here the relative increase of negative ions looked somewhat worrying [25]. Shortly afterwards, ion guides for light-ion and fission reactions were implemented at LIGISOL [26].

1.6 Towards the end of the first IGISOL era

After the first few years of experience with the aforementioned IGISOLs and several others (for example IGISOLs in RIKEN and INS, Japan, in the Heavy Ion Laboratory, Warsaw and in ISN, Grenoble) the following main advantages and limitations of the skimmer-type IGISOLs were apparent:

Pros:

- very fast (≈ 100 μs) on-line mass separation of primary recoil ions from reactions
- comparable efficiencies for volatile and non-volatile elements
- operation at room temperature
- continuously stable operation.

Cons:

- pressure limitation in the extraction chamber of the separator
- acceleration of ions at high pressure resulted in a poor mass resolving power.

To overcome the above limitations, further development led to a next generation of ion guides whereby the skimmer electrode was replaced first by a low voltage squeezer [27, 28] and then by an rf sextupole device, SPIG [29], the latter of which will be described in connection with developments at IGISOL–2 and IGISOL–3 (Section 3).

2 IGISOL–2: Expanding performance and possibilities

2.1 The move to the new Accelerator Laboratory

In 1993, the IGISOL facility was moved to the new Accelerator Laboratory on the Ylistönrinne campus of the University of Jyväskylä. The old low-energy accelerator (up to 20 MeV proton energy) limited the acceleration of beams relevant for nuclear physics studies to the elements hydrogen and helium, and delivered rather modest beam intensities (up to a few 100 nA). The new K = 130 light- and heavy-ion cyclotron together with its appropriate ion sources expanded the available beams tremendously. Proton beams with energies up to about 65 MeV and intensities in the μA range, and Coulomb-barrier energy beams up to krypton became available. Due to a vigorous ion source development and upgrade programme, heavier and more intense ion beams became available throughout the years. Most important for IGISOL was the increase in intensity of proton beams, up to about 50 μA by the late 1990s, especially for the production of mass-separated fission product beams.

The move to the new laboratory was used to perform a major upgrade to the IGISOL facility [30]. At the front end, this mainly consisted of a much improved roots pumping system on the ion guide chamber (resulting in an effective pumping speed of 3000 m^3/h) and a new helium gas feeding system with a liquid-nitrogen-cooled activated charcoal purification stage. The isotope separator was coupled to a "switchyard" allowing for the addition of a new beam line to the left and right of the single central beam line previously available. The central beam line was branched off to a second floor level where groups from the Universities of Manchester and Birmingham installed a collinear laser spectroscopy facility in 1994, thus resulting in a total of 4 beam lines. This proved to be a very important driving force for the IGISOL programme and related development. Apart from resulting in a very successful physics programme, the laser spectroscopy set-up proved invaluable for the IGISOL development as it was in many cases a unique tool for beam diagnostics (measuring e.g. the energy spread and performing ion identification of mass separated beams). The beam quality requirements for optimised laser spectroscopy (small emittance and energy spread and the need for pure beams of atomic ions) in turn helped to push the development of the facility towards the delivery of such beams.

2.2 Installation of an ion beam cooler-buncher

A major step towards the development of higher quality beams commenced in 1998 with the installation of a so-called RFQ cooler-buncher and, after some rearrangement of the beam lines, its coupling to the laser spectroscopy line [31]. It had been known for some time that a drawback in the ion guide method is the relatively large energy spread (10–150 eV) of the ion beam caused mainly by collisions with the rest gas atoms during the extraction and acceleration process [19]. Both the yield of activity and the energy spread (thus the mass resolving power) were found to depend sensitively on the skimmer voltage: the energy spread as well as the yield increases with increasing voltage. The beam quality requirements are particularly important for collinear laser spectroscopy and therefore only by compromising the beam intensity was it possible to run a successful laser programme. Interestingly however, a local maximum in yield was found to exist at low skimmer voltage

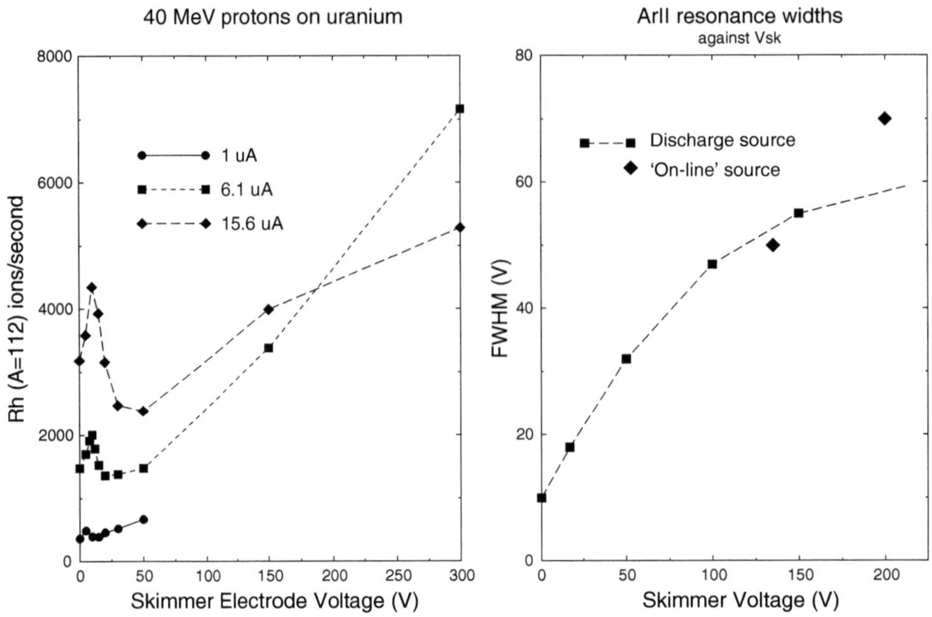

Fig. 14 The left panel indicates the fission ion guide yield of ^{112}Rh as a function of skimmer voltage (primary beam intensity indicated). On the right the graph illustrates the increase in FWHM of the laser resonance of argon ions as a function of skimmer voltage [32]. A compromise had to be reached between energy spread and ionic yield when doing laser spectroscopy

(see Fig. 14) allowing sufficient yields to be produced without significant energy spread [33].

The coupling of the beam cooler to the laser spectroscopy line proved to be immediately fruitful and one of the first measurements involved an investigation of the energy spread of the cooled beam, performed off-line using $^{138}Ba^+$ ions. Figure 15 illustrates the resulting laser-induced resonances using low and high skimmer voltages, −12 V and −250 V, respectively, corresponding to initial energy spreads of $\leq$10 and $\leq$100 eV. Previously, one would have expected to see a change in the resonance FWHM, however no change in the profile lineshape or centroid is observed. In this example the Gaussian content of the fit profile was measured to be compatible with an energy spread of 2–4 eV for both ensembles. The cooler not only reduces the longitudinal energy spread (to below 1 eV) but decouples the cooled beam from the front end parameters, for example variations in gas pressure and skimmer voltage, affording long term stability during experiments.

The improved emittance of the cooled ion beam (estimated to be $\sim 3\pi$-mm-mrad as illustrated in Fig. 16) results in a much better spatial overlap with the counter-propagating laser beam. By achieving a narrower beam waist while maintaining low angular divergence, a lower laser power can be used to provide the same power density, thus contributing less to scattered photon background. The cooler not only improved the ion beam quality however, but provided a unique opportunity to suppress the constant amount of laser scatter. By releasing the ions from the cooler in bunches, only those photons which correspond to the transit of the ion bunch through

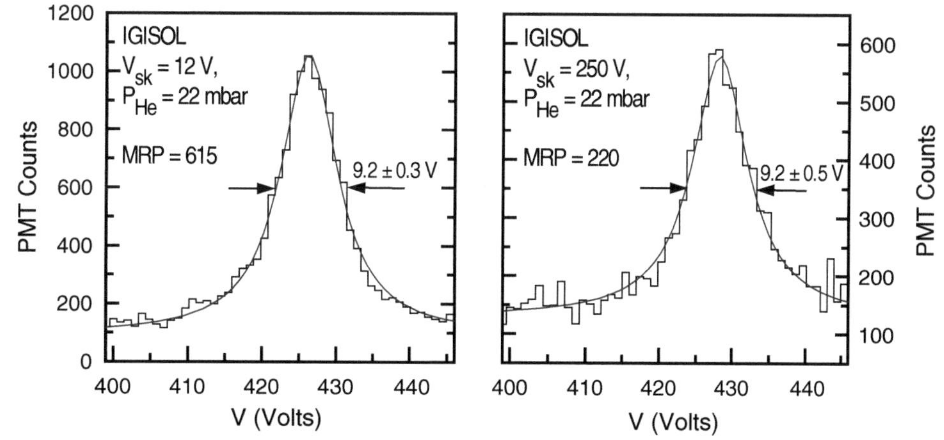

Fig. 15 Laser-induced resonances of cooled $^{138}Ba^{+}$ ions produced off-line. Two different skimmer voltages were used, −12 V (*left*) and −250 V (*right*). The horizontal axis represents the Doppler tuning voltage applied to the light collection region on the collinear laser line. Note the difference in mass resolving power. Post-cooler, no change in the profile centroid or lineshape was observed and the FWHM in both cases is ~9.2 V [31]

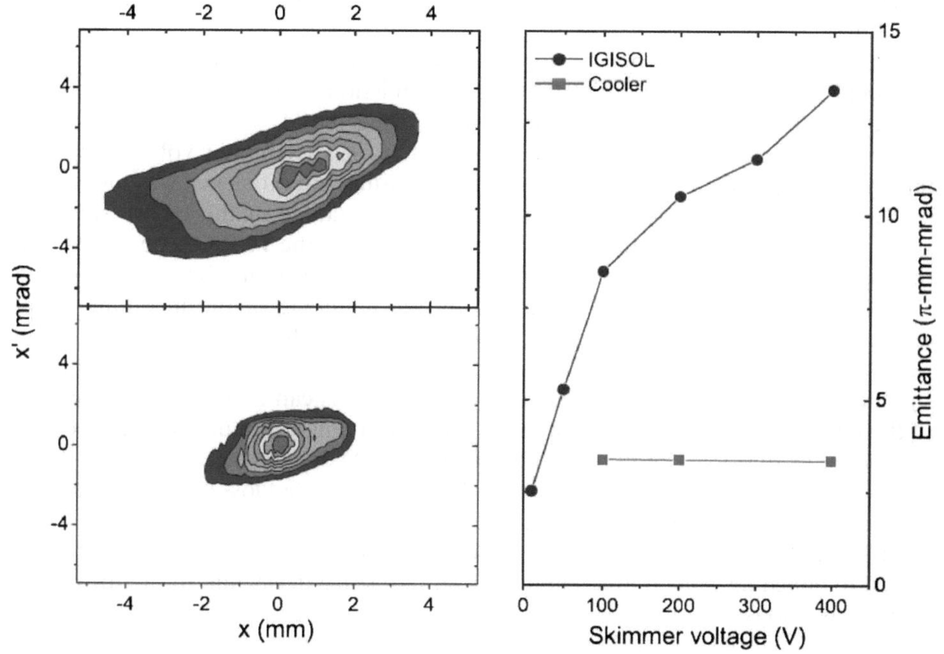

Fig. 16 Measured emittance plots after the IGISOL (*top left*) and the cooler (*bottom left*). These diagrams were measured with a skimmer voltage of 400 V. The *right panel* illustrates the emittance dependence on the skimmer voltage using a $^{84}Kr^{+}$ beam at 1 nA intensity [34]

the photon-ion interaction region are counted [35]. In this manner, the dominant source of background is suppressed by the ratio of the ion accumulation time to the time width of the bunch, typically four orders-of-magnitude.

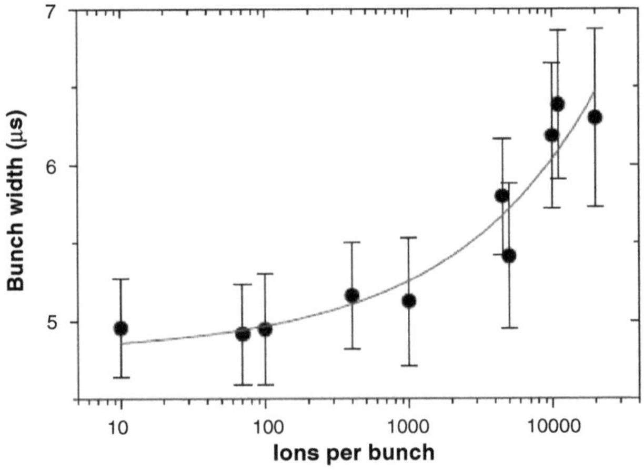

Fig. 17 The cooled ion bunch width as a function of the number of ions accumulated in the cooler, extracted from a measurement of the TOF distribution of the ions [34]

Following the successful application of an ion beam cooler to laser spectroscopy at JYFL, this method has now been adapted at other ISOL facilities, notably ISOLDE CERN [36], and at TRIUMF, Vancouver [37]. The techniques of collinear laser spectroscopy and results achieved at IGISOL are discussed elsewhere in this Portrait [38, 39].

Finally, we note that when performing experiments on rare isotopes produced in minute amounts it is essential that the transmission efficiency of all ion-optical elements is high. An on-line measurement utilizing the $A = 112$ beam produced in the proton-induced fission of ^{238}U showed that for all skimmer voltages up to a maximum of 300 V the corresponding cooler transmission efficiency was more than 60 %. A maximum in transmission of 70 % was obtained at the lowest skimmer voltages corresponding to the lowest energy spread of the IGISOL beam [31]. In principle there should be no reason why the transmission should be much less than 100 % and any limitations are governed by the matching of the incoming beam emittance with the RF quadrupole acceptance. For typical IGISOL radioactive beam intensities of 10^4–10^5 ions/s the effect of space charge in the cooler appears not to be significant. The most sensitive indicator for such effects can be seen in the increase of the ion bunch width as a function of the number of ions accumulated in the cooler, illustrated in Fig. 17. The bunch width shows a $\sqrt[3]{N}$ dependence, where N is the number of ions per bunch. It has not been clear however what losses the space charge effects amount to and up to 10,000 ions per bunch can be easily handled. Most likely though the effect reduces the accumulation time that can be used which in turn is linked to bunching-related losses arising from radiactive decay and to the formation of molecular ions.

2.3 New ion guides for fission and heavy ion reactions

As the light-ion reactions used were basically the same as in the old laboratory, no major design changes to the light-ion reaction ion guide were implemented. However, minor modifications were introduced now and then throughout the years, leading to a more uniform gas flow inside (due to a streamlined shape) and a design with 2 target positions allowing one to interchange between them without breaking the ion guide chamber vacuum [40]. On the other hand, for fission and heavy ion reactions (now available due to the heavy ion beams from the cyclotron), new ion guides were developed.

2.3.1 A major increase in fission yields

The design and performance of the new fission ion guide is described in detail in [41]. By 1997, the mass-separated fission product yields had increased by about 2 orders of magnitude [30] compared to the old IGISOL facility [19], resulting in almost 3000 ions/s per mb cross section. Roughly one order of magnitude of the increase could be attributed to the increase in primary beam intensity, the rest being due to an increased ion guide efficiency (for example to a larger effective target thickness and a larger stopping volume). In subsequent years, it proved difficult to further substantially increase the fission product yields, indicating that the performance limit of the present design had been reached. However, the research possibilities using fission products have continued to expand tremendously via improvements of the separator beam quality. The impetus for such improvements came in the mid 1990's when investigations into neutron-rich isotopes around mass 80 were started. It turned out that these beams were overwhelmed by doubly-charged fission products of twice the mass. Trying to reduce the amount of doubly-charged ions within the ion guide was considered to be a poor strategy as it carried the risk of reducing the amount of singly-charged ions at the same time. Instead, the goal became isobaric separation of the beam.

The chosen technical realisation was the construction of a longitudinal Penning trap [42]. However, although the IGISOL beam has a better emittance than beams from conventional ion sources, the typical energy spread ($\sim$100 eV) is too large for it to be injected efficiently into a Penning trap in a continuous mode. In order to overcome this problem, an RFQ ion beam cooler was first implemented. As discussed earlier, the cooler can provide beams with an energy spread of about 1 eV and an emittance of $\sim$3 π mm mrad with 60 % efficiency, allowing for a more efficient continuous injection. When operated in a bunching mode it is possible to use a pulsed injection into the Penning trap in which the trap potential can be dynamically adjusted to capture the arriving ion cloud. It is interesting to note that the Penning trap for isobaric separation was also coupled to a Penning trap for mass measurements, leading to a very successful research programme [43].

2.3.2 Heavy-ion ion guide development

A heavy-ion ion guide earlier developed in a joint effort by groups from Lyon, Dubna and Grenoble [44] was first implemented at IGISOL (and referred to as the HIGISOL facility) in 1996. Its design and first performance is described in detail in [45, 46]. Given the ion guide stopping length of about 5 cm, a helium pressure of up to $\sim$1 bar is needed to maximize the stopping efficiency of the relatively high-energy heavy ion reaction products. In order to keep the pressure in the extraction region within limits (less than about 2×10^{-4} mbar) when using such high helium pressures, the exit-hole diameter can be reduced or the skimmer-to-exit-hole distance can be increased. The former is not a good option as it slows down the evacuation of the ion guide and thus reduces the extraction efficiency (in particular for the short-living

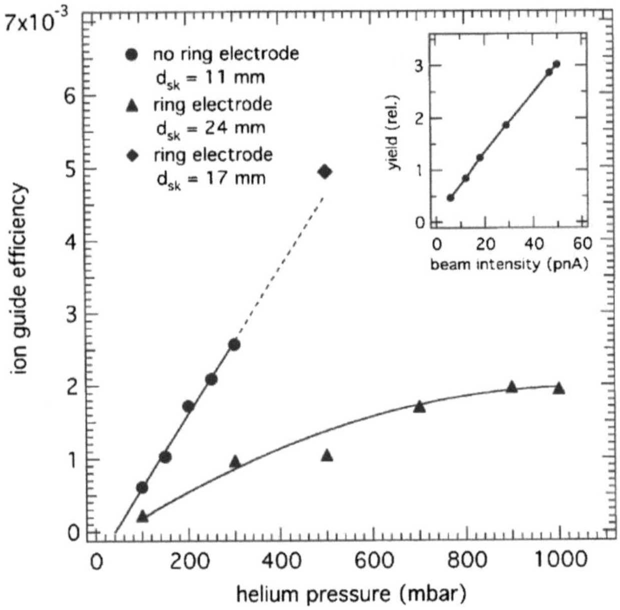

Fig. 18 The effect on the ion guide efficiency of adding a ring electrode between the ion guide and skimmer using the ^{116}Cd(^{40}Ar, 6n)^{150}Dy reaction [45]. Results were taken at three skimmer distances (d_{sk}) with respect to the ion guide. The inset reveals the linearity of the yield as a function of primary beam intensity

radioisotopes). The latter reduces the efficiency of transporting ions through the skimmer.

However, these drawbacks could be mitigated by the introduction of a ring electrode in between the exit-hole and skimmer [45]. Figure 18 summarizes the ion guide efficiency as a function of helium pressure, with and without the ring electrode. The saturation of the data with the ring electrode represents the stopping of the full-energy recoils in the stopping chamber. Without the ring electrode the trend in efficiency is linear as a function of pressure, limited by the pumping capacity in the extraction region. The individual result taken at a skimmer distance of 17 mm follows a careful reoptimization of the IGISOL parameters and shows that the ring electrode allows both the skimmer distance and the helium pressure to be increased without any loss in efficiency. The inset of Fig. 18 indicates the independence of the ion guide efficiency on primary beam intensity within the range studied.

Although the use of the ring electrode allowed the efficiency to be maintained when going to higher helium pressures, it did not improve the notoriously poor mass resolving power (MRP) of the IGISOL technique (typically 250–400 depending on helium pressure and skimmer voltage, even though the theoretical limit of the separator magnet is about 1500). Replacing the skimmer with an RF sextupole (SPIG) increased the MRP to 1100 (see Section 3.6), a value independent of helium pressure [40, 47] (a similar performance was obtained for the fission ion guide [40]). In parallel with improvements in the extraction/differential pumping region, the assembly of beam energy degraders, targets, beam stopper and heavy-ion ion guide chamber was improved on a number of occasions, leading to the system described in [40] with an overall efficiency of ~1 %.

The use of the ring electrode for light-ion induced reactions has been described fully in [40]. In brief, for low proton beam intensities the optimum ring voltage is negative and corresponds to the equipotential voltage between the skimmer and exit

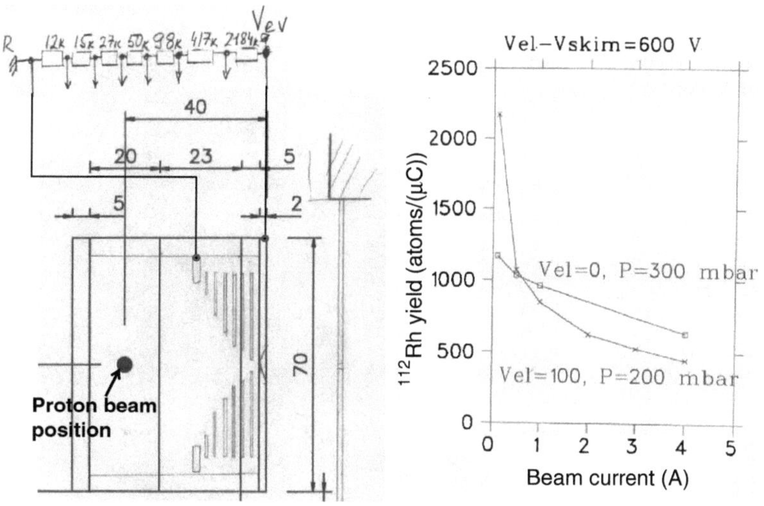

Fig. 19 *Left*: Sketch of the dual-chamber ion guide for fission with internal electrodes. *Right*: Normalized yield values for ^{112}Rh as a function of proton beam intensity for two sets of voltages and pressures [48]. Figure courtesy of A. Popov

hole (so it does not really change anything). For increasing proton beam intensity, the optimum ring voltage increases and eventually becomes positive (up to +30 V) at high intensities. This clearly illustrates some qualitative changes in the conditions of the exit hole-skimmer region with beam intensity. Earlier observations with the light-ion ion guide noted a non-linear trend of the secondary yield vs. the primary beam intensity which was attributed to a plasma effect within the ion guide. However, as the ring-electrode-skimmer system improved the yields significantly when using high-intensity beams (linear trends were seen even up to 35 μA [40]) the earlier observations are likely to have been caused by distortions in the electric field between the ion guide and skimmer caused by the large amount of charge flowing out of the ion guide.

2.4 Implementation of electric fields

Due to the reduction of the survival time of singly-charged ions in high pressure He buffer gas (required to stop the high energy recoils in fission or heavy-ion fusion-evaporation), it is important to reduce the average evacuation time of the reaction products. This motivated the first attempts in 1994 to incorporate static electric field guidance, initially in connection with the fission ion guide. Figure 19 shows a schematic of the ion guide with electrodes whose voltages were selected following simulations. A number of tests were performed [48]. The results of the influence of the beam current intensity on the normalized yield of ^{112}Rh for two sets of electrode voltages and buffer gas pressures are shown in the right panel of Fig. 19. At the lowest beam intensities the electric field results in an almost doubling of the normalized yield, however the subsequent reduction as a function of beam intensity reflects a drop in efficiency, the effect more strongly pronounced for $V_{\mathrm{el}} = 100$ V than for

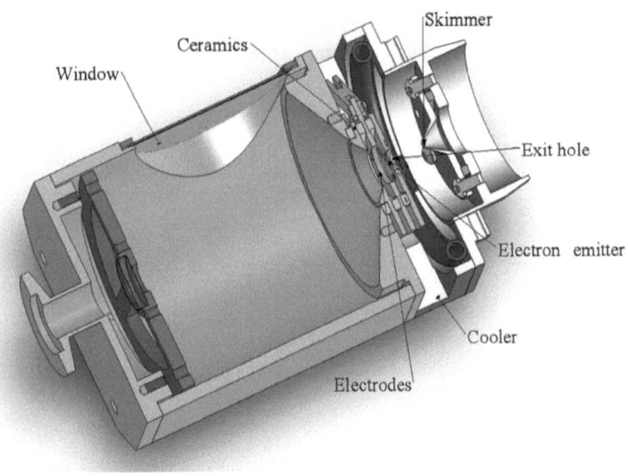

Fig. 20 The heavy-ion ion guide with dc electrodes and electron emitter. Figure courtesy of A. Popov

$V_{el} = 0$ V. The mechanism causing the levelling off is difficult to identify but is likely to be due to the effects of space charge inside the stopping chamber.

In general, the homogenous distribution of the fission fragments and resulting high plasma density inside the stopping volume considerably reduces the effectiveness of static electric fields. In the case of HIGISOL, the reaction recoils and small fraction of primary beam (~1 %) which scatters past the beam stopper create a localized ion-electron density which is concentrated behind the entrance window. Due to the low density of plasma between this localized volume and the exit nozzle, static electric field guidance in this region should be possible.

A modified HIGISOL ion guide chamber was subsequently designed, incorporating a series of electrode plates near the nozzle region to provide a focusing field for a faster transportation of recoiling ions towards the exit hole. Off-line tests using a ^{223}Ra ($\tau_{1/2} = 11.43$ d) α-source attached to the tip of a needle, moveable along the ion guide symmetry axis, revealed that the evacuation efficiency was constant for ions thermalized far from the exit hole, with a drop in efficiency at short distances [47]. The drop can be understood as arising from the difficulty in shaping the field properly near the nozzle. This overall success however has only been demonstrated for cases of very low ionization rates of the buffer gas. On-line, in the presence of beam, the dc field appears to have a negative effect on the extracted yield as a function of electric field strength. The situation is complex but may be understood as arising from a distortion of the electric field caused by the presence of ion-electron pairs created by the scattered primary beam (Popov, private communication).

The injection of additional electrons near the exit hole was discussed as a second approach to create an axis-directed field [40]. Electrodes and an electron emitter (a loop of tungsten wire doped with 1 % thorium) were mounted in the regular HIGISOL chamber on the exit hole plate as illustrated in Fig. 20. With the emitter placed at a negative potential with respect to the ion guide, electrons were emitted into the stopping chamber. The resulting electron cloud creates a negative space charge near the exit hole, generating an electric field which is summed with the field created by the electrodes. As the emitter current is increased, the potential gradient within the guide increases providing faster and more efficient ion extraction. The calculation of the electric field using the particle-in-cell approach suggested that the efficiency could be increased, at least in the presence of low ionization rates (at higher ionization densities the simulations became unstable). Indeed, this was shown to be the case using the ^{223}Ra α-source. Compared with earlier off-line tests in the HIGISOL chamber, the absolute efficiency in 500 mbar helium was about a factor of 10 higher when using the electron emitter and the corresponding delay times were considerably reduced [40]. Importantly, the possible increase of neutralisation due to the high electron density was not observed. However, the question of whether the electric field generated by the electrons would be screened by the plasma created by ionizing radiation in on-line conditions was yet to be answered. Indeed, the HIGISOL chamber was later tested on-line at IGISOL–3 using the reaction ^{32}S+^{nat}Ni at a beam energy of 140 MeV. Although a factor of more than 10 increase was observed in the extracted yield of ^{82}Y, the absolute efficiency of the ion guide was low due to stopping gas impurities. The electron emitter was also successfully implemented in several iterations of gas cell design for the production of low-energy ion beams of daughter products of alpha-emitting isotopes, specifically utilized in the search for the low-lying isomeric state in ^{229}Th [49, 50].

2.5 Handling large beam intensities

From the earliest days of the ion guide development, the fact that ionisation of the buffer gas (by primary beam or reaction products) leads to neutralisation via 3-body recombination between a singly-charged ion, an electron and a neutral helium atom was understood. At IGISOL–2, this neutralisation was expected to be much more severe as far larger beam intensities were available and, relevant in some cases, larger helium pressures could be used due to the increased pumping capacity. To our surprise, a new phenomenon was observed: a peak in mass-separated beam intensity at low skimmer voltages [51] (see left panel of Fig. 14). Earlier, the intensity was observed to increase monotonically with skimmer voltage (see right panel of Fig. 11), reaching saturation because of limitations due to the maximum extraction efficiency and/or to a decreased transmission efficiency through the separator related to an increase in energy spread. This phenomenon may have been previously overlooked; it certainly showed up prominently at high beam intensities. Following some further investigations [33, 51], it was concluded that we were witnessing a decrease in skimmer transmission due to space charge problems. In fact, the current exiting the ion guide could reach ~25 μA. One observation supporting this conclusion is that when using high cyclotron beam intensities, an increase in mass-separated yield of up to a factor of two was observed when mixing a small amount of impurities in the helium buffer gas. The impurities caused a large decrease of the total current coming out of the ion guide which, as concluded, reduced the electric field distortion thus resulting in an increase in skimmer transmission [41, 51].

2.6 Improving the beam quality with a radiofrequency sextupole

The main problem with using a skimmer electrode in the first stage of the differential pumping system is caused by the collisions of the ions (attracted to the skimmer by a field of typically several 100 V/cm) with the relatively high number density of gas atoms between the ion guide and skimmer, resulting in poor emittance beams with a large energy spread. The latter, reaching values of several 10 s of eV on a

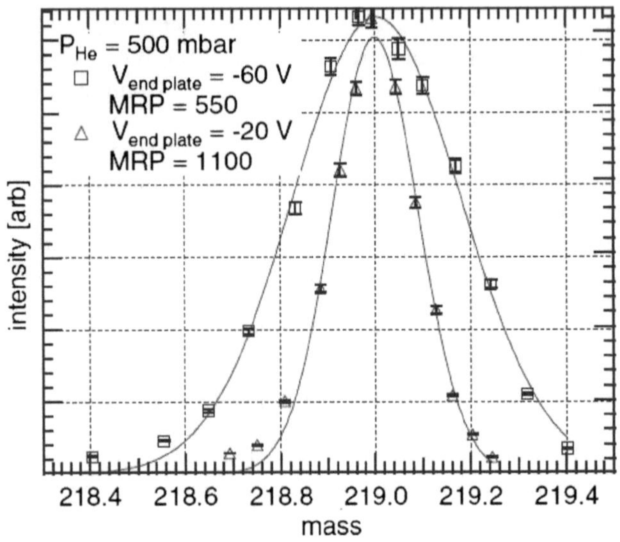

Fig. 21 The mass resolving power of $^{219}Rn^{+}$ ions using a SPIG. The *solid lines* are Gaussian profiles. The difference in MRP is due to the scattering of ions between the SPIG rods and end electrode during acceleration, caused by the vacuum chamber limitations of IGISOL–2 [40]

total energy of 38 keV, is especially devastating for laser spectroscopy as it Doppler-broadens resonance lines. Another problem due to the poor beam quality is the loss in transport efficiency throughout the IGISOL separator (especially at the 7 mm gap of the 30° electrostatic beam bending electrodes in the switchyard).

Ways to address the above issues without losing the secondary yield were sought, with the obvious solution to bring the ions into high vacuum before they are accelerated. One system developed in Jyväskylä was the so-called squeezer ion guide [27, 28]. This device consists of three rings (to which voltages less than 25 V are applied) mounted between the ion guide exit hole and skimmer electrode. The viscous drag of the helium flow, together with the weak electric fields, focus the ions through the skimmer. Ions created in the gas cell can be brought into high vacuum in less than 1 ms without accelerating them to energies higher than approximately 10 eV. In off-line conditions, using α-decay recoils from a ^{227}Ac source, a transport efficiency of about 75 % and an energy spread of 2.5 eV were measured [27]. A squeezer device was also tested on-line at IGISOL in connection with the fission ion guide and showed that the mass resolving power approximately doubled without any losses to the fission yields.

The idea of using a radiofrequency sextupole (SPIG) to transport ions extracted from the ion guide to high vacuum was first tested by Xu et al. in 1993 [29]. A similar system was installed at the Leuven Isotope Separator On-Line (LISOL) facility to guide ions from the ion guide laser ion source [52]. The principal motivation to implement a SPIG at JYFL was that it allows a much larger pumping speed of buffer gas in the extraction region providing lower density and thus less collisions between gas atoms and ions prior to the acceleration stage. This results in a much smaller energy spread (~1 eV) compared to the skimmer system (~100 eV) and the possibility to use higher buffer gas pressure thereby increasing the stopping efficiency of the ion guide. The performance of a SPIG at HIGISOL is described in [47]. As shown in Fig. 21, a mass resolving power of 1100 was obtained in that work, independent of ion guide gas pressure, whereas it was only typically about

400 and pressure dependent with the skimmer electrode. The difference seen in the two measurements of Fig. 21 results from the scattering of ions being accelerated between the SPIG rods and the end plate. This is caused by the relatively high pressure in that region as it was not possible to optimise the differential pumping due to restrictions set by the main vacuum chamber housing the ion guide and SPIG. Importantly, the SPIG significantly improved the mass resolving power on-line and removed the typical feature of a long tail in the mass peak on the low-mass side (i.e. the low-energy side) seen with the skimmer. Although only a few percent of the yield is in the tail it can be significant when performing spectroscopy of exotic nuclei. A noted improvement of a factor of 5 in the yield of ^{112}Rh was seen using the fission ion guide and SPIG compared to that measured with the skimmer system [40]. This trend was also observed with the SPIG developed for the laser ion source discussed in connection with IGISOL–3.

3 IGISOL–3: Improving yields and selectivity and a few new ideas

3.1 Improvements at the front end of the facility

From 1993 onwards, the intense upgrade project of IGISOL–2 both in terms of the coupling to the K = 130 MeV cyclotron and development of ion beam manipulation tools including the rf cooler-buncher and Penning trap facility, soon led to increasing demands for improvements to the front-end of the IGISOL facility. The main goals included increasing the mass-separated beam intensity, a new target chamber with more space for an additional external target-beam degrader system for the HIGISOL facility, as well as addressing the need for chemical selectivity via laser ionization in the ion guide technique. Initially, the improvements were rather technical in nature and were realized during a shutdown period during 2003. Details may be found in [53, 54] however will be briefly summarized in the following.

The full intensity of proton beams from the cyclotron (>50 μA) could not be delivered to IGISOL–2 due to the elevated radiation levels in the adjacent experimental cave. In the upgrade to what was to be termed IGISOL–3, the concrete shielding in the primary beam direction was increased from 1.5 m to 2.5 m. Additionally, in order to handle the expected higher activation of the front-end of the mass separator, all valves and electrodes in the area were transferred to a computer-controlled system. The ion guide mounting platform was also modified such that ion guides could be rather quickly and therefore more safely changed as complete units.

In order to address limitations in the use of the rf sextupole discussed in the context of IGISOL–2 (Section 2.6), namely insufficient differential pumping due to restrictions in the main vacuum chamber, a larger volume chamber was installed providing more space towards the separator. This not only served to improve the pumping, but also afforded extra space for a larger heavy-ion fusion-evaporation ion guide and a more permanent sextupole which was to eventually replace the skimmer electrode for all IGISOL experiments. The extraction region of the separator was also modified to allow for a better evacuation using a more effective diffusion pump (8000 l/s *vs.* 2000 l/s) and wider pumping channels.

Following completion of the upgrade in 2004, the first on-line tests performed with light-ion induced fusion-evaporation reactions saw improvements in the yields

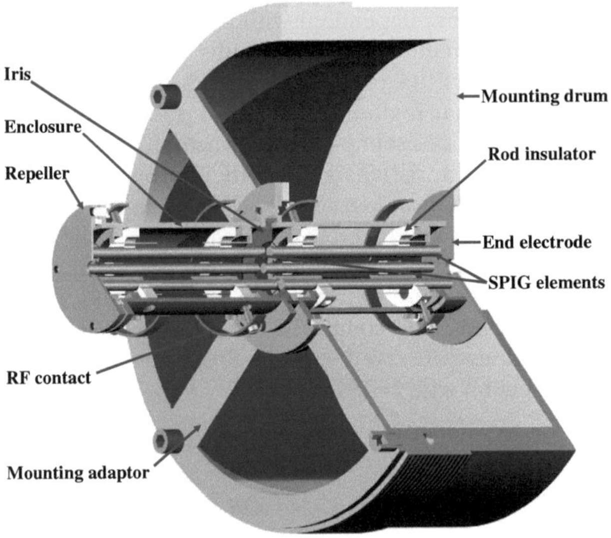

Fig. 22 A cut-out illustration of the radiofrequency sextupole for IGISOL–3 [57]. The ion beam enters from the *left*. Colour on-line

per μC of a factor of 2 to 8 whenever a comparison could be made with IGISOL–2. In the first proton-induced fission experiments, the size and shape of the fission ion guide was exactly the same, only the mounting was different. With 25 μA proton beam intensity, the cumulative yield of ^{112}Rh finally reached the 10^5 ions/second level in June 2004, a new record for the IGISOL facility. The primary reason for the improvement in efficiency appears to have been due to a better transmission through the extraction electrode which, due to the increased pumping efficiency, allowed the use of a 7 mm aperture instead of the previously used 4.2 mm one that may well have been collimating the secondary beam.

3.2 An increase in mass-separated yields with a new radiofrequency sextupole

With the extra space available in the target chamber, a new radiofrequency (rf) sextupole was designed in 2005 in connection with the laser ion source project (to be discussed in Section 3.4). The earlier device, motivated by the wish to reduce the energy spread of the skimmer-ring system, had a very much smaller geometry than the new design. Compared with an inner diameter of 3 mm, 40-mm long rods and an end plate hole of 1.5 mm, the new sextupole, illustrated in Fig. 22, was divided into two elements separated by an adjustable iris aperture, boasting an inner diameter of 5 mm, a total axial length of almost 170 mm, an end electrode aperture of 6 mm and a first electrode, known as the repeller, also with an aperture of 6 mm. The choice of geometry was constrained early on by factors associated with the goal of eventually performing laser ionization in the expanding gas jet immediately downstream from the ion guide [55]. Off-line studies of the gas jet properties revealed that the background pressure in the vicinity of the jet region is an important parameter in collimating the gas jet, improving the future spatial overlap between neutral atoms and counter-propagating laser beams [56]. The first segment of the SPIG was thus enclosed in a steel cylinder in order to increase the local pressure which could be coarsely adjusted with the iris. The buffer gas could then be pumped away through the gaps between the rods of the second element. Later, this design was modified following discharge problems when using argon as a buffer gas. A study of the measured breakdown voltage as a function of argon pressure led to the unobstruction of both segments of the sextupole to improve the pumping capacity [57]. Future collimation of the gas jet would be achieved via optimization of the ion guide exit nozzle shape [58].

It was known from IGISOL–2 and other facilities that the use of an rf sextupole not only improves the beam quality but also the transmission efficiency from the ion guide to the mass separator. The improved beam quality is associated both with buffer gas cooling and low voltage acceleration introduced by the SPIG. The buffer gas cooling effect is not, however, soley able to explain the experimentally observed improvement of the transmission efficiency when compared with the traditional skimmer system. Moreover, a clear discrepancy of beam qualities between the two different SPIG systems tested at IGISOL was noted, even though their effects on transmission efficiency are quite similar. With the new design, mass resolving powers between 400 to 600 were measured, while values almost twice as high were reached with the SPIG reported in [40]. Between 2005 and 2008 a number of experiments were performed with the SPIG covering both non-primary beam-related conditions (using an alpha decay recoil source of ^{223}Ra), as well as a range of on-line reactions including proton-induced fission, light-ion and heavy-ion induced fusion-evaporation. In each case, the experiments were repeated with the skimmer-based IGISOL. In parallel, Monte Carlo simulations were performed, with and without the effect of space charge, to compare ion beam properties such as emittance and transmission efficiency. Earlier indications pointed towards a reduction in skimmer transport efficiency for higher ion guide currents [51], yet this was the first time these studies were performed in more detail.

Details of the experimental results and the first simulations can be found in [59]. Later, the simulations were revised to improve the accuracy of the model for buffer gas flow and to develop the space charge model [57]. Without the addition of space charge, the simulations showed that the skimmer efficiency decreases as a function of background pressure because of the increased scattering of the ions, in particular for lighter ions. The SPIG however retains a good transmission efficiency, partly due to the buffer gas cooling effect and also to the larger geometrical acceptance of the repeller electrode. Losses in transmission efficiency due to the addition of space charge occur mainly in the region between the ion guide and skimmer. A higher efficiency may be obtained by using higher skimmer voltages (in particular for lighter mass ions) but at the cost of a reduced beam quality. The acceptance of the aperture of the repeller electrode is such that the majority of space charge losses occur within the SPIG, not between the SPIG and the ion guide.

The transmission efficiency of the SPIG was experimentally measured as a function of ion guide current using a proton beam of 40 MeV impinging onto a magnesium target. The total output current of the ion guide was measured on a Faraday cup before the SPIG, while a second cup measured the current before the mass separator. The results can be seen in Fig. 23. For comparison, the transmission efficiency of ions with mass $A = 20$ (chosen to reflect the dominance of light masses, of which helium, nitrogen, oxygen and water form a major fraction) was simulated using similar parameters as the experiment. The SPIG can transmit 100 % of the

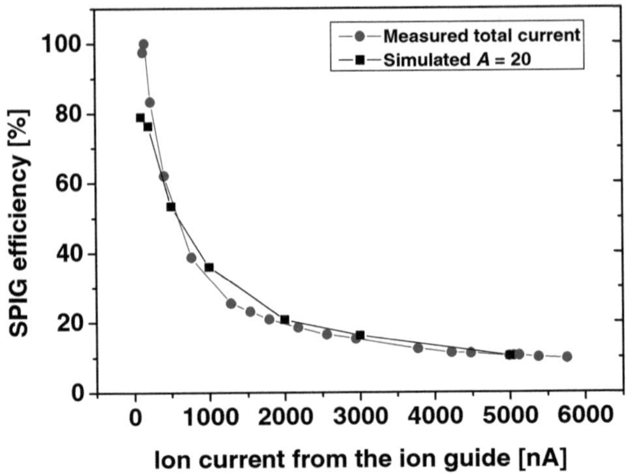

Fig. 23 A comparison between the measured and simulated SPIG efficiency as a function of the beam current from the ion guide [57]

incoming charge density up to currents of ~100 nA (~10^{12} ions s^{-1}) however above this, space charge effects steadily reduce the transmission before tending to saturation after a few μA. A transmission efficiency of 50 % is reached at an ion guide current of ~600 nA. The agreement between simulation and experiment is impressive, particularly when taking into account that the data sets are not normalized to each other and the experimental data was not used to adjust the simulation.

In general, the SPIG has improved the mass-separated yield extracted from fission reactions by a factor of 5 over the skimmer system, in good agreement with that reported in [40]. In light-ion induced fusion-evaporation reactions the SPIG results in an even higher improvement, with a factor of ~8 increase in yield being measured compared to the skimmer. Interestingly, in heavy-ion induced fusion-evaporation reactions no increase in yield was observed, once again supporting earlier results [40]. Similarly, both the SPIG and the skimmer perform equally well under off-line conditions. Following the impressive improvements in the mass-separated yields of the majority of experiments at IGISOL, the skimmer electrode is no longer in use.

3.3 An ion guide for quasi- and deep-inelastic reactions

Proton-induced fission of actinide targets has been and continues to be the tool of choice at IGISOL facilities for the production of beams of neutron-rich nuclei down to about $A = 70$, however below this the production cross sections decrease rapidly. In order to gain access to new regions of the nuclear chart, specifically the neutron rich isotopes of refractory elements between Ca and Ni, it was proposed to merge the IGISOL concept with quasi- and deep-inelastic reactions for the first time. Such reactions are rather complicated for two reasons, firstly due to the high kinetic energy of the projectile-like products and secondly, the energy distribution is rather broad. The conventional IGISOL technique is therefore rather unsuitable as the reaction products thermalize in a relatively small volume filled with rather modest helium gas pressures. For this reason a novel kind of ion guide was developed in collaboration with Lawrence Berkeley National Laboratory following a study of

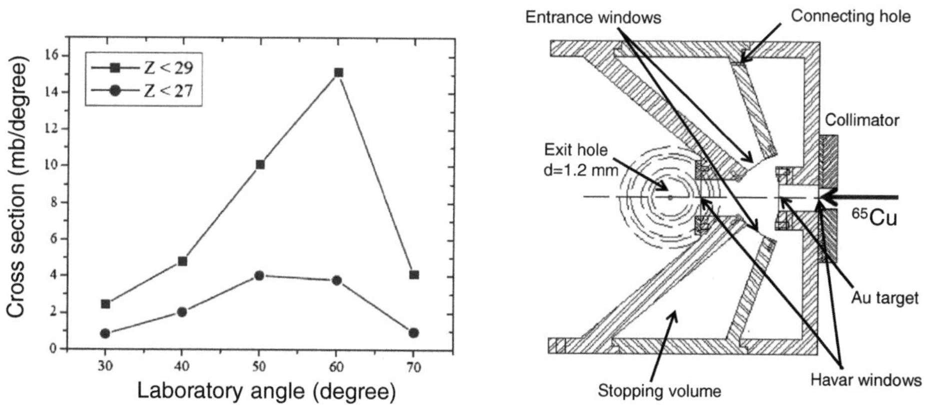

Fig. 24 Angular distributions of projectile-like reaction products $Z < 29$ and $Z < 27$ produced by bombarding a ^{197}Au target with a 403 MeV ^{65}Cu beam (*left*). Based on these results the angular acceptance of the IGISOL stopping chamber should be from 40 to 70 degrees. The resulting chamber design is shown (*right*). Figure courtesy of K. Peräjärvi

the energy and emission angle distributions of the chosen reaction, ^{197}Au(^{65}Cu,X)Y, at the 88″ cyclotron.

The deep-inelastic ion guide design, shown in Fig. 24, was tested in both on- and off-line conditions at IGISOL [60]. Using the same reaction as the Berkley experiment, the yields of mass-separated radioactive projectile-like species were studied. Isotopes including ^{62,63}Co were measured with focal plane rates of about 0.8 ions/s/pnA, corresponding to a total stopping and extraction efficiency of approximately 0.06 %. With a helium pressure of 300 mbar, about 9 % of the Co recoils are thermalized and of this stopped fraction, 0.7 % are converted into mass-separated beams. This intrinsic efficiency is comparable to that reported for the HIGISOL system. Further tests with this new ion guide design were not performed at IGISOL-3 however it was proposed that dramatic improvements in the overall efficiency of the ion guide could be expected in the future via the combination of introducing Ar as a buffer gas and by relying on selective laser re-ionization.

3.4 A move towards a more element-selective approach

The on-going success of the IGISOL facility and the rich programme of physics utilizing laser spectroscopy, mass spectrometry and decay spectroscopy has been a testament to the niche the ion guide technique has offered in comparison to other target-ion source methods. The unique ability to study refractory elements, combined with the speed of extraction from the source in an almost chemically independent manner continued to be exploited throughout the years of IGISOL-3. Nevertheless, it was recognized that in order to push further from stability more stringent requirements would need to be set on the ion guide method. The fact remained that with access to higher primary beam currents and better transmission using the rf sextupole, the ion guide was not only producing the short-lived exotic species of interest, but more often than not such beams would be accompanied or even overwhelmed by longer-lived more abundant isobars. Although the longitudinal

Penning trap can provide isobarically clean beams for experiments located after the trap, the device itself, as well as the rf cooler, can suffer from space charge limitations. The technique of cooling and bunching the ions for laser spectroscopy, pioneered in Jyväskylä, led to dramatic improvements in suppressing the dominant source of photon background. However, isobaric contamination in the bunched beam may also limit what can be achieved, with photons scattering off the bunched beam as it passes through the light collection region.

Therefore, in 2004, following infrastructure funding from the Academy of Finland, the decision was taken to combine the benefits afforded by selective laser ionization with those of the ion guide technique. Resonance photo-ionization with high power pulsed lasers provides in principle isotopically and isobarically pure ion beams, utilizing the unique atomic fingerprint to selectively excite and finally ionize the atom of interest. If one can achieve neutralization, this method may be applied resulting in both high ionization efficiency and high selectivity, while maintaining the fast and universal thermalization of nuclear reaction products in a chemically independent manner. Laser resonance ionization is a mature technique which has played a key role in the selective and efficient production of radioactive beams (RIBs) at on-line isotope separator (ISOL) facilities. The technique has advanced and developed to a point where it is the preferred method for the ionization of nuclear reaction products at ISOLDE [61]. The RILIS ion source at ISOLDE has been used to resonantly ionize 31 elements of the periodic table [62] with ionization efficiencies ranging from 0.1–30 %. Other existing or upcoming facilities coupling laser ionization with a hot cavity approach include TRIUMF [63], GANIL [64], IRIS at Gatchina [65], HRIBF at Oak Ridge [66], ALTO at IPN Orsay [67] and SPES, Legnaro [68].

The development of laser ionization of neutral recoils in an on-line gas cell was pioneered by the University of Leuven (see [69] and references therein). The group demonstrated the power of combining a high-power low-duty cycle laser system with the ion storage capability of a high-pressure gas cell. During the course of many studies, the need for conditions of high purity (sub-ppb) within the gas cell was demonstrated [70]. Traditionally this has been less stringent for the standard ion guide technique which has delay times often shorter than molecular ion formation times. At Leuven, argon rather than helium is used as this provides stronger recombination of reaction products in the gas. Such conditions favour the presence of only a weakly ionized plasma and thus the gas flow to transport photo-ions out of the gas cell can be significantly lower than at IGISOL. This permits the use of higher pressures, leading to better stopping. The consequence however of a smaller gas flow and the time required for neutralization makes the laser ion guide slower than the standard one by about two orders of magnitude. Although the laser ion guide no longer operates as an ion guide in the strictest sense, it would turn out during the course of the developments at IGISOL–3 that many things learned by the Leuven group would need to be applied at IGISOL. We note that such selective ionization techniques will play a key role in the future at fragmentation facilities including S^3 at GANIL [71] and the PArasitic RI-beam by Laser Ion Source (PALIS) facility at RIKEN [72, 73].

The choice of laser system for the new laser ion source facility was motivated by the requirements for highest ionization efficiency and elemental selectivity, combined with a universal coverage of ionization schemes throughout the periodic table. In order to avoid large duty cycle losses, lasers with a relatively high repetition rate, typically between 100 Hz and 10 kHz are used. The elemental selectivity is provided by matching the spectral linewidth of the laser to the spectral width of the transitions investigated in the ion source, typically reaching values of about 1 GHz. In addition, by applying widely tunable laser radiation a maximum in wavelength coverage can be obtained. Finally, reliable long-term operation with negligible maintenance has to be demanded. The Fast Universal Resonant laser IOn Source (FURIOS), developed at JYFL, initially started as a *twin* laser system, combining high-repetition rate (10 kHz) Ti:Sapphire and dye lasers each with independent pump lasers [74]. Over the years the laser system has evolved and developed further, with the dye lasers being superceded by solid-state technology. The copper vapour laser, originally bought for the pumping of the dye lasers, is now primarily used for non-resonant ionization. New resonator designs have been employed to ease the search for new resonance ionization schemes and for a reduction in laser linewidth for spectroscopic applications. Currently, the laser system is capable of covering wavelengths ranging from ~690 nm to 1000 nm and from ~500 nm to 205 nm with higher harmonic generation. Further technical details regarding the laser system may be found elsewhere [75].

From the early stages of implementing the laser ion source at IGISOL, discussions regarding the possible exploration of laser ionization in the expanding supersonic gas jet proceeded. It was understood from the work at Leuven that achieving 100 % neutralization of recoils would be unlikely, therefore the selectivity of the laser ionization process (defined as the count rate with lasers on compared with lasers off) would always be somewhat limited. In the so-called LIST (Laser Ion Source Trap) approach, neutral radioactive atoms are selectively ionized upon exit from the ion guide within the gas jet, captured by the rf field of an rf multipole (quadrupole or sextupole) and transported to the high-vacuum region of the mass separator by a voltage gradient. A positive DC voltage applied to the multipole acts to repel any non-neutral fraction thus ensuring the highest possible beam purity is maintained for subsequent experiments. Such a concept was originally proposed to improve the beam quality from a hot-cavity ion source by addressing the problem of poor selectivity from unwanted surface ionization [76]. The idea of coupling the LIST technique with a gas cell was then suggested at JYFL [55] and provided the motivation for the developed of the rf sextupole, discussed in Section 3.2.

Figure 25 illustrates the original laser system and the LIST scheme of operation. Late in 2004, a dipole magnet was loaned from the old ISOL separator facility at GSI to replace the existing IGISOL magnet which did not have optical access. Unfortunately, after several tests it was discovered that the window on the GSI magnet suffered from high losses and was inaccessible for either cleaning or replacement. A full optimization of the LIST geometry would therefore be impossible to achieve throughout the period of IGISOL–3, and only with the move to the new accelerator hall (IGISOL–4) would the full exploitation of the gas jet for both highly selective laser ionization and spectroscopy be realised. Nevertheless, throughout the years the LIST mode was tested on a number of occasions in a somewhat limited capability, with ionization being performed in the jet following focusing of the laser beams through the exit hole of the ion guide [55]. In addition to demonstrating the feasibility of LIST coupled with an ion guide, the prospect of utilizing the gas jet for laser spectroscopy on short-lived exotic nuclei has become highly attractive due to the substantial reduction in the effects of line broadening mechanisms (Doppler

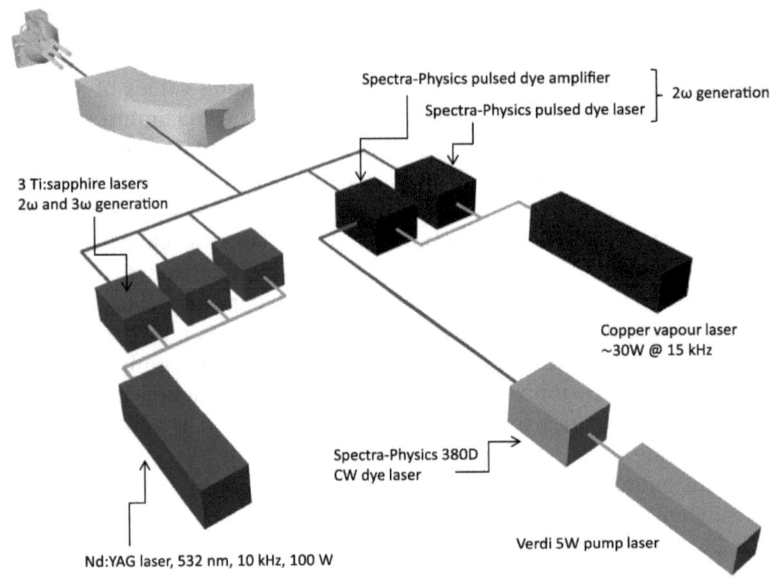

Fig. 25 The layout of the high repetition rate laser system at IGISOL–3 and laser transport through the mass separator towards the rf sextupole ion beam guide. Figure originally made by A. Nieminen

and pressure). The first proof of principal tests at IGISOL–3 were performed on stable Ni isotopes [75]. Although this illustrated the insensitivity to nuclear volume effects in lighter mass systems using broadband pulsed lasers, it motivated a major development programme to reduce the laser linewidth which saw new infrastructure funding being successfully applied for in 2012.

The application of laser ionization at IGISOL was originally focused on opportunities for the study of nuclear structure in three areas: $Z \geq N$ nuclei up to ^{100}Sn; on neutron-rich nuclei available in fission between $A = 70$ and 160; heavy nuclides in the actinide region. Coupling optical light with the fission ion guide was deemed to be rather complicated and therefore modifications were initially made to the existing HIGISOL ion guide, with the goal of improving the rather modest ion guide efficiency of $\sim$1 %. A number of considerations were taken into account when designing the gas cell including efficient evacuation (gas flow transport and nozzle design), water cooling and baking capabilities, and modularity in the construction in order to incorporate filaments and dc electrodes. The recoil distribution in a heavy-ion induced reaction requires a rather large stopping volume and thus efficient evacuation becomes critically important. Gas flow simulations were applied including both the macroscopic motion of the buffer gas as well as a microscopic random diffusion process, closely following earlier work by Peräjärvi [77]. Such models can be combined with evacuation time profile distributions and were found to be extremely useful when disentangling the many complex processes imprinted on an ion of interest as it is evacuated from a gas cell. Further details of the models used and the application to laser ionization may be found in [78].

Figure 26 illustrates the usefulness of gas flow simulations in the design of ion guides. The gas in the original HIGISOL ion guide (shown on the left) was unable

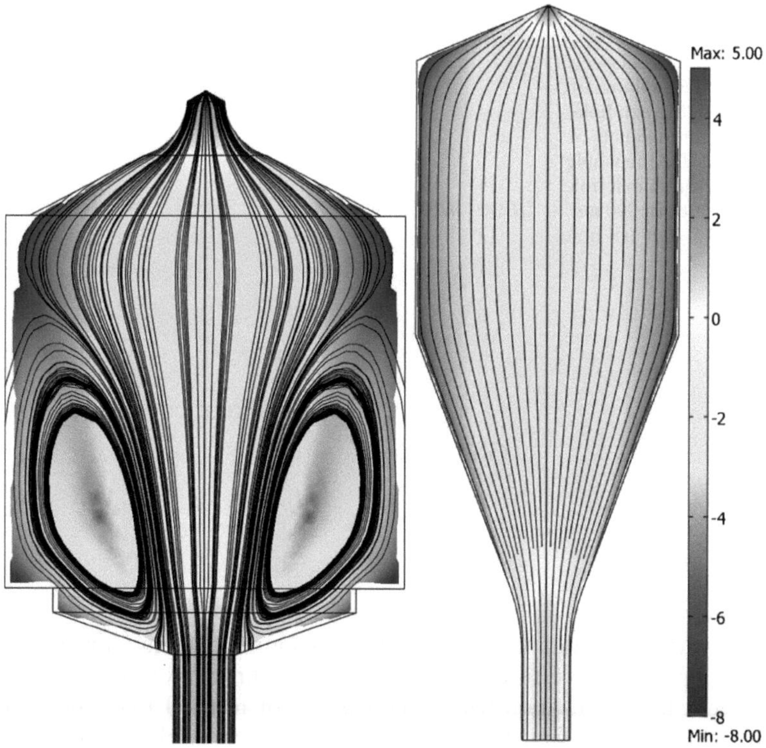

Fig. 26 Gas flow comparison of the old HIGISOL (without gas splitter, *left*) and new HIGISOL ion guide (*right*) for a He gas pressure of 200 mbar and a 0.5 mm exit hole [78]. The colour coding represents the gas velocity (m/s) in logarithmic scale. Colour on-line

to expand fast enough and therefore formed a channel in the centre with vortices in the outer region. This was circumvented by adding a gas splitter to the inlet of the ion guide. This however leads to complications with laser ionization along the extraction axis as a vortex formed immediately behind the splitter thus deteriorating the gas flow. The new HIGISOL design, shown on the right of Fig. 26, featured an extended input side which smoothly diverges allowing the gas flow to expand without formation of vortices. The final technical design, illustrated in Fig. 27, consisted of three separate modules; a gas feeding part, the main body and a removable exit nozzle/hole. The number of components was minimized in comparison with the original HIGISOL design to reduce the open surface area within the guide. Sealing was achieved using indium foil. The gas cell was successfully used in several off-line and on-line laser ionization studies as well as standard HIGISOL experiments.

Finally, we note that the operation of an efficient and reliable laser ion source depends critically on the laser transport path. A notable restriction to performing laser ionization at IGISOL–3 was the severe losses in laser power due to several compromises which had to be made when transporting the laser light to the target area. This is the unfortuate situation when one tries to construct a laser ion source at an existing facility, rather than incorporating plans in the original layout and design. From the early design phase of the new laboratory at JYFL, IGISOL–4 was

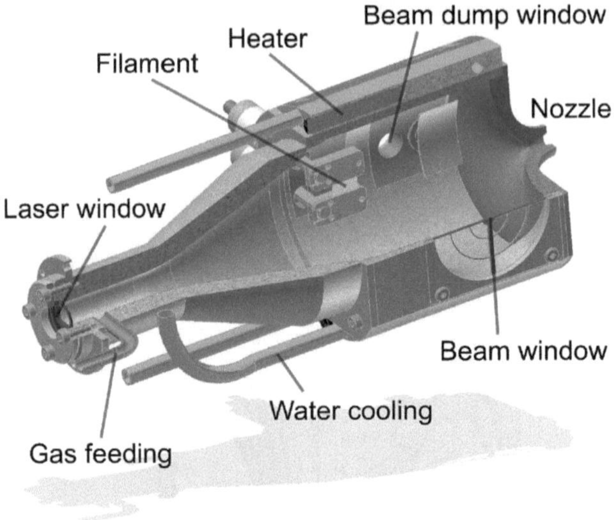

Fig. 27 Design of the new HIGISOL ion guide compatible for use with laser ionization [78]. The volume of the main body to the exit hole is ~250 cm^3, resulting in an evacuation time (for the whole guide) of 2.25 s for an exit hole of 0.5 mm diameter or 390 ms for a standard 1.2 mm exit hole diameter

constructed with the installation of the laser ion source playing a primary role and it is currently possible to efficiently transport up to three laser beams to the target area for either laser ionization within the gas cell or in the LIST mode. The laser ion source at IGISOL–3 was therefore used primarily as a selective probe to study the processes occuring within a gas cell both under off-line and on-line conditions, whereby the laser ionization efficiency was not of immediate concern. This was viewed as an important intermediate step towards a future successful laser ionization programme, particularly as the optimum conditions for the ion guide principle are rather different to those requiring neutralization followed by selective re-ionization.

3.5 Gas phase chemistry and weak plasma studies via laser ionization

The argument for chemical independence of the ion guide technique is that all species have a much lower ionization potential than helium, therefore, after the reaction products have been stopped they will remain ionized during extraction. The presence of impurities in the gas can however lead to losses via molecular formation, in particular for the most chemically active ions which include several refractory elements that can only be produced with the gas cell method. Such losses can be quantified with characteristic timescales—in this instance chemical reaction rate constants which can often differ by orders of magnitude [79]. One can immediately understand the requirement for conditions of high purity (sub-ppb) for both laser ion guides and large gas catchers due to the competition between atomic ion survival and evacuation timescales. Direct measurements of gas purity at this level is not possible and therefore at Leuven such studies were performed by adding a known amount of an impurity gas and observing the change in the mass spectra and ion time profiles [70]. At IGISOL, the requirement of high purity operation had not been strictly adhered to in the past, primarily due to the short extraction times which did not result in severe losses to molecular formation. Nevertheless, with the installation

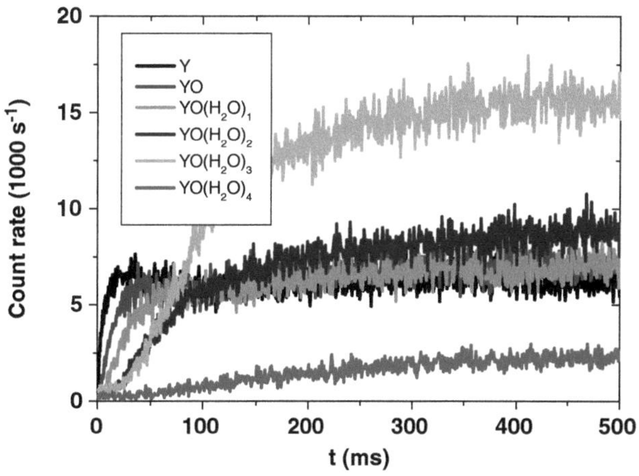

Fig. 28 Time distribution profiles of atomic yttrium ions and associated molecules following a laser on period of the first 90 ms [78]. Only 10 % of the sum of all species survives extraction from the ion guide in atomic form. An impurity level of ~0.1 ppm can be extracted from this data. Colour on-line

of a laser ion source it became possible to selectively probe the effects of molecular formation via the mass-separated time distribution profiles.

A series of measurements was performed at IGISOL during the period between 2006 and 2007, using yttrium as a benchmark due to its refractory nature and to the extremely strong affinity towards binding with the main contaminants in the buffer gas, water and oxygen. Yttrium atoms were introduced into an ion guide from a heated filament and following resonant laser ionization the surviving atomic ions as well as molecular ions were extracted, mass separated and detected downstream from the IGISOL focal plane. An example of the resulting time distribution profiles when the ion guide purity conditions are not controlled is shown in Fig. 28. Such data can yield some rather simple yet fundamental insights. The saturation time of the atomic yttrium profile is related to a combination of survival against losses due to molecular formation and evacuation from the ion guide. An exponential fit to the data yields a value for τ of ~5 ms. Due to an evauation time of several hundreds of ms, such a time constant reflects the dominance of molecular formation.

When combined with the exit-hole conductance, a rather useful quantity can be extracted, the *effective volume* for laser ionization. Additionally, by combining the saturation time with the known reaction rate coefficient an estimate for the impurity level can be made. In the example shown in Fig. 28, the impurity level is ~0.1 ppm. The bare atomic fraction of yttrium is effectively evacuated from a volume of ~0.6 cm^3 around the nozzle region and amounts to 10 % of the sum of all species shown in the figure. The result of a reduction in impurity level is experimentally translated into a larger ratio of atomic to molecular ions, a longer saturation time constant and thus a larger effective volume. Indeed, in order to uniquely determine the effect of chemistry independently from a competing evacuation time requires a rather more complicated analysis of the data whereby a framework of rate equations is used to extract time profiles of single laser shots. Such a model is discussed along with the effect of gas phase chemistry in [80]. With better control over the gas purity, this model was used to show that the impurity level at IGISOL can be reduced to <1 ppb and results in atomic yttrium being extracted as the dominant ion.

The addition of a non-selective ionization mechanism such as the passage of a primary beam through an ion guide or from the extraction of recoils out of the gas cell from a fixed target such as in a fission reaction leads to an additional set of complex processes. One of the most important loss mechanisms for an ion of interest on-line is that of neutralization within the high density of (primarily) buffer gas ion–electron pairs. In a similar manner to the molecular formation times in gas phase chemistry, neutralisation is governed by ion recombination rate constants. Losses due to recombination were understood from the earliest days of IGISOL operation and thus in the ion guide approach the extraction time of the radioactive ions out of the gas cell should be much less than the recombination time constant. On the other hand, the operational principal of a laser ion source is based on the fast ion–electron recombination in the high ion/electron density created by a projectile beam, with subsequent re-ionization being applied in a zone where the density is low enough for a high survival chance of the photo-ions. In the latter case, the evacuation time should be longer than the recombination time constant to obtain neutralisation, but the evacuation time of the laser produced ions should be shorter than the respective recombination time.

The effects associated with the presence of a large density of ion–electron pairs had been carefully studied at the Leuven laser ion source (LISOL) [81, 82]. The LISOL team showed that the ion source efficiency was element and intensity dependent. Even at time scales of approximately 5 ms recombination losses start to become important once the ion–electron density is above 10^7–10^8 pairs/cm^3. The main loss mechanism for the laser re-ionized species is the recombination process with electrons created by the primary beam. The removal of these electrons via a DC/RF electrical field is a possible solution, however, space charge effects start to play a role at high ionization rates. Space charge leads to a reduction in the effectiveness of the external field and thus a reduction in the removal of electrons, finally returning the ion guide to the conditions of a neutral low-density plasma. In order to overcome these limitations one may consider the application of a pulsed primary beam time structure, pulsed electric fields to remove the remaining primary electrons before finally introducing the laser ionization pulse. This however is a rather complicated approach.

At IGISOL, experiments continued with the study of yttrium, looking at the effects of the addition of a primary beam on the filament atoms, molecules and laser-produced ions. Studies of the time profiles provided insight into saturated yields under different conditions as well as dynamic competing processes when the primary beam was operated in a pulsed mode. Many of the conclusions made at LISOL could be verified, however additional understanding was sought via in-depth simulations based on differential equations describing the competition between direct beam ionization, charge exchange, recombination, molecular formation and evacuation. In addition, the role of a plasma-generated electric field was discussed and its effect of suppressing ion transport within the gas cell [83]. This mechanism is rather sensitive to whether an ion of interest is transported in helium or argon buffer gas due to the differences in recombination coefficients of the buffer gas ion–electron pairs as well as the difference in gas flow velocity. Perhaps the most important conclusion from these experiments is that the effective volume for laser ionization in the presence of the primary beam is restricted to the nozzle region, a volume which is evacuated within milliseconds. Although this volume was not probed at different beam intensities, the results support the time scales at which the effect of recombination has been seen at Leuven.

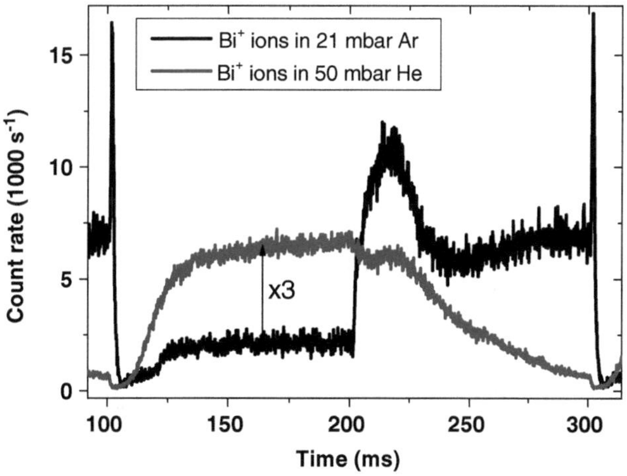

Fig. 29 Time distribution profiles of atomic bismuth ions in argon and helium buffer gases. The cyclotron is pulsed on between 100 and 200 ms in a repeated cycling mode. A strikingly different behaviour in ion survival is observed. More details can be found in [78]

In later beam time periods, the effect of the primary beam was also studied on less chemically active elements including technetium and bismuth. One such experiment used bismuth material evaporated on a tantalum backing foil, subsequently irradiated by a 500 nA beam of 60 MeV protons. Bismuth was thus sputtered from the target providing comparable on-line conditions to that expected for the production of radioactive nuclei. The primary beam was pulsed in a variety of duty cycles in order to probe the dynamics of the Bi time profiles. Figure 29 illustrates the striking differences between the Bi^+ ion time profile in helium and argon buffer gases. It is clear that both the chemical nature of the ion of interest as well as the buffer gas type are important parameters which directly influence the ion survival.

The gas cell is an extremely complex, harsh environment. Earlier experiments at IGISOL did not require a detailed understanding of the competing processes due to the dominance of the short evacuation timescale compared to that associated with losses due to impurities and/or recombination. The development of the laser ion source during IGISOL–3 resulted in a renewed interest in both experiments and simulations to identify the important parameters which determine the survival of the ion of interest. Qualitatively we have been able to simulate the dynamic effects seen in the ion time profiles. Future steps could yet be taken into account for the role of diffusion of the plasma, to incorporate realistic beam and gas flow distributions, and to include the different mobility of the ions of interest and the plasma (buffer gas) ions. The operation of the ion guide as traditionally utilized during the years at IGISOL is very different from the conditions required to operate a successful laser ion source. The work described here may thus be viewed as an intermediate step in the path to improve the efficiency and selectivity of the ion guide technique.

3.6 Development of the dual-chamber gas cell

The complexities associated with trying to perform laser ionization in a gas cell which at the same time includes the influence of a primary accelerator beam, recoils and radioactivity, results in a very limited volume within which one can successfully extract the laser ions before they recombine. At the same time, the application of

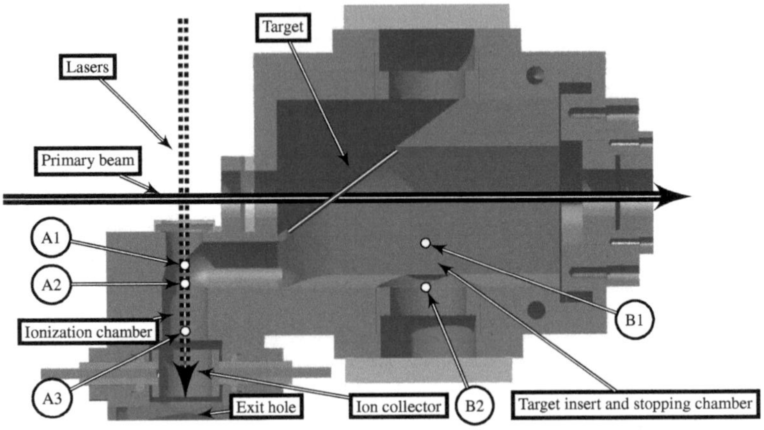

Fig. 30 Schematic view of the dual-chamber gas cell (modified from [85]). ^{223}Ra α-recoil source locations are labeled *A1–A3*, *B1*, and *B2*. Buffer gas enters via a ring slit that homogeneously distributes the gas across the cell (not labeled). The target insert is shown in *red*. A filament (not shown) can be mounted within the main stopping volume of the ion guide to act as a source of atoms. Colour on-line

electrical fields to remove electrons or to reduce the background of non-resonant ions is hindered by the effect of space charge due to the presence of the high density of buffer gas ions. One such method to address these issues is to physically separate the laser ionization zone so that it is not visible to primary beam, trajectories of recoils or scattered particles. This approach was pioneered in a novel design of ion guide, the so-called dual-chamber gas cell, by the Leuven group [84]. The guide consists of two chambers, the stopping and ionization volumes, which are connected by a channel in the shape of an elbow. Gas flow transports recoils from the stopping chamber to the laser ionization volume, which can be illuminated either along the extraction axis or transversely. The latter ionization geometry, close to the exit hole, allows the use of an ion collector, located upstream, to collect non-neutralized ions without disturbing the laser-produced ions. The total selectivity for heavy-ion induced fusion-evaporation reactions was measured at LISOL to be an impressive 2200 [84].

In 2009, based on the LISOL design, a dual-chamber gas cell was constructed at IGISOL. With a fundamentally different geometry to normal ion guides in operation, the efficiency of the gas cell was investigated under off-line conditions using a radioactive ^{223}Ra α-recoil source. Figure 30 schematically illustrates the ion guide as well as the locations of the recoil source (labelled A1 to A3, and B1, B2), probing both regions within the ionization volume as well as the stopping volume. Targets can be installed on the tilted surface of the insert (shown in red) that is fixed in the stopping chamber. We note that the angle between the target surface and the accelerator beam can be changed. The shape of the insert ensures a turbulent-free laminar gas flow towards the elbow. Full details of the source measurements can be found in [85], with the result that efficiencies measured from the ionization chamber (positions A1, A2 and A3) are 10–20 %, dropping by approximately a factor of ten with the source located in the stopping chamber (positions B1 and B2). Ions created close to the chamber walls (position B2 in Fig. 30) suffer due to a negligible gas flow velocity and high diffusion losses. Without the target insert, the efficiency from position B1 drops from ~2.3 % to ~0.9 %. Gas flow simulations performed by T. Sonoda suggest strong effects of turbulence without the target insert and this may well be contributing to the decrease in efficiency.

Prior to the shutdown of the IGISOL–3 facility in June 2010, some initial laser ionization tests on nickel were performed with the dual-chamber cell under both off- and on-line conditions. The ionization chamber was fitted with an extension piece after the ion collector so as to test the ability to collect ions from the stopping chamber, without disturbing laser-produced ions created close to the exit hole. Voltage pulses of different polarity (but equal amplitude) were applied to the electrodes. In the presence of the passage of a primary beam of $^{4}He^{2+}$, non-neutralized ions which reach the ion collector are completely suppressed whereas the Ni atoms laser ionized before the nozzle are undisturbed [85]. Further measurements and first on-line radioactive studies are keenly awaited at IGISOL–4.

3.7 Development of a cryogenic ion guide

High purity (sub-ppb) conditions required for both laser ion guides and for large gas catchers is achievable however technically rather challenging. This is particularly true for the latter devices which can be viewed as a step forward in the evolutionary process of the IGISOL technique; a system with a large stopping volume introduced in the late 1990s at in-flight separators for the conversion of high-energy radioactive beams into low-energy beams required for precision experiments. Traditionally, room temperature devices have been built, however, ultra-high vacuum (UHV) techniques must be used in the design and construction of the catcher. For example, the RIA gas catcher prototype built at Argonne National Laboratory and tested at the Fragment Recoil Separator, GSI, utilized about 7300 components with over 4000 components cleaned to UHV standards [86]. Partly because of these stringent requirements as well as the challenge to minimize the size of the gas catcher, an alternative approach was suggested in Jyväskylä in 2001 in collaboration with and following the work of N. Takahashi [87]; the use of superfluid (SF) liquid helium as a stopping and transport medium for radioactive ions. The main advantage over stopping in a gas is the much larger density (a factor of 800), thus allowing the use of a very small stopping chamber. This work was peformed under the auspices of an EU R&D network, IONCATCHER, the goal of which was to further develop ion deceleration and gas stopping techniques.

Between 2001 and 2003 several off-line experiments using alpha-decay recoils from a ^{223}Ra source were performed at JYFL. This work resulted in the successful transport of recoils through a closed cell, from inside SF helium at the bottom of the cell, across the liquid surface and through the vapour region to a detector at the top of the cell. This was the first observation of the extraction of any positive ion from SF helium and a successful proof-of-principle of this new method [88, 89]. This work continued at KVI, Groningen, with similar experiments performed in 2005 as part of the RIASH (Radioactive Ions and Atoms in Superfluid Helium) project, focussing on the extraction of ions from SF helium at a temperature close to 1 K. By evaporating a thin surface layer of the liquid using so-called second-sound pulses, evidence of an improved extraction efficiency from the surface was obtained [90]. This improvement was unfortunately unable to be reproduced and thus the limitation

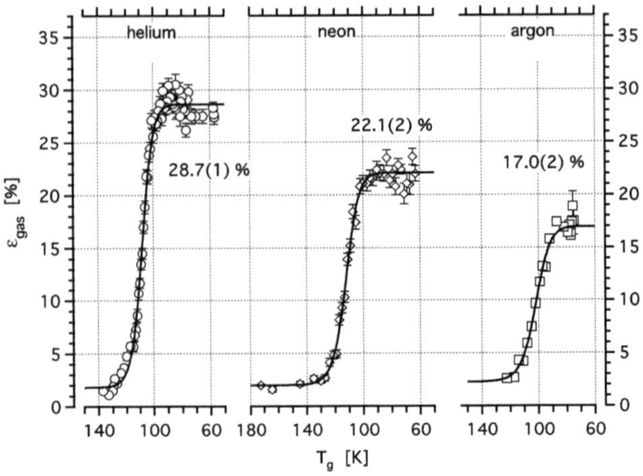

Fig. 31 Measured thermalization and transport efficiency of 100 keV ^{219}Rn ions in 1 bar room temperature equivalent noble gases as a function of gas temperature. Saturated efficiencies are given. This data was obtained in a closed cell [93]

of transporting the ions through the liquid helium surface still remains the curent bottleneck of this technique [91].

In connection with the KVI experiments, the equipment built for the RIASH project was utilized in a second approach to achieve ultra-pure conditions: freezing out the impurities using a cryogenic noble gas. This project, for the JYFL side supported through an EU Design Study, DIRACsecondary-Beams (a GSI-led proposal to address the design and development of scientifically and technically challenging projects for the future Facility for Antiproton and Ion Research, FAIR), was to be the start of a very successful programme of research which would lead to the testing of a cryogenic ion guide at IGISOL and, furthermore, to the development and successful commissioning of a cryogenic gas catcher [92] built for the Fragment Recoil Separator (FRS) at GSI within the framework of a collaboration between GSI and the University of Groningen. We note that this catcher is a prototype of a device to be installed at the Super-FRS at FAIR.

The first experiments with cryogenic gas were performed at KVI in a simple closed cell, once more using the unique identification of alpha spectra from a ^{223}Ra source. A beautiful demonstration of the improvement in ion survival as the temperature of three noble gas media is reduced can be seen in Fig. 31. The combined ion survival and transport efficiency of ^{219}Rn ions starts to improve at about 120 K for all three gases and subsequently saturates below ~90 K to the values given in the figure. The systematic decrease in saturated efficiencies when going from helium towards argon demonstrates that the results are based on intrinsic properties of the noble gases. As discussed in [93], this high and temperature-independent efficiency at low temperature is interpreted as being due to the ultra-pure conditions that are obtained by freezing out the impurities.

In the expected absence of impurities at low temperature the fate of the ions may still be affected by the ionization of the buffer gas. Thus experiments performed at JYFL during 2005–2006 (initially IGISOL was not used) addressed this issue by using a 13 MeV proton beam to ionize the helium gas in the volume in which the α-decay recoils from ^{223}Ra are stopped. The combined efficiency of survival and transport of the recoil ions over several cm was measured as a function of the ionization density, i.e. beam intensity, the electric field strength within the closed chamber, the

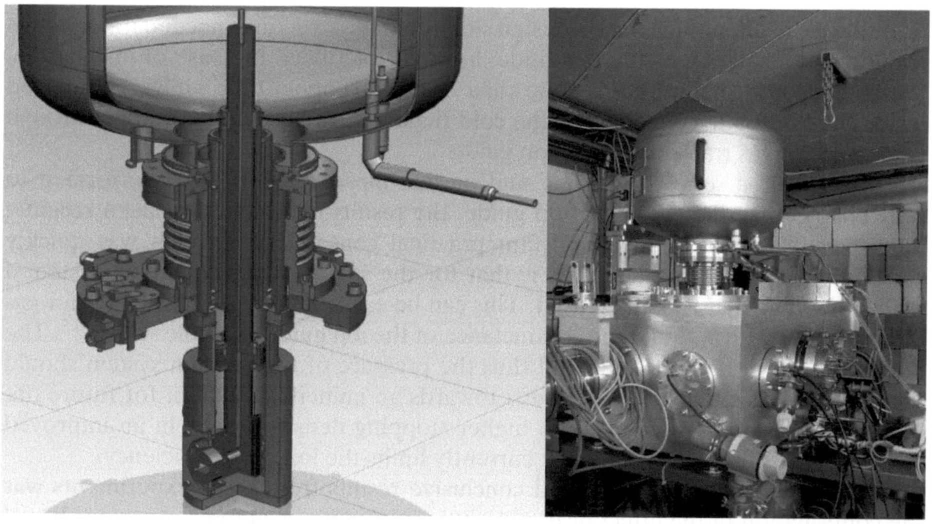

Fig. 32 The cryogenic ion guide setup at IGISOL–3. *Left*: The technical structure of the system. The LN_2 bath cryostat used to cool the cold finger and He gas can be see at the top. The cold finger is surrounded by an insulating vacuum vessel. The cryogenic ion guide, attached to the tip of the cold finger is outside the insulation vacuum. *Right*: The whole system mounted from the top of the IGISOL target chamber [97]

helium pressure and temperature [94, 95]. At low beam intensity or high electric field, the off-line results were confirmed and a saturation efficiency of 30 % was obtained. Above a certain threshold in beam intensity a sharp drop in efficiency was observed. The threshold at which this occured increased with the square of the applied electric field. This behaviour was interpreted as resulting from the shielding of the applied field by the weak plasma created by the proton beam. As discussed in Section 3.5, the ions and electrons are no longer pulled apart under such conditions and recombination sets in.

The saturated efficiency seen at low temperatures and low ionization-rate densities most likely reflects the fraction of ions at thermalization following the charge exchange and stripping cross sections involved in the slowing down process of the ions. Similar tendencies towards saturation of extraction efficiencies from gas cells have been seen at other facilities, discussed further in [96], and suggest a fundamental upper limit for the efficiency of noble gas ion catchers. This upper limit depends not only on the chemical nature of the ion of interest but also on the buffer-gas type.

The difference between these initial studies and the operation of IGISOL is that all aforementioned measurements were performed in a closed cell, ions were not extracted to high vacuum and, importantly, a mass analysis could not be performed. In the next step, a cryogenically cooled ion guide was developed and constructed at KVI, appropriately called the "cryogenic ion guide", CIG, in order to verify whether primary ions and/or secondary nuclear reaction products can be stopped and extracted from a low temperature gas cell. The design of the guide closely followed that of the existing light ion guide, mounted onto a vacuum-isolated cold finger which is cooled by a liquid nitrogen (LN_2) bath. The whole assembly sits vertically on top of the IGISOL target chamber, illustrated in Fig. 32, rather than from the more traditional side mounting. Such a design simplified the structures needed for effective cooling and insulation of the ion guide, however sacrificed the ease of installation. Pre-cooling of the helium gas is done via a 7 m spiral copper tube which is immersed inside the LN_2 bath, connected to the cold finger which has a hole running through the centre for gas transport to the ion guide.

During 2008 a series of off-line and on-line measurements were performed to fully characterize the cryogenic ion guide, the results of which have been recently published [98]. One of the important practical consequences which was quickly verified when using cryogenic gas is that for the same vacuum pumping system a higher stopping density can be used. This can be easily understood as the buffer gas density scales with T^{-1} and the conductance of the ion guide exit hole with $T^{0.5}$. The mass flow out of the ion guide and thus the pressure of the vacuum system should therefore scale with $T^{-0.5}$. This points towards an immediate benefit for future on-line fission reactions at IGISOL: a higher stopping density results in an improved stopping of fission fragments which currently limits the ion guide efficiency.

One of the most interesting and conclusive results from these experiments was the confirmation of the effect of liquid nitrogen cooling in suppressing water-based impurities, complex molecules and hydrocarbons. This was not only seen in mass-separated spectra but also via a reduction of the total current coming out of the ion guide. On the other hand, insufficient pumping arrangements of the cold finger insulating vacuum resulted in a typical operating temperature (with gas load) of $\sim$90 K. Thus, key impurities such as oxygen and nitrogen could not be frozen from the gas. In a rather surprising observation, any impurity not frozen showed an enhancement in yield, in some cases close to a factor of 10 [98]. This most likely indicates an overall reduction in the number of charge exchange reactions between the unfrozen species and heavier complex molecules and/or hydrocarbons which are frozen, improving the ion survival of the former. The most practical means to improve this situation would be to reach temperatures below that of LN_2, possible through the use of a cryocooler.

In general, when decreasing the temperature of the system, ion–molecule reactions tend to become faster. This couples with an increase in the extraction time from the ion guide. One of the features built into the CIG is a quartz window mounted on the cold finger opposite to the placement of the ion guide. In the future, one might utilize laser ionization in connection with the CIG in order to probe the neutral fraction and perhaps gain a better understanding of a possible temperature dependence effect in recombination. The reduction of impurities would be critically important to future HIGISOL reactions which have often suffered from contamination problems in addition to low yields of the nuclei of interest. Laser ionization coupled with a cryogenically cooled heavy ion guide would be a powerful tool for the production and study of $Z \geq N$ nuclei at IGISOL–4.

Finally, and in support of the on-line closed cell measurements, an energetic ion beam (^{58}Ni) was stopped and extracted from the ion guide. The results of the extraction efficiency as a function of ionization rate density are shown in Fig. 33. Experimentally, the ionization density in the gas cell was varied by changing both the helium gas pressure as well as the primary beam intensity which covered a dynamic range of approximately five orders of magnitude, i.e. from about 10^3 to 10^8 particles per second. Athough one cannot directly compare these results with other facilities, the trend differs from room temperature systems in that the extraction efficiency remains rather constant at relatively high ionization rate densities.

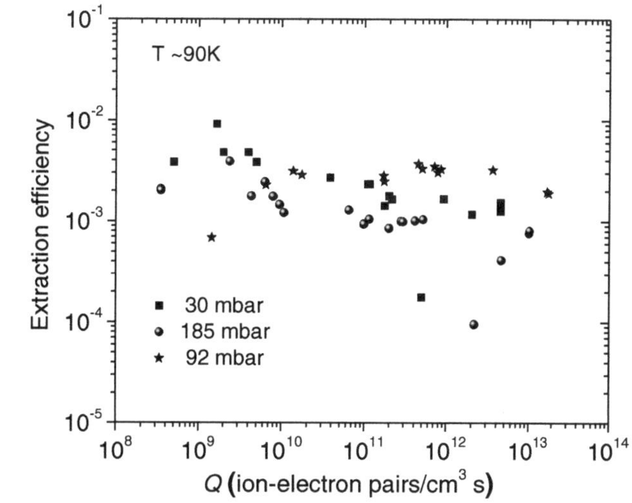

Fig. 33 Extraction efficiency of the cryogenic ion guide for $^{58}Ni^{+}$ ions as a function of ionization rate density (proportional to beam intensity) [98]

4 Outlook towards IGISOL–4

In June 2010, the IGISOL facility reached the end of an era having operated at its current location in the Accelerator Laboratory since 1993. A strong programme of research and development througout the years of IGISOL–3 was motivated by the need to improve selectivity, to increase the level of mass-separated yields and at the same time to reduce isobaric and general contaminants. In parallel, a vigorous scientific programme was pursued, primarily led by the use of the Penning trap and collinear laser spectroscopy stations. The final experiment at IGISOL–3 fittingly used JYFLTRAP, temporarily closing a chapter on an experimental facility which has by now been employed in the measurement of over 250 atomic masses of short-living nuclei, coupled to a variety of spectroscopic setups for beta, gamma and beta-delayed neutron studies. The ground state nuclear structure research of refractory elements at IGISOL is a testament to the on-going developments of not only the ion guide method (and its many variants) but also to the push towards the implementation of novel ion manipulation techniques.

Currently, the IGISOL facility is once again going through an upgrade, one which sees a move to a considerably larger experimental hall, the coupling to not only the K=130 cyclotron but also to a new high intensity 30 MeV proton cyclotron (MCC30), as well as a dramatic improvement in the layout of beam-lines and access for laser light to the target area. The front end of the facility includes a new gas handling system and a full re-design of the ion guide mounting. New modes of operation are planned which will see the application of neutron-induced fission of actinide targets for a variety of nuclear energy-related applications as well as fundamental physics, thus the development of a new gas cell will be required. The facility is currently (Spring 2013) undergoing an intense period of commissioning with first milestones having been reached through implantation experiments, laser ionization both in-gas cell and in-jet, and the first off-line results from the collinear laser spectroscopy station. A current overview of the new IGISOL–4 facility as well as the status of

JYFLTRAP and the FURIOS laser ion source have recently been submitted for publication as part of the EMIS 2012 conference series. Research and development, driven by an outstanding scientific programme, has been a priority at IGISOL over the last three decades. This will continue well into the coming decade and beyond with IGISOL–4.

Acknowledgements The authors wish to thank the Academy of Finland and a variety of EU programmes for providing financial support over the last three decades. The efforts of the cyclotron team, the mechanical and electrical workshops over the years deserve a particular mention. Thanks also to all those who provided discussions and support during the process of writing this manuscript.

References

1. Bohr, N., Lindhard, J.: Electron capture and loss by heavy ions penetrating through matter. Selsk. Mat. Fys. Medd. **28**, 1 (1954)
2. Ghiorso, A., Sikkeland, T., Larsch, A.E., Latimer, R.M.: New element, lawrencium, atomic number 103. Phys. Rev. Lett. **6**, 473 (1961)
3. Stevenson, P.C., Larson, J.T., Leary, J.J.: Proc. Int. Conf. on the Properties of Nuclei Far From the Region of Beta-Stability (CERN 70–30), vol. 1, p. 143 (1970)
4. Nitschke, J.M.: Proc. Int. Conf. on the Properties of Nuclei Far From the Region of Beta-Stability (CERN 70–30), vol. 1, p. 153 (1970)
5. Äystö, J., Valli, K.: Transport efficiency of the helium-jet recoil-transport method with pure helium. Nucl. Instrum. Methods **111**, 531 (1973)
6. Äystö, J., Hillebrand, S., Hellmuth, K.H., Valli, K.: Transport of recoil atoms in a stream of liquid-air-cooled pure helium. Nucl. Instrum. Methods **120**, 163 (1974)
7. Äystö, J., Puumalainen, P., Valli, K.: The carrier-loaded helium-jet transport method. Nucl. Instrum. Methods **115**, 65 (1974)
8. Äystö, J., Rantala, V., Valli, K., Hillebrand, S., Kortelahti, M., Eskola, K., Raunemaa, T.: Efficiency of an on-line isotope separator system employing cooled and NaCl-loaded He-jet methods. Nucl. Instrum. Methods **139**, 325 (1976)
9. Ärje, J., Valli, K.: Helium-jet ion guide for an on-line isotope separator. Nucl. Instrum. Methods **179**, 533 (1981)
10. Ärje, J., Äystö, J., Taskinen, P., Honkanen, J., Valli, K.: Ion guide method for on-line isotope separation. Nucl. Instrum. Methods Phys. Res. B **26**, 384 (1987)
11. Ärje, J.: Charge creation and reset mechanisms in an ion guide isotope separator (IGIS). Phys. Scr. **T3**, 37 (1983)
12. Ärje, J., Äystö, J., Valli, K., Taskinen, P., Honkanen, J., Koponen, V., Hautojärvi, A., Vierinen, K.: JYFL annual report (1982)
13. Ärje, J., Hyvönen, H., Äystö, J., Taskinen, P., Koponen, V., Honkanen, J., Valli, K., Hautojärvi, A., Vierinen, K.: JYFL Annual report (1983–1984)
14. Ärje, J., Äystö, J., Hyvönen, H., Taskinen, P., Koponen, V., Honkanen, J., Hautojärvi, A., Vierinen, K.: Submillisecond on-line mass separation of nonvolatile radioactive elements: an application of charge exchange and thermalization processes of primary recoil ions in helium. Phys. Rev. Lett. **54**, 99 (1985)
15. Ärje, J.: Ion guide method for isotope separator. Ph.D. thesis, University of Jyväskylä (1986)
16. Äystö, J., Ärje, J., Koponen, V., Taskinen, P., Hyvönen, H., Hautojärvi, A., Vierinen, K.: Beta decay of $T_z = -1/2$ nuclides ^{51}Fe and ^{55}Ni: a new approach to on-line isotope separation. Phys. Lett. **B138**, 369 (1984)
17. Lönnroth, T., Äystö, J., Ärje, J., Honkanen, J., Koponen, V., Taskinen, P., Hyvönen, H.: Half-life and configuration of the $1/2^+$ intruder state in ^{203}Bi. Z. Phys. A **319**, 149 (1984)
18. Äystö, J., Taskinen, P., Yoshii, M., Honkanen, J., Jauho, P., Ärje, J., Valli, K.: Separation of fission products by the ion guide fed isotope separator, IGISOL. Nucl. Instrum. Methods Phys. Res. B **26**, 394 (1987)
19. Taskinen, P., Penttilä, H., Äystö, J., Dendooven, P., Jauho, P., Jokinen, A., Yoshii, M.: Efficiency and delay of the fission ion guide for on-line mass separation. Nucl. Instrum. Methods Phys. Res. A **281**, 539 (1989)

20. Äystö, J.: Fast and nonselective on-line mass separation of neutron-rich isotopes produced in proton-induced fission. Nucl. Instrum. Methods Phys. Res. B **40–41**, 489 (1989)
21. Yoshii, M., Hama, H., Takuchi, K., Ishimatsu, T., Shinozuka, T., Fujioka, M., Ärje, J.: The ion-guide isotope separator on-line at the Tohoku University Cyclotron. Nucl. Instrum. Methods Phys. Res. B **26**, 410 (1987)
22. Hama, H., Shinozuka, T., Yoshii, M., Taguchi, K., Kobayashi, K., Suzuki, S., Matsuura, J., Kamiya, T., Fujioka, M.: Precision half-life measurements of $t_Z = 1/2$ nuclei in the $1f_{7/2}$-shell region using IGISOL (1985). Available at http://www.cyric.tohoku.ac.jp/english/report/repo1985/85p07.pdf
23. Deneffe, K., Brijs, B., Huyse, M., Coenen, E., Duppen, P.V., Ärje, J.: LIGISOL: the Leuven ion guide isotope separator on-line at the cyclone cyclotron. I.K.S. Report (1985)
24. Deneffe, K., Brijs, B., Coenen, E., Gentens, J., Huyse, M., Duppen, P.V., Wouters, D.: LIGISOL: the Leuven ion guide isotope separator on-line. Nucl. Instrum. Methods Phys. Res. B **26**, 399 (1987)
25. Deneffe, K., Brijs, B., Huyse, M., Ärje, J., Gentens, J., Duppen, P.V., Wouters, D.: JYFL Annual Report (1985)
26. Huyse, M., Dendooven, P., Gentens, J., Vancraeynest, G., Vandenberghe, P., Duppen, P.V.: Status report of the Leuven ion guide isotope separator on-line (LISOL). Nucl. Instrum. Methods Phys. Res. B **70**, 50 (1992)
27. Iivonen, A., Riikonen, K., Saintola, R., Valli, K., Morita, K.: Focusing ions by viscous drag and weak electric fields in an ion guide. Nucl. Instrum. Methods Phys. Res. A **307**, 69 (1991)
28. Iivonen, A., Kuhalainen, J., Saintola, R., Valli, K., Inamura, T., Koizumi, M., Morita, K., Yoshida, A.: The squeezer ion guide. Nucl. Instrum. Methods Phys. Res. B **70**, 213 (1992)
29. Xu, H., Wada, M., Tanaka, J., Kawakami, H., Katayama, I., Ohtani, S.: A new cooling and focusing device for ion guide. Nucl. Instrum. Methods Phys. Res. A **333**, 274 (1993)
30. Penttilä, H., Dendooven, P., Honkanen, A., Huhta, M., Jauho, P.P., Jokinen, A., Lhersonneau, G., Oinonen, M., Parmonen, J.M., Peräjärvi, K., Äystö, J.: Status report of the Jyväskylä ion guide isotope separator on-line facility. Nucl. Instrum. Methods Phys. Res. B **126**, 213 (1997)
31. Nieminen, A., Huikari, J., Jokinen, A., J.Äystö, Campbell, P., Cochrane, E.C.A.: Beam cooler for low-energy radioactive ions. Nucl. Instrum. Methods Phys. Res. A **469**, 244 (2001)
32. Moore, I.D.: Laser spectroscopic studies at the IGISOL on-line separator. Ph.D. thesis, University of Manchester (2001)
33. Billowes, J., Campbell, P., Cochrane, E.C.A., Cooke, J.L., Dendooven, P., Evans, D.E., Grant, I.S., Griffith, J.A.R., Honkanen, A., Huhta, M., Levins, J.M.G., Liukkonen, E., Oinonen, M., Pearson, M.R., Penttilä, H., Persson, J.R., Richardson, D.S., Tungate, G., Wheeler, P., Zybert, L., Äystö, J.: First collinear laser spectroscopy measurements of radioisotopes from an IGISOL ion source. Nucl. Instrum. Methods Phys. Res. B **126**, 416 (1997)
34. Nieminen, A., Campbell, P., Billowes, J., Forest, D.H., Griffith, J.A.R., Huikari, J., Jokinen, A., Moore, I.D., Moore, R., Tunate, G., Äystö, J.: Cooling and bunching of ion beams for collinear laser spectroscopy. Nucl. Instrum. Methods Phys. Res. B **204**, 563 (2003)
35. Nieminen, A., Campbell, P., Billowes, J., Forest, D.H., Griffith, J.A.R., Huikari, J., Jokinen, A., Moore, I.D., Moore, R., Tungate, G., Äystö, J.: On-line ion cooling and bunching for collinear laser spectroscopy. Phys. Rev. Lett. **88**, 094801 (2002)
36. Mane, E., Billowes, J., Blaum, K., Campbell, P., Cheal, B., Delahaye, P., Flanagan, K.T., Forest, D.H., Franberg, H., Geppert, C., Giles, T., Jokinen, A., Kowalska, M., Neugart, R., Neyens, G., Nörtershäuser, W., Podadera, I., Tungate, G., Vingerhoets, P., Yordanov, D.T.: An ion cooler-buncher for high-sensitivity collinear laser spectroscopy at ISOLDE. Eur. Phys. J. A **42**, 503 (2009)
37. Levy, C.D.P., Pearson, M.R., Morris, G.D., Chow, K.H., Hossain, M.D., Kiefl, R.F., Labbe, R., Lassen, J., MacFarlane, W.A., Parolin, T.J., Saadaoui, H., Smadella, M., Song, Q., Wang, D.: Development of the collinear laser beam line at TRIUMF. Hyperfine Interact. **196**, 287 (2010)
38. Forest, D.H., Cheal, B.: Physics highlights from laser spectroscopy at the IGISOL. Hyperfine Interact. (2012). doi:10.1007/s10751-012-0620-9
39. Cheal, B., Forest, D.H.: Collinear laser spectroscopy techniques at JYFL. Hyperfine Interact. (2012). doi:10.1007/s10751-012-0608-5
40. Huikari, J., Dendooven, P., Jokinen, A., Nieminen, A., Penttilä, H., Peräjärvi, K., Popov, A., Rinta-Antila, S., Äystö, J.: Production of neutron deficient rare isotope beams at IGISOL; on-line and off-line studes. Nucl. Instrum. Methods Phys. Res. B **222**, 632 (2004)
41. Dendooven, P., Hankonen, S., Honkanen, A., Huhta, M., Huikari, J., Jokinen, A., Kolhinen, V.S., Lhersonneau, G., Nieminen, A., Oinonen, M., Penttilä, H., Wang, J.C., Äystö, J.: The fission

ion guide: present performance and future development. In: AIP Conference Proceedings on Nuclear Fission and Fission-Product Spectroscopy: Second International Workshop, vol. 447, p. 135 (1998)
42. Kolhinen, V., Kopecky, S., Eronen, T., Hager, U., Hakala, J., Huikari, J., Jokinen, A., Nieminen, A., Rinta-Antila, S., Szerypo, J., Äystö, J.: JYFLTRAP: a cylindrical penning trap for isobaric beam purification at IGISOL. Nucl. Instrum. Methods Phys. Res. A **528**, 776 (2004)
43. Eronen, T., Kolhinen, V.S., Elomaa, V.V., Gorelov, D., Hager, U., Hakala, J., Jokinen, A., Kankainen, A., Karvonen, P., Kopecky, S., Moore, I.D., Penttilä, H., Rahaman, S., Rinta-Antila, S., Rissanen, J., Saastamoinen, A., Szerypo, J., Weber, C., Äystö, J.: JYFLTRAP: a Penning trap for precision mass spectroscopy and isobaric purification. Eur. Phys. J. A **48**, 46 (2012)
44. Beraud, R., Emsallem, A., Astier, A., Bouvier, R., Duffait, R., Coz, Y.L., Morier, S., Wojtasiewicz, A., Lazarev, Y.A., Shirokovsky, I.V., Izosimov, I.N., Barneoud, D., Genevey, J., Gizon, A., Guglielmini, R., Margotton, G., Vieux-Rochaz, J.L.: Development of a new SARA/IGISOL technique for the study of short-lived products from heavy-ion induced fusion-evaporation reactions. Nucl. Instrum. Methods Phys. Res. A **346**, 196 (1994)
45. Dendooven, P., Beraud, R., Chabanat, E., Emsallem, A., Honkanen, A., Huhta, M., Jicheng, W., Jokinen, A., Lhersonneau, G., Oinonen, M., Penttilä, H., Peräjärvi, K., Äystö, J.: Improved ion guide for heavy-ion fusion-evaporation reactions. Nucl. Instrum. Methods Phys. Res. A **408**, 530 (1998)
46. Oinonen, M., Beraud, R., Canchel, G., Chabanat, E., Dendooven, P., Emsallem, A., Hankonen, S., Honkanen, A., Huikari, J., Jokinen, A., Lhersonneau, G., Miehe, C., Nieminen, A., Novikov, Y., Penttilä, H., Peräjärvi, K., Popov, A., Seliverstov, D.M., Wang, J.C., Äystö, J.: Production of refractory elements close to the Z = N line using the ion-guide technique. Nucl. Instrum. Methods Phys. Res. A **416**, 485 (1998)
47. Beraud, R., Canchel, G., Emsallem, A., Dendooven, P., Huikari, J., Huang, W., Wang, Y., Peräjärvi, K., Rinta-Antila, S., Jokinen, A., Kolhinen, V.S., Nieminen, A., Penttilä, H., Szerypo, J., Äystö, J., Bruyneel, B., Popov, A.: Staus of HIGISOL, a new version equipped with SPIG and electric field guidance. Hyperfine Interact. **132**, 481 (2001)
48. JYFL Internal Report
49. Tordoff, B., Eronen, T., Elomaa, V.V., Gulick, S., Hager, U., Karvonen, P., Kessler, T., Moore, I., Popov, A., Rahaman, S., Rinta-Antila, S., Sonoda, T., Äystö, J.: An ion guide for the production of a low energy ion beam of daughter products of α-emitters. Nucl. Instrum. Methods Phys. Res. B **252**, 347 (2006)
50. Sonnenschein, V., Moore, I.D., Raeder, S., Hakimi, A., Popov, A., Wendt, K.: The search for the existence of ^{229m}Th at IGISOL. Eur. Phys. J. A **48**, 52 (2012)
51. Dendooven, P.: The development and status of the IGISOL technique. Nucl. Instrum. Methods Phys. Res. B **126**, 182 (1997)
52. den Bergh, P.V., Franchoo, S., Gentens, J., Huyse, M., Kudryavtsev, Y.A., Piechaczek, A., Raabe, R., Reusen, I., Duppen, P.V., Vermeeren, L., Wöhr, A.: The SPIG, improvement of the efficiency and beam quality of an ion-guide based on-line isotope separator. Nucl. Instrum. Methods Phys. Res. B **126**, 194 (1996)
53. Karvonen, P., Penttilä, H., Äystö, J., Billowes, J., Campbell, P., Elomaa, V.V., Hager, U., Hakala, J., Jokinen, A., Kessler, T., Kankainen, A., Moore, I.D., Peräjärvi, K., Rahaman, S., Rinta-Antila, S., Rissanen, J., Ronkainen, J., Saastamoinen, A., Sonoda, T., Tordoff, B., Weber, C.: Upgrade and yields of the IGISOL facility. Nucl. Instrum. Methods Phys. Res. B **266**, 4454 (2008)
54. Penttilä, H.: The layout of the IGISOL 3 facility (3 December 2003–29 June 2010). Hyperfine Interact. (2012). doi:10.1007/s10751-012-0607-6
55. Moore, I.D., Billowes, J., Campbell, P., Eronen, T., Geppert, C., Jokinen, A., Karvonen, P., Kessler, T., Marsh, B., Nieminen, A., Penttilä, H., Rinta-Antila, S., Sonoda, T., Tordoff, B., Wendt, K., Äystö, J.: Laser ion source development at IGISOL. AIP Conf. Proc. **831**, 511 (2006)
56. Marsh, B.: In-source laser resonance ionization at ISOL facilities. Ph.D. thesis, University of Manchester (2007)
57. Karvonen, P.: Fission yield studies with SPIG-equipped IGISOL: a novel method for nuclear data measurements. Ph.D. thesis, University of Jyväskylä (2010)
58. Reponen, M., Moore, I.D., Pohjalainen, I., Kessler, T., Karvonen, P., Kurpeta, J., Marsh, B., Piszczek, S., Sonnenschein, V., Äystö, J.: Gas jet studies towards and optimization of the IGISOL LIST method. Nucl. Instrum. Methods Phys. Res. A **A35**, 24 (2011)
59. Karvonen, P., Moore, I.D., Sonoda, T., Kessler, T., Penttilä, H., Peräjärvi, K., Ronkanen, P., Äystö, J.: A sextupole ion beam guide to improve the efficiency and beam quality at IGISOL. Nucl. Instrum. Methods Phys. Res. B **266**, 4794 (2008)

60. Peräjärvi, K., Cerny, J., Hakala, J., Huikari, J., Jokinen, A., Karvonen, P., Kurpeta, J., Lee, D., Moore, I., Penttilä, H., Popov, A., Äystö, J.: New ion-guide for the production of beams of neutron-rich nuclei between Z =20–28. Nucl. Instrum. Methods. Phys. Res. A **546**, 418 (2005)
61. Mishin, V.I., Fedoseyev, V.N., Kluge, H.J., Letokhov, V.S., Ravn, H.L., Scheerer, F., Shirakabe, Y., Sundell, S., Tengblad, O.: Chemically selective laser ion-source for the CERN-ISOLDE on-line mass separator facility. Nucl. Instrum. Methods Phys. Res. B **73**, 550 (1993)
62. Fedosseev, V.N., Berg, L.E., Fedorov, D.V., Fink, D., Launila, O.J., Losito, R., Marsh, B.A., Rossel, R.E., Rothe, S., Seliverstov, M.D., Sjödin, A.M.: Upgrade of the resonance ionization laser ion source at ISOLDE on-line isotope separator facility: new lasers and new ion beams. Rev. Sci. Instrum. **83**, 02A903 (2012)
63. Lassen, J., Bricault, P., Dombsky, M., Lavoie, J.P., Geppert, C., Wendt, K.: Resonant ionization laser ion source project at TRIUMF. Hyperfine Interact. **162**, 69 (2005)
64. Lecesne, N., Alves-Conde, R., Coterreau, E., Olveira, F.D., Duboise, M., Flambard, J.L., Franberg, H., Gottwald, T., Jardin, P., Lassen, J., Blanc, F.L., Leroy, R., Mattolat, C., Olivier, A., Pacquet, J.Y., Pichard, A., Rothe, S., Saint-Laurent, M.G., Wendt, K.: GISELE: a resonant ionization laser ion source for the production of radioactive ions at GANIL. Rev. Sci. Instrum. **81**, 02A910 (2010)
65. Barzakh, A.E., Fedorov, D.V., Ivanov, V.S., Molkanov, P.L., Panteleev, V.N., Volkov, Y.M.: New laser setup for the selective isotope production and investigation in a laser ion source at the IRIS (Investigation of Radioactive Isotopes on Synchrocyclotron) facility. Rev. Sci. Instrum. **83**, 02B306 (2012)
66. Liu, Y., Gottwald, T., Havener, C.C., Howe, J.Y., Kiggans, J., Mattolat, C., Vane, C.R., Wendt, K., Beene, J.R.: Laser ion source development at Holifield radioactive ion beam facility. Rev. Sci. Instrum. **83**, 02A904 (2012)
67. The Tandem-Alto Facility. http://ipnweb.in2p3.fr/tandem-alto/index_E.html
68. Scarpa, D., Vasquez, J., Tomaselli, A., Grassi, D., Biasetto, L., Cavazza, A., Corradetti, S., Manzolaro, M., Montano, J., Andrighetto, A., Prete, G.: Studies for aluminium photoionization in hot cavity for the selective production of exotic species project. Rev. Sci. Instrum. **83**, 02B317 (2012)
69. Kudryavtsev, Y., Andrzejewski, J., Bijnens, N., Franchoo, S., Gentens, J., Huyse, M., Piechaczek, A., Szerypo, J., Reusen, I., Duppen, P.V., den Bergh, P.V., Vermeeren, L., Wauters, J., Wöhr, A.: Beams of short lived nuclei produced by selective laser ionization in a gas cell. Nucl. Instrum. Methods Phys. Res. B **114**, 350 (1996)
70. Kudryavtsev, Y., Bruyneel, B., Huyse, M., Gentens, J., den Bergh, P.V., Duppen, P.V., Vermeeren, L.: A gas cell for thermalizing, storing and transporting radioactive ions and atoms. Part I: off-line studies with a laser ion source. Nucl. Instrum. Methods Phys. Res. B **179**, 412 (2001)
71. Drouart, A., Amthor, A.M., Boutin, D., Delferriere, O., Duval, M., Manikonda, S., Nolen, J., Payet, J., Savajols, H., Stodel, M.H., Uriot, D.: The Super Separator Spectrometer (S^3) for SPIRAL2 stable beams. Nucl. Phys. A **834**, 747c (2010)
72. Wada, M., Takamine, A., Sonoda, T., Schury, P., Okada, K., Collaboration, S.: Precision laser spectroscopy of Be isotopes and prospects for SLOWRI facility at RIKEN. Hyperfine Interact. **196**, 43 (2010)
73. Sonoda, T., Wada, M., Tomita, H., Sakamoto, C., Takatsuka, T., Furukawa, T., Iimura, H., Ito, Y., Kubo, T., Matsuo, Y., Mita, H., Naimi, S., Nakamura, S., Noto, T., Schury, P., Shinozuka, T., Wakui, T., Miyatake, H., Jeong, S., Ishiyama, H., Watanabe, Y.X., Hirayama, Y., Okada, K., Takamine, A.: Development of a resonant laser ionization gas cell for high-energy, short-lived nuclei. Nucl. Instrum. Methods Phys. Res. B **295**, 1 (2013)
74. Moore, I.D., Nieminen, A., Billowes, J., Campbell, P., Geppert, C., Jokinen, A., Kessler, T., Marsh, B., Penttilä, H., Rinta-Antila, S., Tordoff, B., Wendt, K.D.A., Äystö, J.: Development of a laser ion source at IGISOL. J. Phys. G: Nucl. Part. Phys. **31**, S1499 (2005)
75. Reponen, M., Moore, I.D., Kessler, T., Pohjalainen, I., Rothe, S., Sonnenschein, V.: Laser developments and resonance ionization spectroscopy and IGISOL. Eur. Phys. J. A **48**, 45 (2012)
76. Blaum, K., Geppert, C., Kluge, H.J., Mukherjee, M., Schwarz, S., Wendt, K.: A novel scheme for a highly selective laser ion source. Nucl. Instrum. Methods Phys. Res. **B204**, 331 (2003)
77. Peräjärvi, K., Dendooven, P., Huikari, J., Jokinen, A., Kolhinen, V.S., Nieminen, A., Äystö, J.: Transport of ions in ion guides under flow and diffusion. Nucl. Instrum. Methods Phys. Res. A **449**, 427 (2000)
78. Kessler, T.: Development and application of laser technologies at radioactive ion beam facilities. Ph.D. thesis, University of Jyväskylä (2008)

79. Koyanagi, G.K., Caraiman, D., Blagojevic, V., Bohme, D.K.: Gas-phase reactions of transition-metal ions with molecular oxygen: room temperature kinetics and periodicities in reactivity. J. Phys. Chem. A **106**, 4581 (2002)
80. Kessler, T., Moore, I.D., Kudryavtsev, Y., Peräjärvi, K., Popov, A., Ronkanen, P., Sonoda, T., Tordoff, B., Wendt, K.D.A., Äystö, J.: Off-line studies of the laser ionization of yttrium at the IGISOL facility. Nucl. Instrum. Methods Phys. Res. B **266**, 681 (2008)
81. Huyse, M., Facina, M., Kudryavtsev, Y., Duppen, P.V., Collaboration, I.: Intensity limitations of a gas cell for stopping, storing and guiding of radioactive ions. Nucl. Instrum. Methods Phys. Res. B **187**, 535 (2002)
82. Facina, M., Bruyneel, B., Dean, S., Gentens, J., Huyse, M., Kudryavtsev, Y., den Bergh, P.V., Duppen, P.V.: A gas cell for thermalizing, storing and transporting radioactive atoms and ions. Part II: on-line studies with a laser ion source. Nucl. Instrum. Methods Phys. Res. B **226**, 401 (2004)
83. Moore, I.D., Kessler, T., Sonoda, T., Kudryavtsev, Y., Peräjärvi, K., Popov, A., Wendt, K.D.A., Äystö, J.: A study of on-line gas cell processes at IGISOL. Nucl. Instrum. Methods Phys. Res. B **268**, 657 (2010)
84. Kudryavtsev, Y., Cocolios, T.E., Gentens, J., Huyse, M., Ivanov, O., Pauwels, D., Sonoda, T., den Bergh, P.V., Duppen, P.V.: Dual chamber laser ion source at LISOL. Nucl. Instrum. Methods Phys. Res. B **267**, 2908 (2009)
85. Reponen, M.: Resonance laser ionization developments for IGISOL–4. Ph.D. thesis, University of Jyväskylä (2012)
86. Savard, G., Clark, J., Boudreau, C., Buchinger, F., Crawford, J.E., Geissel, H., Greene, J.P., Gulick, S., Heinz, A., Lee, J.K.P., Levand, A., Maier, M., Münzenberg, G., Scheidenberger, C., Seweryniak, D., Sharma, K.S., Sprouse, G., Vaz, J., Wang, J.C., Zabransky, B.J., Zhou, Z., and the S248 collaboration: Development and operation of gas catchers to thermalize fusion-evaporation and fragmentation products. Nucl. Instrum. Methods Phys. Res. B **204**, 582 (2003)
87. Takahashi, N., Shigematsu, T., Shimizu, S., Horie, K., Hirayama, Y., Izumi, H., Shimoda, T.: Impurities in superfluid helium detected via radioactivity. Phys. B **89**, 284 (2000)
88. Huang, W.X., Dendooven, P., Gloos, K., Takahashi, N., Arutyunov, K., Pekola, J.P., Äystö, J.: Transport and extraction of radioactive ions stopped in superfluid helium. Nucl. Instrum. Methods Phys. Res. B **204**, 592 (2003)
89. Huang, W.X., Dendooven, P., Gloos, K., Takahashi, N., Pekola, J.P., Äystö, J.: Extraction of radioactive positive ions across the surface of superfluid helium: A new method to produce cold radioactive nuclear beams. Europhys. Lett. **63**, 687 (2003)
90. Dendooven, P., Purushothaman, S., Gloos, K., Äystö, J., Takahashi, N., Huang, W.X.: Radioactive ions and atoms in superfluid helium. AIP Conf. Proc. **831**, 439 (2006)
91. Purushothaman, S., Peräjärvi, K., Ranjan, M., Saastamoinen, A., Gloos, K., Takahashi, N., Dendooven, P.: Positive ion extraction across the superfluid-vapor helium interface. J. Phys. Conf. Ser. **150**, 032086 (2009)
92. Ranjan, M., Purushothaman, S., Dickel, T., Geissel, H., Plass, W.R., Schaefer, D., Scheidenberger, C., de Walle, J.V., Weick, H., Dendooven, P.: New stopping cell capabilities: RF carpet performance at high gas density and cryogenic operation. Europhys. Lett. **96**, 52001 (2011)
93. Dendooven, P., Purushothaman, S., Gloos, K.: On a cryogenic noble gas ion catcher. Nucl. Instrum. Methods Phys. Res. A **558**, 580 (2006)
94. Purushothaman, S., Dendooven, P., Moore, I., Penttilä, H., Ronkainen, J., Saastamoinen, A., Äystö, J., Peräjärvi, K., Takahashi, N., Gloos, K.: Cryogenic helium as stopping medium for high-energy ions. Nucl. Instrum. Methods Phys. Res. B **266**, 4488 (2008)
95. Purushothaman, S.: Superfluid helium and cryogenic noble gases as stopping media for ion catchers. Ph.D. thesis, University of Groningen (2008)
96. Moore, I.D.: New concepts for the ion guide technique. Nucl. Instrum. Methods Phys. Res. B **266**, 4434 (2008)
97. Saastamoinen, A.: Studies of $T_z = -3/2$ nuclei of astrophysical interest. Ph.D. thesis, University of Jyväskylä (2011)
98. Saastamoinen, A., Moore, I.D., Ranjan, M., Dendooven, P., Penttilä, H., Peräjärvi, K., Popov, A., Äystö, J.: Characterization of a cryogenic ion guide at IGISOL. Nucl. Instrum. Methods Phys. Res. A **685**, 70 (2012)

DOI 10.1007/978-94-007-5555-0_3
Reprinted from *Hyperfine Interactions* Volume 223, Issues 1-3, 2014
and *The European Physical Journal A* Volume 48, Issue 4, 2012

THE EUROPEAN PHYSICAL JOURNAL A

Special Article – Tools for Experiment and Theory

JYFLTRAP: a Penning trap for precision mass spectroscopy and isobaric purification

T. Eronen[a], V.S. Kolhinen, V.-V. Elomaa[b], D. Gorelov, U. Hager[c], J. Hakala, A. Jokinen, A. Kankainen, P. Karvonen, S. Kopecky[d], I.D. Moore, H. Penttilä, S. Rahaman[e], S. Rinta-Antila, J. Rissanen, A. Saastamoinen, J. Szerypo[f], C. Weber[f], and J. Äystö

University of Jyväskylä, P.O. Box 35 (YFL), FI-40014 University of Jyväskylä, Finland

Received: 31 January 2012
Published online: 18 April 2012 – © Società Italiana di Fisica / Springer-Verlag 2012
Communicated by E. De Sanctis

Abstract. In this article a comprehensive description and performance of the double Penning-trap setup JYFLTRAP will be detailed. The setup is designed for atomic mass measurements of both radioactive and stable ions and additionally serves as a very high-resolution mass separator. The setup is coupled to the IGISOL facility at the accelerator laboratory of the University of Jyväskylä. The trap has been online since 2003 and it was shut down in the summer of 2010 for relocation to the upgraded IGISOL facility. Numerous atomic mass and decay energy measurements have been performed using the time-of-flight ion-cyclotron resonance technique. The trap has also been used in several decay spectroscopy experiments as a high-resolution mass filter.

1 Introduction

Ion manipulation and trapping techniques have opened a new powerful way to study ground-state properties of stable and of short-living ions. Penning-trap mass spectrometry has become a routine technique for high-precision mass measurements and one can measure atomic masses of stable ions with a relative uncertainty $\delta m/m < 10^{-10}$ and short-living ions with $\delta m/m < 10^{-8}$ [1]. Uncertainties of this level allows to investigate many physics phenomena through atomic mass [2].

Stable-ion traps such as SMILETRAP in Stockholm [3], the trap setup at Florida State University [4] and the University of Washington Mass spectrometer (now at the Max-Planck-Institute for nuclear physics in Heidelberg, Germany) [5,6] have measured several masses of stable nuclei with precision of better than 10^{-10}. Some of the high-precision results are actually used as reference masses for on-line Penning-trap setups. Also, Penning traps for anti-matter studies (*e.g.*, ATRAP [7] and ASACUSA [8]) have performed high-precision mass measurements of anti-matter ions.

Most of the radioactive ion beam facilities that can provide beams of short-living ions have implemented a trap to be a part of their setup. Presently there are several functioning traps around the world which perform mass measurements with radioactive ions like CPT at ANL [9], ISOLTRAP at CERN [10], LEBIT at MSU [11], MLLTRAP at LMU [12], SHIPTRAP at GSI [13], TITAN at TRIUMF [14], TRIGATRAP at the Mainz research reactor [15] and JYFLTRAP at the University of Jyväskylä [16,17]. These so-called on-line traps are typically very fast (they can access nuclei with short half-lives; see *e.g.*, [18]) and also efficient in terms of very low production rates (see, *e.g.*, [19]).

In this article the JYFLTRAP double Penning-trap setup will be described in detail. In short, the setup is designed to perform high-resolution beam purification and mass measurements with both stable- and radioactive-ion beams created with the Ion Guide Isotope Separator On-Line (IGISOL) technique [20,21].

[a] *Present address*: Max-Planck-Institut für Kernphysik, Saupfercheckweg 1, 69117 Heidelberg, Germany;
e-mail: tommi.eronen@jyu.fi
[b] *Present address*: Turku PET Centre, Accelerator Laboratory, Åbo Akademi University, 20500 Turku, Finland.
[c] *Present address*: Colorado School of Mines, Golden, CO, USA.
[d] *Present address*: European Commission - Joint Research Centre, Institute for Reference Materials and Measurements, Retieseweg 11, B-2440 Geel, Belgium.
[e] *Present address*: Los Alamos National Laboratory, Los Alamos, NM 87545, USA.
[f] *Present address*: Fakultät für Physik, LMU München and Maier-Leibnitz Laboratory, Am Coulombwall 1, 85748 Garching, Germany.

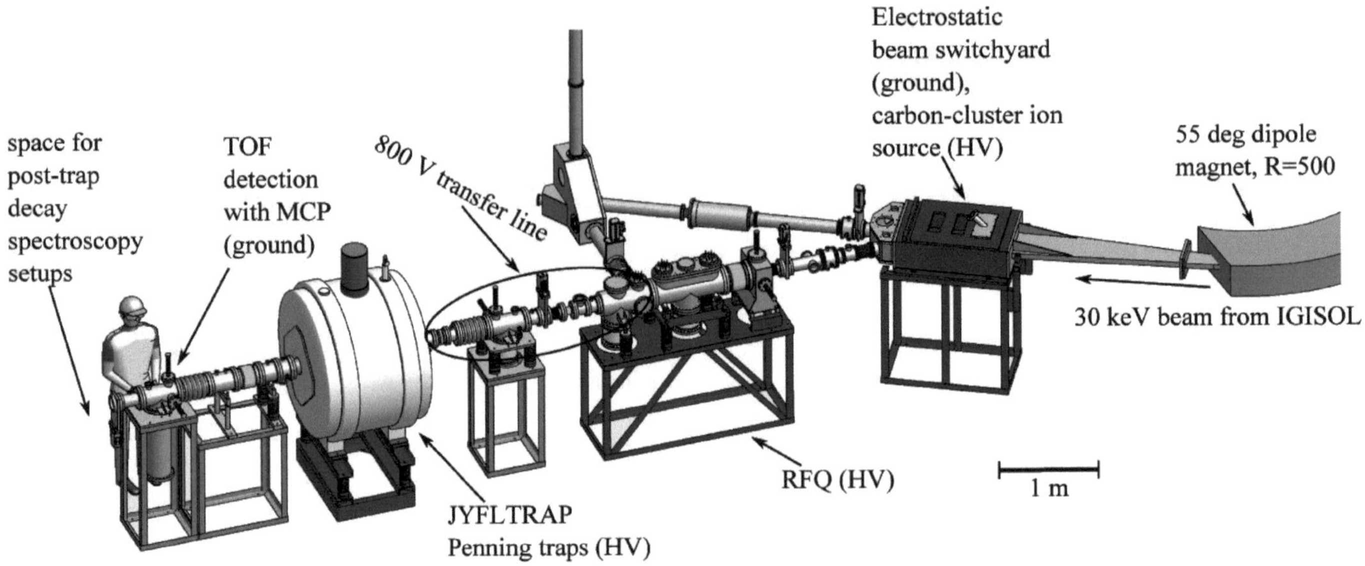

Fig. 1. The experimental area of the IGISOL facility which is mostly occupied by the radiofrequency quadrupole (RFQ) cooler-buncher and the JYFLTRAP Penning-trap setup. Devices operated in high voltage are marked with (HV). The 30 keV ion beam from IGISOL (incident from right) is mass-separated with a 55° dipole magnet and deflected with an electrostatic 30° deflector to left towards the RFQ and JYFLTRAP setups. The setup is described in detail in sect. 3.

2 Basic principles

2.1 Ion beam production and separation

The JYFLTRAP setup receives beams from IGISOL, which is extensively discussed elsewhere in this special issue of the European Physical Journal A. The radioactive ions created either by fission or fusion reactions are stopped in helium gas, extracted by using electric fields and a helium gas jet via a sextupole ion guide (SPIG) [22], and are finally electrostatically accelerated to $30q$ keV of energy, where q is the charge state of ions (usually $q = 1$). Alternatively, an electric discharge ion source can be used to create ions of stable isotopes (see, *e.g.*, ref. [23] for more details). The extracted ion beam is mass-separated with a 55° dipole magnet (see fig. 1) allowing for a mass resolving power ($M/\Delta M$) of about 500.

2.2 Radiofrequency quadrupole cooler-buncher

The mass-separated ion beam from IGISOL is injected into a radiofrequency quadrupole cooler-buncher (RFQ) [24]. The $30q$ keV ion beam is electrostatically decelerated to ~ 100 eV of energy by having the whole RFQ on a high-voltage (HV) platform. The decelerated beam then enters inside the quadrupole rod structure filled with helium buffer gas at low pressure. The ions are cooled by collisions with buffer gas atoms and collected in a potential well. The cooled ions are periodically released to the Penning traps as short, 10–15 μs, bunches.

2.3 Principle of a Penning trap

In general, a Penning trap is a device to confine charged particles to a small volume with a static quadrupolar electric field and a homogeneous magnetic field. The electric potential in cylindrical coordinates (z, ρ) is of the form

$$V(z,\rho) = \frac{U_0}{4d^2}\left(2z^2 - \rho^2\right), \qquad (1)$$

where U_0 is the potential difference between the ring and endcap electrodes and $d = \sqrt{2z_0^2 + r_0^2}$ is the characteristic trap parameter defined by the trap geometry: $2z_0$ is the distance between endcap electrodes and r_0 is the inner radius of the ring electrode.

The trapped charged particles exhibit three different eigenmotions. One of the motions, called the axial motion with frequency ν_z, occurs along the magnetic field lines. The other two motions are in a plane perpendicular to the magnetic field and are commonly called radial motions. These can be distinguished by their frequencies ν_- and ν_+. The magnetron motion with the lower frequency ν_- is almost mass independent. The second motion described by the reduced cyclotron frequency ν_+, exhibits a mass dependence, and the sum of the two frequencies is given by the cyclotron frequency ν_c:

$$\nu_- + \nu_+ = \nu_c = \frac{1}{2\pi}\frac{q}{m}B, \qquad (2)$$

where q and m are the charge and mass of the ion, respectively, and B is the magnetic field. For an ion with $A/q = 100$ trapped in a 7 T magnetic field via a Penning trap (potential depth of the order of 100 V), the reduced

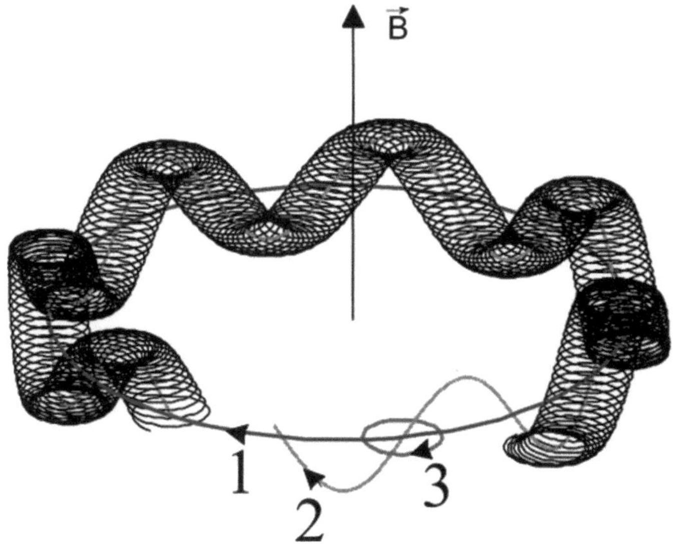

Fig. 2. (Color online) Motion of a positively charged ion in a Penning trap with magnetic field lines pointing up. The figure shows a track of an ion having component of all three motions. The blue and red lines show motion of an ion having only magnetron (1) or reduced cyclotron (3) motion, respectively, while the green track is for an ion with magnetron and axial motion (2). For clarity, the frequencies are not to scale.

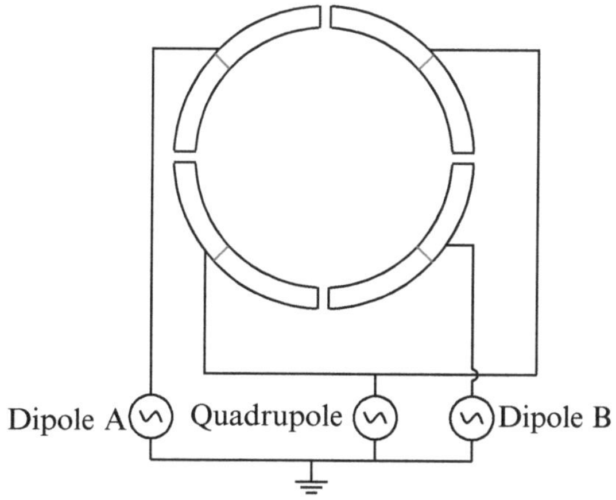

Fig. 3. The four-fold split ring electrode (top view). Each quadrant is formed by interconnecting two octants. At JYFLTRAP, two of the quadrants are used for quadrupole excitation and the two other for dipolar excitations: Dipole A for magnetron (ν_-) and Dipole B for modified cyclotron (ν_+) motion excitation.

cyclotron frequency, the magnetron frequency, and the axial oscillation frequency are of the order of $\nu_+ \approx 1\,\text{MHz}$, $\nu_- \approx 2\,\text{kHz}$, and $\nu_z \approx 50\,\text{kHz}$. It should be noted that eq. (2) is only valid in an ideal case where the quadrupolar electric field is fully harmonic and the magnetic field perfectly homogeneous. In the case of a real Penning trap a more robust relationship is the invariance theorem [25,26]

$$\nu_c^2 = \nu_-^2 + \nu_+^2 + \nu_z^2. \tag{3}$$

In fig. 2 the three eigenmotions in a Penning trap are illustrated.

One important aspect of many on-line Penning traps such as ISOLTRAP, SHIPTRAP, MLLTRAP and JYFLTRAP is ion motion damping in a gas-filled trap. Introducing buffer gas into the Penning trap will lead to a modification of the ion's motion [27]. The interaction between the gas and the ions will decrease the oscillation amplitude in the axial direction and reduce the radius of the modified cyclotron motion. However, the radius of the magnetron motion will increase.

2.4 Excitation of the ion motion in a Penning trap

The operation principle of a Penning trap is based on the manipulation of the trap eigenmotions. This is achieved by applying RF signals to different trap electrodes. In the case of JYFLTRAP, the ring electrodes are eight-fold segmented for exciting the radial eigenmotions. The effect of these excitations to the motion of trapped ions depends on the excitation mode and the applied parameters: the frequency, the amplitude, the duration and the envelope of the RF field. At JYFLTRAP, the RF electric fields are applied either in a dipolar or in a quadrupolar mode.

In the dipolar excitation the frequency is applied to the two opposite segments of the central ring electrode in such a way that the segments have the same frequency and amplitude, but opposite phase. At JYFLTRAP, a simplified excitation scheme is used, where the RF signal is fed only to one quadrant (two neighboring octants) of the ring electrode, while applying a static voltage to the others. Dipolar excitations at eigenfrequencies can be used to excite one particular eigenmotion without affecting the other two. Typically, the dipolar excitation is applied at the mass-independent magnetron frequency ν_- or at the mass-dependent reduced cyclotron frequency ν_+ to enlarge the radius of the corresponding motion, *i.e.*, to move ions away from the trap center.

A quadrupolar excitation can be achieved by using four segments and applying voltages in such a way that opposite segments receive the same frequency, amplitude and phase, while the two other segments, that are in a 90° angle relative to the two first ones, receive the same frequency and amplitude but the opposite phase. Here also a simplified excitation scheme is used at JYFLTRAP by connecting two opposite quadrants (two octants) to the same RF signal, while applying a static voltage to the two other quadrants. This quadrupolar type excitation is usually called as a side-band excitation. When one applies quadrupolar excitation with the mass-dependent cyclotron frequency $\nu_{\text{RF}} = \nu_c = \frac{1}{2\pi}\frac{q}{m}B$ the two radial eigenmotions are bound together and there is a continuous conversion from one motion to the other. Figure 3 illustrates the connections of the ring electrode during dipolar and quadrupolar excitations. A more detailed description of the operation principles of a Penning trap can be found, for example, in ref. [28].

2.5 Time-of-flight ion-cyclotron resonance technique

At JYFLTRAP, the cyclotron frequency (ν_c) is determined with the time-of-flight ion-cyclotron resonance (TOF-ICR) technique [29,30] which is based on probing the cyclotron frequency in the trap and measuring the flight time of the ions from the trap to a microchannel plate (MCP) detector located outside the strong magnetic field, see fig. 1. This technique provides a universal and fast way to perform mass measurements with a relatively low number of ions.

Initially, a small number of ions (ideally just one since larger number induces frequency shifts due to their mutual interaction) is injected into the trap. Next, a dipolar excitation with the magnetron frequency ν_- is applied which leads to an increased magnetron orbit of all ions. Subsequently quadrupolar excitation is then used to mass-selectively convert the magnetron motion into a modified cyclotron motion. Due to this conversion from the low-frequency magnetron motion to the high-frequency modified cyclotron motion (the difference in the frequency can be as large as 10^5), the radial energy E_r of the ions will be increased. Larger radial energy leads to stronger axial acceleration of ions in the gradient of the magnetic field after the extraction according to the equation

$$\boldsymbol{F} = -\boldsymbol{\mu}(\boldsymbol{\nabla} \cdot \boldsymbol{B}) = -\frac{E_r}{B_0}\frac{\partial B(z)}{\partial z}\hat{z}, \quad (4)$$

where $\boldsymbol{\mu} = (E_r/B_0)\hat{z}$ is the magnetic moment of the ion; B_0 is the magnetic field at precision trap and E_r radial energy of the ion. The measurement procedure is then done by scanning the quadrupolar excitation frequency ν_{RF} around the cyclotron frequency ν_c and determining the frequency resulting in the shortest flight time from the trap to the MCP detector. The flight time can be calculated by using the formula

$$T(\omega) = \int_0^{z'} \sqrt{\frac{m}{2(E_0 - qU(z) - \mu B(z))}}\mathrm{d}z, \quad (5)$$

where E_0 is the initial axial kinetic energy of the ion, $U(z)$ is the electrostatic potential and $B(z)$ is the magnetic field along the flight path [30]. Figure 4 shows a typical TOF-ICR spectrum. The conversion from magnetron motion to modified cyclotron motion is periodic and the conversion rate depends on the excitation time T_{RF} and the amplitude U_{RF}. To achieve full conversion after the excitation in resonance the values T_{RF} and U_{RF} must be carefully chosen. Once the values are experimentally found, their product is kept constant:

$$T_{RF}U_{RF} = \mathrm{const.} \quad (6)$$

The value of this constant depends of the trap geometry (see eq. (1)) and has been experimentally determined to be 11.2 mVs for JYFLTRAP. This full conversion happens only at the frequency $\nu_{RF} = \nu_c = \nu_+ + \nu_-$. At other frequencies the conversion is only partial. The line width of the resonance is inversely proportional to the excitation time T_{RF}, *i.e.*, longer excitation time gives a better mass resolving power.

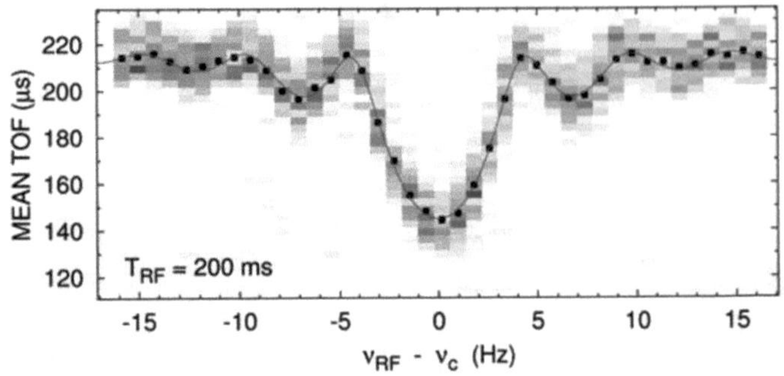

Fig. 4. A time-of-flight ion-cyclotron resonance curve for $^{54}Co^+$ ions ($T_{1/2} \approx 200$ ms). An excitation time T_{RF} of 200 ms was used. The pixels represent detected ions; the shading is proportional to the number of detected ions.

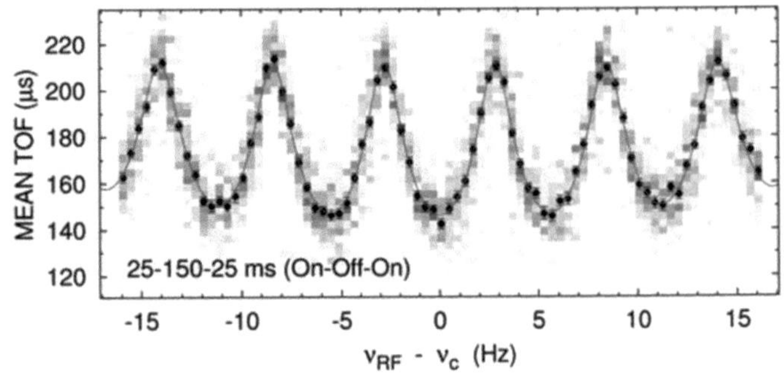

Fig. 5. A time-of-flight ion-cyclotron resonance curve for $^{54}Co^+$ ions ($T_{1/2} \approx 200$ ms) obtained with Ramsey's method of time-separated oscillatory fields. An excitation time pattern of 25-150-25 ms (On-Off-On) was used. It should be noted that this figure has twice the amount of scanned frequency points than fig. 4.

2.5.1 Excitation with time-separated oscillatory fields

Excitation with time-separated oscillatory fields was introduced by N.F. Ramsey [31] and applied in Penning-trap mass spectrometry for the first time at ISOLTRAP [32]. Once the analytical form for the spectrum shape became available [33,34], this method has been routinely used in many trap facilities. Typically the time-separated oscillatory fields method is performed with two equally long RF-on periods interrupted with a certain time duration. It is important to retain phase coherence for the two RF-on periods. With the same statistics typically a factor of 2 to 3 better precision is obtained than with the conventional cyclotron frequency determination described previously. The main peak is 40% narrower but most of the enhancement is due to fact that more frequency scan points lie on high-slope parts of the curve than in the conventional resonance. Typically the quadrupole RF field is switched on for 20–50 ms, then off for 150–750 ms (mostly depending on the half-life of the ion of interest), and finally switched on for another 20–50 ms. Naturally, any excitation scheme must fulfill eq. (6) for "RF-on" periods in order for ions to undergo one full conversion from magnetron to modified cyclotron motion. A typical TOF-ICR curve obtained with time-separated oscillatory fields technique is shown in fig. 5. In order to unambiguously assign the center fringe

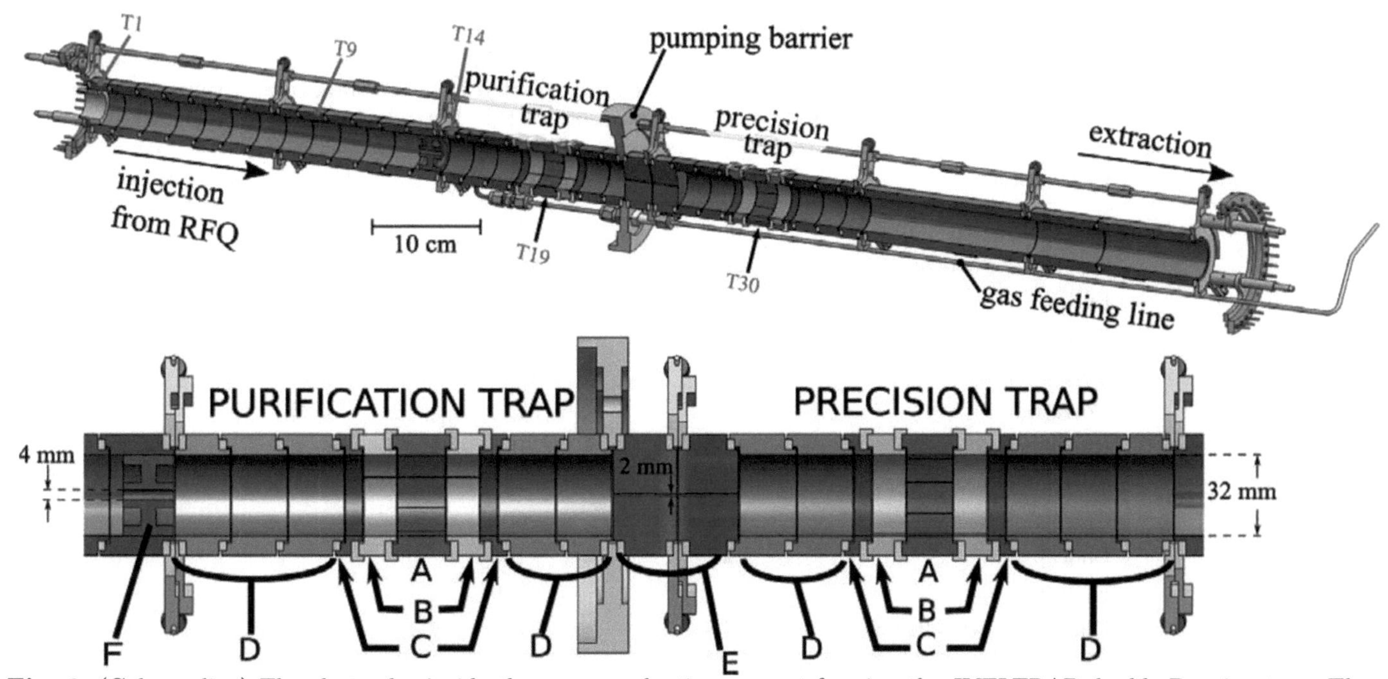

Fig. 6. (Color online) The electrodes inside the superconducting magnet forming the JYFLTRAP double Penning trap. The center of the trap is expanded for clarity. The 8-fold split ring electrodes are marked with A, the two-fold split inner correction electrodes with B, outer correction electrodes with C and the endcap electrodes with D. For reference, the trap electrodes are indexed from left to right with T1, T2, and so on. The ring electrodes are marked T19 and T30. The electrodes restricting the gas flow out are marked with E and F; their inner diameters are 2 mm and 4 mm, respectively.

corresponding to ν_c, a conventional TOF-ICR curve can be recorded.

One important application employing ion motion excitation with time-separated oscillatory fields is the so-called "Ramsey cleaning technique" [35] which is used in high-resolution mass separation. This method has been developed at JYFLTRAP and is described in detail in sect. 5.1.

3 The JYFLTRAP experimental setup

The JYFLTRAP consists of two Penning traps located inside the same superconducting solenoid. The ions are injected from the RFQ (see ref. [24] for more details). The RFQ cooler-buncher and JYFLTRAP are on a single 30 kV high-voltage platform. This way, the ion beam incident from IGISOL can be electrostatically slowed down. The ion bunches from the RFQ are ideal to be injected into the JYFLTRAP setup: they have energy and temporal spreads of less than 1 eV and 15 μs, respectively. The JYFLTRAP setup is shown in fig. 1. Marked with HV are the devices that are operated on a high-voltage platform. The ions are transferred from the RFQ to the Penning traps as an 800 eV ion beam. In the following subsections technical details of the JYFLTRAP setup are given.

3.1 Superconducting magnet

The most visible part of the JYFLTRAP setup is the superconducting magnet that creates the magnetic field of the Penning traps. It is an actively screened 7.0 T superconducting solenoid manufactured by Magnex Scientific Ltd. in the UK. The magnet has a 160 mm diameter warm bore. The generated field is fine-tuned both with superconducting shimming coils and with ferromagnetic metal strips placed around the bore tube in order to create two homogeneous 1 cm^3 field regions 10 cm apart from the center of the magnet. The relative homogeneity of the magnetic field ($\Delta B/B$) has been ≈ 0.4 ppm in both traps after restarting the magnet in 2007. The superconducting solenoid is placed on a high-voltage platform along with all the trap electronics.

Table 1. Typical voltages used at the injection ion optics and at the injection section of the trap. Voltages are given with respect to the purification trap ring electrode potential. Transfer section refers to the beam line section between RFQ and trap and Injection 1...3 to three cylindrical electrodes in front of the first trap electrode T1. See also table 2.

Electrode	Voltage (V)	Beam energy qeV
Transfer section	−800	834
Injection 1	−800	834
Injection 2	−390	423
Injection 3	−350	384
T1-T10	−340	374
T11-T12	−30	64

Table 2. Axial lengths of the electrodes in the purification trap (in addition, there is a 0.5 mm gap between each electrode) together with the operating voltages for the three modes of operation. The "trapping" denotes voltages used when the ions are stored in the trap and "Inj." and "Extr." when ions are being injected or extracted from the trap, respectively. The potentials are given with respect to the ring electrode potential. See fig. 6 for positions of the electrodes.

Electrode	Description	L (mm)	trapping (V)	Inj. (V)	Extr. (V)
T13	Diaphragm ($d = 4$ mm)	24	+100	−20	
T14	Gas feeding endcap	21.5	+100	−20	
T15	Endcap	22	+100	−20	
T16	Endcap	22	+100	+30	
T17	Correction 2	6.7	+66	+20	
T18	Correction 1	12.8	+17	+20	
T19	Ring	18.5	0	0	0
T20	Correction 1	12.8	+17		−1
T21	Correction 2	6.7	+66		−5
T22	Endcap	22	+100		−5
T23	Endcap	22	+100		−5
T24	Diaphragm ($d = 2$ mm)	24.5	+100		−5

The stainless-steel (grade 316L) vacuum tube inside the bore is mounted with adjustable fasteners allowing alignment of the beam tube with respect to the magnetic field axis. The alignment was performed by inserting an electron source to the center of the magnet and guiding the two beams of electrons following the magnetic field lines through a set of narrow collimators placed on both sides of the trap center. The same alignment procedure was later adapted also at MLLTRAP and details of the alignment procedure can be found in ref. [12].

3.2 Trap structure

All vacuum chambers are built of grade 316L non-magnetic stainless steel using the con-flat (CF) flange standard. Two turbomolecular pumps of 880 l/s are located at the injection and at the extraction side of the magnet. The pressure in the vacuum volume of the Penning traps (without any gas load) is well below the lower limit of the Penning pressure gauge (5×10^{-9} mbar).

All electrodes not within the superconducting solenoid are made of either aluminium or of stainless steel. The electrodes within the solenoid forming the trap electric field are made of gold-plated oxygen-free copper and they are electrically isolated from each other by aluminium oxide insulators. In total about 50 gold-plated copper electrodes form a 1046 mm long electrode structure, see fig. 6. For ease of assembly, the electrodes are subdivided to seven segments held together with aluminium rods. Once the segments are interconnected the whole package can be inserted into the vacuum tube as one unit from the extraction side. All trap electrodes are cylindrical. These are much easier to manufacture than hyperbolic electrodes and also better vacuum can be achieved in the precision trap due to open geometry. Most of the electrodes have an inner diameter of 32 mm.

Each of the two cylindrical Penning traps of JYFLTRAP (see fig. 6) consists of an 8-fold split ring electrode (A), two-fold splitted inner correction electrodes B) on each side of the ring electrode, outer correction electrodes (C) and endcap electrodes (D) located next to the correction electrodes. The dimensions of the electrodes have been scaled from the ISOLTRAP purification trap [36–38]. The lengths of the ring electrode, the first and the second correction electrode and the endcap electrode are 18.5, 12.8, 6.7 and 44 mm, respectively, separated by 0.5 mm. The trap nearest to the RFQ is called the purification trap and is filled with dilute helium gas to enable the use of buffer-gas cooling of ions (see sect. 4 for more explanation). Gas flow to other sections of the system is minimized with the use of electrodes having narrow channels (see fig. 6 electrodes E and F). The other trap, commonly referred as the precision trap is located on the extraction side, 20 cm away from the purification trap center.

Table 1 shows the typical voltages used in the injection section of the trap. Tables 2 and 3 show the dimensions of the trap electrodes and trapping and extraction voltages applied in the purification and precision traps, respectively. The extraction side voltages when the trap is used for atomic mass measurements are given in table 4. When ions are extracted to the post-trap decay spectroscopy station, the voltages are somewhat different to optimally transfer ions further out from the trap setup.

The voltages to the trap electrodes are fed into the vacuum tube on both sides of the solenoid with so-called plug chambers. Each plug chamber houses three 500 V 10 pin feedthroughs connected via silver-plated copper wires into 30 evenly distributed sockets on the circumference of an insulating ring made of PEEK. At the ends of the trap structure there are similar rings with pins that fit into the sockets of the plug chambers. The trap electrodes are connected to the pins via silver-plated copper leads.

Table 3. Axial lengths of the electrodes in the precision trap (in addition, there is a 0.5 mm gap between each electrode) together with the operating voltages for three modes of operation. The "Trapping" denotes voltages used when ions are stored in the trap and "Inj." and "Extr." when ions are being injected or extracted from the trap, respectively. The potentials are given with respect to the ring electrode potential. It should be noted that the potential of the precision trap ring electrode is 4.2 V higher than the potential of the purification trap ring electrode. See fig. 6 for positions of the electrodes.

Electrode	Description	L (mm)	Trapping (V)	Inj. (V)	Extr. (V)
T25	Diaphragm ($d = 2$ mm)	24.5	+10	−10	
T26	Endcap	22	+10	−10	
T27	Endcap	22	+10	−10	
T28	Correction 2	6.7	+6.6	−10	
T29	Correction 1	12.8	+1.7	−10	
T30	Ring	18.5	0	0	0
T31	Correction 1	12.8	+1.7		−1
T32	Correction 2	6.7	+6.6		−1
T33	Endcap	22	+10		−1
T34	Endcap	22	+10		−1
T35	Endcap	22	+10		−1

Table 4. Typical voltages used at the extraction section of the trap. The electrode voltages above the single horizontal line are given with respect to the precision trap ring electrode. After "Extraction 2" electrode the ions are accelerated to $30q$ keV. The voltages below the single horizontal line are given with respect to the ground potential.

Electrode	Voltage (V)
T36	−2
T37	−3
T38	−4
Plug chamber el.	−100
Extraction 1	−150
Extraction 2	−1200
Shield grid	−900
Einzel 1	+8000
(Einzel 2	+2500)

This structure allows for a relatively fast way to insert the trap structure into the vacuum tube and to connect the electrical contacts.

3.3 Electronic devices

The electronic devices required to operate the Penning traps include DC power supplies, high-voltage switches and arbitrary waveform generators. In addition, a pulse pattern generator is used to generate 5 V *ON* and *OFF* TTL signals for the switches, waveform generators and other equipment requiring time-dependent operation.

3.3.1 Power supplies

DC power supplies from two manufacturers, ISEG Spezialelektronik GmbH and Spellman High Voltage Electronics Corporation, are used. The ISEG power supplies used in the setup are modular units having their own control electronics with 16 independent outputs and with a voltage range from −500 V to 0 V. All electrodes listed in tables 1, 2, 3, and 4 having voltages less than 500 V are connected to these units. The ISEG modules have better voltage stability than the Spellman ones. Other electrodes requiring more than 500 V are connected to Spellman power supplies. Each of these power supplies have one output whose output level is determined by a 0–10 V control signal. These control signals are provided by digital-to-analog (DAC) modules in modular I/O systems manufactured by WAGO Kontakttechnik GmbH & Co. The ISEG crate and WAGO I/O systems are connected to a computer by a Control Area Network (CAN) fieldbus via fiber optic repeaters. For a summary of the DC power supply modules, see table 5.

3.3.2 High-voltage switches

Ion samples are either injected to, trapped in or extracted from the purification or precision trap by switching the voltages of electrodes from "injection" or "extraction" to "trapping" voltage levels or vice versa. For each switchable electrode two voltage outputs from the ISEG power supply are connected to a high-voltage switch. One ISEG output may supply voltage to multiple electrodes as shown in tables 2 and 3, for example, to the electrodes from T13 to T15 in "open" mode. The switches are controlled by TTL signals.

Two kinds of high-voltage switches have been used, one for the 100 V deep purification trap and one for the

Table 5. Summary of the DC power supply modules in use.

Manufacturer	Model	Outputs	$V_{\max}$ (kV)	I_{max} (mA)
ISEG	EHQ F005n	16	−0.5	0.25
Spellman	MP5N	1	−5	2.0
Spellman	MP10N	1	−10	1.0
Spellman	MP5P	1	5	2.0
Spellman	MP20P	1	20	0.5
Spellman	MP40P	1	40	0.2

Table 6. Arbitrary waveform generators and their usage.

Model	Maximum frequency (MHz)		Connectivity
	Continuous	*Burst start*	
33120A	15	5	GPIB, RS232
33220A	20	6	GPIB, Ethernet
33250A	80	25	GPIB, RS232

ID	Trap	Type	Motion	Model	Mode	Remark
RF1	1	Dipole	ν_-	33120A	*Burst start*	
RF2	1	Quadrupole	ν_c	33250A	*Gated*	
RF3	2	Dipole	ν_-	33220A	*Burst start*	
RF4	2	Quadrupole	ν_c	33250A	*Gated*	Fixed starting phase
					AM	Ramsey excitation
RF5	2	Dipole	ν_+	33120A	*Gated*	Ramsey cleaning
RF6	2	AM		33120A	*Burst start*	AM of RF4

10 V deep precision trap. The switch for the purification trap is based on a Supertex inc. HV20822 16-channel high-voltage analog switch integrated circuit, which has two sets of eight analog switches and two corresponding digital inputs for TTL signals. The switch for the precision trap has been designed using UC2707 dual channel power drivers for switching and HFBR-2524 fiber optical receivers to control switching. OPA445 precision amplifiers have been used to boost the low current outputs from the ISEG power supplies.

3.3.3 Arbitrary waveform generators

The segmented ring electrodes of the two traps are connected to arbitrary waveform generators through AC/DC coupling boxes consisting of one resistor and a capacitor. Three models manufactured by Agilent Technologies are in use: 31120A, 31220A and 33250A (see table 6). The major difference between these models is the maximum frequency in *continuous* and so-called *burst start* mode. A LAN/GPIB gateway made by Agilent Technologies is used to connect the waveform generators to a computer via Ethernet. A pair of Ethernet media converters bridges the high-voltage platform and the ground via a pair of fiber optic cables.

Arbitrary waveform generators have a connector for a TTL input signal that can be used in two ways. In *burst start* mode the TTL signal determines when the generator outputs a predefined number of (sine) waveforms, for example, 17 periods of sine waveform with $\nu = 1.7\,\text{kHz}$ suitable for the 10 ms long magnetron excitation in the purification trap. In *gated* mode the generator repeats a (sine) waveform as long as the TTL signal remains in the *ON* level. In addition, amplitude modulation (*AM*) used with a Ramsey type of excitation is possible via a separate input terminal. In this mode the starting phase of the excitation waveform is random but the phase is continuous for the excitation periods [39].

3.3.4 Pulse pattern generator

The high-voltage switches and RF generators require precise timing signals. A PCI-card named PulseBlaster (model PB24-32k) is installed to a computer in order to provide these TTL signals for timings. It has 24 output channels and a 100 MHz clock which translates to a pulse

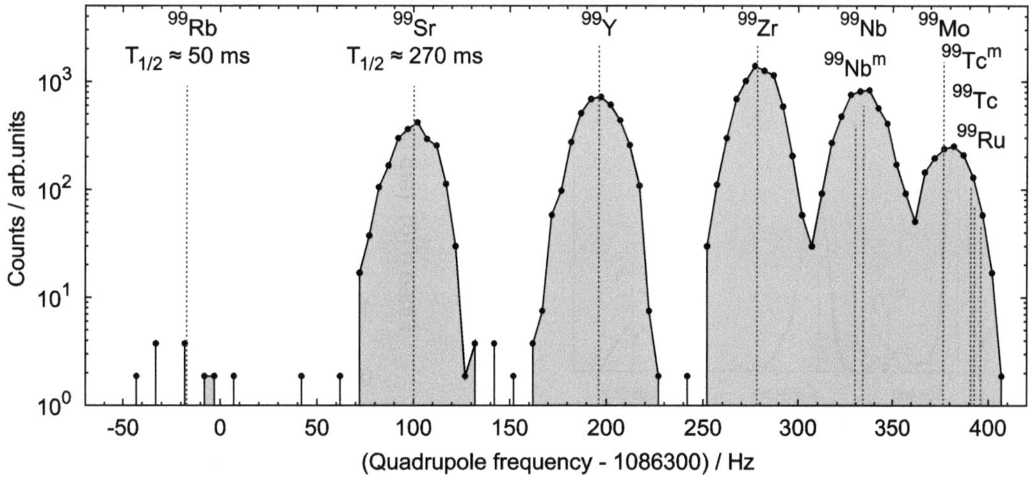

Fig. 7. Quadrupole frequency scan in the purification trap for ions produced in proton-induced fission at $A = 99$. Cyclotron resonance frequencies ν_c for various products have been labeled and marked with dashed vertical lines. The isotopes and isomers close to the stable ^{99}Ru are not fully resolved. A full width at half maximum ($\Delta\nu_{\mathrm{FWHM}}$) of 30 Hz is obtained, corresponding to a mass resolving power $M/\Delta M \approx 30000$.

resolution of 10 ns. The program executed by the card can contain up to 32 768 instructions that can last from 90 ns to 1.4 years each.

Most of the outputs from the card are connected to fiber optic transmitters through a buffer circuit. At the other end of each fiber optic cable is a fiber optic receiver which provides TTL signals for example to a high-voltage switch or to an arbitrary waveform generator.

3.3.5 Time-of-flight measurement

The signal of the amplifier of the MCP detector is recorded by a Multichannel Scaler/Averager (MCS) — Model SR430— that is made by Stanford Research System. It has a built-in discriminator with a typical pulse pair resolution of 10 ns. In the most common scenario, the recording time is 1024 times the 0.64 μs bin width. A TTL timing signal from the pulse pattern generator signals the device when to start counting ions as a function of time. The MCS is connected to a computer via GPIB bus and a LAN/GPIB gateway.

4 Purification trap

The first of the JYFLTRAP Penning traps is called the purification trap. It is filled with helium gas and is used for isobaric purification of ion beam. Trapped ions are manipulated by applying multipole RF fields to the azimuthally split ring electrode. After an initial cooling period without excitation, an azimuthal dipole field with magnetron frequency ν_- is switched on for a short duration (≈ 10 ms). This inreases the magnetron radius of all ions. The amplitude of the RF field is chosen so that no ion, upon extraction, can pass through the narrow channel of electrode E shown in fig. 6. After the azimuthal dipolar excitation, a quadrupole RF field with the cyclotron frequency of the ion of interest (see eq. (2)) is switched on. This excitation causes conversion from magnetron motion to modified cyclotron motion. Due to this, amplitudes of both the magnetron motion starts to decrease and also the modified cyclotron motion decreases in the presence of buffer gas, which will center the ions of interest in the trap. Therefore, a mass-selected beam can be extracted through a diaphragm out of the purification trap. The mass resolving power of the purification depends on the physical dimension of the diaphragm, on the buffer gas pressure and the amplitudes of the applied fields. At JYFLTRAP, a mass resolving power $M/\Delta M$ of the order of 10^5 has been achieved, see fig. 7. More details are given in ref. [16].

5 Precision trap

The precision trap is geometrically almost identical to the purification trap. It is primarily used for high-precision atomic mass measurements employing time-of-flight ion-cyclotron resonance (TOF-ICR) technique [29,30], and secondarily for high-resolution beam purification [35] reaching a mass resolving power of 10^6 or more.

5.1 High-resolution beam purification

If the mass resolving power of the purification trap is not sufficient to prepare monoisobaric (or monoisomeric) ion samples, the precision trap can be utilized to provide even better mass resolution. Typically $\Delta\nu_{\mathrm{FWHM}}$ of 10 . . . 20 Hz (or slightly better with reduced transmission) can be reached with the purification trap. Here we give a brief description of a high-resolution cleaning method employing both the purification and the precision trap providing even better than ~ 1 Hz resolution. A more comprehensive discussion of this so-called "Ramsey cleaning method" is given in ref. [35].

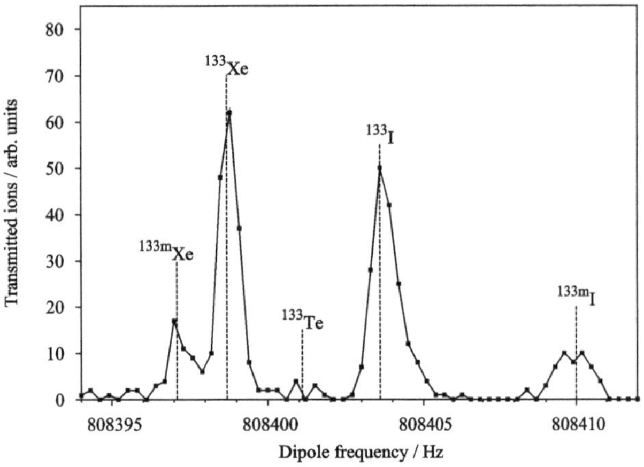

Fig. 8. Dipole frequency scan near the ν_+ frequency of ^{133}Xe using 20-40-20 ms On-Off-On pattern for dipolar RF field. The two states of xenon are clearly separated. Other isotopes are also present due to low resolving power of the purification trap. Resolving power $M/\Delta M \approx 10^6$ is reached.

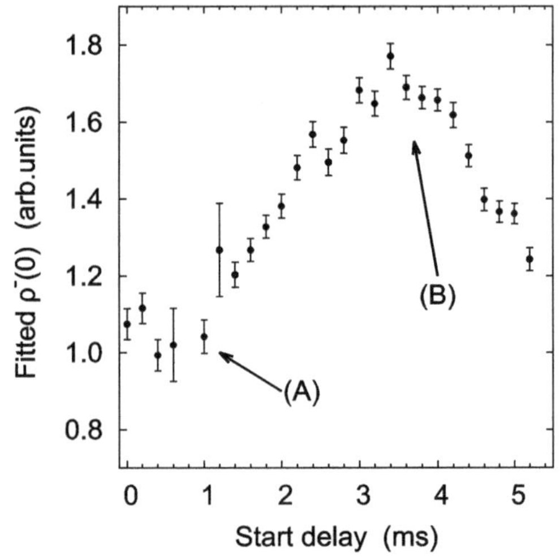

Fig. 9. A magnetron phase-locking scan. The starting time of the dipolar excitation time scanned. On the vertical axis the magnetron orbit radius after RF dipolar excitation ($\rho^-(0)$) is plotted. The radius is obtained from fits to the resonance curves (see fig. 10 where resonances corresponding to delay times A and B are shown). The period of the magnetron motion is $\approx \frac{1}{170\,\mathrm{Hz}} \approx 5.9\,\mathrm{ms}$.

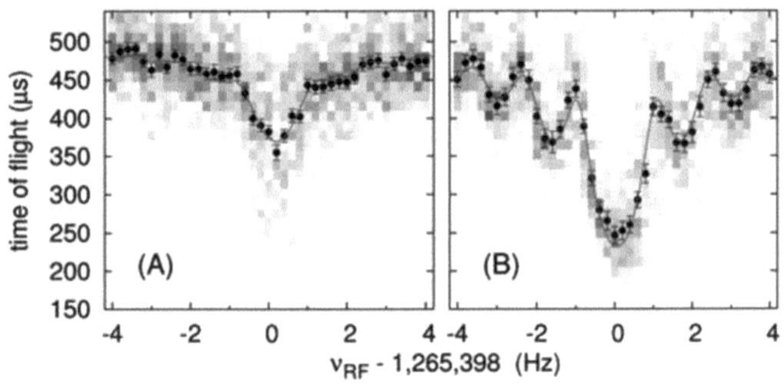

Fig. 10. (Color online) The TOF-ICR curves obtained with magnetron excitation start delays marked in fig. 9 with (A) and (B). The resonance curve will be deeper when the initial magnetron radius ($\rho^-(0)$) is larger.

The high-resolution cleaning is performed only after the pre-purification with the purification trap. The pre-cleaned bunch of ions is captured to the precision trap where contaminant ions are excited to a larger cyclotron orbit while the ion of interest remains mostly unexcited. To boost the performance in both resolution and the required time, the contaminants are excited with a dipolar RF field employing Ramsey's method of time-separated oscillatory fields technique [31]. Here, a short RF pulse (with constant amplitude) is applied. This pulse is followed by a waiting period after which an RF pulse with the same duration is applied again. It is important to preserve phase coherence for the both RF-on periods.

This method is relatively fast. To resolve ions having about 1 Hz cyclotron frequency difference typically less than 100 ms is needed for the excitation procedure in total. The ions of interest are allowed to gain some cyclotron motion in the excitation process since the ion bunch is sent back to the purification trap for recooling and recentering. On the way the contaminants hit the 2 mm electrode and thus will be completely removed. Once the ions have been properly recentered in the purification trap, they are sent back to precision trap for mass measurements or further downstream for decay spectroscopy experiments.

The state-of-the-art example of the purification process is the separation of the isomer in ^{133}Xe at 233 keV, corresponding to 1.7 Hz cyclotron frequency difference at JYFLTRAP [40]. Here only 60 ms was used for the excitation process in the precision trap and all in all about 500 ms was needed to prepare a monoisomeric bunch of ^{133m}Xe ions. A dipolar frequency scan in the precision trap plotted in fig. 8 showing the transmission of different ion species.

Without this cleaning method several experiments would have not been possible, such as the Q-value measurements of ^{54}Co and ^{50}Mn [41] which both have half-lives less than 300 ms. Other methods would have required much more time to separate the states [42].

5.2 Time-of-flight ion-cyclotron resonance measurement

Here the steps required to perform high-precision cyclotron frequency measurement using time-of-flight ion-cyclotron (TOF-ICR) [29,30] are explained.

5.2.1 Magnetron excitation

The first step after capturing the monoisomeric bunch of ions to the precision trap is the magnetron radius expansion. This is done with a short dipole RF pulse at magnetron frequency of 170 Hz. It is important that the magnetron radius expansion prior to the quadrupolar excitation is the same for all ions, and more importantly, for all ion bunches. In order to eliminate bunch-to-bunch variations, the phase of the dipole rf field is fixed to the injection time of the ions [43]. This is called *magnetron phase locking*. Figure 9 shows a scan where the time between the ion injection and the start of the dipolar excitation has been varied. After dipolar excitation, the quadrupolar excitation is switched on to convert the induced magnetron motion to cyclotron motion. Two TOF-ICR curves are shown in fig. 10. One with a delay time of 0.6 ms (marked with A) and one with 3.6 ms (marked with B). For case A, the excitation is in the opposite phase to the magnetron motion of the ions, and, for case B, they are in the same phase.

Thus, to ensure similar magnetron radii for all ion bunches it is important to lock the phase of the excitation. Its absolute value is less important, since a larger resonance effect can be accomplished by increasing the amplitude of the driving RF generator.

5.2.2 Quadrupolar excitation

After the magnetron excitation, the quadrupole excitation is switched on to mass-selectively convert the magnetron motion to cyclotron motion. The excitation duration T_{RF} and the quadrupole RF field amplitude V_{RF} are tuned so that one conversion from magnetron to cyclotron happens only at the resonance frequency $\nu_+ + \nu_-$. With other frequencies the conversion is only partial. The quadrupole RF field frequency (ν_{RF}) is scanned over an interval including the cyclotron frequency ν_c of the ions.

6 Complete atomic mass measurement procedure

Penning-trap mass spectrometry provides cyclotron frequency ratios between the ions of interest and the reference ions. The frequency ratio is converted to a mass ratio using eq. (2). By using a reference with a well-known atomic mass, the mass of the ions of interest can be obtained precisely. The determination of frequency ratios is explained in the following.

6.1 Measurement pattern

An ion bunch goes through the following steps:

1) Accumulation in the RFQ.
2) Purification in the purification trap.
3) High-resolution cleaning in the precision trap (if needed).
4) Recooling and recentering in the purification trap.
5) Ion motion excitations in the precision trap.
6) TOF recording.

These steps are repeated for subsequent bunches using different quadrupolar excitation frequencies ν_{RF}. The full procedure is shown in fig. 11. To maximize the efficiency, the RFQ and the Penning traps are operated in parallel, thus minimizing the measurement time needed. Once enough data are accumulated for the reference ions, the measurement is switched over to the ion species of interest. When enough data are obtained for the ions of interest, the reference ions are scanned again. This procedure is repeated as long as needed.

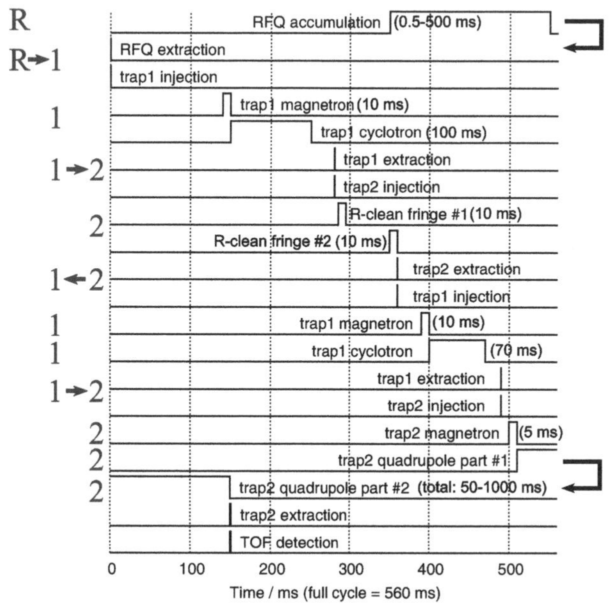

Fig. 11. (Color online) A typical timing pattern used. The blue letters on the left show the location of the ions: R = RFQ, 1 = purification trap, 2 = precision trap. The arrow indicates the transfer of the ions between the traps. From the ions' point of view the steps happen from the first row downwards. It should be noted that each of the traps (RFQ, purification and precision) are run in parallel. The arrows on the right depict the start of the next timing cycle.

6.2 Interleaved scanning

The production rate for the ions of interest can be very poor so that most cycles have no detected ions. It may take several hours to gather sufficient statistics for data analysis. Thus, the time between two reference measurements will be long, and changes in the trap environment may cause shifts in the resonance frequency. The conditions of the measurements will be more similar when they are performed in almost parallel fashion.

Switching between two ion species is a simple operation during the measurements. The parameters requiring a change are limited to the following ones.

- Mass-dependent parameters, such as:
  1) IGISOL separator magnet;
  2) transportation time from the cooler to the purification trap;
  3) transportation time from the purification trap to the precision trap;
  4) the purification trap cyclotron frequency.
- Production-rate–dependent beam gate timing.
- Ramsey-cleaning parameters (if applicable).

Except for the separator magnet, the changes can be performed in less than a second. Changing IGISOL dipole separator magnet to transmit another mass number may take up to 15 seconds. Thus, it is not practical to change between the ion species after every measurement cycle (0.2–5 s). A measurement program, which switches between two ion species after one or more fully completed frequency sweeps (0.5–2 min), has been developed.

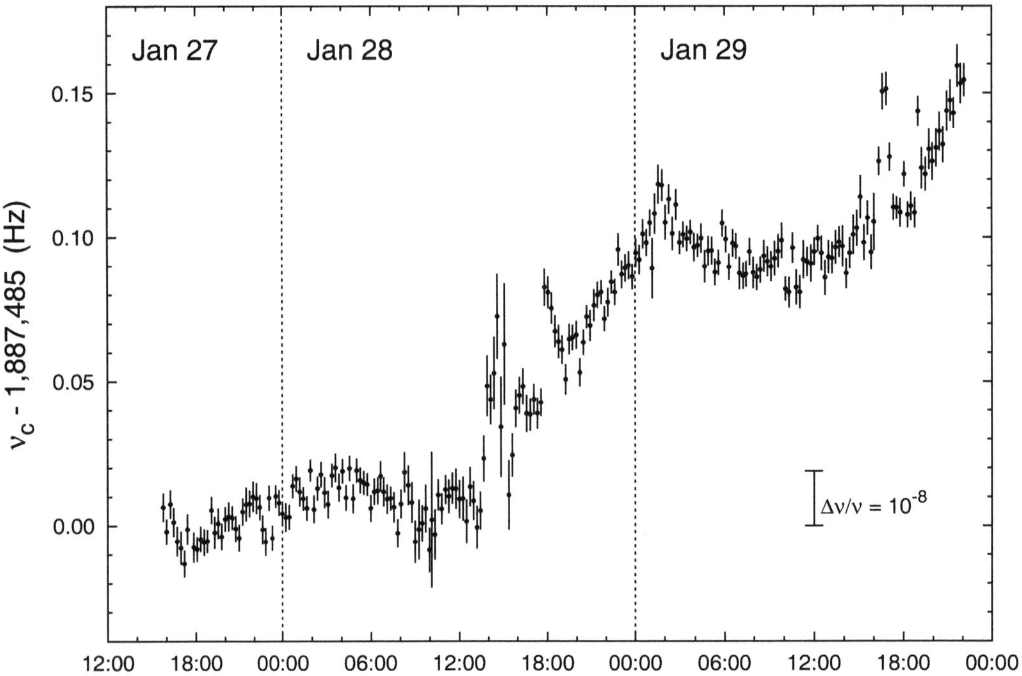

Fig. 12. Cyclotron frequencies of $^{57}Fe^{+}$ ions measured continuously for about 55 hours in January 2009. Every datapoint consist of 22 minutes of collected data. Statistical uncertainty is about 3×10^{-9} for each point. The fluctuations around 14:00–18:00 in January 28 were due to an opening of the entrance doors to outside in the experimental hall (the temperature outside was about $-5\,^{\circ}$C). The origin of the frequency deviation around 17:00 in January 29 is not known.

7 Trap performance and systematic studies

The trap performance has been extensively studied. Fluctuations and inhomogeneity of the magnetic field as well as mass-dependent and residual systematic errors related to JYFLTRAP have been investigated. In this section, different optimization procedures are described and systematic uncertainties quantified.

7.1 Magnetic-field fluctuations

Although the magnetic field of the superconducting magnet is actively screened, there are some fluctuations in the field. These are mostly caused by the varying temperature of the immediate surroundings of the superconducting magnet and pressure of the liquid-helium vessel of the magnet. Some effort has been put into the stabilization of the temperature by covering the high-voltage cage of the trap platform with plastic cover in order to minimize air flow through the setup. Exhaust helium from the magnet cryostat is released to the lab through a differential pressure valve that keeps a small overpressure in the magnet cryostat.

The fluctuations of the magnetic field have been characterized by continuously measuring the cyclotron frequency of the same ion species. For example, fig. 12 shows the cyclotron frequency of $^{57}Fe^{+}$ ions measured continuously for about 55 hours. The observed fluctuations are mostly due to changes in the magnetic field, but can include other effects.

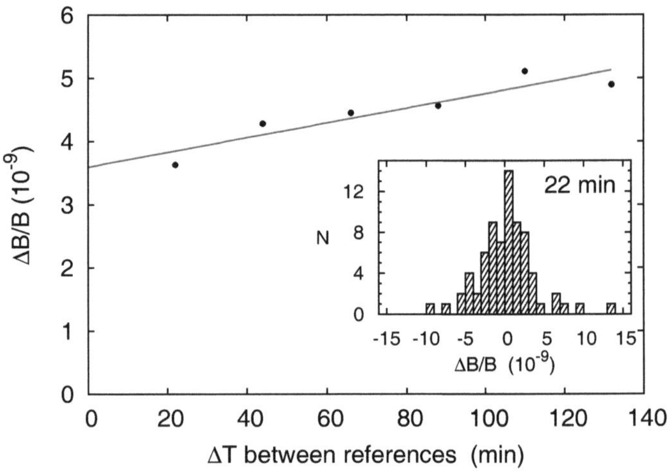

Fig. 13. The standard deviations obtained for different time durations between two consecutive reference scans. A slope of $1.2(2) \times 10^{-11}$ min^{-1} and y-axis crossing of $3.6(2) \times 10^{-9}$ are obtained. The inset shows the distribution of offsets for the first point (22 min). This figure is from ref. [44].

In frequency ratio (or mass) measurements the linear drifts are taken into account by interpolating the reference scans recorded right before and after the ion of interest. To account for short-term fluctuations, a long frequency scan such as shown in fig. 12 is subdivided to 22 min files. A real measurement process is simulated by taking three consecutive files, the first and the third for reference and the intermediate one for the ion of interest. A frequency value is interpolated from the references and compared to

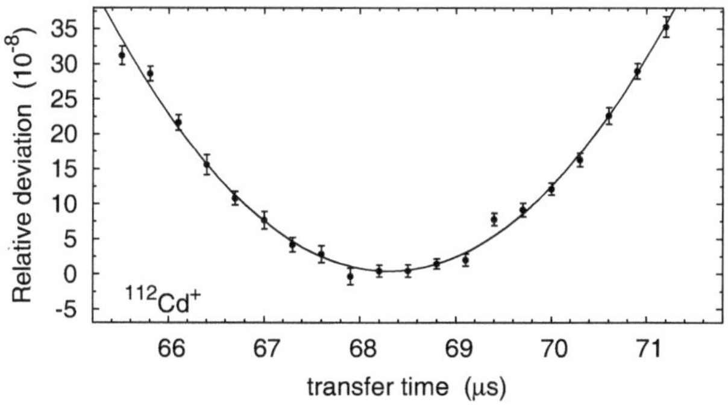

Fig. 14. Measured cyclotron frequency as a function of the transfer time. Here, the transfer time is the time difference between the lowering of the extraction side potential wall of the purification trap and the rising of the injection side potential wall of the precision trap. The frequency is at minimum at the optimum transfer time of 68.3 μs (for $A/q = 112$ ions). TOF-ICR curve fitting also indicates higher energy gain with non-optimal transfer time.

the measured value in order to obtain the offset $B - B_{\text{int}}$. The whole 55 hours of data are treated like this, and a distribution of offsets is obtained. Next, the data interval is increased to 44 minutes and so on. Finally, the standard deviations of the different distributions as a function of time interval are obtained as shown in fig. 13, where a line is fitted through the points.

In the absence of non-linear drifts, the distributions should be constant and correspond to the statistical uncertainty of the individual TOF-ICR curve fits. With non-linear drifts present, a positive slope is expected. The y-axis crossing resembles the statistical uncertainty of the individual fits. The data shown in fig. 13 are from January 2009. A similar measurement was done prior to the quenching of the superconducting magnet in 2007 and reported in ref. [44]. That time the slope was $3.2(2) \times 10^{-11}\,\text{min}^{-1}$, slightly larger than in 2009. To be on the conservative side, this larger value has been used in all JYFLTRAP measurements.

Non-linear drifts of the magnetic field at JYFLTRAP is a factor of two smaller than the ones reported from ISOLTRAP in 2003 [45] (since 2008 their magnet has been stablized to level beyond their measurement sensitivity [46]). Compared to TRIGATRAP, their fluctuations are a factor of two higher than JYFLTRAP [47]. Non-linear drifts at SHIPTRAP were almost 20-fold worse than at JYFLTRAP but with their recent implementation of pressure and temperature stabilization system their fluctuations have set a record low value of $0.13 \times 10^{-11}\,\text{min}^{-1}$ [48]. LEBIT trap has implemented an additional coil to counter the field decay due to the magnet; the remaining non-linear fluctuations are of the order of $0.83 \times 10^{-11}\,\text{min}^{-1}$ [11] and Canadian Penning-trap reports $1.3 \times 10^{-11}\,\text{min}^{-1}$ [49].

If a measurement does not exceed 30 minutes at JYFLTRAP, the contribution from non-linear magnetic field drifts are below 1×10^{-9} level. Since the shifts have no preferred direction, contribution from the magnetic field fluctuations are added to the statistical uncertainty.

7.2 Magnetic- and electric-field inhomogeneities

To minimize the effects of inhomogeneities of the magnetic field the volume in which the ions are confined should be as small as possible. Radially some expansion is needed for the TOF-ICR technique to work, but the amplitude of the axial motion should be minimized. Higher amplitudes result in the ions probing a larger extent of the magnetic field, especially in axial direction, where they are confined only by the electric potential. The axial extent can be minimized by optimizing the ion transfer time between the traps. With a non-optimal transfer time the ions gain energy in the transfer process. This can be observed not only as a shorter time of flight of the ions when extracted from the trap but also in shifts of the measured cyclotron frequency. In fig. 14, the cyclotron frequency as a function of the transfer time is shown. With non-optimal times the frequency increases, indicating that the magnetic field is on average weaker at the center. This coincides with the shimming data of the magnet: The field was measured to be stronger on both sides of the center of the precision trap.

Both electric- and magnetic-field inhomogenities could well be minimized by following the procedure described in [50]. Especially the electric field optimization will be performed in the new JYFLTRAP setup.

7.3 Carbon clusters cross-reference measurements

7.3.1 Background

Mass measurements aim for better and better accuracies. As always in experimental science, this requires a better understanding of the measurement instrument. A detailed description of the device and its features has been given in the previous sections. In addition to this, also better, more accurate reference-mass ions are needed for more accurate mass measurements with Penning traps. Since the atomic mass unit (u) is defined as

$$1\,\text{u} = \frac{1}{12} m\left(^{12}\text{C}\right), \tag{7}$$

the carbon atom or a multiple of carbon atoms, a so-called carbon cluster, is an ideal reference mass ion. Carbon clusters are available in equidistant steps of 12 mass units practically covering the whole chart of nuclei. Therefore, carbon clusters are typically used both for systematic studies of the mass spectrometer and also as reference masses.

To study the performance of Penning-trap devices, a carbon-cluster ion source based on laser ablation has been used already at ISOLTRAP [45], at SHIPTRAP [51] and at TRIGATRAP [47] to quantify mass-dependent systematic effects.

The effort to build a carbon-cluster ion source for JYFLTRAP started around 2005. The first carbon-cluster ions were fired through the trap setup using the first version of the ion source described in ref. [52]. The ions

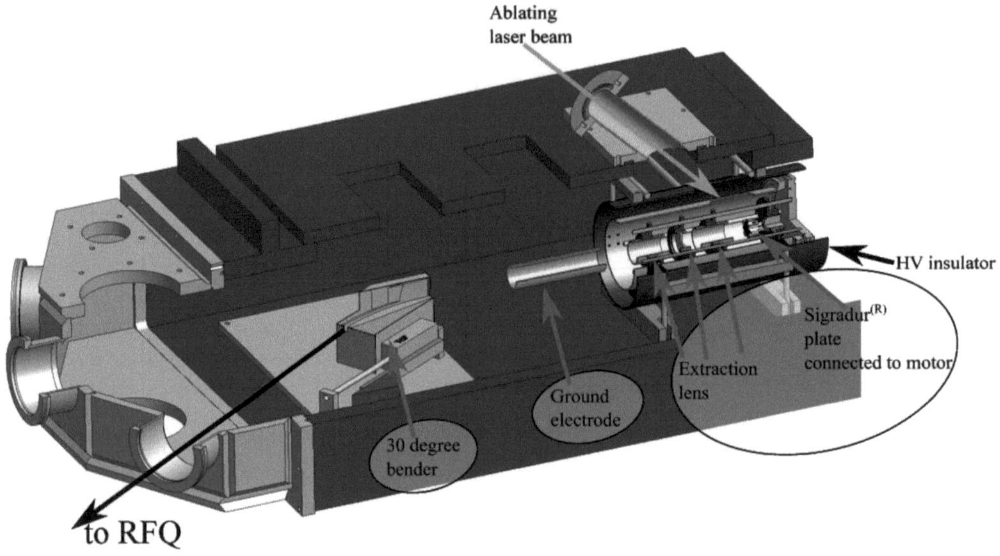

Fig. 15. (Color online) A cut view of the carbon cluster ion source located at the electrostatic switchyard. The carbon plate at 30 kV of potential and extraction electrodes are housed within a high-voltage insulator. The laser pulses hit the Sigradur® plate at 45° angle through a sapphire window on top of the vacuum chamber. The clusters are accelerated to 30q keV of energy, deflected by 30° with an electrostic beam deflector to the JYFLTRAP beamline for the injection to the RFQ.

were produced in a setup situated in an extension of the 90° cross-chamber between the RFQ and Penning traps. The setup was closely related to the SHIPTRAP and ISOLTRAP setups. The main difference to JYFLTRAP was the production of ions at the 30 kV high-voltage platform where the ion source was installed.

This place for the source had some clear advantages: the source could be built inside the same 30 kV platform without having a separate high-voltage installation. Also, switching between the carbon cluster source and the main IGISOL source would be easy by simply pulsing the electrodes of the quadrupole bender. Unfortunately, a good beam quality could not be achieved. The energy spread of the ions was far bigger than the depth of the purification trap (100 V) and carbon cluster ions of broad mass range were captured into the trap. Although this version of the cluster ion source could be used for some tests, it was decided to relocate the source further upstream of the RFQ or even before the IGISOL dipole magnet. This would allow an ideal, cooled reference ion beam identical to the beam of the ions of interest from IGISOL.

The ideal place for the carbon-cluster ion source would have been in the IGISOL cave in place of the main ion source. Here, the 55° dipole magnet could have been used to sieve only the cluster species with certain mass to charge ratio. Since IGISOL was almost in constant use by other experiments, this was not feasible. Finally, the ion source was built inside the electrostatic switchyard (at the focal plane of the IGISOL dipole separator magnet), see figs. 1 and 15. At the time there was only an electrostatic bender and a deflector unit inside the switchyard. Even without the full mass separation of the IGISOL dipole magnet, at least the modest mass-selectivity of the RFQ cooler buncher could be used.

7.3.2 Carbon cluster ion source

In short, the carbon cluster ion source consists of a carbon plate which is impinged by intense enough laser pulses that some clusters of carbon are evaporated and ionized (see fig. 15). The created ions are guided and accelerated with a set of electrodes to create a (bunched) beam. Here, a Q-switched Quantel Brilliant Nd:YAG 532 nm laser was used to create carbon ions from a Sigradur® glassy carbon plate. The laser operates at up to 10 Hz repetition rates. The laser spot is focused down to smaller than 1 mm diameter to achieve high enough energy densities for ablation. The carbon plate is situated on a rotating disk to ensure that the laser does not burn a hole through the plate. A more thorough description of the carbon-cluster ion source can be found in refs. [52,53].

7.3.3 Measurements and results

The cross-reference measurements were performed to quantify the mass-dependent frequency shifts. Since the mass ratios of carbon clusters are known, the shifts in frequency ratios can be quantified. In total more than 200 cyclotron frequency ratios were obtained. Three different sized cluster ions ($^{12}C_7^+$, $^{12}C_{10}^+$ and $^{12}C_{13}^+$) were chosen as reference masses. These were measured against five to eight different sized clusters. To achieve better precision in determining the centre frequency, Ramsey excitation schemes were used. A full description of the measurement parameters can be found in ref. [53].

Two main systematic properties of the JYFLTRAP setup were quantified: the mass-dependent uncertainty and the so-called residual uncertainty. The mass-dependent uncertainty estimates the relative frequency shift due to the difference in mass between the reference ion and the ion of interest. The residual uncertainty contains all the remaining systematic uncertainties of the setup.

The mass-dependent uncertainty was measured to be

$$\sigma_m(r)/r = 7.8(3) \times 10^{-10} \times \frac{\Delta m[\mathrm{u}]}{\mathrm{u}}, \quad (8)$$

and the residual uncertainty

$$\sigma_{res}(r)/r = 1.2 \times 10^{-8}, \quad (9)$$

where r is the measured cyclotron frequency ratio. With $\Delta m \leq 24\,\mathrm{u}$ —which is the case for all JYFLTRAP mass measurements performed so far— the uncertainties are

$$\sigma_m(r)/r = 7.5(4) \times 10^{-10} \times \frac{\Delta m[\mathrm{u}]}{\mathrm{u}}, \quad (10)$$

$$\sigma_{res}(r)/r = 7.9 \times 10^{-9}. \quad (11)$$

7.3.4 Future of carbon clusters at JYFLTRAP

The carbon cluster ion source that has been built and used for the systematic studies of the JYFLTRAP setup worked very well. Because it has to be manually removed from the beam line to inject beam from IGISOL into the trap, it was never used as a reference ion source. Building the high-voltage platform inside a vacuum chamber which was on the ground potential was not ideal. If the usual 30 kV potential was used, occasional sparks occurred between the platform and the ground potential. This was circumvented by using a slightly lowered high voltage than normal. This had no effect on the final systematic measurements and results.

Now that the whole IGISOL setup is moving to a different location, the systematic measurements should be at least partly repeated to confirm the good operation of JYFLTRAP. It is also a great opportunity to improve the cluster ion source itself.

A tentative location for the carbon-cluster ion source in the future is a separate off-line ion source station, located so that the mass-selection power of the dipole magnet can be used.

7.4 Mass doublets

Frequency ratio measurements of ions with same A/q ratio form a special class of JYFLTRAP measurements. Although frequency shifts due to electric field imperfections and tilt between electric and magnetic fields axis are present [54], the effect in the frequency ratio is negligible compared to statistical uncertainty.

Several mass doublet measurements have been done at JYFLTRAP. Mostly these are Q-value measurements of superallowed beta emitters [55] and of rare weak decays (see, *e.g.*, [56,57]). In these measurements no residual systematic uncertainty as described in the previous section has been added, and precisions in the 10^{-9} range have been achieved. The mass differences of several mass doublets that are known to very high precision have been measured to check the reliability. These include $^{26}\mathrm{Al}^m$ with $^{26}\mathrm{Al}$ as reference, $^{34}\mathrm{Cl}^m$ with $^{34}\mathrm{Cl}$ as reference and $^{76}\mathrm{Ge}$ with $^{76}\mathrm{Se}$ as reference [23,58].

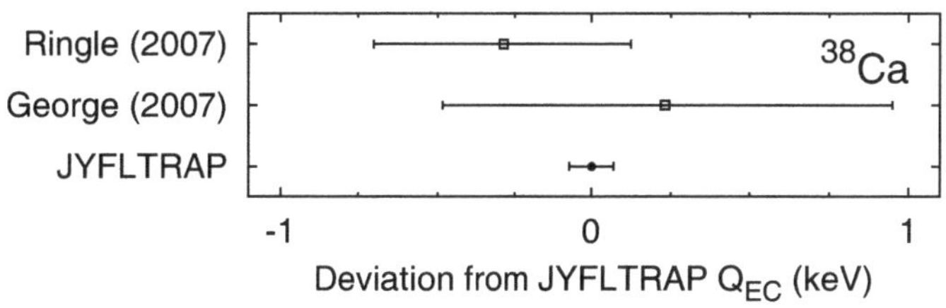

Fig. 16. Comparison of Q-values obtained with different trap setups. All results are in perfect agreement. JYFLTRAP measurement was performed as mass doublets while the other two measurements determined only the mass of the parent.

The doublet measurement technique gives a significant boost compared to "ordinary" mass measurements. In fig. 16 precisions obtained for superallowed β emitter $^{38}\mathrm{Ca}$ in three trap facilities are given. In addition to JYFLTRAP, the Q-value was measured at ISOLTRAP (George 2007 [34]) and at LEBIT (Ringle 2007 [59]). Their measurements concentrated only on $^{38}\mathrm{Ca}$ mass; mass of the daughter, $^{38m}\mathrm{K}$, was taken from the literature. LEBIT measurement was performed with doubly charged ions while ISOLTRAP measurement relied on Ramsey method. As seen, the JYFLTRAP result is at least 5 times more precise. Recently the doublet technique has been utilized at SHIPTRAP to measure Q-values of double-electron-capture candidates (see, *e.g.*, [60,61]).

7.5 Present performance and limits

In terms of IGISOL production yields, the most exotic species so far measured are neutron-rich nuclei such as $^{122}\mathrm{Pd}$, $^{114}\mathrm{Tc}$ and $^{103}\mathrm{Y}$. Their yields were less than 1 ion/s and half-lives about 100 ms [62].

The narrowest peak width achieved with the purification trap has been about 10 Hz ($\Delta\nu_{FWHM}$), which corresponds to a mass resolving power $R = M/\Delta M$ of about 10^5 for singly charged $A = 100$ ions. With a moderate mass resolving power ($\approx 10^4$) the transmission of the whole RFQ and trap line has been about 40%. This was measured using $^{62}\mathrm{Ga}$ ions by measuring their decay rate with a silicon detector before the RFQ and after the Penning trap.

With the high-resolution cleaning method the best achieved FWHM so far is 1 Hz which corresponds to $R = 10^6$ for $A = 100$ ions. The transmission is very low when using high-resolution cleaning method: about 30 ions/bunch was demonstrated in ref. [40] although yield from IGISOL would have allowed much larger bunch sizes. Here the limiting factor is space charge in the precision trap.

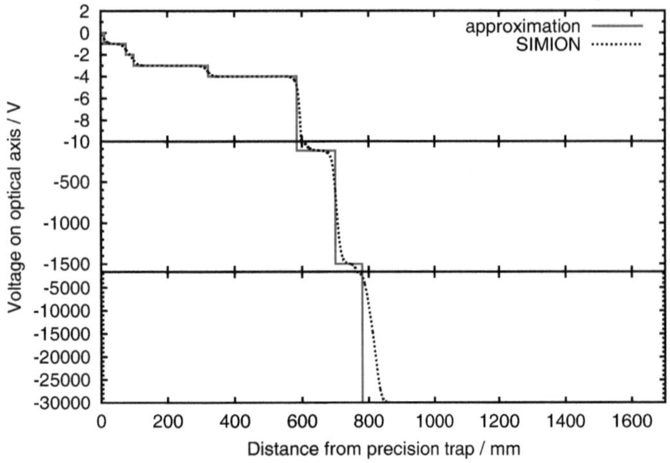

Fig. 17. (Color online) Comparison of the realistic (from SIMION) and the approximated electric potentials. The zero potential is chosen to be the potential of the ring electrode of the precision trap. The earth ground in this scale is then at −30 kV. The MCP detector is located about 1600 mm away from the precision trap. The approximation reproduces the most crucial low voltage part at 0–600 mm very well.

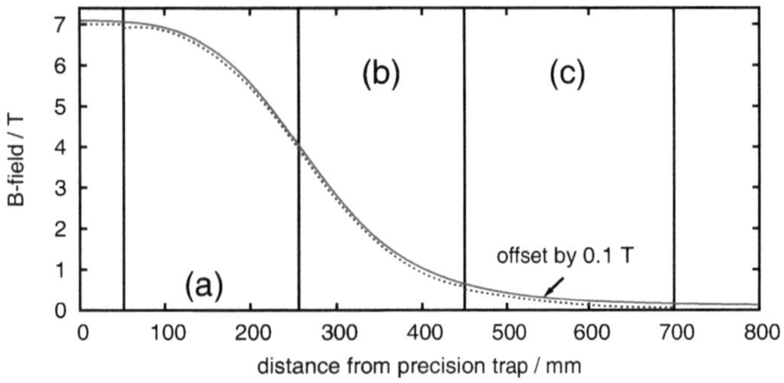

Fig. 18. (Color online) Comparison of the realistic and the approximated magnetic fields. The field is split in five parts, where two sections have constant 7 T and 0 T and the three sections marked (a–c) in between are well described with second order polynomial functions (dashed blue line). The realistic field shape from SIMION (solid red line) is offset by 0.1 T for clarity.

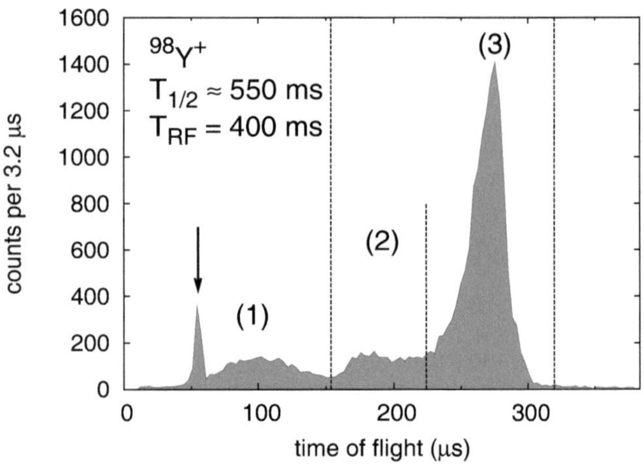

Fig. 19. (Color online) A typical time-of-flight spectrum of short-living β^- decaying ions consisting of all ions regardless of excitation frequency in the trap. The peaks at (2) and (3) mark the ions excited with $\nu_{\mathrm{RF}} \approx \nu_c$ and $\nu_{\mathrm{RF}} \neq \nu_c$, respectively. Peak (1) consists of doubly charged β^- decay daughter ions. Their time of flight is short not only because they are doubly charged but also due to recoil energy gained in the decay process. The sharp peak marked with arrow consists of He^+ ions, formed in collisions of β decay recoils with residual gas atoms. Only the ions detected between the two dashed vertical bars are used for TOF determination. Naturally all detected ions are used for the countrate-class analysis.

8 Data analysis procedure

The recorded TOF-ICR data consist of the time of flight of the ions as a function of the frequency of the exciting RF field. The lineshape of the resonance curve is well described in ref. [30] for a rectangular excitation amplitude shape and in ref. [63] for time-separated oscillatory fields. Many aspects of the data analysis procedure have been elaborated at ISOLTRAP (see, *e.g.*, [45]). More details about data analysis procedures at JYFLTRAP can be found in ref. [64].

8.1 Theoretical TOF determination

After quadrupolar excitation in the precision trap (as described in sect. 5.2.2) the extraction side potential wall of the trap is lowered and the ions are ejected towards the MCP detector, which is located at the ground potential, about 1.6 m downstream from the precision trap (see also fig. 1). The time of flight can be calculated since the magnitudes of the electric and magnetic fields are known from simulations. The electric potential along the geometry axis is shown in fig. 17, and the magnetic field in fig. 18. The fields are extracted from the ion optics simulation program SIMION, which gives the field values along the flight path with grid size of 1 mm. To make this less computer power demanding the fields are averaged over sections which can be described either with a constant number or, at most, with a second order polynomial. The full time of flight can then be calculated with eq. (5).

It is important to extract the ions slowly over the magnetic field gradient so that ions with more radial energy will gain significantly more axial energy when crossing the field gradient. At ≈ 600 mm the ions are electrostatically significantly accelerated. At this point the magnetic field is only about 0.1 T. The ions are then hitting the MCP detector with $30q$ keV of energy.

8.2 Determination of the experimental TOF and its uncertainty

Several bunches of ions are recorded for each frequency. A TOF gate is applied so that only ions within a certain TOF interval are accepted. The gate has to be set so that no ions of interest are left out. This is illustrated in fig. 19,

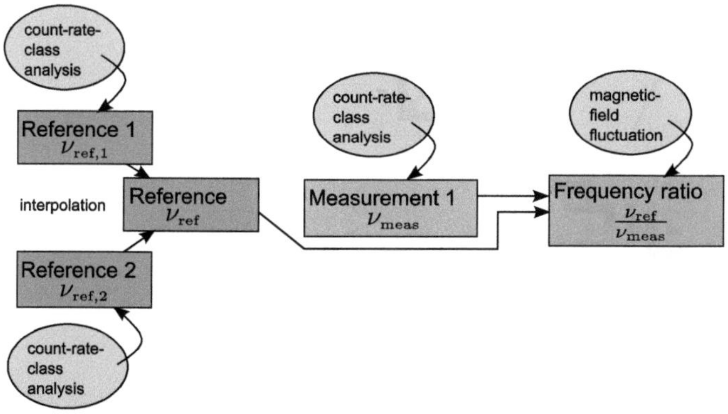

Fig. 20. Determination of a single frequency ratio using two reference ion scans and an ion-of-interest scan. The countrate class analysis is performed for each scan. Uncertainty due to the magnetic field fluctuation is added to frequency ratio.

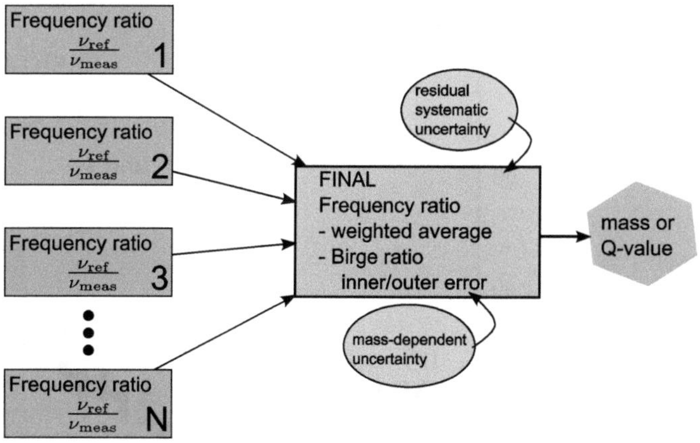

Fig. 21. Determination of the final frequency ratio. Using N individual frequency ratios (as shown in fig. 20) the final frequency ratio is obtained by calculating the weighted average of the individual values. The inner and outer errors are calculated. The mass-dependent and residual systematic uncertainty are added at the very end.

where the TOF gate has been set to be 155–330 μs (depicted with full vertical bars). Ions outside the gate do not contribute to the cyclotron frequency determination. Another gate is imposed by the number of detected ions. Typically, the analysis is performed by dividing the data into classes according to how many ions were detected in a single bunch. This is called countrate class analysis and is well described in ref. [45]. This way, the fitted frequency can be extracted as a function of detected ions per bunch. If contaminants are present, the frequency is expected to shift with increasing number of ions.

To calculate the TOF uncertainty at each frequency a so-called mean-corrected sum-statistics method is used in order to provide uncertainty even for the frequencies where only few ions were detected. In these cases ordinary standard deviation of the mean value would produce unrealistic uncertainties [64]. First, the mean TOF for each frequency point is calculated. Then, the distributions of each frequency are stacked together keeping the mean value as a central value. The standard deviation (σ) of such a distribution is

$$\sigma\left(\overline{T}\right) = \sqrt{\frac{\sum_{m=1}^{M}\left[\sum_{n=1}^{N_m}\left(\overline{T}_m - T_{m,n}\right)^2\right]}{N-1}}, \tag{12}$$

where the inner sum is over all ions (N_m) excited with frequency ν_m. $T_{m,n}$ is the TOF of one ion and $\overline{T}_m$ is their mean TOF. The outer sum is over all excitation frequencies ν_m. The total number of ions is $N = \sum_m N_m$.

For each frequency the standard deviation of the mean is obtained as

$$\sigma\left(\overline{T_k}\right) = \frac{\sigma\left(\overline{T}\right)}{\sqrt{N_k}}, \tag{13}$$

where N_k is the number of ions detected at that frequency.

8.3 Cyclotron frequency ratio determination

The ion-of-interest scans have been obtained consecutively or interleaved as described in sect. 6.2. The TOF-ICR curves are fitted using the countrate class analysis (see ref. [45]) to account for shifts due to contaminating ions. At JYFLTRAP, the MCP efficiency was measured to be about 60% [65] thus, in order to get a frequency value corresponding to one stored ion in the trap, the countrate classed results were extrapolated to a value of 0.6.

With consecutive scanning, where the ion species is switched every 30 minutes or more, the reference frequency needs to be interpolated from adjacent scans to the ion-of-interest scan. From the interpolated frequency and the ion-of-interest frequency the final cyclotron frequency ratio is obtained.

For a single measurement consisting of a reference scan (at time t_0 having frequency ν_0), an ion-of-interest scan (at time t_1 having frequency ν_{meas}) and a reference scan again (at t_2 with ν_2), the interpolated reference ion frequency ν_{ref} at the time of ν_{meas} is

$$\nu_{\mathrm{ref}} = \frac{1}{t_2 - t_0}\left[\nu_0\left(t_2 - t_1\right) + \nu_2\left(t_1 - t_0\right)\right], \tag{14}$$

with uncertainty $\sigma(\nu_{\mathrm{ref}})$

$$\sigma(\nu_{\mathrm{ref}}) = \frac{1}{t_2 - t_0}\sqrt{\left(t_2 - t_1\right)^2\sigma(\nu_0)^2 + \left(t_1 - t_0\right)^2\sigma(\nu_2)^2}. \tag{15}$$

After having both the interpolated reference frequency (ν_{ref}) and the frequency of the ion of interest (ν_{meas}), the frequency ratio r is

$$r = \frac{\nu_{\mathrm{ref}}}{\nu_{\mathrm{meas}}}, \tag{16}$$

with uncertainty

$$\sigma(r) = r\sqrt{\left(\frac{\sigma(\nu_{\mathrm{ref}})}{\nu_{\mathrm{ref}}}\right)^2 + \left(\frac{\sigma(\nu_{\mathrm{meas}})}{\nu_{\mathrm{meas}}}\right)^2}. \tag{17}$$

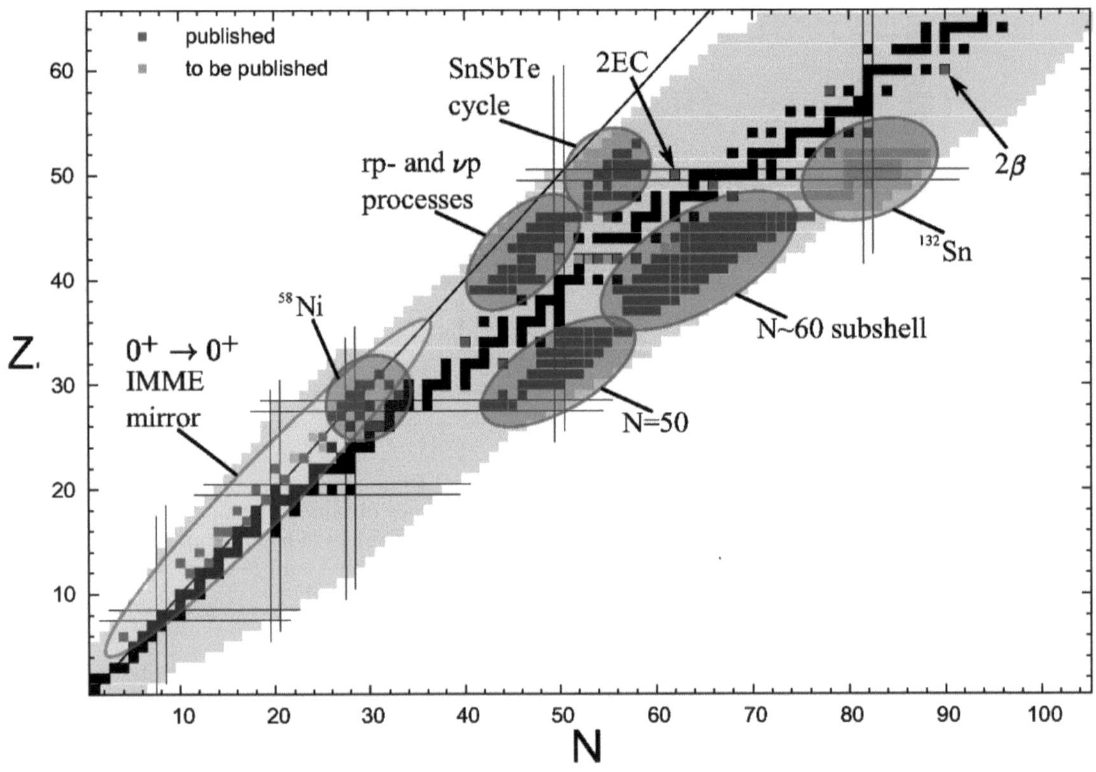

Fig. 22. (Color online) The chart of nuclei, where nuclei whose mass has been determined with JYFLTRAP have been indicated. Red symbols indicate published results and light blue the measured ones (see also tables 7 and 8). Also to which studies the measured nuclei contribute have been marked (see text for more explanation).

With interleaved scanning no interpolation is needed. Uncertainty due to magnetic field fluctuation is quadratically added to the frequency ratio uncertainty. The analysis steps are summarized in fig. 20.

When a chain of data (REF-MEAS-REF-...-MEAS-REF) has been taken, the final frequency ratio is weighted average of the individual frequency ratios. With interleaved scanning procedure a long measurement is split to convenient chunks that consists usually of about 30 minutes of data. The mass-dependent uncertainty and the residual uncertainty are added in this point, as illustrated in fig. 21. So far any data obtained with JYFLTRAP has not been corrected with mass-dependent shift. Instead, this shift is added as an uncertainty. Additionally the Birge ratio [66] needs to be calculated. For this, both the inner error,

$$\sigma_{\text{int}}^2 = \frac{1}{\sum_i \frac{1}{\sigma_i^2}}, \tag{18}$$

and the outer error ,

$$\sigma_{\text{ext}}^2 = \frac{\sum_i \frac{1}{\sigma_i^2}\left(r_i - \overline{r}\right)^2}{(n-1)\sum_i \frac{1}{\sigma_i^2}}, \tag{19}$$

are needed (r_i are the individual frequency ratios and the Birge ratio is $R = \sigma_{\text{ext}}/\sigma_{\text{int}}$ [66]. If both the inner and the outer errors are about equal, the fluctuation around the mean value is purely statistical. Common practise has been that if either of the error is larger, then the larger one is used as the final error. At this stage the mass dependent and residual systematic uncertainties are added (see fig. 21).

8.4 Mass and Q-value

Finally the atomic mass of the ion of interest m_{meas} is obtained (for singly charged ions and omitting binding energies of the missing electrons)

$$m_{\text{meas}} = r \times (m_{\text{ref}} - m_e) + m_e, \tag{20}$$

where m_e is mass of an electron. Mass uncertainty is

$$\sigma\left(m_{\text{meas}}\right) = \sqrt{\sigma(r)^2(m_{\text{ref}} - m_e)^2 + \sigma(m_{\text{ref}})^2 r^2}, \tag{21}$$

where uncertainty of the electron mass as well as binding energies of the missing electrons are neglected.

Mass differences or Q-values are calculated for singly charged ions by using the relation

$$Q = (m_{\text{meas}} - m_{\text{ref}})c^2 = (r-1)(m_{\text{ref}} - m_e)c^2 + \Delta B_e, \tag{22}$$

where ΔB_e term (with singly charged ions typically few eV) arises from the binding-energy difference of the missing atomic electrons. The Q-value uncertainty is

$$\sigma(Q) = \sqrt{\sigma(r)^2(m_{\text{ref}} - m_e)^2 c^4 + \sigma(m_{\text{ref}}c^2)^2(r-1)^2}, \tag{23}$$

Table 7. List of isotopes whose mass has been measured with JYFLTRAP. References to publications have also been given.

Z	Element	Mass numbers	References
6	C	10	[77]
12	Mg	23	[78]
13	Al	23, 26	[78,79]
14	Si	26	[80]
16	S	30, 31	[81,82]
17	Cl	34, 34m	[58]
18	Ar	34	[77]
19	K	38, 38m	[58]
20	Ca	38	[77]
21	Sc	42, 42m	[79]
22	Ti	42	[83]
23	V	46	[79]
25	Mn	50, 50m	[41]
26	Fe	56–57	[84]
27	Co	53, 53m, 54, 56	[41,84]
28	Ni	55–57, 70–73	[84,44]
29	Cu	57–58, 62, 73, 75	[84,79,44]
30	Zn	59–60, 62, 76–80	[84,79,85]
31	Ga	62, 78–83	[79,85]
32	Ge	76, 80–85	[23,85]
33	As	81–87	[85]
34	Se	74, 84–89	[57,85]
35	Br	85–92	[86]
37	Rb	94–97	[86]
38	Sr	95–100	[87]
39	Y	80–84, 95–103	[88–90,62]
40	Zr	83–88, 98–105	[87–89]
41	Nb	85–88, 100–108	[88,90,62]
42	Mo	88–89, 100, 102–111	[89,23,87,62]
43	Tc	88–92, 106–114	[89,91,62]
44	Ru	90–94, 106–116	[89,91,62]
45	Rh	92–95, 108–119	[89,91,62]
46	Pd	94–99, 101, 112–122	[89,92,91,62]
47	Ag	100	[92]
48	Cd	101–105, 116	[92,93]
49	In	102,104, 115	[92,94]
50	Sn	104–108, 112	[95,56]
51	Sb	106–110	[95]
52	Te	108–109, 130	[95,93]
53	I	111	[95]
54	Xe	136	[96]
58	Ce	136	[96]
60	Nd	150	[97]

when neglecting uncertainty of the electron mass and the atomic electron binding energies. Typically contribution from the reference mass uncertainty $\sigma(m_{\text{ref}})$ can also be omitted since for A/q doublets $(r-1) < 10^{-3}$ and thus in most of the cases uncertainty in the Q-value can be simply written as

$$\sigma(Q) \approx \sigma(r) \times (m_{\text{ref}} - m_e). \tag{24}$$

Table 8. List of isotopes whose mass has been measured at JYFLTRAP but not yet published.

Z	Element	Mass numbers
15	P	29
24	Cr	49
25	Mn	49
41	Nb	108
42	Mo	92, 94–98, 100
43	Tc	104–105
48	Cd	121–128
49	In	129, 131
50	Sn	130–135
51	Sb	131–136
52	Te	130, 132–140

9 Summary of JYFLTRAP atomic mass measurements

The JYFLTRAP Penning-trap setup was operational since 2003 until June 2010 when the whole IGISOL facility, including JYFLTRAP, were shut down for relocation. The atomic masses of more than 200 short-living nuclei have been measured and published as summarized in table 7. Nearly 50 more have been measured which are expected to be published shortly, see table 8. An up-to-date list of the measured and published nuclei can be found from the JYFLTRAP website [67].

The atomic masses measured at JYFLTRAP are shown in fig. 22. The main areas of studies are (see for mass measurement references in tables 7 and 8):

- Nuclear structure at the n-rich side [68].
- Nuclear astrophysics (near ^{58}Ni, rp and νp processes, SnSbTe cycle) [69–72].
- Q-values of superallowed $0^+ \to 0^+$ β emitters [55].
- Q-values of mirror decays [73].
- Testing the Isobaric Mass Multiplet Equation (IMME) [74].
- Q-values of neutrinoless double–electron-capture candidates (2EC) [75].
- Q-values for double-beta decay studies (2β) [76].

All Q-value measurements have measured as doublets and typically precision of better than 1 keV (10^{-8}) have been obtained.

The mass measurement program will be continued as soon as the relocation and upgrade of the IGISOL and JYFLTRAP setups are complete in 2012. Upgrades include, for instance, a new off-line ion source which can be operated independently of the IGISOL frontend, and optional beam ports between the RFQ and Penning traps for adding various devices such as a multi-reflection time-of-flight mass separator.

This work has been supported by the European Union (RTD project EXOTRAPS, Contract No. ERBFMGECT980099; RTD project NIPNET, Contract No. HPRI-CT-2001-50034; Fifth Framework Programme "Improving Human Potential-Access to Research Infrastructure", Contract No. HPRI-CT-1999-00044; 6th Framework programme "Integrating Infrastructure Initiative- Transnational Access" Contract No. 506065 (EURONS)) and by the academy of Finland under the Finnish Centre of Excellence Programmes 2000-2005 and 2006-2011.

References

1. H.-J. Kluge, Hyperfine Interact. **196**, 295 (2010).
2. K. Blaum, Phys. Rep. **425**, 1 (2006).
3. I. Bergström *et al.*, Nucl. Instrum. Methods Phys. Res. A **487**, 618 (2002).
4. M. Redshaw, J. McDaniel, E.G. Myers, Phys. Rev. Lett. **100**, 093002 (2008).
5. R.S. Van Dyck *et al.*, Phys. Rev. Lett. **92**, 220802 (2004).
6. C. Diehl *et al.*, Hyperfine Interact. **199**, 291 (2011).
7. G. Gabrielse, Int. J. Mass Spectrom. **251**, 273 (2006).
8. M. Hori *et al.*, Nature **475**, 484 (2011).
9. J. Wang *et al.*, Nucl. Phys. A **746**, 651 (2004).
10. M. Mukherjee *et al.*, Eur. Phys. J. A **35**, 1 (2008).
11. R. Ringle *et al.*, Nucl. Instrum. Methods Phys. Res. A **604**, 536 (2009).
12. V.S. Kolhinen *et al.*, Nucl. Instrum. Methods Phys. Res. A **600**, 391 (2009).
13. G. Sikler *et al.*, Nucl. Instrum. Methods Phys. Res. B **204**, 482 (2003).
14. TITAN Collaboration (J. Dilling), Hyperfine Interact. **196**, 219 (2010).
15. J. Ketelaer *et al.*, Nucl. Instrum. Methods Phys. Res. A **594**, 162 (2008).
16. V.S. Kolhinen *et al.*, Nucl. Instrum. Methods Phys. Res. A **528**, 776 (2004).
17. A. Jokinen *et al.*, Int. J. Mass Spectrom. **251**, 204 (2006).
18. M. Smith *et al.*, Phys. Rev. Lett. **101**, 202501 (2008).
19. M. Block *et al.*, Nature **463**, 785 (2010).
20. J. Äystö, Nucl. Phys. A **693**, 477 (2001).
21. H. Penttilä *et al.*, Eur. Phys. J. A **25**, 745 (2005).
22. P. Karvonen *et al.*, Nucl. Instrum. Methods Phys. Res. B **266**, 4794 (2008).
23. S. Rahaman *et al.*, Phys. Lett. B **662**, 111 (2008).
24. A. Nieminen *et al.*, Nucl. Instrum. Methods Phys. Res. A **469**, 244 (2001).
25. L.S. Brown, G. Gabrielse, Phys. Rev. A **25**, 2423 (1982).
26. G. Gabrielse, Int. J. Mass Spectrom. **279**, 107 (2009).
27. G. Savard *et al.*, Phys. Lett. A **158**, 247 (1991).
28. L.S. Brown, G. Gabrielse, Rev. Mod. Phys. **58**, 233 (1986).
29. G. Gräff, H. Kalinowsky, J. Traut, Z. Phys. A **297**, 35 (1980).
30. M. König *et al.*, Int. J. Mass Spectrom. **142**, 95 (1995).
31. N.F. Ramsey, Rev. Mod. Phys. **62**, 541 (1990).
32. G. Bollen *et al.*, Nucl. Instrum. Methods Phys. Res. B **70**, 490 (1992).
33. M. Kretzschmar, Int. J. Mass Spectrom. **264**, 122 (2007).
34. S. George *et al.*, Phys. Rev. Lett. **98**, 162501 (2007).
35. T. Eronen *et al.*, Nucl. Instrum. Methods Phys. Res. B **266**, 4527 (2008).
36. K. Farrar, Nucl. Instrum. Methods Phys. Res. A **485**, 780 (2002).
37. G. Gabrielse, L. Haarsma, S. Rolston, Int. J. Mass Spectrom. Ion Process. **88**, 319 (1989).
38. H. Raimbault-Hartmann *et al.*, Nucl. Instrum. Methods Phys. Res. B **126**, 378 (1997).
39. M. Eibach *et al.*, Int. J. Mass Spectrom. **303**, 27 (2011).
40. K. Peräjärvi *et al.*, Appl. Radiat. Isot. **68**, 450 (2010).
41. T. Eronen *et al.*, Phys. Rev. Lett. **100**, 132502 (2008).
42. K. Blaum *et al.*, Nucl. Phys. A **752**, 317c (2005).
43. K. Blaum *et al.*, J. Phys. B: At. Mol. Opt. Phys. **36**, 921 (2003).
44. S. Rahaman *et al.*, Eur. Phys. J. A **34**, 5 (2007).
45. A. Kellerbauer *et al.*, Eur. Phys. J. D **22**, 53 (2003).
46. M. Marie-Jeanne *et al.*, Nucl. Instrum. Methods Phys. Res. A **587**, 464 (2008).
47. J. Ketelaer *et al.*, Eur. Phys. J. D **58**, 47 (2010).
48. C. Droese *et al.*, Nucl. Instrum. Methods Phys. Res. A **632**, 157 (2011).
49. G. Savard *et al.*, Phys. Rev. Lett. **95**, 102501 (2005).
50. D. Beck *et al.*, Nucl. Instrum. Methods Phys. Res. A **598**, 635 (2009).
51. A. Chaudhuri *et al.*, Eur. Phys. J. D **45**, 47 (2007).
52. V.-V. Elomaa *et al.*, Nucl. Instrum. Methods Phys. Res. B **266**, 4425 (2008).
53. V.-V. Elomaa *et al.*, Nucl. Instrum. Methods Phys. Res. A **612**, 97 (2009).
54. G. Gabrielse, Phys. Rev. Lett. **102**, 172501 (2009).
55. J.C. Hardy, I.S. Towner, Phys. Rev. C **79**, 055502 (2009).
56. S. Rahaman *et al.*, Phys. Rev. Lett. **103**, 042501 (2009).
57. V.S. Kolhinen *et al.*, Phys. Lett. B **684**, 17 (2010).
58. T. Eronen *et al.*, Phys. Rev. Lett. **103**, 252501 (2009).
59. R. Ringle *et al.*, Phys. Rev. C **75**, 055503 (2007).
60. S. Eliseev *et al.*, Phys. Rev. Lett. **106**, 052504 (2011).
61. S. Eliseev *et al.*, Phys. Rev. C **84**, 012501(R) (2011).
62. J. Hakala *et al.*, Eur. Phys. J. A **47**, 1 (2011).
63. S. George *et al.*, Int. J. Mass Spectrom. **264**, 110 (2007).
64. T. Eronen, Ph.D. thesis, University of Jyväskylä, 2008.
65. H. Penttilä *et al.*, Eur. Phys. J. A **44**, 147 (2010).
66. R.T. Birge, Phys. Rev. **40**, 207 (1932).
67. JYFLTRAP mass database, `http://research.jyu.fi/igisol/JYFLTRAP_masses/`.
68. L. Schweikhard, G. Bollen (Editors), Int. J. Mass Spectrom. **251** (2006).
69. R. Wallace, S. Woosley, Astrophys. J. Suppl. Ser. **45**, 389 (1981).
70. H. Schatz *et al.*, Phys. Rep. **294**, 167 (1998).
71. B.A. Brown *et al.*, Phys. Rev. C **65**, 045802 (2002).
72. H. Schatz *et al.*, Phys. Rev. Lett. **86**, 3471 (2001).
73. O. Naviliat-Cuncic, N. Severijns, Phys. Rev. Lett. **102**, 142302 (2009).
74. E.P. Wigner, in *Robert A. Welch foundation conference on chemical research*, edited by W.O. Millikan, Vol. **1** (Houston, Texas, 1957) p. 67.
75. R.G. Winter, Phys. Rev. **100**, 142 (1955).
76. F.T. Avignone, S.R. Elliott, J. Engel, Rev. Mod. Phys. **80**, 481 (2008).
77. T. Eronen *et al.*, Phys. Rev. C **83**, 055501 (2011).
78. A. Saastamoinen *et al.*, Phys. Rev. C **80**, 044330 (2009).
79. T. Eronen *et al.*, Phys. Rev. Lett. **97**, 232501 (2006).
80. T. Eronen *et al.*, Phys. Rev. C **79**, 032802 (2009).
81. J. Souin *et al.*, Eur. Phys. J. A **47**, 1 (2011).
82. A. Kankainen *et al.*, Phys. Rev. C **82**, 052501 (2010).
83. T. Kurtukian Nieto *et al.*, Phys. Rev. C **80**, 035502 (2009).
84. A. Kankainen *et al.*, Phys. Rev. C **82**, 034311 (2010).

85. J. Hakala *et al.*, Phys. Rev. Lett. **101**, 052502 (2008).
86. S. Rahaman *et al.*, Eur. Phys. J. A **32**, 87 (2007).
87. U. Hager *et al.*, Phys. Rev. Lett. **96**, 042504 (2006).
88. A. Kankainen *et al.*, Eur. Phys. J. A **29**, 271 (2006).
89. C. Weber *et al.*, Phys. Rev. C **78**, 054310 (2008).
90. U. Hager *et al.*, Nucl. Phys. A **793**, 20 (2007).
91. U. Hager *et al.*, Phys. Rev. C **75**, 064302 (2007).
92. V.-V. Elomaa *et al.*, Eur. Phys. J. A **40**, 1 (2009).
93. S. Rahaman *et al.*, Phys. Lett. B **703**, 412 (2011).
94. J.S.E. Wieslander *et al.*, Phys. Rev. Lett. **103**, 122501 (2009).
95. V.-V. Elomaa *et al.*, Phys. Rev. Lett. **102**, 252501 (2009).
96. V.S. Kolhinen *et al.*, Phys. Lett. B **697**, 116 (2011).
97. V.S. Kolhinen *et al.*, Phys. Rev. C **82**, 022501 (2010).

DOI 10.1007/978-94-007-5555-0_4
Reprinted from *Hyperfine Interactions* Volume 223, Issues 1-3, 2014
and *The European Physical Journal A* Volume 48, Issue 4, 2012

Collinear laser spectroscopy techniques at JYFL

B. Cheal · D. H. Forest

Published online: 30 March 2012

Abstract Over the last 15 years, a collinear laser spectroscopy programme has been developed at the University of Jyväskylä Accelerator Laboratory (JYFL), Finland. Continuous technique development and exploitation has taken place to address physics cases inaccessible elsewhere. In particular, the use of ion beams from the IGISOL and the pioneering application of cooled and bunched beams to laser spectroscopy. Many of these advances have additionally now been exported to facilities worldwide.

Keywords Collinear · Laser · Spectroscopy · IGISOL · Cooler

1 Introduction

Laser spectroscopy [1] is the model-independent study of nuclear properties from optical spectra. Collinear laser spectroscopy was first demonstrated more than thirty years ago [2] and has since become the dominant high-resolution variant of the technique. This method has formed the basis of experiments conducted by the collaboration. Following the closure of the UK Daresbury Nuclear Structure Facility, the universities of Birmingham and Manchester formed a new collaboration with the University of Jyväskylä, which has housed the spectroscopy apparatus since 1994.

Hyperfine structure is the splitting and perturbation of atomic levels and spectral lines due to the nucleus. The nuclear magnetic dipole moment, μ, interacts with the magnetic field caused by the electrons and the nuclear spectroscopic quadrupole moment, Q_s, interacts with the electric field gradient of the electron distribution.

B. Cheal (✉)
School of Physics and Astronomy, University of Manchester, M13 9PL, Manchester, UK
e-mail: Bradley.Cheal@manchester.ac.uk

D. H. Forest
School of Physics and Astronomy, University of Birmingham, B15 2TT, Birmingham, UK

Calibration of the atomic factors is achieved by measuring the hyperfine structure for a single isotope where the nuclear moments are already known. The number of hyperfine peaks and the precise pattern of their frequencies is determined by the nuclear and electronic angular momenta. This permits a measurement of the nuclear spin, I, which also manifests itself in the relative intensities of the hyperfine peaks. Since optical transitions (and their atomic factors) are characteristic of the atomic number, laser spectroscopy systematically studies the isotopic chain of an element. The shift of the resonance position, or centroid of the hyperfine structure, between isotopes (or isomeric states) is known as the isotope shift, comprising mass shift and field shift components. The first arises from the nuclear recoil motion in the atomic centre-of-mass frame, the second, from the changing potential felt by the electrons as the nuclear proton distribution changes. Once the mass shift and the electron density at the centre of the nucleus are accounted for, the change in mean–square charge radius, $\delta\langle r^2\rangle$, can be deduced. These four quantities (μ, Q_s, I, $\delta\langle r^2\rangle$) provide a comprehensive description of the nuclear state, enabling exploration of single-particle and collective phenomena. The mean-square charge radius is also sensitive to the surface diffuseness and a complementary probe of deformation. An overview of the physics results which have been obtained from laser spectroscopic studies at JYFL is given in a separate article in this volume.

The Accelerator Laboratory at the University of Jyväskylä had already developed the IGISOL technique [3–5] offering universal isotope production and separation irrespective of physical and chemical properties. Refractory elements not available from conventional ISOL facilities recoil from thin foil targets and are stopped in a noble buffer gas. Sub-millisecond extraction times mean that the technique also gains favour for production of short-lived isotopes. The collaboration has continued to apply collinear laser spectroscopy to such beams while improving the sensitivity and applicability of the technique.

2 The collinear beams geometry

In the spectroscopy apparatus, the ion beam and laser beam are overlapped in an anti-parallel configuration. A photomultiplier tube (PMT, ~10% quantum efficiency) is placed perpendicular to the beams to detect fluorescence photons as a function of frequency. From the classical definition of kinetic energy, a velocity spread is caused by the energy spread of the ion source and leads to a Doppler broadening of the resonances. While the latter remains constant under acceleration through vacuum, the velocity spread in the forward direction, $\Delta v = \Delta E/mv$, is reduced by the beam velocity. A typical beam energy of 30 keV is sufficient to reduce an energy spread of a few eV such that the Doppler broadening is within the natural line width of the resonances. Rather than scan the laser frequency, an additional, tuning voltage is applied to the region of interaction. This acts to Doppler shift the effective frequency seen by the ions over the hyperfine resonances, while the laser is stabilised to a fixed frequency. In a few cases, spectroscopy of the neutral atom has been preferable to the ion, in which case the ions are neutralised while passing through an alkali vapour, and the tuning voltage is applied at this point (rather than the light collection region). The focussing and overlap of the laser and ion or atom beams are ensured using a 0.9 mm diameter removable aperture, while a 6 mm diameter removable aperture placed ~3 m upstream ensures a low angular divergence. An imaging light collection region (using a quartet of fresnel lenses with a solid angle efficiency of 9%) is designed to capture resonant photons while discriminating against non-resonant photons and maintaining the signal-to-noise ratio. The laser beam is produced using a continuous-wave narrow line-width ring dye laser with intra-cavity frequency doubling. Most transitions studied require less than ~1 mW. Applying a Doppler tuning voltage permits the laser to be locked to a single frequency. This is achieved using two interferometers and absorption lines of molecular iodine, achieving long-term stabilisation to within ~2 MHz.

Continuous non-resonant scattering of the laser beam into the PMT dominates the photon background, causing a typical background rate of 200 photons per second (per mW). Prior to the installation of the cooler and adoption of the bunching technique (see Section 4.2), the background was suppressed in all experiments by only accepting photons in delayed coincidence with the corresponding ion [6] (or atom [7]). Position sensitivity along the photon detection region (with a 16-fold horizontally segmented PMT) can enhance the time resolution (to 20 ns) thereby reducing the gate width and enhancing the background suppression factor [8].

3 Application of IGISOL to laser work

At conventional thick target, hot cavity ISOL production facilities, high yields of many elements are available. However, long delays may be incurred while the reaction products diffuse out of the target material—a duration which may be many times the half-life of a short-lived state. Even if sufficient in-target production were available to compensate, the flux may be swamped by isobaric components. Moreover, if the physical or chemical properties are unfavourable, it can be that the yields are substantially reduced, with virtually no production of many refractory elements.

In the IGISOL technique [5], ions from nuclear reactions recoil directly from thin foil targets into a helium buffer gas at typically 100–200 mbar. The ions, which due to impurities below the part-per-billion level remain predominantly in a singly-charged state, leave the ion guide through a nozzle in a supersonic helium gas jet. Different ion guide designs are used for light-ion induced fusion evaporation reactions and for fission reactions. In each case however, the technique is chemically insensitive and achieves sub-millisecond extraction times. The helium gas is then removed from the outer chamber of the IGISOL while electrodes extract the ions into a differential pumping region. An extraction electrode then accelerates the ions into beam line quality vacuum and for mass selection with a dipole magnet. For laser spectroscopy, the beam was delivered from a switchyard via two 90° bends in a twisted periscope arrangement to the collinear interaction apparatus on the floor above.

Originally, a skimmer electrode (1.35 mm hole diameter) was used to guide the ions into the differential pumping region, and both the skimmer and extraction electrodes had longitudinal adjustment. Acceleration of the ions between the ion guide and skimmer occurs in a region of poor vacuum and collisions with helium atoms occur. Higher skimmer voltages produce larger yields but with a corresponding increase in energy spread. Even with the velocity compression of the collinear fast-beam method, a reported energy spread of 100 eV [9] would result in significant Doppler broadening of the optical resonances. However, a local maximum in yield

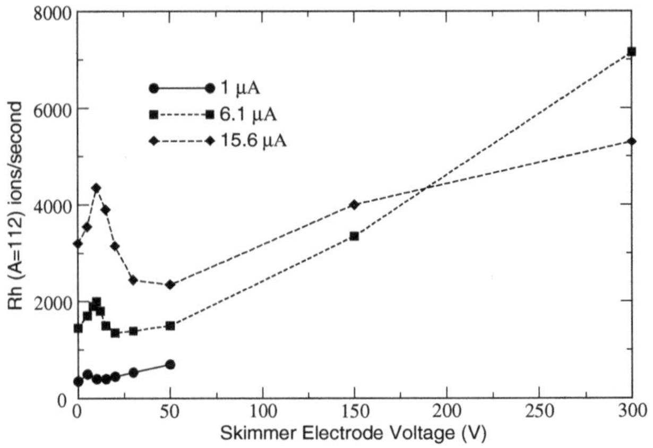

Fig. 1 Fission ion-guide yield of ^{112}Rh as a function of skimmer voltage for 50 MeV H_2^+ on ^{238}U (beam intensity as indicated)

was found to exist at a low skimmer voltage (see Fig. 1) allowing sufficient yields to be produced without significant energy spread (~5 eV).

Collinear laser spectroscopy was first demonstrated for radioactive ^{140,142,144}Ba ions produced from a fission reaction [10, 11]. This early test case was a prelude to the first on-line laser spectroscopy measurements of a refractory element, namely the study of radioactive ^{170,172,173,174}Hf isotopes from fusion evaporation reactions [12].

4 Installation of the cooler-buncher

In order to achieve higher yields from the ion guide, by optimising the skimmer for yields alone, an ion beam cooler [13] was installed in 2000 to reduce the energy spread of the ion beam. This device, a gas-filled linear radio-frequency quadrupole, is mounted on a high voltage platform at ~100 V below the potential of the separator to provide initial deceleration and capture of the incoming beam. Inside, the ions pass through a 40 cm length while confined by an RF-driven electric quadrupole. Through viscous collisions with ~1 mbar of helium buffer gas, the ions lose their residual energies and the energy spread is reduced as the ions thermalise. With the remaining longitudinal motion governed by diffusion alone, a weak axial field is additionally applied (via segmentation of the electrodes) to guide the ions to the exit region within 1 ms. Here, the ions (which may be accumulated with a trapping potential and bunched) are extracted through a miniature quadrupole into an 800 V transfer line before re-acceleration by the platform potential to the experimental set up. In comparison with the IGISOL, the extraction takes place in much better vacuum using a weaker field gradient.

Shortly after the installation of the cooler, the IGISOL area also underwent a major upgrade in 2003 [14]. This included increased IGISOL pumping capability and improved ion guide designs. As a result of this, and the replacement of the skimmer electrode with a SextuPole Ion Guide (SPIG) in 2005 [15], a significant enhancement in yields was seen [14, 15]. The SPIG also removed the practical problems of aligning the small skimmer and ion guide nozzles, particularly with the rotatable dual-target fusion guide [16].

4.1 Cooled beams for laser spectroscopy

Extracted beams from the JYFL cooler have been shown to have an emittance of 3π·mm·mrad and a longitudinal energy spread consistently below 1 eV even with a primary beam energy spread of 100 eV [13, 17]. This allowed optimisation of the IGISOL to produce higher yields while maintaining a much improved emittance and longitudinal energy spread. Additionally, the cooled beam is decoupled from source conditions.

A smaller longitudinal energy spread prevents significant Doppler broadening of the optical resonances. This allows for smaller hyperfine structures to be resolved and a higher spectroscopic efficiency to be achieved. Doppler broadening reduces the number of ions on resonance with the narrow line-width laser, increasing the resonant peak-width at the expense of peak-intensity. An additional increase in the intensity of optical transitions from the ionic ground state is due to the quenching of metastable states inside the cooler.

A smaller ion beam emittance allows the ion and laser beams to be overlapped in the collinear geometry more effectively. By achieving a narrower beam waist while maintaining a low angular divergence, a lower laser power can be used to provide the same power density, contributing less to the photon background.

Laser spectroscopy of the isomeric state ^{130m}Ba was performed using a cooled (continuous) beam flux of 150 isomers/s [18]. This experiment (as all other preceding experiments at JYFL) used the photon–ion coincidence technique. Theoretically, the sensitivity of this method can now be improved with the use of cooled beams, since the narrower beam waist will provide better imaging and a coincidence timing resolution below 5 ns [8].

4.2 Ion beam bunching with fluorescence detection

Although theoretically, the photon–ion coincidence method can potentially achieve a sensitivity required to perform measurements on fluxes of an ion per second, this is prevented in practice by impurities. Isobaric components in the beam, which may be far larger than the isotope under study, trigger "false coincidences". Photons are only accepted within a ~5 ns time window, which is opened on detection of an ion. If contaminant ions hit the micro-channel plates and each create such a window, then many non-resonant photons will be accepted and contribute to the background.

Releasing the ions from the cooler in bunches provides an alternative method of suppressing the constant amount of laser scatter and is less sensitive to lack of beam purity (a problem for all ISOL facilities). Ions are accumulated in the extraction region of the cooler by applying a trapping potential to the end plate, which also acts to seal the gas volume. Every 10–500 ms cycle, an ion bunch is released and only photons detected within a 15 μs gate, corresponding to the transit through the interaction region, are counted. In this way, the dominant source of the background is suppressed by the ratio of these times, typically four orders-of-magnitude. With this method, the hyperfine structure of ^{176m}Yb was measured with a flux of 100 ions/s [19]. Moreover, the improvement over the photon–ion coincidence method is shown in Fig. 2. Despite one-sixth of the ion current from the IGISOL, and one-seventh of the data collection time, the bunching method produces a much clearer resonance. This method was pioneered at JYFL and applied initially to the study of Hf [20] and Zr [21].

Fig. 2 Fluorescence spectra of ^{88}Zr using single-ion coincidence prior to the installation of the cooler (5.25 h of data collection with an ion flux of 12,000 ion/s) and instead using ion bunch gating with the cooler (45 min of data collection with an ion flux of 2,000 ions/s)

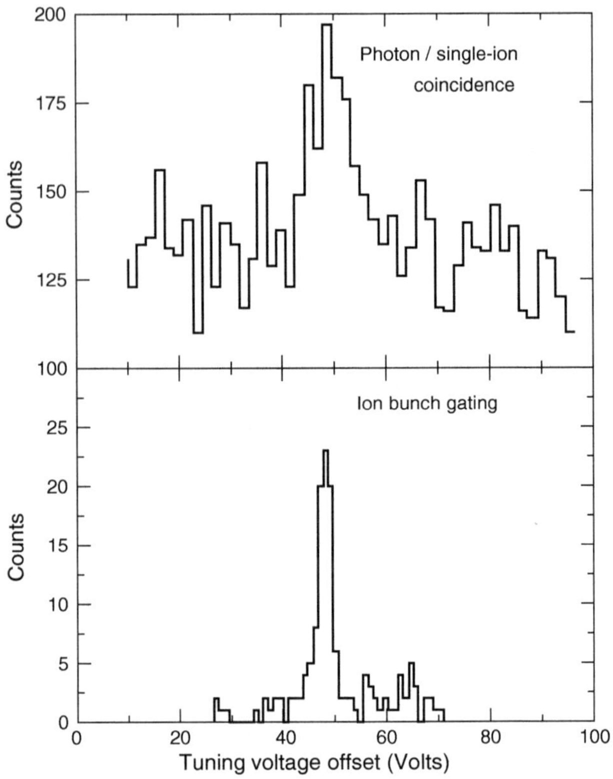

Tests were later performed with an alternative method of releasing the ion bunches from the cooler. The miniature quadrupole mounted inside the end plate was removed leaving a small aperture in the end plate. Ions were released in 10 ms bunches and detected using micro-channel plates at the end of the laser line. Time-of-flight analysis showed that the end plate produced temporally shorter bunches than the miniature quadrupole, as shown in Fig. 3. However, laser measurements revealed that the energy spread was larger with just an end plate and the miniature quadrupole has remained in use.

4.3 Collinear resonance ionisation spectroscopy

In resonance ionisation spectroscopy, most notably used in laser ion sources, atoms are stepwise exited to ionisation [22–25]. By tuning the laser frequency across the lowest (or sometimes other) step in the excitation sequence and counting (deflected) ionised particles rather than fluorescent photons, an optical spectrum is produced. Collinear resonance ionisation spectroscopy (CRIS) is an attempt to combine the high-resolution of the collinear technique [2] with the sensitivity of particle detection (or nuclear decay tagging). High power, and therefore pulsed, lasers are required for ionisation, which leads to duty cycle loses for continuous ion beams. Only with bunched beams, does the opportunity arise to expose all atoms to the laser beams. Preliminary tests were conducted at JYFL [26, 27] and provided a proof of principle of the technique.

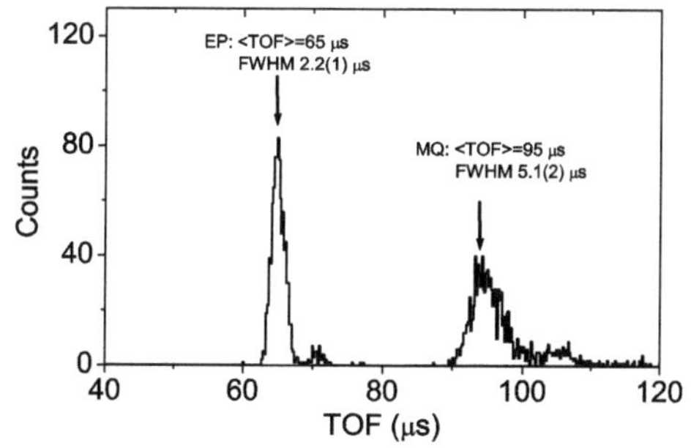

Fig. 3 Time of flight spectrum using two different methods for releasing ion bunches from the cooler. An end plate (EP) skimmer-style extraction produces a temporally shorter bunch than the mini-quadrupole (MQ) but with a larger energy spread (The small secondary peaks in each case are due to delayed ions which each have a water molecule attached, originating from cooler gas impurities.)

4.4 Development of coolers for laser spectroscopy at other facilities

Following the successful application of an ion beam cooler to laser spectroscopy at JYFL, an ion beam cooler ("ISCOOL") was designed and installed [28, 29] immediately downstream of the high resolution separator (HRS) at ISOLDE, CERN. Following commissioning for laser spectroscopy [30], cooled and bunched beams have been delivered and have already led to successful collinear laser fluorescence experiments on isotopes of gallium [31] and copper [32]. In a parallel development, a new beam line dedicated to CRIS experiments (which require bunched beams to avoid laser duty cycle losses) has been constructed and will be used for laser spectroscopy of francium isotopes.

At the TRIUMF TITAN facility, an ion beam cooler has been installed [33]. This is being applied to a new collinear laser spectroscopy programme [34], and has recently enabled the measurement of the neutron-deficient ($N = Z$) isotope ^{74}Rb.

5 Optical pumping in the cooler-buncher

Laser spectroscopy is generally performed using electronic transitions from the ground state due to reasons of population. Many elements chosen for nuclear structure interest have however been unavailable for efficient study due to the unsuitability of such transitions in these cases. Neutralisation with alkali vapour to the atomic state (performed at JYFL using a variety of techniques) can be beset with similar problems or be inefficient due to the spreading of the population over atomic metastable states.

Excitation of ground states within the ion beam cooler-buncher results in optical pumping and the efficient redistribution of the ground state population to a selected metastable state. This enhancement persists on acceleration and delivery of the ion beam to the collinear set-up, where collinear spectroscopy can be performed using transitions from the chosen metastable state in preference to the ground state [35]. Unlike the high-resolution continuous-wave lasers used for collinear spectroscopy, pulsed lasers, which can readily access higher harmonics and a larger wavelength range, are used for the in-cooler optical pumping. In this case, the broad line-width of the pulsed lasers has no effect of spectroscopic resolution and matches the Doppler broadened velocity profile of the ions well.

Fig. 4 Layout of the JYFL IGISOL facility

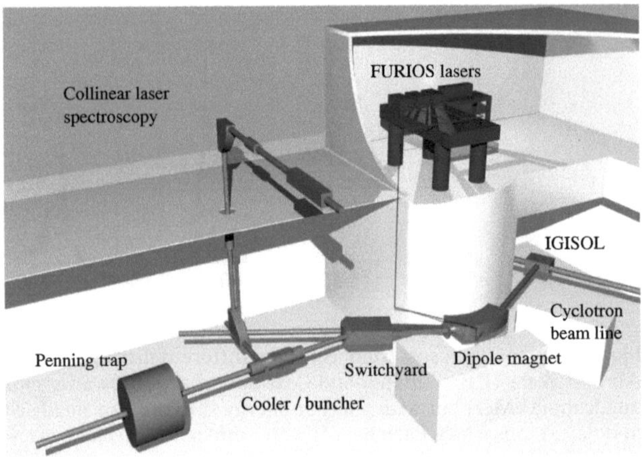

For the study of manganese [36] (and future studies of molybdenum [37]), the wavelengths of transitions from the ground state were too short to be efficiently achieved using a narrow line-width continuous wave laser. Optical pumping was used to enhance the population of a metastable state from which an alternative transition could be efficiently studied. Despite the transition from the ground state having a very weak spectroscopic efficiency, the extended laser–ion interaction time in the cooler permitted efficient excitation [36]. This was also the case for niobium [35]; although the ground state transition was accessible, the weak spectroscopic efficiency was more suited to in-cooler excitation than collinear fast-beam techniques. More importantly, the ionic ground state in this case, and in yttrium [38, 39], had an angular momentum of $J = 0$, limiting all transitions from the ground state to $J = 0 \rightarrow J' = 1$. This limits the hyperfine structure to three peaks and an extraction of the magnetic dipole moment, electric quadrupole moment and mean-square charge radius only in cases where a previous assignment of the nuclear spin has been made. Optical pumping allows transitions from a metastable state of higher-J to be studied, permitting a measurement of the spin also.

Figure 4 shows the the layout of the laboratory at the time of these experiments. Lasers from the FURIOS project [25] (shown here entering the IGISOL in their configuration for the laser ion source) are used for the optical pumping. To gain optical access to the cooler, the Penning trap had to be disconnected and the laser light shone through a quadrupole bend [40] to illuminate the central axis and trapping region of the cooler. In the new IGISOL 4 laboratory (discussed in a separate article in this volume) the design will enable uninterrupted access to the cooler while all experimental stations remain connected.

6 Summary

Application of the collinear technique to beams from an IGISOL has enabled and will continue to enable optical studies of elements not available elsewhere. These include short-lived states and refractory elements—many of which remain completely unexplored. Techniques developed at JYFL such as cooling and bunching of ion beams, and more recently, optical pumping, have greatly improved the sensitivity and applicability of the technique. These improvements in methodology have since been exported to other facilities.

Acknowledgements This work has been supported by the UK Science and Technology Facilities Council (STFC), the EU 7th framework programme "Integrating Activities—Transnational Access", project number: 262010 (ENSAR) and by the Academy of Finland under the Finnish Centre of Excellence Programme 2006–2011 (Nuclear and Accelerator Based Physics Research at JYFL).

References

1. Cheal, B., Flanagan, K.T.: J. Phys. G: Nucl. Part. Phys. **37**(11), 113101 (2010)
2. Wing, W.H., Ruff, G.A., Lamb, W.E., Spezeski, J.J.: Phys. Rev. Lett. **36**(25), 1488 (1976)
3. Ärje, J., Äystö, J., et al.: Nucl. Instrum. Methods Phys. Res., Sect. A **247**(3), 431 (1986)
4. Penttilä, H., et al.: Nucl. Instrum. Methods Phys. Res., Sect. B **126**(1–4), 213 (1997)
5. Äystö, J.: Nucl. Phys. A **693**(1–2), 477 (2001)
6. Eastham, D.A., Walker, P.M., Smith, J.R.H., Griffith, J.A.R., Evans, D.E., Wells, S.A., Fawcett, M.J., Grant, I.S.: Opt. Commun. **60**(5), 293 (1986)
7. Eastham, D.A., Walker, P.M., Smith, J.R.H., Warner, D.D., Griffith, J.A.R., Evans, D.E., Wells, S.A., Fawcett, M.J., Grant, I.S.: Phys. Rev. C **36**(4), 1583 (1987)
8. Eastham, D.A., Gilda, A., Warner, D., Evans, D., Griffiths, J., Billowes, J., Dancey, M., Grant, I.: Opt. Commun. **82**(1–2), 23 (1991)
9. Koizumi, M., et al.: Nucl. Instrum. Methods Phys. Res., Sect. A **313**(1–2), 1 (1992)
10. Billowes, J., et al.: Nucl. Instrum. Methods Phys. Res., Sect. B **126**(1–4), 416 (1997)
11. Cooke, J.L., et al.: J. Phys. G: Nucl. Part. Phys. **23**(11), L97 (1997)
12. Levins, J.M.G., et al.: Phys. Rev. Lett. **82**(12), 2476 (1999)
13. Nieminen, A., Huikari, J., Jokinen, A., Äystö, J., Campbell, P., Cochrane, E.C.A.: Nucl. Instrum. Methods Phys. Res., Sect. A **469**(2), 244 (2001)
14. Karvonen, P., et al.: Nucl. Instrum. Methods Phys. Res., Sect. B **266**(19–20), 4454 (2008)
15. Karvonen, P., Moore, I.D., Sonoda, T., Kessler, T., Penttilä, H., Peräjärvi, K., Ronkanen, P., Äystö, J.: Nucl. Instrum. Methods Phys. Res., Sect. B **266**(21), 4794 (2008)
16. Elomaa, V.V., et al.: Eur. Phys. J. A **40**(1), 1 (2009)
17. Nieminen, A., et al.: Nucl. Instrum. Methods Phys. Res., Sect. B **204**, 563 (2003)
18. Moore, R., et al.: Phys. Lett. B **547**(3–4), 200 (2002)
19. Bissell, M.L., et al.: Phys. Lett. B **645**(4), 330 (2007)
20. Nieminen, A., et al.: Phys. Rev. Lett. **88**(9), 094801 (2002)
21. Campbell, P., et al.: Phys. Rev. Lett. **89**(8), 082501 (2002)
22. Letokhov, V.: Opt. Commun. **7**(1), 59 (1973)
23. Letokhov, V., et al.: In: On-Line in 1985 and Beyond, A Workshop on the ISOLDE Programme, Zinal, CERN (1984)
24. Marsh, B.A., et al.: Hyperfine Interact. **196**, 129 (2010)
25. Moore, I.D., et al.: J. Phys. G: Nucl. Part. Phys. **31**(10), S1499 (2005)
26. Billowes, J.: Nucl. Phys. A **682**(1–4), 206 (2001)
27. Campbell, P., et al.: Eur. Phys. J. A **15**(1–2), 45 (2002)
28. Jokinen, A., Lindroos, M., Molin, E., Petersson, M.: Nucl. Instrum. Methods Phys. Res., Sect. B **204**, 86 (2003)
29. Podadera Aliseda, I., Fritioff, T., Giles, T., Jokinen, A., Lindroos, M., Wenander, F.: Nucl. Phys. A **746**, 647 (2004)
30. Mané, E., et al.: Eur. Phys. J. A **42**(3), 503 (2009)
31. Cheal, B., et al.: Phys. Rev. Lett. **104**(25), 252502 (2010)
32. Flanagan, K.T., et al.: Phys. Rev. Lett. **103**(14), 142501 (2009)
33. Smith, M., Blomeley, L., Delheij, P., Dilling, J.: Hyperfine Interact. **173**(1–3), 171 (2006)
34. Mané, E., et al.: Hyperfine Interact. **199**, 357 (2011)
35. Cheal, B., et al.: Phys. Rev. Lett. **102**(22), 222501 (2009)
36. Charlwood, F.C., et al.: Phys. Lett. B **690**(4), 346 (2010)
37. Charlwood, F.C., et al.: Phys. Lett. B **674**(1), 23 (2009)
38. Cheal, B., et al.: Phys. Lett. B **645**(2–3), 133 (2007)
39. Baczynska, K., et al.: J. Phys. G: Nucl. Part. Phys. **37**(10), 105103 (2010)
40. Charlwood, F.C., et al.: Hyperfine Interact. **196**, 143 (2010)

DOI 10.1007/978-94-007-5555-0_5
Reprinted from *Hyperfine Interactions* Volume 223, Issues 1-3, 2014
and *The European Physical Journal A* Volume 48, Issue 4, 2012

Conversion electron spectroscopy at IGISOL

Sami Rinta-Antila · Wladyslaw Trzaska · Juho Rissanen

Published online: 3 April 2012

Abstract Conversion elecron spectroscopy has been an important part of the nuclear spectrocopy research at the Department of Physics of the University of Jyväskylä since the commissioning of the first cyclotron in the mid 1970s. At the IGISOL facility a specialiced conversion electron spectrometer ELLI was developed in the late 1980s. The first results with ELLI were obtained using the beams from the old MC-20 cyclotron to study newly discovered isotopes of refractory fission products. In the present K130 cyclotron laboratory ELLI has been utilized in many decay-spectroscopy experiments both neutron-deficient and neutron-rich side of the valley of stability. In the early 2000s the new JYFLTRAP ion trap system overthrew ELLI from its permanent place in the IGISOL beamline. Conversion electron spectroscopy has continued with the new Penning trap that has been used in in-trap electron spectroscopy tests and post-trap electron spectroscopy is foreseen.

Keywords Conversion electron spectroscopy · On-line mass separator · Trap-assisted spectroscopy

1 Introduction

Nuclear spectroscopy was the dominant research line at the Department of Physics, University of Jyväskylä (JYFL) already in the early days of the accelerator physics there. In that respect JYFL was basically following the world-wide boom made possible by the availability of commercial germanium detectors. However, what set JYFL apart, already at that time, was the equal research effort devoted to the electron spectroscopy. For instance, during the first full year of operation of the MC-20 cyclotron (1976) 60 of the total of 167 days of beam on target (36%) were

S. Rinta-Antila (✉) · W. Trzaska · J. Rissanen
Department of Physics, University of Jyväskylä, P.O. Box 35 (YFL), 40014 Jyväskylä, Finland
e-mail: sami.rinta-antila@phys.jyu.fi

used for in-beam gamma-ray spectroscopy and 55 days (33%) for in-beam electron spectroscopy. Similar ratio between gamma and electron measurements prevailed throughout the late 1980s. Naturally, it has been self evident that low-energy transitions are heavily converted and that in numerous cases measuring the conversion electron spectra is not just the best way but the only way of looking for exotic excited states or finding the missing transitions. But, to our knowledge, no other laboratory had such a pronounced commitment to the electron spectroscopy as JYFL in the late 1970s and 1980s.

2 Electron spectroscopy at JYFL

Unlike in the case of gamma-ray detection, there are no off-the-shelf solutions for high-resolution electron measurements. The first precision electron spectrometers were developed in Sweden in the mid 1950s. This pioneering role was recognized by the Nobel Committee by awarding in 1981 the Prize in Physics to Kai M. Siegbahn, together with Nicolaas Bloembergen and Arthur L. Schawlow. Indeed, what gave JYFL a considerable head start in electron spectroscopy were the magnetic spectrometers obtained from Sweden [1] and the passion of the late Prof. Juhani Kantele, one of the founders of JYFL and the first chair in experimental nuclear physics in Finland. The main methodological breakthrough that made conversion electron spectroscopy a viable addition to gamma-ray measurements was made in parallel at the tandem laboratory in Uppsala [2] and at JYFL [3]. It was accomplished by combining a magnetic spectrometer operating in a broad-band lens mode with a modern semiconductor detector. The lens mode secured very good background suppression, the broad transmission provided high efficiency, and the excellent resolution was guaranteed by the cooled detector. The most successful of these combination spectrometers was a Siegbahn–Slätnis type intermediate-image magnetic lens beta-ray spectrometer that came to JYFL via Åbo Akademi around 1980. After minor modifications and with a new power supply and a 5 mm thick Ge(HP) in the focal plane this device has established the high-energy record in conversion electron spectroscopy by measuring E0 transitions in ^{208}Pb at above 5 MeV of excitation energy [4, 5].

3 First conversion electron measurements at IGISOL

The invention of IGISOL [6] in 1982 has opened new research opportunities at JYFL. In particular, the possibility to mass separate even the short-lived fission products enabled the production of radioactive ion beams in a very broad mass range. What followed very quickly was the discovery of many new, neutron-rich isotopes, especially in the region $42 \leq Z \leq 47$. The identification of those isotopes was made by well established spectroscopic techniques. At first the use of a plain silicon detector was sufficient to obtain adequate conversion electron spectra. Mass separated sources, the use of tape transport to remove buildup of activities and coincidence requirements yielded adequate background reduction during the first experiments. Already this simple approach resulted in many new discoveries in the mass region $110 \leq A \leq 117$ [7]. Nevertheless, by the end of 1980 it became obvious that IGISOL needed a dedicated electron spectrometer.

By that time a significant progress was achieved in electron spectroscopy. In addition to the bulky Siegbahn-type spectrometers with massive iron yokes and powerful current supplies several new technologies became available. With the arrival of strong, permanent magnets it became possible to assemble compact and self sustained spectrometers consisting of the central rod protecting the silicon detector from the direct radiation from the source and a ring of magnetic blades attached to the central rod to bend the trajectories of electrons towards the detector [8]. These mini-orange spectrometers were often operated in sets as each covered only a limited momentum range. The small size and a care-free operation were at the same time the strongest but also the only assets of mini-oranges.

The second novelty was the supercondicting magnet. The growing industrial and medical demand for such devices significantly reduced the price and made them available to the scientific community at large. As the result many groups world-wide opted for building their conversion electron spectrometers using such magnets. They offered uniform and strong magnetic fields (up to several Tesla) assuring very high transport efficiencies when used in the solenoid mode. The ability to profile the shape of the field by changing the current in the selected coils gave the addition degree of trajectory control. However all these benefits were to some degree compromised by the costs and labor to maintain the coils in the liquid helium temperature.

Finally, the enormous explosion of easily accessible computing power brought by the advent of PCs allowed for creative design of iron-free non-superconducting coils for magnetic spectrometers. After thorough consideration this option was chosen for ELLI.

4 ELLI spectrometer

ELLI stands for Electron Lens for IGISOL. ELLI was the first conversion electron spectrometer that was designed and build for the detection, identification, and study of exotic isotopes separated with IGISOL. To work with mass separated low-intensity radioactive beams the spectrometer had to have high efficiency and incorporate a tape drive to remove the buildup of background activities, see Fig. 1. It was desired to have a provision for up to three detectors allowing for simultaneous measurements of conversion electrons, betas, and gamma- or X-rays. Since oil-free vacuum was still a rarity at that time, the cooled detectors required a cold trap. To tune the separator with a stable ion beam, a provision for a removable Faraday cup had to be provided to place it just in front of the implantation point. Further, compatibility with the existing power sources and detector support systems was required. ELLI is described in details in [9]. As is the case for all solenoid spectrometers, the detection efficiency is a function of the current and electron energy ranging from 15% at 250 keV to 5 % at 2 MeV. The overall background reduction as compared with a plain silicon detector in close geometry is about a factor of 4.

ELLI design included several new solutions; for instance, the abandonment of the axial implantation of the beam. As a rule, all previous designs of similar devices had the beam enter along the symmetry axis through the hole in the detector. Such

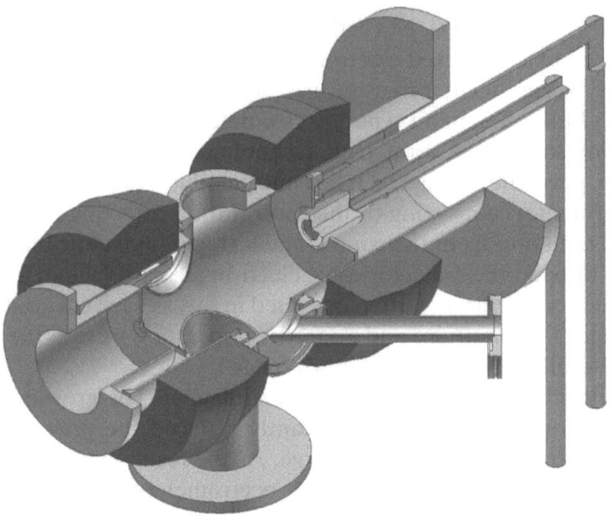

Fig. 1 ELLI spectrometer. The beam enters through the angled beam tube and is implanted in to a movable tape in the centre of the left hand side coil. The cooled Si(Li) detector is facing the implantation point in the centre of the other coil. Surrounding the Si(Li) detector is the cold trap to collect any oil vapour from the vacuum chamber before cooling the detector. Other detectors (Si for β and Ge/LEGe for γ and X-rays) can be inserted into the recessed aluminium cup giving access close to the implantation point. The removable faraday cup for beam tuning is not pictured here but it can be rotated in and out of the beam without breaking the vacuum

annular detectors are very expensive and their performance (resolution and noise level) is often inferior to the regular detectors of the same size. Actually, thanks to the large mass difference between electrons and nuclei there is no need to worry too much about the deflection of the ions in the magnetic fields chosen for bending of electron trajectories. In case of ELLI separated beam enters at a steep 45 degree angle through the gap between the two coils arranged in the Helmholtz configuration. This novel idea, invented for ELLI, of the beam entering at an oblique angle with respect to the symmetry axis of the magnetic field has been subsequently adopted by SACRED [10] and SAGE [11] spectrometers.

5 Measurements with ELLI

The first accelerator at JYFL (the MC-20 cyclotron) has stopped operation in 1991 and the new K=130 became operational in 1993. The tests, commissioning, and the first measurements with ELLI were made still in the old laboratory and continued with the beams from the new cyclotron.

Among the first reported results was the beta decay of ^{114}Ru and Q_β systematics for neutron-rich Ru isotopes [12],the beta decay of ^{113}Rh and the observation of ^{113m}Pd [13], Gamow–Teller decay of ^{118}Pd and the new isotope ^{120}Pd [14], deformation studies in the neutron-rich Zr region [15], and the decay of ^{108}Mo and of the neighboring even Mo-isotopes [16]. At the new cyclotron a new isomer in ^{125}La was discovered [17], isomeric state of ^{80}Y and its role in the astrophysical rp-process was investigated [18], as well as isomers of astrophysical interest in neutron-defficient nuclei at masses $A = 81$, 85 and 86 [19]. ELLI was even used at ISOLDE, CERN in the investigation of non-analog beta decay of ^{74}Rb [20]. The latest experiment using ELLI was an off-line measurement of ^{133m}Xe conversion coefficient from a source produced and implanted at IGISOL [21]. This particular isomer is important in indication of nuclear explosion and is therefore monitored by Comprehensive Nuclear-Test-Ban Treaty Organization (CTBTO) with sampling stations located around the globe.

The installation of the cooler and the Penning trap at the start of the new millennium brought new, unique research possibilities to IGISOL at the same time contributing to the demise of the role of the traditional spectroscopic methods. Nevertheless, after two decades of service ELLI remains in the working condition awaiting new challenges at JYFL or experiments at other facilities.

6 In-trap conversion electron spectroscopy

As mentioned above with the arrival of RFQ cooler and JYFLTRAP Penning trap the main focus of the IGISOL physics program turned towards utilization of ion traps in beam preparation and performing atomic mass measurements. Right from the trap design phase the idea was to harness the high mass resolving power of a Penning trap to benefit the decay spectroscopy both after the trap as well as inside the trap [22].

The concept of an ideal trapped decay source was first developed for precision study needed for search of scalar and tensor currents in weak interaction. In practise the trapped source technique allows for detection of the beta-decay daughter ion recoiling out of the trap. Examples of this kind of ion trap installations are LPC trap at Caen [23] and WITCH trap at ISOLDE [24], the former is a transparent Paul trap and the latter is an open Penning trap. The benefit of a trapped decay source was also recognised for high precision decay spectroscopy. As the trapped ions confined by magnetic and electric fields are emitting conversion electrons the scattering and energy loss in the backing material present in an implanted source are avoided. Furthermore, the strong magnetic field of a Penning trap can be employed to transport electrons efficiently into a small area high resolution detector and thus suppressing γ- or X-ray background.

As described in the previous chapters the conversion electron detection efficiency over γ- or X-ray efficiency is usually increased by introduction of guiding magnetic solenoid field. The effect of the magnetic field is two fold in the case of an implanted source. It increases the electron collection efficiency by capturing and guiding electrons emitted within a certain cone to the detector. On the other hand some of the electrons emitted to larger angles lose considerable portion of their energy in the backing material which shows up in the energy spectrum as a low energy tail of electron peaks. The deterioration of the spectrum becomes more evident with increasing magnetic field and source implantation depth. With a trapped electron source there is not such limitation.

The first feasibility test of in-trap conversion electron spectroscopy was carried out at ISOLDE, CERN using REX-ISOLDE accumulator Penning trap REXTRAP to confine the source ions. A EB10GC-500P detector assembly with a low noise PA1201 preamplifier was acquired from Canberra Semiconductor for the test. The 500 μm

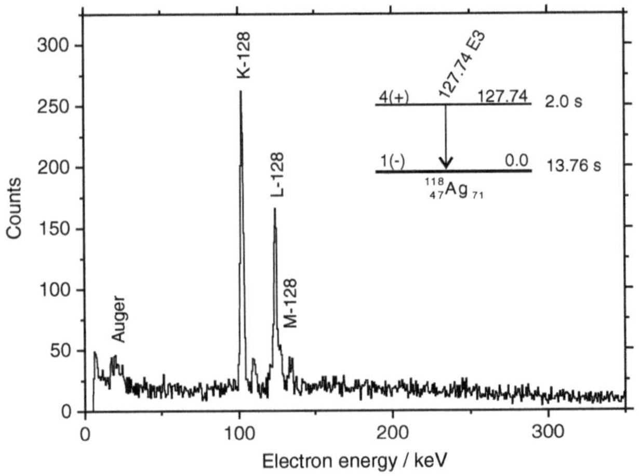

Fig. 2 In-trap electron spectrum recorded for ^{118m}Ag. The K, L, and M conversion electron lines of the 127.7 keV transition are labelled. The *inset* shows the studied isomeric transition in ^{118}Ag

thick detector has 10 mm^2 active area (3.3 mm diameter) and dead layer of 250 Å. The feasibility of the method was demonstrated by measuring isomeric decays of on-line separated ^{116m}In and ^{118m}In sources [25]. The promising results encouraged to bring the detector setup to Jyväskylä where the foreseen physics program and expertise was more suited for further investigation of the method.

So far only one on-line test has been run at JYFLTRAP [26]. In total ten conversion electron emitting isomers of neutron-rich isotopes produced in proton induced fission of ^{238}U were studied. Among these was the same ^{118m_2}In isomer studied earlier at ISOLDE. Compared to the first test some improvement could be seen in the resolution. The showcase of the Jyväskylä test experiment was the detection of conversion electrons at 9.9 keV from short lived ^{117m}Pd isomeric state with half-life of only 19.1 ms. Electron energy spectrum from another measured isomeric decay of ^{118m}Ag is shown in Fig. 2. The largest peaks are K, L, and M conversion electron lines of the 127.0 keV transition at 102.2 keV, 123.9 keV, and 127.0 keV, respectively. The measured peaks exhibit an outstanding energy resolution, taking a 2.2 keV resolution for the 102.2-keV peak as an example. The small peaks at slightly higher energies arise from isomeric decay of ^{118m_2}In that was present in the trap as contamination. At the low end of the spectrum Auger electron peaks are visible. From the spectrum one can see that besides the very good energy resolution the overall background level and the energy threshold are quite low.

Although the first test experiments have been succesfull there are still some problems related to this technique. The confinement of electrons with high magnetic field all the way to the detector enhances the backscattering which is roughly proportional to the square of the electron impact angle.

Compared to the ELLI spectrometer the in trap electron spectroscopy method is still lacking simultaneous gamma-ray detection. Some issues with detector contamination by recoiling highly charged radioactive ions was suspected in the REXTRAP experiment. Furthermore the tuning of the Penning trap becomes challenging when the extraction side is blocked with electron detector. Due to these complications, of which some could be overcome by investing efforts to R&D, the studied physics case should be such that it is impossible or hard to be tackled with traditional techniques. The experimental target could be ultra-high-resolution electron spectroscopy or detection of very low-energy electrons. The problem could also be turned around in a way that conversion electron detection would be used to image the trapped ion cloud and hence to diagnose the trapping process itself.

7 Future of conversion electron spectroscopy at IGISOL

As it has been seen in the decay-spectroscopy experiments concentrating in gamma or neutron detection, the trend has been towards utilizion of the Penning trap as a high-resolution mass filter to provide even isomerically pure sources for experiments. The same is foreseen also for the conversion electron spectroscopy. One option for this kind of experiment would be to install the ELLI spectrometer behind the trap but also other detectors without a magnetic lens could be considered.

For the in-trap technique to be used in more demanding decay studies it probably requires some further development as was outlined in the [26]. A more sophisticated in-trap electron detector with a larger area and segmentation would allow coincidence detection between electrons. An easy insertion and retraction of the detector without breaking the vacuum would be desirable not least for the trap tuning point of view but it also would allow quick checking of emitted conversion electrons in the midst of a post-trap spectroscopy experiment. The data readout and acquisition from the electron detector could be simplified by operating the trap in the ground potential rather than the presently used high-voltage platform. This can be achieved by use of a pulse-down drift-tube in the trap injection line. The ability to detect coincident electromagnetic radiation together with the conversion electrons would enhance the applicability of the technique. A limited choise of gamma-ray detectors can be considered due to the tight space limitation and high magnetic field inside the trap magnet. An option with a reasonable resolution would be an array of LaBr crystals combined with silicon based photomultipliers surrounding the trapping region in a close geometry.

Conversion electron spectroscopy has had an important role throughout the life of decay spectroscopy at IGISOL. It all started with a simple silicon detector and tape transport system, followed by a specialiced ELLI electron spectrometer and has now evolved into a more complex in-trap and post-trap spectroscopy set-ups. Surely in the future the conversion electron spectroscopy will have its place in the decay spectroscopic studies carried out at the IGISOL facility independent of the techniques used for preparing the source or detecting the electrons.

Acknowledgements This work has been supported by the Academy of Finland under the Finnish Centre of Excellence Programme 2000–2005 (Project No. 44875, Nuclear and Condensed Matter Physics Programme at JYFL).

References

1. Kleinheinz, P., Samuelsson, L., Vukanovic, R., Siegbahn, K.: Nucl. Instrum. Methods Phys. Res. **32**, 1 (1965)
2. Westerberg, L., Edvardson, L., Madueme, G., Thun, J.: Nucl. Instrum. Methods Phys. Res. **128**, 61 (1975)
3. Kantele, J., Luontama, M., Passoja, A., Julin, R.: Nucl. Instrum. Methods Phys. Res. **130**, 467 (1975)

4. Julin, R., Kantele, J., Kumpulainen, J., Luontama, M., Passoja, A., Trzaska, W., Verho, E.: Phys. Rev. C **36**, 1129 (1987)
5. Julin, R., Kantele, J., Kumpulainen, J., Luontama, M., Nieminen, V., Passoja, A., Trzaska, W., Verho, E., Blomqvist, J.: Nucl. Instrum. Methods Phys. Res. A **270**, 74 (1988)
6. Ärje, J., Äystö, J., Hyvönen, H., Taskinen, P., Koponen, V. Honkanen, J., Valli, K., Hautojärvi, A., Vierinen, K.: Nucl. Instrum. Methods Phys. Res. A **24**, 774 (1986)
7. Äystö, J., Davids, C., Hattula, J., Honkanen, J., Honkanen, K., Jauho, P., Julin, R., Juutinen, S., Kumpulainen, J., Lönnroth, T., Pakkanen, A. Passoja, A., Penttilä, H., Taskinen, P., Verho, E., Virtanen, A., Yoshii, M.: Nucl. Phys. A **480**, 104 (1988)
8. van Klinken, J., Feenstra, S., Dumont, G.: Nucl. Instrum. Methods **151**, 433 (1978)
9. Parmonen, J., Janas, Z., Trzaska, W., Äystö, J., Kantele, J., Jauho, P., Jokinen, A., Penttilä, H.: Nucl. Instrum. Methods Phys. Res. A **306**, 504 (1991)
10. Kankaanpää, H., Butler, P.A., Greenlees, P.T., Bastin, J.E., Herzberg, R.D., Humphreys, R.D., Jones, G.D., Jones, P., Julin, R., Keenan, A., Kettunen, H., Leino, M., Miettinen, L., Page, T., Rahkila, P., Scholey, C., Uusitalo, J.: Nucl. Instrum. Methods Phys. Res. A **534**, 503 (2004)
11. Papadakis, P., Herzberg, R.-D., Pakarinen, J., Greenlees, P.T., Sorri, J., Butler, P.A., Coleman-Smith, P.J., Cox, D., Cresswell, J.R., Hauschild, K., Jones, P., Julin, R., Lazarus, I.H., Letts, S.C., Parr, E., Peura, P., Pucknell, V.F.E., Rahkila, P., Sampson, J., Sandzelius, M., Seddon, D.A., Simpson, J., Thornhill, J., Wells, D.: J. Phys.: Conf. Ser. **312**, 052017 (2011)
12. Jokinen, A., Äystö, J., Jauho, P.P., Leino, M., Parmonen, J.M., Penttilä, H., Eskola, K., Janas, Z.: Nucl. Phys. A **549**, 420 (1992)
13. Penttilä, H., Enqvist, T., Jauho, P., Jokinen, A., Leino, M., Parmonen, J., Äystö, J., Eskola, K.: Nucl. Phys. A **561**, 416 (1993)
14. Janas, Z., Äystö, J., Eskola, K., Jauho, P.P., Jokinen, A., Kownacki, J., Leino, M., Parmonen, J.M., Penttilä, H., Szerypo, J., Zylicz, J.: Nucl. Phys. A **552**, 340 (1993)
15. Lhersonneau, G., Pfeiffer, B., Kratz, K.L., Enqvist, T., Jauho, P.P., Jokinen, A., Kantele, J., Leino, M., Parmonen, J.M., Penttilä, H., Äystö, J.: Phys. Rev. C **49**, 1379 (1994)
16. Jokinen, A., Enqvist, T., Jauho, P.P., Leino, M., Parmonen, J.M., Penttilä, H. Äystö, J., Eskola, K.: Nucl. Phys. A **584**, 489 (1995)
17. Canchel, G., Béraud, R., Chabanat, E., Emsallem, A., Redon, N., Dendooven, P., Huikari, J., Jokinen, A., Kolhinen, V., Lhersonneau, G., Oinonen, M., Nieminen, A., Penttilä, H., Peräjärvi, K., Wang, J., Äystö, J.: Eur. Phys. J. A **5**, 1 (1999)
18. Novikov, Y., Schatz, H., Dendooven, P., Braud, R., Mieh, C., Popov, A., Seliverstov, D., Vorobjev, G., Baumann, P., Borge, M., Canchel, G., Desagne, P., Emsallem, A., Huang, W., Huikari, J., Jokinen, A., Knipper, A., Kolhinen, V., Nieminen, A., Oinonen, M., Penttilä, H., Peräjärvi, K., Piqueras, I., Rinta-Antila, S., Szerypo, J., Wang, Y., Äystö, J.: Eur. Phys. J. A **11**, 257 (2001)
19. Kankainen, A., Vorobjev, G., Eliseev, S., Huang, W., Huikari, J., Jokinen, A., Nieminen, A., Novikov, Y., H. Penttilä, Popov, A., Rinta-Antila, S., Schatz, H., Seliverstov, D., Suslov, Y., Äystö, J.: Eur. Phys. J. A **25**, 355 (2005)
20. Oinonen, M., Äystö, J., Baumann, P., Cederkäll, J., Courtin, S., Dessagne, P., Franchoo, S., Fynbo, H., Górska, M., Huikari, J., Jokinen, A., Knipper, A., Köster, U., Scornet, G.L., Miehé, C., Nieminen, A., Nilsson, T., Novikov, Y., Peräjärvi, K., Poirier, E., Popov, A., Seliverstov, D., Siiskonen, T., Simon, H., Tengblad, O., Duppen, P.V., Walter, G., Weissman, L., Wilhelmsen-Rolander, K.: Phys. Lett. B **511**, 145 (2001)
21. Peräjärvi, K., Turunen, J., Hakala, J., Jokinen, A., Moore, I.D., Penttilä, H., Saastamoinen, A., Siiskonen, T., Toivonen, H., Äystö, J.: App. Rad. Isotopes **66**, 530 (2008)
22. Szerypo, J., Jokinen, A., Kolhinen, V., Nieminen, A., Rinta-Antila, S., Äystö, J.: Nucl. Phys. A **701**, 588 (2002) (5th International Conference on Radioactive Nuclear Beams)
23. Rodríguez, D., Méry, A., Ban, G., Brégeault, J., Darius, G., Durand, D., Fléchard, X., Herbane, M., Labalme, M., Linard, E., Mauger, F., Merrer, Y., Naviliat-Cuncic, O., Thomas, J., Vandamme, C.: Nucl. Instrum. Methods Phys. Res. A **565**, 876 (2006)
24. Beck, M., Ames, F., Beck, D., Bollen, G., Delauré, B., Golovko, V.V., Kozlov, V.Y., Kraev, I.S., Lindroth, A., Phalet, T., Quint, W., Schuurmans, P., Severijns, N., Vereecke, B., Versyck, S.: Nucl. Instrum. Methods Phys. Res. A **503**, 567 (2003)
25. Weissman, L., Ames, F., Äystö, J., Forstner, O., Reisinger, K., Rinta-Antila, S.: Nucl. Instrum. Methods Phys. Res. A **492**, 451 (2002)
26. Rissanen, J., Elomaa, V., Eronen, T., Hakala, J., Jokinen, A., Rahaman, S., Rinta-Antila, S., Äystö, J.: Eur. Phys. J. A **34**, 113 (2007)

DOI 10.1007/978-94-007-5555-0_6
Reprinted from *Hyperfine Interactions* Volume 223, Issues 1-3, 2014
and *The European Physical Journal A* Volume 48, Issue 4, 2012

THE EUROPEAN PHYSICAL JOURNAL A

Review

Fission yield studies at the IGISOL facility

H. Penttilä[1,a], V.-V. Elomaa[1], T. Eronen[1], J. Hakala[1], A. Jokinen[1], A. Kankainen[1], I.D. Moore[1], S. Rahaman[1,b], S. Rinta-Antila[1], J. Rissanen[1], V. Rubchenya[1,2], A. Saastamoinen[1], C. Weber[1,c], and J. Äystö[1]

[1] Department of Physics, P.O.Box 35 (YFL), FI-40014 University of Jyväskylä, Finland
[2] V.G.Khlopin Radium Institute, 194021, St. Petersburg, Russia

Received: 31 January 2012
Published online: 18 April 2012 –
Communicated by E. De Sanctis

Abstract. Low-energy-particle–induced fission is a cost-effective way to produce neutron-rich nuclei for spectroscopic studies. Fission has been utilized at the IGISOL to produce isotopes for decay and nuclear structure studies, collinear laser spectroscopy and precision mass measurements. The ion guide technique is also very suitable for the fission yield measurements, which can be performed very efficiently by using the Penning trap for fission fragment identification and counting. The proton- and neutron-induced fission yield measurements at the IGISOL are reviewed, and the independent isotopic yields of Zn, Ga, Rb, Sr, Cd and In in 25 MeV deuterium-induced fission are presented for the first time. Moving to a new location next to the high intensity MCC30/15 light-ion cyclotron will allow also the use of the neutron-induced fission to produce the neutron rich nuclei at the IGISOL in the future.

1 Introduction

Fission of the actinides —typically uranium or thorium— has been utilized since the discovery of fission to produce neutron-rich nuclei for nuclear spectrometry. The other possible reactions to reach neutron-rich nuclei — spallation, massive transfer and in-flight fragmentation— demand technically much more from the experimental facilities. Neutron-rich nuclei can be produced even without a reactor or an accelerator via spontaneous fission of a suitable source of radioactive nuclei such as ^{252}Cf. Low-energy-particle–induced fission provides inexpensive means to produce neutron-rich nuclei, suitable in particular for a relatively small laboratory.

When the IGISOL technique was developed in the early 1980s [1–7], the opportunity of proton-induced fission of ^{238}U to produce neutron-rich nuclei was readily observed. In particular, the capability of the IGISOL technique to provide ions of any element made it possible to access the short-lived isotopes of refractory elements, an area in the chart of nuclides that had not been accessible for isotope separators before. The first tests at the IGISOL facility with nuclear fission took place in June 1983. Neutron-rich nuclei up to the limit of the known nuclei and beyond were observed [7].

It was soon realized that the isotropic distribution and high recoil energy of the fission fragments gave an additional degree of freedom in the design of the ion guide. The first fission ion guide with target and stopping volumes separated by a foil was built in 1986 [7]. The first published physics papers were a study of the level structure evolution of even Pd nuclei, studied via the beta decay of Rh isotopes [8] and a study of the new isotopes ^{115,116}Rh [9].

Since then, fission has been routinely used to produce neutron-rich nuclei for nuclear structure studies at the IGISOL. The isotopes of the refractory elements from zirconium to palladium have been a specialty of the IGISOL facility. The thick target ISOL facilities ionize these elements with low efficiency, often not at all. Fragmentation and in-flight fission can produce high-energy beams of exceedingly neutron-rich isotopes of these elements, but the beams cannot be easily decelerated to the low energies needed, for example, in the laser spectroscopy of the isotopes. In fact, the technique to generate a low-energy, pencil-like beam from the energetic fragments in current facilities such as NSCL in Michigan [10] and RIKEN [11], as well as in the future fragmentation facilities FRIB [12] and FAIR [13] is based on the use of gas cells that are direct descendants of the original ion guide.

During the research at IGISOL since 1985, properties of over 150 neutron-rich isotopes have been studied. Thirty of these isotopes were observed for the first time.

[a] e-mail: heikki.t.penttila@jyu.fi
[b] *Present address*: Physics Division, P-23, Mail Stop H803, Los Alamos National Laboratory, Los Alamos, NM 87545, USA.
[c] *Present address*: Faculty of Physics, Experimental Nuclear Physics, Am Coulombwall 1, D-85748 Garching, Germany.

The performed studies include nuclear structure studies via beta-gamma decay spectroscopy, beta delayed neutron studies, total absorption spectroscopy, collinear laser spectroscopy for nuclear shapes, and atomic mass measurements, first via beta end point measurements and after commissioning JYFLTRAP, with much higher precision with the Penning trap [8,9,14–28].

The question about the limits of proton-induced fission as a production method of the neutron-rich nuclei prompted systematic studies of the fission yields of proton-induced fission of ^{238}U and ^{232}Th. These experiments were necessary for estimating the isotopic yields to be able realistically plan nuclear structure experiments, since very little experimental data of the proton-induced fission yields in the region of symmetric proton-induced fission existed. The first peer-reviewed report of proton-induced fission yields was published in 1991 [29]. The collected experimental data soon turned out to be important in helping the development of theoretical models of not only proton-induced fission but the particle-induced fission in general. In [30] the measured yields were compared to the fission model of Rubchenya, whose model has been further developed in close collaboration with the experimental effort at IGISOL —see [31,32].

The later fission yield studies in the 1990s resulted in the discovery of the highly asymmetric or superasymmetric fission mode in the proton-induced fission of ^{238}U [33–35]. Before the upgrade of the IGISOL facility to IGISOL 3 in 2003, experiments of the yields in neutron-induced fission were performed using break-up neutrons of deuterium [36] and evaporation neutrons from ^{13}C target [37]. In 2005 a work was started to develop a new method to determine the fission yields utilizing JYFLTRAP. The method turned out to be successful [38], and fission yields of 25 MeV proton-induced fission of ^{238}U [39] and ^{232}Th [40] as well as fission yields for selected elements in 50 MeV proton and 25 MeV deuteron-induced fission of ^{238}U were measured before IGISOL 3 was closed for the move in summer 2010. These experiments are discussed more in detail below.

In addition to Jyväskylä, the ion guide technique has been utilized to determine fission yields in Lyon, France, for alpha-induced fission of ^{238}U [41], in Louvain-la-Neuve, Belgium, for proton-induced fission of ^{238}U [42–45], and in Sendai, Japan, for proton-induced fission of ^{238}U and ^{232}Th [46–49]. The work in Sendai with 25 MeV protons on ^{238}U target was confirmed in a concurrent study in Jyväskylä [36].

2 Experimental techniques

2.1 Merits and concerns of the ion guide technique

The ion guide technique is well suited to determine the independent fission yields for several reasons:

- The fission yield distribution extends over 40 elements from nickel region to the heaviest lanthanides. All those elements are accessible with the ion guide method.
- With the ion guide technique the fission products are extracted from the target to the mass separated sample in tens of milliseconds that is fast as compared with the majority of half-lives of fission products.
- All the ions in the mass separated beam of fission products come directly from fission. The only build-up effects take place in the mass separated sample, which significantly simplifies the analysis.

The major concerns of the possible limitations of the method are:

- The fission fragments are energetic heavy ions with a large variety of initial energies and different stopping powers. Only a fraction of ions that are produced in fission is stopped in helium and only a fraction of the stopped ones is transported through the separation and detected. Does this sample represent the ensemble of fission fragments?
- Althought the method is free from chemical limitations, in the sense that it can produce ions of any element, it is not totally independent of the electronic structure of ions. Some elements such as yttrium are known to rapidly form oxides, while quite a few other elements are not. Lanthanides are also relatively easily extracted from the ion guide as 2+ ions. How large are efficiency variations between elements? How much do they influence on the deduced fission yields?
- Are there mass-dependent effects? Is the efficiency of the ion guide equal for all isotopes of an element?
- Fission fragment distribution in particle-induced fission at several tens of MeV is not precisely isotropic due to momentum brought in the fissioning system by the inducing particle. Is it necessary to take these effects into account? Is it possible?

In the following, some of the concerns are discussed.

The question of the ensemble can be approached in two ways. The process of slowing down and stopping the fission products can be simulated using the fission ion guide geometry. Such attempts were made already in [50,51]. It is however challenging to take into account all processes taking place in the ion guide. Another approach to solve the dilemma is to compare the measured fission yields to preceding measurements, performed using other techniques. Unfortunately, for proton-induced fission, such earlier measurements are rare.

One of the best sets of data on independent yields in proton-induced fission, which is widely cited, *e.g.*, in theoretical works, is the one by Tracy *et al.* [52]. In the work of Tracy *et al.*, isotopic yields of Rb and Cs isotopes in proton-induced fission of ^{238}U were measured at several proton energies. These energies were higher than typically used at IGISOL measurements. Therefore, in [38] a dedicated yield measurement for 50 MeV proton-induced fission of ^{238}U was performed to be able to make a comparison. The isotopic yields of Rb and Cs obtained using the ion guide technique were found to be in agreement with the results of [52].

A comparison can also be made between IGISOL works and the fission yield measurements performed at Tohoku

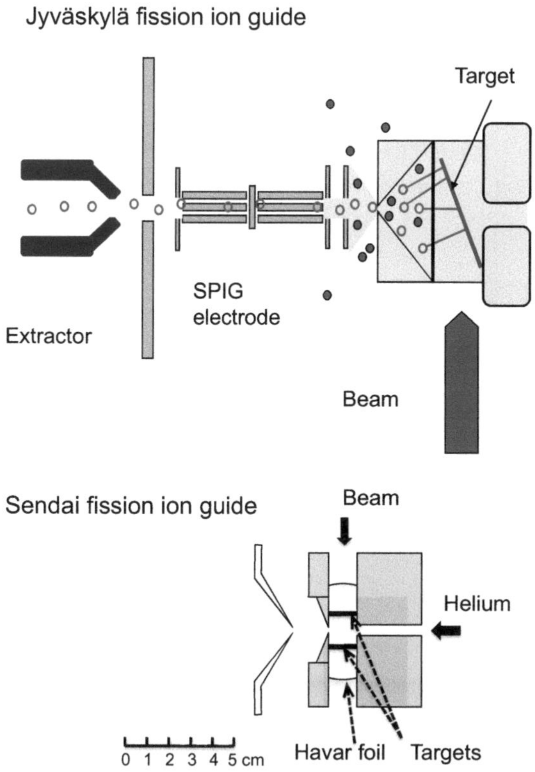

Fig. 1. Schematic views of the double chamber fission ion guide used at the IGISOL and the ion guide used for fission yield studies in Sendai [47]. In the Jyväskylä fission ion guide a thin metal foil separates the target volume from the stopping volume, placed in 90° with respect to the primary beam. In the Sendai ion guide the stopping volume is in between the targets, covering the forward direction of the first fissioning target and the backward direction of the second one. The figure is adapted from Kudo *et al.* [47].

University in Sendai using the local ion guide system [46–49] during the 1990s for proton-induced fission of ^{238}U and ^{232}Th. Though both facilities utilize the ion guide technique, there are differences between the fission ion guides used in Sendai and in Jyväskylä that are worth of pointing out.

Figure 1 shows a schematic view of the fission ion guide used in Jyväskylä having a target tilted at 7° with the respect of beam and a stopping volume separated from the primary beam by a thin nickel foil, and the Sendai fission ion guide with two targets and a stopping volume in between them [47]. In the Sendai ion guide the primary proton beam generates plasma in the evacuated gas volume, which is also so small that only a very small fraction of the fission recoils are stopped. (A double chamber fission ion guide was built also in Sendai [46], but was not used in the yield measurements.) On the other hand, the gas volume is evacuated extremely fast, and its location in between the targets covers equally all the directions that a fission fragment can be emitted, thus compensating the anisotropy of the fission fragment distribution. On the contrary, in the Jyväskylä fission ion guide the stopping volume is located in 90° with respect to the beam direction. There is however no evidence that yield distribution would be different in different angles in anisotropic fission. Furthermore, the ions that are stopped in the buffer gas have very low kinetic energy when they enter in the gas. They have lost their energy in multiple scattering in the target, so the effects due to their original angular distribution can be expected to have vanished.

The 25 MeV proton-induced fission of ^{238}U was studied in both laboratories [36,47]. In [47] the proton energy is given as 24 MeV between the targets (fig. 1). The initial energy of the 25 MeV protons used in [36] is degraded to 24.3 MeV in the half way of the tilted uranium target. In both works the yields were extracted from the cumulative yields of isotopes of a beta decay chain utilizing gamma ray spectroscopy with germanium detectors. The absolute yield cross-sections cannot be extracted from either measurement, but the most probable nuclear charge of fission fragments, Z_p, as function of mass number A was deduced. The results are shown in fig. 2.

The deduced $Z_p(A)$ in these works are in a reasonable agreement, despite the differences between the ion guides. From the result it can be concluded first of all that the possible anisotropy of the fission fragment angular distribution at 25 MeV is too small to have an effect on the measured independent yield distributions. In addition, the factor of about 50 in the stopping volume sizes does not make an observable difference either. In the Jyväskylä fission ion guide, the gas cell is large enough to stop ions whose energy is less than about 2 MeV when they enter in the gas. In the smaller Sendai fission ion guide the respective energy is less than 500 keV. Even so, the collected samples seem to have equal distribution of fission products, which gives reason to believe that the both represent the ensemble of fission products.

2.2 Extracting fission yields using gamma ray spectroscopy

In a typical fission yield measurement [29,30,33–36,42,43,47,51] a mass separated source of fission fragments was implanted on a movable tape in front of a radioactivity counting set-up. In Sendai, also another counting station for long-lived species was used [47]. The set-up consisted of one or more germanium gamma ray detectors. In Jyväskylä, they were typically run in coincidence with a ΔE β-detector, which reduced the background and increased the visibility of the gamma peaks. It is worth of noticing that the beta detection efficiency is function of the β-decay energy: close to the stability the β-efficiency is lower, since larger fraction of the continuous beta particle spectrum lies below the detection threshold. The single gamma ray spectrum of saturated source was thus used in the analysis when applicable. Finally, the tape could be periodically moved to enhance the visibility of the most short-lived activities over the long-lived isobaric background.

To be able to deduce the fission yields from the singles gamma ray spectra it is essential to know both the

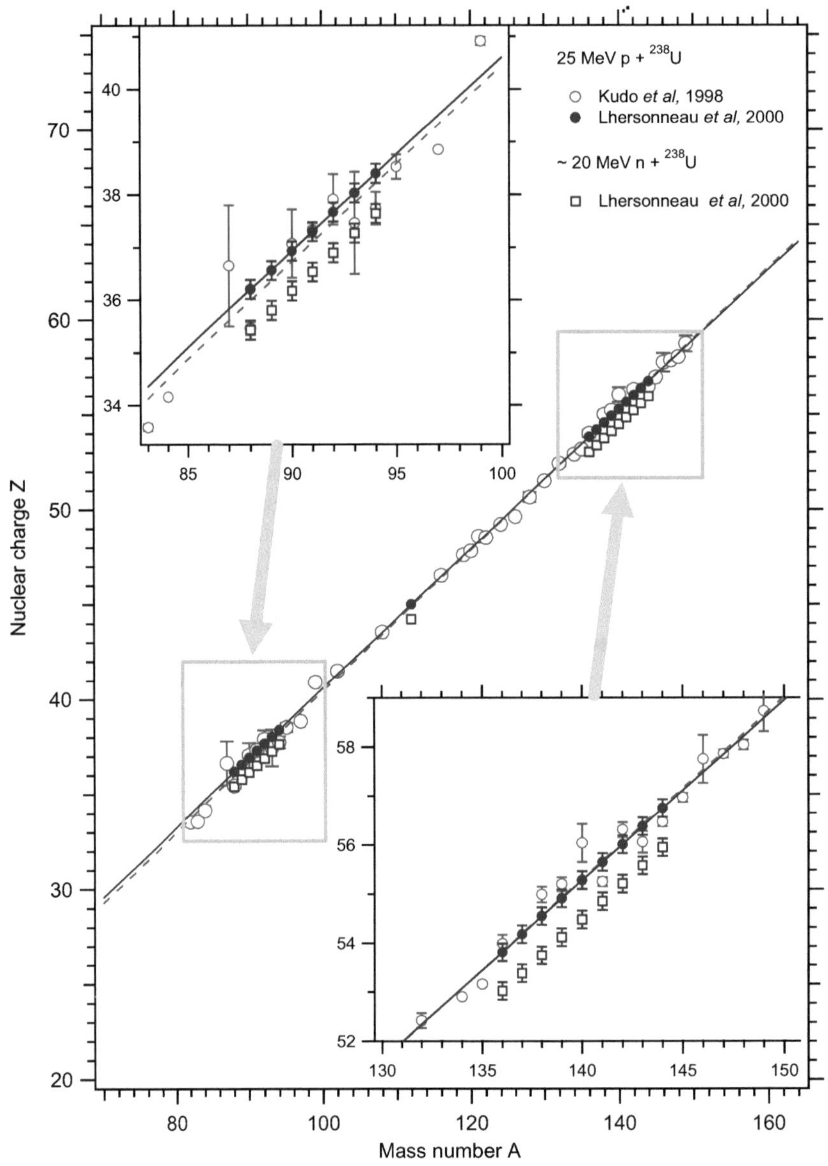

Fig. 2. The most probable nuclear charge Z_p in proton- and neutron-induced fission of ^{238}U. The data are taken from refs. [36, 47]. The linear fits to Z_p of proton-induced fission are shown as solid [36] and dashed [47] lines. The discrepancy between the lines is entirely due to Z_p in masses $A = 82–84$ that are not measured in [36].

detector efficiency and the branching ratios of the decaying isotopes with a good accuracy. In particular, far from stability, the knowledge of the decay schemes is often sparse; specifically the ground-state branchings are not known. A decent detector efficiency calibration is relatively easy to obtain with reasonable effort. The calibration can also be checked using internal calibrations. A notable problem with the detectors is the coincident summing of the gamma rays. Since the rates of gamma rays are low, the detectors are brought to a close geometry to improve the efficiency, which, in turn, leads to true coincidence summing. Making corrections to coincidence summing would again require an accurate knowledge of decay schemes that often does not exist.

If the half-life is so short that the saturation activity is reached, no further information on the half-life is needed: the decay rate equals the cumulative yield. However, as soon as the collected activity is periodically removed, precise half-lives are necessary to calculate the actual yields from the observed activity.

2.3 Direct yield measurements using the Penning trap

All these aforementioned works [29, 30, 33–36, 42, 43, 47, 51] suffered from the limitations of the detection of decay radiations and the limited knowledge of the details of the decays.

The installation of JYFLTRAP [53] allowed to overcome some of the difficulties related to the yield measurements based on decay spectroscopy. JYFLTRAP can be used as a high-resolution mass filter with a mass-resolving power (MRP) of the order of 10^5. This is adequate to separate the isotopes within an isobar A and unambiguously identify the fission products by their mass instead of their decay. The fission products are detected with microchannel plate (MCP) detectors with high efficiency. The fission yields are deduced from the rate of mass separated ions. Since the ions are observed directly, no knowledge of the decay schemes is needed. In fact, since no decay is needed either, also the long-lived and stable isotopes are detected with the same efficiency —and equally fast— as the short-lived ones. There is essentially no background in the recorded mass spectra, which means that almost every observed ion can be identified. Already a very low number of ions is adequate for a yield measurement, which usually more than compensates the fact that the high MRP is reached on the cost of transmission through the Penning trap.

The developed method to extract the fission yields from the high MRP mass spectra obtained with JYFLTRAP is described in detail in [38]. The benefits of the method are obvious. Some limitations of the method that still ask for development in the future are discussed in the following.

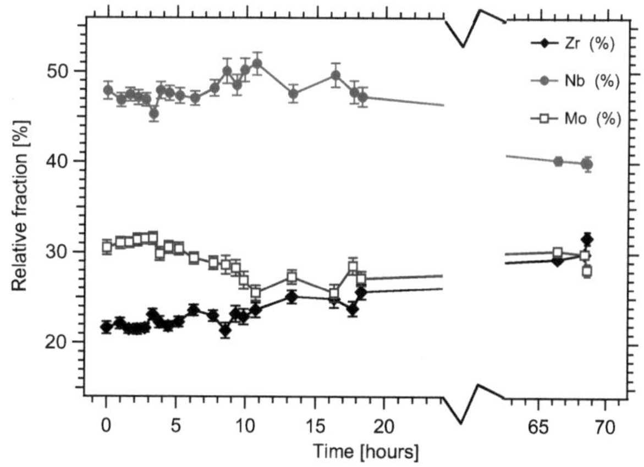

Fig. 3. Evolution of the mass separated yields of Zr, Nb and Mo in a three-day-long experiment. The yield is determined from the mass peaks of singly charged isotopes in $A = 101$, whose mass spectrum was repetitively measured as a reference during the experiment. Since Zr forms a stable monoxide ion easier than Nb or Mo, the change can be interpreted as diminishing of the ZrO fraction as the system —in particular the ion guide— becomes cleaner during the experiment.

Chemistry in the gas-filled traps. The ion guide technique itself is not as such free from chemical effects (*e.g.*, molecule formation, charge distribution dependence on buffer gas purity; see also the discussion of the ion formation and loss rates in [54]). However, the time the ions spend in the fission ion guide is of the order of tens of milliseconds. If necessary, the delay time in the ion guide can be decreased by using smaller ion guide, but as discussed above in sect. 2.1, this seems not currently to be an issue.

The experiments clearly show that the relative yield of elements changes during the measurement. In the fission yield measurements of proton-induced fission of ^{238}U [38] the $A = 101$ isotopes of zirconium, niobium and molybdenum were used as reference isotopes and their yields were repetitively determined over a long period. The evolution of the relative yields is shown in fig. 3. These yield ratios do not carry any information on the absolute yields. It is a well-known tendency that the efficiency of an ion guide improves towards the end of an experiment. This has been interpreted to be due to the purification of the system while heated by the primary ion beam. If that were correct, the change of yield ratios shown in fig. 3 would indicate that the improvement of efficiency due to cleaner conditions is element dependent.

However, the high-resolution mass separation takes place in traps that are also gas-filled devices, albeit in lower pressure. The time ions spent in the gas-filled traps is yet about hundred times longer than the average delay of the ion guide. It is shown [55] that molecule formation takes place in the trap. In principle, the observed effect in fig. 3 can as well be due to changed transmission efficiency of the trap system.

An alpha recoil source has been used to test the ion guides [56]. The alpha decay recoil ions from such a source are stopped in buffer gas, mimicking the ions from induced reactions. The rate of ions can be accurately determined by measuring the alpha activity of the source. The recoil ions have been used to determine the absolute efficiency of the ion guide. In case of fission, a standard fission source such as ^{252}Cf placed inside a fission ion guide could be a way to determine the chemical efficiencies. Such a source has been used, *e.g.*, at LISOL laser ion source in Louvain-la-Neuve to study the extraction efficiency of fission products in argon buffer gas [57]. Calibration measurements with such a source could turn out to be important in the future to improve the accuracy of the yield measurements.

Resolving isomers. Another issue to be properly solved in the fission yield measurements is the determination of the isomeric ratios (the ratio of production rate of the nuclei in the ground state and a possible metastable excited state). Many of the isomeric states can actually be mass separated using the Penning trap. The record-high MRP of over 500000 was reached in separation of ^{133m}Xe from the ground state of ^{133}Xe [58] using special techniques in an experiment concentrating on the separation of this particular isomer only. In a more typical case a MRP of 10^5 or less is achieved. Around mass $A = 100$ this means that the mass peak width is of the order of 1 MeV. In these conditions, mass peaks corresponding isomers that are about 0.5 MeV apart, can be separated by a fit. Most of the isomers that are long-lived enough to be detected at all are however closer to the ground state.

The total intensity of the ground state and the isomer can be normally determined without difficulty. This result must however be corrected for the radioactive decay losses. There are cases where the two half-lives give a vastly different decay correction, and the isomeric ratio should be known. In addition, the isomeric decay seems to be the only case where the daughter of a nucleus decaying in the trap is observed. In beta decay, the daughter will be at least doubly charged because of the change of the nuclear charge. In addition, many electrons are shaken off in the beta decay, resulting in an even more highly charged ion [59]. The multiply charged ions trapped in the RFQ are shown to be quite persistent [35]. The same happens if the isomeric state decays via an internal transition by emitting a conversion electron: the result is a doubly charged ion that will not be observed. However, if the internal transition goes via gamma ray emission, the resulting ion is singly charged and it will be seen in the mass spectrum. This effect is less than a few per cent even in the worst cases, but it needs to be taken into account in the analysis when the precision of the method is improved in the future.

Accumulation of decay products in the trap. The non-observation of the decay products trapped in the Penning trap [38] seems to be incompatible to the experimental results at ISOLTRAP [60]. In the case of EC/β^+ decay of ^{37}K the decay of the trapped singly charged ions leads to a neutral system, and the shake-off of electrons is needed to ionize the decay product in the first place. In the case of fission fragments that undergo β^--decay, the decay product is doubly charged, on top of which comes the electron shake-off. The decay products have not been observed, since they have been multiple charged.

In any case, the next improvement of the method would be the effective means to determine the isomeric ratios in the fission yield measurements. If it is considered sufficient to measure the total yield of the isotope, there are, in fact, only about a dozen isomeric ratios to be measured. In the other cases, the isomeric and the ground-state half-lifes are either similar enough to result in the same decay correction, or both long enough for the correction to be negligible.

3 Fission yield measurements

The fission yields of mostly ^{238}U induced by protons, deuterons and neutrons have been studied at the IGISOL. In the following, some of the most important results are reviewed.

3.1 Superasymmetric fission in intermediate energy proton-induced fission

Among the most important findings in the fission yield studies at the IGISOL is the discovery of pronounced superasymmetric fission mode in intermediate excitation energy [33,34]. The superasymmetric fission mode, which can be inferred to the result from the multimodal nature of nuclear fission, was first obtained in the spontaneous fission of ^{252}Cf [61,62]. Later, the fission mass yields were observed to extend to very low masses in the neutron-induced fission of ^{235}U [63], ^{239}Pu and ^{249}Cf [64]. A shoulder in the mass yield curve of reactor-neutron–induced fission of ^{238}U was observed already earlier [65]. In addition, the superasymmetric fission mode has been seen in the relativistic in-flight fission of ^{238}U [66].

The independent fission yield distribution of the 25 MeV proton-induced fission of ^{238}U was studied at the IGISOL facility in mid 1990s using gamma ray spectroscopy as described above in sect. 2.2 [33,34]. The independent yields of ^{71}Ni, $^{71-74}$Cu, $^{74-79}$Zn, $^{76,78-81}$Ga and $^{78-82}$Ge were determined. From these data the mass yield in the region $A < 80$ could be extracted. It was observed that the mass yield was declining much slower than in thermal neutron-induced fission. This enhancement of yields support the hypothesis that a superasymmetric fission mode connected with $Z = 28$ and $N = 50$ shells plays important role also in the intermediate energy fission.

3.2 Neutron-induced fission

In the research at the IGISOL, the proton-induced fission has been successfully used to produce neutron-rich isotopes for nuclear structure studies. The main goal of the yield studies has often been to explore the production capabilities.

While the capability to produce neutron-rich nuclei is of interest for neutron-induced fission as well, as it is foreseen as a production mechanism for future radioactive beam ISOL facilities such as EURISOL [67], there is also interest for the neutron-induced fission yields. Fast neutron-induced fission is important in designing the Generation 4 (GEN4) nuclear reactors [68].

An attempt to determine neutron-induced fission yields was taken at the IGISOL facility in late 1990s and early 2000s, for which a special ion guide was designed. The results have been published [36,37]. In short, the experiments showed that neutron-induced fission can be combined with the ion guide technique, and that the line of the most probable nuclear charge Z_p moved towards more neutron-rich nuclei about 1.0 unit charges (fig. 2).

3.3 Deuterium-induced fission

Another way to reach further from stability in the production of neutron-rich nuclei is deuterium-induced fission. As compared with the proton-induced fission, deuteron-induced fission can be expected to result in the average to more neutron-rich nuclei, because in addition to (d, *fission*), also the (d,p *fission*) reaction becomes possible.

Deuterium-induced fission of ^{238}U was studied using ion guide technique in Louvain-la-Neuve in the early 1990s with several deuteron energies (18, 25 and 41 MeV) [29, 51]. The nuclear charge distributions were determined for mass numbers $A = 110$, 112 and 114 in the region of symmetric fission and for $A = 80$ in the wing of the lower mass peak. From these data it could be concluded by comparing with the nuclear charge distributions in 20 MeV proton-induced fission that deuteron-induced fission does not shift the nuclear charge distribution towards more neutron-rich nuclei in the region of symmetric fission, indicating that the contribution of (d,p *fission*) reaction is not significant.

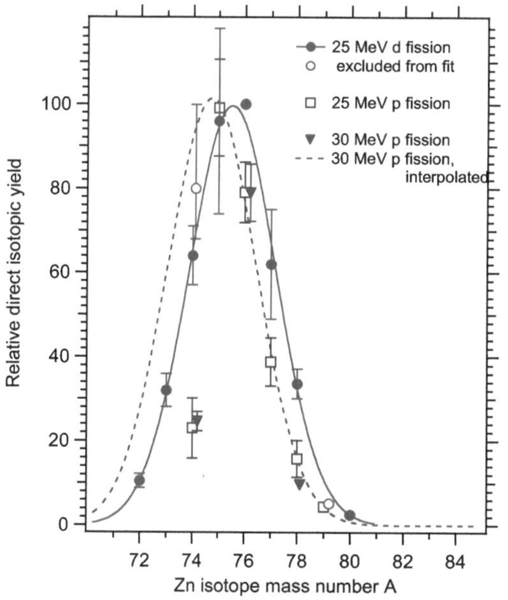

Fig. 4. The independent isotopic yield of neutron-rich Zn isotopes in 25 MeV deuteron-induced fission of ^{238}U. The 25 MeV proton-induced yields from [33] as well as 30 MeV proton-induced yields from [35] are shown for comparison. For an explanation of the interpolated (dashed) curve, see text.

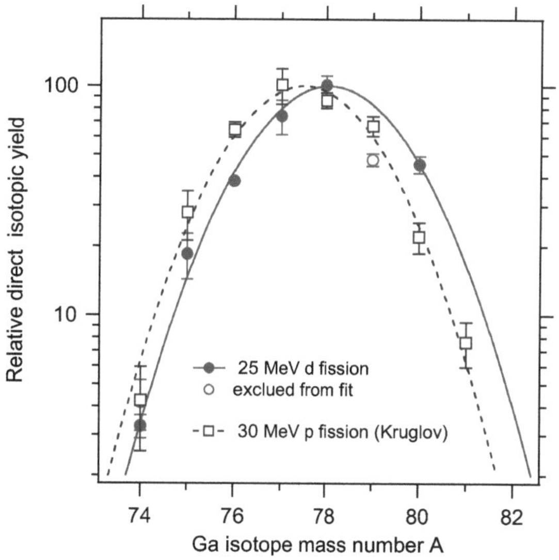

Fig. 5. The independent isotopic yield of neutron-rich Ga isotopes in 25 MeV deuteron-induced fission of ^{238}U (solid circles). The 30 MeV proton-induced fission yields from [42] are shown for comparison (open squares). The curves are Gaussian fits to the data.

Recently, the fission yields in 25 MeV deuteron-induced fission of ^{238}U were studied for a choice of elements at the IGISOL utilising the Penning-trap method [38]. While the earlier work concentrated more in the region of symmetric fission ($A = 110$–114), in this new measurement, the isotopic independent fission yields were determined in the $A < 80$ low mass wing of the distribution (for Zn and Ga), in the lower mass peak ($A = 90$–100, Rb and Sr) and in the heavy end of the symmetric fission ($A = 120$–130, Cd and In). Except for preliminary results for Rb [69] these yields have not been published thus far. The analysis of this experiment has lasted a while because of tedious corrections to be applied for the dipole magnet field fluctuation, see details in [38]. The results of these measurements with comparison to proton-induced fission are presented in figs. 4–7.

Figure 4 shows the independent isotopic yield of neutron-rich Zn isotopes in 25 MeV deuteron-induced fission (solid circles). The line is a Gaussian fit to the yields. Two points have been excluded from the fit: ^{79}Zn, whose yield measurement suffered from the space charge in the Penning trap due to the presence of stable ^{79}Br, and another of the two ^{74}Zn measurements. This rejected ^{74}Zn measurement suffered from extraordinary severe yield fluctuations due to IGISOL dipole magnet instability. Fit is weighed with the experimental error and results in most probable mass $A_p = 75.50 \pm 0.06$ for zinc isotopes.

For comparison, the yield of Zn isotopes in 25 MeV proton-induced fission [33] is shown (open squares). In addition, data points from 30 MeV proton-induced fission are shown [35]. The data from [35] are displayed in such a way that yields for ^{76}Zn are set equal. The choice is arbitrary and made for illustrative purposes only. Both measurements show the same surprisingly low yield for ^{74}Zn, which probably indicates the fallibility of the decay scheme used to determine the yield. This doubt is supported by a yield distribution interpolated from the 30 MeV proton-induced fission data in [42] indicated by a dashed line. In [42] the yield distributions were measured for Co, Ni, Cu and Ga but not for Zn. The centroids of these distributions change linearly as a function of proton number Z. The interpolated value for the average mass number $A_p = 74.88$ for Zn isotopes, and the average FWHM of 3.4 mass units from other yield distributions in [42] are used in the plot. By excluding the suspicious ^{74}Zn yield from the fit and constraining the FWHM to 3.4 mass units one finds from the data in [33] $A_p = 75.27 \pm 0.13$ for 25 MeV proton-induced fission; however, allowing larger FWHM produces smaller χ^2 with $A_p < 75$.

In fig. 5 the independent isotopic fission yield distribution is shown for Ga isotopes in 25 MeV deuteron-induced fission (solid circles). ^{79}Ga yield is excluded from fit because of the numerous stable ^{79}Br ions in the trap impaired the measurement. Since data of 25 MeV proton-induced fission is not available, the comparison is made to 30 MeV proton-induced fission from [42]. The most probable A_p for gallium is 77.50 ± 0.07 and 78.05 ± 0.09 for proton- and deuterium-induced fission, respectively. Deuteron-induced fission thus results in about 0.5 unit masses more neutron-rich isotopes for both Zn and Ga.

In the mass region $A \approx 90$–100, corresponding to the low mass peak in the fission mass yield distribution, the yields of Rb and Sr were measured. Figure 6 shows their independent isotopic yields for both 25 MeV deuteron and 25 MeV proton-induced fission. Both yield distributions have been measured at the IGISOL using the Penning-trap method.

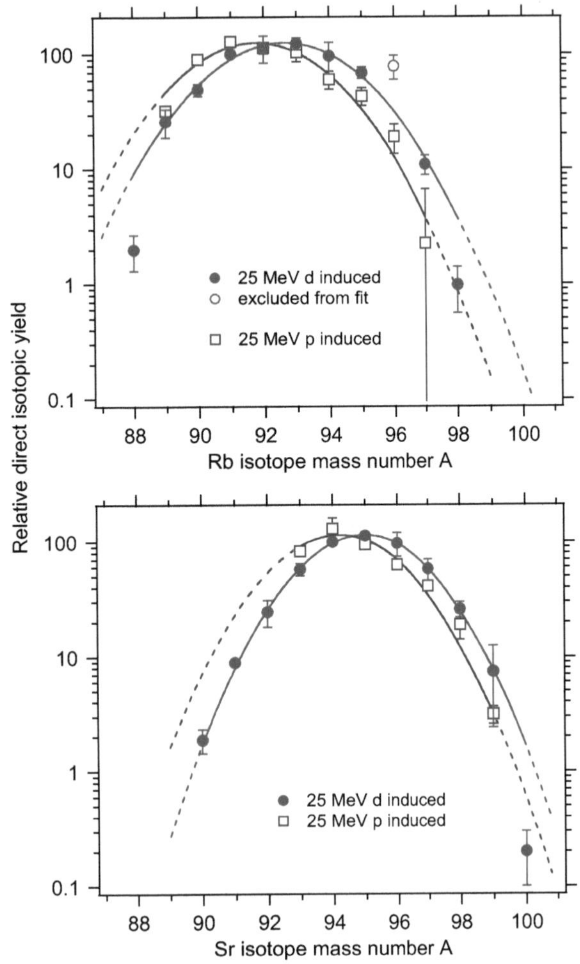

Fig. 6. The top figure shows the independent isotopic yield of neutron-rich Rb isotopes in 25 MeV deuteron-induced fission of ^{238}U (solid circles). The 25 MeV proton-induced yields measured at the IGISOL (open squares) [39] are shown for comparison. The bottom figure is the same for Sr isotopes. The curves are Gaussian fits to the data. Open circle represents a data point that has not been used in the fit.

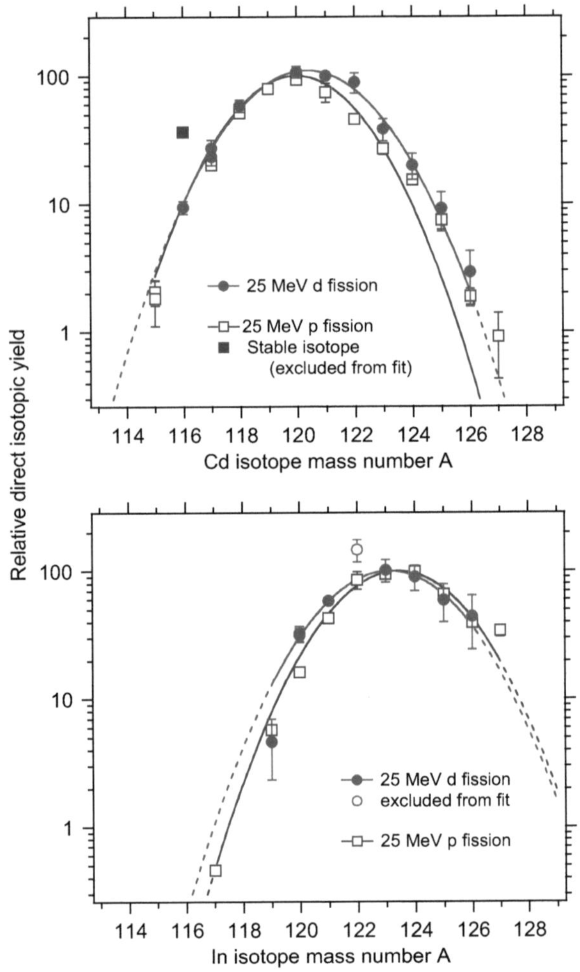

Fig. 7. The independent isotopic yield of neutron rich In and Cd isotopes in 25 MeV deuteron-induced fission of ^{238}U (solid circles). The 25 MeV proton-induced yields from [39] are shown for comparison (open squares). The curves are Gaussian fits to the data. Because of the asymmetry of Cd yield distribution in proton-induced fission, the centroid of the yield distribution of $A_p = 120.2$ cited in the text was determined as a weighed average. An interesting discrepancy for proton- and deuteron-induced fission is observed for the stable isotope ^{116}Cd. For proton-induced fission an excess of ^{116}Cd was observed (as well as stable Cd isotopes in general [38]), while this is not observed for deuterium-induced fission. This difference is certainly instrumental and calls for further investigation.

For Rb and Sr the difference between proton- and deuterium-induced yield is more pronounced than for Zn and Ga. Unweighed Gaussian fits give $A_p = 91.78 \pm 0.14$ and $A_p = 92.65 \pm 0.08$ for Rb, and $A_p = 94.21 \pm 0.23$ and $A_p = 94.99 \pm 0.02$ for Sr in proton- and deuterium-induced fission, respectively. No parameters were restricted in the fit. As can be seen from fig. 6, the distributions are wider (FWHM $\approx$ 4.5) for Rb than for the Sr (FWHM $\approx$ 4.0) in both proton- and neutron-induced fission. A weighed fit that forces the fit closer to the low yields in the distribution wings gives narrower fit for Rb, but the overall agreement with the data is worse. Either way, A_p is independent of the choice between weighed and unweighed fit. In this mass region, the deuteron-induced fission thus seems to result in about 0.8 neutrons more neutron-rich yield distribution than proton-induced fission. In comparison to the $A < 80$ region (Zn,Ga) it is to be noted that in proton-induced fission the 25 MeV protons produce a more neutron-rich distribution than 30 MeV protons that were used for comparison in the low mass region.

Finally, the independent yields of indium and cadmium isotopes in 25 MeV proton- and deuterium-induced fission are shown in fig. 7. The centroids of the distributions are equal or even reversed as compared to the lower masses, centroid of proton-induced fission being more neutron rich ($A_p = 123.44 \pm 0.12$) than the centroid of deuterium-induced fission ($A_p = 123.13 \pm 0.10$) for indium. As can be seen from fig. 7, a Gaussian shape does not properly fit to the 25 MeV proton-induced fission yield distribution for Cd. The curve shown is the best fit to all proton-induced yields of Cd isotopes. The yields of the four most neutron-rich isotopes conrtibute to the fit. The yield distribution seems to have a very asymmetric shape, the neutron-rich side dropping slower than the stable isotope side. It is not quite certain whether this effect is real but is due to difficulties in combining two sets of proton-induced yield measurements with different reference isotope. Namely, the determined isotopic yield distribution for cadmium in deuterium-induced fission is symmetric. The centroid of both cadmium yield distributions can be determined using a weighed average instead of a fit. The centroids determined this way are 120.2 for proton-induced fission and 120.4 for deuterium-induced fission. The latter one agrees with the 120.36 ± 0.08 by a Gaussian fit. The overall result that the independent yield distributions for proton- and deuterium-induced fission do not significantly differ in the region of symmetric fission agrees with the previous measurement in [29, 51].

The result can be interpreted so that in the asymmetric fission the neutron excess of the fissioning $^{2}H + ^{238}U$ system over $^{1}H + ^{238}U$ remains. The attraction of $Z = 50$, $N = 82$ magic shells keeps the heavy mass peak in the same position, and only the low mass peak moves towards more neutron richness. Whether (d,p*fission*) plays an important role here or is the shift of the low mass peak due to more neutron-rich compound nucleus formed in (d,*fission*) cannot be determined from this data. In contrast to asymmetric fission, for symmetric fission the additional neutron in ^{2}H seems to have no effect at all. It would be in fact interesting to measure the deuterium-induced yields over a larger region to be able to follow more in detail the difference between proton- and deuterium-induced fission.

4 Future fission yield studies with MCC30/15 cyclotron

The IGISOL laboratory is currently (as August 2011) being rebuilt to a new location next to new MCC30/15 high current cyclotron [70–72]. The intensive light ion beams provided by the new MCC30/15 cyclotron make it necessary to redesign the present ion guides to fully benefit from the beam intensities available. The performance of the present fission ion guide has not been systematically investigated further than up to $10\,\mu$A [73]. The bottleneck in these experimental tests was not the front end but the radiation level in the vicinity of the switchyard, located in the experimental area, which prevented further increase of proton beam intensity. This will be avoided in the new facility by shielding the entire switchyard. Heat transfer simulations however show that in the present fission ion guide setup, schematically shown in fig. 1, the target temperature exceeds the melting point of uranium with $80\,\mu$A of 30 MeV protons. A practical limit can be expected to be below $50\,\mu$A, while at least twice of that is expected to be available.

A tempting way to fully benefit from the high intense light ion beams would be a neutron converter target. Converting the protons to a few MeV neutrons and using them to induce fission allows, first of all, to study the independent fission yields of neutron-induced fission. These yields are relevant for simulating the advanced nuclear power reactor concepts. The decay spectroscopy of neutron-rich nuclei at the IGISOL would benefit from the availability of neutron-induced fission as well. This happens because the calculated fission yields [31, 32] for the isotopes very far from the stability are much higher for a few MeV neutron-induced fission than for a few tens MeV charged-particle–induced fission. Though the absolute cross-sections are low, they are calculated to be up to two orders of magnitude higher than in proton-induced fission. Furthermore, the isobaric background due to the isotopes closer to stability in neutron-induced fission is orders of magnitude lower than in proton-induced fission, which helps avoiding another bottleneck: the space charge limit of the purification Penning trap. The charge of the collected ions in the Penning trap influences its operation, and even when the trap can tolerate the total charge, the fraction of the ions to be purified should not be less than 10^{-4}.

It is also good to notice that already a very low rate of ions is sufficient for the atomic mass measurements utilising JYFLTRAP. In an optimal atomic mass measurement with a Penning trap, the fraction of ion bunches with exactly one ion in the precision (mass measurement) trap at the time should be as large as possible. Typically, the manipulation of each ion bunch takes 150–400 ms, and therefore the optimal rate of ions is just a few ions per second. For decay spectroscopy, rate of at least 10 ions/s would be favourable.

The earlier experiments [36, 37] have shown that neutron-induced fission can be combined with the ion guide technique. Using neutron-induced fission gives additional freedom to the design of fission ion guide. The ionisation of the buffer gas reduces the performance of the ion guide, and unlike the charged particles, neutrons do not directly ionise the stopping gas. The fission target can thus be placed in the stopping gas volume, while in the case of proton-induced fission, the proton beam and therefore also the target must be separated from the stopping volume. In proton-induced fission, only a relatively narrow cone of all fission fragments can enter the stopping volume in the first place. The gain in this geometrical factor is more than 5. Gain from the improvement of ion guide performance due to the lower ionisation can be expected as well, which, however, it is more difficult to estimate.

The challenging issue is building of the neutron converter target. The proton/neutron conversion rate is relatively low: even in the most favourable case, only a few percent can be expected. In addition, while proton beam can be impinged to a spot of the order of 5–10 mm in diameter, the produced neutron beam is much wider, diverse and subject of severe backscattering, which requires much larger fissioning targets. The most seriously considered options for the neutron conversion target are a heavy stopper such as tungsten with a white neutron spectrum, and a light target such as beryllium or lithium forming a forward-peaked semi-monoenergetic neutron spectrum. In addition to neutron production, the heat deposit and transfer are important issues. The MCC30/15 proton beam can be defocused in an approximately 20 mm in diameter spot, which means heating of the order of $10\,W/mm^2$ with $100\,\mu A$ proton beam, which call for very efficient cooling to keep the target temperature within reason. The converter target is planned to be designed and built in collaboration with the Uppsala University.

Despite the practical obstacles, the design goal is to have a system providing very neutron-rich nuclei with the same rate as proton-induced fission, but with less isobaric background. Since the severest limitation of the fission ion guide is the stopping efficiency, the yields can be improved by building a much larger ion guide than the present one. The survival efficiency of the stopped ions is determined to be about 6% while the stopping efficiency is estimated to be of the order of 10^{-4}. The simulations show that all fission fragments are stopped within a 300 mm radius from the target. Thus, a fission ion guide with a cylindrical symmetry and approximately 400 mm long, 400 mm in diameter would increase the stopping efficiency three orders of magnitude. Such a large gas cell naturally cannot any more operate with the buffer gas flow alone, but a voltage grid is needed to transport the ions to the exit nozzle of the stopping gas cell, similarly to, *e.g.*, the CARIBU ion source in the Argonne National Laboratory [74,75]. An even more advanced option is to use a cryogenic ion guide, where the stopping gas temperature is lowered to below liquid nitrogen temperature. A full-scale prototype of such device for the future heavy ion beam facility FAIR (Facility for Antiproton and Ion Research) has been tested off-line in KVI, University of Groningen. In the first tests the stopped radioactive recoil ions from an internal alpha source have been transported from a distance of about 1 m to the exit nozzle and successfully extracted with an efficiency of about 25% [76].

Use of the neutron-induced fission would help the operation of large fission fragment stopping chamber, since the ionisation of buffer gas is much less than in proton-induced fission. The charge density in the stopping chamber will be smaller, which makes the transportation of ions with electric fields easier.

5 Summary and conclusions

The ion guide technique is very suitable for independent fission yield measurements. Several experiments of the yields of proton, deuterium, alpha and neutron-induced fission of ^{232}Th and ^{238}U have been performed, not only at Jyväskylä IGISOL, but also in Lyon, Louvain-la-Neuve and Sendai.

Recently, a new way to extract fission yields using a Penning trap for the fission product identification was developed at IGISOL. With this technique, the fission yields can be determined from low absolute rate of isotopes.

Thus far, most of the fission yield studies with ion guide have focused on charged-particle–induced fission. The neutron-induced fission is however interesting because of its more direct applications to energy production schemes and nuclear forensics applications, and because of its better prospects to be able produce even more neutron-rich nuclei for research purposes. In the new IGISOL laboratory a possibility to study also neutron-induced fission using a neutron converter target is foreseen in the future.

This work has been supported by the Academy of Finland under project No. 139382 and the Finnish Centre of Excellence Programmme 2006-2011 (Project No. 213503, Nuclear and Accelerator Based Physics Programme at JYFL).

References

1. J. Ärje, K. Valli, Nucl. Instrum. Methods **179**, 533 (1981).
2. J. Ärje, J. Äystö, J. Honkanen, K. Valli, A. Hautojärvi, Nucl. Instrum. Methods **186**, 149 (1981).
3. J. Äystö, J. Ärje, V. Koponen, P. Taskinen, H. Hyvönen, A. Hautojärvi, K. Vierinen, Phys. Lett. B **138**, 369 (1984).
4. J. Ärje, J. Äystö, H. Hyvönen, P. Taskinen, V. Koponen, J. Honkanen, A. Hautojärvi, K. Vierinen, Phys. Rev. Lett. **54**, 99 (1985).
5. J. Ärje, J. Äystö, H. Hyvönen, P. Taskinen, V. Koponen, J. Honkanen, A. Hautojärvi, K. Valli, K. Vierinen, Nucl. Instrum. Methods Phys. Res. A **247**, 431 (1986).
6. J. Ärje, J. Äystö, P. Taskinen, J. Honkanen, K. Valli, Nucl. Instrum. Methods Phys. Res. B **26**, 384 (1987).
7. J. Äystö J., Taskinen P., Yoshii M., Honkanen J., Jauho P., Ärje J., Valli K., Nucl. Instrum. Methods Phys. Res. B **26**, 394 (1987).
8. J. Äystö *et al.*, Nucl. Phys. A **480**, 104 (1988).
9. J. Äystö, P. Taskinen, M. Yoshii, J. Honkanen, P. Jauho, H. Penttilä, C.N. Davids, Phys. Lett. B **201**, 211 (1988).
10. K. Gelbke, Prog. Part. Nucl. Phys. **62**, 307 (2009).
11. M. Wada, Nucl. Instrum. Methods Phys. Res. A **532**, 40 (2004).
12. `http://frib.msu.edu/`, accessed 15 August 2011.
13. `http://www.fair-center.de/index`, accessed 15 August 2011.
14. H. Penttilä, P. Taskinen, P. Jauho, V. Koponen, C.N. Davids, J. Äystö, Phys. Rev. C **38**, 931 (1988).
15. V. Koponen *et al.*, Z. Phys. A **333**, 339 (1989).
16. J. Äystö *et al.*, Nucl. Phys. A **515**, 365 (1990).
17. H. Penttilä, J. Äystö, K. Eskola, Z. Janas, P. Jauho, A. Jokinen, M. Leino, J.-M. Parmonen, P. Taskinen, Z. Phys. A **338**, 291 (1991).
18. H. Penttilä, P. Jauho, J. Äystö, P. Decrock, P. Dendooven, M. Huyse, G. Reusen, P. VanDuppen, J. Wauters, Phys. Rev. C **44**, R935 (1991).
19. A. Jokinen *et al.*, Z. Phys. A **340**, 21 (1991).
20. A. Jokinen, J. Äystö, P. Jauho, M. Leino, J.-M. Parmonen, H. Penttilä, K. Eskola, Z. Janas, Nucl. Phys. A **549**, 420 (1992).
21. J. Äystö *et al.*, Phys. Rev. Lett. **69**, 1167 (1992).
22. H. Penttilä, Ph. D. Thesis, Department of Physics, University of Jyväskylä Research report No. 1/1992, Jyväskylä 1992.
23. Z. Janas *et al.*, Nucl. Phys. A **552**, 340 (1993).
24. H. Penttilä *et al.*, Phys. Rev. C **54**, 2760 (1996).
25. T. Mehren *et al.*, Phys. Rev. Lett. **77**, 458 (1996).
26. J.C. Wang *et al.*, Phys. Lett. B **454**, 1 (1999).
27. A. Jokinen *et al.*, Eur. Phys. J. A **9**, 9 (2000).
28. J. Kurpeta. *et al.*, Eur. Phys. J. A **31**, 263 (2007).

29. M. Leino *et al.*, Phys. Rev. C **44**, 336 (1991).
30. P.P. Jauho, A. Jokinen, M. Leino, J.-M. Parmonen, H. Penttilä, J. Äystö, K. Eskola, V.A. Rubchenya, Phys. Rev. C **49**, 2036 (1994).
31. V.A. Rubchenya, Phys. Rev. C **75**, 054601 (2007).
32. V.A. Rubchenya, J. Äystö, Eur. Phys. J. A **48**, 44 (2012).
33. M. Huhta, P. Dendooven, A. Honkanen, M. Oinonen, H. Penttilä, K. Peräjärvi, V. Rubchenya, J. Äystö, Nucl. Instrum. Methods Phys. Res. B **126**, 201 (1997).
34. M. Huhta *et al.*, Phys. Lett. B **405**, 230 (1997).
35. S. Nummela, P. Heikkinen, J. Huikari, A. Jokinen, S. Rinta-Antila, V. Rubchenya, J. Äystö, Nucl. Instrum. Methods Phys. Res. A **481**, 718 (2002).
36. G. Lhersonneau *et al.*, Eur. Phys. J. A **9**, 385 (2000).
37. L. Stroe, G. Lhersonneau, A. Andrighetto, P. Dendooven, J. Huikari, H. Penttilä, K. Peräjärvi, L. Tecchio, Y. Wang, Eur. Phys. J. A **17**, 57 (2003).
38. H. Penttilä *et al.*, Eur. Phys. J. A **44**, 147 (2010).
39. H. Penttilä *et al.*, in preparation.
40. D. Gorelov *et al.*, in preparation.
41. A. Astier *et al.*, Nucl. Instrum. Methods Phys. Res. B **70**, 233 (1992).
42. K. Kruglov *et al.*, Eur. Phys. J. A **14**, 365 (2002).
43. K. Kruglov *et al.*, Nucl. Phys. A **701**, 145 (2002).
44. S. Franchoo *et al.*, Phys. Rev. Lett. **81**, 3100 (1998).
45. W.F. Mueller *et al.*, Phys. Rev. C **61**, 054308 (2000).
46. H. Kudo, M. Maruyama, M. Tanikawa, M. Fujita, T. Shinozuka, M. Fujioka, Nucl. Instrum. Methods Phys. Res. B **126**, 209 (1997).
47. H. Kudo, M. Maruyama, M. Tanikawa, T. Shinozuka, M. Fujioka, Phys. Rev. C **57**, 178 (1998).
48. S. Goto, D. Kaji, H. Kudo, M. Fujita, T. Shinozuka, M. Fujioka, J. Radioanal. Nucl. Chem. **239**, 109 (1999).
49. Kaji S. Goto, M. Fujita, T. Shinozuka, M. Fujioka, H. Kudo, J. Nucl. Radiochem. Sci. **3**, 7 (2002).
50. P. Taskinen, H. Penttilä, J. Äystö, P. Dendooven, P. Jauho, A. Jokinen, M. Yoshii, Nucl. Instrum. Methods Phys. Res. A **281**, 539 (1989).
51. P.P. Jauho, Ph. D. Thesis, Department of Physics, University of Jyväskylä Research report No. 4/1994, Jyväskylä 1994.
52. Tracy B.L., Chaumont J., Klapisch R., Nitschke J.M., Poskanzer A.M., Roeckl E., Thibault C., Phys. Rev. C **5**, 222 (1972).
53. V. Kolhinen, T. Eronen, U. Hager, J. Hakala, A. Jokinen, S. Kopecky, S. Rinta-Antila, J. Szerypo, J. Äystö, Nucl. Instrum. Methods Phys. Res. A **528**, 776 (2004).
54. I.D. Moore, T. Kessler, T. Sonoda, Y. Kudryavstev, K. Perajarvi, A. Popov, K.D.A. Wendt, J. Aysto, Nucl. Instrum. Methods Phys. Res. B **268**, 657 (2010).
55. S. Rinta-Antila, S. Kopecky, V.S. Kolhinen, J. Hakala, J. Huikari, A. Jokinen, A. Nieminen, J. Äystö, J. Szerypo, Phys. Rev. C **70**, 011301 (2004).
56. J. Huikari, P. Dendooven, A. Jokinen, A. Nieminen, H. Penttilä, K. Peräjärvi, A. Popov, S. Rinta-Antila, J. JÄystö, Nucl. Instrum. Methods Phys. Res. B **222**, 632 (2004).
57. Yu. Kudryatsev *et al.*, Nucl. Instrum. Methods Phys. Res. B **266**, 4368 (2008).
58. K. Peräjärvi *et al.*, Appl. Radiat. Isotopes **68**, 450 (2010).
59. A.H. Snell, F. Pleasonton F., Phys. Rev. **107**, 740 (1957).
60. A. Herlert *et al.*, New J. Phys. **7**, 44 (2005).
61. G. Barreau, A. Sicre, F. Caitucoli, M. Asghar, T.P. Doan, B. Leroux, G. Martinez, T. Benfoughal, Nucl. Phys. A **432**, 411 (1985).
62. C. Budtz-Jorgensen, H.-H. Knitter, Nucl. Phys. A **490**, 307 (1988).
63. J.L. Sida, P. Armbruster, M. Bernas, J.P. Bocquet, R. Brissot, H.R. Faust, Nucl. Phys. A **502**, 233c (1989).
64. R. Hentzschel, H.R. Faust, H.O. Denschlag, B.D. Wilkins, J. Gindler, Nucl. Phys. A **571**, 427 (1994).
65. V.K. Rao, V.K. Bhargava, S.G. Marathe, S.M. Sahakundu, R.H. Iyer, R.H., Phys. Rev. C **19**, 1372 (1979).
66. P. Armbruster *et al.*, Z. Phys. A **335**, 19 (1996).
67. `http://www.eurisol.org/site02/index.php`, accessed 15 August 2011.
68. P. Rullhusen, *ed.*, *Nuclear Data Needs for Generation IV Nuclear Energy Systems, Antwerpen, 2005* (World Scientific Publishing Co. Pte. Ltd., Singapore, 2006).
69. H. Penttilä *et al.*, in *Proceedings of the Fourth International Conference of Fission and Properties of Neutron-Rich Nuclei*, edited by J.H. Hamilton, A.V. Ramayya, H.K. Carter (World Scientific Publishing, Singapore, 2008) p. 410.
70. `https://www.jyu.fi/fysiikka/en/research/accelerator/accelerator/index_html/mcc30` Accessed 30 August 2011.
71. H. Penttilä *et al.*, J. Korean Phys. Soc. **59**, 1589 (2011).
72. H. Penttilä *et al.*, in *Proceedings of NEMEA-6 Wokshop, 25-28 October 2010, Krakow, Poland*, MEA(NSC/DOC(2011)4 (2011), pp. 103-112 .
73. P. Karvonen *et al.*, Nucl. Instrum. Methods Phys. Res. B **266**, 4454 (2008).
74. G. Savard, J. Clark *et al.*, Nucl. Instrum. Methods Phys. Res. B **204**, 582 (2003).
75. G. Savard, S. Baker, C. Davids, A.F. Levand, E.F. Moore, R.C. Pardo, R. Vondrasek, B.J. Zabransky, G. Zinkann, Nucl. Instrum. Methods Phys. Res. B **266**, 4086 (2008).
76. M. Ranjan *et al.*, EPL **96**, 52001 (2011).

DOI 10.1007/978-94-007-5555-0_7
Reprinted from *Hyperfine Interactions* Volume 223, Issues 1-3, 2014
and *The European Physical Journal A* Volume 48, Issue 4, 2012

THE EUROPEAN PHYSICAL JOURNAL A

Regular Article – Experimental Physics

Consistent theoretical model for the description of the neutron-rich fission product yields

V.A. Rubchenya[1,2,a] and J. Äystö[1]

[1] Department of Physics, University of Jyväskylä, POB 35, 40351-Jyväskylä, Finland
[2] V.G. Khlopin Radium Institute, St.-Petersburg 194021, Russia

Received: 30 January 2012
Published online: 18 April 2012 –
Communicated by E. De Sanctis

Abstract. The consistent model for the description of the independent fission product formation cross-section at light projectile energies up to about 100 MeV is described. Pre-compound nucleon emission is described in the framework of the two-component exciton model using the Monte Carlo method, which allows one to incorporate a time duration criterion for the pre-equilibrium stage of the reaction. The decay of the excited compound nuclei, formed after the pre-equilibrium neutron and proton emission, is treated within the time-dependent statistical model with the inclusion of the main dynamical effects of nuclear friction on the fission width and saddle-to-scission descent time. For each member of the compound nucleus ensemble at scission point, the primary fragment isobaric chain yields are calculated using the multimodal approach with the inclusion two superasymmetric fission modes. The charge distribution of the primary fragment isobaric chains was considered as a results of frozen quantal fluctuations of the isovector nuclear matter density at the finite scission neck radius. The calculated fission product formation cross-sections in the neutron, proton, and γ-rays induced fission of the heavy actinides are presented.

1 Introduction

Investigation of nuclear structure and searching new nuclear phenomena predicted to appear when approaching the neutron drip line are some of the most important topics of nuclear physics in the future [1]. Nuclear fission is a promising source for discovering, producing, and investigating exotic nuclei with high neutron excess [2]. The fission process is considered to be very promising for the production of neutron-rich nuclides in different Radioactive Nuclear Beam (RNB) projects [3–8].

Comparative investigations of neutron-rich fission product yields in the fission of heavy nuclei induced by deuterons, protons, neutrons and photons in a wide range of energies are important for the development of RNB facilities. Reliable predictions of fission product yields are also important for the technical application of nuclear reactions at intermediate energy for the energy production and transmutation of nuclear waste in hybrid reactors [9] and for developing advanced new-generation nuclear GEN-IV reactors [10].

During the last years, systematic studies on charge and mass distributions of fragments in proton, deuteron, and neutron-induced fission of ^{238}U at intermediate energy have been carried out at the Accelerator Laboratory of the University of Jyväskylä [11–15]. To analyze experimental results, a new version of the theoretical model for the calculation of the light-particle–induced fission characteristics has been developed. This model is a combination of the new version of the two-component exciton model [16] and a time-dependent statistical model for the fusion-fission process with the inclusion of dynamical effects [17] for accurate calculations of nucleon composition and excitation energy of the fissioning nucleus at scission point. The model for the calculation of fission product yields proposed earlier [18–20] was upgraded by including a dynamical approach for isobaric width and by calculating the parameters of charge polarization at scission point.

In this paper, the description of the model for the calculation of fission product cross-sections in the fission of heavy nuclei induced by light particles with energies up to 100 MeV is presented. The comparison of the neutron-rich fission product yields in the different combination of the projectiles and targets will be done. The structure of the paper is as follows. In the second section, a short description of the theoretical model is given paying special attention to the dynamical effect in the fission process, the extended fission modes model, and the quantal fluctuations in the charge equilibration mode. The applications of the theoretical model for the prediction of the fission product yields are presented and discussed in sect. 3. The last section gives a brief summary of the present study.

[a] e-mail: rubchen@phys.jyu.fi

2 Theoretical model

The detailed theoretical calculation of the fission product formation cross-sections consists of the description of three reaction stages: i) modeling the reaction mechanism to calculate mass, charge, and excitation energy distributions of compound nuclei, ii) modeling the fission process itself and the primary fission fragment formation and iii) modeling the de-excitation process of the heated primary fragments.

2.1 Precompound stage of reaction

In the reaction between target with spin I and the projectile with spin s the partial composite-nucleus formation cross-section is approximately equal to the partial reaction cross-section and can be obtained from the optical model

$$\sigma^{C}_{ljIJ}(A_C, Z_C, E^C_{ex}) = \frac{\pi \cdot \lambda\!\!\!^{-2}}{2I+1}(2J+1)T^J_{ljI}, \qquad (1)$$

where A_C, Z_C, and E^C_{ex} are initial mass and charge numbers and excitation energy of the composite nucleus, $\vec{J} = \vec{I} + \vec{j}$ and $\vec{j} = \vec{l} + \vec{s}$ are the channel projectile spin, and T^J_{ljI} is the projectile transmission coefficients. At a projectile energy higher than 10 MeV per nucleon, the thermally equilibrated compound nuclei are formed after the pre-equilibrium emission of neutrons, protons and other light-charged particles. For each partial entrance channel there is some probability $P^{pre\text{-}eq}_{lJ}(\Delta N^{pre\text{-}eq}, \Delta Z^{pre\text{-}eq}, \Delta E^{pre\text{-}eq}_{ex}, \Delta J^{pre\text{-}eq})$ of the pre-equilibrium emission of $\Delta N^{pre\text{-}eq}$ neutrons and $\Delta Z^{pre\text{-}eq}$ protons, which remove the excitation energy $\Delta E^{pre\text{-}eq}_{ex}$ and the angular momentum $\Delta J^{pre\text{-}eq}$. In this case, the partial compound nucleus formation cross-section for each member of the compound nucleus ensemble is defined by expressions

$$\begin{aligned}
&\sigma^{CN}_{ljIJ_{CN}}(A_{CN}, Z_{CN}, E^{CN}_{ex}) = \\
&\sigma^{C}_{ljIJ} P^{pre\text{-}eq}_{lJ}(\Delta N^{pre\text{-}eq}, \Delta Z^{pre\text{-}eq}, \Delta E^{pre\text{-}eq}_{ex}, \Delta J^{pre\text{-}eq}),\\
&A_{CN} = A_C - \Delta N^{pre\text{-}eq} - \Delta Z^{pre\text{-}eq},\\
&Z_{CN} = Z_C - \Delta Z^{pre\text{-}eq},\\
&E^{CN}_{ex} = E^C_{ex} - \Delta E^{pre\text{-}eq}_{ex}, \quad J_{CN} = J - \Delta J^{pre\text{-}eq}. \qquad (2)
\end{aligned}$$

The pre-equilibrium particle emission process is described in the frame of the two-component exciton model [16,21, 22].

The two-component exciton model considers the time evolution of a nuclear reaction in terms of time-dependent population of exciton states, which are characterized by the proton (p_π) and neutron (p_ν) particle numbers and the proton (h_π) and neutron (h_ν) hole numbers. The exciton states are described by four numbers ($p_\pi, h_\pi, p_\nu, h_\nu$) with the proton exciton number $n_\pi = p_\pi + h_\pi$, the neutron exciton number $n_\nu = p_\nu + h_\nu$, and the exciton number $n = n_\pi + n_\nu$. The initial exciton states depend on the projectile and, in the case of the proton- and neutron-induced reactions, they are (1, 0, 0, 0) and (0, 0, 1, 0), respectively. From these initial states there is no particle emission because that process is included in the elastic scattering of the incident particles. Two-body interactions generate transitions to other states with the selection rule $\Delta n = 0$, 2 neglecting the decay of exciton states into less complex states by pair annihilation. That is known historically in the one-component model as the "never come back" approximation. The nucleon emission generates new exciton states ($p_\pi - 1, h_\pi, p_\nu, h_\nu$), $\Delta n = \Delta n_\pi = -1$ (proton emission) or ($p_\pi, h_\pi, p_\nu - 1, h_\nu$), $\Delta n = \Delta n_\nu = -1$ (neutron emission) with the corresponding transition rates which are proportional to the two-component particle-hole density of daughter nuclei (for details, see [16]). The emission of more complex particles is not included because their probabilities for heavy targets are much lower in comparison with nucleon emission.

2.2 Statistical prescission light-particle emission

After ending the pre-equilibrium stage for every partial wave in the entrance channel, the consideration of the decay of compound nuclei states starts. It is supposed that the duration of the pre-equilibrium process is two orders of magnitude shorter than the average statistical decay time of the initial composite nucleus. Therefore, formation and de-excitation stages of the compound nuclei are decoupled. The light particles are emitted near the equilibrium deformation of the compound nucleus and at descent from saddle to scission point. The emission time is strongly influenced by the nuclear friction [23].

A compound nucleus ensemble with some distributions over mass number A_{CN}, charge number Z_{CN}, excitation energy E^{CN}_{ex}, and spin J_{CN} are formed during the pre-equilibrium stage. To describe the light-particle prescission emission, the time-dependent statistical model [17] with the inclusion of nuclear dissipation effects has been used. The emission of light particles and γ-rays are assumed to start with full statistical decay widths at time $t = 0$ (just after the end of the pre-equilibrium stage). The stationary fission probability flow over the fission barrier approaches an asymptotic value after a delay time τ_d and the time-dependent fission width is determined by the expression [24,25]

$$\Gamma_f(t) = \Gamma^{BW}_f \cdot [1 - \exp(-t/\tau_d)] \cdot \left(\sqrt{1+\gamma^2} - \gamma\right). \qquad (3)$$

Here Γ^{BW}_f is the statistical Bohr-Wheeler fission width and γ is a nuclear friction coefficient

$$\gamma = \frac{\beta}{2\omega_b}, \qquad (4)$$

where β denotes the reduced dissipation coefficient and ω_b describes the potential curvature at fission saddle point. The fission delay time parameter τ_d depends on the fission barrier height, the nuclear friction coefficient and the potential energy curvature at the equilibrium state, and is approximated by the formula

$$\tau_d = \left(\frac{1}{2\gamma\omega_{gs}} + \frac{\gamma}{\omega_{gs}}\right) \cdot \ln\left(\frac{10 \cdot B^{LDM}_f}{T}\right), \qquad (5)$$

where B_f^{LDM} is the liquid-drop–model fission barrier, T is the nuclear temperature, and ω_{gs} stands for the collective frequency at the equilibrium shape. If the fission barrier is too small or the temperature is too high, leading to $\frac{B_f^{LDM}}{T} < 0.1$, the delay time becomes negligibly small, $\tau_d = 0$.

The light-particle emission widths for compound nucleus are calculated according to the expression

$$\Gamma_p(E_{ex}^{CN}, J_{CN}) = \frac{2s_p+1}{2\pi\rho_{CN}(E_{ex}^{CN}, J_{CN})} \sum_{l_p} \sum_{I=|J_{CN}-l_p|}^{I=|J_{CN}+l_p|} \int \mathrm{d}\varepsilon \cdot \rho_d\left(E_{ex}^{CN} - B_p - \varepsilon, I\right) \cdot T_{l_p}(\varepsilon). \quad (6)$$

Here, s_p and B_p are the spin and binding energy of the particle, ρ_{CN} and ρ_d are the level densities of compound and daughter nuclei, respectively, and $T_{l_p}(\varepsilon)$ are the transmission coefficients for particles emitted with angular momentum l_p and kinetic energy ε.

The Bohr-Wheeler width is proportional to the number of the transitional states at the fission barrier

$$\Gamma_f^{BW}(E_{ex}^{CN}, J_{CN}) = \frac{1}{2\pi\rho_{CN}(E_{ex}^{CN}, J_{CN})} \int_0^{E_{ex}^{CN}-B_f(J_{CN})} \mathrm{d}\varepsilon \cdot \rho_b\left(E_{ex}^{CN} - B_f(J_{CN}) - \varepsilon, J_{CN}\right) \cdot T_f(\varepsilon, J_{CN}), \quad (7)$$

where ρ_b is the level density at saddle point, T_f is the fission barrier transmission coefficient.

The fission barrier of the heated rotating nucleus was calculated using the approximation

$$B_f(E_{ex}^{CN}, J_{CN}) = C_b B_f^{LDM}(E_{ex}^{CN}, J_{CN}) - \delta U_{gs}(E_{ex}^{CN}), \quad (8)$$

where B_f^{LDM} is the liquid-drop model fission barrier, C_b is the scale parameter, and δU_{gs} is the shell correction at the equilibrium deformation state of a compound nucleus [26]. The liquid-drop potential energy of the compound nucleus was calculated taking into account the temperature dependence of the surface and Coulomb energies

$$V^{LDM}(E_{ex}^{CN}, J_{CN}) = E_{surf} \cdot (1-\alpha T^2) + E_{Coul} \cdot (1-\eta T^2) + \frac{\hbar^2 J_{CN}(J_{CN}+1)}{2\Im_{sph}} B_{rot}. \quad (9)$$

Here E_{surf} and E_{Coul} are the surface and Coulomb energies of the deformed nucleus, respectively, and $\Im_{sph}$ is the rigid-body momentum of inertia of spherical nucleus. The temperature dependence of the liquid-drop energy was introduced within the Tomas-Fermi model [27] and the parameter values $\alpha = 0.012\,\mathrm{MeV}^{-2}$ and $\eta = 0.001\,\mathrm{MeV}^{-2}$ have been used. The rotation function is $B_{rot} = \Im_{sph}/\Im_\perp$, where $\Im_\perp$ is the rigid-body momentum of inertia associated with the rotation axis perpendicular to the symmetry axis of the deformed nucleus. The inertia momenta were calculated taking into account the diffuseness of the nuclear matter distribution, and the nucleus shape parameterization by Cassini ovals has been used [28].

Gamma-decay width of the dipole transitions was computed as follows:

$$\Gamma_\gamma(E_{ex}^{CN}, J_{CN}) = \frac{4e^2}{3\pi\hbar mc^3} \frac{N_{CN}Z_{CN}}{A_{CN}} \frac{1}{\rho_{CN}(E_{ex}^{CN}, J_{CN})} \cdot \sum_{I=|J_{CN}-1|}^{I=|J_{CN}+1|} \int S_\gamma(E_\gamma)\rho_d(E_{ex}^{CN} - E_\gamma, I)\mathrm{d}E_\gamma. \quad (10)$$

Here, the giant dipole resonance (GDR) strength function for the axially symmetric deformed nucleus has two components

$$S_\gamma(E_\gamma) = w^{\|} \frac{\Gamma^{\|} E_\gamma^4}{(E_\gamma^2 - E_{GDR}^{\|2})^2 + (\Gamma^{\|} E_\gamma)^2} + w^{\perp} \frac{\Gamma^{\perp} E_\gamma^4}{(E_\gamma^2 - E_{GDR}^{\perp 2})^2 + (\Gamma^{\perp} E_\gamma)^2}, \quad (11)$$

where $(w^{\|}, \Gamma^{\|}, E_{GDR}^{\|})$ and $(w^{\perp}, \Gamma^{\perp}, E_{GDR}^{\perp})$ stand for weight, width and position of parallel and perpendicular components of GDR. For the prolate shapes of the heavy nuclei the GDR parameters have been used in accordance with the results of ref. [29].

The change of the angular momentum after the emission of particles or γ quanta with energy ε was taken into account on average as

$$I^f = I^i - \bar{l}_p(\varepsilon). \quad (12)$$

Here, $\bar{l}_p = 1$ for GDR γ quanta, and $\bar{l}_p \propto \sqrt{\varepsilon}$ for particles.

The level density of the rotating nucleus at total excitation energy E^* and spin I is calculated in the approximation

$$\rho(E^*, I) = \rho(E^* - E_{rot}, I_{\text{int}}), \qquad E_{rot} = \frac{\hbar^2}{2\Im_\perp} R^2, \qquad R^2 = I(I+1) - I_{\text{int}}^2, \quad (13)$$

where I_{int} is the internal angular momentum.

After passing the saddle point the light particles and γ-rays are emitted during the saddle-to-scission time, which is altered due to the nuclear dissipation [30]

$$\tau_{ssc} = \tau_{ssc}(\gamma = 0) \cdot \left(\sqrt{1+\gamma^2} + \gamma\right). \quad (14)$$

In numerical calculations the value $\tau_{ssc}(\gamma = 0) = 5 \cdot 10^{-21}$ s has been used.

The initial excitation energy of the compound nucleus at the saddle-to-scission descend stage is assumed to be an average value of the excitation energies at the saddle and scission points, *i.e.*,

$$E_{sdsc}^{CN}(J_{CN}) = E_{sd}^{CN}(J_{CN}) + 0.5(V_{sd}(J_{CN}) - V_{sc}(J_{CN})). \quad (15)$$

Here, E_{sd}^{CN} is the total excitation energy of a compound nucleus just after passing through the fission barrier, V_{sd}

is a potential energy at saddle point. The potential energy at scission point V_{sc} is calculated in the framework of the scission point fission model. The momentum of inertia of the composite system at scission point is

$$\Im_{\perp}^{sc} = \Im_{\perp}^{H} + \Im_{\perp}^{L} + \frac{A(A_{CN} - A)}{A_{CN}} R_{HL}^2, \qquad (16)$$

where $\Im_{\perp}^{H}$ and $\Im_{\perp}^{L}$ are momenta of inertia of heavy and light fragments, R_{HL} is the distance between fragments centers, A_{CN} is the compound nucleus mass at the scission point, and A is the mass number one of the fragments. At scission point, the spin of the compound nucleus is divided between fission fragments and the relative motion degree of freedom according to the relation

$$J_{CN} = I_H + I_L + L, \quad I_H = J_{CN} \frac{\Im_{\perp}^{H}}{\Im_{\perp}^{sc}}, \quad I_L = J_{CN} \frac{\Im_{\perp}^{L}}{\Im_{\perp}^{sc}}. \qquad (17)$$

The kinetic energy of the fragments is the sum of the Coulomb interaction energy, kinetic energy at scission point, and rotational energy

$$E_{kin} = V_{Coul} + A_{rot}^{rel} L(L+1) + E_{kin}^{sc}, \qquad (18)$$

where A_{rot}^{rel} is the rotational constant of the two-fragment system. The kinetic energy of the fragments at scission point consists of two parts

$$E_{kin}^{sc} = E_{kin\_d}^{sc} + E_{kin\_t}^{sc}, \qquad (19)$$

where $E_{kin\_d}^{sc}$ is a dynamical part, which is considered as a parameter (it is usually assumed to be $E_{kin\_d}^{sc} =$ 0–10 MeV), and a thermal part $E_{kin\_t}^{sc} = 3/2T_{sc}$. The temperature at scission point T_{sc} is determined by the conditions at the scission point which will be considered in subsect. 2.5.

The shape and temperature dependences of the nuclear friction strength are an important problem in fission dynamics and, at present, no definite conclusions can be made about their behavior [23]. It was found that the energy dependence of the dissipation strength reveals a threshold character and the dissipation sets up rather rapidly at nuclear excitation energy around 40 MeV [31]. In this study, we used the ansatz for the energy dependence of the reduced friction coefficient:

$$\beta(E^*) = \begin{cases} 0 & \text{at } E^* < E_{th}^* \\ \beta_0(1 + d_1 T + d_2 T^2) & \text{at } E^* \geq E_{th}^* \end{cases}, \qquad (20)$$

where E_{th}^* is the threshold excitation energy value, and β_0, d_1, d_2 are additional parameters describing the temperature dependence of the reduced friction coefficient.

The Monte Carlo simulation method was used to calculate a prescission neutron multiplicity and spectra using the compound nuclei parameters formed in the result of the pre-equilibrium neutron and proton emission. The time-dependent statistical approach was applied for the description of particle emission during the fission process up to scission point to take into consideration the fission, the delay time and the finite descent time from saddle to scission [17]. After prescission particle emission, the compound nucleus arrives at scission point with some distributions W_{sc} of the excitation energy E_{CN}^{sc}, mass A_{CN}^{sc} and charge Z_{CN}^{sc} numbers and spin J_{CN}^{sc}.

2.3 Primary fission fragment distribution

For each member of the ensemble $W_{sc}(A_{CN}^{sc}, Z_{CN}^{sc}, E_{CN}^{sc})$, the primary fission fragment mass and charge distributions are calculated. At low compound nucleus excitation energy, the primary fission fragment mass and charge distributions exhibit odd-even staggering. These distributions can be presented in the factorized form for the preneutron emission mass distribution

$$Y_{pre}(A) = \tilde{Y}_{pre}(A) F_{oe}(A), \qquad (21)$$

and for the global charge distribution

$$P_{pre}(Z) = \tilde{P}_{pre}(Z) F_{oe}(Z). \qquad (22)$$

Here $\tilde{P}_{pre}(Z)$ and $\tilde{Y}_{pre}(A)$ are smoothed distributions, and the functions $F_{oe}(Z)$ and $F_{oe}(A)$ describe the odd-even staggering in the charge and mass distributions.

The parameterization of smoothed mass distribution is based on the multimodal nature of nuclear fission [32], depicting the influence of the nuclear-shell structure on the potential-energy surface (PES) of the fissioning nucleus. The fission process is most probably guided by the valleys and bifurcation points of the PES from equilibrium shape to scission point. For heavy actinides (from Th to Cf), the so-called standard fission modes (symmetric, spherical ^{132}Sn, and deformed $N = 86$–90 shells) have been used. To proceed into a very asymmetric fragment mass region, these fission modes have to be extended with two additional superasymmetric fission modes. The five fission modes: symmetric (SY), standard-I (SI), standard-II (SII), superasymmetric-I (SAI) and superasymmetric-II ($SAII$) were taken into consideration to approximate the smoothed primary mass distribution [33]:

$$\begin{aligned} \tilde{Y}_{pre}(A) = \sum_{m=1}^{5} C_m(A_{CN}^{sc}, Z_{CN}^{sc}, E_{CN}^{sc}) \cdot Y_m(A) = \\ C_{SY} \cdot Y_{SY}(A) + C_{SI} \cdot Y_{SI}(A) \\ + C_{SII} \cdot Y_{SII}(A) + C_{SAI} \cdot Y_{SAI}(A) \\ + C_{SAII} \cdot Y_{SAII}(A). \end{aligned} \qquad (23)$$

The symmetric component $Y_{SY}(A)$ corresponds to a global liquid-drop valley on the potential energy surface in the multidimensional deformation parameter space. Nuclear shells modulate the potential surface creating additional valleys which disappear at high compound nucleus excitation energy. The component Y_{SI} is connected with $Z = 50$ and $N = 82$ nuclear shells in heavy fragments (^{132}Sn-mode), and Y_{SII} component is influenced by a "deformed" nuclear shell at $N = 86$–90. The superasymmetric-I mode, Y_{SAI}, and the superasymmetric-II mode, Y_{SAII}, are connected with the neutron shell $N = 50$ and $Z = 28$, respectively, in light fragments. Each mode is normalized to 2, *i.e.*

$$\sum_A Y_m(A) = 2, \qquad \text{and} \qquad \sum_m C_m = 1. \qquad (24)$$

The symmetric component in (23) has the Gaussian form

$$Y_{SY} = \frac{2}{\sigma_{SY}\sqrt{2\pi}} \exp\{-(A - A^{sc}_{CN}/2)^2/2\sigma^2_{SY}\}. \qquad (25)$$

Each of the asymmetric components consists of two Gaussians representing the heavy and light fragment mass groups:

$$\begin{aligned}
Y_{SI} &= \frac{1}{\sigma_{SI}\sqrt{2\pi}}\{\exp\{-(A - \bar{A}_{SI})^2/2\sigma^2_{SI}\}\\
&\quad + \exp\{-(A - A^{sc}_{CN} + \bar{A}_{SI})^2/2\sigma^2_{SI}\}\},\\
Y_{SII} &= \frac{1}{\sigma_{SII}\sqrt{2\pi}}\{\exp\{-(A - \bar{A}_{SII})^2/2\sigma^2_{SII}\}\\
&\quad + \exp\{-(A - A^{sc}_{CN} + \bar{A}_{SII})^2/2\sigma^2_{SII}\}\}, \qquad (26)\\
Y_{SAI} &= \frac{1}{\sigma_{SAI}\sqrt{2\pi}}\{\exp\{-(A - \bar{A}_{SAI})^2/2\sigma^2_{SAI}\}\\
&\quad + \exp\{-(A - A^{sc}_{CN} + \bar{A}_{SAI})^2/2\sigma^2_{SAI}\}\},\\
Y_{SAII} &= \frac{1}{\sigma_{SAII}\sqrt{2\pi}}\{\exp\{-(A - \bar{A}_{SAII})^2/2\sigma^2_{SAII}\}\\
&\quad + \exp\{-(A - A^{sc}_{CN} + \bar{A}_{SAII})^2/2\sigma^2_{SAII}\}\}.
\end{aligned}$$

Our analysis of experimental data for the thermal neutron-induced fission has shown that an approximation by Gaussian distribution is not correct at a large deviation from the peak center. The Gaussian distribution corresponds to the harmonic approximation of the free energy near the bottom of the fission valley of the potential energy surface [33]. To take into account the anharmonicity correction, the mass dependence of dispersions at low excitation energy, $E^{sc}_{CN} < 10\,\text{MeV}$, was introduced for two asymmetric fission modes (standard-I and standard-II) in the form

$$\begin{aligned}
\sigma_{SI}(A) &= \sigma_{SI}(1 - c^{<}_{SI}|A - \overline{A_{SI}}|) && \text{at} \quad A < \overline{A_{SI}};\\
\sigma_{SII}(A) &= \sigma_{SII}(1 - c^{<}_{SII}|A - \overline{A_{SII}}|) && \text{at} \quad A < \overline{A_{SII}};\\
\sigma_{SI}(A) &= \sigma_{SI}(1 - c^{>}_{SI}|A - \overline{A_{SI}}|) && \text{at} \quad A > \overline{A_{SI}};\\
\sigma_{SII}(A) &= \sigma_{SII}(1 - c^{>}_{SII}|A - \overline{A_{SII}}|) && \text{at} \quad A > \overline{A_{SII}}.
\end{aligned} \qquad (27)$$

The parameters of the fission modes and their weights in expressions (23)–(27) as functions of mass, charge and excitation energy of the composite nucleus at scission point have been obtained by fitting experimental mass distributions in the neutron- and proton-induced fission of heavy nuclei from Th to Cf.

The smoothed primary isobaric chain charge distribution is approximated by a Gaussian distribution. Therefore, the odd-even staggering can be described by a parameter defined as the third difference of the natural logarithms of the fractional yields [34]. The proton and neutron odd-even effects are considered independently and for odd-even correction in (22) one can write

$$F_{oe}(Z) \propto \exp\left((\Pi^H_Z + \Pi^L_Z)\delta_Z(A_{CN}, Z_{CN}, E_{CN})\right), \quad (28)$$

where Π^H_Z and Π^L_Z are parities of the proton number in heavy and light primary fragments:

$$\begin{aligned}
\Pi^{H(L)}_Z &= 1 \quad \text{if } Z \text{ is even},\\
\Pi^{H(L)}_Z &= -1 \quad \text{if } Z \text{ is odd}.
\end{aligned}$$

The proton odd-even difference parameter $\delta_Z(A_{CN}, Z_{CN}, E_{CN})$ is parameterized in accordance with experimental data [35]

$$\delta_Z(A_{CN}, Z_{CN}, E_{CN}) = \frac{\delta_Z(A_{CN}, Z_{CN}, 0)}{1 + \exp((E_{CN} - 10)/2)}, \quad (29)$$

$$\delta_Z(A_{CN}, Z_{CN}, 0) =$$
$$\begin{cases} 1 - 0.1(Z^2_{CN}/A_{CN} - 35.22)^2 & \text{at } Z^2_{CN}/A_{CN} > 35.22\\ 1 & \text{at } Z^2_{CN}/A_{CN} \le 35.22\\ 0 & \text{if } \delta_Z < 0. \end{cases} \qquad (30)$$

The odd-even correction for the primary fragment mass distribution is defined as

$$\begin{aligned}
F_{oe}(A) \propto \exp((\Pi^H_Z + \Pi^L_Z)\delta_Z(A_{CN}, Z_{CN}, E_{CN})\\
+ (\Pi^H_N + \Pi^L_N)\delta_N(A_{CN}, Z_{CN}, E_{CN})), \qquad (31)
\end{aligned}$$

where Π^H_N and Π^L_N are parities of the neutron number in heavy and light fragments and the neutron odd-even parameter δ_N is supposed to be proportional to the proton one

$$\delta_N(A_{CN}, Z_{CN}, E_{CN}) = 0.5 \cdot \delta_Z(A_{CN}, Z_{CN}, E_{CN}). \quad (32)$$

Here, the integer Z value is defined as the nearest integer of an average charge of isobaric chain with a mass number A. Introducing the odd-even correction violates an initial normalization. Therefore, mass and charge distributions have to be normalized again.

2.4 Charge distribution of the primary fission fragment isobaric chain

The smoothed charge distribution of the primary isobaric chain is approximated by a Gaussian function

$$\tilde{P}_{pre}(Z/A) = \frac{1}{\sigma_Z(A)\sqrt{2\pi}} \exp\left[-\frac{(Z - \bar{Z}(A))^2}{2\sigma^2_Z(A)}\right]. \quad (33)$$

The averaged charge of the primary isobaric chain deviates from the unchanged charge density distribution value

$$\begin{aligned}
&\bar{Z}(A, A^{sc}_{CN}, Z^{sc}_{CN}, E^{sc}_{CN}) =\\
&A\frac{Z^{sc}_{CN}}{A^{sc}_{CN}} + \delta\bar{Z}(A, A^{sc}_{CN}, Z^{sc}_{CN}, E^{sc}_{CN}), \qquad (34)
\end{aligned}$$

where the deviation $\delta\bar{Z}$ or charge polarization is determined by global liquid-drop properties of the potential energy surface near scission point and influenced by nuclear-shell effects.

To calculate $\sigma_Z(A)$ we shall consider the isobaric charge width as a result of a frozen quantal fluctuation at scission point. [36]. The fissioning nucleus at scission point is described by two slightly overlapping fragments. The radius of the aperture (neck radius) through which the two fragments may exchange nucleons is equal to r_{neck}. It is assumed that the isovector nuclear density degree-of-freedom oscillation is much faster than the deformation and the mass ones. Therefore, it is possible to fix fragment masses and study the variation of the fragment charge Z (mass and charge of the second nascent fragment are $A^{sc}_{CN} - A, Z^{sc}_{CN} - Z$) alone.

We suppose that the motion along the collective coordinate Z is harmonic with the smoothed potential

$$\tilde{V}(Z) = V(\bar{Z}) + \frac{1}{2}C_{ZZ}(Z - \bar{Z})^2, \tag{35}$$

where C_{ZZ} is the stiffness constant and $\bar{Z}$ the most probable charge. To calculate the inertia parameter M_{ZZ} an expression derived in ref. [37] was used

$$M_{ZZ} = \frac{16}{9} r_0^3 m \frac{A^{sc}_{CN}}{Z^{sc}_{CN} N^{sc}_{CN}} \frac{L + 2r_{neck}}{r^2_{neck}}. \tag{36}$$

Here, r_0 is the nuclear radius parameter ($r_0 = 1.16$ fm), m is the nucleon mass, and L is the neck length ($L = 2$–4 fm). The estimation for the radius neck is $r_{neck} = 2$ fm or $r_{neck} = 0.5(A(A^{sc}_{CN} - A))^{1/6}$ fm.

The wave functions in the oscillator potential are

$$\Psi_n = \left(\frac{1}{\sigma_{ZZ}\sqrt{\pi}2^n n!}\right)^{\frac{1}{2}} \times H_n\left(\frac{Z - \bar{Z}}{\sigma_{ZZ}}\right) \exp\left\{-(Z - \bar{Z})^2/2\sigma^2_{ZZ}\right\}, \tag{37}$$

where H_n are Hermite polynomials which can be calculated from the recursion relations

$$2xH_n(x) = 2nH_{n-1}(x) + H_{n+1}(x) \tag{38}$$

at $n \geq 1$ with $H_0(x) = 1$ and $H_1(x) = 2x$.

The energy of the n-quantum state is equal

$$E_n = (n + 1/2)\hbar\omega_Z, \quad \hbar\omega_Z = \hbar\sqrt{\frac{C_{ZZ}}{M_{ZZ}}}. \tag{39}$$

At low excitation energies ($T_{sc} \ll \hbar\omega_Z$), the charge width is determined by the ground-state wave function of the oscillator,

$$|\Psi_0|^2 = \frac{1}{\sigma_{ZZ}\sqrt{\pi}} \exp\left\{-(Z - \bar{Z})^2/\sigma^2_{ZZ}\right\}, \tag{40}$$

and the standard deviation of the charge distribution of the isobaric chain is defined by the expression

$$\sigma_Z(A) = \frac{1}{2}\frac{\hbar}{\sqrt{M_{ZZ}C_{ZZ}}}. \tag{41}$$

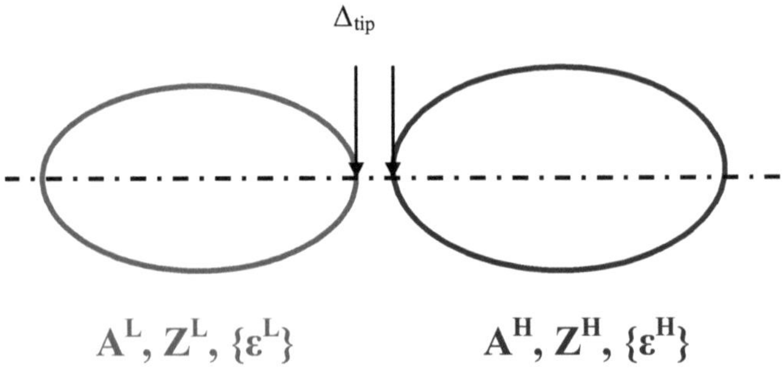

Fig. 1. (Color online) The scission point configuration: $\{\varepsilon^L\}$ and $\{\varepsilon^H\}$ are the deformation parameter sets of a light and a heavy fragment, and Δ_{tip} is the distance between the nearest tips of the fragments.

At high temperature ($T_{sc} \geq \hbar\omega_Z$), the dispersion of the collective function has been calculated by summing higher excited phonon states:

$$|\Psi_{coll}(Z)|^2 = \sum_n |\Psi^2_n|(\exp(E_n/T_{sc}) - 1)^{-1}. \tag{42}$$

The stiffness $C_{ZZ}(A)$ and the equilibrium charge $\bar{Z}(A)$ parameters have been calculated in the framework of the scission point fission model [38,39]. The scission point configuration is approximated by the two nascent deformed fragments as shown in fig. 1.

Potential energy at scission point consists of the interaction energy between the fragments and the deformation energy of both fragments

$$V_{pot} = V_{Coul} + V_{nucl} + E^L_{def}(\{\varepsilon^L\}) + E^H_{def}(\{\varepsilon^H\}), \tag{43}$$

where V_{Coul} and V_{nucl} are the Coulomb and nuclear interaction energies, and E^L_{def} and E^H_{def} are the deformation energies of the fragments. The deformation energy has been calculated using the Strutinsky macroscopic-microscopic method [40]

$$E_{def} = E^{LDM}_{def} + \delta U + \delta E_{pair}. \tag{44}$$

Here, E^{LDM}_{def} is the liquid-drop deformation energy, δU is the shell correction connected with the shell structure of proton and neutron single-particle spectra, and δE_{pair} is the pairing energy correction calculated in the BCS model.

The nuclear shape was described in the lemniscates coordinate system where the Cassini ovals are used as basic figures [28]. The single-particle spectra have been calculated in the axially deformed Woods-Saxon potential [28] using the "universal" nuclear-potential parameters proposed in ref. [41]. For a given compound nucleus after the minimization of the potential relative to the deformation parameters of both fragments, the two-dimensional function of $V^{\min}_{pot}(A, Z)$ as a function of fragment mass and charge was computed. This function at fixed fragment mass was approximated by the parabolic dependence and parameters $\bar{Z}$ and the stiffness parameters C_{ZZ} were determined. An example of the calculated potential energy as a function of fragment charge at $A = 70$ for ^{230}Th, ^{236}U, ^{240}Pu, and ^{248}Cm is shown in fig. 2, where the points present the computed values and the curves are the approximating functions. One can see from fig. 2 that the

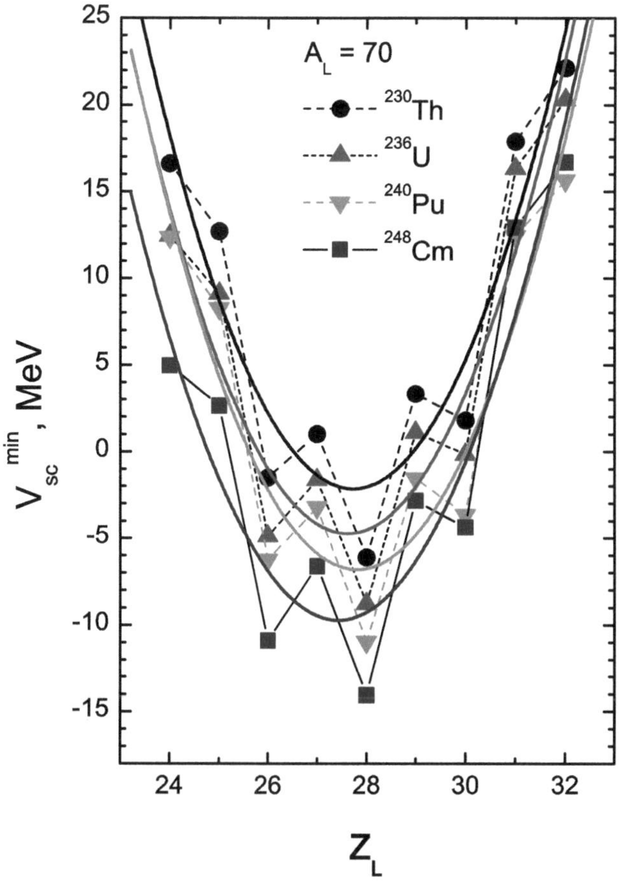

Fig. 2. (Color online) Charge dependences of the potential energy for the isobaric chain $A = 70$ calculated for ^{230}Th, ^{236}U, ^{240}Pu, and ^{248}Cm. The parabolic approximation of the calculated values is shown by curves.

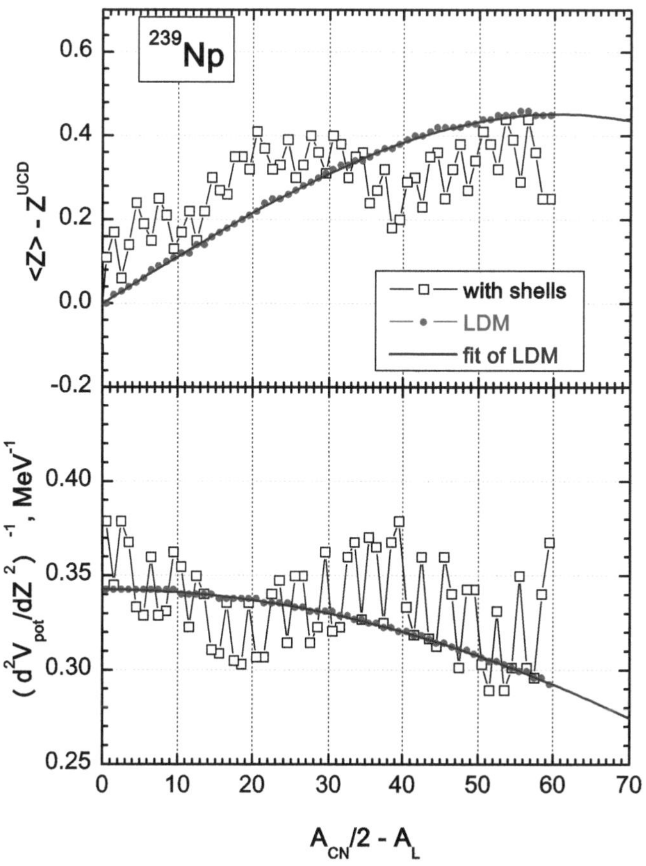

Fig. 3. (Color online) The calculated dependence of $\delta\bar{Z}(A)$ and $C_{ZZ}^{-1}(A)$ for the ^{239}Np nucleus with the inclusion of shell and pairing effects (squares), and in the liquid-drop model (circles). Fitting of the liquid-drop values is presented by the curves.

nuclear shell $Z = 28$ stabilizes the potential energy minimum for a wide range of compound nuclei.

The odd-even staggering and nuclear-shell effect manifest in the dependence of $\bar{Z}(A)$ and $C_{ZZ}(A)$ relative to the global dependencies predicted by the liquid-drop model. Calculated isobaric chain charge distribution parameters as functions of the light fragment mass in the spontaneous fission of ^{239}Np with the inclusion of shell and pairing effects are shown in fig. 3 (squares). These dependences have also been calculated numerically in the framework of the liquid-drop model (points in fig. 3). The liquid-drop values can be calculated using the following analytical approximations:

$$\begin{aligned}\delta\bar{Z}^{LDM}(A) &= (0.1297 - 0.0069Z_{CN}^2/A_{CN}\\ &+1.001\times 10^{-4}\cdot(Z_{CN}^2/A_{CN})^2)\cdot(A_{CN}/2-A)\\ &+10^{-7}\cdot(-144.63 + 6.8186Z_{CN}^2/A_{CN}\\ &-0.08558\cdot(Z_{CN}^2/A_{CN})^2)\cdot(A_{CN}/2-A)^3, \end{aligned} \quad (45)$$

$$\begin{aligned}1/C_{ZZ}^{LDM} &= (-1.161 + 0.07409Z_{CN}^2/A_{CN}\\ &-0.000896862\cdot(Z_{CN}^2/A_{CN})^2)\\ &+10^{-5}\cdot(-7.102 + 0.27956\cdot(Z_{CN}^2/A_{CN})\\ &-0.00336\cdot(Z_{CN}^2/A_{CN})^2)\cdot(A_{CN}/2-A)^2. \end{aligned} \quad (46)$$

To take into account the washing out shell and the pairing effects with the increasing temperature at scission point, the attenuation factor was used as

$$\begin{aligned}\delta\bar{Z}(A, E_{CN}^{sc}) &= \delta\bar{Z}^{LDM}\\ &+\left(\delta\bar{Z}(A,0) - \delta\bar{Z}^{LDM}(A)\right)\frac{1-\exp(-0.1E_{CN}^{sc})}{0.1E_{CN}^{sc}}, \end{aligned} \quad (47)$$

$$\begin{aligned}C_{ZZ} &= C_{ZZ}^{LDM}\\ &+\left(C_{ZZ}(A,0) - C_{ZZ}^{LDM}(A)\right)\frac{1-\exp(-0.1E_{CN}^{sc})}{0.1E_{CN}^{sc}}, \end{aligned} \quad (48)$$

where $\delta\bar{Z}(A,0)$ and $C_{ZZ}(A,0)$ are calculated for the spontaneous fission taking into account shell and pairing effects.

2.5 De-excitation of the primary fission fragments

The fission product independent yields are formed in the process of the light-particle emission from the excited primary fragments. At the composite nucleus excitation energies up to 100 MeV the light-charge particle emission is very small compared with the neutron emission, therefore one can write

$$Y_{ind}(A,Z) = \sum_{n=0}^{n_{max}} Y_{pre}(A+n)\cdot P_{pre}(Z|(A+n))\cdot P_n(A+n,Z). \quad (49)$$

Here, $Y_{pre}(A+n)$ is the primary isobaric chain yield, $P_{pre}(Z|(A+n))$ is the charge distribution of the primary $(A+n)$-isobaric chain, and $P_n(A+n,Z)$ is the n-neutron emission probability from the fragment ^{A+n}Z. The maximal number, n_{max}, in summing is defined by the available fragment excitation energy. The prompt neutron multiplicity distributions for the primary fragments may be calculated using the Hauser-Feschbach statistical theory [42] but this method is too time consuming for the task under consideration. Here, a simplified method was used, where the prompt neutron multiplicity distribution is approximated by the normal one

$$P_n(A,Z,A^{sc}_{CN},Z^{sc}_{CN},E^{sc}_{CN}) = \frac{1}{\sigma_\nu(A,Z)\sqrt{2\pi}}\exp\left\{-\frac{(n-\bar{\nu}(A,Z))^2}{2\sigma^2_\nu(A,Z)}\right\}. \quad (50)$$

Standard deviation is proportional to the averaged prompt neutron multiplicity

$$\sigma_\nu(A,Z,A^{sc}_{CN},Z^{sc}_{CN},E^{sc}_{CN}) = 0.75 + 0.21\cdot\bar{\nu}(A,Z,A^{sc}_{CN},Z^{sc}_{CN},E^{sc}_{CN}). \quad (51)$$

The averaged prompt neutron multiplicities of isobaric chains $\bar{\nu}(A)$ at the most probable charges were calculated within standard statistical model. The parameters of the GDR strength function for the spherical heated fragments were taken from [43] and the level density systematics from ref. [44] was used.

The mean excitation energies of fragments have been calculated in the framework of the fission scission point model. The scission point configuration is approximated by two nascent spheroidal-shaped fragments with a distance between the tips of the fragments $\Delta_{\rm tip} = 2$–$3\,{\rm fm}$ (see fig. 1). Full calculations of the properties of scission point configuration using the macroscopic-microscopic approach are too time consuming. In practical calculations, we used the simple version of the model.

The potential energy at scission point consists of the Coulomb interaction energy between the fragments and the deformation energy of both fragments

$$V_{sc} = V^{Coul}_{sc} + E^L_{def} + E^H_{def}, \quad (52)$$

where V^{Coul}_{sc} is the Coulomb energy, and E^L_{def} and E^H_{def} are deformation energies of the ellipsoidal fragments. The deformation energy has been computed in the parabolic approach

$$E_{def} = C_d(b-R_0)^2, \quad R_0 = r_0A^{1/3}, \quad r_0 = 1.35\,{\rm fm}, \quad (53)$$

where b is the large semi-axis of a deformed fragment. The stiffness coefficient C_d depends on the shell correction δU and the temperature according to the formula

$$C_d = C^{LDM}_d\frac{K-\delta U}{K+\delta U}, \qquad K = 8\,{\rm MeV}. \quad (54)$$

The temperature dependence of shell corrections was described as

$$\delta U(T) = \delta U(0)\frac{2}{1+\exp(t)}, \qquad t = T\frac{2\pi^2}{41A^{-1/3}}. \quad (55)$$

The stiffness coefficient in the liquid-drop model approximation is equal to

$$C^{LDM}_d = \frac{0.16\pi}{r_0^2}\left[a_2\left(1-k_s\frac{(N-Z)^2}{A^2}\right)-\frac{1}{2}c_3\frac{Z^2}{A}\right], \quad (56)$$

where $a_2 = 18.56\,{\rm MeV}$, $k_s = 1.79$, $c_3 = 0.717\,{\rm MeV}$.

The fragment temperature at scission point is determined from the energy balance

$$Q_f(A,Z,A^{sc}_{CN},Z^{sc}_{CN}) + E^{sc}_{CN} = V_{sc} + E^*_{sc} + E^{sc}_{kin} + A^{rel}_{rot}L(L+1). \quad (57)$$

Here, $Q_f(A^{sc}_{CN},Z^{sc}_{CN},A,Z)$ is the fission Q-value and the relative angular momentum is defined from expressions (17). The thermal energy at scission point $E^*_{sc} = E^{*H}_{sc} + E^{*L}_{sc}$ is divided between heavy and light fragments according to the thermal equilibrium condition

$$\frac{E^{*H}_{sc}}{E^{*L}_{sc}} = \frac{a^H}{a^L}, \quad (58)$$

with the level density parameters $a^{L(H)}$ of the deformed fragments. The averaged excitation energy of the primary fragment is the sum of energy deformation and thermal energy calculated at the minimum of the potential energy at scission point.

The shell structure of the fission fragment at scission point plays a decisive role in the fragment mass dependence of the mean neutron multiplicity $\bar{\nu}(A)$ and its temperature dependence. Effects of that shell structure have been taken into account by the determination of the shell corrections $\delta U(A)$ for reference compound nuclei ^{236}U and ^{252}Cf using experimental data of average neutron multiplicities in the spontaneous fission of ^{252}Cf and thermal neutron-induced fission of ^{235}U. The linear interpolation procedure to the compound nucleus mass numbers between these reference shell correction curves $\delta U(A)$ were used to obtain the shell correction of fragments at scission point in fission of other compound nuclei.

Using the calculated values of $\bar{\nu}(A)$ at the most probable charge the multiplicities for the other members of the isobaric chain can be calculated from the relation

$$\bar{\nu}(A,Z) = \frac{\bar{\nu}(A)(Q_f(A,Z,A^{sc}_{CN},Z^{sc}_{CN}) - cZ(Z^{sc}_{CN}-Z))}{\bar{Q}_f(A,A^{sc}_{CN},Z^{sc}_{CN}) - c(Z^{sc}_{CN}\bar{Z}-\bar{Z}^2-\sigma^2_Z)},$$
$$c = \frac{1.44}{1.25(A^{1/3}+(A^{sc}_{CN}-A)^{1/3})+2.5}, \quad (59)$$

where $\bar{Q}_f$ is the averaged fission Q-value for isobaric chain.

3 Results and discussion

The computer code **FIPRODY** (**FI**ssion **PROD**uct **Y**ield), based on the results of the previous section, was developed to calculate the characteristics of the light-particles–induced fission and photo fission. Codes provide calculations of the following characteristics related to the prediction of the fission product yields:

- pre-equilibrium emission mean multiplicity of neutron and protons;

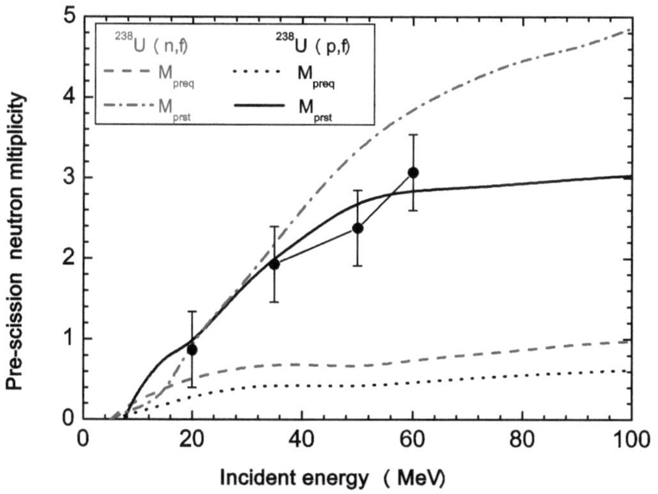

Fig. 4. (Color online) Calculated pre-equilibrium and prescission statistical neutron multiplicities in the neutron (red dash and dot-dashed lines) and proton (black dot and solid lines) induced fission of ^{238}U as function of the incident particle energy. The points with error bars show the experimental prescission statistical neutron multiplicities [20].

- prompt fission neutron spectra;
- pre and postscission multiplicities of light particles;
- preneutron emission fission fragment mass distributions;
- fission product isobaric chain yields;
- fission product independent yields.

The theoretical model parameters were determined by the comparison between calculated and experimental characteristics of the proton- and neutron-induced fission of the heavy actinides from Th to Cf. The main results of that comparisons and the corresponding theoretical calculations will be considered in the following subsections.

3.1 Neutron emission

An accurate calculation of neutron emission during the fission process is very important for the estimation of the formation cross-section of extremely neutron-rich fission products. Prescission neutron emission is formed by a pre-equilibrium emission mechanism and an evaporation from the thermal equilibrated nucleus. The calculated pre-equilibrium (M^n_{preq}) and statistically emitted prescission (M^n_{prst}) neutron multiplicities in the proton- and neutron-induced fission of ^{238}U are presented in fig. 4. The experimental M^n_{prst} values [20] are shown by the point with error bars. The pre-equilibrium neutrons and protons emission substantially reduces the excitation energy of compound nucleus at scission point which defines the postscission neutron multiplicity.

Neutron emission from the primary fission fragments plays a crucial role in the fission product yield formation. The precise calculation of the postscission neutron multiplicity and its variance as a function of the fragment mass and excitation energy at scission point is a challenging problem of the nuclear theory. As it was mentioned in sect. 2.5, the semi-empirical approach was used to get the fragment shell corrections at scission point using the experimental postscission neutron multiplicities in the thermal neutron-induced fission of ^{235}U and the ^{252}Cf spontaneous fission. In the low energy fission of heavy actinides, the fragment mass dependence of the postscission neutron multiplicity $\bar{\nu}(A)$ has the universal sawtooth shape with a deep minimum at $A \approx$ 130–132. Increasing the compound nucleus excitation energy, the $\bar{\nu}(A)$ function will be smoothed (see fig. 5 in ref. [16]) due to the washing out of nuclear shell corrections.

Table 1. Fission mode parameters for the ^{246}Cm compound nucleus obtained from the measured mass yields in ^{245}Cm(n_{th}, f).

Mode	$\overline{A_m}$	σ_m	$c_m^<$	$c_m^>$	C_m,%
SY	123.0	8.94	–	–	2.2
SI	132.32	3.56	0.0	0.60	13.4
SII	141.04	9.73	0.08	0.30	84.0
SAI	82.0	1.17	–	–	0.34
SAII	70.0	0.6	–	–	10^{-4}

3.2 Mass yields

The primary fission fragment mass distributions are described by the fission mode expansion equation (23) with the parameters of eqs. (25)–(27), which were adjusted by comparison with the experimental data. The precise experimental data on the thermal neutron-induced fission obtained at the Lohengrin mass separator at the Institute Laue-Langevin have been used [33]. The manifestation of the superasymmetric-II mode at $A \approx 70$ was observed in the thermal neutron-induced fission of ^{235}U, ^{239}Pu, ^{242m}Am, and ^{245}Cm at the low yield level around 10^{-5}–10^{-4}%. As an example, the fission mode parameters in the ^{245}Cm(n_{th}, f) reaction are listed in table 1.

The dependence of the fission mode parameters on the compound excitation energy has been parameterized using the experimental data in the proton-induced fission of ^{232}Th, ^{238}U, and ^{242}Pu measured in Jyväskylä [11, 15,18,20,45,46]. The prediction for the excitation energy dependences of the fission mode weights of ^{239}U compound nucleus are shown in fig. 5. In the case of the fast-particle–induced fission, the fragment mass yields are formed by the contribution from different compound nuclei at scission point which have slightly different energy dependences of C_m.

The comparison between the theoretical calculations of the primary fragment mass distributions and the preliminary experimental data [46] in the proton-induced fission of ^{232}Th at E_p = 13, 20, 40, and 55 MeV is presented in fig. 6.

3.3 Fission product independent yields

The model and the code allow to predict the independent fission product yields for different entrance channels starting with the thermal neutron-induced fission of heavy

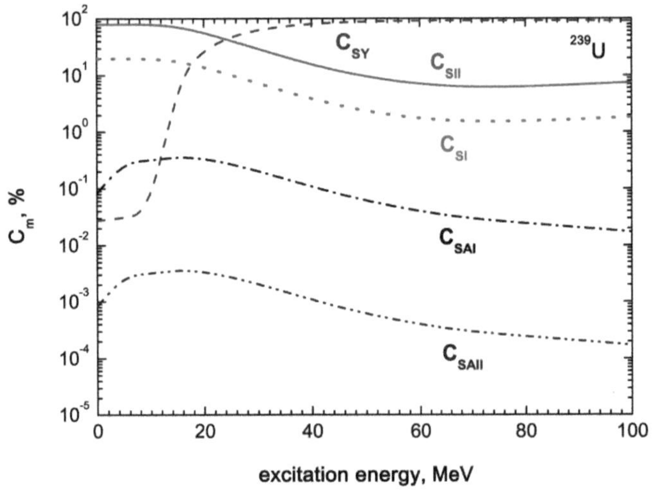

Fig. 5. (Color online) The dependences of the fission mode weights on the excitation energy of the ^{239}U compound nucleus.

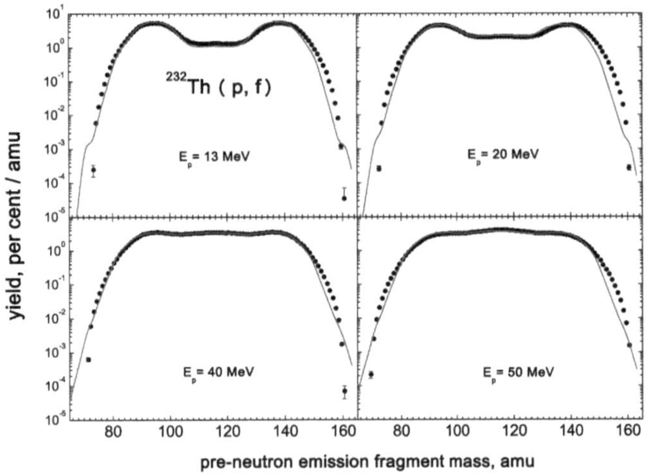

Fig. 6. (Color online) Calculated and experimental primary fragment distributions in the proton-induced fission of ^{232}Th at $E_p = 13$, 20, 40, and 55 MeV [46].

actinides. The latter was demonstrated in the case of ^{243}Cm and ^{245}Cm targets [33]. During the last years, the new powerful method of measurement the fission product yields was developed using the IGISOL-TRAP facility [47,15] which provides the possibility to study extremely neutron-rich fission products in the proton- and neutron-induced fission.

The comparison between experimental [34] and theoretical Cs isotope independent yields in the ^{238}U(p, f) reaction at $E_p = 40$ and 50 MeV is shown in fig. 7. There is good agreement between the theoretical and the experimentally independent fission product formation cross-section at $E_p = 40$ MeV. But, at $E_p = 50$ MeV, the calculated cross-section values systematically underestimates the experimental yields for the neutron-deficient mass region at $A < 137$.

The calculated independent fission product yields in the proton-induced fission of ^{232}Th at $E_p = 25$ MeV in the two-dimensional A-Z plot are presented in fig. 8.

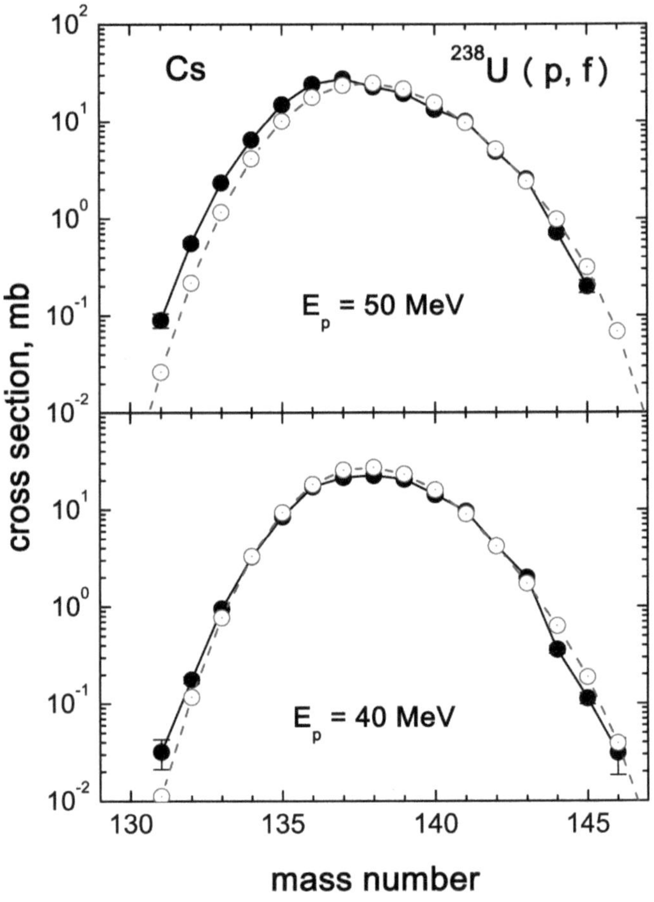

Fig. 7. (Color online) Comparison of the calculated (open circles) and experimental [34] Cs product formation cross-sections in the proton-induced fission of ^{238}U at $E_p = 40$ and 50 MeV.

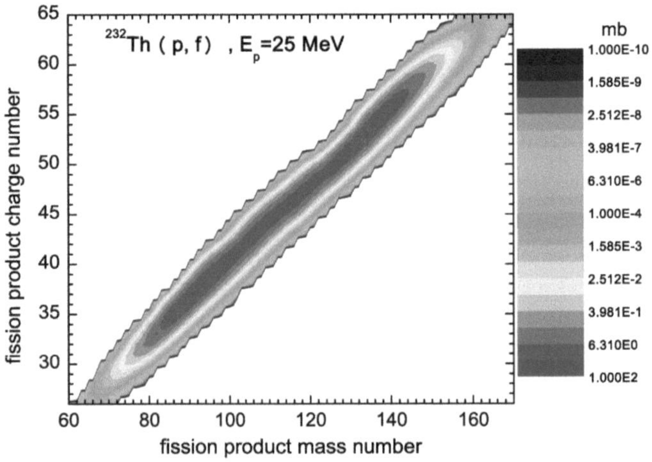

Fig. 8. (Color online) Calculated fission product formation cross-sections in the proton-induced fission of ^{232}Th at $E_p = 25$ MeV.

The calculated isotopic independent fission product distributions for all elements from Ni up to Tb in the ^{238}U(n, f) reaction at $E_n = 5$ MeV are shown in fig. 9. The neutron-induced fission is a promising reaction for the production of very neutron-rich fission products compared with any other bombarding particles.

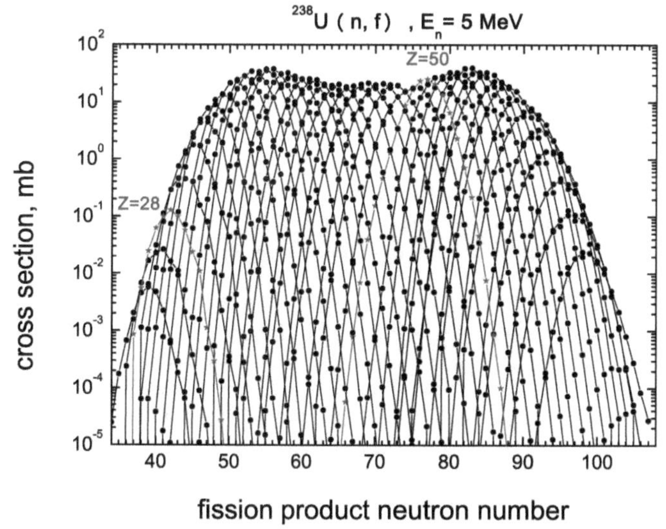

Fig. 9. (Color online) Calculated independent fission product yields for all elemental chains $Z = 26$–65 in the neutron-induced fission of ^{238}U at $E_n = 5\,\mathrm{MeV}$. The isotopic distributions of Ni and Sn chains are marked in red.

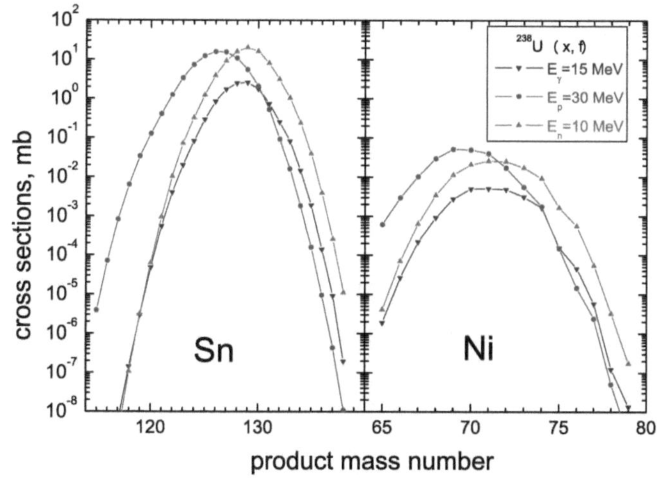

Fig. 10. (Color online) Calculated Sn and Ni isotopes production cross-sections in the protons ($E_p = 30\,\mathrm{MeV}$), neutron ($E_n = 10\,\mathrm{MeV}$) and gammas ($E_\gamma = 15\,\mathrm{MeV}$) induced fission of ^{238}U.

Different RNB facilities are based on using fission by non-relativistic neutrons, protons and gamma-rays. The maximum photon-fission cross-section is at the position of the GDR resonance around 15 MeV, and the proton-induced fission cross-section reaches its maximum value at around 30 MeV proton energy. In the case of the neutron-induced fission, the secondary neutron flux with the wide neutron energy spectrum and the averaged value about a few MeV can be used. To compare these three methods for the production of neutron-rich fission products, the calculated Ni and Sn isotopes production cross-sections in the protons ($E_p = 30\,\mathrm{MeV}$), neutron ($E_n = 10\,\mathrm{MeV}$), and gammas ($E_\gamma = 15\,\mathrm{MeV}$) induced fission of ^{238}U are presented in fig. 10. One can clearly see that the yields of neutron-rich isotopes for the neutron entrance channel are substantially higher in comparison with the photon- or proton-induced fission.

4 Conclusion

In conclusion, the new consistent model for the calculations of the independent fission product formation cross-sections in the proton-, neutron-, and γ-rays–induced fission is formulated and analyzed. Precompound nucleon emission is described in the framework of the two-component exciton model using the Monte Carlo method, which allows one to incorporate a time duration criterion for the pre-equilibrium stage of the reaction. The decay of the excited compound nuclei, formed after the pre-equilibrium neutron and proton emission, is treated within the time-dependent statistical model with the inclusion of the main dynamical effects of the nuclear friction on the fission width and saddle-to-scission descent time. This allows to define more precisely the compound nucleus ensemble at scission point. For each member of that ensemble, the primary fragment mass yields are calculated using the multimodal approach to the fission process which was expanded by adding two superasymmetric fission modes. The charge distribution of the primary fragment isobaric chains was considered as a results of the frozen quantal fluctuations of the isovector nuclear matter density at the scission configuration with the finite scission neck radius. To calculate the primary fission fragment excitation and kinetic energies, the scission point fission model with the semi-phenomenological shell corrections and its temperature dependence was used. The postscission neutron multiplicities, the spectra from the individual primary fission fragment at the average excitation and the kinetic energies were calculated within the standard statistical approach.

The present model was tested using experimental data in the proton- and neutron-induced fission of heavy actinides at projectile energies from the thermal neutrons up to about 100 MeV. The model and computer code `FIPRODY` provide a powerful tool for the experimental data analysis and practically important relevant nuclear data predictions. The calculated fission product formation cross-sections in the neutron- and proton-induced fission of ^{238}U may be found in ref. [48].

The present calculation method has a special importance for optimizing the experimental method of extremely neutron-rich nuclide production. One can conclude that the most neutron-rich fission products were produced in the reactions which lead to the more neutron-rich and less heated compound fissioning nuclei at scission point. The low energy neutron-induced fission of the heavy actinides ^{238}U, ^{244}Pu, ^{247}Cm, and others, is the most promising method for this objective. Multinucleon transfer reactions with neutron-rich radioactive ion beams may also open new possibilities for the production of the neutron-rich fission product in the fission of the very neutron-rich heavy target-like nuclei at a moderate excitation energy.

This work was supported by the Finnish Center of Excellence Programme 2000-2005 (Project No. 44875, Nuclear and Condensed Matter Physics Programme at JYFL), the Finnish Center of Excellence Programme 2006-2011 (Project No. 213503, Nuclear and Accelerator Based Physics Programme at JYFL). The authors acknowledge the financial support of the European Community under the FP6 "Research Infrastructure Action - Structuring the European Research Area" EURISOL DS Project Contract no. 515768 RIDS.

References

1. NuPECC Long range plan 2010, `http://www.nupecc.org/lrp2010/Documents/lrp2010_final.pdf`.
2. J. Äystö, V. Rubchenya, Eur. Phys. J. A **13**, 109 (2002).
3. SPIRAL II, `http://www.ganil-spiral2.eu/spiral2`.
4. SPES, `http://spes.lnl.infn.it`.
5. CARIBU, `http://www.phy.anl.gov/atlas/caribu.html`.
6. ARIEL-TRIUMF, `http://www.triumf.ca/ariel`.
7. D. Habs *et al.*, in *Proceedings of the International Workshop on Research with Fission Fragments*, edited by T.V. Edigy *et al.*, (World Scientific, Singapore, 1997) p. 18.
8. Final Report of the EURISOL Design Study (2005-2009) Nov. 2009, `http://www.eurisol.org/eurisol_design_study_final_report-1156.html`.
9. C.D. Bowman *et al.*, Nucl. Instrum. Methods A **320**, 336 (1992).
10. J. Bouchard, in *Proceedings of the International Conference on Nuclear Data for Science and Technology, April 22-27, 2007, Nice, France, 2008*, edited by O. Bersillon, F. Gunsing, E. Bauge, R. Jacqmin, S. Leray (EDP Sciences, 2008) DOI: 10.1051/ndata:07718.
11. M. Huhta *et al.*, Phys. Lett. B **405**, 230 (1997).
12. G. Lhersonneau, P. Dendooven, G. Canchel, J. Huikari, P. Jardin, A. Jokinen, V. Kolhinen, C. Lau, L. Lebreton, A.C. Mueller, A. Nieminen, S. Nummela, H. Penttilä, K. Peräjärvi, Z. Radivojevic, V. Rubchenya, M.-G. Saint-Laurent, W.H. Trzaska, D. Vakhtin, J. Vervier, A.C.C. Villari, J.C. Wang, J. Äystö, Eur. Phys. J. A **9**, 385 (2000).
13. U. Hager, V.-V. Elomaa, T. Eronen, J. Hakala, A. Jokinen, A. Kankainen, S. Rahaman, S. Rinta-Antila, A. Saastamoinen, T. Sonoda, J. Äystö, Phys. Rev. C **75**, 064302 (2007).
14. J. Hakala, S. Rahaman, V.-V. Elomaa, T. Eronen, U. Hager, A. Jokinen, A. Kankainen, I. D. Moore, H. Penttilä, S. Rinta-Antila, J. Rissanen, A. Saastamoinen, T. Sonoda, C. Weber, J. Äystö, Phys. Rev. Lett. **101**, 052502 (2008).
15. H. Penttilä, P. Karvonen, T. Eronenl, V.-V. Elomaa, U. Hager, J. Hakala, A. Jokinen, A. Kankainen, I.D. Moore, K. Peräjärvi, S. Rahaman, S. Rinta-Antila, V. Rubchenya, A. Saastamoinen, T. Sonoda, J. Äystö, Eur. Phys. J. A **44**, 147 (2010).
16. V.A. Rubchenya, Phys. Rev. C **75**, 054610 (2007).
17. V.A. Rubchenya, A.V. Kuznetsov, W.W. Trzaska, D.N. Vakhtin, A.A. Alexandrov, J. Äystö, S.V. Khlebnikov, V.G. Lyapin, O.I. Osetrov, Yu.E. Penionzhkevich, Yu.V. Pyatkov, G.P. Tiourin, Phys. Rev. C **58**, 1587 (1998).
18. P.P. Jauho, A. Jokinen, M. Leino, J.M. Parmonen, H. Penttilä, J.Äystö, K. Eskola, V.A. Rubchenya, Phys. Rev. C **49**, 2036 (1994).
19. V.A. Rubchenya, J. Äystö, Nucl. Phys. A **701**, 127 (2002).
20. V.A. Rubchenya, W.H. Trzaska, D.N. Vakhtin, J. Äystö, P. Dendooven, S. Hankonen, A. Jokinen, Z. Radivojevich, J.C. Wang, I.D. Alkhazov, A.V. Evsenin, S.V. Khlebnikov, A.V. Kuznetsov, V.G. Lyapin, O.I. Osetrov, G.P. Tiourin, Yu. E. Penionzhkevich, Nucl. Instrum. Methods Phys. Res. A **463**, 653 (2001).
21. E. Gadioli, P.E. Hodgson, *Pre-equilibrium Nuclear Reaction* (Clarendon Press, Oxford, 1992).
22. A.J. Koning, M.C. Duijvestijn, Nucl. Phys. A **744**, 15 (2004).
23. D. Hilscher, H. Rossner, Ann. Phys. (Paris) **17**, 471 (1992).
24. H.A. Kramers, Physika **7**, 284 (1940).
25. P. Grangé, S. Hassani, H.A. Weidenmüller, A. Gavron, J.R. Nix, A.J. Sierk, Phys. Rev. C **34**, 209 (1986).
26. W.D. Myers, W.J. Swiatecki, Nucl. Phys. **81**, 1 (1966).
27. X. Campi, S. Stringari, Z. Phys. A **309**, 239 (1983).
28. V.V. Pashkevich, Nucl. Phys. A **169**, 275 (1971).
29. M. Thoennessen, D.R. Chakrabarty, M.G. Herman, R. Butsch, P. Paul, Phys. Rev. Lett. **59**, 2860 (1987).
30. H. Hofman, J.R. Nix, Phys. Lett. B **122**, 117 (1983).
31. P. Paul, M. Thoennessen, Ann. Rev. Nucl. Part. Sci. **44**, 65 (1994).
32. U. Brosa, S. Grossmann, A. Müller, Phys. Rep. **197**, 167 (1990).
33. I. Tsekhanovich, N. Varapai, V. Rubchenya, D. Rochman, G.S. Simpson, V. Sokolov, I. Al Mahamid, Phys. Rev. C **70**, 044610 (2004).
34. B.L. Tracy, J. Chaumont, R. Klapish, J.M. Nitschke, A.M. Poskanzer, E. Roeckl, C. Thibault, Phys. Rev. C **5**, 222 (1972).
35. J.P. Bocquet, R. Brissot, Nucl. Phys. A **502**, 213 (1989).
36. H. Nifenecker, J. Phys. Lett. (France) **41**, L47 (1980).
37. E.S. Hernandez *et al.*, Nucl. Phys. A **361**, 483 (1981).
38. V.A. Rubchenya, Sov. J. Nucl. Phys. **9**, 697 (1969) (Yad. Fiz. **9**, 1192 (1969)).
39. B. Wilkins, E.P. Steinberg, R.R. Chasman, Phys. Rev. C **14**, 1832 (1976).
40. V.M. Strutinsky, Nucl. Phys. A **122**, 1 (1968).
41. S. Cwiok, J. Dudek, W. Nazarewicz, J. Skalski, T. Werner, Comput. Phys. Commun. **46**, 379 (1987).
42. B.F. Gerasimenko, V.A. Rubchenya, At. Energy **56**, 893 (1985) (At. Energ. **56**, 335 (1985)).
43. M. Thoennessen, D.R. Chakrabarty, M.G. Herman, R. Butsch, P. Paul, Phys. Rev. Lett. **59**, 2860 (1987).
44. A.V. Ignatyuk, G.N. Smirenkin, A.S. Tishin, Sov. J. Nucl. Phys. **21**, 255 (1975).
45. V.A. Rubchenya, W.H. Trzhaska, I.M. Itkis, M.G. Itkis, J. Kliman, G.N. Knyazheva, N.A. Kondratiev, E.M. Kozulin, L. Kruppa, I.V. Pokrovski, V.M. Voskressenski, F. Hanappe, T. Materna, O. Dorvaux, L. Stuttge, G. Chubarian, S.V. Khlebnikov, D.N. Vakhtin, V.G. Lyapin, Nucl. Phys. A **734**, 253 (2004).
46. E.M. Kozulin, J. Äystö, A.A. Bogachev, S. Yamaletdinov, M.G. Itkis, F. Hanappe, O. Dorvaux, S.V. Khlebnikov, J. Kliman, G.N. Knyazheva, L. Krupa, V. Lyapin, M. Mutterer, V.A. Rubchenya, M. Sillanpää, A.M. Stefanini, W.H. Trzaska, E. Vardaci, AIP Conf. Proc. **853**, 336 (2006).
47. H. Penttilä, J. Äystö, V.-V. Elomaa, T. Eronen, D. Gorelov, U. Hager, J. Hakala, A. Jokinen, A. Kankainen, P. Karvonen, T. Kessler, I. Moore, S. Rahaman, S. Rinta-Antila, V. Rubchenya, T. Sonoda, Eur. Phys. J. ST **150**, 317 (2007).
48. `http://www-w2k.gsi.de/eurisol-t11/`.

DOI 10.1007/978-94-007-5555-0_8
Reprinted from *Hyperfine Interactions* Volume 223, Issues 1-3, 2014
and *The European Physical Journal A* Volume 48, Issue 4, 2012

Beta-delayed (multi-)particle decay studies

O. Tengblad · C. Aa. Diget

Published online: 5 April 2012

Abstract Here we report on beta-delayed multi-particle-breakup studies made at IGISOL to determine the nuclear structure and eventual alpha clustering in light nuclei. The experimental studies were performed in full kinematics and we will dedicate part of this article to technical details of the experiments. In particular we will discuss the breakup of excited states in ^{12}C fed by beta decay from ^{12}B and ^{12}N, the study of the beta decay of ^{13}O, as well as studies determining the neutrino spectrum of ^{8}B.

Keywords IGISOL · Nuclear structure · Beta-decay · Multi-particle breakup · ^{12}C · ^{13}O · ^{8}B · Segmented detectors

1 Introduction

The long way to create heavy elements from protons and neutrons passes through the gap between mass numbers A = 5 and A = 8. To bridge this gap of unbound systems, light nuclei such as ^{9}Be and ^{12}C have to play an important role. From a structural point of view huge progress has been made recently in ab-initio calculations, which now are reaching mass number A = 12 [1, 2]. Thus it is becoming a challenge for the experimentalists to gain complete information of excited states in light nuclei. Furthermore, the break-up of nuclear states into more than two fragments has attracted much attention, since unlike the two-body breakup where the mechanism is fully determined by energy and momentum conservation, several different mechanisms for the multi-particle breakup are possible. Due to the stability of the α-particle and

O. Tengblad (✉)
Instituto de Estructura de la Materia CSIC, Serrano 113 bis, 28006 Madrid, Spain
e-mail: olof.tengblad@csic.es

C. Aa. Diget
Department of Physics, University of York, Heslington, York, YO10 5DD, UK

the fact that ^{8}Be is unbound, multi-particle breakup dominates the decay of these very light nuclear systems.

The process of β-delayed particle emission is a subject of much study as it allows, if the final state is known, to uniquely determine the decay-pattern from the energy of the emitted particle. As the decay pattern depends on the structure of the states involved, the decay properties are an important tool to investigate nuclear structure. Beta-delayed multi-particle emission occurs when the nucleus breaks up into more than two particles and therefore the mechanism of the break-up is not fully determined by energy and momentum conservation. Either the break-up proceeds sequentially through sub-system states or the beta-daughter breaks up directly into the three-body continuum. The sequential three-body breakup process is described by the R-matrix theory, which depends on the barrier penetrability and on the reduced width. The case where resonances of width comparable to the energy available in any of the two-body subsystems has been discussed by Korsheninnikov [3] and is described by the use of hyper-spherical harmonics. As the mechanisms involved are so different the fact that both theoretical approaches are able to describe most of the data is still unexplained. It has, however, been shown that kinematically complete studies where all particles are detected and their energy and relative angles measured can in some cases disentangle the decay mechanism.

At IGISOL [4, 5] and at ISOLDE [6, 7] we have developed experimental equipment, analysis procedures and performed several studies, aiming both to disentangle the decay mechanism and to obtain data of high relevance for astrophysical scenarios, see the review by Fynbo and Diget within this journal. In the present contribution we will especially concentrate on the technical advances in this research using the experiments performed at IGISOL as examples.

2 Physics case

Light nuclei are an interesting playground for β-decay experiments. It is a region where we can produce and populate isotopes over the full isobar from drip-line to drip-line. It is also a region where it starts to be possible to make exact ab-initio calculations (one is at the moment approaching A = 12 [2]), and thus an increase in high precision data is needed. At IGISOL we have been studying part of this region, see Fig. 1, and here we will present related experiments and data on ^{12}C, ^{13}O and ^{8}B.

The isotope ^{12}C is the fourth most abundant nuclear species observed in the universe, but in the energy region just above the triple α-threshold broad overlapping resonances make spectroscopy very challenging, and significant uncertainties still remain on the nuclear spectroscopic properties of this isotope. The properties of this region are closely linked to the understanding of the cluster structure of individual ^{12}C states. Different configurations of the three α-particles are predicted to have different rotational bands, thus experimental knowledge of the position of specific states can be used to infer the underlying structure of ^{12}C. The properties of states in ^{12}C above the triple-α threshold are of high current interest for nuclear astrophysics and for the nuclear many-body problem in general. It is well known that the 7.65 MeV state in ^{12}C plays a critical role in astrophysics by enhancing the reaction rate of the triple-α process in red-giant stars by orders of magnitude [8, 9]. The European

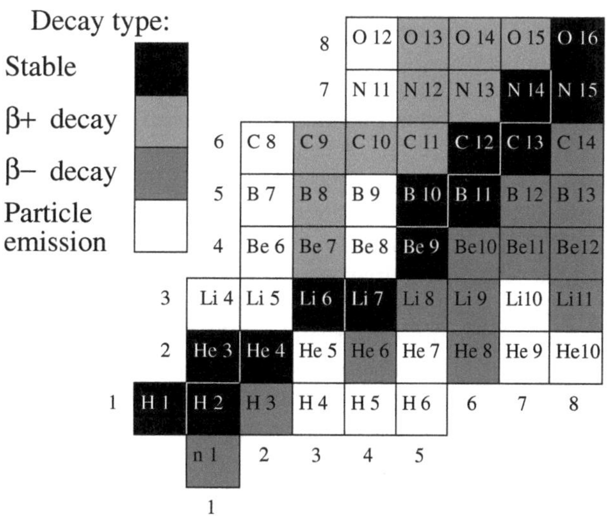

Fig. 1 Light nuclei are an interesting laboratory to study nuclear structure. The absence of bound systems at A = 5 and A = 8 and the stability of the α particle results in a dominating multi-particle breakup of these very light nuclear systems. In the mass 6–15 region shown above, β-delayed α emission has therefore already been experimentally verified for 11 of the 21 β-decaying isotopes, with delayed protons and neutrons having been observed for an additional two isotopes

compilation of astrophysical relevant reaction rates NACRE [10] assumes in addition a 2^+ state at 9.2 MeV, which at temperatures above $4{\cdot}10^9$ K significantly speeds up this process. This state has, however, not been observed in experiments and its existence is deduced only from cluster type calculations [11]. A broad 10.3 MeV resonance, however, exists, which for a long time has been assumed to be this rotational excitation of the Hoyle state [12, 13]. Recently it has been shown to be dominated by a 0^+ state interfering with the high-energy tail of the Hoyle state [14–17]. The interference between states causes further challenges in relation to the triple-α-reaction rate, not only in relation to the 0^+ interference, but also with regard to the 2^+ and 3^- states. See the article of H.O.U. Fynbo and C.Aa. Diget, in this journal, and references therein for further details.

The A = 13 isobar has attracted interest as part of the discussion of states consisting of neutrons occupying molecular-like orbits around α-particle clusters. This has been most successfully demonstrated in the Be isotopes where rotational bands have been identified and compared to detailed theoretical calculations, see, e.g., [18–20] and references therein. The cluster structure suggested for these excited states should also influence their decay to the n + ^{12}C or 3α + n final states. These decay properties are not known in detail for the relevant states. Further, breaking of isospin symmetry has been suggested as a indicator for molecular-like states as the protons, because of Coulomb repulsion, are not expected to occupy the molecular-like orbits [21]. Hence, in the A = 13 isobar one should expect isospin asymmetries between the β decays of ^{13}B and ^{13}O that feed excited states in ^{13}C and ^{13}N respectively. Work on the A = 9 isobar [22–24] has provided the largest asymmetry between mirror β transitions ever measured. Presently no theoretical understanding of this large asymmetry exists, but differences in (cluster) structure of the ^{9}Be/^{9}B daughters could play a role. Because the A = 13 system can be thought of as an α particle added to A = 9 independent of whether molecular structure is present, it would clearly be interesting to search for similar asymmetries within this system. Further, the β decays of ^{13}O and ^{13}B provide a test of the predictions of molecular orbits in ^{13}C: these decays populate $1/2^-$, $3/2^-$, and $5/2^-$ states in the daughters ^{13}N and ^{13}C, and hence spin assignments can be extracted from how the states are being populated. Large isospin asymmetries could indicate the presence of molecular orbits in the neutron-rich part of the isobar.

Knowledge of the energy spectrum of the neutrinos emitted in the β-decay of ^{8}B in the Sun is needed to interpret the neutrino spectrum measured on Earth. Although, the neutrinos from the ^{8}B-decay only constitute a small fraction of the total solar-neutrino flux [25] they play a central role in the interpretation of solar neutrino measurements due to their high energies, reaching up to 17 MeV. Apart from the tiny contribution made by the neutrinos from the ^{3}He + p proton capture, ^{8}B provides the only source of solar neutrinos above 2 MeV and hence the only source of detectable solar neutrinos for the main experiments; Super–Kamiokande (SK) and Sudbury Neutrino Observatory (SNO), which both have a detection threshold of 4 MeV. The ^{8}B neutrino spectrum cannot be derived theoretically because nuclear theory is unable to give a reliable prediction of the distribution of excitation energies populated in the decay. Experimentally, however, the ^{8}B neutrino spectrum may be extracted from a measurement of the β-delayed α–α breakup. A measurement of the total energy of the two α particles thereby provides a direct and hence more reliable determination of the level distribution. Such measurements have only recently become feasible thanks to advances in detector technology that we will explain in the following sections.

In all the cases mentioned here similar techniques and experimental set-ups have been used. In the following sections we will enter in more detail on the experimental procedures, including radioactive ion-beam production (Section 3), multi-particle detection (Section 4), and data analysis (Section 5). In continuation, results from the experimental campaigns are summarised (Section 6), and some future prospects for β-delayed multi-particle-breakup studies at IGISOL are outlined (Section 7).

3 Beam development at IGISOL to facilitate the above

Over the last decade there has been a rapid improvement in techniques for handling radioactive beams, in the push for producing and studying rare nuclei far away from the valley of stability. In the Isotope Separation On-line (ISOL) approach short-lived isotopes are quickly extracted from the target and ion-source, to be mass separated and directed to the experimental set-up for direct measurement. This technique can be applied to species with half-lives as short as milliseconds and is therefore ideally suited for new improved experiments on the β-decays of ^{12}N and ^{12}B (half-lives 11.0 ms and 20.2 ms respectively [26]). The development of new experimental equipment such as highly-segmented detectors and more advanced data-acquisition systems (DAQs) allows for efficient detection in coincidence of all emitted particles in the break-up process and thus experiments can be performed in full kinematics.

We first initiated a campaign aimed at providing new experimental information about this interesting region from the β-decays of ^{12}N and ^{12}B. The advantage in using β-decay here is that only 0^{+}, 1^{+} and 2^{+} states can be fed in allowed transitions. As a further improvement we produced ^{12}N and ^{12}B with the ISOL method, which permits stopping the radioactive beams in thin carbon foils (in the region of 30 μg/cm^{2}) surrounded by thin segmented detectors. This allows the final states to be detected in complete kinematics with negligible contamination from

β-particles and much reduced energy loss of the delayed α-particles compared to in-beam methods.

The first result from this campaign was the development of a new ISOL beam of ^{12}N [27]. The combination of the chemical reactivity of atomic nitrogen, the short lifetime of ^{12}N, and the low production cross sections, however, makes the ISOL production of this isotope extremely challenging. To overcome this, the ^{12}N beam was produced in the ^{12}C(p,n)^{12}N reaction using the IGISOL Ion Guide Isotope Separator On-Line. Here, the recoiling reaction products are thermalised (typically as 1^+ ions) in an inert gas flow, through which they are transported to the accelerating electrodes [4, 5]. In the first experiment (Feb. 2001) we achieved an average yield of 50 ions/s. From this experiment followed the development of an indirect production of ^{12}B at ISOLDE [6, 7]. As in the case of nitrogen the reactivity of boron greatly hinders its extraction from traditional ISOL targets. Instead, ^{12}Be was produced in a tantalum-foil target coupled to the ISOLDE–RILIS laser ion source [28]. ^{12}B was thereby produced in the β decay of ^{12}Be after delivery to our setup [29]. As the decay of ^{12}Be induces no further multi-particle breakup, no additional background was introduced this way. During the experiment (October 2002) the ion-beam intensity achieved was a few thousand ions per second, which was, however, less than expected. The results of the two experiments are detailed in [14, 15, 30, 31].

The second stage of this campaign was therefore launched at IGISOL with an experiment performed in Feb. 2004, for which further beam development was undertaken yielding the highest ISOL beam intensities of ^{12}N and ^{12}B so far. Based on the previous success at IGISOL in producing ^{12}N, as well as previous experience in the production of ^{12}B [4], the beams were produced using the ^{12}C$(p,n)^{12}$N and ^{11}B$(d,p)^{12}$B reactions in thin targets of natural carbon and boron (1.4 mg/cm^2 and 0.5 mg/cm^2 respectively). The energies of the proton (28 MeV) and deuteron (10 MeV) beams were optimised and delivered from the K130 cyclotron at intensities of 25 μA and 10 μA respectively, yielding average yields of 300 ^{12}N ions per second and 4000 ^{12}B ions per second throughout the experiment [16].

Having developed an efficient and highly segmented experimental set-up for this type of experiments, we followed up with an experiment on ^{13}O (Mar. 2004) [32]. For this purpose, an ^{13}O ISOL beam was developed for the first time. Previous experiments on the decay of ^{13}O all used the reaction ^{14}N(p, 2n)^{13}O with N_2 gas targets and a proton energy in the range of 43–51 MeV (threshold 31.2 MeV). The produced radioactivity was then either directly measured from within the target or transported with a helium gas flow system to a counting area [33–35]. The clear disadvantage of this approach is that the detectors see a very extended source, and particles emitted in the decay have to pass through the gas and possibly windows before being detected. To improve the detection conditions at IGISOL we used the same reaction but a solid boron-nitride target (1 mg/cm2), which was placed in front of the IGISOL gas cell. The proton beam energy from the K130 cyclotron was varied between 40 and 50 MeV and the intensity between 10 and 30 μA. The variations resulted in only modest variations in the yield of ^{13}O. The produced ions recoiled out of the target and stopped in the helium buffer gas, and the singly charged ions were then carried out by the helium flow and skimmed from the carrier gas before being injected into the mass separator by a 25 kV voltage, mass separated and directed to the thin catcher foil with a beam spot of only a few mm; see [5] for a full description of the method. This development led to the first ever production of an ISOL beam of the drip-line nucleus ^{13}O and yielded an average beam rate of 4–5 ions per second. Results of this experiment can be found in [36] and are outlined in Section 6.2.

Lastly, a ^{8}B beam has been developed (Feb./Aug. 2006) for studies of the neutrino spectrum, investigated through a detailed study of the ^{8}B β-delayed α–α breakup [37]. The radioactive beam was produced through the ^{6}Li(^{3}He, n)^{8}B reaction with a 5–7 MeV helium-3 beam impinging on a 1.95-mg/cm^{2} 95 % enriched ^{6}LiF target, evaporated on a 3.2 mg/cm^{2} Al backing. Also the ^{12}C$(p, \alpha n)^{8}$B reaction was investigated as a possible production mechanism (proton energies of 40 MeV) but no mass eight production was observed. An acceleration voltage of 20 kV was used, resulting in the ^{8}B ions, on average, being implanted midway into a carbon foil of thickness 26 μg/cm^{2}. The experiment was carried out in Jan. 2008, with an average implantation rate during the 72 hours of measurement of about 200 ions per second utilising a typical ^{3}He beam intensity of 0.3 μA. Results of this experiment may be found in [38, 39], some of which are discussed in Section 6.3.

4 Experimental Setup

The study of excited states of unbound light nuclei includes the simultaneous detection of several charged particles emitted with very low energy. This puts several constraints on the detection system to be used. For the detectors, the following is necessary: high segmentation to be able to simultaneously detect several coincident particles without summing; very thin dead layers in order to reduce the cut-off energy; in combination with thin detectors to minimise sensitivity to beta and neutral particles. Further, heavy particles are easily stopped in the ΔE detector where in many cases identification of mass has to be done by kinematical considerations off-line or pulse-shape analysis on-line.

4.1 Detectors

The ISOL technique followed by β-decay studies is an excellent laboratory, where the experiments discussed here can be performed with very compact experimental set-ups. β-decay experiments require: production of short-lived activity, and transport of it to the experiment where you let it land on a thin catcher-foil for subsequent decay. In order to kinematically reconstruct the event the point like source is surrounded by highly segmented detectors that simultaneously can detect the energy and angle of emission of several particles, e.g. as shown in Fig. 2 with a set-up of 3 DSSD telescopes.

In β decay experiments the isotope of interest is stopped in a catcher foil, where it decays at rest. This means that contrary to the case of reaction experiments, there is no additional energy available due to the impact of an incoming beam. The consequence is that the energy loss in the catcher foil and the entrance window of the detector have to be reduced to a minimum in order to lower the detection threshold. In Fig. 3 it is illustrated how the dead layers experienced by the particle depend on the angle of incidence. To the right is shown the reduction in the corrections needed to be made due to different incident angles when using a detector with less dead layer. In Section 5.1 it is further detailed how the peak-position of α particles is adjusted due to these effects.

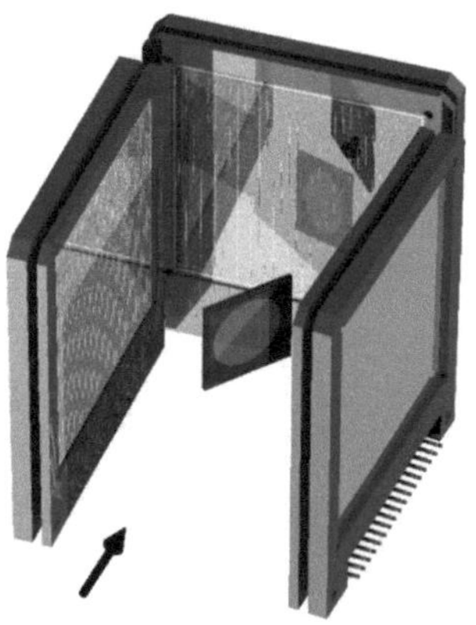

Fig. 2 Compact-geometry detector setup. Here shown for three thin DSSDs backed by thicker beta pad-detectors, all surrounding a thin carbon foil in which the radioactive nuclei, ^{8}B, ^{12}B, ^{12}N, or ^{13}O are implanted. The *arrow* illustrates the direction of the incoming beam

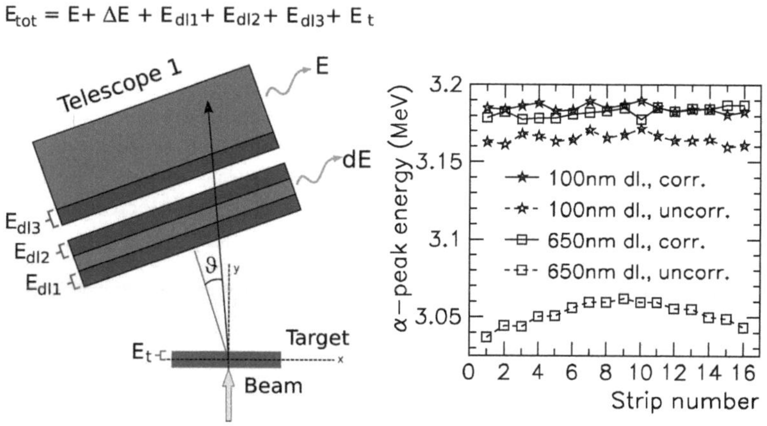

Fig. 3 Illustration of the different regions where energy losses must be taken into account. The energy detected in the ΔE and E detectors does not give the total energy of the incoming particle. To obtain the total energy, the energy loss in the different inactive areas must be added. This includes the energy loss which occurs in the target, as well as the energy loss from the dead layers of the Si detectors. The angle of entry is used to calculate the effective thickness of the different dead layers. To the *right* is shown the detector performances; comparing the needed dead-layer correction for the new 100 nm dead-layer design compared to the standard 600–700 nm, to obtain the correct peak position at 3180 keV α energy (^{148}Gd)

A novel design to achieve an ultra-thin dead layer for large-area Si strip detectors was made [40] in order to lower the detection threshold and to improve upon the corrections needed to be made as a function of incident angle. The traditional contact layer making up the strip is here replaced by a grid covering only 4% of the strip area (this new value has been adopted, see [39]). The dead layer is thus (over 96% of the active surface) reduced to become the implantation depth only. Furthermore, the implantation depth was reduced from the standard design of 400 to 100 nm, Fig. 4.

4.2 Electronics

As explained in the case of the detectors, high segmentation is needed to be able to detect several coincident particles without an exponential drop in efficiency. The

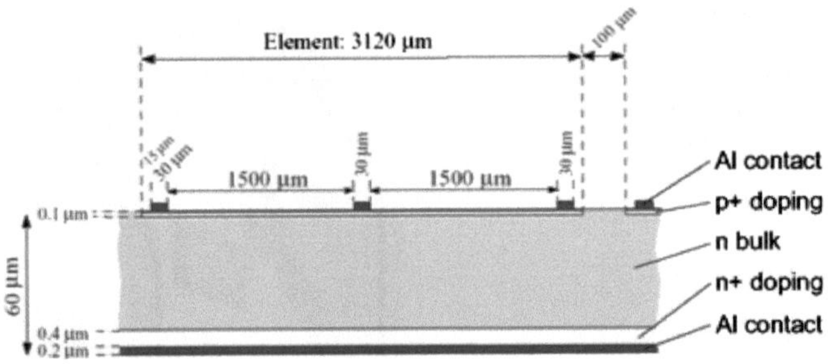

Fig. 4 The thin-window design of the DSSD (Micron Semiconductor Ltd.) is shown as a sketch of the detector profile. The design is with 16 3 × 50 mm^2 strips implanted on the front face, using a 100 nm p$^+$-doping depth, and an Al contact grid covering 4% of the active surface and, orthogonally on the back face are 16 n$^+$-strips implanted with traditional 400 nm, depth and 200 nm Al contact

high segmentation of the detectors leads to experiments with an increased number of electronic channels. For very dedicated experiments integrated electronic chips can be prepared, but in many cases where the detector set-up is frequently changed, one still has to rely on more traditional analogue electronic circuits. The DSSDs are very cost effective as with 16 × 16 strips we have 256 pixel detectors of 3 × 3 mm^2 and this with only 32 readout channels. In these experiments, however, one reaches rapidly 200 electronic channels and the use of standard 1 unit NIM electronics is not possible. We have therefore opted to use integrated units in multiples of 16 channels from MESYTEC [41]. The shaped and amplified signal is read out through 32-channel CAEN V785 peak-sensing ADCs using the Daresbury Multi Instance Data Acquisition System (MIDAS) [42].

5 Analysis Methods

The data analysis of multi-particle breakup data, as described below, is separated into three stages. Firstly, the handling of the raw data set, where energies and angles of all emitted particles are evaluated; secondly, the multi-particle analysis in which the important physical distributions are extracted and the sensitivity of these on experimental geometry and thresholds are evaluated; and thirdly, the interpretation of the derived distributions in terms of the populated states in the β-decay and their break-up to the relevant continuum channels. The handling of raw data as well as multi-particle analysis is given in the present section whereas the interpretations specific to each decay study are discussed in Section 6.

5.1 Calibrations

A range of different α sources were used in the calibration of the detectors to ensure that all systematics were under control. These include plated sources of ^{148}Gd, ^{239}Pu, ^{241}Am, and ^{244}Cm. In addition to the offline sources, online sources of ^{20}Na β-delayed α particles and ^{23}Al β-delayed protons were used for testing consistency of the calibration, the setup geometry, and all energy-loss corrections applied to the

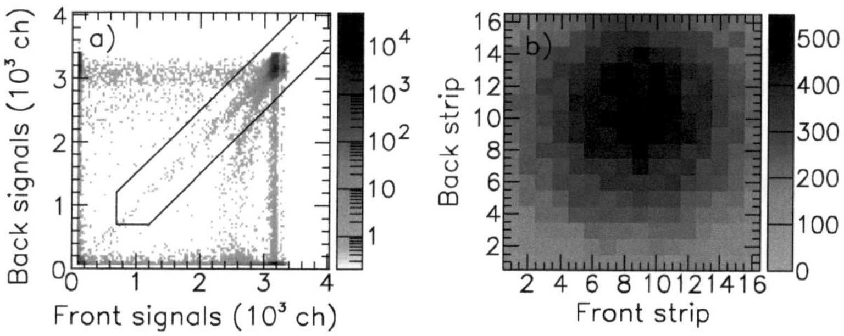

Fig. 5 ^{241}Am data: back-strip energy channel vs. front-strip energy channel (**a**). Cutoff for raw front–back identification shown. With this, a grid of number of events per pixel is deduced (**b**)

data. These sources are available only at ISOL facilities, and yield very strict cross-checks for data analysis of multi-particle breakup.

5.1.1 Detector geometry

Since the gains for all silicon signals are very similar, the detector pixels hit by α particles can easily be identified, even before a detailed calibration of the detectors. This can be seen in Fig. 5a, where the uncalibrated back (ohmic) strip signal is plotted against the front (junction) strip signal for ^{241}Am calibration data. To exclude low-energy noise and to identify the pixel hit by the α particle a cut is placed around the diagonal, accepting only similar-energy front–back pairs as shown. For these identified hits, the pixel distribution is shown in Fig. 5b. This distribution can be described by assuming an isotropic emittance from a point source somewhere in front of the detector taking into account the solid angle the individual pixel covers as seen from the point source. This solid angle is proportional to z/r^3 where z is the distance from the detector plane to the source and r is the distance from the pixel to the point source under the approximation that pixel dimensions are small compared to z. By fitting this to the measured distribution using a Poisson–Likelihood χ^2, the position of the point source relative to the detector is identified as described in [43]. For the present 3 mm detector segmentation and the source being placed 2–4 cm from the face of the detector, detector positions relative to the source are typically determined to within 0.1–0.3 mm.

5.1.2 Energy-loss corrections

The detector dead layers have, as described in Section 4, been minimised to a thickness of 100 nm yielding small dead-layer corrections, typically in the region of 25 keV for α particles in the 2–5 MeV energy range. The multi-particle-breakup studies described here are, however, dependent on the detection of very low energy alpha particles, and an appropriate account of dead layer corrections is therefore critical. Furthermore, traditional detectors have in some cases been used in conjunction with thin window detectors. In addition to this, a similar energy loss is experienced when the decaying isotope is implanted in the foil, yielding a foil energy loss of about 25 keV, with similar variations as those of the dead-layer corrections. In addition to the minimisation of these effects by reduction of dead layer and foil thickness, two

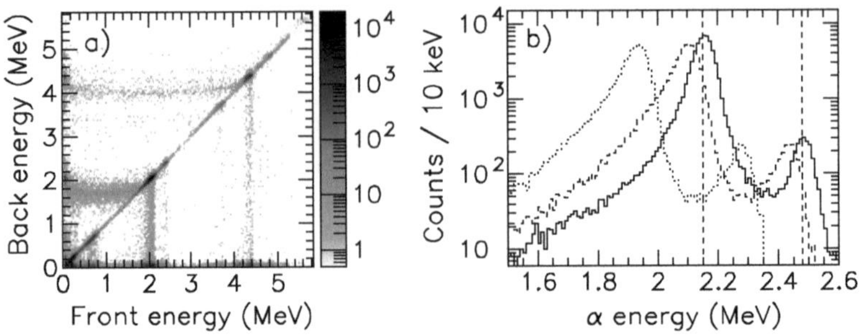

Fig. 6 Calibrated energies for ^{20}Na data. Shown is **a**: front–back energy-identification plot, with groups in the 2–4 MeV range along the diagonal being the major alpha-decay lines from the ^{20}Na decay. Groups below 1 MeV are ^{16}O nuclei emitted back-to-back with an α particle. Frame **b** shows: detected α energy (*dotted*), dead-layer corrected α energy (*dashed*), and reconstructed ^{20}Na-decay α-energy as corrected for detector dead-layer and implantation-foil energy loss. Tabulated energies [45] are shown as *vertical lines*

different strategies are applied to correct for these effects, one for the offline sources and one for online source and multi-particle decay data.

From the derived source position, the impact angle of the α particle on the detector is found, yielding the effective dead layer of the pixel as seen from the source [40, 43]. From Srim [44] calculations, the stopping of the α particles in the dead layer is found, yielding an effective peak position for each pixel. By averaging over all 16 pixels, weighted by their solid-angle coverage as seen from the source, an effective peak position is found for each strip, and used in the calibration in place of the α energy as emitted from the source.

5.1.3 The use of On-line beams

This calibration, is now used in the analysis of the online ^{20}Na-source data and multi-particle data, to deduce the impact energy from the energy deposited in the active region of the strip. This is evaluated on an event-by-event basis based on the identification of the pixel hit by the particle (Section 5.1.1). Similarly, from the impact point of the individual particle on the detector the angle of the emitted particle with respect to the foil can be deduced. From this and the energy at impact, also the foil energy loss can be deduced.

The effect of the corrections is illustrated in Fig. 6, in which the measured 2 MeV ^{20}Na α-energy peak is compared to the energy-loss-corrected energies and the tabulated energy (the data shown here is for a detector with the traditional 600–700 μm dead-layer). Since the effective thickness of both foil and dead layer is different for the individual pixels in one strip, the event-by-event correction not only removes the energy-loss bias in the raw data but improves resolution as well. Using the strongest α groups from the ^{20}Na decay [39, 45, 46], covering the 2–4 MeV energy range, the systematic effects of dead layers, foil-energy loss, and geometry are thereby evaluated. Through this analysis, it is confirmed that the detection efficiency and energy resolution are independent of energy, that the detector geometry is as expected, and the beam-spot position on the stopping foil is determined. The latter can be determined as the thin ^{12}C implantation foil allows not only α particles, but also the ^{16}O recoil to escape. As a result, α-^{16}O coincidences must be detected back-to-back. The events should furthermore obey momentum conservation if corrected for: dead-layer energy loss in the detectors; stopping in the ^{12}C implantation foil; and pulse-height defects in the ^{16}O silicon response [47]. In addition, the combination of thin-window detectors and coincident detection of α particles and ^{16}O recoils yields a strong suppression of low-energy background allowing the identification of extremely weak low-energy α-decay groups [39].

5.2 Kinematics

With single-particle detection under control, the next stage in the analysis process is identifying the β-delayed multi-particle breakup under investigation. The detection of the complete kinematics of the breakup with all emitted particles detected has marked advantages over experiments where one of the break-up particles remains undetected. In the latter case, the remaining particle must be reconstructed from an evaluation of missing momentum in the break-up [14]. When, on the other hand, the full break-up is detected, the total momentum can be used as a strong indicator for the authenticity of the break-up event [16], when compared to random coincidences or low-energy β signals. To minimise background for low-intensity break-up channels, we therefore focus on full-kinematics multi-particle detection. This necessity has been the primary driver for the development of the compact, segmented setup described in Section 4. The acceptance of this setup covers the majority of the multi-particle phase space for up to three emitted particles, thereby allowing the simultaneous study of break-up channels that span significant regions of phase space.

5.2.1 Total momentum

The evaluation of the measured total momentum has the further advantage of allowing a direct test of geometry, similarly to that of the back-to-back α-^{16}O coincidence measurements used to identify the ^{20}Na beam spot position relative to the detectors. For multi-particle detection in full kinematics, the individual projections of the three-particle sum momentum p_x, p_y onto the plane of the implantation foil and p_z perpendicular to the plane yields the same type of information.

This is discussed in [14, 48], where the marked sensitivity to position and size of the beam-spot is demonstrated for both two- and three-particle coincidence data in detailed comparisons to Monte-Carlo simulations.

With the completion of all such geometrical and energy-loss corrections, true multi-particle coincidences can be identified by requiring that the total momentum is sufficiently small. The remaining contributions to the total-momentum resolution are therefore: energy resolution, angular resolution, and beta-decay recoil. All of which are typically in the range of 3–20 MeV/c, depending on the specific multi-particle breakup under investigation.

5.2.2 Two-particle center-of-mass-energy reconstruction

Multi-particle breakup of light nuclei very often yield one or more α particles in the exit channel(s) because of the strong binding of the α particle. Furthermore, because of the strong α clustering in the slightly unbound ^{8}Be 0^+ ground state, multi-particle breakup often proceed via ^{8}Be states [14, 16, 22–24, 31, 39, 49]. For

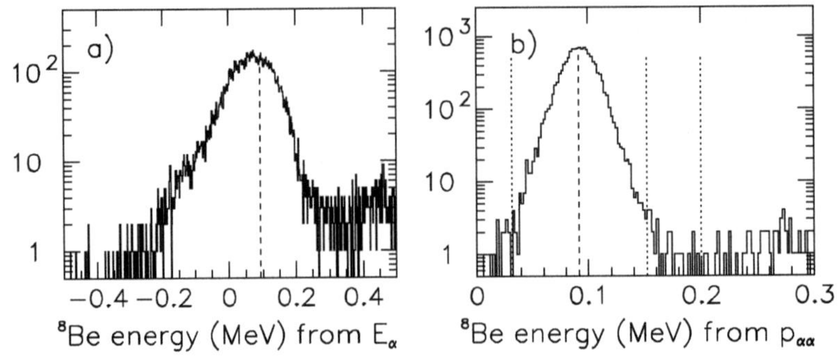

Fig. 7 Frame **a** shows the reconstructed ^{8}Be energy as found from the subtraction of the ^{8}Be recoil energy from the combined energy of the two α particles from the ^{8}Be breakup. The energies in **b** are reconstructed from the α–α relative momentum. The true energy of the ^{8}Be ground state is indicated as a *dashed line*. Gates used to distinguish between break-up channels (being either the ^{8}Be ground state or energies above 200 keV) are shown in **b** as *dotted lines*

identification of exit channels it is therefore useful to reconstruct the intermediate ^{8}Be energy from the data. This energy can be found from the two measured α energies as $E_{^8\mathrm{Be}} = E_{\alpha_1} + E_{\alpha_2} - E_{\mathrm{rec}}$, where E_{rec} is the kinetic energy of the recoiling ^{8}Be nucleus. Alternatively, the resonance energy can be found from the relative momentum of the two alpha particles in the ^{8}Be centre-of-mass frame.

The uncertainty in the reconstructed energy again depends on the specific multi-particle breakup under investigation. For high-precision studies of the ^{8}B β decay, careful analysis has shown that the former method is the most precise of the two, in spite of the β-decay recoil energy not being measured in the experiment [39]. For the β-delayed triple-α-breakup studies, on the other hand, the recoil energy originates from the first emitted α-particle, and is therefore large compared to the α–α-breakup energy. With this, the resolution in reconstructed energy is worsened, even when all three α-particle energies are measured, and the relative momentum yields a better evaluation of the resonance energy. This is illustrated in Fig. 7, where the two methods are compared for triple-α-breakup data.

5.2.3 Angular correlations and break-up dynamics

Triple-α-coincidence data from the β decay of ^{12}B is shown in Fig. 8. The displayed data set includes both the ^{8}Be ground state channel and the ^{8}Be excited state channel. This type of plot was first used in [50], and is very illuminating when studying a break-up where two or more break-up channels are possible. This stems from the way kinematics are displayed. The plot is a scatter plot of the deduced total energy plotted against the individual particle energies.

In the general case, this is most useful when displayed separately for each particle type. However, for the triple-α breakup, shown, only a single plot is required. Each detected coincidence event will in this case yield three dots on a horizontal line in the figure. If the break-up is mediated by a narrow intermediate state (such as the ^{8}Be ground state), a linear relation is observed between the individual-particle energy and the total energy, in this case effectively the ^{12}C excitation energy: $E_\alpha = \frac{2}{3}\,(E_{^{12}\mathrm{C}^*} - E_{^8\mathrm{Be}})$. This is clearly visible in the data as a line with slope 3/2 crossing the

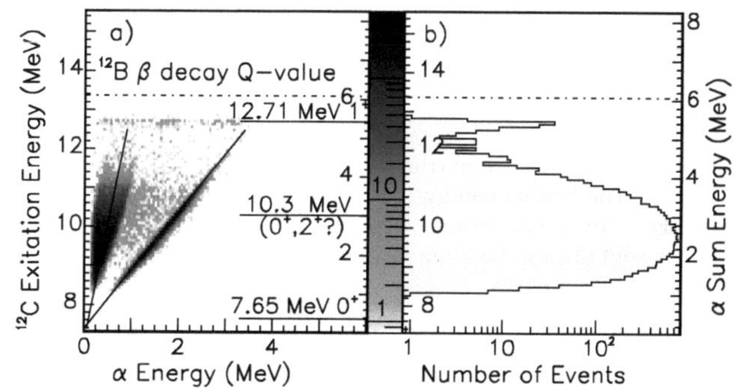

Fig. 8 *Left frame* plot of the ^{12}B β-delayed triple-α-breakup distribution with the total triple-α sum energy plotted against single-α particle energy. *Right frame*: projection of the same data onto the E_{sum} axis.

E_{sum} axis at 91.8 keV. The two remaining α particles will here share the remaining energy, which is visible as well, centred around the line with slope 6.

In these scatter plots, many of the physical properties examined in the experiment can be seen at first glance. One aspect is the very pronounced structure around 10 MeV in ^{12}C clearly decaying through the ^{8}Be ground state channel and in addition to this the 12.7 MeV 1^{+} state in ^{12}C. The fact that it is a 1^{+} state is immediately visible, since it cannot decay through the ^{8}Be 0^{+} ground state (conservation of spin and parity), but must decay through another channel, seen by the completely different sharing of the energy between the three alphas. A third group of break-up events is seen between the two sloped ^{8}Be ground state contributions around 10 MeV in ^{12}C. This can be interpreted as break-up through higher energies of ^{8}Be. The 7.65 MeV Hoyle state is not seen in this plot because of low-energy cutoffs.

6 Some Results

We have so far discussed in detail the experimental procedures in order to study the multi-particle breakup following β decay of light exotic nuclei. As was outlined in Section 2 we have at IGISOL performed experiments on ^{12}C, ^{13}O and ^{8}B. This has produced several interesting outputs and several articles have been published. We will in the following summarise some of the results.

6.1 The case of ^{12}C

The results from the ^{12}C campaign are presented in a dedicated article by Fynbo and Diget, within this journal, which gives a detailed report on all the results obtained. Here it is sufficient to say, that the impact of the ^{12}C campaign can be summarised under three headings: the direct measurement of the ^{12}C continuum states, their β-decay feeding and break-up to the triple-α continuum [16, 17, 51, 52]; the impact on theoretical modelling of three-body dynamics through unprecedented observation of the detailed break-up and the challenge of existing models [31]; and lastly, the input to the evaluation of the astrophysical triple-α reaction rate [14, 15].

6.2 The case of ^{13}O

The β-delayed proton spectrum obtained from ^{13}O can be observed in Fig. 9. The spectrum extends to both lower and higher energy and has slightly better statistics

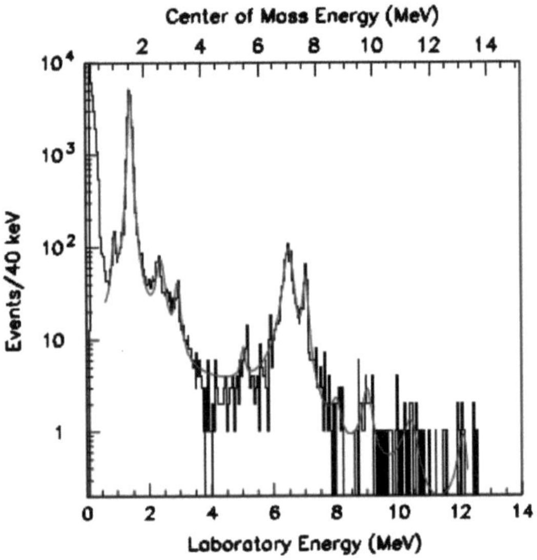

Fig. 9 β-delayed proton spectrum from the β decay of ^{13}O. The *full curve* is a fit using Breit–Wigner peak shapes modified by the integrated Fermi function (the f factor). The *bottom* energy scale is the observed proton energy with the *top* denoting center-of-mass energy

than the previous measurements [33–35]. The peak at 1 MeV center-of-mass energy and everything above 9 MeV is seen for the first time in these IGISOL data. The full curve is a fit to the proton spectrum using Breit–Wigner peak shape multiplied by the integrated Fermi function to take into account that the low-energy sides of (broad) levels populated in β decay are enhanced. In this way energy, width and relative branching ratios could be determined [36]. We can conclude that there is a significant feeding to the region above 10 MeV. A new decay scheme based on the obtained data could be produced and compared to that of ^{13}B. The mirror asymmetries are all consistent with zero within the experimental error, except for the ground-state transitions, where the difference is 2 standard deviations. This asymmetry is in agreement with the estimates of Smirnova and Volpe [53]. However, the obtained data, do not confirm the findings of Milin et al. [21]. In Fig. 10 a photo of the experimental set-up is shown.

6.3 The case of ^{8}B

The ^{8}B experiment was performed in January 2008 at the IGISOL facility. This was aimed at improving the determination of the neutrino spectrum from the ^{8}B β-decay compared to the presently accepted distributions [54, 55] through a utilisation of our highly-segmented detector set-up described in Section 4. In this, the two α particles were measured in coincidence in separate very thin detectors (60 μm) facing a thin carbon foil in which the ^{8}B activity was implanted at 20 keV. Using thin highly-segmented detectors where the β response is negligible reduces enormously the distortion of the spectrum due to β summing and unwanted background from β–α coincidences. Furthermore, at IGISOL we could produce and use the β-delayed α

Fig. 10 Displayed is a photo taken into the chamber of the set-up used at IGISOL for the ^{13}O experiment. The radioactive beam passes through a semi-sphere of Si detectors (36 × 4 quadrants) [56], and is stopped in a C foil surrounded on three sides by the DSSD telescopes as shown in Fig. 2

emitter ^{20}Na for on-line energy calibration and thus reduce the systematic uncertainties from energy loss corrections, as described in Section 5.1.3. Through back-to-back coincidence measurements of α-^{16}O pairs in the calibration, the effect of pulse-height defects from ^{16}O was further avoided. These improvements should be seen in light of the fact that the uncertainty in the measured ^{8}Be excitation-energy spectrum and deduced neutrino-energy distribution is dominated by systematic uncertainties at most energies.

The resulting ^{8}Be spectrum, as given in [38] yields a downward shift of the peak position of 20–25 keV with an uncertainty of 6 keV. This corresponds to a slight increase in the high-energy tail of the ^{8}B neutrino spectrum. Publications detailing the results are presently in preparation [39].

7 Future studies

The ^{12}C states in the triple-α continuum have been subject to the scrutiny of a number of experiments over the past decades. These have utilised either β decay of ^{12}N and ^{12}B [13–17, 31, 51, 52, 57–59] or reactions such as proton, α or ^{12}C scattering, or transfer reactions [60–67] to populate the states.

The second excited 2^+ state in ^{12}C has proven particularly elusive. This state is a key element in our understanding of the entire triple-α continuum, and has for decades been theoretically predicted. In the simplest picture the 2^+ state would be the rotational excitation of the Hoyle state [8, 9], which in the earliest works was described as a rigid triple-α chain [12, 68]. The 2^+ state would thereby reveal the structure of the Hoyle state, and be a cornerstone in the theoretical understanding of the triple-α continuum. More recently, complex theoretical descriptions of the continuum states have been developed [11, 69–71], but experimentally only suggestions of the 2^+ state have been observed and a coherent picture from experiment is yet to be seen. These experimental difficulties stem from the following:

- The broad nature of the states: the states in question, in particular low-lying 0^+ and 2^+ states in the triple-α continuum, couple very strongly to the triple-α continuum, coupling not only through the ^{8}Be ground state but in several cases also through the ^{8}Be excited 2^+ state. This is a considerable challenge to experiments as both the states and their break-up channels are broadened, inhibiting a strictly sequential description of the break-up of the states.
- The presence of background states: firstly, a strong 0^+ strength is present in the energy region where the 2^+ state is expected to be; and secondly, the 3^- state (9.6 MeV) populated strongly in inelastic scattering is located at exactly the suggested 2^+ energy [26, 65]. This reduces the sensitivity to the 2^+ state significantly in reaction studies.
- The weak population in β decay: though β-decay studies offer a unique suppression of all states apart from the 0^+, 1^+ and 2^+ states, the second excited 2^+ state in ^{12}C has been suggested theoretically to be populated very weakly in the β decay of ^{12}N and ^{12}B [69]. As a result, the shape of the combined 0^+, 2^+ strength is insufficient to be sensitive to the suggested 2^+ state.

The possible identifications of the 2^+ state have therefore all been from reaction studies which have suggested a state in the 9.5–11.5 MeV region [60–65]. All, however, suffer from significant backgrounds and none of them consider their measurement a firm confirmation of the 2^+ state position and properties.

In addition to the fundamental interest in the structure of the triple-α continuum states, the mentioned 2^+ state is particularly important because it, if present at the suggested energy, accounts for the single most important contribution to the triple-α reaction rate apart from the Hoyle state [10, 15, 65].

7.1 Facility developments

With the new upgrade of the radioactive isotope facility in Jyväskylä, some of the above challenges will be addressed. The very significant increase in beam intensities will, in combination with recent detector upgrades allow high-statistics measurements of β-triple-α coincidences. With this, the broad nature of the states and break-up channels will be probed in detail for the full β-decay Q-value window; the 3^- state background is eliminated througn the β-decay selectivity; and the discrimination between the 0^+ and 2^+ strength greatly improved through identification of β-triple$-\alpha$ coincidences.

Using the new MCC30/15 cyclotron at JYFL [72], primary beams of 100 μA and 50 μA are planned for proton and deuteron beams respectively, corresponding to an increase in primary beam intensities of factors of 4–5. Furthermore, the recent IGISOL ion-guide upgrade has yielded enhancement of factors up to 10. Even in a conservative evaluation, this is expected to yield at least a factor 20 improvement in the beam intensities for beams of ^{12}N as well as ^{12}B.

7.2 Detection of β-triple-α coincidences

The detector setup to be used for the future multi-particle-breakup studies at IGISOL consists primarily of the Silicon-Cube array [73]. This array is built from six double-sided silicon strip detectors (DSSDs) in a cubic close-packed formation. In the centre of the setup, the new high-intensity IGISOL beams of ^{12}N and ^{12}B will be stopped in a thin carbon foil. The new setup will through the doubling of the spatial coverage, provided by the six close-packed detectors, yield efficiency improvements of: a factor of two for single-α detection; a factor of four for triple-α detection of the back-to-back breakup through the ^{8}Be ground state; and a factor of ten for triple-α coincidence detection of the weakest break-up through higher energies in ^{8}Be. This corresponds to an efficiency of approximately 15% for triple-coincidence detection, regardless of the break-up kinematics.

At least one of the six detectors will be a thick (approximately 1 mm) segmented detector for detection of β particles from the decay, and all of the segmented detectors will be backed by silicon pad detectors. This yields a β efficiency of 10%, and thereby a total efficiency for the quadruple β-triple-α coincidences of 1.5%. In addition to the Silicon Cube detectors, a single HPGe detector will be used in conjunction with the β-detector to obtain an absolute normalisation of the beam intensity from the known decay to the first excited 2^+ state of ^{12}C.

In addition to the two orders of magnitude gain in statistics offered by the combined improvements in beam intensity and detection efficiency, we will in future studies obtain a significantly improved sensitivity to the 2_2^+ state through measurement of β-3α coincidences. This is based on the fact that the 2^+ state, relative to the direction of the emitted β particle, is produced in a highly polarised state [74]. If the 2^+ state suggested in [61, 65] in the region between 9.5 MeV and 10.0 MeV is the 2^+ state calculated by Kanada En'yo [69], the future IGISOL studies should be sensitive to the state, and thereby yield a direct observation of this elusive state.

8 Summary

In this contribution we have discussed β delayed multi-particle-breakup studies of light nuclei performed at the IGISOL facility over the last decade. The interest in these nuclei comes both from the need to understand the nuclear synthesis in the universe as well as to obtain a more detailed understanding of the nuclear structure. To disentangle these systems, kinematically complete studies are needed in which all particles are detected with precise energies and relative emission angles. These types of experiments have recently become possible to perform due to the advances in ISOL beams, highly segmented and compact detector arrays and integrated electronics to deal with the increase of readout channels. We have concentrated a major part of the article explaining these technical advances and the subsequent analysis that also have advanced with the increased computer performance and the possibility to compare obtained data with Monte-Carlo simulations. The techniques have been applied to study the β-delayed break-up of three key nuclei in this light region; ^{12}C (^{12}N and ^{12}B decays), ^{13}O and ^{8}B, for which some of the obtained results have been summarised.

Acknowledgements This work was partly financed by the Spanish Ministry of Science and Technology under project FPA2007-62170, FPA2009-07387 and partly by the European Union under contract EURONS JR4/DLEP (contract no. 506065). We would like to acknowledge The Knut and Alice Wallenberg Foundation, Sweden, for funding of experimental equipment.

References

1. Pieper, S.C., Wiringa, R.B.: Annu. Rev. Nucl. Part. Sci. **51**, 53 (2001)
2. Pieper, S.C.: Nucl. Phys. A **751**, 516 (2005)
3. Korsheninnikov, A.A.: Sov. J. Nucl. Phys. **52**, 827 (1990)
4. Ärje, J., Äystö, J., Hyvönen, H., Taskinen, P., Koponen, V., Honkanen, J., Valli, K.: Nucl. Instrum. Methods A **247**, 431 (1986)
5. Äystö, J.: Nucl. Phys. A **693**, 477 (2001)
6. Kugler, E.: Hyperfine Interact. **129**, 23 (2000)
7. Jonson, B., Riisager, K.: Scholarpedia **5**(7), 9742 (2010). http://www.scholarpedia.org/article/The_ISOLDE_facility
8. Hoyle, F., Dunbar, D.N.F., Wenzel, W.A., Whaling, W.: Phys. Rev. **92**, 1095c (1953)
9. Hoyle, F.: Astrophys. J. Suppl. Ser. **1**, 121 (1954)
10. Angulo, C., Arnould, M., Rayet, M., Descouvemont, P., Baye, D., Leclercq-Willain, C., Coc, A., Barhoumi, S., Aguer, P., Rolfs, C., Kunz, R., Hammer, J.W., Mayer, A., Paradellis, T., Kossionides, S., Chronidou, C., Spyrou, K., Degl'Innocenti, S., Fiorentini, G., Ricci, B., Zavatarelli, S., Providencia, C., Wolters, H., Soares, J., Grama, C., Rahighi, J., Shotter, A., Lamehi Rachti, M.: Nucl. Phys. A **656**, 3 (1999)
11. Descouvemont, P., Baye, D.: Phys. Rev. C **36**, 54 (1987)
12. Morinaga, H.: Phys. Lett. **21**, 78 (1966)
13. Schwalm, D., Povh, B.: Nucl. Phys. **89**, 401 (1966)
14. Diget, C.Aa., Barker, F.C., Borge, M.J.G., Cederkäll, J., Fedosseev, V.N., Fraile, L.M., Fulton, B.R., Fynbo, H.O.U., Jeppesen, H.B., Jonson, B., Köster, U., Meister, M., Nilsson, T., Nyman, G., Prezado, Y., Riisager, K., Rinta-Antila, S., Tengblad, O., Turrion, M., Wilhelmsen, K., Äystö, J.: Nucl. Phys. A **760**, 3 (2005)
15. Fynbo, H.O.U., Diget, C.Aa., Bergmann, U.C., Borge, M.J.G., Cederkäll, J., Dendooven, P., Fraile, L.M., Franchoo, S., Fedosseev, V.N., Fulton, B.R., Huang, W., Huikari, J., Jeppesen, H.B., Jokinen, A.S., Jones, P., Jonson, B., Köster, U., Langanke, K., Meister, M., Nilsson, T., Nyman, G., Prezado, Y., Riisager, K., Rinta-Antila, S., Tengblad, O., Turrion, M., Wang, Y., Weissman, L., Wilhelmsen, K., Äystö, J., the ISOLDE Collaboration: Nature **433**, 136 (2005)
16. Diget, C.Aa., Barker, F.C., Borge, M.J.G., Boutami, R., Dendooven, P., Eronen, T., Fox, S.P., Fulton, B.R., Fynbo, H.O.U., Huikari, J., Hyldegaard, S., Jeppesen, H.B., Jokinen, A., Jonson, B., Kankainen, A., Moore, I., Nieminen, A., Nyman, G., Penttila, H., Pucknell, V.F.E., Riisager, K., Rinta-Antila, S., Tengblad, O., Wang, Y., Wilhelmsen, K., Aysto, J.: Phys. Rev. C **80**, 034316 (2009)
17. Hyldegaard, S., Alcorta, M., Bastin, B., Borge, M.J.G., Boutami, R., Brandenburg, S., Büscher, J., Dendooven, P., Diget, C.Aa., Van Duppen, P., Eronen, T., Fox, S.P., Fraile, L.M., Fulton, B.R., Fynbo, H.O.U., Huikari, J., Huyse, M., Jeppesen, H.B., Jokinen, A.S., Jonson, B., Jungmann, K., Kankainen, A., Kirsebom, O.S., Madurga, M., Moore, I., Nieminen, A., Nilsson, T.: Phys. Rev. C **81**, 024303 (2010)
18. von Oertzen, W., Bohlen, H.G.: C. R. Phys. **4**, 465 (2003)
19. Freer, M.: C. R. Phys. **4**, 475 (2003)
20. von Oertzen, W., Freer, M., Kanada-En'yo, Y.: Phys. Rep. **432**, 43 (2006)
21. Milin, M., von Oertzen, W.: Eur. Phys. J. A **14**, 295 (2002)
22. Prezado, Y., Borge, M.J.G., C.Aa. Diget, Fraile, L.M., Fulton, B.R., Fynbo, H.O.U., Jeppesen, H.B., Jonson, B., Meister, M., Nilsson, T., Nyman, G., Riisager, K., Tengblad, O., Wilhelmsen, K.: Phys. Lett. B **618**, 43 (2005)
23. Prezado, Y., Bergmann, U.C., Borge, M.J.G., Cederkäll, J., Diget, C.Aa., Fraile, L.M., Fynbo, H.O.U., Jeppesen, H., Jonson, B., Meister, M., Nilsson, T., Nyman, G., Riisager, K., Tengblad, O., Weissmann, L., Wilhelmsen Rolander, K., Collaboration, I.: Phys. Lett. B **576**, 55 (2003)
24. Bergmann, U.C., Borge, M.J.G., Boutamib, R., Fraile, L.M., Fynbo, H.O.U., Hornshøj, P., Jonson, B., Markenroth, K., Martel, I., Mukha, I., Nilsson, T., Nyman, G., Oberstedt, A.,

Prezado Alonso, Y., Riisager, K., Simon, H., Tengblad, O., Wenander, F., Wilhelmsen Rolander, K., the ISOLDE collaboration: Nucl. Phys. A **692**, 427 (2001)
25. Bahcall, J.N., Serenelli, A.M., Basu, S.: Astrophys. J. Lett. **621**, L85 (2005)
26. Ajzenberg-Selove, F.: Nucl. Phys. A **506**, 1 (1990)
27. Fynbo, H.O.U., Axelsson, L., Äystö, J., Bergmann, U.C., Borge, M.J.G., Dendooven, P., Fraile, L.M., Forssén, C., Hansper, V.Y., Huikari, J. Jeppesen, H., Jokinen, A., Jonson, B., Markenroth, K., Meister, M., Nilsson, T., Nyman, G., Oinonen, M., Prezado, Y., Riisager, K., Rinta-Antila, S., Siiskonen, T., Tengblad, O., Wenander, F., Wang, Y., Wilhelmsen-Rolander, K.: Study of the multi-particle breakup in the β-decay of ^{12}N (2000). Proposal IGISOL-63 to the March 2000 JYFL PAC
28. Fedoseyev, V.N., Huber, G., Köster, U., Lettry, J., Mishin, V.I., Ravn, H., Sebastian, V.: Hyperfine Interaction. **127**, 409 (2000)
29. Fynbo, H.O.U., Äystö, J., Bergmann, U.C., Borge, M.J.G., Cederkäll, J., Fraile, L.M., Franchoo, S., Fulton, B.R., Jeppesen, H., Jokinen, A., Jonson, B., Köster, U., Meister, M., Nilsson, T., Nyman, G., Prezado, Y., Riisager, K., Tengblad, O., Weissman, L., Wilhelmsen-Rolander, K.: Study of the β-decay of ^{12}B (2002). Proposal IS404 to the January 2002 ISOLDE and Neutron Time-of-Flight Experiments Committee (INTC)
30. Fynbo, H.O.U., Prezado, Y., Äystö, J., Bergmann, U.C., Borge, M.J.G., Dendooven, P., Huang, W., Huikari, J., Jeppesen, H., Jones, P., Jonson, B., Meister, M., Nyman, G., Oinonen, M., Riisager, K., Tengblad, O., Vogelius, I.S., Wang, Y., Weissman, L., Rolander, K.W.: Eur. Phys. J. A **15**, 135 (2002)
31. Fynbo, H.O.U., Prezado, Y., Bergmann, U.C., Borge, M.J.G., Dendooven, P., Huang, W.X., Huikari, J., Jeppesen, H., Jones, P., Jonson, B., Meister, M., Nyman, G., Riisager, K., Tengblad, O., Vogelius, I.S., Wang, Y., Weissman, L., Wilhelmsen Rolander, K., Äystö, J.: Phys. Rev. Lett. **91**, 082502 (2003)
32. Fynbo, H.O.U., Fraile, L.M., Bergmann, U.C., Borge, M.J.G., Cederkäll, J., Dendooven, P., Diget, C.Aa., Franberg, H., Fox, S.P., Fulton, B.R., Groombridge, D., Huikari, J., Jeppesen, H., Jokinen, A., Jonson, B., Meister, M., Nilsson, T., Nyman, G., Riisager, K., Simon, H., Tengblad, O., Turrion, M., Äystö, J.: Study of the β-decay of ^{13}O and ^{13}B at IGISOL (2003). Proposal IGISOL-91 to the September 2003 JYFL PAC
33. McPherson, R., Esterlund, R.A., Poskanzer, A.M., Reeder, P.L.: Phys. Rev. **140**(6B), B1513 (1965)
34. Esterl, J., Hardy, J., Sextro, R., Cerny, J.: Phys. Lett. B **33**, 287 (1970)
35. Asahi, K., Matsuta, K., Takeyama, K., Tanaka, K.H., Nojiri, Y., Minamisono, T.: Phys. Rev. C **41**, 358 (1990)
36. Knudsen, H.H., Fynbo, H.O.U., Borge, M.J.G., Boutami, R., Dendooven, P., Diget, C.Aa., Eronen, T., Fox, S., Fraile, L.M., Fulton, B., Huikary, J., Jeppesen, H.B., Jokinen, A.S., Jonson, B., Kankainen, A., Moore, I., Nieminen, A., Nyman, G., Penttilä, H., Riisager, K., Rinta-Antila, S., Tengblad, O., Wang, Y., Wilhelmsen, K., Äystö, J.: Phys. Rev. C **72**, 044312 (2005)
37. Fulton, B.R., Fynbo, H.O.U., Borge, M.J.G., Dendooven, P., Diget, C.Aa., Fraile, L.M., Hansper, V.Y., Huikari, J., Jeppesen, H.B., Jokinen, A., Jonson, B., Kirsebom, O., Knudsen, H., Moore, I., Nilsson, T., Nyman, G., Penttilä, H., Rinta-Antila, S., Riisager, K., Saastamoinen, A., Tengblad, O., Äystö, J.: ^{8}B β-decay study at igisol (2006). Proposal IGISOL-123 to the JYFL PAC
38. Kirsebom, O., Fynbo, H.O.U., Knudsen, H.H., Riisager, K., Fulton, B.R., Laird, A., Fox, S.P., Borge, M.J.G., Madurga, M., Tengblad, O., Alcorta, M., Jonson, B., Nyman, G., Hultgren, H., Raabe, R., Büscher, J., Saastamoinen, A., Jokinen, A., Moore, I.D., Äystö, J.: Proceedings of the International Symposium on Nuclear Astrophysics—Nuclei in the Cosmos—XI, p. 16 (2010)
39. Kirsebom, O.S., Hyldegaard, S., Alcorta, M., Borge, M.J.G., Buscher, J., Eronen, T., Fox, S., Fulton, B.R., Fynbo, H.O.U., Hultgren, H., Jokinen, A., Jonson, B., Kankainen, A., Karvonen, P., Kessler, T., Laird, A., Madurga, M., Moore, I., Nyman, G., Penttila, H., Rahaman, S., Reponen, M., Riisager, K., Roger, T., Ronkainen, J., Saastamoinen, A., Tengblad, O., Aysto, J.: Precise and accurate determination of the 8B decay spectrum. Phys. Rev. **C83**, 065802 (2011)
40. Tengblad, O., Bergmann, U.C., Fraile, L.M., Fynbo, H.O.U., Walsh, S.: Nucl. Instrum. Methods A **525**, 458 (2004)
41. MESYTEC: STM-16+, 16 fold spectroscopy amplifier with discriminators and multiplicity trigger. http://www.mesytec.com/
42. Pucknell, V.F.E., Letts, S.: The multi instance data acquisition system (MIDAS) (2010). http://npg.dl.ac.uk/MIDAS/
43. Bergmann, U.C., Fynbo, H.O.U., Tengblad, O.: Nucl. Instrum. Methods A **515**, 657 (2003)

44. Ziegler, J.F., Biersack, J.P., Littmark, U.: The Stopping and Range of Ions in Solids. Pergamon Press, New York (2003)
45. Clifford, E.T.H., Hagberg, E., Hardy, J.C., Schmeing, H., Azuma, R.E., Evans, H.C., Koslowsky, V.T., Schrewe, U.J., Sharma, K.S., Towner, I.S.: Nucl. Phys. A **493**, 293 (1989)
46. Huang, W., Xu, X., Ma, R., Hu, Z., Guo, J., Guo, Y., Liu, H., Xu, L.: Sci. China A: Math. **40**, 638 (1997)
47. Knoll, G.F.: Radiation Detection and Measurement. John Wiley, New York (1979)
48. Diget, C.Aa.: Beta delayed particle emission—probing the triple alpha continuum. Ph.D. thesis, University of Aarhus, Denmark (2006)
49. Gete, E., Buchmann, L., Azuma, R.E., Anthony, D., Bateman, N., Chow, J.C., Auria, J.M.D., Dombsky, M., Giesen, U., Iliadis, C., Jackson, K.P., King, J.D., Measday, D.F., Morton, A.C.: Phys. Rev. C **61**, 064310 (2000)
50. Fynbo, H.O.U., Borge, M.J.G., Axelsson, L., Äystö, J., Bergmann, U.C., Fraile, L.M., Honkanen, A., Hornshøj, P., Jading, Y., Jokinen, A., Jonson, B., Martel, I., Mukha, I., Nilsson, T., Nyman, G., Oinonen, M., Piqueras, I., Riisager, K., Siiskonen, T., Smedberg, M.H., Tengblad, O., Thaysen, J., Wenander, F., The ISOLDE Collaboration: Nucl. Phys. A **677**, 38 (2000)
51. Hyldegaard, S., Forssén, C., Diget, C.Aa., Alcorta, M., Barker, F.C., Bastin, B., Borge, M.J.G., Boutami, R., Brandenburg, S., Büscher, J., Dendooven, P., Duppen, P.V., Eronen, T., Fox, S., Fulton, B.R., Fynbo, H.O.U., Huikari, J., Huyse, M., Jeppesen, H.B., Jokinen, A., Jonson, B., Jungmann, K., Kankainen, A., Kirsebom, O., Madurga, M., Moore, I., Navrátil, P., Nilsson, T., Nyman, G., Onderwater, G.J.G., Penttilä, H., Peräjärvi, K., Raabe, R., Riisager, K., Rinta-Antila, S., Rogachevskiy, A., Saastamoinen, A., Sohani, M., Tengblad, O., Traykov, E., Vary, J., Wang, Y., Wilhelmsen, K., Wilschut, H., Äystö, J.: Phys. Lett. B **678**, 459 (2009)
52. Hyldegaard, S., Diget, C.Aa., Borge, M.J.G., Boutami, R., Dendooven, P., Eronen, T., Fox, S.P., Fraile, L.M., Fulton, B.R., Fynbo, H.O.U., Huikari, J., Jeppesen, H.B., Jokinen, A.S., Jonson, B., Kankainen, A., Moore, I., Nyman, G., Penttila, H., Perajarvi, K., Riisager, K., Rinta-Antila, S., Tengblad, O., Wang, Y., Wilhelmsen, K., Aysto, J.: Phys. Rev. C **80**, 044304 (2009)
53. Smirnova, N.A., Volpe, C.: Nucl. Phys. A **714**, 441 (2003)
54. Winter, W.T., Freedman, S.J., Rehm, K.E., Schiffer, J.P.: Phys. Rev. C **73**, 025503 (2006)
55. Bhattacharya, M., Adelberger, E.G., Swanson, H.E.: Phys. Rev. C **73**, 055802 (2006)
56. Fraile, L.M., Äystö, J.: Nucl. Instrum. Methods A **513**, 287 (2003)
57. Dunbar, D.N.F., Pixley, R.E., Wenzel, W.A., Whaling, W.: Phys. Rev. **92**, 649 (1953)
58. Cook, C.W., Fowler, W.A., Lauritsen, C.C., Lauritsen, T.: Phys. Rev. **107**, 508 (1957)
59. Cook, C.W., Fowler, W.A., Lauritsen, C.C., Lauritsen, T.: Phys. Rev. **111**, 567 (1958)
60. John, B, Tokimoto, Y., Lui, Y.W., Clark, H.L., Chen, X., Youngblood, D.H.: Phys. Rev. C **68**, 014305 (2003)
61. Itoh, M., Akimune, H., Fujiwara, M., Garg, U., Hashimoto, H., Kawabata, T., Kawase, K., Kishi, S., Murakami, T., Nakanishi, K., Nakatsugawa, Y., Nayak, B.K., Okumura, S., Sakaguchi, H., Takeda, H., Terashima, S., Uchida, M., Yasuda, Y., Yosoi, M., Zenihiro, J.: Nucl. Phys. A **738**, 268 (2004)
62. Itoh, M.: Mod. Phys. Lett. A **21**, 2359 (2006)
63. Tamii, A., Adachi, T., Fujita, K., Hatanaka, K., Hashimoto, H., Itoh, M., Matsubara, H., Nakanishi, K., Sakemi, Y., Shimbara, Y., Shimizu, Y., Tameshige, Y., Yosoi, M., Fujita, Y., Sakaguchi, H., Zenihiro, J., Kawabata, T., Sasamoto, Y., Dozono, M., Carter, J., Fujita, H., Rubio, B., Perez, A.: Mod. Phys. Lett. A **21**, 2367 (2006)
64. Freer, M., Boztosun, I., Bremner, C.A., Chappell, S.P.G., Cowin, R.L., Dillon, G.K., Fulton, B.R., Greenhalgh, B.J., Munoz-Britton, T., Nicoli, M.P., Rae, W.D.M., Singer, S.M., Sparks, N., Watson, D.L., Weisser, D.C., Phys. Rev. C **76**, 034320 (2007)
65. Freer, M., Fujita, H., Buthelezi, Z., Carter, J., Fearick, R.W., Fortsch, S.V., Neveling, R., Perez, S.M., Papka, P., Smit, F.D., Swartz, J.A., Usman, I.: Phys. Rev. C **80**, 041303 (2009)
66. Kirsebom, O.S., Alcorta, M., Borge, M.J.G., Cubero, M., Diget, C.Aa., Dominguez-Reyes, R., Fraile, L., Fulton, B.R., Fynbo, H.O.U., Galaviz, D.: Phys. Lett. B **680**, 44 (2009)
67. Kirsebom, O.S., Alcorta, M., Borge, M.J.G., Cubero, M., Diget, C.Aa., Dominguez-Reyes, R., Fraile, L.M., Fulton, B.R., Fynbo, H.O.U., Hyldegaard, S., Jonson, B., Madurga, M., Muñoz Martin, A., Nilsson, T., Nyman, G., Perea, A., Riisager, K., Tengblad, O.: Phys. Rev. C **81**, 064313 (2010)
68. Morinaga, H.: Phys. Rev. **101**, 254 (1956)
69. Kanada-En'yo, Y.: Prog. Theor. Phys. **117**, 655 (2007)
70. Álvarez-Rodríguez, R., Garrido, E., Jensen, A.S., Fedorov, D.V., Fynbo, H.O.U.: Eur. Phys. J. A **31**, 303 (2007)

71. Álvarez-Rodríguez, R., Jensen, A.S., Garrido, E., Fedorov, D.V., Fynbo, H.O.U.: Phys. Rev. C **77**, 064305 (2008)
72. Heikkinen, P.: Cyclotrons and Their Applications 2007, Eighteenth International Conference, pp. 128–130 (2007)
73. Matea, I., Adimi, N., Blank, B., Canchel, G., Giovinazzo, J., Borge, M., Domínguez-Reyes, R., Tengblad, O., Thomas, J.C.: Nucl. Instrum. Methods A **607**, 576 (2009)
74. Blin-Stoyle, R.J.: Phys. Rev. **120**, 181 (1960)

DOI 10.1007/978-94-007-5555-0_9
Reprinted from *Hyperfine Interactions* Volume 223, Issues 1-3, 2014
and *The European Physical Journal A* Volume 48, Issue 4, 2012

Structure of ^{12}C and the triple-α process

H. O. U. Fynbo · C. Aa. Diget

Published online: 13 June 2012

Abstract We review experiments at IGISOL and associated work at ISOLDE and KVI on the use of the β-decays of ^{12}N and ^{12}B for the study of the structure of ^{12}C and the consequences for the rate of the triple-α reaction in stars.

Keywords Triple-alpha reaction rate, deduced spin and parity of levels in 12C · Triple-alpha break-up of levels in 12C, 12N and 12B beta-decays

1 Introduction

The aim of the present paper is to review the work performed at the IGISOL facility of the Jyväskylä accelerator laboratory (JYFL) on the β-decays of ^{12}N and ^{12}B. The physics goals of this work are: (1) to study the breakup mechanism of populated resonances in ^{12}C into the final state of three α-particles, in particular for the 1^+ state at 12.71 MeV; (2) to understand the cluster structure of ^{12}C by resolving and identifying 0^+ and 2^+ resonances in ^{12}C; and (3) to establish how resonances in ^{12}C determine the reaction rate of the astrophysical triple-α process. Figure 1 shows a schematic decay scheme of ^{12}N and ^{12}B and for the unbound states the subsequent breakup of ^{12}C to the 3α-final state. Because ^{12}N and ^{12}B have 1^+ ground states, only 0^+, 1^+ and 2^+ states are populated in allowed β-decay transitions, which makes these decays ideally suited for studying these questions.

The rest of the paper is outlined as follows. In Section 2 we summarise the state of knowledge prior to the IGISOL work. Then we give a detailed description of the

H. O. U. Fynbo (✉)
Department of Physics and Astronomy, Aarhus University, Ny Munkegade 120, Aarhus C, 8000 Aarhus, Denmark
e-mail: hans.fynbo@gmail.com

C. Aa. Diget
Department of Physics, University of York, Heslington, York, YO10 5DD, UK

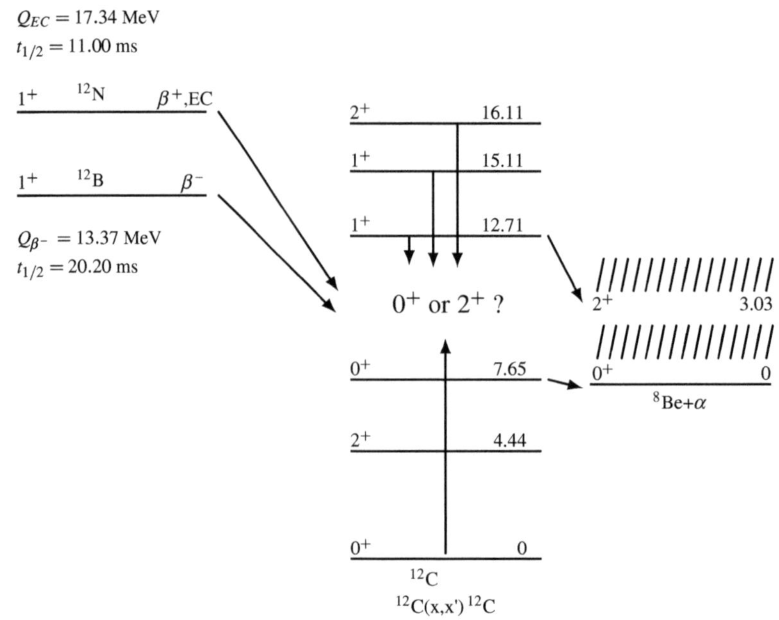

Fig. 1 Schematic level scheme of ^{12}C including methods of populating the states. 3α breakup via intermediate states in ^{8}Be is illustrated. The *arrows* indicate different ways of searching for 0^+ and 2^+ state in the low energy continuum of ^{12}C, which are discussed in the text

results obtained in the series of experiments conducted at IGISOL and associated follow up work in Section 3. The discussion of the possible influence of the experimental findings on the rate of the 3α-reaction rate is given in Section 4. Section 5 concludes the paper by a discussion and outlook.

2 Background

The breakup of states into three or more particles is a general problem in physics with examples in both molecular-, atomic-, nuclear- and particle-physics. In nuclear physics the discussion is mainly centered on the decay of isotopes close to the proton drip line where β-delayed two-proton emission or even ground state two-proton emission can occur, see [1, 2] for recent reviews. These studies are at the limit of what is technically feasible today and the low statistics hinders detailed comparisons to theory. The decay of states in ^{12}C above the 3α threshold provide cases of three-particle breakup which are complimentary to the drip line work. The 0^+ spin of the α-particle removes spin correlations, and the fact that all final state particles have the same mass is also a simplification over the two-proton decay case. The 1^+ state at 12.71 MeV is particularly interesting since it is a narrow state of unnatural parity, which cannot decay sequentially via the ground state of ^{8}Be due to parity conservation. Data [3, 4] prior to our work had been taken as evidence for both direct decay [3, 5] and sequential decay [6–8]. β-decay as a path to populate this state has the advantage of equal population of the magnetic sub-states for an unpolarised source, whereas in reactions this population can not a priori be assumed equal and must be determined experimentally.

The structure of ^{12}C is one of the longest standing problems of nuclear physics. Our main interest here is the question of the nature of the second excited state at 7.65 MeV, which early on was suggested to be a strongly deformed α-cluster state [9]. It was also quickly understood that this state cannot be explained by the shell-model [10], which confirmed the picture of a strongly clustered state. However, this qualitative picture predicts a rotational band built upon the 7.65 MeV state including a low lying 2^+ excited state, which has not been clearly identified in the intervening 50 years despite considerable effort. A broad state found in the β-decays of ^{12}N and ^{12}B [11–13] was suggested as a candidate early on [13, 14], and experimental support for this assignment came shortly after from a study of the breakup of this state [15]. In the evaluations by Ajzenberg-Selove and co-workers [16] this state was assigned as (0^+) because the large width of the state required a very large reduced width for the 2^+ assignment as already argued in [11]. Most theoretical studies supported the 2^+ assignment, including the so-called microscopic cluster calculations [17–19]. These calculations changed the interpretation of the 7.65 MeV state from a crystal-like chain of α-particles to a Bose-gas like state, but maintained that a low-lying 2^+ state should exist. The discussion of the gas-like nature of this state has been taken up again more recently [20].

The 7.65 MeV state, and the associated question of the existence of a 2^+ excitation of it, is also relevant for astrophysics because the natural parity states above the 3α-threshold determine the rate of the triple-α reaction rate. The history of how F. Hoyle predicted the existence of the 7.65 MeV state in an attempt to explain the observed abundances of ^4He, ^{12}C, ^{16}O and ^{20}Ne has recently been critically examined [21]. The tabulation of astrophysical reaction rates by Caughlan and Fowler [22] include only the 7.65 MeV state and a 3^- state at 9.6 MeV, whereas the European compilation of reaction rates NACRE [23] also includes a low lying 2^+ state at 9.1 MeV motivated by the theoretical calculation [24]. The idea that such a low lying state should be searched for in the decays of ^{12}N and ^{12}B using modern radioactive beam methods was advanced in [25], however, not by using the ISOL method as described below, but by using recoil mass separators in inverse kinematics.

Prior to our work at IGISOL there were only three published works aimed at measuring the β-delayed α-particles from ^{12}N and ^{12}B [11–13]. These experiments were performed before the effective methods of studying exotic nuclei, we know today, were developed. Hence these experiments could not provide complete kinematics information on the breakup. The initial motivation for our work was to measure these decays in triple coincidence to resolve the breakup mechanism of the 1^+ state and to identify and understand the nature of the 0^+ and 2^+ states in ^{12}C.

3 Experiments and results

We have performed three sets of experiments on the β-decays of ^{12}N and ^{12}B; in the first we studied ^{12}N at IGISOL and ^{12}B at CERN during 2001–2002, in the second we studied both decays at IGISOL in 2004 and finally in 2006 we again studied both decays at the Kernfysisch Versneller Instituut (KVI), Groningen. We first review the first experiments and then the last two taken together.

3.1 First attempts 2001–2002

The first step to achieve complete kinematics reconstruction of the three α-particle breakup was the use of double sided Silicon strip detectors (DSSSDs). These detectors are ideal for measuring high multiplicity events from weak samples because large solid angles can be covered without suffering from $\beta\alpha$- or $\alpha\alpha$-summing and pile-up. In addition, the detectors can be made relatively thin such that the response to β-particles can be small. The second, equally important step was the use of the isotope separation on-line (ISOL) method for producing ^{12}N and ^{12}B. In the ISOL method the production and detection stages can be especially separated by extracting the produced isotopes and implanting them with low energy in a thin catcher foil, which can be surrounded by an efficient detection system. The thin source reduces the energy loss of the particles emitted in the decay compared to in-beam methods.

The setup used for the first experiment on ^{12}N decay at IGISOL in 2001 consisted of two DSSSDs of thickness 300 μm and 60 μm. The 300-μm detector did not in itself allow to separate the low energy α-particles from the much more abundant β-particles; this could only be achieved for events with three α-particles in coincidence, where the condition of energy and momentum conservation could be used to select the true 3α events. During the data analysis it became clear that the unusually low energy α-particles emitted in these decays suffer significant energy loss in the Carbon collection foil and detector dead-layers. The effect of these energy losses and methods for dealing with them is fully discussed in [26]. To reduce these effects we initiated the development of a new DSSSD design with a reduction of the dead-layer from effectively 600 to 200 nm Si [27]. These detectors were first used in the measurement of ^{12}B decay at ISOLDE in 2002. The 2001 IGISOL experiment achieved for the first time the production of an ISOL beam of ^{12}N with a yield of the order of 100 ions/s, whereas at ISOLDE ^{12}B was produced at a rate of the order of 10^4 ions/s (indirectly via ^{12}Be).

In [28] complete kinematics data from the breakup of the 1^+ state at 12.71 MeV populated in the decay of ^{12}N was compared to R-matrix based sequential models [7] with and without symmetrisation effects included, and with the direct breakup model [5], which neglects all interactions and only take into account constraints from conservation of angular momentum and symmetry. The sequential breakup model including interference [7] clearly gave the best description of the data. The small deviations between this model and the data could be interpreted as being due to final state Coulomb repulsion between the first emitted α-particle and the secondary α-particles from the breakup of ^{8}Be, which pointed to the need for a full three-body calculation including both nuclear and Coulomb interactions in order to fully reproduce this breakup. In our ^{12}B experiment identification of the 1^+ state was achieved for the first time, but with a very small branching ratio of the order of 10^{-6} [29]. In Section 5 we shall have more to say about the problem of the breakup of the 1^+ state.

The two experiments confirmed that very broad states in ^{12}C are populated in the decays of ^{12}N and ^{12}B. In [11, 13] this was analysed using an R-matrix parametrisation as a single 0^+ or 2^+ state with an energy 10.1-10.3 MeV and a full width at half maximum of 2.5-3 MeV. Both [11, 13] could also place an upper limit of 4 % on the fraction of the state which decays via the ^{8}Be 2^+ state. The triple-coincidence data from our measurements [30, 31] corrected for detection efficiency and $\beta\nu$-phase space

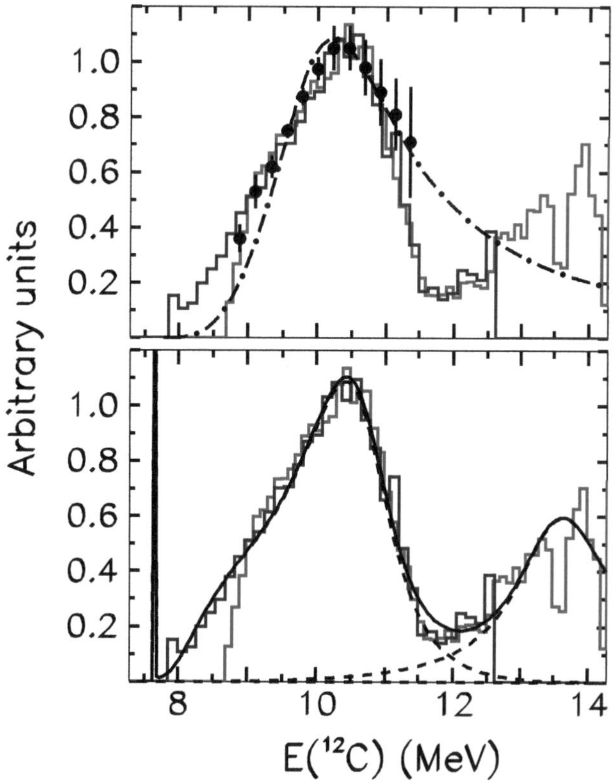

Fig. 2 Excitation energy spectra in ^{12}C corrected for $\beta\nu$-phase space (f-factor), and for detection efficiencies (arbitrary units on vertical axis). Data from ^{12}N/^{12}B decay are shown as the *red/blue histograms*. *Upper plot*: Comparison to the result of [13] (*filled circles with error bars*). The *dashed-dotted curve* shows the 10-MeV resonance using the literature values before our experiment. *Lower plot*: Fit to new data (*solid curve*), where the 7.65-MeV 0^+ resonance, and a 2^+ resonance at high energy, are added (the individual components are shown with the *dashed curves*)

(f-factor) revealed a distribution stretching from the lower detection limit of 8 MeV (700 keV above the 3α threshold) up to the highest energies of more than 14 MeV with a deep minimum around 12 MeV, see Fig. 2. No low lying (and therefore narrow) resonance was found near 9 MeV allowing an upper limit of the feeding of such a state to be placed. The structure below 12 MeV agreed well with the earlier measurement [13] except for the highest data points, however the single-level R-matrix description completely failed to reproduce the new data. Reproduction of the data was achieved by including interference with the high-energy tail (so-called ghost) of the 7.65 MeV state, thereby determing the spin of the high-lying state as 0^+. In a two-level interference description there can be either constructive or descructive interference between the two states and correspondingly descructive and constructive interference outside the two states; with these particular entrance and exit channels (β-decay and α-breakup), the first possibility applies, which explains the steeper decline above 10 MeV and the enhancement below 10 MeV compared to a one-level R-matrix description. The structure above 12 MeV was tentatively assigned as a 2^+ state round 13.5 MeV with a width around 1 MeV.

A re-evaluation of the rate of the triple-α reaction with these findings taken into account will be discussed separately in Section 4.

These first two experiments suffered from two shortcomings which resulted in the remeasurements discussed in Section 3.2. First, they did not provide a way to

Table 1 Branching ratios to ^{12}C from complete kinematics (JYFL) and implantation (KVI) experiments

^{12}C	^{12}N			^{12}B		
Energy (MeV)	Lit. (%)	JYFL (%)	KVI (%)	Lit. (%)	JYFL (%)	KVI (%)
g.s.	94.6(6)*	–	95.96(5)	97.4(3)*	–	98.02(4)
7.65	2.7(4)	–	1.44(3)	1.2(3)*	–	0.59(2)
9–12	0.46(15)	0.38(5)	0.416(9)	0.08(2)	0.060(7)	0.070(3)
12.7	0.28(8)*	0.11(2)	0.123(3)	–	$3.5(7) \cdot 10^{-4}$	$2.8(2) \cdot 10^{-4}$

Literature values for comparison are from [16]. *Updated literature values (see [35, 36])

normalise the branching ratios of the identified states, and hence the predicted ratio between the branching ratios to the interfering 7.65 MeV state and the higher lying state could not be tested. Second, the setups used in the two experiments had small efficiencies for detecting breakup events where the α-particles are emitted with large relative angles between all pairs, such as for breakup via the 2^+ state in ^{8}Be. Important strength might therefore have been left undetected.

3.2 IGISOL 2004 and KVI 2006–2007

The shortcomings of the first measurements were overcome in two complimentary ways in two new experiments at IGISOL (2004) and KVI (2006). In the first approach we remeasured both ^{12}B and ^{12}N at IGISOL with a more efficient setup consisting of three DSSSDs in a horse-shoe configuration combined with the ISOLDE Silicon ball [32]. By including also a Germanium detector for detecting the 4.44 MeV γ-ray from the first excited state of ^{12}C, for which the the branching ratio is well measured, the branching ratio of the α-decaying branches could be determined. However, using this approach it is impossible to measure the 3α-breakup of the 7.65 MeV state since the α-particles from the breakup of the ground state of ^{8}Be are so low in energy that they cannot escape from the Carbon collection foil. Hence, the branching ratio to this state cannot be measured in this way. The second approach consisted of implanting ^{12}B and ^{12}N, produced in inverse kinematics at KVI [33], in a thin DSSSD [34] and measure the summed energy of the α-particles directly with high efficiency (this is similar to what was proposed in [25]). Normalisation is in this approach simply achieved by comparing the number of implantations with the number of subsequent decays in the same detector segment. This approach is nearly optimal in terms of detection efficiency (nearly 100%) and threshold (ca. 2–300 keV), but this is achieved at the expence of loosing all information of the breakup kinematics. Detection of the 7.65 MeV state is possible because the energies of the three α-particles are summed, and the particles suffer no energy loss before being detected.

Table 1 gives some of the branching ratios extracted from the two experiments [35, 36], and the corresponding literature values. The branching ratios resulting from the two methods are consistent with each other in the regions of overlap. The combined result has reduced the uncertainties on the branching ratios by factors of 10–20, and the branching ratios to the 7.65 MeV and 12.71 MeV states are reduced by a factor 2 compared to the literature. The ground state branching ratios are also modified because they are deduced as everything not going to excited states. Figure 3

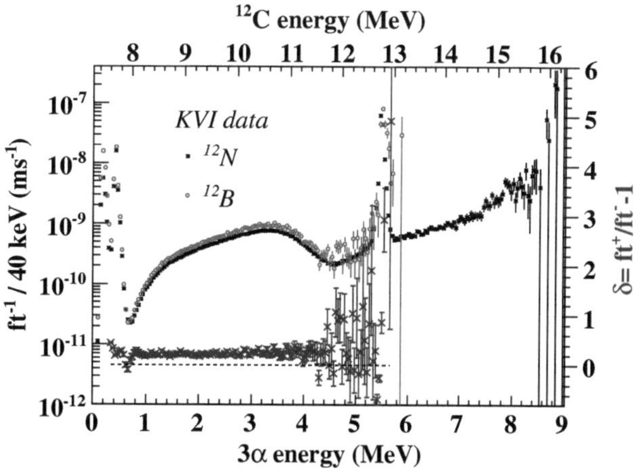

Fig. 3 Spectra of inverse ft value per energy bin for ^{12}N and ^{12}B decays (*left ordinate*). The isospin asymmetry, δ, is also shown (*right ordinate*)

shows the inverse ft values per energy bin for the ^{12}N and ^{12}B decays measured at KVI (left ordinate); this is equivalent to the ordinate in Fig. 2, but now on an absolute scale. The right ordinate shows the β-decay asymmetry $\delta = ft(\beta^+)/ft(\beta^-) - 1$. A small constant positive shift is seen in favor of β^- decay, which is similar to the asymmetry observed for the bound states [16]. The energy independence confirms that the origin of the asymmetry is mainly nuclear structure [37, 38] as a second-class-currents explanation infers an energy-dependent asymmetry [37]. The branching ratios in Table 1 can be converted into Gamow-Teller matrix elements, which can be directly compared to theory. In [35] a comparison to no-core shell-model calculations is presented.

A detailed analysis of the kinematics of the 3α-breakup from the IGISOL experiment is given in [39]. The main idea behind this analysis is similar to that previously mentioned [15], where the breakup pattern was used to provide evidence for a 2^+ assignment of the broad state at 10.3 MeV in ^{12}C. In [39] the energy region 9-16 MeV was divided into 11 bins, and for each bin Dalitz plots defined as $x = \sqrt{3}(E_1 - E_3)/(E_1 + E_2 + E_3)$ and $y = (2E_2 - E_1 - E_3)/(E_1 + E_2 + E_3)$ were projected. Figure 4 shows three examples of these Dalitz plots. Data points close to the tip of the region (left) are events with an equal sharing of the energy between the three emitted particles, whereas the points close to the upper right corner correspond to $E_3 \simeq 0$ and the lower right corner have $E_1 = \frac{2}{3}(E_1 + E_2 + E_3)$, the highest possible energy for an individual α-particle. These Dalitz plots are compared to Monte-Carlo simulations based on the same R-matrix based sequential model used for the 1^+ state [7, 28]. This model is dependent on the spin in ^{12}C and ^{8}Be (both 0^+ or 2^+), and on the orbital angular momentum between the first emitted α-particle and ^{8}Be. By comparing bin by bin data to simulations the contribution of 0^+ and 2^+ states in ^{12}C can be constrained. The result confirmed the conclusion from [30, 31] that below 12 MeV 0^+ dominates, and above 12 MeV 2^+ dominates. Note, this conclusion is opposite to that reached in [15], the reason being that the model used by [15] assumed a specific ratio between the 0^+ and 2^+ channels in ^{8}Be, which is unjustified.

A detailed comparison of Figs. 2 and 3 shows that the structure at 8–12 MeV is consistent in the two experiments, but that the structure at 13–16 MeV can now be

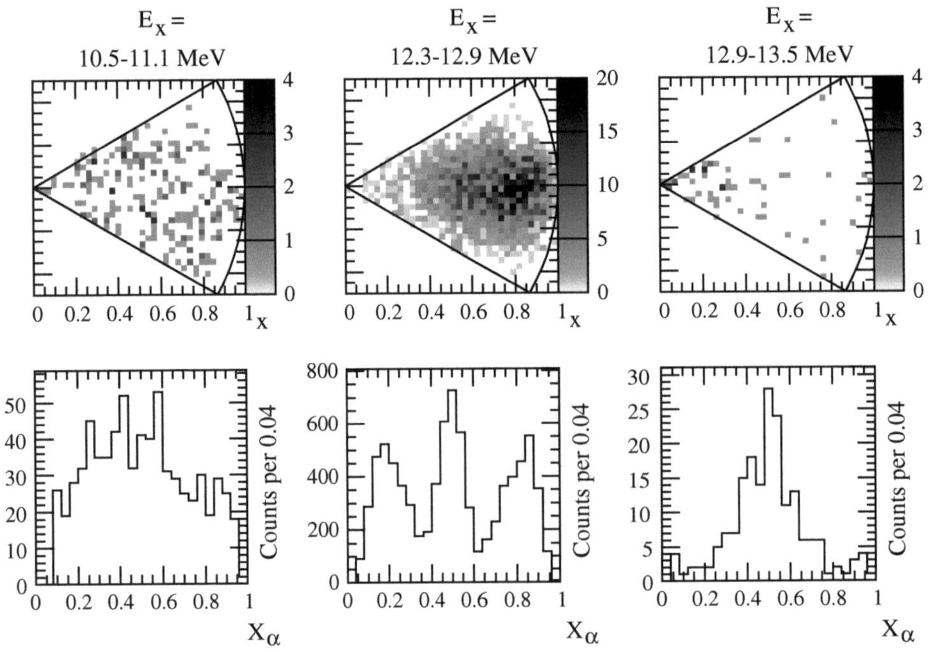

Fig. 4 ^{12}N β-delayed 3α breakup data. Dalitz plots (x, y) defined as described in the text are shown along with individual α-energies (X_α) scaled event by event relative to the maximal kinematically allowed α-energy. Data is shown for three representative ^{12}C excitation energy bins

seen to continue beyond 14 MeV, and can clearly no longer be described as a single state at 13.5 MeV. Also, the fit shown in the lower part of Fig. 2 is not consistent with the branching ratios given in Table 1 because it predicts a ratio between the 7.65 MeV state and the 9–12 MeV region four times higher than observed.

The detailed discussion of how to find a correct description of the data is given in [40]. The procedure is simply to see what combination of 0^+ and 2^+ states in ^{12}C will give the best description of the data. The combined data from the IGISOL and KVI experiments is used in the fits, with IGISOL data providing the relative contribution of ^{8}Be ground state and 2^+ state as a functions of energy in ^{12}C. Each fit will predict a relative contribution of 0^+ and 2^+ in each ^{12}C energy bin, which can be compared to the results from the Dalitz-plot analysis [39]. The Gamow–Teller sum rule and the Wigner-limit provide constraints on the β-feeding parameters and reduced widths respectively. It can be anticipated from Fig. 3 that the result is more complicated than the simple model in Fig. 2 because of the unusual structure above 12 MeV, which seems to be the low-energy tail of something extending beyond the Q_β-window. The result of this analysis is that one 0^+ and one 2^+ state above the 7.65 MeV state are no longer enough to reproduce the data. Our fits give evidence for one broad 0^+ and 2^+ state in the 10.5–12 MeV region. Above the 12.71 MeV peak the increasing β-strength seen in Fig. 3 is probably caused by additional 0^+ and 2^+ components at high energy. Figure 5 is an example of a fit including three 0^+ (including the 7.65 MeV state) and two 2^+ states.

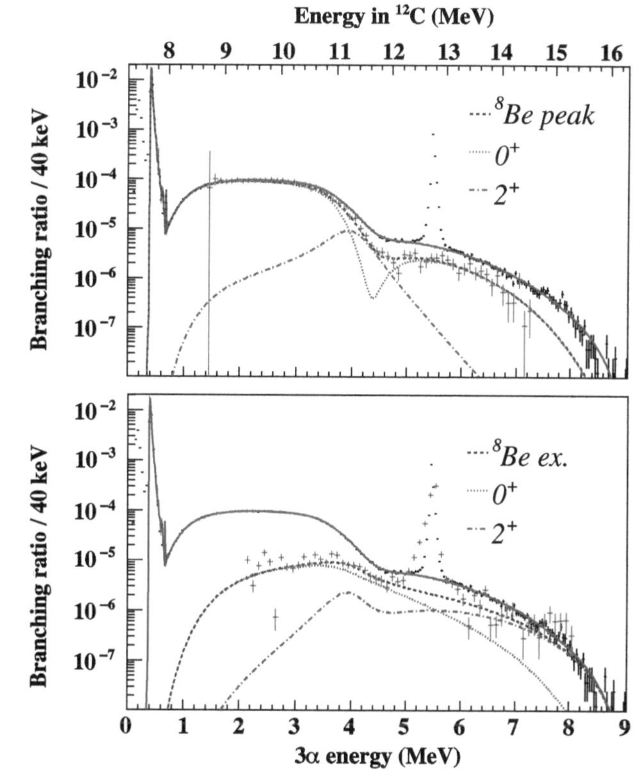

Fig. 5 Fit including three 0^+ and two 2^+ states. Only the ^{12}N spectra are shown. The KVI spectra (*black point with error bars*) are shown in both plots compared to the *solid line* showing the total fit summed over all channels. JYFL spectra (*grey points with error bars*) for the ^8Be peak (*top*) and excited states (*bottom*) channels are compared to the dashed curves, which are the fit components for the respective channels. The *dotted* and *dot-dashed lines* are contributions to each decay channel from 0^+ and 2^+ states in ^{12}C, respectively

4 The rate of the triple-α-reaction

The triple-α reaction rate in stellar environments has been a long-standing challenge in nuclear astrophysics since the identification of its significance more than 50 years ago. The dynamics of the three-body reaction was first investigated and described by Öpik [41] and Salpeter [42], where the latter furthermore included the significant ^8Be 0^+ ground-state resonance. After this the ^{12}C resonance at approximately 7.7 MeV was predicted by Hoyle (the state is now known as the Hoyle state). It was subsequently identified, and it's impact on the triple-α reaction rate presented [43–45], with the first comprehensive evaluation of the reaction rate presented by Burbidge, Burbidge, Fowler and Hoyle (B^2FH [46]). For further historical background, see refs. [21, 47, 48].

In the B^2FH compilation the reaction-rate calculation is based on the reaction proceeding resonantly through a thermal-equilibrium population of the Hoyle state. The reaction rate as a function of temperature is therefore proportional to [46]:

$$\frac{\Gamma_{\text{rad}}}{T^3}\exp(-E_{\text{r}}/k_{\text{B}}T),$$

where E_{r} is the resonance energy for the Hoyle state relative to the triple-α threshold and Γ_{rad} is the radiative width of the state.

In modern calculations of the reaction rate not only the resonant contribution is included but also that of the low-energy tails of the resonances (the ^8Be ground state, and the Hoyle state). These contributions were for the first time taken into account by Nomoto et al. [49] and subsequently in a modified version by Langanke et al. [50]. All later evaluations have to some degree followed the methods laid out in these two papers, for which the low-temperature reaction rate is sensitive not only to the resonance energy and radiative width of the Hoyle state, but also to the ^{8}Be resonance energy and the energy dependent α widths of both resonances.

For the Hoyle state, the critical parameters are therefore: the resonance energy, the radiative width, and (for the non-resonant component) the α-decay width. The previous measurements of these parameters are summarised by Ajzenberg–Selove [16]. The Q-value is here dominated by the measurement of Nolen and Austin [51], and the widths are evaluated through the combined measurement of the radiative branch ($\Gamma_{\text{rad}}/\Gamma$), the pair-decay branch (Γ_π/Γ), and the pair-decay width (Γ_π) [52–54]. For the ^{8}Be 0^+ state, the relevant data (the resonance energy and the width) has been evaluated by Tilley et al. [55]. The data available for the two states has for the most part remained untouched since Nomoto's paper was written, with two notable exceptions: firstly the improvement in the resonance parameters for the ^{8}Be 0^+ state [56] where the α width of the state was reduced by approximately 20%, the uncertainty in the width was improved by a factor seven, and the value for the resonance energy was lowered slightly. Secondly, all available electron-scattering data has been re-evaluated and an updated pair width has recently been deduced [57]. It is here concluded that previous data had led to an approximately 15% too high evaluation of the electron-scattring cross section, and that therefore the value for the pair width should be lowered accordingly. This corresponds to a reduction in the pair width by 2σ. However, a new study [58] has concluded that the previously accepted value was indeed correct, and has improved the uncertainty to yield: $\Gamma_\pi = 62.3(20)$ μeV. This study included firstly: new data from the Darmstadt S-DALINAC; and secondly: the data was analysed using the Distorted-Wave Born Approximation, and it is shown that for the previous Fourier–Bessel analysis [57] to yield the same results, data up to $q = 3.7$ fm^{-1} would have had to be included.

In Fig. 4 the effect of the above developments on the low-temperature reaction rate is summarised. Here, four different reaction rates are compared to the rate of the 1999 NACRE compilation [23]. These rates are: the resonant reaction rate (similar to B^{2}FH [46], but with updated resonance parameters); the two compilations of Caughlan and Fowler from 1985 and 1988 respectively (CF-85 [59] and CF-88 [59]); and our rate from 2005 (FD-05 [30]). The dramatic effect of Nomoto's off-resonance description is clear when comparing the remaining four rates to the resonant rate, which plummets relative to the more recent calculations around $T_9 = 0.08$. The difference between the other rates at the lowest temperatures is presently unresolved, though it should be noted that both calculations by Caughlan and Fowler were performed using the ^{8}Be 0^+ data from ref. [60], i.e. without the more recent update [56]. These compilations might therefore be expected to differ at the lowest temperatures, where the energy and width of the ^{8}Be 0^+ state become increasingly important. Furthermore, though CF-85 was considered to be incorrect in the light of Nomoto's developments and therefore corrected at low energies in CF-88, the resulting reaction rate in the $T_9 = 0.03$–0.08 region happens to agree better with both the NACRE compilation and FD-05 at these temperatures. This coincidence arises from the difference in both the method used for the calculation and the resonance parameters now used. Because of this coincidental overlap, it is not surprising that comparisons between CF-85, NACRE, and FD-05 show little astrophysical impact for this temperature range [61]. This might, however, be the case for the updated CF-88, and the use of either NACRE or FD-05 is therefore strongly recommended for scenarios that are sensitive to the triple-α reaction rate at low temperatures.

5 Discussion and outlook

5.1 Triple-α breakup mechanism

The model used to describe the breakup of the 12.71 MeV state was a phenomenological R-matrix model. As previously mentioned, small deviations between data and model could be interpreted as Coulomb effects in the final state. In the direct decay model [5] no Coulomb effects are included and so one could speculate that this model when corrected for Coulomb effects might also describe the data. A complete quantum mechanical three-body calculation [62], providing a unified description where the interaction is not split into initial and final state effects, gives a reasonable reproduction of the data. This breakup is there called "direct", but the distinction between direct breakup including interactions and sequential breakup including symmetrisation and final-state interaction might be semantic. A conceptual clarification of why most models qualitatively reproduce the decay of the 12.71 MeV state was given in [63]. Here it is shown that symmetry for the decay of a 1^+ state into three 0^+ α-particles requires large areas of the Dalitz plot to vanish for any model. To better test models of three-body breakup it would be desirable to find cases with less stringent symmetry requirements. A step in this direction is taken in [64] where Dalitz plots for all low lying unbound states in ^{12}C, using the same theoretical approach as [62], are provided. Experimental tests of these predictions for the 11.83 MeV (2^-), 13.35 MeV (4^-) and again the 12.71 MeV (1^+) state are published in [65] using instead of β-decay a new method based on high Q-value reactions in complete kinematics [66]. This method is less selective than the β-decay approach, but allows more states to be studied at the expence of more background from other reaction channels. Neither the phenomenological R-matrix model nor the full calculation [64] were able to reproduce the detailed shape of the distributions from the 2^- and 4^- states. The R-matrix model modified to accommodate final-state Coulomb repulsion gives the best fit to the data, but the level of agreement is less than for the 1^+ state, which may be understood as a result of the relaxed constraints from symmetry leaving more room for the dynamics of the breakup process to affect the final energy distribution of the α-particles.

5.2 Identification of 0^+ and 2^+ states in ^{12}C

The experiments discussed above have provided a much better picture of the population of unbound states in the β-decays of ^{12}B and ^{12}N. The analysis of the data in terms of 0^+ and 2^+ states has provided a quite complicated result with several broad overlapping states contributing. This fit is very sensitive to the branching ratio to the 7.65 MeV state because of the contribution of the "ghost" in the 9–12 MeV region. Since this branching ratio was reduced by a factor 2 in our measurent compared to the literature it would be desirable to confirm this result by an independent method.

One approach would be to measure the γ-decay of the 7.65 MeV state when it is populated in β-decay since the relative γ-decay branching ratio is well measured.

In the past decade there has been a large number of both experimental and theoretical studies related to the question of low lying 0^+ and 2^+ resonances in ^{12}C. Experiments using scattering of either protons [67, 68] or α-particles [69–71] on ^{12}C consistently give evidence for 0^+ strength in the region 9–12 MeV, but for the location of 2^+ resonances no consistent answer has yet emerged from this approach. 2^+ states have been suggested at 11.46(2) MeV [69], 9.9(3), 15.4 and 18.2 MeV [70, 71], and 9.6(1) MeV [68]. Coincident detection of the scattered proton or α-particle and the subsequent breakup α-particles is used in [71] to get further constraints, but the correlations did not confirm the 2^+ assignment for the 2^+ resonance at 9.9 MeV. This approach is similar to that used by [15] and [39]. The $^{12}C(^{12}C,3\alpha)^{12}C$ reaction has also been used to sesarch for a low-lying 2^+ resonance [72, 73]. The first of these studies suggested a 2^+ state at 11.16 MeV, while the latter found no evidence for new states between 9 and 11 MeV. The 2^+ resonance around 10 MeV supported by some of these measurements is observed as a weak perturbation of the high energy tail of the 9.6 MeV 3^- state. A potential problem with this is that interference with higher lying broad 3^- states, such as suggested by [72], will affect the shape of the lower lying 3^- state.

In an alternative approach a 2^+ state is searched for using an Optical Readout Time Projection Chamber (O-TPC) operating with a mixture of CO_2 and N_2 with γ-beams from the HIγS facility of TUNL at Duke University. This method has the advantage that it is blind to 0^+ contributions. In the first report no evidence for a 2^+ resonance was found [74], but in later reports evidence for a 2^+ state in the 9–10 MeV range is advanced [75]. As is the case for inelastic scattering experiments, a combination of angular distributions and energy peak shapes must be applied to discriminate between the overlapping 2^+ and 3^- states.

For a detailed review of recent theoretical studies related to the question of unbound 0^+ and 2^+ states in ^{12}C we refer to [76]. The no-core shell-model [35] is unable to find low lying 0^+ and 2^+ resonances, whereas microscopic cluster models [77–79], and cluster models with inert α-particles [76, 80, 81] all find one or more 0^+ and 2^+ states within the first several MeV of the continuum. There is still general agreement among the cluster models that a 2^+ resonance should exist within the first 2–3 MeV of the 3α continuum. We would like to encourage that the theoretical studies extend the predictions to include not only just energies and widths of suggested resonances, but also predictions for how these resonances decay to the 3α final state.

5.3 Future investigation of 0^+ and 2^+ states in ^{12}C

New technical developments at IGISOL due to be completed during 2012 will provide significantly increased production yields of ^{12}B and ^{12}N: (1) A new dedicated cyclotron for the IGISOL beam line will give increased primary beam currents by a factor 5; (2) the replacement of the skimmer extraction system with an RF sextupole ion guide has already improved yields by a factor 10–20 since our last experiment in 2004; (3) the new dedicated cyclotron will make it possible to ask for longer beamtimes; (4) we will build a new detection setup, which is even more efficient than that previously used. The combined effect of these improvements is an increase in the expected total statistics of more than a factor 100. We will therefore repeat our β-decay measurements at IGISOL with a substantial improvement in sensitivity. An independent measurement of the ^{12}B and ^{12}N decays is furthermore underway using a ionisation chamber [82], which might provide independent confirmation of our results.

The γ-decay of the 1^+ states at 15.1 and 12.71 MeV in ^{12}C to 3α decaying states has recently been studied [83] by using complete kinematics detection of the α-particles to identify events with intermediate γ-rays. Using the same method we plan to use the γ-decay of the $T = 1$ 16.11 MeV (2^+) and 17.77 MeV (0^+) states to search for lower lying 2^+ states, see Fig. 1. For the former, isovector M1 transitions to 2^+ states are strongly favoured over isovector E2 transitions to 0^+ states, while for the latter, feeding of 0^+ states is completely avoided. Therefore this provides a selective method for identifying a 2^+ resonance in the 9–12 MeV region where 0^+ strength is known to exist. If a 2^+ state is found, this approach could also provide the first clean measurement of its decay into three α-particles.

A picture is emerging with overlapping 0^+ and 2^+ states in the 10–12 MeV region, and additional very broad 0^+ and 2^+ states in the 13–18 MeV region. A crucial future goal is therefore to relate the structures seen in our β-decay experiments to states seen in other experimental approaches. In order for this to be meaningful, and to extract consistent values for the energies and widths of contributing states, different data sets must be analysed with similar parametrisations, which as a minimum must contain leading physical effects. Barker has made a consistent analysis of 0^+ and 2^+ states in ^{8}Be as observed in various transfer reactions, scattering, and β-decay based on R-matrix theory [84–86]. Such an analysis has not yet been attempted for ^{12}C. As discussed in the preceding sections the R-matrix formalism has been used to analyse the β-decay data since the first measurements in 1958 [11], but data from other approaches are most often analysed with less sophisticated parametrisations.

It is an interesting question whether it is reasonable to expect a consistent determination of resonance properties of broad states as seen in different experimental approaches using R-matrix theory. One argument against this expectation is that there might be non-resonant contributions to the spectra which could be quite different for different experimental probes, and for states as broad as the 0^+ and 2^+ states in ^{8}Be and ^{12}C it is difficult to separate resonance and non-resonant contributions. This issue may play a role also in the discussion of low-lying intruder states in ^{8}Be [87].

Assuming a consistent experimental understanding emerges it will be the task of theory to explain the strength or weakness with which the identified states are observed in different experimental approaches. This pattern should provide additional clues to the structure of these elusive states in ^{12}C.

5.4 The rate of the triple-α-reaction

Several experimental programmes are presently under way to address the outstanding questions in relation to the rate of the triple-α process. The first group of these focuses on reducing the present 10–12% uncertainty in the resonant reaction rate to the 5% level for the $T_9 = 0.1$–1 temperature range where the reaction rate is determined entirely by the Hoyle state properties. Sensitivity studies for the triple-α rate have been performed in two contexts: (1) an investigation of the helium-flash driven dredge-up in AGB stars [88], and (2) systematic studies of the influence on the

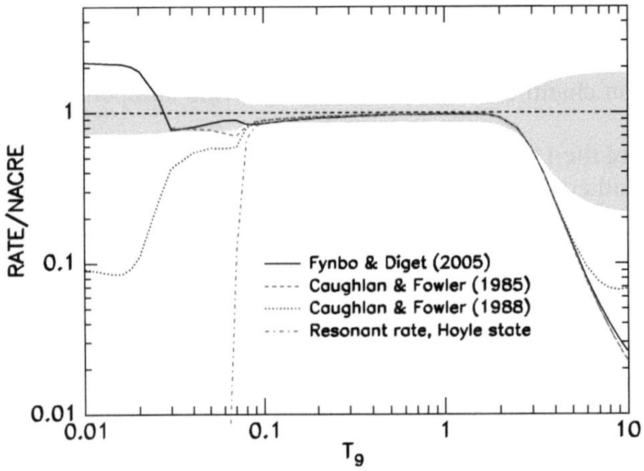

Fig. 6 Four triple-α reaction rates are shown relative to the rate from the NACRE compilation [23], T_9 is the temperature in 10^9 K. Shown are: uncertainty band for the NACRE compilation, not including the uncertainty in the position of the assumed 2_2^+ state (*grey band*) [23]; the FD-05 rate, including only the Hoyle state contribution (*solid line*) [30]; the CF-85 rate (*dashed line*) [59]; the CF-88 rate, with updated calculation method, inclusion of the 3^- state, and extended below $T_9 = 0.03$ (*dotted line*) [22]; and the Hoyle-state resonant reaction rate as calculated by B^2FH [46], but with updated parameter values (*dot-dashed line*)

iron-core size prior to core-collapse supernovae and the following production of elements up to the iron peak [89–91]. To pin down the reaction rate in this temperature region, a new and very challenging measurement of the pair branch (Γ_π/Γ) is underway, with a first attempt at this presented in ref. [92]. Furthermore, a measurement of the radiative width is under development [93]. This will be performed through the measurement of the ratio between the E0 and E2 pair branches ($\Gamma_\pi[E2]/\Gamma_\pi[E0]$), which would yield a measurement of the radiative width that is independent of both the pair branch as well as the radiative branch measurements.

The second type of experimental programmes focuses on the rate outside the temperature regime dominated by the resonant reaction, in particular the contribution from the suggested 2_2^+ state in the 9–11 MeV region. These are spurred by recent indications of the presence of this state described in Section 5.2, and primarily influence the high-temperature reaction rate. Presently, at least two such programmes are underway: For the mentioned γ-induced breakup, $^{12}C(\gamma, 3\alpha)$, data analysis is presently ongoing [74, 75]. Furthermore, the programme of β-delayed-breakup studies at IGISOL is beeing intensified with the present JYFL upgrades. These are, in combination with detector developments [94], expected to yield at least a factor 100 improved statistics for triple-α breakup channels. The high statistics data with simultaneous detection of the emitted β-particles will allow a, for the β-decay measurements, unprecedented discrimination between 0^+ and 2^+ strength in ^{12}C. Through this, the upcoming β-decay IGISOL programme, will be sensitive to the low 2_2^+ β-decay feeding. In the following analysis of the 2^+ strength in the 9–11 MeV region it will, for both the γ-induced and β-delayed breakup programmes, be critical to describe the shape of the 2_2^+ state in a full multi-channel R-Matrix formalism [40]. This is because of the strongly energy-dependent α-width as well as the significant interference with the higher-lying 2_3^+ state as described in the preceding.

Lastly, the reaction rate has also been calculated using theoretical three-body calculations which in recent years have been performed by several groups [95–99]. These have been followed up by evaluations of the potential effect of the rate changes on neutron stars [100], white dwarfs [101], and low-mass stars [102]. It is worth noting, however, that the latter evaluation stresses the significant discrepancy between the first of these calculations [95] and observations of low-mass star evolution. In addition to the above fundamental studies of the triple-α reaction-rate calculations, the traditional reaction-rate calculations on and off resonance will undergo further scrutiny as the resonance parameters are being pinned down. Of particular significance in this contest is the small discrepancy in the 1–2 $\times$ 10^8 K region between the reaction rates presently in use (see Fig. 6). This discrepancy is presently within the uncertainty introduced by the parameters, but will become significant with the expected improvement in the radiative width of the Hoyle-state of a factor of two.

Acknowledgements We would like to thank all collaborators in the experiments discussed here, and in particular the IGISOL and JYFL groups and staff for their great hospitality and for providing such an excellent venue for research.

References

1. Blank, B., Borge, M.J.G.: Prog. Part. Nucl. Phys. **60**(2), 403 (2008)
2. Blank, B., Ploszajczak, M.: Rep. Prog. Phys. **71**(4), 046301 (2008)
3. Olsen, W.C., Dawson, W.K., Neilson, G.C., Sample, J.T.: Nucl. Phys. **61**(4), 625 (1965)
4. Waggoner, M.A., Etter, J.E., Holmgren, H.D., Moazed, C.: Nucl. Phys. **88**(1), 81 (1966)
5. Korsheninnikov, A.A.: Sov. J. Nucl. Phys. **52**(5), 827 (1990)
6. Prats, F., Salyers, A.: Phys. Rev. Lett. **19**(11), 661 (1967)
7. Balamuth, D.P., Zurmühle, R.W., Tabor, S.L.: Phys. Rev. C **10**(3), 975 (1974)
8. Takahashi, T.: Phys. Rev. C **16**(2), 529 (1977)
9. Morinaga, H.: Phys. Rev. **101**(1), 254 (1956)
10. Cohen, S., Kurath, D.: Nucl. Phys. **73**(1), 1 (1965)
11. Cook, C.W., Fowler, W.A., Lauritsen, C.C., Lauritsen, T.: Phys. Rev. **111**(2), 567 (1958)
12. Wilkinson, D.H., Alburger, D.E., Gallmann, A., Donovan, P.F.: Phys. Rev. **130**(5), 1953 (1963)
13. Schwalm, D., Povh, B.: Nucl. Phys. **89**(2), 401 (1966)
14. Morinaga, H.: Phys. Lett. **21**(1), 78 (1966)
15. Jacquot, C., Sakamoto, Y., Jung, M., Girardin, L.: Nucl. Phys. A **201**(2), 247 (1973)
16. Ajzenberg-Selove, F.: Nucl. Phys. A **506**, 1 (1990)
17. Uegaki, E., Okabe, S., Abe, Y., Tanaka, H.: Prog. Theor. Phys. **57**(4), 1262 (1977)
18. Uegaki, E., Abe, Y., Okabe, S., Tanaka, H.: Prog. Theor. Phys. **62**(6), 1621 (1979)
19. Kamimura, M.: Nucl. Phys. A **351**(3), 456 (1981)
20. Tohsaki, A., Horiuchi, H., Schuck, P., Röpke, G.: Phys. Rev. Lett. **87**(19), 192591 (2001)
21. Kragh, H.: Arch. Hist. Exact Sci. **64**(6), 721 (2010)
22. Caughlan, G., Fowler, W.: At. Data Nucl. Data Tables **40**, 283 (1988)
23. Angulo, C., Arnould, M., Rayet, M., Descouvemont, P., Baye, D., Leclercq-Willain, C., Coc, A., Barhoumi, S., Aguer, P., Rolfs, C., Kunz, R., Hammer, J.W., Mayer, A., Paradellis, T., Kossionides, S., Chronidou, C., Spyrou, K., Degl'Innocenti, S., Fiorentini, G., Ricci, B., Zavatarelli, S., Providencia, C., Wolters, H., Soares, J., Grama, C., Rahighi, J., Shotter, A., Rachti, M.L.: Nucl. Phys. A **656**, 3 (1999)
24. Descouvemont, P., Baye, D.: Phys. Rev. C **36**(1), 54 (1987)
25. Betts, R.: Il Nuovo Cimento A (1971–1996) **110**(9), 975 (1997)
26. Bergmann, U.C., Fynbo, H.O.U., Tengblad, O.: Nucl. Instrum. Methods A **515**, 657 (2003)
27. Tengblad, O., Bergmann, U.C., Fraile, L.M., Fynbo, H.O.U., Walsh, S.: Nucl. Instrum. Methods Phys. Res., Sect. A **525**(3), 458 (2004)
28. Fynbo, H.O.U., Prezado, Y., Bergmann, U.C., Borge, M.J.G., Dendooven, P., Huang, W.X., Huikari, J., Jeppesen, H., Jones, P., Jonson, B., Meister, M., Nyman, G., Riisager, K., Tengblad,

O., Vogelius, I.S., Wang, Y., Weissman, L., Rolander, K.W., Äystö, J.: Phys. Rev. Lett. **91**(8), 082502 (2003)
29. Fynbo, H.O.U., Diget, C.A., Prezado, Y., Äystö, J., Bergmann, U.C., Cederkll, J., Dendooven, P., Fraile, L.M., Franchoo, S., Fulton, B.R., Huang, W., Huikari, J., Jeppesen, H., Jokinen, A., Jonson, B., Jones, P., Käster, U., Meister, M., Nilsson, T., Nyman, G., Borge, M.J.G., Riisager, K., Rinta-Antila, S., Storgaard, Vogelius, I., Tengblad, O., Turrion, M., Wang, Y., Weissman, L., Wilhelmsen, K.: Nucl. Phys. A **738**, 59 (2004)
30. Fynbo, H.O.U., Diget, C.A., Bergmann, U.C., Borge, M.J.G., Cederkall, J., Dendooven, P., Fraile, L.M., Franchoo, S., Fedosseev, V.N., Fulton, B.R., Huang, W., Huikari, J., Jeppesen, H.B., Jokinen, A.S., Jones, P., Jonson, B., Koster, U., Langanke, K., Meister, M., Nilsson, T., Nyman, G., Prezado, Y., Riisager, K., Rinta-Antila, S., Tengblad, O., Turrion, M., Wang, Y., Weissman, L., Wilhelmsen, K., Aysto, J., T.I. Collaboration: Nature **433**(7022), 136 (2005)
31. Diget, C.A., Barker, F.C., Borge, M.J.G., Cederkll, J., Fedosseev, V.N., Fraile, L.M., Fulton, B.R., Fynbo, H.O.U., Jeppesen, H.B., Jonson, B., Käster, U., Meister, M., Nilsson, T., Nyman, G., Prezado, Y., Riisager, K., Rinta-Antila, S., Tengblad, O., Turrion, M., Wilhelmsen, K., Äystö, J.: Nucl. Phys. A **760**(1–2), 3 (2005)
32. Fraile, L.M., Äystö, J.: Nucl. Instrum. Methods Phys. Res., Sect A **513**(1–2), 287 (2003)
33. Traykov, E., Rogachevskiy, A., Bosswell, M., Dammalapati, U., Dendooven, P., Dermois, O.C., Jungmann, K., Onderwater, C.J.G., Sohani, M., Willmann, L., Wilschut, H.W., Young, A.R.: Nucl. Instrum. Methods Phys. Res., Sect. A **572**(2), 580 (2007)
34. Smirnov, D., Aksouh, F., Dean, S., De, Witte, H., Huyse, M., Ivanov, O., Mayet, P., Mukha, I., Raabe, R., Thomas, J.C., Van Duppen, P., Angulo, C., Cabrera, J., Ninane, A., Davinson, T.: Nucl. Instrum. Methods Phys. Res., Sect. A **547**(2–3), 480 (2005)
35. Hyldegaard, S., Forssèn, C., Diget, C.A., Alcorta, M., Barker, F.C., Bastin, B., Borge, M.J.G., Boutami, R., Brandenburg, S., Büscher, J., Dendooven, P., Van Duppen, P., Eronen, T., Fox, S., Fulton, B.R., Fynbo, H.O.U., Huikari, J., Huyse, M., Jeppesen, H.B., Jokinen, A., Jonson, B., Jungmann, K., Kankainen, A., Kirsebom, O., Madurga, M., Moore, I., Navrtil, P., Nilsson, T., Nyman, G., Onderwater, G.J.G., Penttilä, H., Peräjärvi, K., Raabe, R., Riisager, K., Rinta-Antila, S., Rogachevskiy, A., Saastamoinen, A., Sohani, M., Tengblad, O., Traykov, E., Vary, J.P., Wang, Y., Wilhelmsen, K., Wilschut, H.W., Äystö, J.: Phys. Lett. B **678**(5), 459 (2009)
36. Hyldegaard, S., Diget, C.A., Borge, M.J.G., Boutami, R., Dendooven, P., Eronen, T., Fox, S.P., Fraile, L.M., Fulton, B.R., Fynbo, H.O.U., Huikari, J., Jeppesen, H.B., Jokinen, A.S., Jonson, B., Kankainen, A., Moore, I., Nyman, G., Penttilä, H., Peräjärvi, K., Riisager, K., Rinta-Antila, S., Tengblad, O., Wang, Y., Wilhelmsen, K., Äystö, J.: Phys. Rev. C **80**(4), 044304 (2009)
37. Wilkinson, D.H.: Eur. Phys. J. A **V7**(3), 307 (2000)
38. Wilkinson, D.H., Alburger, D.E.: Phys. Rev. Lett. **26**(18), 1018 (1971)
39. Diget, C.A., Barker, F.C., Borge, M.J.G., Boutami, R., Dendooven, P., Eronen, T., Fox, S.P., Fulton, B.R., Fynbo, H.O.U., Huikari, J., Hyldegaard, S., Jeppesen, H.B., Jokinen, A., Jonson, B., Kankainen, A., Moore, I., Nieminen, A., Nyman, G., Penttila, H., Pucknell, V.F.E., Riisager, K., Rinta-Antila, S., Tengblad, O., Wang, Y., Wilhelmsen, K., Aysto, J.: Phys. Rev. C (Nucl. Phys.) **80**(3), 034316 (2009)
40. Hyldegaard, S., Alcorta, M., Bastin, B., Borge, M.J.G., Boutami, R., Brandenburg, S., Büscher, J., Dendooven, P., Diget, C.A., Van Duppen, P., Eronen, T., Fox, S.P., Fraile, L.M., Fulton, B.R., Fynbo, H.O.U., Huikari, J., Huyse, M., Jeppesen, H.B., Jokinen, A.S., Jonson, B., Jungmann, K., Kankainen, A., Kirsebom, O.S., Madurga, M., Moore, I., Nieminen, A., Nilsson, T.: Phys. Rev. C **81**(2), 024303 (2010)
41. Öpik, E.J.: Proc. R. Irish Acad. A **54**, 49 (1951)
42. Salpeter, E.E.: Astrophys. J. **115**, 326 (1952)
43. Hoyle, F., Dunbar, D.N.F., Wenzel, W.A., Whaling, W.: Phys. Rev. **92**, 1095 (1953)
44. Dunbar, D.N.F., Pixley, R.E., Wenzel, W.A., Whaling, W.: Phys. Rev. **92**, 649 (1953)
45. Hoyle, F.: Astrophys. J. Suppl. Ser. **1**, 121 (1954)
46. Burbidge, E.M., Burbidge, G.R., Fowler, W.A., Hoyle, F.: Rev. Mod. Phys. **29**(4), 547 (1957)
47. Fowler, W.A.: In: Ekspång, G. (ed.) Nobel Lectures, Physics 1981–1990. World Scientific, Singapore (1993)
48. Salpeter, E.E.: Publ. Astron. Soc. Aust. **25**(1), 1 (2008)
49. Nomoto, K., Thielemann, F.K., Miyaji, S.: Astron. Astrophys. **149**, 239 (1985)
50. Langanke, K., Wiescher, M., Thielemann, F.K.: Zeitsch. Phys. A **324**, 147 (1986)
51. Nolen, J.A., Austin, S.M.: Phys. Rev. C **13**(5), 1773 (1976)
52. Markham, R.G., Austin, S.M., Shahabuddin, A.M.: Nucl. Phys. A **270**, 489 (1976)

53. Alburger, D.E.: Phys. Rev. C **16**, 2394 (1977)
54. Strehl, P.: Zeitsch. Phys. A **234**(5), 416 (1970)
55. Tilley, D., Kelley, J., Godwin, J., Millener, D., Purcell, J., Sheu, C., Weller, H.: Nucl. Phys. A **745**, 155 (2004)
56. Warner, R.E., Okihana, A., Fujiwara, M., Matsuoka, N., Kakigi, S., Hayashi, S., Fukunaga, K., Kasagi, J., Tosaki, M.: Phys. Rev. C **45**(5), 2328 (1992)
57. Crannell, H., Jiang, X., J.T. O'Brien, Sober, D.I., Offermann, E.: Nucl. Phys. A **758**, 399 (2005)
58. Chernykh, M., Feldmeier, H., Neff, T., von Neumann-Cosel, P., Richter, A.: Phys. Rev. Lett. **105**(2), 022501 (2010)
59. Caughlan, G.R., Fowler, W.A., Harris, M.J., Zimmerman, B.A.: At. Data Nucl. Data Tables **32**(2), 197 (1985)
60. Benn, J., Dally, E.B., Müller, H.H., Pixley, R.E., Staub, H.H., Winkler, H.: Nucl. Phys. A **106**(2), 296 (1967)
61. Weiss, A., Serenelli, A., Kitsikis, A., Schlattl, H., Christensen-Dalsgaard, J.: Astron. Astrophys. **441**, 1129 (2005)
62. Alvarez-Rodriguez, R., Jensen, A.S., Fedorov, D.V., Fynbo, H.O.U., Garrido, E.: Phys. Rev. Lett. **99**(7), 072503 (2007)
63. Fynbo, H.O.U., Álvarez Rodrìguez, R., Jensen, A.S., Kirsebom, O.S., Fedorov, D.V., Garrido, E.: Phys. Rev. C **79**(5), 054009 (2009)
64. Alvarez-Rodriguez, R., Jensen, A.S., Garrido, E., Fedorov, D.V., Fynbo, H.O.U.: Phys. Rev. C (Nucl. Phys.) **77**(6), 064305 (2008)
65. Kirsebom, O.S., Alcorta, M., Borge, M.J.G., Cubero, M., Diget, C.A., Dominguez-Reyes, R., Fraile, L.M., Fulton, B.R., Fynbo, H.O.U., Hyldegaard, S., Jonson, B., Madurga, M., Muñoz Martin, A., Nilsson, T., Nyman, G., Perea, A., Riisager, K., Tengblad, O.: Phys. Rev. C **81**(6), 064313 (2010)
66. Alcorta, M., Kirsebom, O., Borge, M.J.G., Fynbo, H.O.U., Riisager, K., Tengblad, O.: Nucl. Instrum. Methods Phys. Res., Sect. A **605**(3), 318 (2009)
67. Tamii, A., Adachi, T., Fujita, K., Hatanaka, K., Hashimoto, H., Itoh, M., Matsubara, H., Nakanishi, K., Sakemi, Y., Shimbara, Y., Shimizu, Y., Tameshige, Y., Yosoi, M., Fujita, Y., Sakaguchi, H., Zenihiro, J., Kawabata, T., Sasamoto, Y., Dozono, M., Carter, J., Fujita, H., Rubio, B., Perez, A.: Mod. Phys. Lett. A **21**(31–33), 2367 (2006)
68. Freer, M., Fujita, H., Buthelezi, Z., Carter, J., Fearick, R.W., Förtsch, S.V., Neveling, R., Perez, S.M., Papka, P., Smit, F.D., Swartz, J.A., Usman, I.: Phys. Rev. C **80**(4), 041303(R) (2009)
69. John, B., Tokimoto, Y., Lui, Y.W., Clark, H.L., Chen, X., Youngblood, D.H.: Phys. Rev. C **68**(1), 014305 (2003)
70. Itoh, M., Akimune, H., Fujiwara, M., Garg, U., Hashimoto, H., Kawabata, T., Kawase, K., Kishi, S., Murakami, T., Nakanishi, K., Nakatsugawa, Y., Nayak, B.K., Okumura, S., Sakaguchi, H., Takeda, H., Terashima, S., Uchida, M., Yasuda, Y., Yosoi, M., Zenihiro, J.: Nucl. Phys. A **738**, 268 (2004)
71. Itoh, M.: Mod. Phys. Lett. A **21**(31–33), 2359 (2006)
72. Freer, M., Boztosun, I., Bremner, C.A., Chappell, S.P.G., Cowin, R.L., Dillon, G.K., Fulton, B.R., Greenhalgh, B.J., Munoz-Britton, T., Nicoli, M.P., Rae, W.D.M., Singer, S.M., Sparks, N., Watson, D.L., Weisser, D.C.: Phys. Rev. C (Nucl. Phys.) **76**(3), 034320 (2007)
73. Muñoz-Britton, T, et al.: J. Phys. G Nucl. Part. Phys. **37**(10), 105104 (2010)
74. Gai, M., the Uconn-Yale-Duke-Weizmann-Ptb-Ucl Collaboration: J. Phys. Conf. Ser. **202**(1), 012016 (2010)
75. Gai, M.: In: Zakopane Conference on Nuclear Physics, 2010, Acta Phys. Pol. B **42**, 775 (2011)
76. Descouvemont, P.: J. Phys. G Nucl. Part. Phys. **37**(6), 064010 (2010)
77. Arai, K.: Phys. Rev. C (Nucl. Phys.) **74**(6), 064311 (2006)
78. Kanada-En'yo, Y.: Prog. Theor. Phys. **117**(4), 655 (2007)
79. Chernykh, M., Feldmeier, H., Neff, T., von Neumann-Cosel, P., Richter, A.: Phys. Rev. Lett. **98**(3), 032501 (2007)
80. Álvarez-Rodríguez, R., Garrido, E., Jensen, A., Fedorov, D., Fynbo, H.: Eur. Phys. J. A Hadrons Nuclei **31**(3), 303 (2007)
81. Kurokawa, C., Kato, K.: Nucl. Phys. A **792**(1–2), 87 (2007)
82. Patel, N.R., Greife, U., Rehm, K.E., Deibel, C.M., Greene, J., Henderson, D., Jiang, C.L., Kay, B.P., Lee, H.Y., Marley, S.T., Notani, M., Pardo, R., Tang, X.D., Teh, K.: AIP Conf. Proc. **1098**(1), 181 (2009)
83. Kirsebom, O.S., Alcorta, M., Borge, M.J.G., Cubero, M., Diget, C.A., Dominguez-Reyes, R., Fraile, L., Fulton, B.R., Fynbo, H.O.U., Galaviz, D., Garcia, G., Hyldegaard, S., Jeppesen, H.B.,

Jonson, B., Joshi, P., Madurga, M., Maira, A., Muñoz, A., Nilsson, T., Nyman, G., Obradors, D., Perea, A., Riisager, K., Tengblad, O., Turrion, M.: Phys. Lett. B **680**(1), 44 (2009)
84. Barker, F.C., Hay, H.J., Treacy, P.B.: Aust. J. Phys. **21**, 239 (1967)
85. Barker, F.C.: Aust. J. Phys. **22**, 293 (1969)
86. Barker, F.C.: Phys. Rev. C **62**(4), 044607 (2000)
87. Warburton, E.K.: Phys. Rev. C **33**, 303 (1986)
88. Herwig, F., Austin, S.M., Lattanzio, J.C.: Phys. Rev. C **73**(2), 025802 (2006)
89. Woosley, S.E., Heger, A., Rauscher, T., Hoffman, R.D.: Nucl. Phys. A **718**, 3 (2003)
90. Tur, C., Heger, A., Austin, S.M.: Astrophys. J. **671**(1), 821 (2007)
91. Tur, C., Heger, A., Austin, S.M.: Astrophys. J. **718**(1), 357 (2010)
92. Tur, C., Wuosmaa, A.H., Austin, S.M., Starosta, K., Yurkon, J., Estrade, A., Goodman, N., Lighthall, J.C., Lorusso, G., Marley, S.T., Snyder, J.: Nucl. Instrum. Methods Phys. Res., Sect. A **594**(1), 66 (2008)
93. Kibedi, T., Stuchbery, A.E., Dracoulis, G.D., Devlin, A., Teh, A., Robertson, K.: AIP Conf. Proc.**1109**(1), 66 (2009)
94. Matea, I., Adimi, N., Blank, B., Canchel, G., Giovinazzo, J., Borge, M.J.G., Domìnguez-Reyes, R., Tengblad, O., Thomas, J.C.: Nucl. Instrum. Methods Phys. Res., Sect. A **607**(3), 576 (2009)
95. Ogata, K., Kan, M., Kamimura, M.: Prog. Theor. Phys. **122**(4), 1055 (2009)
96. de Diego, R., et al.: EPL (Europhys. Lett.) **90**(5), 52001 (2010)
97. de Diego, R., Garrido, E., Fedorov, D.V., Jensen, A.S.: Phys. Lett. B **695**(1–4), 324 (2011)
98. Umar, A.S., Maruhn, J.A., Itagaki, N., Oberacker, V.E.: Phys. Rev. Lett. **104**(21), 212503 (2010)
99. Kat, K.: Mod. Phys. Lett. A **25**(21–23), 1819 (2010)
100. Peng, F., Ott, C.D.: Astrophys. J. **725**(1), 309 (2010)
101. Saruwatari, M., Hashimoto, M.-a.: Prog. Theor. Phys. **124**(5), 925 (2010). doi:10.1143/PTP.124.925
102. Dotter, A., Paxton, B.: Astron. Astrophys. **507**(3), 1617 (2009). doi:10.1051/0004-6361/200912998

DOI 10.1007/978-94-007-5555-0_10
Reprinted from *Hyperfine Interactions* Volume 223, Issues 1-3, 2014
and *The European Physical Journal A* Volume 48, Issue 4, 2012

Decays of $T_Z = -3/2$ nuclei ^{23}Al, ^{31}Cl, and ^{41}Ti

A. Kankainen · A. Honkanen · K. Peräjärvi · A. Saastamoinen

Published online: 24 April 2012

Abstract This article gives an overview on the decay spectroscopy of $T_Z = -3/2$ nuclei ^{23}Al, ^{31}Cl, and ^{41}Ti performed at the Ion Guide Isotope Separator On-Line (IGISOL) facility. The results of the IGISOL experiments are compared to the experimental results that have been published since. The isobaric multiplet mass equation (IMME) has been studied for the $T = 3/2$ quartets at $A = 23$ and $A = 31$. For ^{41}Ti, a detailed comparison to the Gamow–Teller strengths obtained for the analog transitions via charge-exchange reactions has been done. Further improvements in the experimental instrumentation and methods and possible implementations for studying $T_Z = -3/2$ nuclei at the new IGISOL facility are discussed.

Keywords IGISOL · Beta decay · Beta-delayed protons

1 Introduction

Nuclei having an isospin component $T_Z = (N - Z)/2 = -3/2$ offer an interesting possibility to study beta-decay strength over broad energy range. Since the main part of the beta-decay strength is concentrated on states above the proton separation energy S_p, beta-delayed proton spectrum is essential in order to study the strength distribution. A large fraction of the beta-decay strength is within the Q_{EC} window allowing a detailed comparison to shell-model calculations or to the strength of

A. Kankainen (✉) · A. Saastamoinen
Department of Physics, University of Jyväskylä, P.O. Box 35,
40014 University of Jyväskylä, Jyväskylä, Finland
e-mail: anu.k.kankainen@jyu.fi

A. Honkanen
Philips Healthcare, 01511 Vantaa, Finland

K. Peräjärvi
STUK Radiation and Nuclear Safety Authority, 00881 Helsinki, Finland

analogous transitions studied via charge-exchange reactions, such as in the case of the $T = 3/2$ quartet at $A = 41$ [1].

The relative β-delayed proton emission window $(1 - S_p/Q_{EC})$ of the $T_Z = -3/2$ nuclei depends strongly on whether they belong to the $A = 4n + 3$ (odd Z) or to the $A = 4n + 1$ series (even Z) [2]. Generally for the $T_Z = -3/2$ nuclei with odd Z, such as ^{23}Al and ^{31}Cl, the β-delayed proton emission window is between 0.3 and 0.45, whereas for even-Z nuclei, like ^{41}Ti, it is between 0.8 and 1. This gives a rough indication of the expected β-delayed proton versus γ-ray emission probability. As the β-decay phase space favors the decay to the low-energy states, the γ decay by slow electromagnetic interaction is especially pronounced for the $A = 4n + 3$ series β-decay precursors. Also due to Coulomb and angular momentum barriers, γ decay starts to compete with isospin-forbidden proton decay for low-lying proton unbound states. Therefore, in addition to protons, detecting γ rays with good sensitivity is essential.

According to the Isobaric Multiplet Mass Equation (IMME) [3–5], masses of the members of an isobaric multiplet (T) should lie along a parabola $M(T_Z) = a + bT_Z + cT_Z^2$. In order to study the IMME, masses and excitation energies of the isobaric analog states (IAS) belonging to the T multiplet should be precisely measured. Previously, masses were determined from beta-decay energies, but nowadays Penning trap mass spectrometry offers a more accurate method for measuring the ground state mass excesses directly. Beta-decay studies of $T_Z = -3/2$ nuclei offer information on the excitation energies of the IAS in the $T_Z = -1/2$ nuclei. In addition, isospin mixing in the wavefunctions of the $T = 3/2$ IAS, manifested in isospin-forbidden proton decays of these IAS, can be systematically studied in these nuclei.

The studied $T_Z = -3/2$ nuclei ^{23}Al, ^{31}Cl, and ^{41}Ti lie at the path of the rapid proton capture (rp) process [6, 7]. In particular, beta-decay studies of ^{23}Al and ^{31}Cl yield information on the states in the daughter nuclei, ^{23}Mg and ^{31}S, whose properties are relevant for the modeling the nucleosynthesis in ONe novae. Namely, the reactions ^{22}Na(p, γ)^{23}Mg and ^{22}Mg(p, γ)^{23}Al have to be well-known in order to model the production of ^{22}Na in ONe novae [8]. The non-observation of 1275-keV γ-rays from ^{22}Na with COMPTEL telescope [9] has drawn attention to the production mechanisms of ^{22}Na and constrained nova models during the last years. However, there are recent papers on possible observation of these γ-rays based on long-term COMPTEL observations [10, 11]. Iyudin et al. [10] reports possible detection from a diffused source in the galactic bulge, explaining it to originate most likely from photo-activation of ^{22}Ne by cosmic rays whereas [11] claims a more localized source from a very slow Nova Cassiopeia 1995. The reaction ^{30}P(p, γ)^{31}S is important since ^{30}P is a mandatory passing point in ONe novae and it will stop further nucleosynthesis unless proton captures on ^{30}P ($T_{1/2} = 2.5$ min) are fast enough [12].

2 Experimental details

2.1 Production of $T_Z = -3/2$ nuclei at IGISOL

At IGISOL [13], the ions of interest with $T_Z = -3/2$ have been produced in light ion fusion evaporation reactions with a proton or ^{3}He beam from the K-130 cyclotron on a thin (few mg/cm^2) target. The recoiling products from the target are thermalized in the gas cell where they undergo charge-exchange reactions and a good fraction of the ions end up at a charge state 1^+. The ions are extracted from the gas cell with the help of differential pumping and a skimmer electrode. During recent years, the skimmer electrode has been replaced by a sextupole ion beam guide SPIG [14]. After the gas cell, the ions are accelerated typically to 40q kV energy and mass-separated by a 55° dipole magnet providing a mass resolving power of $M/\Delta M \approx 500$. The mass-separated beam is implanted on a thin carbon foil for the studies of beta-delayed protons and/or on a movable tape for beta-delayed γ-rays at the experimental station of the IGISOL facility. This on-line mass separation provides much cleaner spectra of the studied nuclides compared to experiments performed with the He-jet technique [15]. Experimental setups and production reactions used for ^{23}Al, ^{31}Cl, and ^{41}Ti are summarized below.

^{23}Al was produced at IGISOL via ^{24}Mg(p,2n)^{23}Al reactions with a 7–10 μA, 40-MeV proton beam on ^{nat}Mg target [16]. The observed yields for mass-separated ^{23}Al and ^{23}Mg were ≤ 20 atoms/s and about 4000 atoms/s, respectively. ^{31}Cl was produced with the same type of reaction, ^{32}S(p,2n)^{31}Cl, using a 10–20 μA, 40 or 45-MeV proton beam on a thin ZnS target [17]. The yields for ^{31}Cl and ^{31}S at 40 MeV were around 14 atoms/s and 20000 atoms/s, respectively [17]. For the production of ^{41}Ti, a 1–7 eμA, 40-MeV ^{3}He beam on ^{nat}Ca target was used to produce the ions of interest via ^{40}Ca(^{3}He,2n)^{41}Ti [18]. The production rate of ^{41}Ti was about 1 atom/s with the highest beam intensity [18].

2.2 Detector setups for observing β-delayed γ-rays and protons

Identification of beta-delayed protons from the beta-particle background is a key issue in beta-delayed proton and gamma spectroscopy. Therefore, different $\Delta E - E$ detectors have been used, since the energy deposit in a ΔE detector depends on the type of the particle. With the same initial energy, protons leave more energy than beta particles but less than alpha particles. A $\Delta E_{gas} - E_{Si}$ detector telescope [19] was used to detect beta-delayed protons of ^{23}Al [16] and ^{41}Ti [18]. The telescope consisted of an E detector which was an Ortec Ultra series silicon detector with an active area of 300 mm^2 and a thickness of 300 μm [19]. A proportional counter mode was applied to the CF_4 gas ΔE detector in order to reach a large enough signal-to-noise ratio. A lower detection limit of 155 keV and an energy resolution of 20 keV was achieved for the telescope [19].

In addition to the $\Delta E_{gas} - E_{Si}$ detector telescope for detecting protons, a 1-mm-thick plastic ΔE_β detector and a HPGe (37.5% relative efficiency) were used in the initial IGISOL ^{23}Al experiment for detecting beta particles and γ-rays, respectively. The 40 keV ^{23}Al$^+$ beam was implanted on a 40 μg/cm^2-thick carbon foil surrounded by the three detectors. This setup is illustrated in Fig. 1. In the experiment on the beta decay of ^{41}Ti, two different measurement setups were used. For detecting beta-delayed protons, the $\Delta E_{gas} - E_{Si}$ detector telescope was used behind a 40 μg/cm^2-thick carbon foil into which the mass-separated beam was implanted. Beta- and proton-delayed γ-rays were observed with a setup consisting of a 0.9-mm-thick plastic scintillator for detecting beta particles, a 50% HPGe detector for γ-rays, and an ion-implanted silicon detector for both beta detection and proton energy measurements [18]. There, the 40 keV ^{41}Ti$^+$ beam was implanted into aluminized mylar tape.

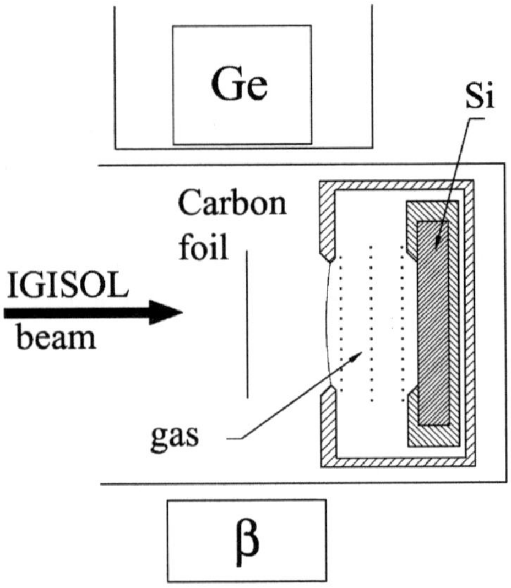

Fig. 1 Schematic presentation of the setup in the IGISOL central beam-line for ^{23}Al β-decay experiment reported in [16, 23]. The 40-keV beam was implanted into 40 μg/cm^2 carbon foil surrounded by 37.5% HPGe detector, 1-mm-thick plastic β-detector and the $\Delta E_{gas} - E_{Si}$ detector telescope

The mass-separated 25-keV $^{31}Cl^+$ was implanted into a 30 μg/cm^2-thick carbon foil surrounded by a novel state-of-the-art silicon detector assembly consisting of the ISOLDE Silicon Ball [20], three double-sided silicon strip detectors (DSSSDs) [21, 22] backed with three thick silicon detectors, and a 70% HPGe detector [17]. The DSSSDs were about 60 μm thick, and therefore, beta particles left very little energy in them. This made DSSSDs ideal for detecting beta-delayed protons from ^{31}Cl. The ISOLDE Silicon Ball and three thick Si detectors were used for detecting betas. The total beta efficiency was measured as 24.9(19)%. Protons below 700 keV could not be observed due to noise and β-tail caused mainly by ^{31}S in the spectrum.

3 Beta-delayed gamma and proton spectroscopy of $T_Z = -3/2$ nuclei at IGISOL

3.1 ^{23}Al

β-delayed proton emitter ^{23}Al can be used as a tool to probe astrophysically interesting states just above the proton separation threshold in ^{23}Mg. These states are relevant for understanding resonant proton capture ^{22}Na(p,γ)^{23}Mg and the amount of ^{22}Na ejected in ONe novae. Measuring this reaction directly is challenging because the need for radioactive ^{22}Na targets complicates the measurement [24, 25]. ^{23}Al was first produced in the early 1970's in Berkeley with the He-jet technique where a single proton group with an energy of 870(30) keV and a half-life of 470(30) ms was discovered [26]. In the mid-90's Tighe et al. [27] extended the study to lower energies and found a low-energy proton group with high intensity, which was assigned to originate from the $T = 3/2$ isobaric analog state (IAS) of the ground state of ^{23}Al. This was interpreted to occur due to extremely strong isospin-mixing as proton decay from the IAS does not conserve isospin.

Due to the astrophysical importance, and in order to confirm the results of Tighe et al. [27], a project to investigate the decay of ^{23}Al was initiated in Jyväskylä in

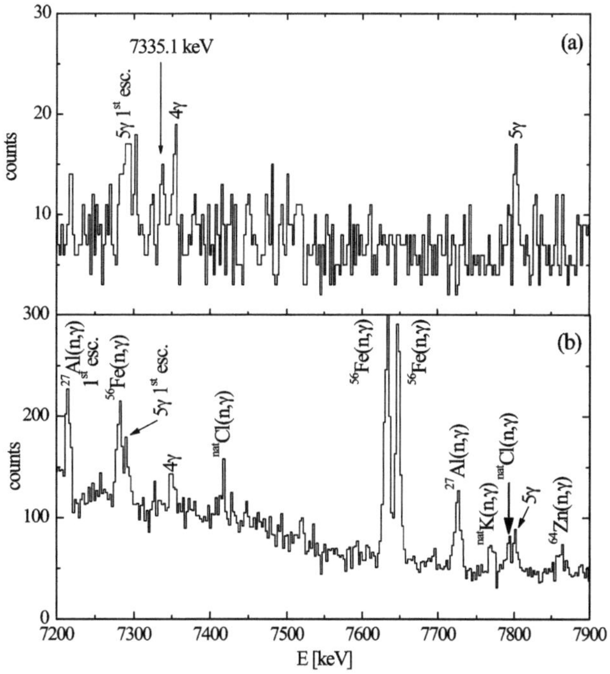

Fig. 2 The high energy part of the γ-spectrum from β-decay of ^{23}Al as reported in [16]

the late 1990's. An experimental setup capable of detecting γ-rays, electrons and heavier charged particles was installed to the end of the central beam line at IGISOL, see Fig. 1. Experimental details related to the ^{23}Al production are summarized in Section 2.1. This was the first ever study of β-decay of ^{23}Al to have a capability for simultaneous observation of both protons and γ-rays.

As the spectroscopy setup was also capable of detecting γ-rays, a γ-decay of the IAS to the ground and first excited state of ^{23}Mg was reported in [16] for the first time. These observations were confirmed later with much higher statistics by Iacob et al. [28]. Figure 2 presents Fig. 2a and b of [16], where peak labelled with 5γ is related to the transition from the IAS to the ground state and 4γ from the IAS to the first excited state.

From the decay spectroscopy point of view, it turned out to be that this initial IGISOL experiment [16] did not confirm the low energy part (below 400 keV) of the proton spectrum reported in [27]. At these energies significantly less intensity was observed at IGISOL [16] compared to [27]. Iacob et al. [28] suggested that this difference was caused by the higher cut-off energy in the $\Delta E_{gas} - E_{Si}$ telescope at IGISOL [16]. However, in a recent experiment at Texas A&M, β-delayed proton spectrum with high statistics [29, 30] was found out to be closer to the IGISOL spectrum [16] than the Berkeley spectrum [27]. A comparison of these spectra is illustrated in Fig. 3.

^{23}Mg was known to have a level with $J^{\pi} \geq 3/2^{+}$ at 17 keV below the IAS already in [16]. Using shell-model calculations and the data obtained in [16] it was concluded that the spin and the parity of the state is not $3/2^{+}$, $5/2^{+}$ or $7/2^{+}$, i.e., allowed beta-decay is not populating it. This conclusion also meant that all observed 200-keV protons were associated to the decay of the IAS. This was, however, a wrong

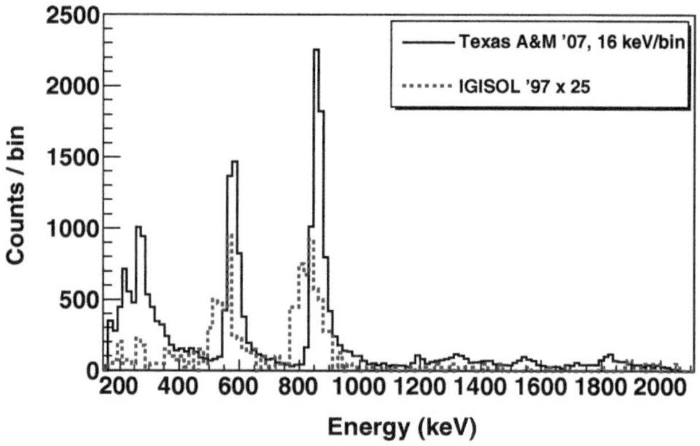

Fig. 3 Comparison of the proton spectra from IGISOL and Texas A&M. The IGISOL spectrum is multiplied ×25 to scale with statistics of the Texas A&M experiment. Note that the IGISOL spectrum has its energy recorded as proton energy in lab, while the Texas A&M spectrum is the recorded total decay energy

conclusion made due to the low statistics of the experiment. Namely, recent studies [28, 31] have assigned this state to be $(7/2)^+$ and to receive a β-feeding of 4.89(25)%. As also correctly predicted by shell-model calculations [16], this $(7/2)^+$ state primarily decays to the first excited state of ^{23}Mg at 450.71(15) keV. γ-rays with an energy of 7335.1 keV are associated to this transition [28]. The location of this line is marked into Fig. 2a. As shown, a small peak, almost at the level of the surrounding background, is visible, but it basically vanishes into the background when the full statistics is employed (Fig. 2b). Notice also, that due to the higher statistics and better energy resolution the more recent proton experiment (Fig. 3) assigned the lowest proton group to the 7787-keV state instead of the IAS.

Using a setup of four DSSSD—thick Si-pad telescopes, data from β-decay of ^{23}Al was measured at IGISOL during the calibration phases of another experiment in 2008. Improvements in the front-end after the first ^{23}Al experiment at IGISOL [16] increased the yield of ^{23}Al by a factor of around 35. This fact and the high efficiency of the setup enabled collection of sufficient statistics to observe 13 proton groups above the two main groups at 554 and 839 keV in only about 28 hours [32, 33]. One of these groups was observed in the earlier experiment [16]. β-tail in the spectrum caused by the isobars at $A = 23$ extended all the way up to 700 keV, preventing the studies of low-energy proton groups.

Soon after this experiment a high-precision mass measurement of ^{23}Al and ^{23}Mg ground states was done with JYFLTRAP. The resulting masses, reported in [34], were used to test the IMME at $A = 23$. The quadratic form of the IMME was found to hold well for this $T = 3/2$ quartet [34]. The new mass values also confirmed the Q_{EC} value deduced in [16].

3.2 ^{31}Cl

Beta-decay of ^{31}Cl has been previously studied with the He-jet technique at the MC-35 cyclotron of the University of Oslo [35] and at the Lawrence Berkeley Laboratory 88-inch Cyclotron [36, 37] by using proton beam with an energy of 34 or 45 MeV on ZnS targets, respectively. In these studies, eight proton peaks in the energy range of 845 to 2204 keV and a half-life of 150(25) ms [35] have been observed. However, these He-jet experiments suffer from contaminants with different mass numbers, such as ^{32}Cl [35–37] and ^{25}Si [37]. Therefore, it was important to study the beta decay of ^{31}Cl at IGISOL where on-line mass-separation is used.

At IGISOL, γ-rays with energies of 1249.1(14) keV and 2234.5(8) keV corresponding to the de-excitations of the first two excited states in ^{31}S were measured [17]. In addition, weak γ peaks at 3536(2) and 4045(2) keV were observed. These γ transitions could correspond to the transitions between the $3/2^+$ IAS at 6268(10) keV [38, 39] and from the $5/2^+$ state at 5777(5) keV [38, 39] to the $5/2^+$ state at 2235.6(4) keV [38, 39]. Thus, the deduced excitation energies of the IAS and the $5/2^+$ state are 6280(2) keV and 5772(2) keV, respectively [17]. In the proton spectrum, previously controversial proton peaks were confirmed to belong to ^{31}Cl and a new proton group with an energy of 762(14) keV was found [17]. Proton captures to this state at 6921(15) keV in ^{31}S contribute to the total reaction rate of ^{30}P(p, γ)^{31}S in ONe novae.

Whereas the previous studies of the beta decay of ^{31}Cl have aimed for determination of experimental Gamow–Teller strength distribution and its comparison to the shell-model predictions [36, 37], the recent study has also been motivated by nuclear astrophysics. Namely, the reaction rate of ^{30}P(p, γ)^{31}S plays a key role in ONe novae. ^{30}P is a mandatory passing point to ^{32}S via ^{30}P(p, γ)^{31}S(p, γ)^{32}Cl(β^+)^{32}S or via ^{30}P(p, γ)^{31}S(β^+)^{31}P(p, γ)^{32}S [12]. With a half-life of about 2.5 min, ^{30}P stops further nucleosynthesis unless the proton captures are fast enough. In addition, the knowledge of the final abundance pattern is important for example in the identification of the possible nova origin of presolar grains [40, 41]. Large excesses in ^{30}Si/^{28}Si and close-to-solar ^{29}Si/^{28}Si abundance ratios have been measured in these grains. In fact, the calculations based on several hydrodynamic nova simulations show that if the ^{30}P(p, γ) rate is reduced by a factor of 100, an enhancement in ^{30}Si is obtained [12, 42].

Since the direct study of this reaction rate is currently limited by the difficulty of producing an intense ^{30}P beam for studies in inverse kinematics, most of the experimental effort has been concentrated on the studies of the excitation energies of the states in ^{31}S either via β^+ decay [17] or different reactions, such as ^{12}C(^{20}Ne, n)^{31}S [43], ^{32}S(p, d)^{31}S [44], or ^{31}P(^{3}He, t)^{31}S [45–47]. Previously, the calculated reaction rates were based on statistical Hauser–Feshbach [48] calculations, which can have uncertainties as high as a factor of 100 up or down. In [43], a first evaluation of the ^{30}P(p, γ)^{31}S astrophysical reaction rate based on calculated resonant reaction rates of 13 experimentally determined states in ^{31}S was performed.

In the study of ^{32}S(p, d)^{31}S reactions [44], a total of 26 states including five new states in ^{31}S were observed and altogether 66 states were used for calculating a new ^{30}P(p, γ)^{31}S reaction rate. The new rate was found to differ from Hauser–Feshbach estimates by a factor of 10 and be comparable to previous estimates [43] in typical nova peak temperatures. Ma et al. [44] observed the states at 5781(5), 7728(4), 7912(5), 8049(6), and 8517(13) keV corresponding to the states at 5772(2), 7705(21), 7878(23), 8019(24), and 8509(30) keV measured via beta-decay at IGISOL [17]. The IAS at 6280(2) keV [17] was not populated in ^{32}S(p, d)^{31}S reactions [44].

The IAS ($T = 3/2$) state at 6280(2) keV [17] was confirmed in [45] where an excitation energy of 6283(2) keV was measured from the triton spectrum of the ^{31}P(^{3}He, t)^{31}S reaction. In a later publication on ^{31}P(^{3}He, t)^{31}S reactions measured at Yale University's Wright Nuclear Structure Laboratory [46], 17 new levels and 5 new tentative levels were found in ^{31}S. The obtained reaction rate for ^{30}P(p, γ)^{31}S [46] was about a factor of two larger than that from [43, 44] for $0.4 \leq T \leq 1.9$ GK. Above 1 GK the reaction rate calculated in [46] is a factor of seven larger than in [44].

All states observed in the beta decay of ^{31}Cl at IGISOL [17], were confirmed in [46], except the very weakly populated state at 8669(40) keV [17]. This peak was already marked as a smaller than a 3σ peak [17], and thus, it can be due to random statistical fluctuations. The beta-delayed proton peak observed at 1174(14) keV [17, 36] was assumed to correspond to the proton decay to the ^{30}P ground state, and an excitation energy of 7347(14) keV [17] was deduced for the state in ^{31}S. This was assumed to correspond to a known state at 7311(11) keV in ^{31}S. However, in Jenkins et al. [43, 49], an excitation energy of 7302.8(7) keV and a spin $11/2^+$ was determined for the 7311(11) keV state. Thus, this state cannot be strongly fed in the beta decay of ^{31}Cl having a ground-state spin of $3/2^+$. Wrede et al. [46] have found that several states in ^{31}S proton decay also to states at 677.01(3) keV (0^+) and 708.70(3) keV (1^+) [50] in ^{30}P. They have also suggested that the 1174(14)-keV beta-delayed protons [17, 36] would result from a proton decay to the 0^+ state at 677.01(3) keV in ^{30}P. Then, the corresponding excitation energy would be 8024(14) keV, close to the excitation energy of 8021(16) keV deduced from the 1826(15)-keV proton peak [17]. In future, coincidences of the beta-delayed protons with the γ-rays in ^{30}P should be searched for with better statistics.

The most recent study on ^{31}P(^{3}He, t)^{31}S reactions [47] performed at Maier–Leibnitz–Laboratorium (MLL) in Garching, Germany confirmed the results from previous (^{3}He, t) experiments. However, no evidence for the suggested doublets [46] at 6835 keV or 7030 keV was found. In addition, the angular distributions were analyzed and spin constraints were obtained for almost all critical states [47]. The hydrodynamic nova simulations performed in [47] show that the remaining uncertainties in the spin values and in the relevant proton spectroscopic factors may still lead to a factor of up to 20 variation in the proton capture rate on ^{30}P and a factor of up to four in the nova yields in the Si-Ar region.

The recent ^{31}P(^{3}He, t)^{31}S studies [45–47] have improved the precision of the IAS in ^{31}S. A weighted mean of 6283.2(8) keV [17, 45–47] for the IAS can be utilized in a more precise IMME fit to the $A = 31$ $T = 3/2$ quartet. At JYFLTRAP, the ground state mass of ^{31}S has been measured very precisely, $-19042.55(24)$ keV [51], and found to deviate from the Atomic Mass Evaluation 2003 (AME03) [52] by more than 1σ. In this paper, we have adopted the mass of ^{31}S from [51], the mass of ^{31}P from [53], the latest value for the IAS in ^{31}S, and used otherwise the same, most recent results for the $T = 3/2$ quartet at $A = 31$, as in [54], and performed a new IMME fit. The resulting mass excess value for ^{31}Cl is $-7048(5)$ keV, which is 10 keV higher than the value obtained in [54] and about 20 keV higher than in the AME03 [52]. In future, JYFLTRAP aims to measure the mass of ^{31}Cl produced via ^{32}S(p, 2n) reactions.

β-decay of ^{31}Cl has been studied also at Texas A&M in two experiments with a similar setup as in the case of ^{23}Al [29, 30]. These experiments confirm the results of the IGISOL experiment [17] for the proton spectrum up to about 2 MeV, even though suffering from ^{29}S impurities and the fact that the higher-energy protons escaped the implantation detector. Final analysis of the proton data from these experiments is an ongoing process. The first experiment yielded a good statistics for $\beta\gamma$-data, see [55–57] for the preliminary results. In the second experiment, the

γ-data were extended further by acquiring $\beta\gamma\gamma$-coincidences to allow building up a more complete level scheme [57].

A more precise proton separation energy of ^{31}S, $S_p = 6130.95(39)$ keV, could be determined from the mass excess of ^{31}S measured at JYFLTRAP [51]. This new value deviates from the adopted AME03 value [52] by more than 1σ [51]. Taking into account the new proton separation energy and the new excitation energy for the $T = 3/2$ IAS in ^{31}S [17, 45–47], β-delayed protons from the IAS should have laboratory energies of about 147.3(9) keV.

3.3 ^{41}Ti

Beta-decay of ^{41}Ti offers a possibility to study beta-decay strength over a wide energy range and compare the strength and its sum to the shell-model calculations in the crossing region of *sd* and *fp* shells. In addition, isospin mixing in the wave function of the $T = 3/2$ isobaric analog state (IAS), which makes the isospin-forbidden proton decay from the IAS possible, can be investigated.

Delayed proton emission following the beta-decay of ^{41}Ti was observed for the first time at Brookhaven National Laboratory [58, 59]. Seventeen beta-delayed proton peaks of ^{41}Ti were observed and a half-life of 88(1) ms was determined [59]. Later, a high-resolution study of the ^{41}Ti decay was performed at Berkeley with the He-jet technique and $\Delta E - E$ counter telescopes [60, 61]. A half-life of 80(2) ms and 27 proton peaks belonging to ^{41}Ti beta decay were observed [61]. Most of the peaks agreed with the peaks observed in [59]. The location of the lowest $T = 3/2$ in ^{41}Sc was determined as 5935(8)keV [60] disagreeing with the old ^{40}Ca(p, p) resonance measurements. Studies at Berkeley continued in the 1980s and six new beta-delayed proton peaks were found in [62]. ^{41}Ti was also used for calibration purposes in experiments at GSI and GANIL where half-lives of 81(4) ms [63] and 80.1(9) ms [64], respectively, were measured for ^{41}Ti.

In the experiment conducted at IGISOL, no beta- or proton-delayed γ-rays were observed [18]. Therefore, the intensities of the observed 25 proton peaks were directly proportional to the beta-decay transition intensities. For three peaks, beta-proton summing was taken into account in the analysis of peak intensities. The main differences from [61] were that the intensity of the 986-keV proton peak was about half of the one given in [61], the proton group just below 1.6 MeV was found to be a double peak, and the 2063-keV proton group was not detected at IGISOL. On the other hand, proton peaks with energies of 754(12), 1586(11), 4298, and 4684(11) keV not observed in [61, 62] were observed at IGISOL [18]. Of these peaks, the 4684(11)-keV protons originated from a new level at an energy of 5886 keV [18].

Extensive shell-model calculations of the ^{41}Ti beta decay were performed with the code OXBASH [18]. The observed experimental decay strength was found to be 0.64 times the theoretical strength corresponding to a quenching factor of $\sqrt{0.64} = 0.80$. Theoretical Fermi strength of the β^+ decay to the IAS is $B(F)^{theor} = T(T+1) - T_Z(T_Z+1) = 3$ for $T = 3/2$ and $T_Z = -3/2$. Shell-model calculations give a Gamow–Teller strength of $B(GT)^{theor} = 0.302$ to the IAS, which yields an estimate of $B(GT)^{exp} = 0.18$ for the experimental decay strength after correction due to the quenching. The experimentally observed total decay strength to the IAS was $B(F)^{exp} + B(GT)^{exp} = 2.87(22)$ [18], thus yielding a value of $B(F)^{exp} = 2.87(22) - 0.18 \approx 2.69(22)$ for the experimental Fermi strength. This can be explained by the isospin impurity of the IAS. Namely, if there is a state with $J^\pi = \frac{3}{2}^+$, $T = \frac{1}{2}$ near the IAS ($J^\pi = \frac{3}{2}^+$, $T = \frac{3}{2}$), the isospins $T = \frac{3}{2}$ and $T = \frac{1}{2}$ of these near-lying states can mix with each other with an amplitude b according to a simple perturbation theory. Then, the perturbed IAS has a structure $a|T = \frac{3}{2}\rangle + b|T = \frac{1}{2}\rangle$ and the other state is $a|T = \frac{1}{2}\rangle - b|T = \frac{3}{2}\rangle$, where $|b| = \sqrt{1-a^2}$. In order to explain the observed Fermi strength, a fraction $b^2 = 10(8)\%$ of the Fermi strength of the IAS ($B(F)^{theor} = 3$) is shifted to this other state.

After the experiment performed at IGISOL, beta decay of ^{41}Ti has been studied at GSI [65]. The isotopically separated ^{41}Ti beam from FRS was implanted into a silicon detector stack consisting of eight 300 μm thick, 30-mm diameter Si detectors [65]. An array of 14 large-volume Crystal Ball NaI detectors were used to measure emitted γ rays [65]. The energy resolution of around 40 – 100 keV [65] was worse than at IGISOL, and some of the peaks observed at IGISOL [18] could not be resolved. In [65], the total number of implanted ^{41}Ti ions could be deduced from the ΔE versus A/q data after correction for the intensity loss due to secondary reactions in the energy degrader at F_4 [65]. In addition, proton-γ coincidence data were collected and five transitions were assigned to populate the 3904-keV state in ^{40}Ca [65]. With the absolute counting of ^{41}Ti ions and proton-γ coincidence data, Liu et al. [65] obtain significantly larger B(GT) values at high excitation energies in ^{41}Sc than at IGISOL [18] where the branching ratios were based on the total number of observed protons. The total experimental B(GT) of 3.6(5) determined at IGISOL [18] is also smaller than the value obtained at GSI, 4.83(29) [65].

The GT strength obtained for the transitions in the beta decay of ^{41}Ti ($T_Z = -3/2$) to the states in ^{41}Sc ($T_Z = -1/2$), $B(GT)_\beta$, can be interestingly compared to the strength $B(GT)_{CE}$ obtained for analogous transitions from ^{41}K ($T_Z = +3/2$) to ^{41}Ca($T_Z = +1/2$) via ^{41}K(^{3}He, t)^{41}Ca charge-exchange reactions at 0° [1]. Assuming isospin symmetry, these transitions should have equal energies and strengths. The 35-keV energy resolution of the observed tritons at the Grand Raiden spectrometer at the Research Center for Nuclear Physics (RCNP) allowed to resolve states of ^{41}Ca up to an excitation energy of $E_x = 10$ MeV [1]. The beta-decay study of ^{41}Ti conducted at IGISOL had a similar energy resolution of around 30 keV and states up to around 8 MeV in ^{41}Sc were populated, which offered a possibility to compare the observed Gamow–Teller strengths in detail. The determination of the Gamow–Teller strength in the charge-exchange reactions relies on the approximate proportionality of the reaction cross sections measured at the scattering angle 0° and their B(GT) values. In order to normalize this proportionality, a B(GT) value known well from beta-decay studies is typically used. For the A = 41 system, such well-known beta transition does not exist. Therefore, the total sum of the $B(GT)_{CE}$ was normalized to the total sum of $B(GT)_\beta$, which was obtained as an average from [18, 65]. The transition to the IAS in ^{41}Ca contains both Fermi and GT components, and the $B(GT)_{CE}$ was estimated to be 0.24(4). If a similar isospin impurity of 10(8)% [18] is assumed as for ^{41}Sc, the $B(GT)_{CE}$ value would be increased by 0.055(45).

The $B(GT)_{CE}$ distribution [1] is quite similar to the $B(GT)_\beta$ distribution obtained at IGISOL [18] (see Fig. 4). Almost one-to-one correspondence of the observed states and GT transition strengths to them is observed up to around 6 MeV. The energy resolution of around 40 – 100 keV in [65] did not allow a good comparison to the charge-exchange reactions. In addition, the observed rather strong GT strength to states around 7 MeV in ^{41}Sc [65] was not observed via charge-exchange reactions

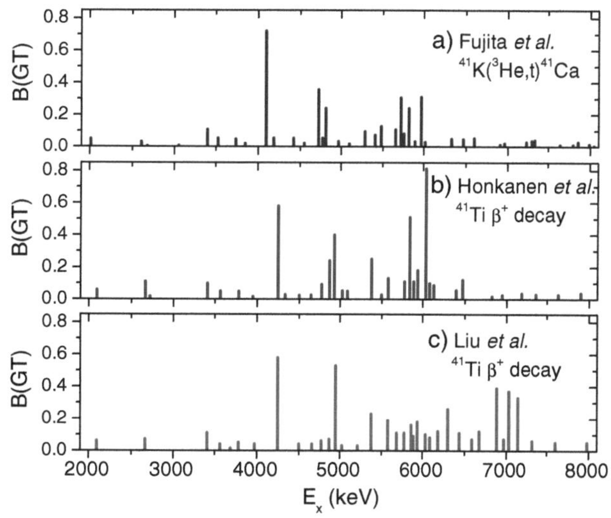

Fig. 4 Experimental B(GT) distributions for $T_Z = \pm 3/2 \rightarrow T_Z = \pm 1/2$ isospin mirror transitions from [1, 18, 65]

to ^{41}Ca [1] nor at IGISOL [18]. Most probable J^π values were deduced for each analog pair, which lead to a confirmation of $J^\pi = 1/2^+$ for the states at 3951(14) and 6038(25) keV, $J^\pi = 3/2^+$ for the states at 3562.6(3), 5576(4) and 5939(4) keV, and $J^\pi = 5/2^+$ for the states at 4928(5), 5774(4), 5840(5) and 5886(12) keV suggested as $1/2^+, 3/2^+, 5/2^+$ states in the beta-decay study at IGISOL [18]. The main part of the GT strength was found to the $5/2^+$ states in charge-exchange reactions [1].

Mass of ^{41}Ti has been directly measured at the FRS-ESR facility at GSI [66]. However, the obtained precision for the mass excess, −15090(360) keV, is rather poor compared to Penning trap measurements. In future, the mass of ^{41}Ti could be measured with the purification trap of JYFLTRAP (as was done e.g. for ^{97}Kr at ISOLTRAP [67]) if the half-life is too short for precision trap measurements.

4 Discussion

Various experiments on beta decays of other $T_Z = -3/2$, $A = 4n + 1$ nuclei than ^{41}Ti have been performed in the past: ^{9}C [68], ^{13}O [69], ^{17}Ne [70], ^{21}Mg [62], ^{25}Si [71, 72], ^{29}S[62], ^{33}Ar [73], ^{37}Ca [74], ^{49}Fe [63], ^{53}Ni [75], and ^{57}Zn [76] (only the most recent references given). In addition to ^{23}Al and ^{31}Cl, $T_Z = -3/2$, $A = 4n + 3$ nuclei such as ^{27}P [37], ^{35}K [77, 78], ^{43}V [79], and ^{47}Mn [63] have been studied. At IGISOL, some of these beta decays could be investigated in future. JYFLTRAP mass spectrometer could be applied for example in the direct mass measurement of ^{31}Cl. Future measurements on these nuclei are also motivated by possible one-proton halos in ^{17}Ne, ^{23}Al, ^{31}Cl, and ^{35}K, and two-proton halos in ^{9}C and ^{13}O [80].

Straightforward experimental interpretation of the low-energy part of the ^{23}Al β-delayed proton spectrum, around 200 keV, is a challenging task. To distinguish unambiguously protons from the IAS and the neighboring 7787 keV state a detection setup that has a particle identification capability, excellent proton-energy resolution and a thin dead-layer is required. The challenge is similar in the case of β-decay of

^{31}Cl, where the β-delayed protons from the IAS have an energy of about 147 keV as discussed in Section 3.2.

This is a non-trivial experimental challenge. For example, with the telescope used in the studies of ^{23}Al and ^{41}Ti decays at IGISOL, the lower energy limit was about 160 keV. Even the high-end DSSSDs with ultra-thin dead layers have the detection thresholds around 150-200 keV, mostly due to the electronics noise in the measurement area. In addition, reducing the β-background requires extremely pure source, preferably with capability of removing the activity from the following daughter decays. In future, these requirements could possibly be fulfilled at IGISOL by combing a Penning trap and a micro-calorimeter detector [81] that employs digital electronics.

Novel detector assemblies could provide cleaner spectra and better energy resolution in future. An energy resolution of 1.06 keV, excellent compared to modern silicon detectors with resolutions of around 8.8 keV, has been achieved for 5.3-MeV alpha particles with a cryogenic microcalorimeter detector [82]. A microcalorimeter is based on the conversion of the particle's kinetic energy into thermal excitations in a tin absorber. The temperature change is measured with a superconducting transition edge sensor (TES) at its superconducting transition temperature of 140 mK where a small change in temperature results in a large change in resistance. An advantage of this technique is that it does not suffer from the surface dead layer effect but the disadvantage is the required low temperature. Typically, an adiabatic demagnetization refrigerator at 80 mK is used for a copper mount, and the TES is heated into its resistive transition by electrical bias [82]. This method requires that the sample is cooled and maintained at 80 mK in the same vacuum as the microcalorimeter. Therefore, it is not suitable for on-line experiments where the sample should be implanted and moved in short time periods unless a clever way to maintain the low temperature for TES is invented. Microcalorimeters can also be used for detecting γ-rays: a 47-eV resolution was achieved for a 103-keV γ-peak [83]. Major drawback of this method is that these microcalorimeters are small (≈ 1 mm^2) and slow (50 – 100 counts/s), and a large array of these microcalorimeters would be required for decent efficiency.

After the IGISOL upgrade in 2002-2003 [84] (including e.g. higher pumping efficiency, better radiation shielding to fully gain the high-intensity light ion beams, and improved gas purification control) and after the replacement of the skimmer by SPIG [14], the yields have increased a lot. For example, with a 8–10 μA, 40-MeV proton beam on ^{nat}Mg, yields of 700 atoms/s and 150000 atoms/s were observed for ^{23}Al and ^{23}Mg, respectively [34]. Thus, the production rates of ^{23}Al and ^{23}Mg were about 35 times higher than with the skimmer at the old IGISOL facility. Similar improvement for the production of ^{41}Ti can also be expected. ^{31}Cl was measured already after the IGISOL upgrade using the skimmer. The introduction of SPIG should increase the chlorine yields with a factor of about 4 – 10, as reported in [14] .

The production of ^{31}Cl has been challenging at IGISOL. This is seen for example from the production rates of ^{23}Al and ^{31}Cl produced via similar $(p, 2n)$ reactions at IGISOL: the yield of ^{23}Al has been higher than for ^{31}Cl (see Section 2.1). Chlorine is a halogen and it has a high electron affinity of 3.612724(27) eV [85] (compare to 0.43283(5) eV [86] for Al). ^{31}Cl is, unfortunately, in the mass region where major molecular contaminant beams are present in form of different nitrogen and oxide compounds. These molecular beams, e.g. ^{15}N^{16}O, arise mostly from either dirty

He-gas or leaks in the gas feeding system. In the past these background beams have been so intense that they have overloaded the Penning trap and thus prevented even purification of the samples for mass measurements. Improvements for the gas-feeding system and He-gas purification for the new IGISOL facility are in the high priority, thus making measurements around $A = 30$ region with the JYFLTRAP a more viable venture. In future, it would be interesting to study the fraction of negative chlorine ions produced at IGISOL and to search for possibilities to produce a beam of these negative ions.

The experiments performed at IGISOL providing mass-separated, thin sources of short-lived nuclei have demonstrated the potential of high-resolution detection technique for β-delayed protons and γ-rays. Taking into account the expected improvement in the yields, these experiments also pave the way for the studies of more exotic nuclei, such as ^{22}Al and ^{40}Ti. ^{22}Al has a very exotic and interesting β-delayed decay spectrum from γ-rays to two-proton and alpha emissions. The decay of ^{40}Ti has significance in determining the detection efficiency of the ICARUS detector for the study of solar neutrinos by utilizing the inverse β-decay of ^{40}Ar. These two nuclei have earlier been studied using the He-Jet technique [87, 88], fragment separator at GANIL [75, 89, 90] and FRS at GSI [65]. The new IGISOL facility may offer an alternative way for studying these nuclei.

Acknowledgements This work has been supported by the Academy of Finland under the Finnish Centre of Excellence Programme 2006-2011 (Nuclear and Accelerator Based Physics Programme at JYFL). A.K. acknowledges the support from the Academy of Finland under the project 127301.

References

1. Fujita, Y., et al.: Phys. Rev. C **70**, 054311 (2004)
2. Cerny, J., Hardy, J.C.: Ann. Rev. Nucl. Sci. **27**, 333 (1977).
3. Wigner, E.P.: In: W.O. Millikan (ed.) Proceedings of the Robert A. Welch Foundation Conferences on Chemical Research, vol. 1, p. 67. Robert A. Welch Foundation, Houston (1958)
4. Weinberg, S., Treiman, S.: Phys. Rev. **116**, 465 (1959)
5. Jänecke, J.: Phys. Rev. **147**, 735 (1966)
6. Wallace, R.K., Woosley, S.E.: Astrophys. J. Suppl. Ser. **45**, 389 (1981)
7. Schatz, H., et al.: Phys. Rep. **294**, 167 (1998)
8. José, J., Hernanz, M.: J. Phys. G **34**, R431 (2007)
9. Iyudin, A.F., et al.: Astron. Astrophys. **300**, 422 (1995)
10. Iyudin, A.F., et al.: Astron. Astrophys. **443**, 477 (2005)
11. Iyudin, A.: Astronomy Reports **54**, 611 (2010)
12. José, J., Coc, A., Hernanz, M.: Astrophys. J. **560**, 897 (2001)
13. Äystö, J.: Nucl. Phys. A **693**, 477 (2001)
14. Karvonen, P., et al.: Nucl. Instrum. Methods Phys. Res. B **266**, 4794 (2008)
15. Wollnik, H.: Nucl. Instrum. Methods **139**, 311 (1976)
16. Peräjärvi, K., et al.: Phys. Lett. B **492**, 1 (2000)
17. Kankainen, A., et al.: Eur. Phys. J. A **27**, 67 (2006)
18. Honkanen, A., et al.: Nucl. Phys. A **621**, 689 (1997)
19. Honkanen, A., et al.: Nucl. Instrum. Methods Phys. Res. A **395**, 217 (1997)
20. Fraile, L.M., Äystö, J.: Nucl. Instrum. Methods Phys. Res. A **513**, 287 (2003)
21. Bergmann, U.C., Fynbo, H.O.U., Tengblad, O.: Nucl. Instrum. Methods Phys. Res. A **515**, 657 (2003)
22. Tengblad, O., et al.: Nucl. Instrum. Methods Phys. Res. A **525**, 458 (2004)
23. Peräjärvi, K.: Ph.D. thesis. Department of Physics, University of Jyväskylä (2001)
24. Stegmüller, F., et al.: Nucl. Phys. A **601**, 168 (1996)

25. Sallaska, A.L., et al.: Phys. Rev. Lett. **105**, 152501 (2010)
26. Gough, R.A., Sextro, R.G., Cerny, J.: Phys. Rev. Lett. **28**, 510 (1972)
27. Tighe, R.J., et al.: Phys. Rev. C **52**, R2298 (1995)
28. Iacob, V.E., et al.: Phys. Rev. C **74**, 045810 (2006)
29. Saastamoinen, A., et al.: J. Phys.: Conf. Ser. **202**, 012010 (2010)
30. Saastamoinen, A., et al.: Phys. Rev. C **83**, 045808 (2011)
31. Jenkins, D.G., et al.: Phys. Rev. Lett. **92**, 031101 (2004)
32. Kirsebom, O.: Ph.D. thesis. Department of Physics and Astronomy, Aarhus University (2010)
33. Kirsebom, O.S., et al.: Eur. Phys. J. A **47**, 130 (2011)
34. Saastamoinen, A., et al.: Phys. Rev. C **80**, 044330 (2009)
35. Äystö, J., et al.: Phys. Scr. **1983**, 193 (1983)
36. Äystö, J., et al.: Phys. Rev. C **32**, 1700 (1985)
37. Ognibene, T.J., et al.: Phys. Rev. C **54**, 1098 (1996)
38. Endt, P.M.: Nucl. Phys. A **521**, 1 (1990)
39. Endt, P.M.: Nucl. Phys. A **633**, 1 (1998)
40. Amari, S., et al.: Astrophys. J. **551**, 1065 (2001)
41. José, J., et al.: Astrophys. J. **612**, 414 (2004)
42. Iliadis, C., et al.: Astrophys. J. Suppl. Ser. **142**, 105 (2002)
43. Jenkins, D.G., et al.: Phys. Rev. C **73**, 065802 (2006)
44. Ma, Z., et al.: Phys. Rev. C **76**, 015803 (2007)
45. Wrede, C., et al.: Phys. Rev. C **76**, 052802 (2007)
46. Wrede, C., et al.: Phys. Rev. C **79**, 045803 (2009)
47. Parikh, A., et al.: Phys. Rev. C **83**, 045806 (2011)
48. Rauscher, T., Thielemann, F.-K.: At. Data Nucl. Data Tables **75**, 1 (2000)
49. Jenkins, D.G., et al.: Phys. Rev. C **72**, 031303 (2005)
50. Basunia, M.S.: Nucl. Data Sheets **111**, 2331 (2010)
51. Kankainen, A., et al.: Phys. Rev. C **82**, 052501 (2010)
52. Audi, G., Wapstra, A.H., Thibault, C.: Nucl. Phys. A **729**, 337 (2003)
53. Redshaw, M., McDaniel, J., Myers, E.G.: Phys. Rev. Lett. **100**, 093002 (2008)
54. Wrede, C., et al.: Phys. Rev. C **79**, 045808 (2009)
55. Trache, L., et al.: PoS(NIC X) 163. 10th Symposium on Nuclei in the Cosmos (2008)
56. Saastamoinen, A., et al.: AIP Conf. Proc. **1409**, 71 (2011)
57. Saastamoinen, A.: Ph.D. thesis. University of Jyväskylä (2011)
58. Reeder, P.L., Poskanzer, A.M., Esterlund, R.A.: Phys. Rev. Lett. **13**, 767 (1964)
59. Poskanzer, A.M., McPherson, R., Esterlund, R.A., Reeder, P.L.: Phys. Rev. **152**, 995 (1966)
60. Gough, R.A., Sextro, R.G., Cerny, J.: Phys. Lett. B **43**, 33 (1973)
61. Sextro, R.G., Gough, R., Cerny, J.: Nucl. Phys. A **234**, 130 (1974)
62. Zhou, Z.Y., et al.: Phys. Rev. C **31**, 1941 (1985)
63. Faux, L., et al.: Nucl. Phys. A **602**, 167 (1996)
64. Trinder, W., et al.: Phys. Lett. B **415**, 211 (1997)
65. Liu, W., et al.: Phys. Rev. C **58**, 2677 (1998)
66. Stadlmann, J., et al.: Phys. Lett. B **586**, 27 (2004)
67. Naimi, S., et al.: Phys. Rev. Lett. **105**, 032502 (2010)
68. Millener, D.: Eur. Phys. J. A **25**, 97 (2005)
69. Knudsen, H.H., et al.: Phys. Rev. C **72**, 044312 (2005)
70. Chow. J.C., et al.: Phys. Rev. C **66**, 064316 (2002)
71. Robertson, J.D., et al.: Phys. Rev. C **47**, 1455 (1993)
72. Thomas, J.-C., et al.: Eur. Phys. J. A **21**, 419 (2004)
73. Adimi, N., et al.: Phys. Rev. C **81**, 024311 (2010)
74. Trinder, W., et al.: Nucl. Phys. A **620**, 191 (1997)
75. Dossat, C., et al.: Nucl. Phys. A **792**, 18 (2007)
76. Jokinen, A., et al.: Eur. Phys. J. Direct **4**, 1 (2002)
77. Ewan, G.T., et al.: Nucl. Phys. A **343**, 109 (1980)
78. Äystö, J., et al.: Phys. Rev. Lett. **55**, 1384 (1985)
79. Borrel, V., Anne, R., Bazin, D., Borcea, C., Chubarian, G.G., Del Moral, R., Détraz, C., Dogny, S., Dufour, J.P., Faux, L., Fleury, A., Fifield, L.K., Guillemaud-Mueller, D., Hubert, F., Kashy, E., Lewitowicz, M., Marchand, C., Mueller, A.C., Pougheon, F., Pravikoff, M.S., Saint-Laurent, M.G., Sorlin, O.: The decay modes of proton drip-line nuclei with A between 42 and 47. Z. Phys. A **344**(2), 135–144 (1992). doi:10.1007/BF01291696
80. Lunney, D.: Int. J. Mod. Phys. E **18**, 2077 (2009)

81. Horansky, R.D., et al.: J. Appl. Phys. **107**, 044512 (2010)
82. Horansky, R.D., et al.: Appl. Phys. Lett. **93**, 123504 (2008)
83. Doriese, W.B., et al.: Appl. Phys. Lett. **90**, 193508 (2007)
84. Penttilä, H., et al.: Eur. Phys. J. A **25**, 745 (2005)
85. Berzinsh, U., et al.: Phys. Rev. A **51**, 231 (1995)
86. Scheer, M., Bilodeau, R.C., Thøgersen, J., Haugen, H.K.: Phys. Rev. A **57**, R1493 (1998)
87. Cable, M.D., et al.: Phys. Rev. C **26**, 1778 (1982)
88. Cable, M.D., et al.: Phys. Rev. Lett. **50**, 404 (1983)
89. Blank, B., et al.: Nucl. Phys. A **615**, 52 (1997)
90. Achouri, N., et al.: Eur. Phys. J. A **27**, 287 (2006)

DOI 10.1007/978-94-007-5555-0_11
Reprinted from *Hyperfine Interactions* Volume 223, Issues 1-3, 2014
and *The European Physical Journal A* Volume 48, Issue 4, 2012

Decay spectroscopy of neutron-rich nuclei with A $\simeq$ 100

G. Lhersonneau · B. Pfeiffer · K.-L. Kratz

Published online: 11 July 2012

Abstract We review structure data obtained by decay spectroscopy of neutron-rich nuclei of mass close to 100. Emphasis is put on the contribution of experiments at IGISOL in the nineties. They confirmed the earlier postulated shape coexistence in the fast shape-transition region between N = 58 (spherical ground states and low collectivity) and N = 60 (strong axial deformation). A detailed spectroscopic study of the A = 99 chain established the upper-Z limit of the N = 56 shell closure region with ^{99}Nb, owing to striking similarities with ^{97}Y. A consequence of the N = 56 closure is that the $s_{1/2}$ odd-neutron becomes the ground state of the most neutron-rich N = 57 isotones, starting with ^{99}Mo, instead of the degenerated $d_{5/2}$ and $g_{7/2}$ subshells familiar in the tin region. Consequences on the change of spin on astrophysical r-process calculations are briefly discussed. Finally, we say a few words about neutron-rich rhodium and palladium isotopes near the neutron midshell where regular and intruder states coexist very close to each other.

Keywords Gamma transitions and level energies · $90 \leq A \leq 149$ · PACS Nuclear structure models and methods

G. Lhersonneau (✉)
GANIL, Caen, France
e-mail: gerard.lhersonneau@ganil.fr

B. Pfeiffer · K.-L. Kratz
Mainz University, Mainz, Germany

B. Pfeiffer
e-mail: B.Pfeiffer@gsi.de

K.-L. Kratz
e-mail: klk@uni-mainz.de

1 Introduction

In the early seventies the first gamma-spectroscopy experiments of fission fragments using germanium detectors have been carried out. Remarkable were those at Berkeley, using spontaneous fission sources like ^{252}Cf. Cheifetz and his collaborators observed cascades in excited fragments that they assigned within small uncertainty to a nucleus via coincidences with X-rays and time-of-flight of the recoiling fragments. Soon after, decay experiments were implemented at research reactors to benefit from the high thermal-neutron flux to induce fission of ^{235}U. These experiments used off-line radiochemistry after fast transport of the target out of the reactor or on-line mass separation. The German community was involved in running several facilities, the ISOL-type OSTIS and recoil separator LOHENGRIN at ILL-Grenoble, France, and the gas-filled JOSEF separator at Jülich, Germany. LOHENGRIN has survived the shutdown wave of reactor-based facilities, underwent upgrades and is currently the place of an active program on the decay of μ-second isomers, e.g. by G. Simpson and W. Urban.

An interesting picture emerged in the systematics of the 2^+ level energies. In the second half of the Z = 28–50 shell, the energy trend is rather smooth as one would expect being far from magic numbers. In contrast, near the middle of the proton shell, the 2^+ energies in $_{38}$Sr and $_{40}$Zr neutron-rich isotopes exhibited a different behaviour. The 2^+ energies of strontium isotopes with N$<$ 60 were remarkably constant around 800 keV and, therefore, quite higher than the average at higher Z. Even more surprising were the very high-lying 2^+ states at 1751 keV and 1223 keV in ^{96}Zr and ^{98}Zr. It was especially striking that the 2^+ energy of their N = 60 isotones dropped dramatically to 144 keV (^{98}Sr) and 213 keV (^{100}Zr), while the ground-state bands indicated large deformation. This unique transition of structure remained unaccounted for by models for a long time. Nevertheless, very low-lying 0^+ states at 215 keV (^{98}Sr) and 331 keV (^{100}Zr) were soon found and suggested shape coexistence. An updated systematics of 2^+states of nuclei in the light-fission group is shown in Fig. 1.

A key contribution to that problem has been the study of the odd-Z $_{39}$Y isotopes at JOSEF. The JOSEF separator had some features similar to IGISOL in slowing down the fission fragments in helium. The ionised fragments, however, were not thermalised, but their trajectory was bent by a magnetic field while they were slowed down in the gas. JOSEF could deliver fragments of less than 1 μs half-life and, since concentrating the various charge states of the recoiling ions, had a relatively high transmission allowing γ-γ-time measurements. The moderate resolving power made spectroscopy a rather difficult task. Yet, before the advent of prompt spectroscopy of fission fragments with large Ge-arrays initiated in 1989 using EUROGAM [1] and later pursued by EUROBALL (see e.g. papers by J. Durell and W. Urban) and GAMMASPHERE (see e.g. J. Hamilton) in the US, the long bands fed by K-isomers observed at JOSEF were the only ones known in that region. A band on an excited state at 496 keV in ^{98}Y was the first one discovered, even during the testing of a forerunner of JOSEF, but was understood only 20 years later as a $K^\pi = 4^-$ band [2]. The $5/2^+$ ground-state band in ^{99}Y was the playing ground for the rotor + particle coupling model [3]. The $\beta \simeq 0.4$ deformation postulated from the band analysis was later confirmed by a fast timing experiment by Mach et al. [4], who succeeded to measure the enhancement of the E2 component in the

Fig. 1 Energies of 2^+ levels in the light fission group. *Solid lines* mark the Z and N = 50 magic numbers limiting the plot and some numbers of special interest in this presentation, namely the (Z = 40, N = 56) closures crossing at ^{96}Zr and the N = 60 line where the ground states of Sr and Zr isotones are strongly deformed ($\beta = 0.40$ for ^{98}Sr). *Full circles* mark nuclei discussed here

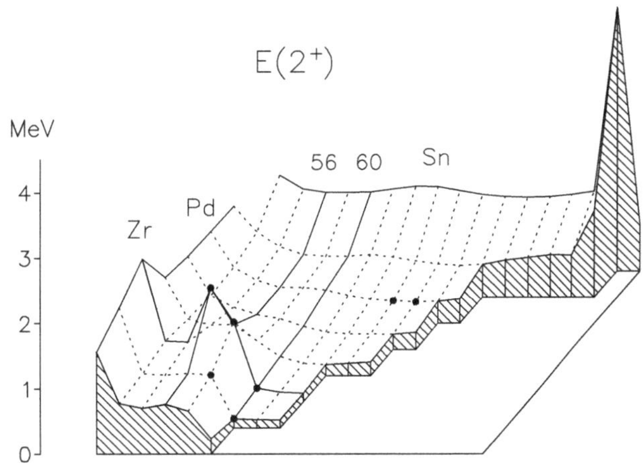

(K+1, 7/2)→(K, 5/2) transition. The modest resolution of JOSEF was the reason for the accidental discovery of another isomer in $^{97}Y_{58}$, which, according to its decay and in agreement with its neutron number, was interpreted in the context of few valence particles outside the N = 56 closure [5]. The isomer half-life of 0.14 s allows its study at IGISOL. It has been studied in detail by γ-spectroscopy (see later in this contribution), but also by collinear laser spectroscopy (see contribution by P. Campbell). The very complex but appealing odd-odd ^{98}Y with neutron number at the transition of shape has been further studied at IGISOL and by prompt fission for searching for more isomers (see papers by W. Urban) and at LOHENGRIN for measuring the transition probabilities using new type $LaBr_3$ fast-timing detectors (see. G. Simpson).

The last experiment at OSTIS (1986) before its shutdown was focused on lifetimes of levels in ^{97}Sr and ^{98}Sr. In addition, a detailed investigation of the levels in ^{97}Sr revealed a strongly enhanced E2 component in a transition from a level at 687 keV [6]. This was the first hint for a deformed structure in a N = 59 isotone of Sr, Y and Zr. The GAMMASPHERE group later confirmed the existence of a band by gating on that transition [7].

The OSTIS measurement did not solve the issue it was aimed at, namely the structure of the first excited state in ^{97}Sr. It, however, opened a Pandora box because of the inconsistencies of the newly measured deformation of ^{98}Sr and the one reported for ^{100}Sr. New measurements were carried out over several years at CERN-ISOLDE by Rb on-line mass separation. There, the very deformed $^{98-100}$Sr were investigated in detail, see e.g. [8], while several levels in ^{101}Sr [9] and the 2^+ state in ^{102}Sr [10] were identified for the first time. These measurements showed quadrupole deformations of $\beta = 0.4$, very rigid bands and several pairs of transitions of identical energies in neighbouring even and odd nuclei. The deformations of Sr isotopes had not been thought to be so large. The older lifetime measurements in order to get the B(E2, $2^+ \rightarrow 0^+$) values had been biased by the existence of at that time unknown high-lying high-K isomers in ^{98}Sr [11] and ^{100}Sr [12]. Laser spectroscopy by the Mainz Physics group later confirmed the large deformations [13, 14]. Evidence for such different structures as nearly magic character in ^{96}Zr and $\beta = 0.4$ axial deformation after addition of a neutron pair was even more challenging as ever.

2 Decay spectroscopy near A = 100 at IGISOL

A key question still remained the observation of deformed structures in other N = 59 isotopes and whether any would be observable in the N = 58 isotopes. The very first experiment performed in this context at IGISOL in the old laboratory turned out to give key information. Later on, a number of experiments has been carried out after IGISOL was moved to the nowadays accelerator laboratory. They, in particular, established the high-Z border of the closed shell region and explored the decays of refractory elements, further pushing spectroscopy to very neutron-rich nuclei towards the r-process path. All of them cannot be described here, but a selection will be presented.

2.1 Excited deformed levels in spherical N = 58 and 59 isotones

In our first experiment at IGISOL at the old Physics Department we searched for a short-lived isomer in ^{98}Y. This isomer had been proposed by the Braunschweig group on the basis of end-point energies of β spectra they had observed with a plastic telescope. In addition to γ-rays, β's and conversion electrons were recorded. The ELLI Electron Lens system provided high-quality spectra without interference of γ-rays. Evidence for the postulated isomer could not be seen in the spectra but new E0 transitions were found in ^{98}Zr [15]. The large $\rho^2 = 0.075$ for the 0_3^+(1436 keV) to 0_2^+ (854 keV, below the 2^+ at 1223 keV) is only possible if the connected states have very different values of their deformation parameters β^2. This suggested the existence of a strongly deformed 0^+ state at 1436 keV. A similar level with a large ρ^2 and very close in energy (1465 keV) had been earlier reported in ^{96}Sr at OSTIS. A convincing band built on it was later observed with GAMMASPHERE [7]. Although a clear band has not been found to be built on the ^{98}Zr level, it is tempting to believe both have the same nature.

In that same experiment data of the mass A = 99 were also recorded. It was attempted to find deformed levels in ^{99}Zr, isotone of ^{97}Sr, in which an enhanced E2 had been found shortly before. In spite of huge effort, and even later in renewed attempts with the new IGISOL and much better experimental conditions, the candidate levels remained rather speculative [16]. Unlike strontium, zirconium does not develop a clean band structure at same N. Moreover, the β-decay selection rules limit the spins of the levels observable in decay studies to rather low values. All together, in the absence of a high spin K-isomer, prompt-fission is a better tool for the identification of rotational bands. The existence of deformed levels was thus proved by series of analysis of prompt data at EUROGAM and EUROBALL, see e.g. [17]. We finally note that the long-seeked [404]9/2 Nilsson level was found in ^{97}Sr too, quite recently at GAMMASPHERE [18]. This orbital has a very important role in stabilizing the large prolate deformation once N reaches 60. Its strong up-sloping in the Nilsson diagram counteracts the deformation driving of the low-K down-sloping orbitals of $h_{11/2}$ origin which come to the Fermi surface.

Whereas reasonable evidence for excited deformed 0^+ states in even-even ^{96}Sr and ^{98}Zr was growing, hardly anything pointed to a deformed level in their intermediate isotone ^{97}Y. Such a level being presumably a $5/2^+$ state by analogy with the ground state of ^{99}Y, and being around 1.5 MeV, would be very difficult to identify as such. Neither β decay of ^{97}Sr nor isomeric decay of ^{97}Y (see next paragraph) led to such a level. The only candidate is a level at 1428 keV, basing the arguments for the assignment to the otherwise very close level correspondence in the isotones ^{97}Y and ^{99}Nb [19].

To summarize this section, there had been a hint about a well-deformed excited state in ^{96}Sr, a nucleus with 2 neutrons added to the closed N = 56 core, which remained an isolated case for a long time. An experiment at IGISOL added a similar level in ^{98}Zr. In addition, an enhanced E2 component was found in a (K+1$\rightarrow$K = $3/2^+$) transition in ^{97}Sr, a nucleus with one more neutron but still with a spherical ground state. These observations led to the picture of a transition of ground-state shape due to the exchange of the energies of the spherical and deformed potential minima. The ground-state shape transition in strontium isotopes is shown in Fig. 2. Some of the levels shown there have not been discussed so far and deserve a short comment. The 0_2^+ in ^{96}Sr is a probable spherical intruder due to promotion of proton pairs from $p_{1/2}$ to $g_{9/2}$, following a mechanism described by K. Heyde and J. Wood. The lowest-lying $3/2^+$ in ^{97}Sr is not the $d_{3/2}$ orbit but a high-seniority state found near the $s_{1/2}$ and $g_{7/2}$ single-particle states in other odd-A Sr and Zr isotopes with N = 57 and 59. The next $3/2^+$ at 585 keV is the [422] Nilsson level, on which there exist band structure, while the $3/2^-$ does not form the head of a clear band due to the large Coriolis mixing usual to low-K orbitals of high j origin (here $h_{11/2}$). In ^{98}Sr, a 2^+ state has been associated with the first excited 0^+ to form a band similar to the ^{96}Sr g.s. band. The analogy further suggests the alone standing 0^+ could be a particle-hole intruder like the one just mentioned in ^{96}Sr. Even if there are levels of perhaps suitable (but uncertain) spin allowing to extend the postulated bands, this interpretation obviously remains tentative. Note also the two-quasineutron band in ^{98}Sr discovered at ISOLDE, built on the [404]9/2$\otimes$[411]3/2 broken neutron pair.

2.2 Nuclei with few valence particles outside N = 56; ^{97}Y,^{97}Zr,^{99}Nb,^{99}Mo

Decay experiments at IGISOL were commonly based on singles and coincidences with 2 or 3 Ge-detectors and a thin plastic scintillator to trigger the acquisition on β particles. Two experiments used a modified TESSA array featuring twelve Compton-suppressed Ge-detectors. The array was brought from UK by the Liverpool group after closure of the Daresbury Laboratory. These experiments involved the JYFL gamma-spectroscopy group headed by R. Julin too. Owing to Compton suppression and higher efficiency it became possible to see transitions below one percent of the strongest one in almost each decay scheme.

The A = 97 experiment was still performed shortly after IGISOL had been moved to the Physics Department and before it had reached its optimal performance, which it did few years later. The main goal of the experiment was the study of decay of the high-spin isomer in ^{97}Y discovered at JOSEF. New levels in ^{97}Y were indeed found. The high-spin levels in ^{97}Y were later analysed in terms of the interacting Boson + Fermion + broken pair model developed in Zagreb [20]. They result from the odd $g_{9/2}$ proton being coupled with a broken neutron pair. The postulated $27/2^-$ spin

Fig. 2 Selected levels in the shape transition region observed by decay spectroscopy at IGISOL and elsewhere. The lowest spherical and deformed levels are connected by *dotted lines* to emphasize the exchange of ground-state shapes. Some levels are commented in the text

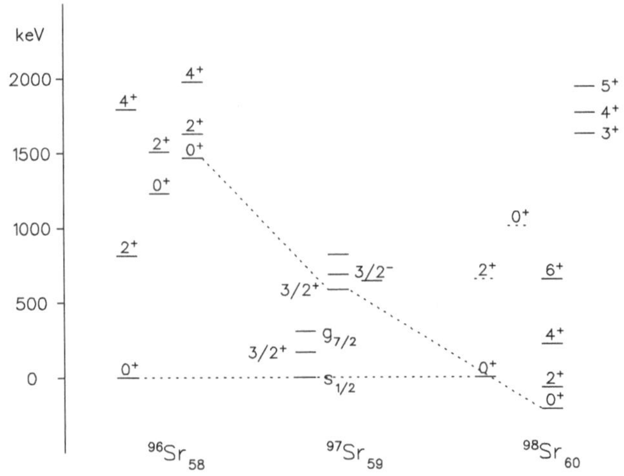

and parity of the isomer appeared likely in view of that analysis, the neutron pair being in the aligned $(d_{5/2} \otimes h_{11/2})_{8^-}$ configuration. The collinear laser spectroscopy at IGISOL performed quite recently did not contradict that interpretation [21]. But, perhaps even more exciting was the discovery of a weak beta branching from the high-spin isomer to ^{97}Zr. It made it possible to observe high-spin levels in that single-neutron nucleus, with the $s_{1/2}$, $g_{7/2}$ and $h_{11/2}$ states and some of their couplings to the core [22]. The high-spin scheme was later confirmed and extended by on-line spectroscopy [23].

The experiment on A = 99 was performed in excellent conditions regarding the IGISOL performance. By analogy with ^{97}Y it was searched for a high-spin isomer in ^{99}Nb (Z = 41). In spite of high statistics it could not be found. Possibly, other decay modes than the E3 isomeric transition are opening since the core states for Nb are lower than those for Y. The very high quality of the data nevertheless allowed major extensions of the schemes of isobars, including Zr, Nb and Mo. As already said, the data on ^{99}Zr did not reveal definite evidence for new bands, but those for Nb and Mo turned out to be very interesting. The level scheme of ^{99}Nb$_{58}$ has been considerably extended with respect to the older works at OSTIS [19]. A IBFBPM calculation including a proton and a broken neutron pair outside the N = 56 closure was carried out. The good agreement with experiment showed that the N = 56 shell gap is not destroyed by the proton outside Z = 40. This contrasts with its full eradication with three protons in Technetium. As discussed below, the N = 56 closure is still relevant for molybdenum, although weaker. We further note that levels in ^{100}Nb have been well reproduced in the spherical basis [24]. This establishes Z = 41 as the upper bound of the clearly defined closed-shell region.

The decay scheme of ^{99}Nb to the odd-neutron nucleus ^{99}Mo$_{57}$ has been considerably extended too. The most salient feature is the discontinuity of levels in the systematics of N = 57 isotones. Closer to stability, the N = 57 ground state is the $d_{5/2}$ neutron orbital, while $g_{7/2}$ is almost degenerated with $d_{5/2}$. Going away from stability, removing the $g_{9/2}$ protons, the interaction of protons and neutrons in overlapping orbitals decreases and moves the $d_{5/2}$ and $g_{7/2}$ neutron upwards with respect to $s_{1/2}$. The ground state of ^{99}Mo, still with two $g_{9/2}$ protons, is that $s_{1/2}$ neutron. When the

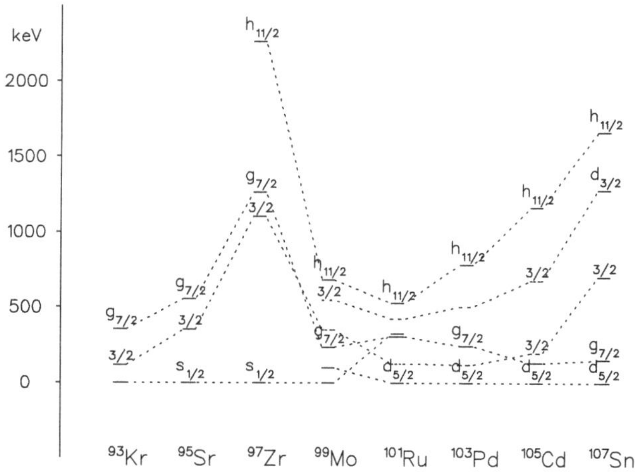

Fig. 3 Levels in N = 57 isotones observed by decay spectroscopy at IGISOL and elsewhere. The N = 56 gap closes at Molybdenum where the occupation of $g_{9/2}$ by a proton pair lowers the $g_{7/2}$ neutron. The $3/2^+$ state in the closed-shell region is not the $d_{3/2}$ shell but a high-seniority state. The connection to the $d^3_{5/2}$ state in ^{107}Sn, established by transfer reactions, is tentative

$g_{9/2}$ protons are fully removed, the vanishing interaction between spin-orbit partners creates the N = 56 closure. The most neutron-rich N = 57 isotones have a $1/2^+$ ground state, with an excited $7/2^+$. The first excited state, a $3/2^+$, does not fit in the picture. It has been reproduced as having a complex structure. The systematics of levels is shown in Fig. 3. The lower Z bound of the closed-shell region is not known. Yet, a decay study of ^{93}Br suggests that the ^{93}Kr scheme presents a good analogy with the scheme of the spherical ^{95}Sr [25]. It has therefore been included as such in Fig. 3. We also note that the recently identified 2^+ in ^{96}Kr$_{60}$ is remarkably low-lying at 241 keV [26] (compare with 213 keV in ^{100}Zr). This all is suggesting that an evolution like in strontium and zirconium from shell closure towards strongly deformed shape is probable.

The β-decay patterns of ^{99}Zr to ^{99}Nb and of ^{99}Nb to ^{99}Mo have been calculated with the Microscopic Quasiparticle Phonon Model (MQPM) developed by J. Suhonen at JYFL. The main features are well reproduced. An obvious application of the model is a prediction of the β-decay branches, and consequently of the decay half-lives and P_n-values (the integrated probability of β-delayed neutrons) of nuclei too far to be experimentally reached. This is the opportunity to say a few words about the astrophysical implications of the measurements.

2.3 A few words about the connection with the r-process

The mechanism for synthesis of elements has to be modeled so that it reproduces the measured abundances of nuclei. Neutron-rich nuclei heavier than iron are created by succession of neutron captures and β decays during exposure to the intense flux of fast neutrons following explosion of a type-II supernova. This is the so-called r-process. Apart from few exceptions, the so-called waiting point nuclei with a neutron closed shell, the nuclei involved are still far from those that presently can be studied in some detail. Measurements of gross properties ($T_{1/2}$, P_n) of nuclei at the border of the part of the nuclide chart within experimental reach can improve extrapolations to the r-process nuclei. In this context, half-lives and P_n-values have been measured at IGISOL for a number of nuclei of refractory nature and therefore not available as beams at ISOL systems and compared with global-model predictions [27, 28]. The change of ground state-spin in N = 57 isotones caused by the N = 56 shell closure far from stability is an example that illustrates the importance of more detailed calculations, such as e.g. the MQPM model. The decay half-life and β-delayed neutron emission depend on a number of factors having their origin in nuclear structure. Only to come back to the data presented here, the decay pattern of a $s_{1/2}$ neutron is different from the one of a $d_{5/2}$ neutron as a consequence of the rather strict β-decay selection rules, and the different excitation energies of the levels that can be reached in the daughter. Another example is the striking difference of decay patterns in spherical and deformed nuclei. In the daughter nucleus there can be two-quasiparticle states in the decay energy window with an internal structure making them easily populated, e.g. via a Gamow-Teller spin-flip transition. In spherical nuclei, this can lead to a strong single decay branch, in which case the β end-point energy is low and the half-life is long. In deformed nuclei, since K-states are a mixture of j-states there is a broad distribution of the pattern, resulting in large β-branches to low-lying states, and consequently a shorter half-life and a low P_n-value. It is therefore obvious that the shape transition at N = 60 in the Sr and Zr isotopes deserves dedicated treatment.

2.4 The challenging level scheme of ^{103}Zr

Mostly high-spin levels had been seen in prompt fission by Hotchkis et al. [1]. It was interesting to search for low-spin levels in ^{103}Zr using β decay. At ISOLDE this had required on-line separation of ^{103}Y, which is hardly coming out of the target, or alternatively of a more exotic parent (Rb, Sr) of too short half-life or/and too low cross-section. This decay was thus observed at IGISOL for the first time [29]. The experimental conditions actually reached the limit of observation. The few identified transitions turned out to fit well in a scheme interpolated between those of ^{101}Sr [9] and ^{105}Mo [30]. It suggests that deformation, after an increase smoother than in strontium, reaches a maximum at N = 63 in Sr, Zr and molybdenum. Since then, the N = 63 levels have been revisited in prompt fission [31].

3 Intruder states near the N = 66 neutron midshell

The N = 66 neutron midshell was only reachable for higher-Z elements. The odd-Z rhodium (Z = 45) and silver (Z = 47) isotopes had been systematically studied at the TRIGA reactor at the Institute for Nuclear Chemistry in Mainz, by decay after on-line chemical separation by N. Kaffrell, J. Rogowski and their collaborators. The even Z palladium (Z = 46) isotopes had been studied in the early days of IGISOL by Äystö et al., while Cd (Z = 48) isotopes have been extensively studied at JYFL via on-line γ-spectroscopy by R. Julin and his group. These works established coexisting structures too. The evolution of band-head energies versus neutron number shows the characteristic V-shape with a sharp turn rather than a smooth variation near the minimum. In cadmium the minimum energy is, like expected, at neutron midshell. In contrast, the K = 1/2 band in rhodium isotopes is the lowest in ^{109}Rh (with the $3/2^+$ level at 226 keV being lower than $1/2^+$), and the excited 0^+ state, assigned as the intruder in Pd isotopes is in ^{110}Pd (947 keV). Remarkably, both nuclei are N = 64 isotones. At midshell, the upward going trend postulated by the Mainz group in ^{111}Rh was confirmed [32], with 395 keV for the 3/2$^+$ level of the K = 1/2 band. The position of the 0$^+$ intruder in ^{112}Pd has been revised higher to 1126 keV [33], which is surprisingly close to the two-phonon 0$^+$ state at 1140 keV. Thus, there is a shift of intruder energies with respect to the $N_\pi \otimes N_\nu$ scheme of the IBM-2. Perhaps the shift would even evolve to lower N-values at lower Z to connect with the equal deformations in the ground states of $^{98}Sr_{60}$ and $^{100}Sr_{62}$. The studies at IGISOL continued with the N = 68 isotones ^{113}Rh [34] and ^{114}Pd [35], and even further, see contributions by J. Kurpeta (Rh) and Y.B. Wang (even-even nuclei).

4 Summary

The data presented here display the richness and challenging aspects of nuclear structure in the region of neutron-rich nuclei near mass 100–110. It is necessarily a partial review, since a lot of complementary data have been collected elsewhere, e.g. at ISOL facilities and by on-line spectroscopy of fission fragments with large Ge-arrays. Data presented here span an about 10 years period, starting with IGISOL at the old Physics Department and ending in year 2000. They necessarily are a selection reflecting personal taste of the authors. Surely, the coexistence in the A $\simeq$ 100 region of a non-standard shell closure and of some of the largest deformations known at low spin and low energy is one of the highlights of nuclear structure. Numerous works have been performed later on by others, including important ones at IGISOL reported elsewhere in this book. The author thus apologizes for the truncated reference list. This contribution is the opportunity to acknowledge the merits of IGISOL as a unique instrument for spectroscopy of exotic nuclei.

Acknowledgements The results presented here had not been obtained without the reliability of the K-130 cyclotron beam, the skillful running of IGISOL by its team and the gracious help by some other JYFL physicists, not always formally members of the IGISOL group. The list is too long to be written here without a risk to forget someone having contributed to this adventure. We thank also the support offered by the German DAAD and the Finnish Academy.
One of the authors (G.L.) dedicates this report to the memory of his friend Slobodan Brant who passed away unexpectedly much too early and whose contribution to the study of nuclei in the A = 100 region and elsewhere has so much helped us in the understanding of nuclear structure.

References

1. Hotchkis, M.A.C., et al.: Nucl. Phys. A **530**, 111 (1991)
2. Brant, S., Lhersonneau, G., Sistemich, K.: Phys. Rev. C **69**, 034327 (2005)
3. Meyer, R.A., et al.: Nucl. Phys. A **439**, 510 (1985)
4. Mach, H., et al.: Phys. Rev. C **41**, 1141 (1990)
5. Lhersonneau, G., et al.: Z. Phys. A **323**, 59 (1986)
6. Lhersonneau, G., et al.: Z. Phys. A **337**, 149 (1990)
7. Hamilton, J.H., et al.: Progr. Part. Nucl. Phys. **35**, 635 (1995)
8. Lhersonneau, G., et al.: Phys. Rev. C **63**, 054302 (2001)
9. Lhersonneau, G., et al.: Z. Phys. A **352**, 293 (1995)
10. Lhersonneau, G., et al.: Z. Phys. A **351**, 357 (1995)
11. Lhersonneau, G., et al.: Phys. Rev. C **65**, 024318 (2002)
12. Pfeiffer, B., et al.: Z. Phys. A **353**, 1 (1995)
13. Buchinger, F., et al.: Phys. Rev. C **41**, 2883 (1990)
14. Buchinger, F., et al.: Phys. Rev. C **42**, 2754 (1990) (erratum)
15. Lhersonneau, G., et al.: Phys. Rev. C **49**, 1379 (1994)

16. Lhersonneau, G., et al.: Phys. Rev. C **56**, 2445 (1997)
17. Urban, W., et al.: Eur. Phys. J. A **16**, 11 (2003)
18. Hwang, J. K., et al.: Phys. Rev. C **67**, 054304 (2003)
19. Lhersonneau, G., et al.: Phys. Rev. C **57**, 2974 (1998)
20. Lhersonneau, G., et al.: Phys. Rev. C **57**, 681 (1998)
21. Campbell, P.: Hyperfine Interact. **171**, 143 (2006)
22. Lhersonneau, G., et al.: Phys. Rev. C **54**, 1117 (1996)
23. Matejska-Minda, M., et al.: Phys. Rev. C **80**, 017302 (2009)
24. Lhersonneau, G., Brant, S., Paar, V.: Phys. Rev. C **62**, 044304 (2000)
25. Lhersonneau, G., et al.: Phys. Rev. C **63**, 034316 (2001)
26. Marginean, N., et al.: Phys. Rev. C **80**, 021301 (2009)
27. Mehren, T., et al.: Phys. Rev. Lett. **77**, 458 (1996)
28. Wang, J.C., et al.: Phys. Lett. **454B**, 1 (1999)
29. Lhersonneau, G., et al.: Phys. Rev. C **54**, 1592 (1996)
30. Liang, M., et al.: Z. Phys. A **351**, 13 (1995)
31. Urban, W., et al.: Phys. Rev. C **79**, 067301 (2009)
32. Lhersonneau, G., et al.: Eur. Phys. J. A **1**, 285 (1998)
33. Lhersonneau, G., et al.: Phys. Rev. C **60**, 014315 (1999)
34. Kurpeta, J., et al.: Eur. Phys. J. A **33**, 307 (2007)
35. Lhersonneau, G., et al.: Phys. Rev. C **67**, 024303 (2003)

DOI 10.1007/978-94-007-5555-0_12
Reprinted from *Hyperfine Interactions* Volume 223, Issues 1-3, 2014
and *The European Physical Journal A* Volume 48, Issue 4, 2012

Fast life-time measurements on fission products

Henryk Mach · Luis Mario Fraile for the IGISOL Fast Timing Collaboration

Published online: 4 May 2012

Abstract Ultra Fast Timing has been applied at the IGISOL facility since more than a decade ago with the aim to systematically study nano- and subnanosecond lifetimes in the neutron-rich nuclei from the A∼110 region. Over this period two generations of crystals and photomultipliers have been introduced, which allowed to study more complex level schemes populated in β decay. The IGISOL facility provides unique capabilities to study the A∼110 region not matched elsewhere in the world.

Keywords Ultra Fast Timing · β decay · Level half-lives · Transition matrix elements · Fast scintillators · BaF_2 · $LaBr_3(Ce)$

1 Introduction

Absolute electromagnetic transition probabilities between nuclear states provide a unique insight into nuclear structure and allow critical comparison of experimental data to model predictions. Electromagnetic γ radiation is emitted by an atomic nucleus as a result of the rearrangement of protons and neutrons within the nuclear orbitals. The interaction of the electromagnetic field with the nucleons can be expanded into an infinite series of electric and magnetic terms, out of which only the lowest moments (dipole, quadrupole and hexadecapole) provide significant contributions. In most cases the electromagnetic radiation has mixed contribution from only the two lowest multipoles.

H. Mach (✉)
Medical and Scientific Time Imaging Consulting, Fruängsgatan 56 E, 611 30 Nyköping, Sweden
e-mail: henryk.mach@masticon.se

H. Mach · L. M. Fraile
Grupo de Física Nuclear, Facultad de CC. Físicas, Universidad Complutense, 28040 Madrid, Spain

L. M. Fraile
e-mail: fraile@nuc2.fis.ucm.es

The direct measurement of lifetimes of excited nuclear states provides a model-independent method of determining electromagnetic transition rates and is therefore of the outmost importance. Lifetimes of nuclear levels span many orders of magnitude, from femtoseconds to days, while the energy of emitted γ-rays may be as low as a few keV or as high as several MeV. The mathematical formalism of the emitted electromagnetic radiation caused by the oscillating electric and magnetic moments is well understood. Exact formulas describe the reduced transition probability in terms of a partial mean-life of a level and account for the large energy factor. The total transition rate for a electric or magnetic ($X = E$ or M) transition of multipolarity λ from a nuclear state J_i to a final nuclear state J_f is given by [1]:

$$\tau_\gamma^{-1} = P_\gamma \left(X\lambda; J_i \to J_f\right) = \frac{8\pi\,(\lambda+1)}{\lambda\,[(2\lambda+1)!!]^2\,\hbar} \left(\frac{E_\gamma}{\hbar c}\right)^{2\lambda+1} B\left(X\lambda; J_i \to J_f\right), \quad (1)$$

where $B(X\lambda)$ is the *reduced transition probability* defined as

$$B\left(X\lambda; J_i \to J_f\right) = (2J_i+1)^{-1} \,|<\psi_f||M\,(X\lambda)\,||\psi_i>|^2, \quad (2)$$

and obtained by averaging over the initial magnetic substates M_i and summing over the final state substates M_f and the third component of the electromagnetic multipole λ.

Therefore for a defined magnetic or electric operator of a given multipolarity λ

$$B\left(X\lambda; J_i \to J_f\right) = \frac{\lambda\,[(2\lambda+1)!!]^2\,\hbar}{8\pi\,(\lambda+1)} \left(\frac{\hbar c}{E_\gamma}\right)^{2\lambda+1} P_\gamma\left(X\lambda; J_i \to J_f\right), \quad (3)$$

and the measurement of the partial mean life—which is the inverse of the rate P_γ—provides direct and model independent access to the reduced transition probability. In particular for $E2$ transitions (E_γ in MeV and τ_γ in s)

$$B\left(E_2; I_i \to I_f\right) = 8.161 \times 10^{-10} \cdot E_\gamma{}^{-5} \cdot \tau_\gamma^{-1}\, e^2\, fm^4. \quad (4)$$

Reduced transition probabilities allow for comparison of rates for different transitions of the same type within the same nucleus. Single Particle Weisskopf estimates provide a convenient—albeit model-dependent—benchmark to intercompare transition rates between different nuclei. For instance, for $E2$ transitions

$$B(E2)_W = 0.0594 \cdot A^{4/3}\, e^2\, fm^4. \quad (5)$$

When expressing transition rates in Weisskopf units one obtains an indication of enhancement of a transition rate caused by collective nuclear effects, for example due to the quadrupole or octupole degrees of freedom.

2 Ultra Fast Timing method

At the IGISOL facility we have measured lifetimes of levels in neutron-rich nuclei populated in β decay. These lifetimes are typically in the fs to μs range. Lifetimes down to a few ns are generally measured by time-delay methods using Ge detectors. Shorter lifetimes in the ns and sub-ns range were measured by the Ultra Fast Timing Method developed by Mach, Moszynski and Gill [2–4]. The method involves three

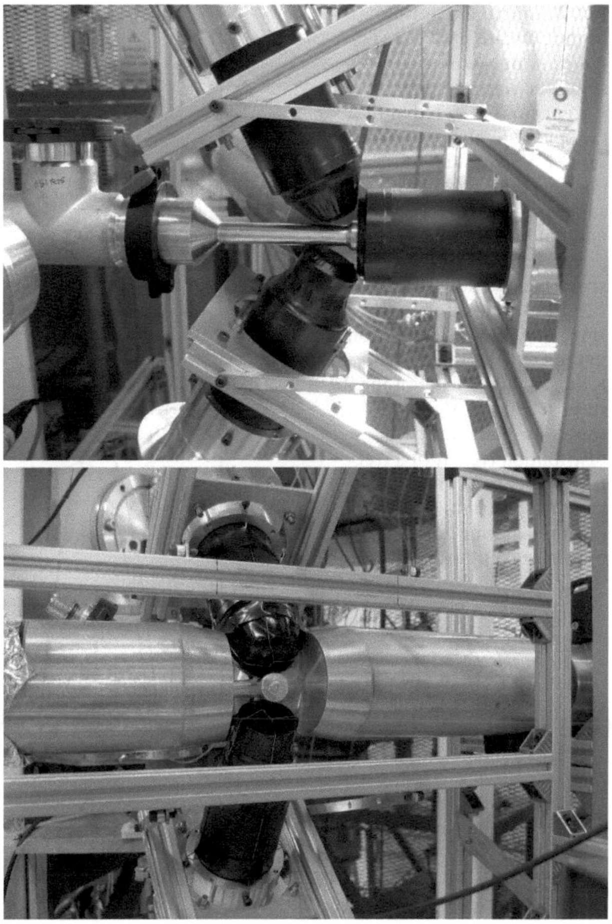

Fig. 1 *Top figure*: A close geometry arrangement of five detectors around the beam deposition spot at the IGISOL facility. The beam comes from the *left*. The β detector is positioned just behind the source separated from it by a thin Al foil which ends the beam pipe and seals the vacuum inside. Two fast timing γ detectors are positioned *above* and *below*, respectively. Two Ge detectors (one removed for the picture) are positioned horizontally. *Bottom figure*: Similar arrangement of detectors during another timing experiment at IGISOL using two $LaBr_3(Ce)$ γ detectors. For this picture the β detector was removed

detectors in triple coincidence. Each detector is of different type and provides a unique contribution. Fast timing information is obtained from coincidences between two fast scintillators, one for β rays and one for γ rays, while a Ge detector allows for a precise selection of the decay path. In order to increase efficiency we use two Ge and two fast γ detectors, as shown in Fig. 1. A thin β detector gives time response largely independent on the energies of the detected β-rays. The residual time dependence on the β-ray energy is determined and corrected for in the off-line analysis.

The Ultra Fast Timing method is capable of measuring lifetimes down to about 60 ps by the slope method and in the low-ps region via the centroid shift method. The latter involves precise calibration procedures that include determination of time response for prompt events where the full energy of γ-rays has been absorbed in the crystal (Full-Energy-Peak curve) and a second curve due to Compton events where part of the energy escaped from the crystal in the form of a Compton-scattered photon [4]. For very precision demanding experiments, a third correction is applied to account for the time dependence of Compton events at a given

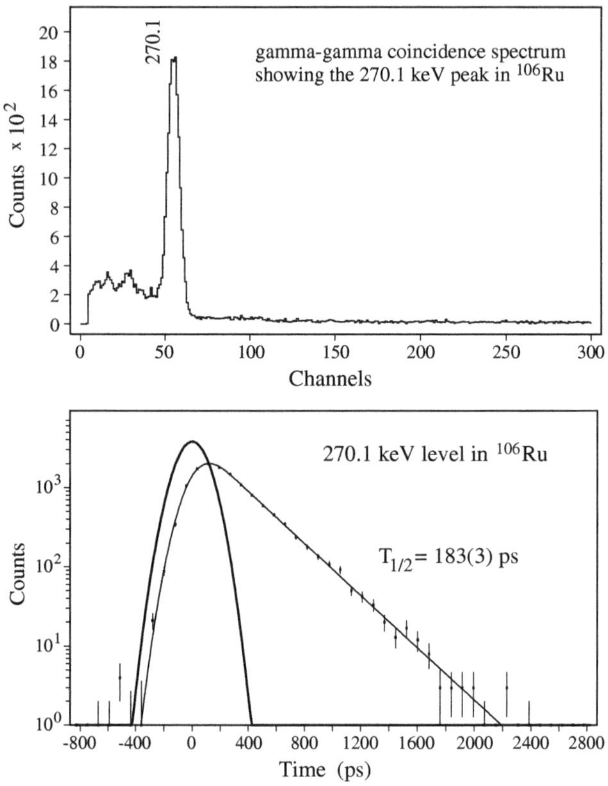

Fig. 2 Lifetime determination of the 2_1^+ state in ^{106}Ru. *Top panel*: A BaF_2 coincidence sum spectrum gated by several γ transitions selected in the Ge spectrum above the energy of 1 MeV. Note the full energy peak at 270.1 keV superimposed on a very low and flat background. *Bottom panel*: Time delayed (β-BaF_2) spectrum gated by the 270.1 keV transition in the BaF_2 and selected transitions in the Ge detector. The slope is due to the half-life of the 2_1^+ state at 270.1 keV. A *thin line* shows the Gaussian distribution (result of a free fit) which approximates the prompt spectrum. It is centered at time zero. This figure is taken from [5]

energy originating from primary photons of different energy; see discussion of the "Compton correction" in [4].

2.1 Choice of crystals

A thin plastic scintillator NE111A is used for the β detector. One can select different scintillator thickness depending on the requirement of the experiment. A standard value is 3 mm since it provides a compromise between the requirement of a strong signal for good timing response and the need for the scintillator to be as thin and small as possible in order to minimize the γ sensitivity. This sensitivity can be measured. For γ-rays of energy ~60 keV, the ratio of events recorded in a reversed arrangement, namely γ-rays recorded in NE111A scintillator and β-rays recorded on the γ-ray scintillator, is about 2% of those recorded in the 'normal' arrangement. The sensitivity of the β detector to γ-rays of higher energies is much smaller. The time resolution of the β detector is about 60–70 ps FWHM.

For the first series of experiments at IGISOL we have used a BaF_2 crystal as fast scintillator to detect γ-rays, see Fig. 2. It is still the fastest commercially produced crystal with the time resolution of ~80 ps FWHM for small, ~1–2 cm^3, crystals and up to 155 ps FWHM for large crystals of our Studvik design (see below). Pioneering Ultra Fast Timing experiments at the TRISTAN separator were carried out using very small truncated cone BaF_2 crystals with diameter of 25 mm at the base and

19 mm at the top, and height of 13 mm. The first round of measurements at the OSIRIS, ISOLDE and IGISOL separators were done using similar truncated cone crystals, but with height of 25 mm. In some later measurements where the efficiency of detection of γ-rays in the 1–3 MeV range was critical we have used semi-conical BaF_2 crystals of one of the two "Studsvik design" versions. Their diameter at the base was 40 mm followed by a cylindrical shape for the first 7 or 17 mm (depending on the version) and then for the next 30 mm a conical shape with the diameter of 20 mm on the top. Longer crystals were not considered due to the deterioration of the crystal quality (self-absorption, internal scattering of light and internal activity) that comes with larger crystal size.

We have carefully tested about 20 large BaF_2 crystals of the Studsvik design. Their quality strongly vary even for those coming from the same production batch. The energy resolution for the 661.7 keV line is 9.0% for the best crystals and about 10.0–10.5% for most of them. In two exceptional cases the energy resolution is as bad as 12.9%. For the experiments at IGISOL we have selected the best crystals.

In more recent experiments at IGISOL we have used a second generation of fast crystals, $LaBr_3(Ce)$. These crystals are characterized by much better energy resolution, $\sim$3% at 661 keV, but about 10–15% worse time resolution in comparison to BaF_2 crystals of the same size or shape. We note here that time resolution of $LaBr_3(Ce)$ crystals depends on Ce doping [6]. Standard crystals currently available have 5% doping. The energy resolution is of premium for complex decay schemes where many transitions have similar energies. Such is the case in the odd-nuclei in the A$\sim$110 mass region. Moreover, better energy resolution means higher peak-to-Compton background ratio and a smaller correction due to the Compton part below the full-energy-peak, see Figs. 4 and 5.

In the future we will replace the cylindrical shape crystals with crystals of conical shape which recently became available. In general conical shape gives about 15–25% improvement in time resolution when compared to the cylindrical shape. However, the improvement depends on the details of the shape. Similarly to the case of BaF_2 crystals, $LaBr_3(Ce)$ crystals of a given shape may show about 10% variation in the energy resolution and time resolution for batches produced at different times.

Crystals are shielded as much as possible, but not completely due to close geometry, against Compton-scattered γ-rays from surrounding detectors, mainly from the Ge detectors. As shielding we mainly use 2–4 mm Pb sheets occasionally aided by 1 mm copper sheet next to the crystal.

2.2 Choice of photomultipliers

The BaF_2 crystals were attached to the quartz-window XP2020URQ photomultiplier tubes (PMTs) using a standard silicon grease. The voltage divider was model S563/04. As all ultrafast phototubes they were assembled by hand and produced in small batches. Since their quality varied strongly, we have used phototubes selected by the manufacturer for the highest sensitivity and lowest noise. These 12-stage tubes had too high gain for the BaF_2 crystals, thus they were operated at voltages of 2100–2400 Volts with the nominal gain of $\sim 5 \times 10^5$. These voltages are well below the design value of 3000 V (and the expected gain of 6×10^6) for the best time response. Before 2005 ultrafast 50 mm PMT's with lower number of dynodes were not available.

$LaBr_3(Ce)$ crystals yield about 74,000 photons/MeV (see for instance [6]) in comparison to about 2 500 photons/MeV for BaF_2 [3]. The phototube designed by Photonis to match $LaBr_3(Ce)$ is XP20D0, with 8 dynodes, and the base 184K/T. However, the front end did not follow the superior design used in the XP2020UR PMT, but instead used a simplified one. To optimize time resolution, these tubes are equipped with double anode. The XP20D0 has the rise time of 1.6 ns and the Transit Time Spread (TTS) of 520 ps FWHM (based on Time jitter, $\sigma = 220$ ps [7]), which are worse parameters than 1.4 ns and 350 ps, respectively, for the XP2020UR tubes.

We operate the XP20D0 tubes at a maximum reasonable gain at about 950 V with the expected amplification of $1–2 \times 10^5$. Due to short duration of the light flash in the crystal and high photon yield per MeV, there is a space-charge effect for electrons moving along the path in the phototube. The higher the charge the larger fraction of electrons will be missed at the last dynode causing nonlinearity. At this voltage we accept about 15% nonlinearity in energy.

Photonis also started the development of 6-stage phototubes XP20E0, which had expected better characteristics than the XP20D0 PMT. However, in 2009 Photonis stopped all production of phototubes.

The Hamamatsu R9779, with the PMT and base assembly model H10570 MOD, is the Hamamatsu 50 mm PMT designed for $LaBr_3(Ce)$. This 8-stage tube has the timing characteristics similar if not slightly better than XP20D0 with the anode pulse rise time of 1.8 ns and the TTS of 250 ps.

3 Advantages of the IGISOL separator

The IGISOL separator offers excellent beam quality for fast timing studies. The beam implantation spot is very small with the area of only a few mm^2. This allows placement of a few detectors in close geometry. The best time response is obtained for a fast γ detector when γ rays from the source illuminate the crystal along its axis of symmetry. Moreover, truncated cone crystals with sharply inclined walls can be used in γ detectors. Typical sequences of γ-rays populated in β decay are very short with about 2–4 cascading γ-rays in most cases.

Narrow beam allows using a narrow lead collimator about 15 cm upstream the deposition spot, which in such geometry is highly effective. The beam is practically a DC beam, which provides a steady counting rate in the detectors and allows for a steady operation of the photomultiplier tubes. With high beam fluctuations, like encountered at the ISOLDE facility at CERN, the instantaneous strong fluctuations in the beam intensity causes deterioration in the energy and time resolutions of the PMT-based detectors.

The IGISOL separator provides full isobaric separation. The drifts of the beam position and contamination from neighboring masses were observed only in exceptional cases on two occasions and they lasted over a very short period of time. The possibility of contamination is monitored on-line and since we collect data in list mode, the contaminated section of a data string can be identified and eliminated during the off-line analysis.

Finally the yields of isotopes in the mass range of interest fit very well the intensity desired for timing studies. Although measurements can be done with lower beam intensities, the saturated source strength at IGISOL was typically between 15 to 70

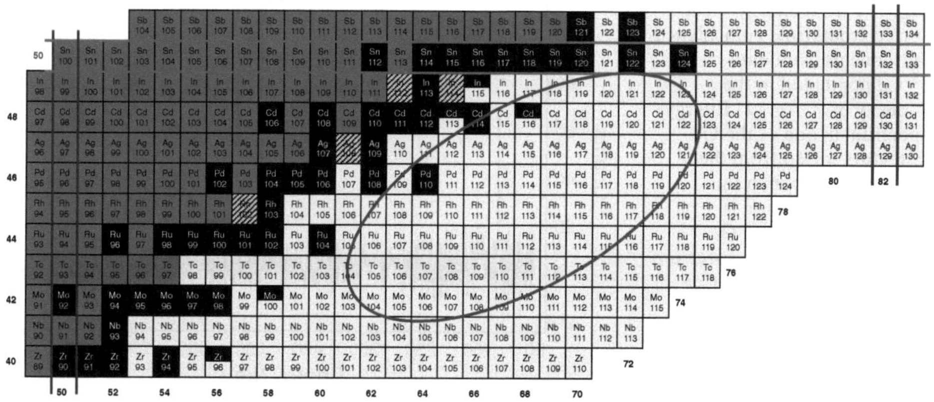

Fig. 3 The region of interest for fast timing studies with mass around A∼110

kcps during the timing studies. The fast timing detectors cannot take higher count rate than about 10 kcps for γ detector and 25 kcps for β detector. Since the β detector has 40% efficiency the maximum source strength can be ∼62 kcps. With higher rates the quality of signal response of timing detectors deteriorates significantly.

4 The physics case

At IGISOL, our region of interest for the fast timing studies in the subnanosecond range is the mass region of $100 \leq A \leq 120$, as shown in Fig. 3. There are a few reasons why this region is of special interest. Firstly, it is within a wider region marked by three doubly magic regions of ^{78}Ni, ^{100}Sn and ^{132}Sn which are all in part amenable to investigation in, although ^{78}Ni and ^{100}Sn themselves are on the edge of what can be studied at present. There is a strong experimental effort going on in this region at various facilities and using many different and complementary techniques.

The region of neutron-rich nuclei at A∼100–120 is located just below the first r-process abundance peak at A∼130. There is a strong effort to theoretically reproduce the exact location of this maximum and the shape of the abundances distribution. Of particular interest is the modification of the shell structure when moving away from stability. The r-process path comes close to the known nuclei in the $A = 120$ region. Nuclear structure properties—including gross properties such as masses, ground state half lives and β-n branches, but also level energies, spin-parity assignments and transition rates—are of the greatest importance to determine the role of these nuclides on the r-process path.

We have studied level lifetimes in the nuclei in the even mass chains from 106 to 114 using a small BaF_2 crystal for γ detection, and the even mass chains 108 to 122 using two small fast γ detectors with $LaBr_3$(Ce) and BaF_2 crystals, respectively. In these measurements only triple coincident $\beta\gamma\gamma$(t) events, involving β-BaF_2-Ge, β-$LaBr_3$-Ge or β-Ge-Ge detectors were collected.

The odd mass chains from 105 to 111 were studied using two $LaBr_3$(Ce) detectors. In these measurements, besides triple coincidences, other events involving an array

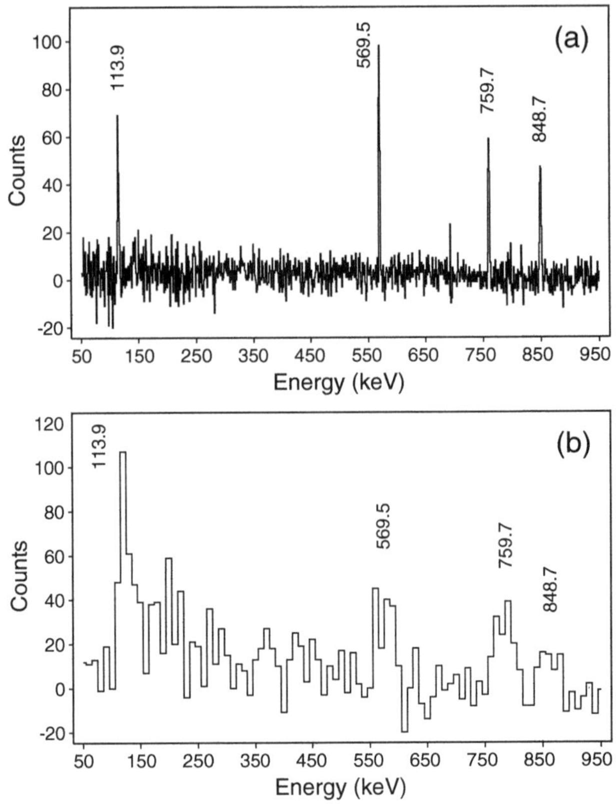

Fig. 4 Partial energy spectra sorted from triple coincidence $\beta\gamma\gamma$ events from data on ^{122}Cd populated in the β-decay of ^{122}Ag. (**a**) Energy spectrum in one of the Ge detectors sorted from the β-Ge-Ge data when the 324.7-keV transition is selected in the other Ge detector. (**b**) Energy spectrum recorded in the $LaBr_3$(Ce) detector sorted from the β-Ge-$LaBr_3$ data when the same 324.7-keV transition was selected in Ge detector. Energy spectra shown in the **a** and **b** panels are equivalent and illustrate the difference in the energy resolutions of the Ge and $LaBr_3$(Ce) detectors. This figure is taken from [8]

of five detectors: β, two Ge and two $LaBr_3$(Ce) detectors placed in a close geometry were also collected, see Fig. 1. The energy resolution provided by $LaBr_3$(Ce) crystals was vital in the data analysis of the odd-A nuclei were many transitions are very close in energy as shown in Figs. 4 and 5.

All measurements were made with the radioactive source in saturation, which means the beam was continuously deposited. This was done to maximize the count rate. All nuclei populated in a given mass chain were of interest since in our run their lifetimes in the subnanosecond range were measured for the first time. Without a tape transport system we have a higher solid angle efficiency for the detector array. Moreover, when a tape system is used to suppress the longer-lived activities, there are also losses due to the tape movement and the process of build up and decay of the source. However, in order to study the most exotic nuclei in this region we plan to implement a tape transport system in the future.

Studies at IGISOL are complementary to those performed by other techniques at other facilities, including the present measurements at FRS at GSI, and the future ones at FAIR. Although at FAIR one will be able to study nuclei very far from stability, yet the size of the beam spot for stopped beams at DESPEC of 10 × 10 cm imply that it will be very challenging to achieve very high precision for subnanosecond lifetime measurements, since the geometrical corrections for the time-of-flight of γ rays from the source to the detectors will vary strongly, and need

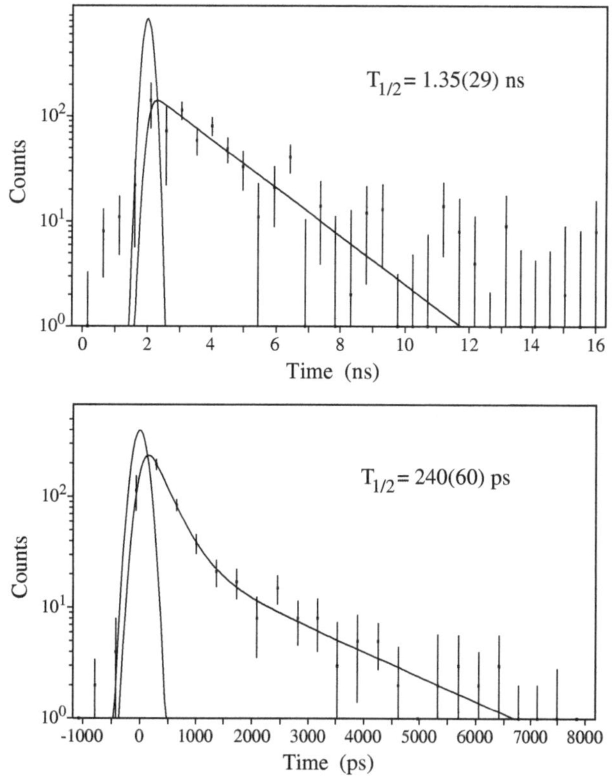

Fig. 5 Time-delayed spectra due to the lifetimes of the 2616.6-keV (*top panel*) and the 2502.7-keV levels (*bottom panel*), from [8]. The first spectrum is due to only one lifetime component, while the second one has two components: one due to the 1.35 ns lifetime from indirect feeding from the 2616.6-keV level and one due to the 2502.7 keV level alone, 0.24 ns, that is directly β fed, see [8] for details. Figures 3 and 4 show that higher resolution and better peak-to-background for the γ peaks in $LaBr_3(Ce)$ allows to resolve complex cases involving multiple lifetimes and weaker transition intensities

to be taken into account. Note that a difference of 1 cm in flight path results in a difference of 33 ps in time.

5 Outlook for the future

The new IGISOL facility will have higher yields of exotic nuclei. We intend to study nuclei further away from stability using a tape transport system. Furthermore, the IGISOL separator can be coupled to the JYFLTRAP Penning trap in order to select specific ions out of the the isobaric beam coming form IGISOL [9]. This allows to perform trap-assisted studies and opens up vast possibilities for fast-timing experiments where the tape transport system is used in combination with the trap, which provides isobar and even isomer purification. Moreover, with this approach the trap will allow to study lifetimes of levels populated from the μs isomers.

There is a steady development of new scintillation crystals with higher energy resolution and higher time resolution, and read-out systems with faster response. This development is driven mainly by the needs of Time-Of-Flight Positron-Emission-Tomography (TOF-PET) applications in medical imaging. One expects that within the next few years there will be a new generation of fast detectors for Ultra Fast Timing.

Over the last few years many different groups have started fast timing measurements, bringing new ideas and calibration methods, and further developing fast timing methodology [10].

Acknowledgements The authors would like to acknowledge support and cooperation of the IGISOL Fast Timing Collaboration, excellent operation of the IGISOL facility by the IGISOL group, hospitality of the Jyväskylä Laboratory staff during experiments at IGISOL and EU support for members of the Fast Timing Collaboration during measurements. The authors acknowledge support by the Autonomous Region of Madrid under project ARTEMIS S2009/DPI-1802 and by the Spanish Ministry of Science and Innovation (MICINN) under projects FPA2010-17142 and Consolider-Ingenio2010 CSD2007-00042.

References

1. Bohr, A., Mottelson, B.R.: Nuclear Structure, pp. 379–394. World Scientific, Singapore (1999)
2. Mach, H., Gill, R.L., Moszynski, M.: Nucl. Instrum. Methods A **280**, 49 (1989)
3. Moszynski, M., Mach, H.: Nucl. Instrum. Methods A **277**, 407 (1989)
4. Mach, H., et al.: Nucl. Phys. A **253**, 197 (1991)
5. Sanchez-Vega, M., et al.: Eur. Phys. J. A **35**, 159 (2008)
6. Glodo, J., et al.: IEEE Trans. Nucl. Sci. **52**, 1805 (2005)
7. Moszynski, M., et al.: Nucl. Instrum. Methods A **567**, 31 (2006)
8. Smith, D.L., et al.: Phys. Rev. C **77**, 014309 (2008)
9. Kurpeta, J., et al.: Eur. Phys. J. A **31**, 263 (2007)
10. Régis, J.-M., et al.: Nucl. Instrum. Methods A **622**, 83 (2010)

DOI 10.1007/978-94-007-5555-0_13
Reprinted from *Hyperfine Interactions* Volume 223, Issues 1-3, 2014
and *The European Physical Journal A* Volume 48, Issue 4, 2012

High-precision half-life measurements of atomic nuclei

**B. Blank · A. Bacquias · G. Canchel · J. Giovinazzo ·
T. Kurtukian Nieto · I. Matea · J. Souin**

Published online: 11 April 2012

Abstract In a series of measurements performed at the IGISOL facility at the Accelerator Laboratory of the University of Jyväskylä, the half-lives of ^{26}Si, ^{29}P, ^{30,31}S, ^{42}Ti and ^{62}Ga were measured with high precision. These half-lives are important ingredients for tests of the weak-interaction standard model and allow, together with other experimental inputs and theoretical correction terms, the determination of the vector coupling constant of the weak interaction. The IGISOL method used in Jyväkylä is an ideal tool for these measurements as it allows to access all elements without distinction of their chemical properties.

Keywords High-precision half-life measurements · Weak interaction · Conserved vector current

1 Introduction

The atomic nucleus is a fantastic laboratory for the study of fundamental interactions. Three of the four fundamental interactions have to be taken into account when describing the properties of the atomic nucleus. Only the action of gravity can be neglected. In particular, the radioactive decay of instable nuclei allows us to perform high-precision studies of the fundamental interactions, be it the strong or the weak interaction.

B. Blank (✉) · A. Bacquias · G. Canchel · J. Giovinazzo · T. Kurtukian Nieto · I. Matea · J. Souin
Centre d'Études Nucléaires de Bordeaux Gradignan - Université Bordeaux 1 - UMR 5797 CNRS/IN2P3, Chemin du Solarium, BP 120, 33175 Gradignan Cedex, France
e-mail: blank@cenbg.in2p3.fr

Present Address:
I. Matea
Institut de Physique Nucléaire, Université Paris-Sud, IN2P3-CNRS, 91406 Orsay Cedex, France

The present article focuses on studies performed in the frame work of the weak-interaction standard model where the super-allowed β decay of nuclei can be used to determine the reduced transition strength ft which depends on the decay heat of the radioactive decay, the so-called Q value, on the branching ratio of a particular decay branch and the total β-decay half-life of the nucleus. After corrections which take into account that, in β^+ decay, the decay of a proton does not take place in an isolated environment, but in a nucleus where the strong and the electromagnetic interactions are also acting, a universal $\mathcal{F}$t value can be determined which depends only on the isospin of the decaying nucleus:

$$\mathcal{F}t = ft \times (1 - \delta_C + \delta_{\mathrm{NS}}) \times (1 + \delta'_R) = \frac{K}{g_V^2 \times \langle M_F \rangle^2 \times (1 + \Delta_R)} \tag{1}$$

In this equation, δ_C, δ_{NS}, δ'_R and Δ_R are the correction factors which are necessary due to isospin impurities of the nuclear states, nuclear structure differences impacting on radiative corrections as well as nucleus dependent and nucleus independent radiative corrections. K is a constant. The Fermi matrix element is $M_F = \sqrt{T(T+1) - T_{zi}T_{zf}}$ and g_V is the vector coupling constant. Details about this formula as well as the corrections can be found in references [1, 2].

The observed constancy [3] of these corrected $\mathcal{F}t$ values is a proof that, at least at the level of precision reached today, the vector current is conserved and does not need to be renormalized in the nuclear medium. Once this is established, the average $\mathcal{F}t$ value can be used to determined the vector coupling constant g_V^2 and ultimately the up-down quark mixing matrix element of the Cabibbo–Kobayashi–Maskawa matrix:

$$V_{\mathrm{ud}}^2 = \frac{g_V^2}{g_V^{\mu\,2}} = \frac{K}{2 g_V^{\mu\,2}(1 + \Delta_R)\mathcal{F}t} \tag{2}$$

g_V^{μ} is the weak vector coupling constant for the purely leptonic μ decay.

A recent review of super-allowed $0^+ \rightarrow 0^+$ Fermi transitions reported such measurements in 20 nuclei with $(T; T_z) = (1; -1, 0)$ [3]. Thirteen nuclei have a precision close to or better than 10^{-3} for the experimental ingredients needed and can be used to determine $\mathcal{F}t$ with a precision close to 10^{-4}. The reported average value is 3072.08 ± 0.79 s [3]. This yields a V_{ud} value of 0.97425(22).

Similar measurements with nuclear mirror β decays allow to arrive at the same final quantity V_{ud}, however, with reduced precision. A value of 0.9719(17) was recently determined [4] from a compilation of the relevant data from literature [5]. Whereas dedicated studies of $0^+ \rightarrow 0^+$ decays are performed for several decades now, the potential of mirror transitions was only rediscovered recently. Therefore, it can be expected that important progress is possible with high-precision studies of different mirror β decays.

The purpose of continued measurements of $0^+ \rightarrow 0^+$ and mirror decays is to test the weakest element of the procedure just laid out, the nuclear-structure dependent corrections δ_C and δ_{NS}. These can be best tested by high-precision measurements with nuclei where these corrections are predicted to be important or in cases where a high precision can be reached for the theoretical calculations as well as for the experimental observables.

In a long-standing program started in 2000 at the Accelerator Laboratory of the University of Jyväskylä, high-precision half-life measurements have been performed at the IGISOL facility for the nuclei of ^{26}Si, ^{29}P, ^{30,31}S, ^{42}Ti and ^{62}Ga. Elements like silicon, phosphorus, and titanium are rather difficult if not impossible to obtain as beams at standard ISOL facilites, as they hardly diffuse out of the production target. Therefore, the IGISOL technique which is not sujected to these limitations is an ideal tool for the production of these nuclei and isotopes of other elements. For most of the studies, the preparation of ultra-pure samples of the radioactive species with the double-Penning trap system JYFLTRAP, the so-called trap-assisted decay spectroscopy, was a key ingredient to the success of the measurements.

In the next chapter, the experimental setup and the experimental procedure will be described. This will be followed by a description of the individual experiments and their results. These results will be put in the international context and experimental results prior and after the JYFL measurements will be presented shortly.

2 Experimental setup and experimental procedure

The isotopes of interest were produced by proton induced fusion reactions on appropriate targets. Only in the case of ^{42}Ti, a ^{3}He induced reaction was used. The fusion evaporation residues were thermalised in the helium gas of the IGISOL target chamber from where they were extracted and mass-analysed by the IGISOL magnet separator. The measurements of the half-lives of ^{26}Si, ^{29}P, and ^{42}Ti were completely performed with trap assistance, whereas the measurements performed for ^{31}S used only for part of the data taking this option. The ^{30}S and ^{62}Ga measurements were fully performed without JYFLTRAP.

The reason for not using the Penning trap separation for a full experiment or for part of the data taking was simply that depending on the mass of the nucleus of interest the Penning trap is more or less filled not only with the ions of interest but also with mainly stable ions or molecules. If the ratio between the isotope of interest and the unwanted species is too unfavourable, the transmission of JYFLTRAP drops and most of the isotopes of interest are lost thus not allowing to reach the statistics needed. The drawback of not using trap-assisted decay spectroscopy is that unwanted radioactive species produced in the same reaction contaminate the IGISOL beam and may prevent a high-precision measurement to be performed. Therefore, a careful analysis of the beam content is needed and the contaminants have to be taken into account in the analysis procedure.

The procedure of purification with JYFLTRAP applied in most experiments was as follows: The fusion-evaporation products were mass-analysed in the IGISOL magnet and collected in the radiofrequency cooler and buncher of JYFLTRAP. After the collection time, these bunches of radioactive and stable ions were then injected into the purification trap of JYFLTRAP where the unwanted species were cleaned away by (1) exciting all ions to larger orbits in the Penning trap, (2) selectively recentering the ions of interest and (3) ejecting the sample through a 2 mm diaphragm placed between the purification and the precision trap of JYFLTRAP. In order to increase the sample size, the purified sample could be reflected in the precision trap and recaptured in the purification trap where a new sample from the

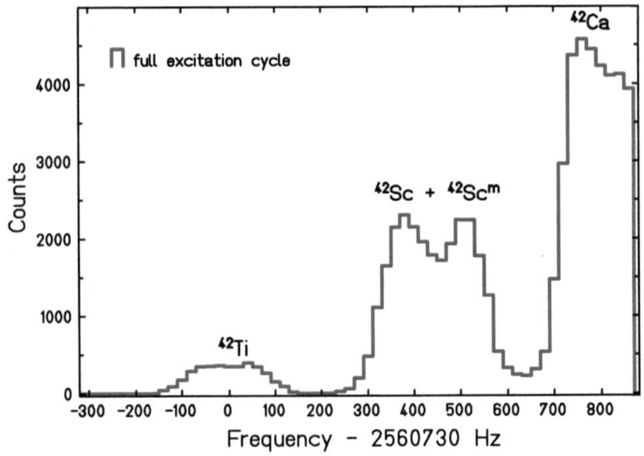

Fig. 1 Frequency scan of the purification trap of JYFLTRAP in order to test the separation power of the trap. As a function of the excitation frequency different mass $A = 42$ isobars are selected by the trap and detected by a MCP detector at the exit of the JYFLTRAP. The Sc peak contains the ground state and the isomeric state. The dip in this peak is not the separation of the two states, but rather a saturation effect

RFQ cooler and buncher was overlaid with the purified sample and a new cleaning process was started.

As this purification procedure takes some time (minimum 100 ms), the number of samples that can be overlaid depends on the half-life of the species of interest. In the case of ^{26}Si, eight samples were accumulated reaching a final activity of about 6–7 times the activity of a single sample. In the case of the much shorter-lived isotope ^{42}Ti, saturation was reached already after one accumulation.

Figure 1 shows the action of the purification procedure. In the case of ^{42}Ti, the contaminants ^{42}Sc, ^{42}Sc^{m}, and ^{42}Ca are produced in the same reaction and these reaction products are not filtered out by the IGISOL magnet (resolution m/Δm $\approx$ 300). However, by fixing the cleaning frequency of the purification trap at the value of ^{42}Ti, all other isotopes are efficiently cleaned away.

The radioactive beam, with or without cleaning, is finally sent to the experimental setup and implanted in a mylar tape in the center of a close to 4π plastic scintillator to detect the β particles from the radioactive decay. A γ-ray detection setup consisting of three germanium detectors completed the setups (see Fig. 2). The germanium detectors were used to either determine a low-precision value for the super-allowed branching ratio which came for free or to check the purity of the beam.

The measurements were always structured in cycles of primary beam ON for production and accumulation of the species in the RFQ cooler and buncher and of beam OFF for the purification and decay-measurement sequence. At the end of such a cycle the remaining activity was removed from the experimental setup by a tape move and a new cycle could start.

The data were registered by two (or three in the case of ^{62}Ga) different and independent data acquisition systems. One of them was optimised for a basically dead-time free time stamping of the decay events. After a cycle, these data were written on disk and cleared. This data acquisition has up to eight independent channels that can e.g. be used with different fixed front-end electronics dead times. We typically used two channels with dead times of 2 and 8 μs. The second data acquisition is of the listmode type storing each event individually first in a buffer and then on disk. This data acquisition was run with a fixed dead time of 100 μs and registered the time of an event, the energy signal from the photomultipliers

Fig. 2 Setup used in the present series of experiments. The β particles are observed with a plastic scintillator coupled to two photomultipliers (PM 1 and PM 2). Three germanium detectors (Ge 1, Ge 2, Ge 3) are used to measure the branching ratios and to test the sample for contamination. The beam comes from the back of the picture

coupled to the 4π plastic scintillator and the energy signals from the germanium detectors. Offline the cycle structure was restored and the data were analysed in a similar manner as for the first data acquisition.

The data acquisition and the front end electronics parameters were routinely changed during the experiments to search for possible bias of the experimental result on one of these parameters. In all cases, such a possible bias was found to be much smaller than the statistical uncertainty.

3 Analysis procedure

The data analysis started with a selection of valid cycles. The criterion for accepting a cycle were that (1) the number of counts in a cycle had to be larger than a user chosen limit, (2) the fit of the experimental decay time spectrum converged and (3) yielded a reduced χ^2 lower than a user chosen value.

The accepted cycles with identical experimental conditions were then dead-time corrected and grouped in runs. Figure 3 shows such a run from the ^{26}Si experiment decomposed into its different contributions. The fit of these runs with the theoretical curve yielded the final half-life of a run. The half-life results from all runs were finally averaged to yield the final result with its statistical uncertainty.

The final step of the analysis is to search for bias on the half-life either from experimental parameters or from analysis parameters. Thus the half-life is investigated as a function of both the experimental and the analysis parameters which are varied within reasonable limits to check their influence on the half-life. From these considerations, the systematic uncertainties are deduced and added quadratically to the statistical uncertainty yielding the final experimental result.

For each step of the analysis, the errors are increased if a fit does not yield a normalised χ^2 of unity or smaller. Thus after increasing the uncertainties by the square root of the normalised χ^2, a new fit yields the result with a χ^2 of unity. This procedure is applied when fitting the experimental distributions and when averaging the different runs.

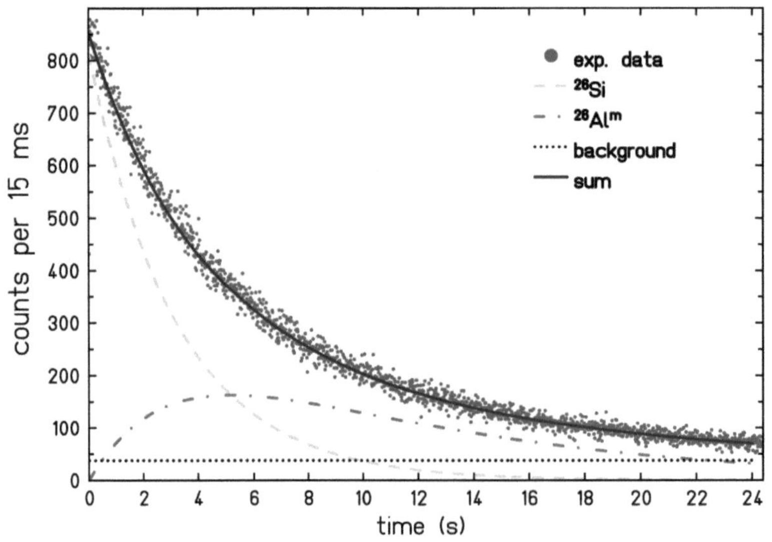

Fig. 3 Time distribution obtained for a single run from the ^{26}Si experiment. The *full line* is the result of the fit. The contributions from ^{26}Si, ^{26}Al^m and the background are represented separately. A half-life of 2.229(11) s was obtained for this run

4 Results

In the following, we will first discuss the $0^+ \to 0^+$ Fermi transitions before turning our attention to the mirror decays.

4.1 $0^+ \to 0^+$ transitions

The half-lives of the decays of ^{26}Si, ^{29}P, ^{42}Ti and ^{62}Ga were measured with high precision. All these nuclei decay mainly by a super-allowed $0^+ \to 0^+$ transition. In the case of ^{62}Ga, we performed also a high-precision measurement of the super-allowed branching ratio.

4.1.1 ^{26}Si

The ions were produced in proton induced fusion-evaporation reactions with a 35 MeV proton beam with an average intensity of 45 μA on a 2.3 mg/cm^2-thick ^{nat}Al target [6]. A master cycle started with a 500 ms accumulation time in the RFQ followed by eight trap cycles and a decay measurement period. In the master cycle, the eighth trap cycle was followed by a final cleaning of the accumulated bunch and by a 24.4 s decay measurement. Data were effectively taken over a period of 68 hours and a total of $3.559(2) \cdot 10^6$ ^{26}Si ions were accumulated.

The measurement yielded a value of 2.2283(27) s which is in agreement with previous measurements but has a precision better by a factor of 4. The error is dominated by the statistical uncertainty (2.5 ms), whereas the systematic errors amount to 1.1 ms.

Table 1 Half-life values published for ^{26}Si

	Hardy et al. [7]	Wilson et al. [8]	Matea et al. [6]	Iacob et al. [9]
$T_{1/2}$ (ms)	2210 ± 21	2240 ± 10	**2228.3 ± 2.7**	2245.3 ± 0.7

Our value is given in bold face. Our value is in contradiction with the latest experimental value (see text)

Table 2 Half-life values published for ^{30}S

	Barker et al. [11]	Moss et al. [12]	Wilson et al. [8]	Souin et al. [10]
$T_{1/2}$ (ms)	1180 ± 40	1220 ± 30	1178.3 ± 4.8	**1175.9 ± 1.7**

Our value is given in bold face. It is a factor of three more precise than any previous value

In the mean time, another high-precision measurement of the half-life of ^{26}Si has been performed [9] which revealed that in our analysis a potentially important correction was not applied. This correction is needed as the detection efficiency of the ^{26}Si parent decay is somewhat higher than the detection efficiency of the daughter decay $^{26}Al^m$. This difference is only of the order of a few per mille. However, for measurements with a precision of the order of 10^{-3}, even such a small difference affects the final result significantly. Our data are presently revised in order to include this correction. Table 1 summarises the existing results for the half-life of ^{26}Si.

4.1.2 ^{30}S

A 35 MeV proton beam impinged on a 2 mg/cm^2 Ni_2P target with a 2.4 mg/cm^2 nickel backing foil to produce ^{30}S [10]. The ^{30}S evaporation residues recoiled out of the target and were stopped in the helium gas of the ion guide. After mass separation, they were implanted on the tape inside the plastic scintillator for the half-life measurement.

Each half-life measurement started with a 2 s measurement of the background. Then, after a 4 s accumulation period of the $A = 30$ products, the proton beam was switched off and the IGISOL beam was deflected and a 12 s decay period started. In total, 3.4×10^6 ^{30}S decays were detected. The final result is $T_{1/2} = 1175.9(17)$ ms. This results is strongly dominated by the statistical error bar of 1.7 ms. The systematic error of 0.3 ms is negligible. Table 2 gives all existing experimental results for the half-life of ^{30}S.

4.1.3 ^{42}Ti

^{42}Ti was produced by means of a fusion-evaporation reaction of a ^3He beam at 17 MeV impinging on a 1.5 mg/cm^2 thick ^{nat}Ca target [13]. The reaction products were thermalized in the helium gas of the IGISOL target chamber, extracted, accelerated to 30 keV, and $A/q = 42$ selected by the IGISOL magnet. Pure samples of ^{42}Ti were prepared by means of the purification Penning trap and were then sent to the experimental setup for half-life measurements.

We accumulated about $1.6*10^6$ ^{42}Ti decays. The resulting experimental half-life value for the ^{42}Ti ground state is $T_{1/2} = (208.14 \pm 0.45)$ ms. This value contains a small systematic error of 0.04 ms.

Table 3 Half-life values published for ^{42}Ti

	Aldridge et al. [14]	Gallmann et al. [15]	Nicholas et al. [16]	Kurtukian Nieto et al. [13]
$T_{1/2}$ (ms)	173 ± 14	202 ± 5	200 ± 20	**208.14 ± 0.45**

Our value is given in bold face. It is a factor of ten more precise than any individual previous value

Table 4 Experimental results of half-life measurements for ^{62}Ga

	Alburger [18]	Chiba et al. [19]	Davids et al. [20]	Hyman et al. [21]
$T_{1/2}$ (ms)	115.95(30) ms	116.4(15) ms	116.34(35) ms	115.84(25) ms
	Blank et al. [22]	Canchel et al. [17]	Hyland et al. [23]	Grinyer et al. [24]
$T_{1/2}$ (ms)	116.19(4) ms	**116.09(17)ms**	116.01(19) ms	116.100(25) ms

Our result is given in bold face

This final result is compared to previous measurements in Table 3. Our result is by far the most precise one, such that the other results basically count no longer in the weighted average.

4.1.4 ^{62}Ga

Historically, the measurement of the half-life of ^{62}Ga was the first high-precision half-life measurement performed by our collaboration in Jyväskylä in 2002 [17]. The ^{62}Ga activity was produced via the ^{64}Zn(p,3n) fusion-evaporation reaction. The 3 mg/cm^2 ^{64}Zn target was used as the entrance window of the light-ion ion guide and was bombarded with a 48 MeV proton beam (20 μA) from the $K = 130$ Jyväskylä cyclotron. The β decay of ^{62}Ga was studied after mass separation only.

Approximatively 10^7 β decay events of ^{62}Ga were recorded. Our experimental result for the half-life of ^{62}Ga of $T_{1/2} = 116.09(17)$ ms is in nice agreement with all previous half-life measurements (see Table 4). At the moment of the publication, it was the second most precise value. Only a measurement performed by members of the collaboration and others at the GSI on-line separator had a smaller error bar.

Since these measurements, two other half-life measurements were performed at TRIUMF with an improved precision. All these measurements make of ^{62}Ga one of the most precisely measured half-life of exotic $N = Z$ odd-odd nuclei. A comparison of these half-life values is given in Table 4.

In another measurement, we could determine the super-allowed branching ratio of ^{62}Ga to be BR = 99.893(24)% [25]. The precisely measured half-life and the branching ratio (see also [26]) allow to determine the partial half-life of the super-allowed branch.

4.2 Mirror transitions

For the moment, only two mirror transitions have been measured. The experiments on ^{29}P and ^{31}S are the start-up of a longer program to get high-precision half-lives and branching ratios for the mirror β decays.

4.2.1 ^{29}P and ^{31}S

During the experiment of ^{30}S, the half-lives of the mirror β decays of ^{29}P and ^{31}S were also measured. The decay of ^{29}P was measured partly with and partly without JYFLTRAP, thus the decay of clean and mixed samples of ^{29}P were studied. For the study of ^{31}S, the measurements were performed with JYFLTRAP.

Both data sets are presently analysed. The expected precision for both isotopes is of the order 10^{-3}.

5 Summary and conclusions

The half-life measurements performed at the Accelerator Laboratory of the University of Jyväskylä reached precisions close to or better than 10^{-3}. This is typically the precision to reach in order to include an isotope in the list of precisely measured ft values. However, the half-life is only one experimental input needed to determine the ft value. The Q value has also been measured for all these isotopes with the necessary precision. The experimental quantity which still has not the required precision for almost all these isotopes, except for ^{62}Ga, is the branching ratio. During our half-life measurements we have measured some of the branching ratios needed with a low precision. These branching ratios came as a by-product of the half-life measurements.

In order to measure these branching ratios with high precision, the group at CEN Bordeaux-Gradignan is presently calibrating a germanium detector in efficiency with high precision. Once this calibration is done, the branching ratios can be measured with sufficient precision. However, it will not be possible to determine these branching ratios with an uncertainty of 0.1% or below. Therefore, these branching ratios will certainly stay for quite a while the limiting experimental factor, at the same level as the theoretical corrections.

With the half-life measurements summarised here, the half-lives of all $T_z = -1$ nuclei up to ^{42}Ti are now known with good precision. As for the $N = Z$, odd-odd nuclei which decay by a super-allowed β decay an even better precision is reached, future half-life measurements will certainly mainly concentrate on mirror β decay, where significant progress can be achieved.

Acknowledgements The present experimental series was performed in collaborations of the CEN Bordeaux-Gradignan, JYFL, GANIL, LPC Caen, CSIC Madrid, ISOLDE CERN, Los Alamos National Laboratory, University of Liverpool and KVI Groningen.

References

1. Towner, I.S., Hardy, J.C.: Phys. Rev. C **66**, 035501 (2002)
2. Towner, I.S., Hardy, J.C.: Phys. Rev. C **77**, 025501 (2008)
3. Hardy, J.C., Towner, I.S.: Phys. Rev. C **79**, 055502 (2009)
4. Naviliat-Cuncic, O., Severijns, N.: Phys. Rev. Lett. **102**, 142302 (2009)
5. Severijns, N., Tandecki, M., Phalet, T., Towner, I.S.: Phys. Rev. C **78**, 055501 (2008)
6. Matea, I., Souin, J., Äystö, J., Blank, B., Delahaye, P., Elomaa, V.V., Eronen, T., Giovinazzo, J., Hager, U., Hakala, J., Huikari, J., Jokinen, A., Kankainen, A., Moore, I.D., Pedroza, J.L., Rahaman, S., Rissanen, J., Ronkainen, J., Saastamoinen, A., Sonoda, T., Weber, C.: Eur. Phys. J. A **37**, 151 (2008)

7. Hardy, J., Schmeing, H., Geiger, J., Graham, R., Towner, I.: Nucl. Phys. **A246**, 61 (1975)
8. Wilson, H.S., Kavanagh, R.W., Mann, F.M.: Phys. Rev. C **22**, 1696 (1980)
9. Iacob, V.E., Hardy, J.C., Banu, A., Chen, L., Golovko, V.V., Goodwin, J., Horvat, V., Nica, N., Park, H.I., Trache, L., Tribble, R.E.: Phys. Rev. C **82**, 035502 (2010)
10. Souin, J., Eronen, T., Ascher, P., Audirac, L., Äystö, J., Blank, B. Elomaa, V.V., Giovinazzo, J., Hakala, J., Jokinen, A., Kolhinen, V.S., Karvonen, P., Moore, I.D., Rahaman, S., Rissanen, J., Saastamoinen, A., Thomas, J.: Eur. Phys. J. A **47**, 40 (2011)
11. Barker, P.H., Drysdale, N., Phillips, W.R.: Proc. Phys. Soc. (London) **91**, 587 (1967)
12. Moss, C.E., Detraz, C., Zaidins, C.S.: Nucl. Phys. A **174**, 408 (1971)
13. Kurtukian Nieto, T., Souin, J., Eronen, T., Audirac, L., Äystö, J., Blank, B., Elomaa, V.V., Giovinazzo, J., Hager, U., Hakala, J., Jokinen, A., Kankainen, A., Karvonen, P., Kessler, T., Moore, I.D., Penttilä, H., Rahaman, S., Reponen, M., Rinta-Antila, S., Rissanen, J., Saastamoinen, A., Sonoda, T., Weber, C.: Phys. Rev. C **80**, 035502 (2009)
14. Aldridge, A.M., Kemper, K.W., Plendl, H.S.: Phys. Lett. B **30**, 165 (1969)
15. Gallmann, A., Aslanides, E., Jundt, F., Jacobs, E.: Phys. Rev. **186**, 1160 (1969)
16. Nicholas, F.M., Lawley, N., Main, J.G., Thomas, M.F., Twin, P.J.: Nucl. Phys. A **124**, 97 (1969)
17. Canchel, G., Blank, B., Chartier, M., Delalee, F., Dendooven, P., Dossat, C., Giovinazzo, J., Huikari, J., Lalleman, A.S., Jiménez, M.J.L., Madec, V., Pedroza, J.L., Penttilä, H., Thomas, J.C.: Eur. Phys. J. A **29**, 409 (2005)
18. Alburger, D.E.: Phys. Rev. C **18**, 1875 (1978)
19. Chiba, R., Shibasaki, S., Numao, T., Yokota, H., Yamada, S., Kotajima, K., Itagaki, S., Iwasaki, S., Takeda, T., Shinozuka, T.: Phys. Rev. C **17**, 2219 (1978)
20. Davids, C.N., Gagliardi, C.A., Murphy, M.J., Norman, E.B.: Phys. Rev. C **19**, 1463 (1979)
21. Hyman, B.C., Iacob, V., Azhari, A., Gagliardi, C.A., Hardy, J.C., Mayes, V., Neilson, R., Sanchez-Vega, M., Tang, X., Trache, L., Tribble, R.: Phys. Rev. C **68**, 015501 (2003)
22. Blank, B., Savard, G., Döring, J., Blazhev, A., Canchel, G., Chartier, M., Henderson, D., Janas, Z., Kirchner, R., Mukha, I., Roeckl, E., Schmidt, K., Zylicz, J.: Phys. Rev. C **69**, 015502 (2004)
23. Hyland, B., Melconian, D., Ball, G.C., Leslie, J.R., Svensson, C.E., Bricault, P., Cunningham, E., Dombsky, M., Grinyer, G.F., Hackman, G., Koopmans, K., Sarazin, F., Schumaker, M.A., Scraggs, H.C., Smith, M.B., Walker, P.M.: J. Phys. G **31**, S1885 (2005)
24. Grinyer, G.F., Finlay, P., Svensson, C.E., Ball, G.C., Leslie, J.R., Austin, R.A.E., Bandyopadhyay, D., Chaffey, A., Chakrawarthy, R.S., Garrett, P.E., Hackman, G., Hyland, B., Kanungo, R., Leach, K.G., Mattoon, C.M., Morton, A.C., Pearson, C.J., Phillips, A.A., Ressler, J.J., Sarazin, F., Savajols, H., Schumaker, M.A., Wong, J.: Phys. Rev. C **77**, 015501 (2008)
25. Bey, A., Blank, B., Canchel, G., Dossat, C., Giovinazzo, J., Matea, I., Elomaa, V.V., Eronen, T., Hager, U., Jokinen, J.H.A., Kankainen, A., Moore, I., Penttila, H., Rinta-Antila, S., Saastamoinen, A., Sonoda, T., Aysto, J., Adimi, N., de France, G., Thomas, J.C., Voltolini, G., Chaventre, T.: Eur. Phys. J. A **36**, 121 (2008)
26. Hyland, B., Svensson, C.E., Ball, G.C., Leslie, J.R., Achtzehn, T., Albers, D., Andreoiu, C., Bricault, P., Churchman, R., Cross, D., Dombsky, M., Finlay, P., Garrett, P.E., Geppert, C., Grinyer, G.F., Hackman, G., Hanemaayer, V., Lassen, J., Lavoie, J.P., Melconian, D., Morton, A.C., Pearson, C.J., Pearson, M.R., Phillips, A.A., Schumaker, M.A., Smith, M.B., Towner, I.S., Valiente-Dobon, J.J., Wendt, K., Zganjar, E.F.: Phys. Rev. Lett. **97**, 102501 (2006)

DOI 10.1007/978-94-007-5555-0_14
Reprinted from *Hyperfine Interactions* Volume 223, Issues 1-3, 2014
and *The European Physical Journal A* Volume 48, Issue 4, 2012

Collective properties of neutron-rich Ru, Pd, and Cd isotopes

Y. B. Wang · J. Rissanen

Published online: 3 April 2012

Abstract Collective properties of neutron-rich Ru, Pd, and Cd isotopes are reviewed, combining the original results from the IGISOL β-decay experiments with recent experimental and theoretical progress. The transitional nature of Ru and Pd nuclei is discussed via the low-lying level systematics, including the low-lying 0^+ states. Although the role of an anharmonic quadrupole vibrator in Cd nuclei was recently questioned, level systematics for the three-phonon quintuplet in ^{116,118,120}Cd are presented, and an outlook of the spectroscopic methods for the level lifetime or $B(E2)$ values is given.

Keywords Collective structure · Ru, Pd and Cd nuclei · IGISOL · β decay

1 Introduction

The structure of neutron-rich Ru, Pd, and Cd isotopes has been the subject of many experimental and theoretical works in recent years. While Ru and Pd nuclei have been predicted to show transitional features from an anharmonic vibrator to a γ-soft rotor when approaching the neutron midshell [1], the general character of Cd nuclei has long been regarded as a good example of an anharmonic quadrupole vibrator with the observation of candidates for complete sets of multi-phonon states [2, 3]. For neutron-rich Ru and Pd nuclei, traditional high-spin studies by prompt γ spectroscopy were carried out by several groups using spontaneous or

Y. B. Wang (✉)
China Institute of Atomic Energy, PO Box 275(10), Beijing 102413,
People's Republic of China
e-mail: ybwang@ciae.ac.cn

J. Rissanen
Department of Physics, University of Jyväskylä, PO Box 35(YFL),
40014 Jyväskylä, Finland

heavy-ion-induced fission, for example, $^{108-112}$Ru in [4–12], ^{114}Ru in [4, 13] as well as ^{112}Pd in [14], ^{114,116}Pd in [9, 15], ^{118}Pd in [16], and ^{120}Pd in [17]. These investigations highlight the nuclear shape degree of freedom of neutron-rich Ru and Pd nuclei by analyzing the rotational band structures, and a possible shape transition from a triaxial prolate to a triaxial oblate has been intensively explored. The absolute transition probabilities for the first 2^+ states in ^{110,114}Pd have also been measured using the recoil distance Doppler shift technique following projectile Coulomb excitation at intermediate beam energies [18], and a sizable deformation of $\beta = 0.24(1)$ in ^{114}Pd has been determined, consistent with systematic Interacting Boson Model (IBM) calculations [1]. Towards very neutron-rich Ru and Pd nuclei relevant to the astrophysical rapid neutron capture process, β-decay half-lives and β-delayed neutron emission branchings for $^{120-122}$Pd isotopes and half-lives for $^{114-118}$Ru isotopes have recently been measured [19].

The Cd nuclei span a full shell from $N = 50$ to $N = 82$, which provides an ideal test-bed for structure evolution in the context of varying valence neutron number. Therefore, numerous experimental investigations have been carried out on even-mass Cd nuclei by using various spectroscopic methods. A striking feature revealed in these studies has been the coexistence of the anharmonic quadrupole vibrator and intruder excitations at low excitation energies [20]. Experimental methods on stable and neutron-rich Cd nuclei include neutron inelastic scattering [21–24], high-spin studies [17, 25–27], and β decay [28–34]. These experimental works have generated a wealth of data on detailed level schemes and transition probabilities in even-mass Cd nuclei.

At IGISOL, systematic studies of neutron-rich even-mass Pd isotopes by β decay started at its inauguration in the 1980's [35, 36], followed soon by similar studies of Ru isotopes [37]. Later on, the constant improvement of the ion guide technique enabled more detailed studies of ^{110}Ru [38] and $^{110-118}$Pd [39–44] isotopes. These experimental studies resulted in the identification of many new γ transitions and excited levels in the daughter Pd and Cd nuclei. The systematics of the low-lying collective and two-quasineutron levels in even Pd nuclei was also extended up to ^{118}Pd. In the experiments of neutron-rich Rh decay studies, the Ag isotopes from the same mass chain were delivered to the spectroscopic station in large quantities and therefore the β decays of neutron-rich Ag to Cd nuclei were studied simultaneously. This led to more detailed decay schemes of $^{116-120}$Ag isotopes and to the identification of many new low-lying collective levels. The complete three-phonon quadrupole vibrational quintuplet was proposed for the ^{116}Cd nucleus for the first time [45], and later extended to ^{120}Cd nuclei [46]. Recently, the utilization of the JYFLTRAP Penning trap setup as a high-resolution mass filter enabled the first study of the ^{114}Tc β decay to ^{114}Ru.

2 IGISOL experiments

The IGISOL facility utilizes a universal technique for the production of radioactive ion beams with extraction times as short as sub-ms. The use of a fast helium-filled ion guide combined with proton-induced fission provides very effective access to various fission fragments in their ground and isomeric states regardless of their chemical properties. In the following series of β-decay works, very similar β and

γ spectroscopic setups were used. The mass separated beam was implanted into a collection tape, which was periodically moved at a preset time interval to reduce the background radiation emitted from the long-lived isobaric contaminants. The ions-of-interest were implanted in the center of a cylindrical 2-mm-thick BC408 plastic scintillator, which was viewed by several Ge detectors placed in close geometry. This kind of setup enabled conventional β-γ and γ-γ coincidence studies. The time stamping of events defined by the period of the tape movement also enabled half-life analyses. In most cases, separate decay schemes were built for the ground state and high-spin isomeric states, respectively. The β feedings and log *ft* values were computed according to the constructed level schemes and γ intensities.

3 Results and discussion

The original results obtained from IGISOL β-decay experiments have been reported in several publications. Here we present some selected level systematics for discussion.

3.1 Ru nuclei

The Ru isotopes show evidence of undergoing a shape transition from a spherical to a γ-soft prolate shape around $N \approx 60$–70 [37, 38, 47–52] and therefore they have been described as transitional nuclei in the Interacting Boson Approximation (IBA) model [47, 48, 51] with increasing rigidity when going towards more neutron-rich species [49–51]. Another shape transition, from γ-soft prolate to γ-soft oblate has been predicted to take place around $N \approx 70$ [53]. Therefore the Ru isotopes have been under detailed investigation, both experimentally and theoretically. Despite many studies on this topic, the exact location of the second shape transition has not been unambiguously identified and more work is certainly needed.

The level structures of the even $^{110-114}$Ru isotopes have been studied at IGISOL via β decays of $^{110-114}$Tc isotopes in three different investigations. In the first one, the ^{110,112}Ru nuclei were studied [37] resulting in new β-decay schemes for the mother nuclides and emphasizing the importance of triaxiality in interpreting the experimental results. The second study concentrated on the level structure of ^{110}Ru resulting in a remarkably extended decay scheme [38]. The experimental results were also compared to a variety of nuclear models. In the third study the β decay of very neutron-rich ^{114}Tc was studied with the help of a Penning trap as a mass selective filter. The level structure of ^{114}Ru populated by the β decay revealed the existence of two beta-decaying states in ^{114}Tc.

The level systematics of even Ru isotopes is presented in Fig. 1. One can observe how the energies of the first 2^+ states are decreasing as a function of neutron number up to $A = 108$, which indicates an increase of collectivity for these nuclei. Near the neutron mid-shell ($N = 66$) the decreasing trend levels off and the energies of the lowest 2^+ states in $^{108-112}$Ru isotopes are rather similar, which is the case also for the other ground state band members. This behaviour has been suggested to correlate with the negative-parity excitations in odd-A Ru nuclei [54]. The energies of the second 2^+ states also have a decreasing trend reaching a clear minimum at $A = 112$, two neutrons above the mid-shell. This trend is the same as in the Pd isotopes, where

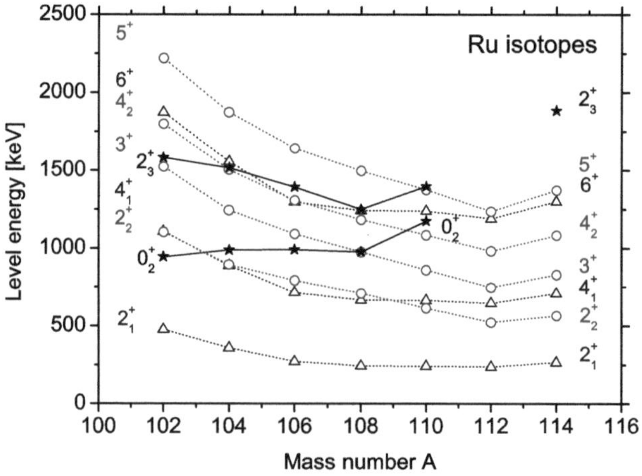

Fig. 1 Systematics of low-lying excited levels in even-mass Ru isotopes. The levels in the ground-state band (*open triangles, blue*) and in the γ-band (*open circles, red*) are connected with *dotted lines*. The first excited 0^+ levels and the third 2^+ levels (*solid stars, black*) are connected by *solid lines*. See text for more details

the energies of the 2_2^+ levels have a minimum at $N = 68$. The level energies in the ground state and gamma bands start to increase again at $A = 114$. The sum rule of $2_1^+ + 2_2^+ \approx 3_1^+$ as well as having the second 2^+ state below the first 4^+ state and the characteristic gamma band indicates that the Ru isotopes have triaxial shapes, which has been supported also by several theoretical calculations [50, 55, 56].

The energies of the first excited 0^+ states are increasing whereas those of the 2_3^+ states are decreasing smoothly as a function of mass number up to $A = 108$. The sudden kink at $A = 110$ suggests a possible change in the nature of the 0^+ states from a β-band-like structure to an intruder excitation. These structures exhibit different behaviours in the level systematics of odd Pd isotopes, see Fig. 2 for a comparison. Unfortunately, the excited 0^+ states have not been observed for the most neutron-rich Ru isotopes ($A = 112, 114$).

3.2 Pd nuclei

The systematics of low-lying levels in neutron-rich doubly-even Pd nuclei is shown in Fig. 2. The series of IGISOL β-decay experiments enabled identification of a large number of new levels in $^{110-118}$Pd isotopes that substantially extend the level systematics for these isotopes. Low-spin levels or band-heads of different origins have been observed enriching the spectroscopic information of Pd nuclei. As already mentioned, Pd nuclei are predicted to exhibit transitional features of triaxiality and γ softness when approaching the neutron midshell [1]. While vibrational degrees of freedom are important for lower mass ^{106,108}Pd isotopes [57], the gradual decrease of excitation energies of yrast states indicates the increase of deformation up to ^{114}Pd followed by a smooth transition towards diminishing deformation in ^{116}Pd and ^{118}Pd isotopes. The systematics of $B(E2, 0_1^+ \rightarrow 2_1^+)$ values in Pd nuclei is presented in [18], which shows a smooth development of collectivity when approaching to the neutron midshell. A $B(E2, 0_1^+ \rightarrow 2_1^+) = 51(9)$ W.u. in ^{114}Pd is close to the value in ^{110}Pd. A relatively large deformation $\beta = 0.24(1)$ in ^{114}Pd indicates that the rotational collectivity reaches its maximum around the neutron midshell, which is consistent with the theoretical predictions [1].

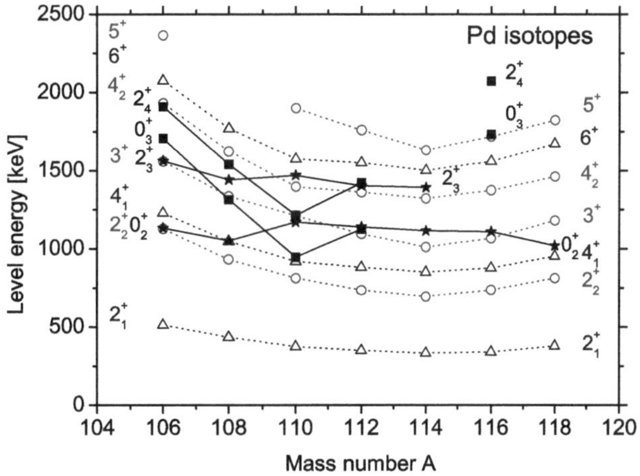

Fig. 2 Systematics of low-lying excited levels in even-mass Pd isotopes with the ground-state band (*open triangles, blue*) and the γ-band (*open circles, red*). Two pairs of 0^+ and 2^+ levels that are indicative of the β band (*solid stars, black*) and the intruder band (*solid squares, black*) are also shown

There are two sets of low-lying 0^+, 2^+ pairs in neutron-rich Pd nuclei, which are populated in the decays of 1^+ states of Rh isotopes. One pair of 0^+, 2^+ excited states exhibits very smooth energy systematics in the 1.0–1.6 MeV range. The 2^+-0^+ energy difference is only slightly lower than the first 2^+ level energies and shows the same decreasing trend with increasing neutron number. Therefore, this pair of 0^+-2^+ levels can probably be associated with a β-band-like structure. The energy systematics of the other pair of 0^+, 2^+ excited levels forms a V-shape versus neutron number, which is characteristic of an intruder excitation. The energy of this $(0^+, 2^+)$ pair is lowest in ^{110}Pd and then moves upwards with larger N. It is interesting to note that the energy minimum occurs at $N = 64$, two neutrons before the midshell. On the other hand, the gamma band energies have a minimum at $N = 68$, two neutrons above the mid-shell. The V-shape behaviour has also been observed for the $K = 1/2$ band in odd-mass Rh isotopes [58, 59], and it has been interpreted as originating from the $\pi 1/2^+[431]$ Nilsson orbital with a prolate deformation.

3.3 Cd nuclei

A three-phonon multiplet had been reported in even-even $^{108-114,118}$Cd isotopes before the IGISOL β-decay experiments. However, information on ^{116}Cd and ^{120}Cd isotopes were lacking in order to fill in the gap in the three-phonon multiplet systematics. Based on the β-decay works at IGISOL, we identified a new 4_2^+ level at 1869.7 keV in ^{116}Cd, and proposed candidates for the complete three-phonon quintuplet along with other four known levels near 2 MeV. The relative $B(E2)$ branching ratios were deduced, which showed strongly preferred decays to the two-phonon triplet [45]. The three-phonon multiplet in ^{118}Cd was previously studied by Aprahamian et al. [28] and the IGISOL experiment modified their interpretation by the substitution of the 1915.8 keV 2_3^+ state with a newly identified 2023.0 keV (2_4^+) state in accordance with the intruder systematics [26, 30]. In ^{120}Cd, we pointed out that the 1899.0-keV (3^+), 1920.5-keV (2_3^+), 1997.9-keV (4_2^+), and 2032.8-keV 6^+ levels can be candidates for the three-phonon multiplet members. The relative $B(E2)$ branching ratios were studied whenever possible to verify the preferred decay to the

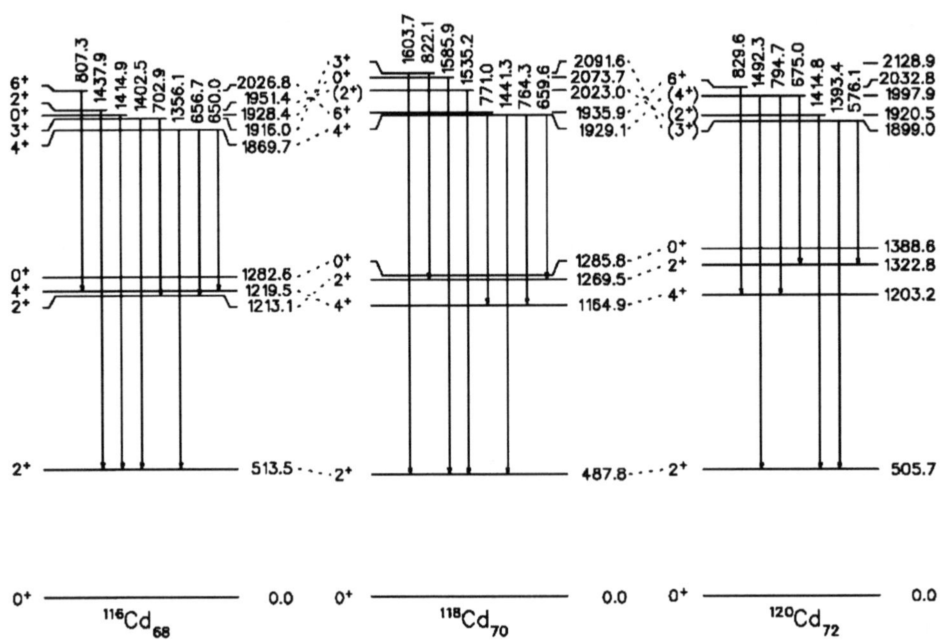

Fig. 3 Comparison of the three-phonon states in ^{116,118,120}Cd isotopes

$N = 2$ states. The comparison of the suggested three-phonon states in ^{116,118,120}Cd isotopes is shown in Fig. 3.

Recently, Garrett et al. made a systematic study of the $^{110-116}$Cd isotopes by using the available experimental data and the IBM calculations incorporating both normal vibrational levels and intruder excitations [60, 61]. For the two-phonon levels, the calculations appear to reproduce the data reasonably well in $^{110-114}$Cd nuclei. However, systematic deviations in low-spin states at the three-phonon levels have been found. A quasi-rotational structure, with the $K = 0$ and $K = 2$ bands built on the ground state has been suggested for $^{110-116}$Cd isotopes [61].

4 Summary and outlook

Many studies with various spectroscopic methods have provided detailed information on even-mass Ru, Pd, and Cd nuclei, thus rendering the possibility to check the mixing of different excitation modes at low excitation energies. However, the level lifetime or $B(E2)$ values that allow for the insight into the intrinsic quadrupole moment are usually available only for stable Pd and Cd nuclei around the neutron midshell. Therefore, it is crucial to apply various spectroscopic studies to more neutron-rich species to extend the $B(E2)$ systematics.

For less neutron-rich Pd and Cd nuclei, fast timing studies with a $\beta - \gamma - \gamma$ coincidence technique have been intensively performed by Mach et al. at ISOL facilities with sources produced by neutron- or proton-induced fission. As shown in the case of ^{116,118,120}Cd isotopes [29], the $B(E2)$ values are very useful to distinguish between the vibrational or intruder nature of narrowly separated 0_2^+ and 0_3^+ states. A better precision may be achieved for such kind of fast timing measurements with the use of high-resolution $LaBr_3$ scintillation detectors [62]. On the other hand, since the very neutron-rich isotopes can be produced now by projectile fragmentation at intermediate beam energies [19, 63], measurements of the Coulomb excitation cross sections or level lifetimes by the Doppler shift technique are promising ways to determine the absolute transition probabilities in the near future [18].

Acknowledgements This work was supported by the Academy of Finland under the Finnish Center of Excellence Programmes 2000–2005 (Project No. 44875, Nuclear and Condensed Matter Program at JYFL) and 2006–2011 (Nuclear and Accelerator Based Physics Programme at JYFL). Y.-B. Wang gratefully acknowledges the financial support from the National Basic Research Program of China under Grant No. 2007CB815003, and the National Natural Science Foundation of China under Grant Nos. 11021504,11175261 and 10875173.

References

1. Kim, K.-H., et al.: Nucl. Phys. A **604**, 163 (1996)
2. Bohr, A., Mottelson, B.R.: Nuclear Structure, vol 2, p. 33. Benjamin, Reading, Massachusetts (1975)
3. Kern, J., Garrett, P., Jolie, J., Lehmann, H.: Nucl. Phys. A **593**, 21 (1995)
4. Shannon, J.A., et al.: Phys. Lett. B **336**, 136 (1994)
5. Lu, Q.H., et al.: Phys. Rev. C **52**, 1348 (1995)
6. Che, X.L., et al.: Chin. Phys. Lett. **21**, 1904 (2004)
7. Jiang, Z., et al.: Chin. Phys. Lett. **20**, 350 (2003)
8. Che, X.L., et al.: Chin. Phys. Lett. **23**, 328 (2006)
9. Hua, H., et al.: Phys. Lett. B **562**, 201 (2003)
10. Wu, C.Y., et al.: Phys. Rev. C **73**, 034312 (2006)
11. Luo, Y., et al.: Int. J. Mod. Phys. E **18**, 1697 (2009)
12. Zhu, S., et al.: Int. J. Mod. Phys. E **18**, 1717 (2009)
13. Yeoh, E.Y., et al.: Phys. Rev. C **83**, 054317 (2011)
14. Krücken, R., et al.: Eur. Phys. J. A **10**, 151 (2001)
15. Butler-Moore, K., et al.: J. Phys. G: Nucl. Part. Phys. **25**, 2253 (1999)
16. Zhang, X.Q., et al.: Phys. Rev. C **63**, 027302 (2001)
17. Stoyer, M., et al.: Nucl. Phys. A **787**, 455c (2007)
18. Dewald, A., et al.: Phys. Rev. C **78**, 051302(R) (2008)
19. Montes, F., et al.: Phys. Rev. C **73**, 035801 (2006)
20. Wood, J., et al.: Phys. Rep. **215**, 101 (1992)
21. Corminboeuf, F., et al.: Phys. Rev. Lett. **84**, 4060 (2000)
22. Kadi, M., et al.: Phys. Rev. C **68**, 031306 (2003)
23. Garrett, P.E., et al.: Phys. Rev. C **75**, 054310 (2007)
24. Bandyopadhyay, D., et al.: Phys. Rev. C **76**, 054308 (2007)
25. Kumpulainen, J., et al.: Phys. Rev. C **45**, 640 (1992)
26. Juutinen, S., et al.: Phys. Lett. B **386**, 80 (1996)
27. Fotiades, N., et al.: Phys. Scr. **T88**, 127 (2000)
28. Aprahamian, A., et al.: Phys. Rev. Lett. **59**, 535 (1987)
29. Mach, H., et al.: Phys. Rev. Lett. **63**, 143 (1989)
30. Zamfir, N.V., et al.: Phys. Rev. C **51**, 98 (1995)
31. Kautzsch, T., et al.: Phys. Rev. C **54**, R2811 (1996)
32. Kautzsch, T., et al.: Eur. Phys. J. **9**, 201 (2000)
33. Batchelder, J.C., et al.: Phys. Rev. C **72**, 044306 (2005)
34. Batchelder, J.C., et al.: Phys. Rev. C **80**, 054318 (2009)
35. Äystö, J., et al.: Nucl. Phys. A **480**, 104 (1988)
36. Äystö, J., et al.: Phys. Lett. B **201**, 211 (1988)
37. Äystö, J., et al.: Nucl. Phys. A **515**, 365 (1990)
38. Wang, J.C., et al.: Phys. Rev. C **61**, 044308 (2000)
39. Lhersonneau, G., et al.: Eur. Phys. J. **2**, 25 (1998)

40. Lhersonneau, G., et al.: Phys. Rev. C **60**, 014315 (1999)
41. Jokinen, A., et al.: Eur. Phys. J. **9**, 9 (2000)
42. Wang, Y.B., et al.: Phys. Rev. C **63**, 024309 (2001)
43. Lhersonneau, G., et al.: Phys. Rev. C **67**, 024303 (2003)
44. Wang, Y.B., et al.: Chin. Phys. Lett. **23**, 808 (2006)
45. Wang, Y.B., et al.: Phys. Rev. C **64**, 054315 (2001)
46. Wang, Y.B., et al. Phys. Rev. C **67**, 064303 (2003)
47. Stachel, J., et al.: Nucl. Phys. A **383**, 429 (1982)
48. Stachel, J., et al.: Z. Phys. A **316**, 105 (1984)
49. Troltenier, D., et al.: Nucl. Phys. A **601**, 56 (1996)
50. Zajac, K., et al.: Nucl. Phys. A **653**, 71 (1999)
51. Stefanescu, I., et al.: Nucl. Phys. A **789**, 125 (2007)
52. Luo, Y., et al.: Phys. Lett. B **670**, 307 (2009)
53. Xu, F.R., Walker, P.M., Wyss, R.: Phys. Rev. C **65**, 021303 (2002)
54. Kurpeta, J., et al.: Phys. Rev. C **82**, 027306 (2010)
55. Faisal, J.Q., et al.: Phys. Rev. C **82**, 014321 (2010)
56. Hakala, J., et al.: Eur. Phys. J. A **47**, 129 (2011)
57. Svensson, L., et al.: Nucl. Phys. A **584**, 547 (1995)
58. Lhersonneau, G., et al.: Eur. Phys. J. A **1**, 285 (1998)
59. Kurpeta, J., et al.: Eur. Phys. J. A **13**, 449 (2002)
60. Garrett, P.E., Green, K.L., Wood, J.L.: Phys. Rev. C **78**, 044307 (2008)
61. Garrett, P.E., Wood, J.L.: J. Phys. G: Nucl. Part. Phys. **37**, 064028 (2010)
62. Deloncle, I., et al.: J. Phys.: Conf. Ser. **205**, 012044 (2010)
63. Nishimura, S., et al.: Phys. Rev. Lett. **106**, 052502 (2011)

DOI 10.1007/978-94-007-5555-0_15
Reprinted from *Hyperfine Interactions* Volume 223, Issues 1-3, 2014
and *The European Physical Journal A* Volume 48, Issue 4, 2012

Trap-assisted studies of odd, neutron-rich isotopes from Tc to Pd

J. Kurpeta · A. Jokinen · H. Penttilä · A. Płochocki · J. Rissanen · W. Urban · J. Äystö

Published online: 4 April 2012

Abstract We review the present and future of trap-assisted structure studies of odd, neutron-rich Tc, Ru, Rh and Pd isotopes at the limits of present experimental techniques. These nuclei of refractory elements are produced in light-particle induced fission and filtered by their mass number with the IGISOL mass separator. Further mass separation with the JYFLTRAP Penning trap system provides a clean, monoisotopic beam perfectly suited for precise nuclear spectroscopy. Connecting the IGISOL and the JYFLTRAP facilities to the recently installed MCC30/15 cyclotron opens new prospects for post-trap spectroscopy of very exotic, neutron-rich nuclei.

Keywords Nuclear spectroscopy · Fission fragments · Mass separation · Ion trap

1 Introduction

For many years the IGISOL (Ion Guide Isotope Separator On-Line) technique [1] has been used to produce beams of neutron-rich, exotic nuclei from Tc to Pd. These elements are difficult to study due to their high melting points. With its stopping and

J. Kurpeta (✉) · A. Płochocki · W. Urban
Faculty of Physics, University of Warsaw, ul. Hoża 69, 00-681 Warsaw, Poland
e-mail: jkurpeta@mimuw.edu.pl

A. Jokinen · H. Penttilä · J. Rissanen · J. Äystö
Department of Physics, University of Jyväskylä, P.O.Box. 35, 40014 Jyväskylä, Finland

W. Urban
Institut Laue-Langevin, 6 rue J. Horowitz, 38042 Grenoble, France

transportation of fission products in a gas stream IGISOL provides ionized beams of Tc, Ru, Rh, Pd and other refractory elements. A vast amount of data gathered at the front line of neutron-rich nuclei over past years with the IGISOL method is reviewed in this volume. A recent use of the JYFLTRAP tandem Penning trap system for producing isobarically purified ion beams considerably enhances the sensitivity and selectivity of nuclear spectroscopy studies.

The nuclei discussed here are created as light-particle-induced fission products. Typically beams of 20–30 MeV protons or deuterons are used to induce fission process of natural uranium or thorium target. Fission products of interest are instantly transported by the IGISOL system to a dipole magnet which works as a mass separator. Its mass resolving power is suitable to separate nuclei of one mass number A from all the other nuclear species produced in fission. These isobaric beams have successfully been used to study many exotic, neutron-rich nuclei of refractory elements, which were poorly known before the IGISOL technique. Many odd-mass, neutron-rich nuclei from Tc to Pd investigated for the first time with IGISOL were reported in [2].

After passing through a radio-frequency cooler and buncher [3], the isobaric beam from the IGISOL separator is injected into the purification trap of JYFLTRAP [4]. For fission products with $A = 100$ mass resolving power of the trap reaches 10^5 which is sufficient to separate a single isotope out of the isobaric beam. As a result, the Penning trap delivers a beam of one desired nuclide, here and after to be called a monoisotopic beam. For nuclear structure studies, the most interesting nuclei in each isobaric chain are the ones furthest away from the valley of beta stability. For the isobaric chains of neutron-rich technetium to palladium nuclei, the cross section usually drops by about one order of magnitude when going by one isobar away from stability. The more abundant, less exotic isobars result in many relatively strong lines in the γ spectra. Their presence produce a lot of background events (mainly by Compton scattering) and sometimes they overlap with the usually weak lines of the exotic nuclei of interest. A huge background may cover weak lines of interest and overlapping background transitions make it hard to identify the peaks of interest as well as their intensities and half-lives. A monoisotopic beam of exotic nuclei is free from the dominating background of reaction-produced less exotic isobars. Thus, it is excellent for precise nuclear spectroscopy studies. For general review on JYFLTRAP, see [5] and a separate article in this special issue.

The combination of the ion-guide method and a Penning trap delivers monoisotopic samples of exotic nuclei which were not available before or were available only together with a huge amount of less interesting nuclides. Trap-assisted measurements investigate these very exotic nuclei with the well-known methods of γ and β coincidence spectroscopy providing reliable information on the nuclear structure.

2 Recent achievements

The neutron-rich nuclei from technetium to palladium ($43 \leq Z \leq 46$) in the mass region $A \approx 110$ are located below the closed $Z = 50$ proton shell and just above the $N = 66$ neutron mid-shell. Nuclear shape in this area of the nuclide chart changes from strongly deformed in Sr and Zr around $N \geq 59$ to spherical closed shell at $Z = 50$ in Sn isotopes. Shape coexistence phenomena, stable oblate deformation,

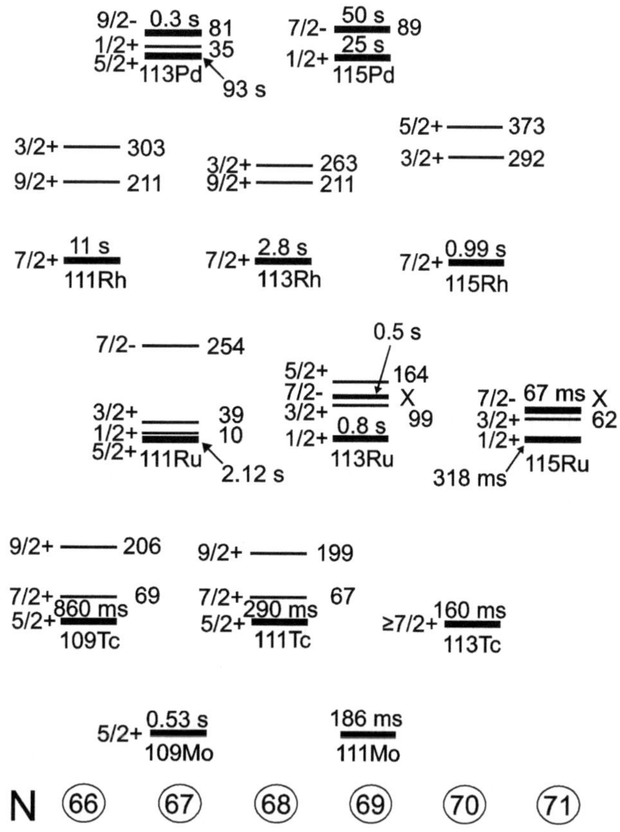

Fig. 1 Systematics of the ground and low-energy excited states in odd, neutron-rich nuclei from Mo to Pd. Monoisotopic beams of ^{109,111}Mo, ^{111,113}Tc and ^{115}Ru were produced using the combination of IGISOL and JYFLTRAP. The spin and parity assignments are tentative and brackets are omitted for better readability. The level energies are given in keV, the X sign for ^{113,115}Ru indicates that the exact energy of the isomer is unknown. The data are based on refs. [2, 9–11, 15–29]. The estimation of the ^{111}Mo half-life is from [14]

triaxial shapes and presence of isomeric states are the topics of current theoretical and experimental investigations in the region. Results of these studies are crucial for the development of nuclear structure models as well as for the modeling of the rapid neutron capture process (*r*-process).

The JYFLTRAP Penning trap was used for nuclear spectroscopy for the first time in the β-decay experiment of even zirconium isotopes [6]. Soon after, the β-decay of ^{115}Ru monoisotopic samples [7] started the trap-assisted spectroscopy studies of Tc, Ru, Rh and Pd isotopes. Further development of the isobaric beam purification in the Penning trap resulted in a preliminary extension of the ^{115}Rh level scheme fed by β decay [8], discovery of an isomeric state in ^{115}Ru and finding a new half-life of its ground state β-decay [9], see Fig. 1. It was also possible to study excited states in ^{115}Pd, populated as a granddaughter in the β^- decay chain of the trap-purified ^{115}Ru [10]. Monoisotopic samples of ^{113}Tc were used to determine its β decay half-life by observing the time decay pattern of the 98.5 keV γ transition in ^{113}Ru, see Fig. 2. The experimental half-life value of 154(6) ms is in accord with the previous results obtained with the IGISOL separator: 130(50) ms [11], 170(20) ms [12], and with in-flight fragment separator 160^{+50}_{-40} ms [13]. Recent trap-assisted studies of ^{109,111}Mo β decays provided new data on the structure of low-lying states in the daughter Tc isotopes, in particular signatures of oblate deformation in ^{111}Tc were found [14]. The results for ^{109}Tc will be published soon.

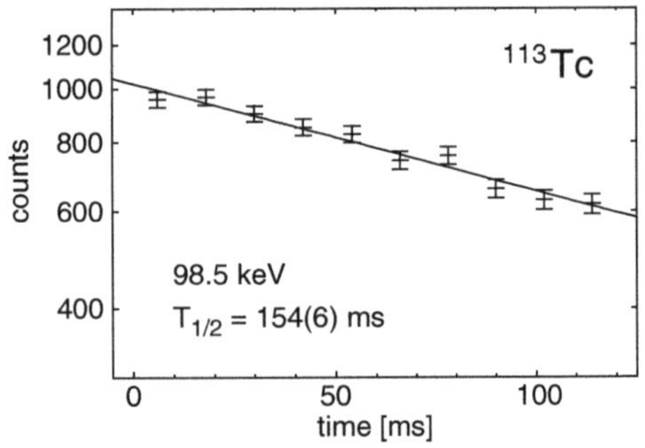

Fig. 2 Decay of 98.5 keV gamma line fed by β^- decay of ^{113}Tc monoisotopic samples

3 Future of trap-assisted spectroscopy

Our study of Tc, Ru, Rh and Pd isotopes is specifically motivated by the question of the structure of very neutron-rich isotopes in the region located close to the path of the astrophysical r-process. While the r-process nuclei are not yet accessible experimentally, the study of somewhat lighter isotopes may help to predict their properties, which are presently estimated only theoretically. For instance, the theory suggests that around the neutron number $N = 70$, nuclei in this region develop an oblate shape, which becomes the ground-state configuration at $N > 70$ [30]. On the other hand, the available experimental data for neutron-rich Pd isotopes are not in favor of such a picture. A recent theoretical work [31] on very neutron-rich Zr and Mo isotopes points to the important role of deformation and predicts almost equal potential minima for prolate and oblate shapes. Since the location of the r-process path depends on the nuclear deformation, it is now important to gather more experimental data in the critical region around $N = 70$.

The theory [30] points to Zr isotopic chain as the one where the prolate-oblate transition should be the most pronounced. However, zirconium nuclei with $N \sim 70$ are presently unreachable for spectroscopy. The Tc, Ru and Rh isotopic chains are the nearest to Zr, where one hopes to access the $N = 70$ line in experimental spectroscopic studies. A possible way to verify the nuclear shape is to examine the single-particle structure of odd-A and odd-odd nuclei. Such a structure in a prolate-deformed nuclear potential is different from that in an oblate one.

Below we present several ideas of improving and broadening the trap-assisted decay spectroscopy studies on the way towards more exotic nuclei located closer to the r-process path. The proposed enhancements are mainly based on the use of isobarically purified (so monoisotopic) ion beams in combination with various spectroscopic and mass separation methods. The monoisotopic beams are to be provided by the IGISOL mass separator coupled to JYFLTRAP system installed at the beam line of the new MCC30/15 cyclotron designed to deliver proton and deuteron beams with currents of 100 and 50 μA, respectively.

Combining β decay and prompt γ data Detailed experimental investigation of nuclear structure is greatly improved when more than one experimental technique is available. The odd nuclei from technetium to palladium have been intensively studied in β^- decay experiments at IGISOL. Over the past decade a new tool has been developed, namely the measurements of multiple-γ coincidences of prompt

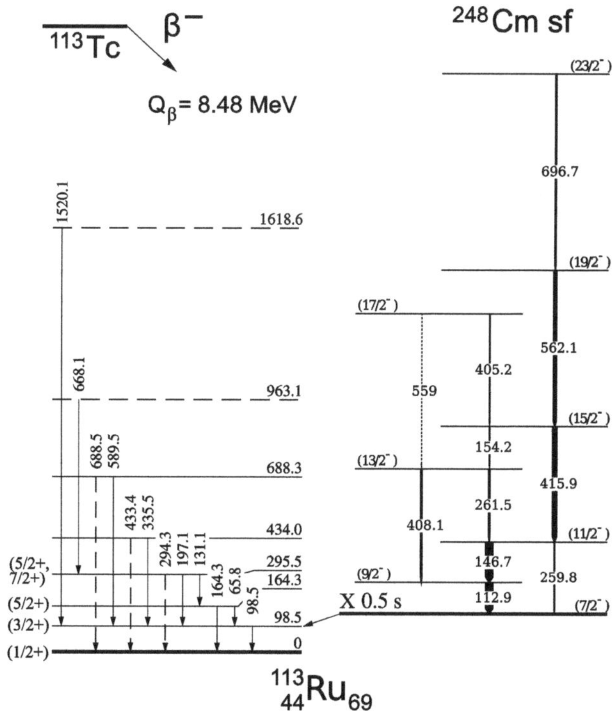

Fig. 3 Excited states in ^{113}Ru as an example of very different nuclear structures populated in the same nucleus by β decay (*left*) and spontaneous fission (*right*). Data on ^{113}Tc β^- decay comes from IGISOL [17, 32] and on spontaneous fission of ^{248}Cm from prompt gamma-ray spectroscopy with the EUROGAM 2 array [21]

γ-rays following fission, collected using large arrays of germanium spectrometers. The structure of excited levels in a given nucleus fed by β-decay and spontaneous fission differs significantly. The spin values of the levels populated in β-decay are similar to the spin of the mother which is usually low. Sometimes a β-decaying isomer populates high-spin excited states, for example the 0.5 s isomer in ^{113}Ru [17]. High-energy levels populated in the daughter by β-decay depopulate either by many weak low-energy γ transitions or a few high-energy ones. Both types are hardly detectable with germanium spectrometers due to their limited efficiency. Most often low-spin and low-energy nuclear levels are populated in the daughter nucleus via beta decay whereas spontaneous fission fragments are produced in high-spin, high-energy states which usually de-excite by long cascades of γ transitions. The latter reveals band structures of nuclear excited levels. The β decay data, among others, provides information on the band heads, especially their location relative to the ground state which is often hard to determine from prompt γ-ray studies. Furthermore, β decay spectroscopy is sensitive to phenomena at the timescale of milliseconds, like some isomeric states, which cannot be measured by prompt γ spectroscopy.

Combining both methods proved to be very successful as they provide complementary information while there is still some overlap for consistency checks. The IGISOL results on ^{113}Tc and ^{113}Ru decays [17, 32] with complementing ^{248}Cm prompt γ spectroscopy data enabled to identify a new band in ^{113}Ru [21]. Thus, a revision of spins and parities for the states in ^{113}Ru and its daughter ^{113}Rh was possible, see Fig. 3.

Half-life measurements with monoisotopic beams The half-life values of the exotic, neutron-rich nuclei are directly useful in the models of astrophysical *r*-process. The use of monoisotopic beam highly reduces background counts in the γ spectra thus considerably improving the peak to background ratio. Such conditions permit half-lives to be measured with a relatively low uncertainty as in the above mentioned example of ^{113}Tc β^- decay (Fig. 2).

To determine the β-decay half-life one usually measures the half-lives of a few (or at least one) of the most intensive γ lines in the daughter. Such decay measurements are usually done after a mass separation of the reaction products. A typical mass separator delivers a mixture of the desired nuclide and a few others which may be considered as background (although often more intensive than the wanted nuclide). In case of the IGISOL facility the background is formed by the less exotic isobars more easily produced in fission.

It may accidentally happen that the γ line used for the half-life estimation contains an unwanted component (of a very similar energy unresolvable in a Ge spectrometer) emitted by one of the "background" isobars. As the half-lives of the unwanted isobars are always longer than of the most exotic one in the isobaric chain such a contamination makes the measured half-life longer. An illustrative example is the case of the ^{115}Ru β^- decay half-life. For the first time, it was measured as 740(80) ms [11] with the isobaric beam from the IGISOL separator. Later, reinvestigation of the same decay with a monoisotopic beam from the trap, showed a shorter value of 318(19) ms [9]. The longer half-life in the first measurement was most probably caused by an overlapping γ line emitted in the β^- decay of ^{115}Pd, naturally present in the IGISOL isobaric beam.

When doing trap-assisted spectroscopy study, the daughter activities in the chain of consecutive β decays are produced only by the decaying monoisotopic mother source. Therefore, any possible overlapping γ lines coming from the isobars are of low intensity as compared to the mother activity delivered by the trap.

Moreover, with the trap one may always unambiguously distinguish the γ lines fed by the most exotic β decay from the ones fed by the less exotic isobars. By setting the trap to deliver a beam of the daughter of the activity of interest, disturbing γ-lines constituting the unwanted isobaric background in the gamma spectra are recorded. The method is also very useful for constructing the decay schemes.

Angular correlations and polarization of γ-rays In order to get relevant physical information one must complement the level schemes with information on transition multipolarities and mixing ratios. The multipole order of a gamma transition can be found by measuring its angular distribution. The simplest setup of three germanium detectors arranged at 90° to each other, usually used for post-trap spectroscopy, can distinguish a dipole from a quadrupole transition. To achieve a higher resolution and sensitivity it is desirable to use a larger array of germanium detectors. In the case of odd nuclei it is especially important to use germanium detectors of high efficiency in the low-energy range, like those equipped with thin beryllium windows.

The experiments using the γ–γ angular correlation technique were already successfully carried out at the IGISOL facility for palladium isotopes [33, 34]. This technique will benefit from the purity achieved with JYFLTRAP.

The angular distribution measurements cannot distinguish between the electric and magnetic multipoles which is possible via a determination of the linear polarization of the γ radiation. Moreover, linear polarization measurements make it possible to unambiguously find multipole mixing ratios. The basic technique for finding polarization of γ-rays is Compton scattering. In odd nuclei one expects a high density of γ lines at low energies. Thus, a good resolution in the low-energy range is an important parameter. An example of a germanium Compton polarimeter with a high resolution at low energies is described in [35]. It is a planar device with electrically defined sectors, compactly assembled in one cryostat and capable of completing the polarization measurements without changing its orientation.

Measurements of conversion electrons The β-decay process often populates excited states in the daughter nucleus which de-excite by γ transitions or via the competing process of internal conversion. For a chosen nucleus, the internal conversion probability increases with the multipole order of the transition and decreases strongly with the energy (inversely proportional to the 3rd or 4th power). The internal conversion coefficient (ICC), defined as the probability of electron emission relative to γ emission, carries information on the parity and multipole order of the transition.

Low-energy and high-multipolarity transitions are often associated with isomers. Some isomeric states in odd palladium and ruthenium isotopes are shown in Fig. 1. For highly-converted transitions, the ICC can be determined using only γ spectrometers with the intensity balance and the fluorescence methods if there exists a γ line in coincidence with the isomeric transition. Both methods depend on the details of the decay scheme and some other factors like coincidence efficiency. Additionally the fluorescence method depends on the observation of x-rays. For Tc, Ru, Rh and Pd nuclei these are in practice K_α x-rays of energies around 18–21 keV.

A more universal method for determining ICCs is to record an electron spectrum with a dedicated detector. To find the transition multipolarity one determines experimentally the intensity ratio of the K and L electron peaks, and γ spectrum is not necessary. The traditional conversion electron spectroscopy can be considerably improved by measuring conversion electrons from monoisotopic samples inside a Penning trap. A radioactive source in a form of an ion cloud trapped by the magnetic field in vacuum does not suffer from interactions of emitted electrons with any surrounding material. The magnetic field of the trap transports the electrons to a detector placed far away from the source. The feasibility of such in-trap conversion electron spectroscopy has already been demonstrated at JYFLTRAP [36]. Among others, conversion electron spectra were measured for a short living, $T_{1/2} = 19.1(7)$ ms, isomeric state in ^{117}Pd [36]. It is important to continue these in-trap conversion electron studies to fix the multipolarities of the transitions in the exotic nuclei accessible at IGISOL.

Identification of isomeric transitions Isobarically purified ion samples are very useful for the identification of new isomeric transitions. The lack of contamination from unwanted reaction products and a high intensity of the nuclei of interest let us observe new isomeric transitions directly in the singles spectra. When using a germanium detector with a high efficiency at low energies one can observe both the isomeric γ line and the conversion induced x-rays. By comparing the β-gated and the single γ spectra one can immediately identify the isomeric γ line as it is not present when a coincidence with a β particle is required. As an example, the single and

β-gated γ-ray spectra for the case of the 62 keV isomeric transitions in ^{115}Ru are shown in Fig. 5 of [9].

Ground-state branching in β decay There is no direct way to determine the ground state to ground state beta decay in a typical decay spectroscopy measurement using only γ spectrometers and thin β counters (for β-γ coincidence). There is no γ-ray to be associated with such a transition and the energy cannot be determined with a transmission β detector. Furthermore, it is difficult to reliably find the efficiency of such a detector at high β energies typical for ground-state transitions of very exotic nuclei with Q_β values close to 10 MeV. At JYFLTRAP the number of ions released from the trap can be counted with a movable microchannel plate (MCP) detector inserted into the beam. When the MCP is lifted up and the beam is implanted at the spectroscopy station the beam intensity is indirectly monitored by the number of γ and β events in the germanium and scintillation detectors, respectively. The fraction of the ions lost on the way from the MCP to the spectroscopy station is estimated by numerical simulations of ion trajectories. As the beam is purely monoisotopic, the total β-decay intensity can be measured with a combination of the MCP ion counts and the γ and β decay intensities measured at the spectroscopy station. Consequently, there is an experimental base to estimate the ground-state branching relevant for the determination of β feedings and log ft values to the other populated levels.

Besides the above discussed ideas there are numerous ways to use the traps of the JYFLTRAP to complement and improve the quality of decay studies of refractory elements. The precision Penning trap is capable of determining nuclear masses and Q_β values with a high accuracy. A new method of preparing isomerically clean samples of short-lived nuclei at JYFLTRAP [37] opens a field of trap-assisted spectroscopy of pure isomeric decays. Using a neutron detector after the trap for studying β-delayed neutron emission is of growing interest as the Q_β values increase when going further from the stability. Last but not least, trap-assisted spectroscopy at IGISOL may offer new findings for the nuclides studied long time ago with less selective methods like chemical separation.

It is interesting to estimate the post-trap yields expected with the new MCC30/15 cyclotron providing proton beam currents up to 100 μA for the new IGISOL facility. Among the odd, neutron-rich Tc to Pd nuclei most of the trap-assisted measurements were done for the $A = 115$ isobars. There, the last nuclide accessible for post-trap spectroscopy was ^{115}Ru and the yield for the more exotic ^{115}Tc was 2.1 ions/s for a 4 μA proton beam and a 100 ms trap cycle. At higher cyclotron beam intensities, the IGISOL yield scales rather with the square root of the beam intensity [1]. Thus, increasing the proton beam current from 4 to 100 μA should increase the IGISOL yield by a factor of 5. Assuming that the Penning trap can fully accept the increased separator beam, a yield of 10.5 ions/s sufficient for post-trap spectroscopy studies of ^{115}Tc is expected. As the ^{115}Tc half-life is 73^{+32}_{-22} ms [38], using a trap cycle shorter than 100 ms would decrease the decay losses and provide higher yields. A decrease in the mass resolving power due to a shorter trap cycle is not so critical as the mass differences between neighboring isobars increase when going towards more exotic nuclei. Therefore, the most exotic isobars can be well separated even with a lower mass resolving power related to a shorter trap cycle. In conclusion, trap-assisted spectroscopy experiments at the new IGISOL facility may reach neutron-rich nuclei of interest located one isobar further away from the stability.

References

1. Äystö, J.: Nucl. Phys. A **693**, 477 (2001)
2. Penttilä, H.: PhD Thesis, Dept. of Physics, University of Jyväskylä, Research Report No. 1/1992
3. Nieminen, A., Huikari, J., Jokinen, A., Äystö, J., Campbell, P., Cochrane, E.C.A., EXOTRAPS Collaboration: Nucl. Instrum. Methods A **469**, 244 (2001)
4. Kolhinen, V., Eronen, T., Hager, U., Hakala, J., Jokinen, A., Kopecky, S., Rinta-Antila, S., Szerypo, J., Äystö, J.: Nucl. Instrum. Methods A **528**, 776 (2004)
5. Jokinen, A., Eronen, T., Hager, U., Moore, I., Penttilä, H., Rinta-Antila, S., Äystö, J.: Int. J. Mass Spectrom. **251**, 204 (2006)
6. Rinta-Antila, S., Eronen, T., Elomaa, V.-V., Hager, U., Hakala, J., Jokinen, A., Karvonen, P., Penttilä, H., Rissanen, J., Sonoda, T., Saastamoinen, A., Äystö, J.: Eur. Phys. J. A **31** 1 (2007)
7. Kurpeta, J., Elomaa, V.-V., Eronen, T., Hakala, J., Jokinen, A., Karvonen, P., Moore, I., Penttilä, H., Płochocki, A., Rahaman, S., Rinta-Antila, S., Rissanen, J., Ronkainen, J., Saastamoinen, A., Sonoda, T., Urban, W., Weber, Ch., Äystö, J.: Eur. Phys. J A **31**, 263 (2007)
8. Kurpeta, J., Rissanen, J., Elomaa, V.-V., Eronen, T., Hakala, J., Jokinen, A., Karvonen, P., Moore, I.D., Penttilä, H., Płochocki, A., Rahaman, S., Rinta-Antila, S., Ronkainen, J., Saastamoinen, A., Sonoda, T., Szerypo, J., Urban, W., Weber, Ch., Äystö, J.: Act. Phys. Pol. B **41**, 469 (2010)
9. Kurpeta, J., Rissanen, J., Płochocki, A., Urban, W., Elomaa, V.-V., Eronen, T., Hakala, J., Jokinen, A., Kankainen, A., Karvonen, P., Małkiewicz, T., Moore, I.D., Penttilä, H., Saastamoinen, A., Simpson, G.S., Weber, C., Äystö, J.: Phys. Rev. C **82**, 064318 (2010)
10. Kurpeta, J., Urban, W., Płochocki, A., Rissanen, J., Elomaa, V.-V., Eronen, T., Hakala, J., Jokinen, A., Kankainen, A., Karvonen, P., Moore, I.D., Penttilä, H., S. Rahaman, Saastamoinen, A., Sonoda, T., Szerypo, J., Weber, C., Äystö, J.: Phys. Rev. C **82**, 027306 (2010)
11. Äystö, J., Astier, A., Enqvist, T., Eskola, K., Janas, Z., Jokinen, A., Kratz, K.-L., Leino, M., Penttilä, H., Pfeiffer, B., Żylicz, J.: Phys. Rev. Lett. **69**, 1167 (1992)
12. Wang, J.C., Dendooven, P., Hannawald, M., Honkanen, A., Huhta, M., Jokinen, A., Kratz, K.-L., Lhersonneau, G., Oinonen, M., Penttilä, H., Peräjärvi, K., Pfeiffer, B., Äystö, J.: Phys. Lett. B **454**, 1 (1999)
13. Pereira, J., Hennrich, S., Aprahamian, A., Arndt, O., Becerril, A., Elliot, T., Estrade, A., Galaviz, D., Kessler, R., Kratz, K.-L., Lorusso, G., Mantica, P.F., Matos, M., Möller, P., Montes, F., Pfeiffer, B., Schatz, H., Schertz, F., Schnorrenberger, L., Smith, E., Stolz, A., Quinn, M., Walters, W.B., Wöhr, A.: Phys. Rev. C **79**, 035806 (2009)
14. Kurpeta, J., Urban, W., Płochocki, A., Rissanen, J., Pinston, J.A., Elomaa, V.-V., Eronen, T., Hakala, J., Jokinen, A., Kankainen, A., Karvonen, P., Moore, I.D., Penttilä, H., Saastamoinen, A., Weber, C., Äystö, J.: Phys. Rev. C **84**, 044304 (2011)
15. Fong, D., Hwang, J.K., Ramayya, A.V., Hamilton, J.H., Luo, Y.X., Gore, P.M., Jones, E.F., Walters, W.B., Rasmussen, J.O., Stoyer, M.A., Zhu, S.J., Lee, I.Y., Macchiavelli, A.O., Wu, S.C., Daniel, A.V., Ter-Akopian, G.M., Oganessian, Yu.Ts., Cole, J.D., Donangelo, R., Ma, W.C.: Phys. Rev. C **72**, 014315 (2005)
16. Penttilä, H., Enqvist, T., Jauho, P.P., Jokinen, A., Leino, M., Parmonen, J.M., Äystö, J., Eskola, K.: Nucl. Phys. A **561**, 416 (1993)
17. Kurpeta, J., Lhersonneau, G., Płochocki, A., Wang, J.C., Dendooven, P., Honkanen, A., Huhta, M., Oinonen, M., Penttilä, H., Peräjärvi, K., Persson, J.R., Äystö, J.: Eur. Phys. J. A **13**, 449 (2002)
18. Äystö, J., Taskinen, P., Yoshii, M., Honkanen, J., Jauho, P., Penttilä, H., Davids, C.N.: Phys. Lett. **B201**, 211 (1988)
19. Franz, G., Hermann, G., J. Inorg. Nucl. Chem. **40**, 945 (1978)
20. Lhersonneau, G., Pfeiffer, B., Alstad, J., Dendooven, P., Eberhardt, K., Hankonen, S., Klöckl, I., Kratz, K.-L., Nähler, A., Malmbeck, R., Omtvedt, J.P., Penttilä, H., Schoedder, S., Skarnemark, G., Trautmann, N., Äystö, J.: Eur. Phys. J. A **1**, 285 (1998)

21. Kurpeta, J., Urban, W., Droste, Ch., Płochocki, A., Rohoziński, S.G., Rząca-Urban, T., Morek, T., Próchniak, L., Starosta, K., Äystö, J., Penttilä, H., Durell, J.L., Smith, A.G., Lhersonneau, G., Ahmad, I.: Eur. Phys. J. A **33**, 307 (2007)
22. Pfeiffer, B., Lhersonneau, G., Dendooven, P., Honkanen, A., Huhta, M., Klöckl, I., Oinonen, M., Penttilä, H., Persson, J.R., Peräjärvi, K., Wang, J.C., Kratz, K.-L., Äystö, J.: Eur. Phys. J. A **2**, 17 (1998)
23. Urban, W., Rząca-Urban, T., Droste, Ch., Rohoziński, S.G., Durell, J.L., Phillips, W.R., Smith, A.G., Varley, B.J., Schulz, N., Ahmad, I., Pinston, J.A.: Eur. Phys. J. A **22**, 231 (2004)
24. Penttilä, H., Taskinen, P., Jauho, P., Koponen, V., Davids, C.N., Äystö, J.: Phys. Rev. C **38**, 931 (1988)
25. Mehren, T., Pfeiffer, B., Schoedder, S., Kratz, K.-L.: Phys. Rev Lett. **77**, 458 (1996)
26. Urban, W., Rząca-Urban, T., Durell, J.L., Smith, A.G., Ahmad, I.: Eur. Phys. J. A **24**, 161 (2005)
27. Luo, Y.X., Hamilton, J.H., Rasmussen, J.O., Ramayya, A.V., Stefanescu, I., Hwang, J.K., Che, X.L., Zhu, S.J., Gore, P.M., Jones, E.F., Fong, D., Wu, S.C., Lee, I.Y., Ginter, T.N., Ma, W.C., Ter-Akopian, G.M., Daniel, A.V., Stoyer, M.A., Donangelo, R., Gelberg, A.: Phys. Rev. C **74**, 024308 (2006)
28. Luo, Y.X., Rasmussen, J.O., Hamilton, J.H., Ramayya, A.V., Hwang, J.K., Zhu, S.J., Gore, P.M., Wu, S.C., Lee, I.Y., Fallon, P., Ginter, T.N., Ter-Akopian, G.M., Daniel, A.V., Stoyer, M.A., Donangelo, R., Gelberg, A.: Phys. Rev. C **70**, 044310 (2004)
29. Urban, W., Droste, Ch., Rząca-Urban, T., Złomaniec, A., Durell, J.L., Smith, A.G., Varley, B.J., Ahmad, I.: Phys. Rev. C **73**, 037302 (2006)
30. Xu, F.R., Walker, P.M., Wyss, R.: Phys. Rev. C **65**, 021303 (2002)
31. Sarriguren, P., Pereira, J.: Phys. Rev. C **81**, 064314 (2010)
32. Kurpeta, J., Lhersonneau, G., Wang, J.C., Dendooven, P., Honkanen, A., Huhta, M., Oinonen, M., Penttilä, H., Peräjärvi, K., Persson, J.R., Płochocki, A., Äystö, J.: Eur. Phys. J. A **2**, 241 (1998)
33. Lhersonneau, G., Wang, J.C., Hankonen, S., Dendooven, P., Jones, P., Julin, R., Äystö, J.: Eur. Phys. J. A **2**, 25 (1998)
34. Lhersonneau, G., Wang, J.C., Hankonen, S., Dendooven, P., Jones, P., Julin, R., Äystö, J.: Phys. Rev. C **60**, 014315 (1999)
35. Sareen, R.A., Urban, W., Barnett, A.R., Varley, B.J.: Rev. Sci. Instrum. **66**, 3653 (1995)
36. Rissanen, J., Elomaa, V.-V., Eronen, T., Hakala, J., Jokinen, A., Rahaman, S., Rinta-Antila, S., Äystö, J.: Eur. Phys. J. A **34**, 113 (2007)
37. Eronen, T., Elomaa, V.-V., Hager, U., Hakala, J., Jokinen, A., Kankainen, A., Rahaman, S., Rissanen, J., Weber, C., Äystö, J.: Nucl. Instrum. Methods B **266**, 4527 (2008)
38. Montes, F., Estrade, A., Hosmer, P.T., Liddick, S.N., Mantica, P.F., Morton, A.C., Mueller, W.F., Ouellette, M., Pellegrini, E., Santi, P., Schatz, H., Stolz, A., Tomlin, B.E., Arndt, O., Kratz, K.-L., Pfeiffer, B., Reeder, P., Walters, W.B., Aprahamian, A., Wöhr, A.: Phys. Rev. C **73**, 035801 (2006)

DOI 10.1007/978-94-007-5555-0_16
Reprinted from *Hyperfine Interactions* Volume 223, Issues 1-3, 2014
and *The European Physical Journal A* Volume 48, Issue 4, 2012

β-delayed neutron emission studies

M. B. Gómez-Hornillos · J. Rissanen · J. L. Taín · A. Algora · K. L. Kratz · G. Lhersonneau · B. Pfeiffer · J. Agramunt · D. Cano-Ott · V. Gorlychev · R. Caballero-Folch · T. Martínez · L. Achouri · F. Calvino · G. Cortés · T. Eronen · A. García · M. Parlog · Z. Podolyak · C. Pretel · E. Valencia · IGISOL Collaboration

Published online: 2 May 2012

Abstract The study of β-delayed neutron emission plays a major role in different fields such as nuclear technology, nuclear astrophysics and nuclear structure. However the quality of the existing experimental data nowadays is not sufficient for the various technical and scientific applications and new high precision measurements are necessary to improve the data bases. One key aspect to the success of these high precission measurements is the use of a very pure ion beam that ensures that only the

M. B. Gómez-Hornillos (✉) · V. Gorlychev · R. Caballero-Folch · F. Calvino · G. Cortés · C. Pretel
Secció d'Enginyeria Nuclear, Universitat Politecnica de Catalunya, Barcelona, Spain
e-mail: belen.gomez@upc.edu

J. Rissanen · T. Eronen · IGISOL Collaboration
JYFL, Jyväskylä, Finland

J. L. Taín · A. Algora · J. Agramunt · E. Valencia
Instituto de Física Corpuscular, CSIC-Universidad de Valencia, Valencia, Spain

K. L.Kratz
MPI für Chemie, Otto-Hahn-Institut, Mainz, Germany

G. Lhersonneau
GANIL, Caen, France

B. Pfeiffer
GSI, Darmstadt, Germany

D. Cano-Ott · T. Martínez · A. García
CIEMAT, Madrid, Spain

L. Achouri · M. Parlog
LPC, Caen, France

Z. Podolyak
University of Surrey, Guildford, UK

ion of interest is produced. The combination of the IGISOL mass separator with the JYFLTRAP Penning trap is an excellent tool for this type of measurement because of the ability to deliver isobarically and even isomerically clean beams. Another key feature of the installation is the non-chemical selectivity of the IGISOL ion source which allows measurements in the important region of refractory elements. This paper summarises the β-delayed neutron emission studies that have been carried out at the IGISOL facility with two different neutron detectors based on ^{3}He counters in a polyethylene moderator: the Mainz neutron detector and the BEta deLayEd Neutron detector.

Keywords Beta delayed • Neutron emission • ^{3}He counter

1 Introduction

β-delayed neutron emission takes place when a precursor nucleus β-decays and the resulting daughter-nucleus emits a neutron. This neutron emission is energetically allowed if the excitation energy of the state populated in the β-decay is larger than the neutron separation energy of the daughter nucleus, S_n.

The study of β-delayed neutron emission probabilities, P_n, is of interest in different fields, such as nuclear technology applications, nuclear astrophysics and nuclear structure.

The technological interest of this type of study is related to nuclear power generation. In nuclear fission β-delayed neutron emission plays an essential role in safely controlling the sustainability of the fission reaction. Research into such nuclei is, therefore, fundamental for the design of safer and more efficient nuclear reactors. The Nuclear Energy Agency (NEA) highlights the importance of experimental measurements and data evaluation of delayed neutron emission in its working group 6, "Delayed neutron data" [1]. Furthermore, recently the International Atomic Energy Agency (IAEA) has held a consultants meeting on beta delayed neutron emission in order to boost further work on this field [2].

In nuclear astrophysics, the delayed neutron emission modulates the element abundance curve of stellar nucleosynthesis [3, 4]. Improved experimental data from delayed neutron emission represents an important input to r-process model calculations since properties of nuclei on the expected r-process path can be predicted by extrapolation on the basis of systematics of experimental $T_{1/2}$ and P_n values.

Furthermore, in nuclear structure, β-delayed neutron emission constitutes an important probe for the structure of neutron-rich nuclei far away from the valley of stability where other measurements are not yet possible [5–7]. The probability of neutron emission after β-decay, P_n, carries information on the β-strength just above the neutron separation energy, S_n.

However the quality of the existing experimental data nowadays is not sufficient for the various technical and scientific applications and it is necessary to perform new high precision measurements. This paper presents the studies of neutron emission probability after β-decay, P_n, carried out at IGISOL. The work with two different detectors, the Mainz neutron detector and the BEta deLayEd Neutron detector, BELEN, is summarised here. Both detectors consist of ^{3}He counters embedded in a polyethylene matrix. In the case of the Mainz neutron detector, experiments were carried out in the late 90s before the Penning trap was available and therefore the ion beams delivered by IGISOL were mass separated with relatively low mass resolution, typically M/δM of about several hundred. In the experiments with BELEN the Penning trap was fully functional and it allowed additional Z separation thus delivering a pure beam of the ion of interest.

The high purity of the ion beam is a key feature for this type of neutron detector since this measurement technique with ^{3}He cannot discriminate contaminant neutrons from those originating from the ion of interest. Neither does this technique allow discrimination of background neutrons. Another key advantage of IGISOL is the non-chemical selectivity of the ion-guide method which allows the production of radioactive beams of any element including the refractory elements which are very difficult to produce at other mass separators equipped with more conventional ion sources.

2 Mainz 4π neutron detector at IGISOL

Two experiments have been performed at the IGISOL facility to measure β-delayed neutron emission with the Mainz 4π neutron long counter [8]. This detector is formed by 42 ^{3}He tubes embedded in a polyethylene matrix and arranged in two crowns around the beam hole. The efficiency of this counter was determined to be (24.9 $\pm$ 0.2)% using a combination of two calibrated sources, Am/Li and ^{252}Cf, and online with the well-known neutron emitter ^{95}Rb [9].

The experimental technique at IGISOL was as follows. A beam of mass-separated short-lived ions was implanted directly onto a collection tape placed inside the neutron long counter. At the implantation position there was also a plastic scintillator for the detection of the β-particles and a Germanium detector that was used for the identification of the implanted ions via their corresponding γ-rays. The collection tape was moved periodically in order to remove the contamination from longer lived species.

2.1 P_n-values of very neutron rich Y to Tc isotopes

A first measurement was carried out by Mehren et al. [10] and it aimed at the measurement of very neutron rich isotopes from yttrium to technetium.

The ions were produced in fusion-fission reactions by bombarding a uranium target with a 50 MeV beam of H_2^+ of 6–8 μA intensity from the K = 130 MeV cyclotron at the University of Jyväskylä. The IGISOL separator was used to produce and separate isobarically pure beams of ions with a delay in the order of milliseconds. This experiment was carried out before the Penning trap was available, therefore there was no Z-separation of the different species.

The data was recorded as delayed coincidences between the neutron detection and the β-decay [6] in order to minimize the background. In this technique the coincidence rate was measured by an overlap technique in which the β-signal was stretched up to 40 μs by a gate-delay generator. In order to measure the random coincidences from the background a second coincidence was performed with the same stretched β-signal and a neutron signal which had been delayed by 45 μs. The use of this technique removes the dependence on the β-detection efficiency which is usually a source of error.

Table 1 Experimental values from this work [10, 11] compared to QRPA predictions [12]

Nuclide	$T_{1/2}$ (s)		P_n (%)	
	Experiment	Theory	Experiment	Theory
^{94}Kr	0.3 ± 0.1	0.44		0.55
^{99}Y	1.48 ± 0.02	0.93	2.5 ± 0.5	3.7
^{100g}Y	0.71 ± 0.03	0.29	1.8 ± 0.6	0.9
^{101}Y	0.40 ± 0.02	0.14	1.5 ± 0.5	1.4
^{102g}Y	0.29 ± 0.02	0.18	4.0 ± 1.5	1.6
^{103}Y	0.23 ± 0.02	0.08	8 ± 3	4.6
^{104}Y	0.18 ± 0.06	0.035		
^{105}Zr	0.6 ± 0.1	0.095		0
^{104g}Nb	5.0 ± 0.4	2.07	0.06 ± 0.03	0
^{105}Nb	2.8 ± 0.1	2.73	1.7 ± 0.9	0.26
^{106}Nb	0.9 ± 0.02	0.15	4.5 ± 0.3	0.27
^{107}Nb	0.30 ± 0.03	0.45	6.0 ± 1.5	3.7
^{108}Nb	0.19 ± 0.02	0.23	6.2 ± 0.5	11
^{109}Nb	0.19 ± 0.03	0.28	31 ± 5	26
^{110}Nb	0.17 ± 0.02	0.19	40 ± 8	20
^{109}Tc	0.8 ± 0.1	0.42	0.08 ± 0.02	0.02
^{110}Tc	0.78 ± 0.15	0.25	0.04 ± 0.02	0.075
^{111}Tc	0.29 ± 0.02	0.235	0.85 ± 0.2	0.53
^{112}Tc	0.29 ± 0.02	0.26	1.5 ± 0.2	1.1
^{113}Tc	0.17 ± 0.02	0.13	2.1 ± 0.3	
^{114}Tc	0.15 ± 0.03	0.08	1.3 ± 0.4	

The QRPA predictions for ^{94}Kr and ^{99}Y take into account the shape coexistance in these nuclei

If the neutron efficiency of the detector ϵ_n is known, the probability of β-delayed neutron emission P_n, can be expressed in terms of the number of β-neutron coincidences $n_{\beta n}$ and the number of β decays of the precursor, n_β, using the following equation,

$$P_n = \frac{n_{\beta n}}{\epsilon_n \cdot n_\beta} \tag{1}$$

The data from this experiment was also used to obtain the half-lives, $T_{1/2}$, of the β-delayed neutron precursors via a fit to the growth and decay curves. The β-decay half-lives of ^{103}Y, $^{108-110}$Nb were reported for the first time. For the isotopes for which previous experimental values existed there was good agreement with the values obtained in this experiment.

P_n values were obtained for ^{99}Y, ^{100g}Y, ^{101}Y, ^{102g}Y, ^{103}Y, ^{104g}Nb, ^{104m}Nb, ^{105}Nb, ^{106}Nb, ^{107}Nb, ^{108}Nb, ^{109}Nb, ^{110}Nb, ^{109}Tc, ^{110}Tc, ^{111}Tc and ^{112}Tc, out of which there were 13 new P_n values and several isotopes of yttrium whose P_n values, already known, were remeasured with improved statistics and higher reliability. The results are presented in Table 1. The higher reliability in the measurement is due to the fact that previously the yttrium isotopes had been obtained as decay products of rubidium or strontium, which introduced some uncertainty. However at IGISOL, refractory elements, such as yttrium can be obtained directly.

This experiment produced isotopes in the mass range from A = 99(Y) to A = 112(Tc) which allowed the authors to obtain the isotopic yield curves for fission products which can be used to test and improve theoretical models.

2.2 β-delayed neutron decay of ^{104}Y and $^{112-114}Tc$

A second measurement with the Mainz neutron detector was carried out by Wang et al. [11] and used the same experimental setup described above.

This experiment was a continuation of the systematic investigation started in the previous measurement. In this later experiment more neutron-rich nuclei were studied with a significant yield increase due to improvements in the ion guide and to an increased accelerator beam intensity resulting in isobaric yields of the order of 10^5 ions/s for medium mass fission-product beams.

The ions were produced in 25 MeV proton-induced fission of ^{238}U and were mass-separated by the IGISOL separator magnet. Signals from the γ-rays and from the observation of neutrons and β-particles were time-stamped within the measurement. These time-gated spectra were used to obtain the half-lives and production rates from the growth and decay curves.

The half-lives of the β-delayed neutron precursors were determined from the neutron time spectra by fitting the growth and decay periods. Only one component and a constant background were required for a satisfactory fit. The half-lives of ^{104}Y and ^{114}Tc were obtained for the first time. For ^{112}Tc and ^{113}Tc there existed previous experimental values that were in good agreement with the values found by Wang et al.

The β-delayed neutron branching ratios of ^{112}Tc, ^{113}Tc and ^{114}Tc were determined from the ratio of the neutron- and β-intensities obtained from the growth and decay curves. The low counting rate of ^{104}Y did not allow for an extraction of the P_n value. The values for ^{113}Tc and ^{114}Tc were reported for the first time in this work. The good statistics for ^{112}Tc resulted in a more precise value than in the previous attempt [10].

This data set also provided values for production yields of neutron-rich nuclei in proton-induced fission of ^{238}U.

2.3 Discussion of experimental results

The experimental values obtained by Mehren [10] and Wang [11] are presented in Table 1. The $T_{1/2}$ and P_n values were compared to predictions of an unpublished QRPA model [12] based on the FRDM masses. This QRPA calculation included Gamow–Teller and First Forbidden transitions. According to this model a trend of decreasing $T_{1/2}$ and increasing P_n value with increasing neutron number is expected. In general the P_n-values were larger than expected which could be due to a more abrupt decrease of the S_n values for $N \geq 60$ than predicted by the models. The large P_n values obtained indicate the importance of β-transitions to high-lying levels with large neutron excess. The fluctuations of the QRPA predictions from the experimental values can be explained taking into account the rapid changes in energies and ordering of the single particle orbitals that give rise to deformations and shape coexistance phenomena. For example, the β-decay daughters of ^{94}Kr and ^{99}Y have spherical-prolate shape coexistence. Furthermore $^{106-108}Nb$ and $^{109-111}Tc$ are predicted by Möller et al. [13] to be triaxial. While in the QRPA used in this calculation it is possible to account for shape coexistence, the model cannot account for triaxial shapes, therefore, discrepancies with respect to experimental data are expected in the triaxal nuclei.

3 BEta deLayEd Neutron detector at IGISOL

A prototype version of the BEta deLayEd Neutron detector, which is being developed for the FAIR/DESPEC experiment, has been used for the first time in an experiment at JYFL. This detector is based on ^{3}He counters and its first run was primarily intended to commission the detector and verify the working principles for future experiments. A new triggerless data acquisition has been developed for these measurements. This DACQ time-stamps the events and allows complete flexibility to construct correlations offline.

In the version employed at JYFL the detector consisted of 20 ^{3}He proportional counters with an effective length of 60 cm, a diameter of 2.54 cm and a gas pressure of 20 atm. The counters were embedded in a polyethylene block of dimensions 90 × 90 × 80 cm^{3} and placed in two concentric crowns around the beam hole.

There are two polyethylene matrices for BELEN-20 whose difference is the radial position of the crowns and the radius of the beam hole. In the first version the counters are placed at a radius of 11 cm (8 counters) and 20 cm (12 counters) around a central longitudinal hole of 10 cm diameter. The detection efficiency for this version calculated with MCNPX (https://mcnpx.lanl.gov/) simulations is nearly constant below 1 MeV reaching 30% and decreases for higher energies (24% at 5 MeV). In the second version the first crown of detectors is placed at a radius of 9.5 cm and the second crown at a radius of 14.5 cm around a beam hole of 11 cm diameter. The average detection efficiency according to MCNPX simulations is 46% in the range from 1 keV to 1 MeV and it decreases to 29% at 5 MeV. Further details about the simulation work performed for this detector can be found in [14].

An isotopically pure beam was obtained using the JYFLTRAP Penning trap setup at the IGISOL facility and it was implanted on a movable tape placed in the centre of the BELEN-20 detector.

In this experiment, as well as with the Mainz detector, a measurement of the coincidence of a β-decay of the precursor and the neutron is required. However the main difficulty is the long moderation time of the neutron in the polyethylene (around 200 μs) which in a conventionally triggered system will require opening a long time window for the correlation. Such a long correlation window would cause a large dead time in the system. Therefore the BELEN-20 detector works with a purpose-built triggerless DACQ [15] where for energy signals above a certain threshold, time-energy pairs are registered independently for every channel. The GasificTL DACQ software builds the β-neutron coincidence with the desired correlation time. With this system the probability of neutron emission after a β-decay can be obtained with (1) and it is not subject to the uncertainty on the β-detection efficiency.

Two different experimental runs have been performed at IGISOL with the BELEN-20 detector. Each experiment was carried out with a different version of the polyethylene matrix.

3.1 First measurement with the BELEN detector at JYFL

In this measurement the first version of the BELEN-20 neutron detector was used and the known delayed neutron emitters of interest for nuclear power generation ^{88}Br, ^{94,95}Rb and ^{138}I were studied [16].

Fig. 1 BELEN neutron detector and Ge detector at JYFL

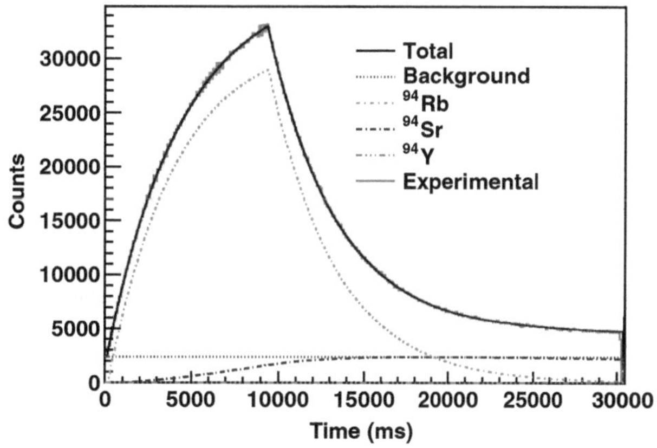

Fig. 2 Spectrum of the Si detector. Implantation and decay curve for ^{94}Rb and the fit of the Bateman equation

The radioactive species were produced at IGISOL by deuteron-induced fission ($E_d = 30$ MeV) on a uranium target. The IGISOL separator magnet was used for mass separation of the IGISOL beam which was then introduced into the Penning trap. The Penning trap was used as a very high resolution mass separator which removed all isotopes other than the one of interest. The isotopically pure beams extracted from the trap were delivered via a vacuum tube to the centre of the detector, where they were implanted on a movable tape. Two collimators were used to define the implantation position. The trap and tape transport cycles were adjusted in such a way that the radioactivity was accumulated during a period of $3 \cdot T_{1/2}$, while the measurement period was extended up to $10 \cdot T_{1/2}$ before removing the activity. A Si detector (0.9 mm thick, 25.2 mm diameter) for the detection of β-particles was placed closely behind the implantation position (distance 3 mm) also in vacuum. The instrumentation was complemented with an 80% efficiency HPGe detector for the detection of γ-rays, situated inside the central hole of the counter at a distance of 9 cm from the tape. The picture of the setup is presented in Fig. 1.

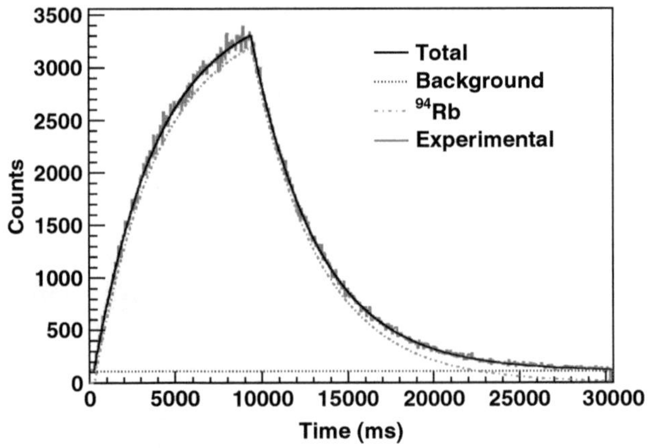

Fig. 3 Spectrum of the neutron detector. Implantation and decay curve for ^{94}Rb and the fit of the Bateman equation

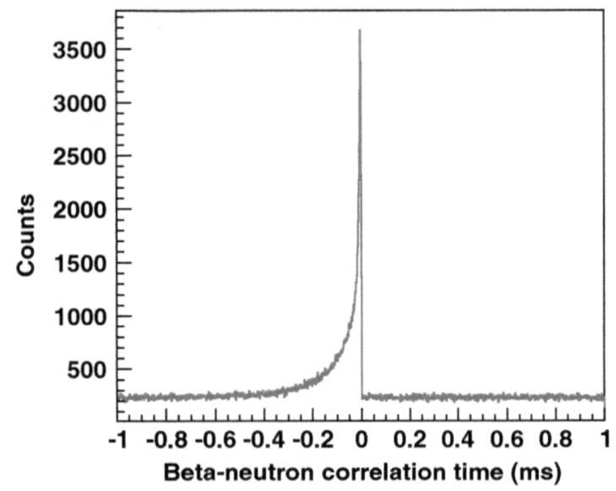

Fig. 4 Beta-neutron correlation spectrum

The measurements were performed for known delayed neutron emitters ^{88}Br, ^{94,95}Rb and ^{138}I. These emitters, with well known P_n values, together with measurements using a ^{252}Cf source were used to obtain the counter detection efficiency and to tune the Monte Carlo simulations.

In the preliminary analysis the growth and decay curves for the β-particles and the neutrons were fitted with the Bateman equation [17], see Figs. 2 and 3. These growth and decay curves were obtained by plotting the counts corresponding to the beta particles from the Si spectrum with respect to the cycle time and similarly with the neutron counts. The fit to the Bateman equation disentangles the contribution from each nucleus within the decay chain from the background and singles out the counts of the nucleus of interest.

The β-neutron coincidence spectrum, presented in Fig. 4, was built from the coincidences between the neutrons and all the beta particles in a 1 ms time window, forward and backward from the detection of a neutron. The true coincidences are the counts on left half of the spectrum on top of the flat background of random coincidences which is defined by the counts on the right half of the spectrum.

The P_n values of ^{88}Br and ^{95}Rb were used as references to calculate the efficiency of BELEN-20 since their values in references [9] and [18] have good agreement and small uncertainties. From these two values the average detection efficiency for BELEN-20 is $(27.1 \pm 0.8)\%$ (see Table 2).

Table 2 BELEN-20 detection efficiency obtained using ^{88}Br and ^{95}Rb [9] as calibration

Isotope	P_n (%) [9]	N_β	$N_{\beta n}$	ϵ_n
^{88}Br	6.58 ± 0.18	867,701	14,350	27.6 ± 0.7
^{95}Rb	8.73 ± 0.20	588,116	13,301	26.6 ± 0.8

Table 3 P_n values for ^{94}Rb and ^{138}I obtained in this work, compared to other authors [9] and [18]

Isotope	N_β	$N_{\beta n}$	P_n (%)	Author
^{94}Rb	3,005,635	83,768	10.28 ± 0.31	This work
			10.01 ± 0.23	Rudstam [9]
			9.1 ± 1.1	Pfeiffer [18]
^{138}I	343,890	4,955	5.32 ± 0.2	This work
			5.46 ± 0.18	Rudstam [9]
			5.17 ± 0.36	Pfeiffer [18]

Using the above average efficiency and (1), the P_n values were obtained for ^{94}Rb and ^{138}I. They were in good agreement with the values reported in [9] and [18] (see Table 3).

3.2 Current and future work

Further measurements have been planned at the IGISOL facility with the motivations of technological applications, nuclear structure and astrophysics.

Some of the nuclei that we plan to measure in the future are fission fragments that have a major contribution to the number of delayed neutrons in nuclear power reactors [19]. We will focus on the new data requirements for reactor technology because of the use of high burn-up fuels or the burning of minor actinides where the fission product inventory differs from that of conventional reactors.

These and other nuclei that we plan to study are also important from the nuclear structure point of view. This will be the case for nuclei around the doubly magic ^{78}Ni and ^{132}Sn where the P_n values are sensitive to differences in the beta strength distribution. The shell reordering for these very neutron rich nuclei is actively discussed and information can be gained from our new measurements since the P_n values are sensitive to the structure of the daughter and parent nuclei.

Finally we will also target nuclei along the *r*-process path, where the decay towards the line of stability happens by a series of β-decays, accompanied by the emission of delayed neutrons in the case of the most neutron-rich isotopes. This is also the case of the above mentioned regions around ^{78}Ni and ^{132}Sn.

Some of these nuclei were already measured in June 2010. A primary beam of protons at 25 MeV with an intensity of 7 μA produced fission products from a ^{232}Th target. The BELEN-20 detector had the second version of the polyethylene matrix. The rest of the experimental setup and acquisition system was identical to the one described in the previous section.

The isotopes measured were ^{85}Ge, ^{86}Ge, ^{85}As, ^{91}Br and ^{137}I. ^{88}Br and ^{95}Rb were also measured in order to obtain the neutron detector efficiency. The data analysis process is being carried out and it is similar to that described in Section 3.1. Results will be published in the future.

Acknowledgement This work was partially supported by Spanish FPA2008-04972-C03-03 grant.

References

1. Rudstam, G., et al.: Delayed neutron data for major actinides. http://www.oecd-nea.org/science/wpec/volume6/volume6.pdf(2002). Accessed 27 March 2012
2. Albiola, D., et al.: Beta delayed neutron emission evaluation. IAEA Report, INDC(NDS) 0599 (2011)
3. Kratz, K.-L., et al.: Isotopic r-process abundances and nuclear structure far from stability—implications for the r-process mechanism. Astrophys. J. **403**, 216 (1993)
4. Farouqi, K., et al.: Charged-particle and neutron-capture processes in the high-entropy wind of the core-collapse supernovae. Astrophys. J. **712**, 1359 (2010)
5. Kratz, K.-L., et al.: Beta-minus strength function phenomena of exotic nuclei—a critical examination of the signature of nuclear model predictions. Nucl. Phys. A **417**, 447 (1984)
6. Reeder, P.L., et al.: P_n measurements at Tristan by a β-n coincidence technique. In: Proceedings of the Specialist Meeting on Delayed Neutron Properties, p. 37. Univ. of Birmingham, Birmingham, England. ISBN 07044 0926 7 (1986)
7. Möller, P., Pfeiffer, B., Kratz, K.-L.: New calculations of gross β-decay properties for astrophysical applications: speeding-up the classical r-process. Phys. Rev. C **67**, 055802 (2003)
8. Reeder, P.L., et al: Average neutron energies from separated delayed-neutron precursors. Phys. Rev. C **15**(6), 2108 (1977)
9. Rudstam, G., et al.: Delayed neutron branching ratios of precursors in the fission product region. Atom. Data Nucl. Data Tab. **53**, 1 (1993)
10. Mehren, T., et al.: β-decay half-lives and neutron-emission probabilities of very neutron-rich Y to Tc isotopes. Phys. Rev. Lett. **77**, 3 (1996)
11. Wang, J.C., et al.: β-delayed neutron decay of ^{104}Y, ^{112}Tc, ^{113}Tc and ^{114}Tc: test of half-life predictions for neutron-rich isotopes of refractory elements. Phys. Lett. B **454**, 1 (1999)
12. Möller, P., Kratz, K.-L.: private communication (2006)
13. Möller, P., et al.: Axial and reflection asymmetry of the nuclear ground state. Atom. Data Nucl. Data Tab. **94**, 758 (2008)
14. Gómez-Hornillos, M.B., et al.: First Measurements with the BEta deLayEd Neutron detector (BELEN-20) at the JYFL penning trap. In: Proceedings of the International Nuclear Physics Conference 2010, Vancouver, Canada. Journal of Physics: Conference Series, vol. 312, p. 052008 (2011)
15. Agramunt, J., et al.: A triggerless DACQ system for nuclear decay experiments. Nucl. Instrum. Methods A (in preparation)
16. Gómez-Hornillos, M.B., et al.: Monte Carlo simulations for the study of a moderated neutron detector. In: Proceeding of the Int Conf on Nuclear Data for Science and Technology 2010, Korea. Journal of the Korean Physics Society, vol. 59, p. 1573 (2011)
17. Skrable, K.W., et al.: A general equation for the kinetics of linear first order phenomena and suggested applications. Health Phys. **27**, 155 (1974)
18. Pfeiffer, B. et al.: Status of delayed neutron precursor data: half lives and neutron emission probabilities. Prog. Nucl. Energy **41**, 39 (2002)
19. D'Angelo, A.: Overview of the delayed neutron data activities and results monitored by the NEA/WPEC subgroup 6. Prog. Nucl. Energy **41**, 5 (2002)

DOI 10.1007/978-94-007-5555-0_17
Reprinted from *Hyperfine Interactions* Volume 223, Issues 1-3, 2014
and *The European Physical Journal A* Volume 48, Issue 4, 2012

THE EUROPEAN PHYSICAL JOURNAL A

Review

Isomer and decay studies for the rp process at IGISOL

A. Kankainen[1,a], Yu.N. Novikov[2], M. Oinonen[3], L. Batist[2], V.-V. Elomaa[1,b], T. Eronen[1,c], J. Hakala[1], A. Jokinen[1], P. Karvonen[1], M. Reponen[1], J. Rissanen[1], A. Saastamoinen[1], G. Vorobjev[4], C. Weber[1d], and J. Äystö[1]

[1] Department of Physics, University of Jyväskylä, P.O. Box 35, FI-40014 University of Jyväskylä, Jyväskylä, Finland
[2] Petersburg Nuclear Physics Institute, 188300 Gatchina, Russia
[3] Dating Laboratory, P.O. Box 64, FI-00014 University of Helsinki, Finland
[4] Gesellschaft für Schwerionenforschung mbH, Planckstrasse 1, D-64291 Darmstadt, Germany

Received: 31 January 2012
Published online: 19 April 2012 –
Communicated by E. De Sanctis

Abstract. This article reviews the decay studies of neutron-deficient nuclei within the mass region $A =$ 56–100 performed at the Ion-Guide Isotope Separator On-Line (IGISOL) facility in the University of Jyväskylä over last 25 years. Development from He-jet measurements to on-line mass spectrometry, and eventually to atomic mass measurements and post-trap spectroscopy at IGISOL, has yielded studies of around 100 neutron-deficient nuclei over the years. The studies form a solid foundation to astrophysical *rp*-process path modelling. The focus is on isomers studied either via spectroscopy or via Penning-trap mass measurements. The review is complemented with recent results on the ground and isomeric states of ^{90}Tc. The excitation energy of the low-spin isomer in ^{90}Tc has been measured as $E_x = 144.1(17)$ keV with JYFLTRAP double Penning trap and the ground state of ^{90}Tc has been confirmed to be the (8^+) state with a half-life of $T_{1/2} = 49.2(4)$ s. Finally, the mass-excess results for the spin-gap isomers ^{53}Co^m and ^{95}Pd^m and implications from the JYFLTRAP mass measurements for the (21^+) isomer in ^{94}Ag are discussed.

1 Introduction

Decay studies of neutron-deficient nuclides have a long tradition in the IGISOL group in Jyväskylä. Even before the era of the ion-guide method, several studies had been performed with the helium-jet technique at the MC-20 cyclotron in Jyväskylä. Those studies included the $T_Z = -1$ beta-delayed proton/alpha emitters ^{20}Na [1], ^{24}Al [2], ^{24}Al^m [2], ^{28}P [3], ^{32}Cl [3], ^{36}K [4], ^{40}Sc [5] as well as the mirror nuclei ^{49}Mn [6], ^{53}Co [6], and ^{59}Zn [7]. After the invention of the new ion-guide method [8], these decay studies were continued at IGISOL and focused on the $T_Z = -1/2$ mirror nuclei ^{43}Ti [6], ^{51}Fe [9], and ^{55}Ni [9].

After commissioning of the K-130 cyclotron, another milestone in the decay studies of neutron-deficient nuclides has been the application of the heavy-ion ion-guide HIGISOL in the production of these nuclides [10,11]. Previously, the ions of interest had been produced by light ion (p or ^{3}He) fusion-evaporation reactions, which limited the studies close to the stability line. With HIGISOL, the heavy-ion fusion-evaporation reactions became available, and a broader range of radioactive isotopes could be studied in the heavier mass region.

The third major step forward has been the application of the JYFLTRAP Penning-trap mass spectrometer for the mass measurements of neutron-deficient nuclides at IGISOL. This has also made it possible to measure the excitation energies of the isomers very precisely. JYFLTRAP has also been used for isobarical, and in some cases, isomerical purification of the IGISOL beams. The mass measurements of neutron-deficient nuclides are reviewed in the following paper of this special issue of the European Physical Journal A.

We hope that this contribution provides a view to the history, presence and opportunities of the ion-guide–related methodology in studying neutron-deficient nuclei beyond the doubly magic ^{56}Ni. In the following sections, decay studies of mirror nuclei and the isospin triplet at $A = 58$ will be discussed at first. Then, the mass region around $A =$ 80–90 rich of isomers is discussed with the main focus on the isomers. In addition, the three spin-gap isomers studied at JYFLTRAP are reviewed in sect. 4. The investigated nuclides lie on the path of the astrophysical rapid proton capture (rp) process, a sequence of

[a] e-mail: anu.k.kankainen@jyu.fi
[b] *Present address*: Turku PET Centre, Accelerator Laboratory, Åbo Akademi University, FI-20500 Turku, Finland.
[c] *Present address*: Max-Planck-Institut für Kernphysik, Saupfercheckweg 1, D-69117 Heidelberg, Germany.
[d] *Present address*: Department of Physics, Ludwig-Maximilians-Universität München, D-85748 Garching, Germany.

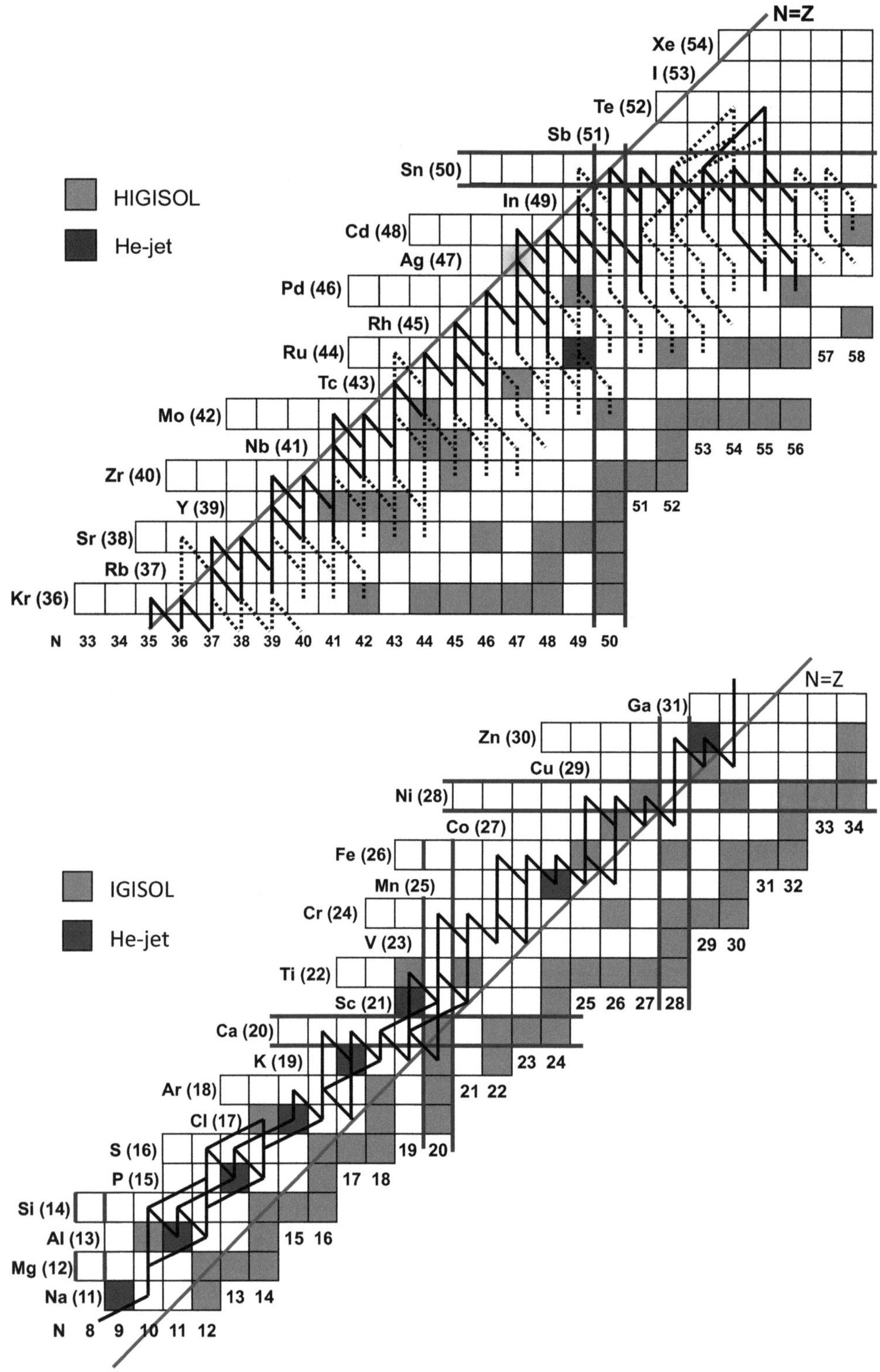

Fig. 1. Nuclides studied via beta-decay spectroscopy with He-jet, IGISOL and HIGISOL techniques in Jyväskylä. For the lower mass region, the main path of the *rp* process is drawn for one-zone X-ray burst model according to ref. [12]. For the higher mass region, the time-integrated reaction flow has been plotted for conditions where the thermonuclear burning proceeds in steady state [13]. There, the solid line represents the major path (more than 10% of the reaction flow through 3α reaction) and the dashed line the minor path (1–10% of the reaction flow through 3α reaction). Recent mass measurements at JYFLTRAP [14] have shown that the SnSbTe cycle is not so strong as plotted here. ^{53}Co has been studied both with the He-jet technique [6] and with JYFLTRAP at IGISOL [15]. The energy of the spin-gap isomer in ^{94}Ag has been estimated based on JYFLTRAP mass measurements. Decay studies of ^{23}Al, ^{31}Cl, and ^{41}Ti are discussed in a separate article [16].

proton captures and β^+ decays occurring at high temperatures and hydrogen densities, such as in X-ray bursts [17, 18] (see fig. 1). The *rp* process has been one motivation for these studies as the detailed knowledge of beta decays, half-lives, energy levels and isomers of the involved nuclides is required for an accurate modeling of the process. Long-lived isomers could play a role by increasing the reaction flow and thus reducing the timescale for the *rp*-process nucleosynthesis during the cooling phase [19].

2 Decay studies of isospin doublets and the A = 58 triplet

2.1 $T_Z = -1/2$ mirror nuclei

Mirror nuclei offer an interesting possibility to study the Gamow-Teller (GT) strength in fast mixed Fermi and GT transitions between the ground states of these nuclei. Since the Fermi strength is constant in these mirror transitions, the Gamow-Teller matrix element of the transition can be extracted and compared to the strength obtained from shell-model calculations. A persistent quenching of the GT strength has been observed and quenching factors of 0.744(15) [20] and 0.77(2) [21] have been determined for the *pf* and *sd* shells, respectively.

At IGISOL, beta-decay study of ^{55}Ni was one of the first experiments performed with the ion-guide method [9]. There, a 27 MeV ^{3}He beam on enriched ^{54}Fe target was applied from the old MC-20 cyclotron to produce ^{55}Ni. With a beta telescope, an endpoint energy of 7.70(18) MeV and a half-life of 208(5) ms were determined [9]. The γ spectrum was recorded with a 19% Ge(Li) detector but no feeding to other states than the ground state was observed. Also the beta decay of ^{51}Fe produced with the same beam on $^{50}Cr_2O_3$ target was studied with the same setup. The obtained half-life and the endpoint energy were 310(5) ms and 7.26(15) MeV [9], respectively. A beta feeding of 5.0(13)% to a $7/2^-$ state at 237 keV was determined from the intensity ratio to the annihilation peak [9]. The ion-guide method was also used for a half-life determination of ^{43}Ti with a beta telescope [6]. The ions of interest were produced via ^{40}Ca$(\alpha, n)^{43}$Ti reactions with an 18 MeV alpha beam from the MC-20 cyclotron. The measured half-life was 509(5) ms [6].

2.2 T = 1 triplet at A = 58

In addition to beta decays, Gamow-Teller strength can be studied via charge-exchange (CE) reactions. The CE reactions, such as ^{58}Ni(^{3}He, t)^{58}Cu [22], are not limited by the Q value as beta-decay strength studies, and thus, GT strengths over a broader energy range can be investigated. However, the charge-exchange reactions rely on the proportionality between the cross-section near 0° scattering angle and the GT strength. This proportionality has to be normalized, for example, at $A = 58$ with the GT strength value from the beta decay of ^{58}Cu.

In order to provide a more precise value for the normalization, beta decay of ^{58}Cu was measured at IGISOL. About 4000 ions/s of ^{58}Cu were produced at IGISOL with an 18 MeV, 15 μA proton beam from the K-130 cyclotron impinging on an enriched ^{58}Ni target [23]. At the implantation point, the yield of ^{58}Cu and corresponding γ transitions were monitored by two HPGe detectors and a plastic ΔE_β detector. The tape was moved to the measurement position, where the ratio of the intensities of the 511 keV annihilation radiation to the 1454 keV γ-transition in ^{58}Ni was precisely measured. A branching ratio of 80.8(7)% to the ^{58}Ni ground state was obtained [23].

Recently, a Q_{EC} value for the beta decay of ^{58}Cu has been precisely determined at JYFLTRAP [15]. When this new Q_{EC} value is combined with the branching ratio of 81.1(4)% (from the values of 80.8(7)% [23], 81.2(5)% [24] and 82(3)% [25]) and a half-life of 3.204(7) s [26] for ^{58}Cu, a Gamow-Teller strength $B(GT) = 0.08285(46)$ and a squared Gamow-Teller matrix element $\langle\sigma\tau\rangle^2 = 0.05141(33)$ are obtained. The values are little higher and more precise than previously (cf. $B(GT) = 0.0821(7)$ and $\langle\sigma\tau\rangle^2 = 0.0512(5)$ in ref. [23]).

The $T = 1$ triplet at $A = 58$ offers a possibility to study the isospin symmetry of transitions. Namely, the Gamow-Teller strength to the states in ^{58}Cu obtained from the (^{3}He, t) charge-exchange reactions on ^{58}Ni ($T_Z = +1$) [22] can be compared with the strenghts of the analogous transitions from the beta decay of ^{58}Zn ($T_Z = -1$). For this purpose, beta-decay of ^{58}Zn has been studied with a beta-delayed gamma and proton setups at ISOLDE and at IGISOL [27]. At ISOLDE, the earlier results [28] concerning the half-life and two lowest γ-transitions were confirmed. No beta-delayed protons were observed with the ISOLDE Silicon Ball detector [29] similarly to the previous work [28]. Therefore, the experiment was also tried at IGISOL with a 50 MeV, 1 μA ^{3}He beam on a ^{nat}Ni target. The estimated yield of ^{58}Zn at IGISOL was about 0.6 atoms/s based on the 203 keV γ-transition [30]. Unfortunately, no beta-delayed protons were observed at IGISOL either.

3 Studies around A = 80–90 for the rp process

3.1 Yttrium isotopes: $^{80}Y^m$, ^{81}Y, and ^{82}Y

$^{80}Y^m$

^{80}Y has a long-lived isomeric state at 228.5 keV relevant for the astrophysical *rp* process. Since the spin assignment and internal conversion coefficient were not certain, an experiment aiming for the spin identification of this isomer was performed at HIGISOL [10,11] by employing a 150 MeV $^{32}S^{7+}$ beam on a 2.7 mg/cm^2 thick ^{54}Fe target [31]. The isomeric transition was studied with a magnetic conversion electron transporter spectrometer ELLI [32] and a low-energy Ge detector (LeGe). ELLI transported electrons from the implantation point to a cooled Si(Au) surface barrier detector in a remote detection area, which helped to reduce the background. The observed 228.5 keV γ-rays were not in coincidence with the beta particles. Moreover, the coincidence of yttrium K_α X-rays with the 228.5 keV transition confirmed the isomeric character. The internal conversion coefficient for the isomeric transition was determined from the ratio of the conversion electrons to the corresponding γ-rays as well as from the ratio of yttrium K X-rays to the 228.5 keV γ-rays. The contribution from the beta decay of ^{80}Zr to the yttrium X-rays is negligible since the production rate of ^{80}Zr is very small. The obtained values $\alpha_K = 0.51(9)$ [31] and $\alpha_K = 0.48(13)$ [31] agree well with each other and with the result of ref. [33], $\alpha_K = 0.47(15)$. The average value of $\alpha_K = 0.50(7)$ [31] gives $M3 + (E4)$ multipolarity for the 228.5 keV transition and a spin assignment 1^- for the isomer.

The measured half-life of the isomeric transition, $T_{1/2} = 5.0(5)$ s [31], is in agreement with the previous result of $T_{1/2} = 4.7(3)$ s [34]. The weighted mean for the half-life of the 228.5 keV isomeric transition is $T_{1/2} = 4.8(3)$ s, which converts into a bare half-life of $T_{1/2} = 6.8(5)$ s [31] corresponding the fully ionized ^{80}Y^m in the stellar plasma of an X-ray burst. In the *rp* process, ^{80}Y^m is produced via beta decay of ^{80}Zr which dominantly populates the (1^+) states at 623 keV and at higher energies [35]. De-excitations of these states lead finally to the 228.5 keV 1^- isomeric state. About 19(2)% of the isomeric decays proceed via beta decay and the rest 81(2)% via internal transition to the ^{80}Y ground state [36]. Since the decay of the isomeric state is many orders of magnitudes slower than the proton captures on it during most of the *rp* process, the proton captures will predominantly occur on the isomeric state [31] and not on the ground state of ^{80}Y as previously assumed. This is a clear example how isomers can play a role in the *rp* process and should be known for an accurate modeling.

^{81}Y

^{81}Y was produced at HIGISOL with a 150–170 MeV ^{32}S^{7+} beam impinging on an enriched ^{54}Fe (1.8 mg/cm^2) target [37]. The beta-decay of ^{81}Y was studied at HIGISOL with a setup consisting of two HPGe detectors, a LeGe detector and the magnetic conversion-electron transporter spectrometer ELLI [32]. Before this HIGISOL experiment [37], the beta-decay of ^{81}Y had been studied already in refs. [38–40]. All known γ-transitions [41] following the beta-decay of ^{81}Y could be observed at HIGISOL except the 216.6 keV transition. Conversion electrons from the 43.2, 79.23, 101.05, 115.39, 124.16, and 155.20 keV transitions in ^{81}Sr were measured. The internal conversion coefficient for the 43.2 keV transition was determined as $\alpha_K = 1.5(3)$, which supports an $M1+E2$ multipolarity for the transition [37]. The obtained low conversion coefficient ($\alpha_K = 0.03(1)$) for the 124 keV transition supports an $E1$ multipolarity for the transition. The measured conversion coefficient for the 79.23 keV transition, $\alpha_K = 2.3(1)$, agrees nicely with the literature value of 2.3(6) [39], and supports an $E2$ multipolarity. Figure 2 shows an example of conversion electron spectrum measured with ELLI at $A = 81$.

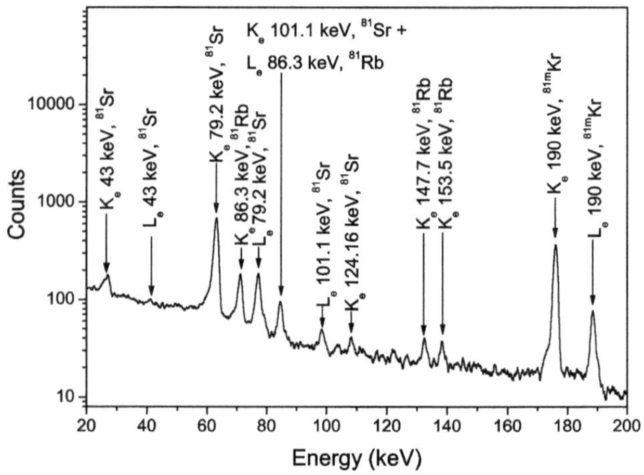

Fig. 2. Conversion-electron spectrum measured at $A = 81$ with the magnetic conversion-electron transporter spectrometer ELLI [32] at IGISOL.

^{82}Y

^{82}Y was produced at HIGISOL using the reaction 165 MeV ^{32}S + ^{nat}Ni [10]. A 37% Ge detector was used for detecting γ-rays, a LeGe detector for X-rays and a 1 mm thick plastic scintillator for detecting beta particles [10]. The beta-decay of ^{82}Y had been studied via ^{60}Ni(^{24}Mg, pn)^{82}Y and ^{60}Ni(^{28}Si, αpn)^{82}Y reactions with beam energies of 90 and 110 MeV [38] and via ^{54}Fe(^{32}S, $3pn$)^{82}Y reaction at an energy of 123 MeV [39]. The measured half-lives of 9.5(4) s [38] and 9.5(5) s [39] agree with each other but the values for the relative intensity of the 602 keV γ-transition, 41% [38] and 13(1)% [39], differ a lot. At HIGISOL, the half-life was determined from the beta particles, annihilation γ-rays, as well as from the 573, 602, 737, 1176 and 1291 keV γ-transitions following the beta-decay of ^{82}Y [10]. The shorter-lived ^{82}Zr might distort the half-lives obtained from the beta particles and from the annihilation radiation, and since the measured half-life values of ^{82}Zr vary a lot, the value of 8.30(20) s based on the most intensive, 573 keV γ-transition was adopted for ^{82}Y [10]. The statistics of the other γ-transitions were too poor for precise half-life fitting. The half-life measured at HIGISOL disagrees with the former values of [38,39] but is much more precise.

The obtained relative intensity of 4.4(4)% [10] for the 602 keV transition deviates from the previous He-jet studies [38,39]. In the He-jet studies no mass separation is used and contamination for example from the 602 keV γ-transitions following the beta-decay of ^{84}Y, (6^+) state, cannot be excluded. Therefore, the obtained relative intensities for the 602 keV transition in [38,39] can be higher than at HIGISOL, where a clear spectrum at mass $A = 82$ was obtained.

The mass of ^{82}Y has been measured precisely with the JYFLTRAP Penning-trap mass spectrometer [42]. The difference from the Atomic Mass Evaluation 2003

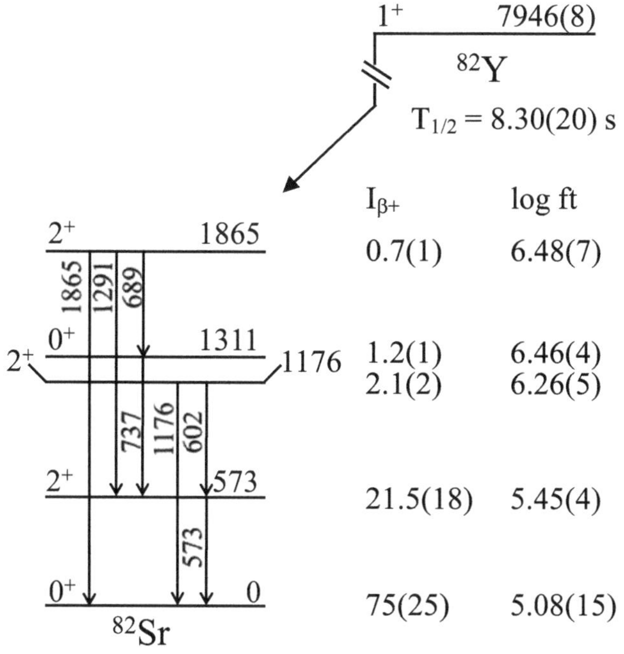

Fig. 3. Proposed decay scheme of ref. [10] revised with the Q_{EC} value from ref. [42].

(AME03) [43] value is 127(100) keV. The updated Q_{EC} value for ^{82}Y is 7946(8) keV, which is 130 keV higher and 13 times more precise than in the AME03. The updated log ft values of the ^{82}Y beta-decay are little higher than previously (see fig. 3).

Neutron-deficient yttrium isotopes have typically isomers. No clear evidence for an existence of an isomer in ^{82}Y with a significantly different half-life than the 1^+ ground state was observed in ref. [10]. On the other hand, for example the production of ^{84}Y was found to be almost totally concentrated on the 5^-, 39.5 min isomeric state: only about 1% of the ^{84}Y production was due to the 1^+ ground state based on the intensities of the 793 keV and 974 keV γ-transitions [10].

3.2 Niobium isotopes: ^{85}Nb^m and ^{86}Nb

^{85}Nb^m

A 37% Ge detector for detecting γ-rays, a LeGe detector for X-rays and a 1 mm thick plastic scintillator for detecting beta particles were employed in a spectroscopic survey at $A = 85$ carried out at HIGISOL by using 165 MeV ^{32}S + ^{nat}Ni reactions [10]. Several unknown γ-rays were found with energies of 166, 272, 423, 434, 484, 532, 538, 590, 610, 660, 709, and 759 keV at $A = 85$. The most intense, 759 keV γ transition had a half-life of 12(5) s agreeing with the beta-decay half-life of ^{85}Zr^m ($T_{1/2} = 10.9(3)$ s [44]), and disagreeing with the half-lives of ^{85}Mo ($T_{1/2} = 3.2(2)$ s [45]) and ^{85}Nb ($T_{1/2} = 20.9(7)$ s [46]). However, the transition could not be associated with the beta-decay of ^{85}Zr^m since such an intense transition should have been observed in the detailed beta-decay study of ref. [47]. At HIGISOL, only the two most intensive peaks at 416 and 454 keV following the ^{85}Zr beta decay were observed. The half-life for the unknown 484 keV transition was determined as $T_{1/2} = 4^{+8}_{-2}$ s [10] consistent with the reported half-life of ^{85}Mo. This transition could not be associated with the beta decays of ^{85}Nb and ^{85}Zr due to the different half-lives. Similarly to the 759 keV transition, the 484 keV transition should have been detected in ref. [47] if it was associated with ^{85}Zr^m. As a summary, the 759 keV, 12(5) s transition follows either the beta-decay of ^{85}Nb^m or ^{85}Mo^m. In addition to the isomer in Nb or Mo, the 484 keV transition could originate from the beta decay of ^{85}Mo.

Unknown γ-rays at energies of 272, 484, 530, 536, 590, 660, 709, and 758 keV were also observed with two HPGe detectors at the implantation point in a run employing a 150–170 MeV ^{32}S^{7+} beam on a ^{nat}Ni target at HIGISOL [37]. After 15 or 40 seconds, the activity was moved to the LeGe detector station. There, none of these γ-rays could be observed supporting short half-lives for these transitions. It should also be noted that about half of the observed 12 unknown γ-transitions in ref. [10] at $A = 85$ have an energy difference of 50 keV: $484 - 434 = 50$ keV, $660 - 610 = 50$ keV and $759 - 709 = 50$ keV. In addition, the differences $532 - 484 = 48$ keV and $590 - 538 = 52$ keV are quite close to 50 keV which corresponds to the energy of a well-known 50 keV transition in ^{85}Zr. One possibility is that these γ-rays follow the beta decay of an isomer in ^{85}Nb to the possible states at 484 keV, 660 keV and 759 keV (and possibly at 590 keV), which de-excite to the $(7/2^+)$ ground state and to the $(7/2^+, 9/2^+)$ state at 50 keV in ^{85}Zr. However, no coincidences have been checked between these γ-rays or between the Zr, Nb or Mo X-rays and these transitions. Thus, we cannot conclude to which nucleus these transitions belong. Trap-assisted spectroscopy at $A = 85$ would enlighten the situation provided the yields are high enough.

In the latter run at $A = 85$ [37], conversion electrons at 32.1 and 50 keV were observed with ELLI. The 32.1 keV conversion electrons were in coincidence with Zr K X-rays, thus confiming that they belong to the 50 keV transition in ^{85}Zr. The determined internal conversion coefficient for the 50 keV transition, $\alpha_K = 1.7(2)$, suggests a mixed $M1 + E2$ multipolarity for the transition. The half-life determined for this 50 keV transition fed by the beta decay of ^{85}Nb was 17(2) s, which is little shorter than $T_{1/2} = 20.9(7)$ s given for ^{85}Nb in ref. [46].

The 50 keV conversion electrons were in coincidence with the Nb K X-rays and had a half-life of 3.3(9) s. Since the electrons were not in coincidence with beta particles and because the production rate of ^{85}Mo is much smaller than for ^{85}Nb, we identify this transition as a 69 keV isomeric transition in ^{85}Nb. Based on the internal conversion coefficients, the preferred multipolarity for the transition would be $E2$ or $M2$, which would mean a much shorter half-life than 3.3 s. An indication of beta-decay of this isomer to the 292 keV isomeric state in ^{85}Zr is observed in the time behaviour of the 292 keV γ-line [37].

At JYFLTRAP, the Q_{EC} value of ^{85}Nb has been measured as 6898(9) keV [42] which is 900(200) keV higher than the previous value [46]. There, the beta-endpoint energy was determined from the beta spectrum in coincidence with the 50 keV transition in ^{85}Zr. If the 50 keV transition is fed by de-excitations from possible states at 484 keV, 660 keV and 759 keV in ^{85}Zr, this would increase the deduced Q_{EC} value of ref. [46] and move it closer to the JYFLTRAP value. Although the beta-decay experiments tend to underestimate Q_{EC} values, such a big difference suggests that it should be verified for which state the mass of ^{85}Nb has been measured. For example, a half-life measurement using the isobarically purified beta emitters after JYFLTRAP would allow us to distinguish between the 20.9 s ground state and the suggested isomers with half-lives of 3.3(9) s [37] and 12(5) s [10].

^{86}Nb

^{86}Nb has a (6^+) ground state with a half-life of 88(1) s and a suggested high-spin isomer with a half-life of 56(8) s [48]. In the HIGISOL run employing a 150–170 MeV $^{32}S^{7+}$ beam on a ^{nat}Ni target [37], accumulation times of 40 s and 200 s were used for $A = 86$. There, γ peaks at energies of 47.7, 50.1, 97.8 and 186.8 keV following the beta-decay of the 0^+ ground state of ^{86}Mo were observed [37]. These low-spin states in ^{86}Nb are located at the energies of E_0, $E_0 + 50.1$ keV, $E_0 + 97.8$ keV, and $E_0 + 236.9$ keV. Multipolarities for the γ-transitions have been determined and they are most likely $E1$ except $M1$ for the 47.7 keV transition. A half-life of 19.1(3) s for ^{86}Mo was determined from the time behaviour of γ, Nb K X-ray, and electron peaks [37]. This is in agreement with the value of 19.6(11) s obtained in ref. [48]. No evidence for the 56 s isomer was found in ref. [37]. No converted low-energy transitions with an energy E_0 were observed. Thus, the lowest low-spin state E_0 fed by the ^{86}Mo beta-decay, should be either highly excited or decay mainly via beta-decay to ^{86}Zr.

At JYFLTRAP, the mass-excess value of ^{86}Nb [37] was found to be 700(90) keV higher than in the AME03 [43]. Here, again, the ground-state nature of the measured mass should be verified for example via half-life determination after the trap. In addition, the production ratio of the isomer to ground state at HIGISOL can shed light on which mass has been measured.

3.3 ^{90}Tc^m

The beta decay of ^{90}Tc has been studied in refs. [49] and [50]. A 1^+ state with a half-life of 8.7(2) s and a high-spin state (6^+) with a half-life of 49.2(4) s were found in ref. [50]. The corresponding excitation energy of the suggested (6^+) isomer has been estimated to be 124(390) keV [51]. In ref. [52], the ground state was suggested to be an 8^+ state based on the systematics of all neighbouring $N = 47, 49$ nuclei. In a study of the ^{90}Ru 0^+ ground-state beta-decay [53], feeding to three low-spin states in ^{90}Tc was found. There, a high-spin 8^+ ground state was obtained, when the low-spin and high-spin results of shell-model calculations were compared to the experimental results.

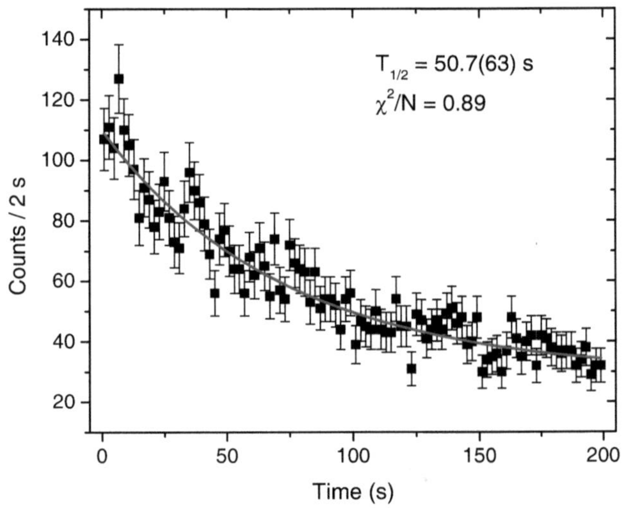

Fig. 4. One-component fit on the time behaviour of the beta particles at $A = 90$ measured after the IGISOL dipole magnet.

^{90}Tc has been studied previously at HIGISOL via a JYFLTRAP mass measurement [51]. There, the measured mass-excess value was assigned for the (8^+) ground state. Since the identification of the ground or isomeric state is uncertain, a new experiment was performed with a $^{40}Ca^{8+}$ beam on a ^{nat}Ni target at 210 MeV. In that experiment, a half-life for the mass $A = 90$ was determined by measuring the time behaviour of beta particles with a silicon detector after the IGISOL dipole magnet. A half-life of $T_{1/2} = 50.7(63)$ s was obtained with a one-component fit (see fig. 4). This is in agreement with the half-life of the high-spin state (8^+), $T_{1/2} = 49.2(4)$ s [50] and shows that the contribution from the low-spin 1^+, 8.7(2) s state [50] is small. The half-lives of ^{90}Nb (14.60(5) h [54]) and ^{90}Mo (5.56(9) h [54]) can be considered to be constant within the time scale of 200 s. ^{90}Ru ($T_{1/2} = 11(3)$ s) is much less produced than ^{90}Tc and thus, does not contribute much to the beta-particle time behaviour. The Mo and Nb isomers decay via internal transition and do not contribute to the beta spectrum.

The isomer in ^{90}Tc was also observed with JYFLTRAP in the experiment employing $^{40}Ca^{8+} + {}^{nat}$Ni at HIGISOL, as shown in fig. 5. Three measurements of the ^{90}Tc ground state and five measurements of the isomeric state against the ^{86}Kr reference were performed with the time-of-flight ion-cyclotron resonance (TOF-ICR) method [55,56]. In the precision trap, a quadrupolar RF excitation time of 800 ms was applied for the ions of interest. In total, there were 14331 ions of ^{90}Tc and 4229 ions of ^{90}Tc^m in the analysis. The count-rate class analysis [57] was performed for the ground state with three classes. For the isomer, no count-rate class analysis could be performed. Therefore, the ion number was limited to 1–2 ions per bunch in the analysis of the isomeric state. The uncertainty of the cyclotron frequency of the isomer was multiplied by a factor

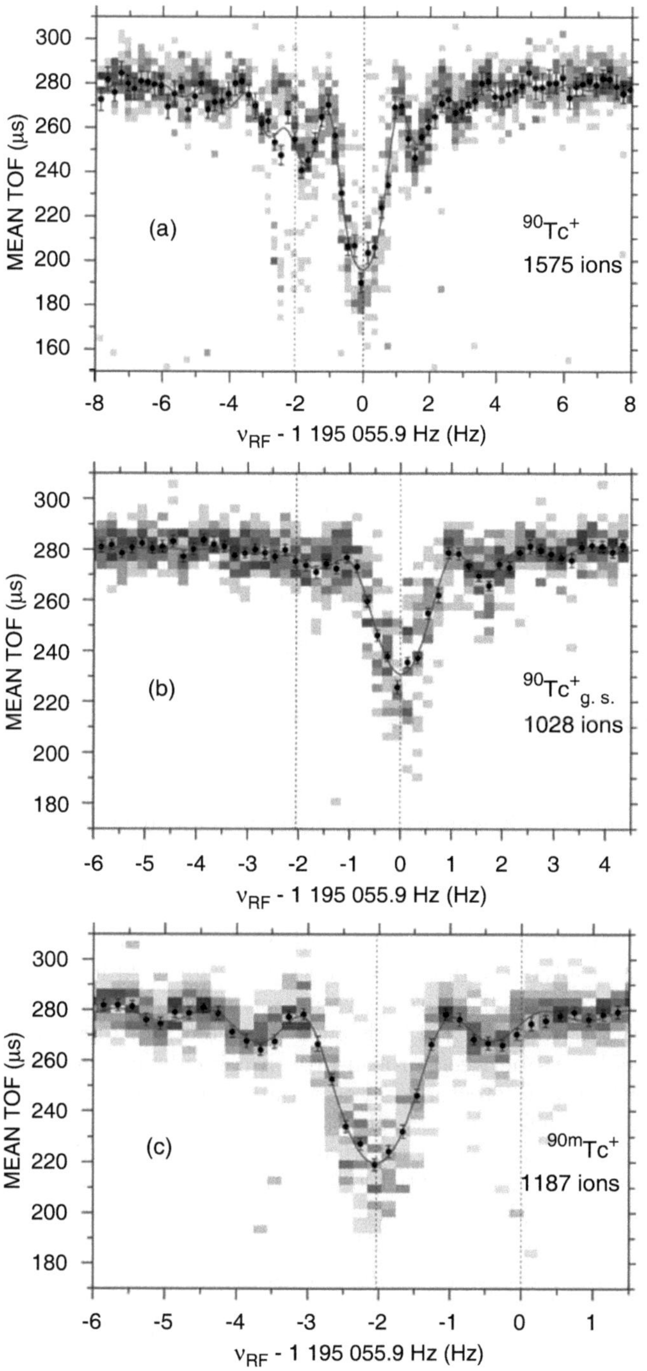

Fig. 5. (Colour on-line) Time-of-flight spectra after 800 ms quadrupolar RF excitation for (a) both the ground and the isomeric state in ^{90}Tc, (b) the ground state of ^{90}Tc and (c) the isomeric state ^{90}Tc^m. Blue shading around the data points indicates the number of ions in each time-of-flight bin: the darker the bin, the more ions. The isomer contribution is clearly seen in panel (a). The dashed lines show the positions of the ground and isomeric states obtained from panels (b) and (c). The TOF spectra shown here are based on single files, where the number of ions has been limited to 1–2 ions per bunch. The selected time window has been 147.2–307.2 μs for the panel (a) and 179.2–307.2 μs for (b) and (c). The cleaning method [60] based on electric dipole excitation with time-separated oscillatory fields has been applied in panels (b) and (c).

Table 1. Measured frequency ratios and mass-excess (ME) values of ^{90}Tc and ^{90}Tc^m obtained at JYFLTRAP in this work using ^{86}Kr$^+$ as a reference. The result for the ground-state mass agrees with the previous experiments at JYFLTRAP (ME = −70723.6(39) keV) [51] and SHIPTRAP (ME = −70724.1(74) keV) [51], which employed ^{85}Rb$^+$ ions as a reference.

Nuclide	r	ME (keV)	E_x (keV)
^{90}Tc	1.046717034(13)	−70724.7(11)	0
^{90}Tc^m	1.046718834(16)	−70580.6(13)	144.1(17)

of two. This was based on a comparison of the non-Z-classed results to the Z-classed results of the ground state and the reference. The weighted mean of the frequency ratios and the internal and external errors [58] were calculated. The internal errors were bigger than the external errors showing that the fluctuations of the frequency ratios were purely statistical. The mass-dependent and residual uncertainties [59] were added to the measured frequency ratios quadratically. The resulting frequency ratios and mass-excess values are given in table 1.

When the time-of-flight spectrum for ^{90}Tc is fitted with two peaks (see fig. 5(a)), the fractions of the lower-mass state and the higher-mass state are 89(5)% and 11(5)%, respectively. A very similar result is obtained if a two-component fit is performed to the beta-particle time behaviour at $A = 90$: half-lives of 60(21) s and 8(14) s are obtained with fractions of 84(19)% and 16(16)%, respectively. This indicates that the low-spin state is much less produced than the high-spin state at HIGISOL. Thus, we confirm that the previously measured mass-excess value of ^{90}Tc belongs to the more favorably produced (8^+) ground state with a half-life of $T_{1/2} = 49.2(4)$ s.

The excitation energy obtained for the isomer, E_x = 144.1(17) keV, fits well with the shell-model prediction of a 2^+ state (143 keV [52], 160 keV [53]). However, $\log ft$ values of around 5–6 to the 0^+ ground state and to the 2^+ state at 948.1 keV in ^{90}Mo [49,50] support a 1^+ assignment. The 2^+ assignment is possible if the ground-state feeding has not been correctly determined.

3.4 ^{93}Ru^m

The beta-decay of ^{93}Ru^m ($T_{1/2} = 10.8(3)$ s, $J^\pi = (1/2^-)$) has been studied via a 24 MeV ^{3}He beam from the University of Jyväskylä MC-20 cyclotron on an isotopically enriched ^{92}Mo target employing the He-jet technique [61]. The total decay energy of the isomer was measured as 7070(85) keV and the proton binding energy of ^{93}Tc as 4086.5(10) keV [61]. The excitation energy of the isomer is 734.4(1) keV [62]. The transition was suggested as an isomeric transition in ^{93}Ru since it was not in coincidence with any other γ-ray or annihilation radiation. Later, in ref. [53], the beta-decay of ^{93}Rh was studied via the reaction ^{58}Ni(^{40}Ar, $p4n$)^{93}Ru. There, no evidence for the $(1/2)^-$ isomeric state in ^{93}Ru was found. Thus, the low-spin isomeric state is not populated at all or only weakly populated in these kind of reactions.

The mass of ^{93}Ru has been studied at JYFLTRAP [51] and it agrees with the values of ref. [61]. The ground state should be well resolved from the isomeric state at 734.4.(1) keV. In future, it would be interesting to measure the mass of the isomeric state by using similar reaction as in ref. [62] at IGISOL. These measurements could have implications for the *rp* process, because such long-lived isomers can play a role in it.

4 High-spin isomers

High-lying, high-spin isomers with relatively long half-lives have been observed close to magic shell closures. Examples of high-spin isomers are ^{52}Fe^m (12$^+$) [63], ^{53}Fe^m (19/2$^-$) [64], and ^{53}Co^m (19/2$^-$) [65] near $N = Z = 28$, ^{94}Ag^m (7$^+$ and 21$^+$) [66,67], ^{95}Ag^m [68], and ^{95}Pd^m (21/2$^+$) [69] near $N = Z = 50$ and ^{212}Po^m [70] near $Z = 82$, $N = 126$. These high-spin isomers are called spin-gap isomers since they lie lower in energy than the intermediate spin states, thus forcing the isomer to decay either via beta-decay, particle emission or a high-multipolarity internal transition. The excitation energies of these high-spin states are lower due to the effective proton-neutron interaction, which favors an aligned proton-neutron pair. In this section, studies on ^{53}Co^m (19/2$^-$), ^{94}Ag^m (21$^+$) and ^{95}Pd^m (21/2$^+$) performed at IGISOL will be reviewed.

4.1 ^{53}Co^m

The first observation of proton radioactivity was from the high-spin (19/2$^-$) isomer in ^{53}Co [65,71,72]. This state with a half-life of 247(12) ms [72] is the isobaric analogue state of the (19/2$^-$) isomeric state in its mirror nucleus ^{53}Fe [64]. The excitation energy of ^{53}Co^m as determined from the energy of the observed protons, is 3197(29) keV [71,72]. The isomeric state decays mainly via beta-decay to its analogue state in ^{53}Fe —only about 1.5% decays to ^{52}Fe via proton decay [72].

In an experiment at the JYFLTRAP mass spectrometer, both the ground and the isomeric state of ^{53}Co could be produced with a 40 MeV proton beam impinging on an enriched 2 mg/cm^2 thick ^{54}Fe target [15]. At JYFLTRAP, a Ramsey excitation with a time pattern of 25–150 (wait)–25 ms was applied for ^{53}Co and its reference ^{53}Fe. The ground state and the isomeric state were measured against each other and against the ^{53}Fe ground state. The obtained mass-excess values were −42657.3(15) and −39482.9(16) keV for the ground and isomeric state, respectively [15]. The mass-excess value for the ground state agrees well with the earlier results [43,73]. The mass-excess value of the isomer agrees with the latter proton-decay experiment [72] but deviates from the first experiment [71]. The excitation energy for the (19/2$^-$) isomeric state obtained as a weighted average from the frequency ratio ^{53}Co^m–^{53}Co and from the frequency ratios of ^{53}Co^m–^{53}Fe and ^{53}Fe–^{53}Co, was 3174.3(10) keV [15]. This new value agrees well with the previously adopted value 3197(29) keV [43] but is 29 times more precise. With the ^{52}Fe mass from ref. [43], this would correspond to a proton energy of $E_{p,lab} = 1530(7)$ keV [15].

Since ^{53}Co and ^{53}Co^m were measured against their beta-decay daughter ^{53}Fe at JYFLTRAP, also Q_{EC} values could be determined precisely. The obtained Q_{EC} values from the ground and isomeric state of ^{53}Co to the ground state of ^{53}Fe were 8288.12(45) keV and 11462.2(12) keV, respectively [15]. These precisely measured mirror decay Q_{EC} values are important for an accurate determination of the corrected ft values. In addition, a Coulomb energy difference of 133.9(10) keV [15] was obtained for the (19/2$^-$) states in ^{53}Fe and ^{53}Co, which improves the adopted value 157(29) keV [43] a lot.

In future, a direct measurement of the proton separation energy of ^{53}Co at JYFLTRAP would yield a sub-keV value for this separation energy. This, in turn, would give a precise calibration value for proton spectroscopy experiments. In addition, precise mass measurements of ^{53}Fe, ^{53}Fe^m and ^{52}Fe^m could be considered in future at JYFLTRAP.

4.2 ^{94}Ag^m

Two high-spin states, (7$^+$) and (21$^+$) have been observed in the $N = Z$ nucleus ^{94}Ag. After beta-delayed γ-ray and proton studies of the isomeric states in ^{94}Ag performed at GSI [66,67,74,75], direct one-proton decay from the (21$^+$) state to the high-spin states in ^{93}Pd was observed by detecting protons in coincidence with γ-γ correlations and applying γ gates based on the known ^{93}Pd levels [76]. A year later also direct two-proton radioactivity was observed from the (21$^+$) isomer by detecting fourfold coincidences between proton-proton coincidences measured with the Si detectors and γ-γ coincidences measured with the Ge detectors at GSI [77]. The γ-γ gates were set to two γ-transitions in the daughter nucleus ^{92}Rh. A two-proton energy of 1900(100) keV [77], an excitation energy of 5780(30) keV [76], and a half-life of 0.39(4) s [75] has been obtained for this (21$^+$) isomer.

At JYFLTRAP, we have measured the mass excesses of ^{92}Rh and ^{94}Pd [51], the two-proton decay daughter of the (21$^+$) isomer, and the beta-decay daughter of ^{94}Ag, respectively. The mass-excess value of ^{94}Ag was not directly measured but it was extrapolated to −53330(360) keV based on the linear behaviour of the Coulomb displacement energies of odd-odd $N = Z$ nuclei [78]. The extrapolated mass of ^{94}Ag yields a two-proton separation energy $S_{2p} = -\Delta(^{94}\mathrm{Ag}) + \Delta(^{92}\mathrm{Rh}) + 2\Delta(^{1}\mathrm{H}) = 4910(360)$ keV [78], where Δ refers to the mass excess. If this value is now combined with the two-proton decay data [77], an excitation energy of 8360(370) keV [78] is obtained for the (21$^+$) state.

The one-proton decay daughter ^{93}Pd was not measured directly at JYFLTRAP. Instead, two-proton separation energies of several $N = 47$ isotones have been determined at JYFLTRAP. A parabolic fit to the smooth behaviour of these S_{2p} values yields $S_{2p} = 5780(160)$ keV [78] for ^{93}Pd. Since the mass excess of ^{91}Ru has been measured at JYFLTRAP, we obtain from the $S_{2p}(^{93}\mathrm{Pd})$ value a mass excess of −59440(160) keV for ^{93}Pd. Finally, we can calculate the proton separation energy of ^{93}Pd with the mass of ^{92}Rh measured at JYFLTRAP, $S_p(^{93}\mathrm{Pd})$ = 3730(160) keV [78]. From the mass-excess values of ^{93}Pd and ^{94}Ag, a proton separation energy of 1180(390) keV [78] is obtained for ^{94}Ag. If we now combine this value with the one-proton decay data [76], we obtain an excitation energy of 6960(400) keV [78] for the (21^+) isomer. This disagrees with the value based on the two-proton decay data and suggests that either the one-proton or two-proton decay data of the (21^+) state in ^{94}Ag needs revision.

The two-proton decay of the (21^+) isomer in ^{94}Ag has raised a lot of discussion recently. Shell-model calculations using a *gds* model space have shown that the additional binding energy due to the large attractive *pn* interaction in the $0g_{9/2}$ orbit lowers the energy of the 21^+ state primarily but the level inversion is caused eventually by the mixing with the $1d_{5/2}$ configurations [79]. The *gds* shell-model calculations [79] do not support strong deformation for the 21^+ state in contrast to ref. [77], where strong prolate deformation was suggested to explain the unexpectedly high two-proton decay probability [77]. In addition, recent calculations based on statistical theory of hot rotating nucleus combined with the macroscopic-microscopic approach do not support a strong prolate deformation [80].

The direct two-proton decay has been questioned in many papers. For example, the level scheme of ^{92}Rh has been improved from ref. [81]: the tentative 575 keV γ-transition [81] was not observed, the 307 keV transition was assigned as the $20^{(-)} \to 19^{(-)}$ transition, the 632 keV γ-rays as the $19^{(-)} \to 18^{(-)}$ transition and the order of the 939 and 1034 keV transitions was reversed [82]. No γ-rays at energies of 565 and 833 keV in coincidence with the known γ-transitions in ^{92}Rh were observed in ref. [82] in contrast to ref. [77]. The data of ref. [82] suggest that the spin difference between the two-proton decay mother and daughter should be greater than 10 and that the isomer should be even more deformed. When the in-beam data of ref. [82] is combined with the (extrapolated) mass-excess data of AME03 [43], the two-proton decay scenario seems to be impossible. Later, these considerations have been discussed in refs. [83,84].

Protons could not be unambigously identified in ref. [77] since two 1 mm thick Si detectors, which are also sensitive to electrons and positrons, were used. Detailed discussion in ref. [85] shows that the 1.9 MeV two-proton peak could originate from Compton-scattered γ-rays following the beta-decay of the 7^+ isomer in ^{94}Ag and annihilation radiation, and not from ^{92}Rh. However, these considerations are based on the assumption that Compton scattering events between adjacent Ge crystals had not been reduced in ref. [77]. In a recent paper [86], it is clarified that the Compton scattering effects in γ-γ coincidence events were reduced by excluding double hits in adjacent Ge crystals while they were accepted in the other crystals [67,86]. In addition, the coincidence events with the sum energy of the two γ-rays corresponding to 511 ± 1.5 keV were excluded for all crystals [86].

To identify the protons from the 21^+ isomer in ^{94}Ag, a measurement employing an array of 24 $\Delta E1$(gas)-$\Delta E2$(gas)-E(Si) detectors was performed at Lawrence Berkeley National Laboratory [87]. There, the total two-proton coincidence yield for the used reaction of 197 MeV ^{40}Ca beam on ^{58}Ni target should have been comparable to the yield at GSI. However, only the lower energy one-proton decay group at 0.79(3) MeV with its branching ratio of 1.9(5)% was confirmed [87]. No evidence for the direct two-proton decay was found [87]. Thus, ref. [87] suggests that the one-proton decay data should be correct and we should adopt the excitation energy of 6960(400) keV [78] for the (21^+) isomer.

At IGISOL, developments to produce ^{94}Ag continue [88,89]. The hot cavity laser ion source has been successfully commissioned at IGISOL and the laser ionization has been found to be fully saturated [89]. If the tests with stable $^{107}\mathrm{Ag}^{21+}$ beam from the K-130 cyclotron will be successfull, the on-line experiments with ^{94}Ag will be carried out at IGISOL. In future, the mass measurements of the ground and isomeric states of ^{94}Ag with JYFLTRAP Penning-trap mass spectrometer would solve the two-proton decay energy puzzle. The measurement of the hyperfine structure of the isomeric states would experimentally determine the spectroscopic quadrupole moment, and thus, also the shape of the isomer.

4.3 ^{95}Pd^m

The $(21/2^+)$ isomer in ^{95}Pd with a half-life of 13.3(3) s [90] was first observed at Munich MP tandem via beta-delayed proton and gamma spectroscopy [69]. The total beta branching to proton-emitting states was determined as 0.74(19)%. The final state to be populated after the beta decay was found to be 8^+ in ^{94}Ru. The most strongly populated level in the beta decay of $^{95}\mathrm{Pd}^m$ was the $(21/2^+)$ level at 2449 keV in ^{94}Rh. Soon after this experiment, the known properties of ^{95}Pd were confirmed at GSI [90]. Shell-model calculations [91] suggest that this isomer lies at 1.90 MeV and is a spin-gap isomer in nature. Other shell-model calculations have predicted excitation energies of 1803 keV [92,93] and 1973 keV [94] for this spin-gap isomer. The extrapolated value in the AME03 is 1860(500)# keV [43]. The energy of the isomeric state was experimentally determined as 1876 keV [68] based on the γ-transitions connecting the states built on the ground and isomeric states of ^{95}Pd.

At JYFLTRAP, a 170 MeV ^{40}Ca beam on a ^{nat}Ni target was used for the production of the ground and isomeric states of ^{95}Pd [51]. For these rather long-lived states with an energy difference of about 2 MeV, a quadrupole RF excitation time of 800 ms in the precision trap allowed separate mass measurements for the ground and isomeric state. The reference ion used in these measurements was ^{94}Mo. The measured mass-excess values for the ground and isomeric state of ^{95}Pd were −69961.6(4.8) keV and

−68086.2(4.7) keV [51], respectively. This yields an excitation energy of 1875.4(6.7) keV [51] for the $(21/2^+)$ isomer in ^{95}Pd in agreement with the result of ref. [68]. It further confirms the spin-gap character of the transition as the $21/2^+$ state lies lower in energy than the $15/2^+$ (1973 keV) and $17/2^+$ (1879 keV) states [68]. A recent mass measurement of the reference ^{94}Mo against ^{85}Rb has shown that the mass excess of ^{94}Mo is off by −3.0(21) keV. This has an effect on the mass-excess values of the ground and isomeric states but does not change the value for the excitation energy of the isomer.

The obtained Q_{EC} value for the isomer, 10256.1(6.3) keV [51], together with the branching ratio of 35.8% [90] for the beta decay to the $(21/2^+)$ state at 2449 keV in ^{94}Rh, yields a $\log ft = 5.5$ for this most dominant beta-transition. This confirms the allowed character of the transition. The measurement of $^{95}\mathrm{Pd}^m$ was the first direct Penning-trap mass measurement of such a high-spin isomer at JYFLTRAP. It proved that JYFLTRAP can also be used for the studies of spin-gap isomers.

5 Conclusions and outlook

The studies of neutron-deficient nuclides at the *rp*-process path have been an important part of the physics performed at IGISOL. Decay studies and Q_{EC}-value measurements of the mirror beta decays as well as the studies of the $T = 1$ triplet at $A = 58$ with the light-ion ion guide have yielded information necessary for precise and corrected ft values and for the studies of the isospin symmetry of transitions. The HIGISOL method has shown its strength in the production of Y, Nb and Zr isotopes. JYFLTRAP Penning-trap mass spectrometer has been employed for Q_{EC}-value measurements as well as for measuring the masses or excitation energies of the long-lived isomers, such as $^{53}\mathrm{Co}^m$ and $^{95}\mathrm{Pd}^m$.

By combining the latest state-of-the-art detectors for spectroscopy experiments and employing JYFLTRAP for isobaric purification, many of the older experiments performed with He-jet or IGISOL techniques could be superseded in future. Other future challenges are, for example, the identification of isomeric and ground states in the mass $A = 80$–90 region via post-trap spectroscopy. Since the states are populated differently in the heavy-ion and light-ion fusion-evaporation reactions, some isomeric states, such as $^{93}\mathrm{Ru}^m$ or $^{91}\mathrm{Tc}^m$, could also be produced with the light-ion ion-guide method similarly to the earliest studies of these isomers. There are still many open questions in this mass region, such as the observed γ-rays at $A = 85$, and the excitation energies of $^{85}\mathrm{Nb}^m$, $^{86}\mathrm{Nb}^m$ and $^{94}\mathrm{Ag}^m$. For the production of $^{94}\mathrm{Ag}^m$ (21^+) isomer, the development and testing of the hot-cavity laser ion source at IGISOL are essential.

Spectroscopic experiments provide information on the ground, excited and isomeric states (spins, energies, half-lives), and possible decay modes of the studied nuclides. These data are relevant for the *rp* and recently proposed νp process [95] network calculations, which should also take into account the contribution from the isomers. Close collaboration with nuclear astrophysicists modeling the *rp* and νp processes will continue in future. It would be beneficial to establish a large campaign to measure the interesting spectroscopic details within the relevant mass region at the new IGISOL4 facility in collaboration with the *rp* and νp process modelers. In this way, the impact of new data on the modeling of the *rp* or νp processes would be seen immediately.

This work has been supported by the Academy of Finland under the Finnish Centre of Excellence Programme 2006-2011 (Nuclear and Accelerator Based Physics Programme at JYFL). The support via the Finnish-Russian Interacademy Agreement (project No. 8) is acknowledged. AK acknowledges the support from the Academy of Finland under the project 127301.

References

1. J. Honkanen, Ph.D. thesis, Department of Physics, University of Jyväskylä (1981).
2. J. Honkanen *et al.*, Phys. Scr. **19**, 239 (1979).
3. J. Honkanen *et al.*, Nucl. Phys. A **330**, 429 (1979).
4. K. Eskola *et al.*, Nucl. Phys. A **341**, 365 (1980).
5. J. Honkanen *et al.*, Nucl. Phys. A **380**, 410 (1982).
6. J. Honkanen *et al.*, Nucl. Phys. A **471**, 489 (1987).
7. J. Honkanen, M. Kortelahti, K. Eskola, K. Vierinen, Nucl. Phys. A **366**, 109 (1981).
8. J. Ärje *et al.*, Phys. Rev. Lett. **54**, 99 (1985).
9. J. Äystö *et al.*, Phys. Lett. B **138**, 369 (1984).
10. M. Oinonen *et al.*, Nucl. Instrum. Methods Phys. Res. A **416**, 485 (1998).
11. J. Huikari *et al.*, Nucl. Instrum. Methods Phys. Res. B **222**, 632 (2004).
12. H. Schatz, K. Rehm, Nucl. Phys. A **777**, 601 (2006).
13. H. Schatz *et al.*, Phys. Rev. Lett. **86**, 3471 (2001).
14. V.-V. Elomaa *et al.*, Phys. Rev. Lett. **102**, 252501 (2009).
15. A. Kankainen *et al.*, Phys. Rev. C **82**, 034311 (2010).
16. A. Kankainen, A. Honkanen, K. Peräjärvi, A. Saastamoinen, to be published in Hyperfine Interact. (2012).
17. R. Wallace, S.E. Woosley, Astrophys. J. Suppl. Ser. **45**, 389 (1981).
18. H. Schatz *et al.*, Phys. Rep. **294**, 167 (1998).
19. Y. Sun, M. Wiescher, A. Aprahamian, J. Fisker, Nucl. Phys. A **758**, 765 (2005).
20. G. Martínez-Pinedo, A. Poves, E. Caurier, A.P. Zuker, Phys. Rev. C **53**, R2602 (1996).
21. B.H. Wildenthal, M.S. Curtin, B.A. Brown, Phys. Rev. C **28**, 1343 (1983).
22. H. Fujita *et al.*, Phys. Rev. C **75**, 034310 (2007).
23. K. Peräjärvi *et al.*, Nucl. Phys. A **696**, 233 (2001).
24. Z. Janas *et al.*, Eur. Phys. J. A **12**, 143 (2001).
25. H. Jongsma, A.D. Silva, J. Bron, H. Verheul, Nucl. Phys. A **179**, 554 (1972).
26. J.M. Freeman *et al.*, Nucl. Phys. **69**, 433 (1965).
27. A. Kankainen *et al.*, Eur. Phys. J. A **25**, 129 (2005).
28. A. Jokinen *et al.*, Eur. Phys. J. A **3**, 271 (1998).
29. L.M. Fraile, J. Äystö, Nucl. Instrum. Methods Phys. Res. A **513**, 287 (2003).
30. A. Kankainen, Ph.D. thesis, Department of Physics, University of Jyväskylä (2006).
31. Y. Novikov *et al.*, Eur. Phys. J. A **11**, 257 (2001).

32. J. Parmonen *et al.*, Nucl. Instrum. Methods Phys. Res. A **306**, 504 (1991).
33. A. Piechaczek *et al.*, Phys. Rev. C **61**, 047306 (2000).
34. J. Döring *et al.*, Phys. Rev. C **57**, 1159 (1998).
35. J.J. Ressler *et al.*, Phys. Rev. Lett. **84**, 2104 (2000).
36. J. Döring *et al.*, Phys. Rev. C **59**, 59 (1999).
37. A. Kankainen *et al.*, Eur. Phys. J. A **25**, 355 (2005).
38. C.J. Lister, P.E. Haustein, D.E. Alburger, J.W. Olness, Phys. Rev. C **24**, 260 (1981).
39. S. Della Negra, H. Gauvin, D. Jacquet, Y. Le Beyec, Z. Phys. A **307**, 305 (1982).
40. S. Mitarai *et al.*, Nucl. Phys. A **557**, 381 (1993).
41. C.M. Baglin, Nucl. Data Sheets **109**, 2257 (2008).
42. A. Kankainen *et al.*, Eur. Phys. J. A **29**, 271 (2006).
43. G. Audi, A.H. Wapstra, C. Thibault, Nucl. Phys. A **729**, 337 (2003).
44. R. Iafigliola, J.K.P. Lee, Phys. Rev. C **13**, 2075 (1976).
45. W.X. Huang *et al.*, Phys. Rev. C **59**, 2402 (1999).
46. T. Kuroyanagi *et al.*, Nucl. Phys. A **484**, 264 (1988).
47. D. Bucurescu *et al.*, Z. Phys. A **342**, 403 (1992).
48. T. Shizuma *et al.*, Z. Phys. A **348**, 25 (1994).
49. R. Iafigliola, S.C. Gujrathi, B.L. Tracy, J.K.P. Lee, Can. J. Phys. **52**, 96 (1974).
50. K. Oxorn, S.K. Mark, Z. Phys. A **303**, 63 (1981).
51. C. Weber *et al.*, Phys. Rev. C **78**, 054310 (2008).
52. D. Rudolph *et al.*, Phys. Rev. C **47**, 2574 (1993).
53. S. Dean *et al.*, Eur. Phys. J. A **21**, 243 (2004).
54. E. Browne, Nucl. Data Sheets **82**, 379 (1997).
55. G. Gräff, H. Kalinowsky, J. Traut, Z. Phys. A **297**, 35 (1980).
56. M. König *et al.*, Int. J. Mass Spectrom. Ion Process. **142**, 95 (1995).
57. A. Kellerbauer *et al.*, Eur. Phys. J. D **22**, 53 (2003).
58. R.T. Birge, Phys. Rev. **40**, 207 (1932).
59. V.-V. Elomaa *et al.*, Nucl. Instrum. Methods Phys. Res. A **612**, 97 (2009).
60. T. Eronen *et al.*, Phys. Rev. Lett. **100**, 132502 (2008).
61. J. Äystö *et al.*, Nucl. Phys. A **404**, 1 (1983).
62. J.C. de Lange *et al.*, Z. Phys. A **279**, 79 (1976).
63. D.F. Geesaman *et al.*, Phys. Rev. Lett. **34**, 326 (1975).
64. K. Eskola, Phys. Lett. **23**, 471 (1966).
65. K. Jackson *et al.*, Phys. Lett. B **33**, 281 (1970).
66. M.L. Commara *et al.*, Nucl. Phys. A **708**, 167 (2002).
67. C. Plettner *et al.*, Nucl. Phys. A **733**, 20 (2004).
68. N. Mărginean *et al.*, Phys. Rev. C **67**, 061301 (2003).
69. E. Nolte, H. Hick, Z. Phys. A **305**, 289 (1982).
70. I. Perlman *et al.*, Phys. Rev. **127**, 917 (1962).
71. J. Cerny, J. Esterl, R. Gough, R. Sextro, Phys. Lett. B **33**, 284 (1970).
72. J. Cerny, R.A. Gough, R.G. Sextro, J.E. Esterl, Nucl. Phys. A **188**, 666 (1972).
73. D. Mueller, E. Kashy, W. Benenson, H. Nann, Phys. Rev. C **12**, 51 (1975).
74. K. Schmidt *et al.*, Z. Phys. A **350**, 99 (1994).
75. I. Mukha *et al.*, Phys. Rev. C **70**, 044311 (2004).
76. I. Mukha *et al.*, Phys. Rev. Lett. **95**, 022501 (2005).
77. I. Mukha *et al.*, Nature **439**, 298 (2006).
78. A. Kankainen *et al.*, Phys. Rev. Lett. **101**, 142503 (2008).
79. K. Kaneko, Y. Sun, M. Hasegawa, T. Mizusaki, Phys. Rev. C **77**, 064304 (2008).
80. M. Aggarwal, Phys. Lett. B **693**, 489 (2010).
81. D. Kast *et al.*, Z. Phys. A **356**, 363 (1997).
82. O.L. Pechenaya *et al.*, Phys. Rev. C **76**, 011304 (2007).
83. I. Mukha, H. Grawe, E. Roeckl, S. Tabor, Phys. Rev. C **78**, 039803 (2008).
84. O.L. Pechenaya *et al.*, Phys. Rev. C **78**, 039804 (2008).
85. D.G. Jenkins, Phys. Rev. C **80**, 054303 (2009).
86. I. Mukha, E. Roeckl, H. Grawe, S. Tabor, arXiv: 1008.5346v1 [nucl-ex] (2010).
87. J. Cerny *et al.*, Phys. Rev. Lett. **103**, 152502 (2009).
88. T. Kessler *et al.*, Nucl. Instrum. Methods Phys. Res. B **266**, 4420 (2008).
89. M. Reponen *et al.*, Eur. Phys. J. A **42**, 509 (2009).
90. W. Kurcewicz *et al.*, Z. Phys. A **308**, 21 (1982).
91. K. Ogawa, Phys. Rev. C **28**, 958 (1983).
92. S.E. Arnell *et al.*, Phys. Rev. C **49**, 51 (1994).
93. J. Sinatkas, L.D. Skouras, D. Strottman, J.D. Vergados, J. Phys. G **18**, 1401 (1992).
94. K. Schmidt *et al.*, Nucl. Phys. A **624**, 185 (1997).
95. C. Fröhlich *et al.*, Phys. Rev. Lett. **96**, 142502 (2006).

DOI 10.1007/978-94-007-5555-0_18
Reprinted from *Hyperfine Interactions* Volume 223, Issues 1-3, 2014
and *The European Physical Journal A* Volume 48, Issue 4, 2012

Double-β decay studies with JYFLTRAP

V. S. Kolhinen · S. Rahaman · J. Suhonen

Published online: 28 March 2012

Abstract We have applied JYFLTRAP double Penning trap mass spectrometer to study several double-β decay Q values that could be useful for the search of neutrinoless double-β decay and, hence for the determination of neutrino properties.

Keywords Double-β decay · Q value · Neutrino mass

1 Introduction

The neutrino is one of least well known particles in modern physics—one does not even know its mass. Present-day high-precision neutrino-oscillation experiments performed with solar and reactor neutrinos have shown evidence of nonzero neutrino mass [1–3]. This data has allowed to extract the difference of the squared neutrino masses and the mixing angles, but not the absolute mass scale of the neutrinos, which remains have to be determined by other means. Another important question to be answered is whether neutrinos are Dirac or Majorana particles. A Dirac neutrino has a distinct antineutrino but a Majorana neutrino does not since it is its own antiparticle.

Neutrinoless double-β decay experiments provide the best means to probe these questions [4, 5]. The existence of $0\nu\beta\beta$ decay would mean that neutrino is a Majorana particle. Since this decay violates the lepton-number conservation, its existence would also be evidence of physics beyond the Standard Model. The mass of a Majorana neutrino is then inversely proportional to the double-β decay half-life [6].

V. S. Kolhinen (✉) · J. Suhonen
Department of Physics, University of Jyväskylä,
Post Office Box 35(YFL), 40014 Jyväskylä, Finland
e-mail: veli.kolhinen@jyu.fi

S. Rahaman
Physics Division, P-25, Los Alamos National Lab,
Los Alamos, NM 87545, USA

The widest interest in the search of a neutrinoless double-β decay is focused in $0\nu\beta^-\beta^-$ decays since they usually have relatively large Q values that are above most of the background β-decay energies. Experiments that use this decay mode for neutrino research need to know the decay Q value as accurately as possible to be able to narrow the energy window in the two electron spectrum around the small peak coming from the $0\nu\beta^-\beta^-$ decay.

Another interesting case is the double-electron-capture decay, 0νECEC, which can occur as a resonant decay [7] or as a radiative process without resonance condition [8]. The resonance condition, where the atomic ground state of the mother has the same energy as the excited state of the decay daughter, can enhance the decay rate by a factor of 10^6. This would make the experimental testing of the Majorana nature of the neutrinos possible if one could find a good candidate for such a resonant 0νECEC decay. Thefore, there is a strong motivation to measure the Q values of the possible candidates for the resonant double-electron-capture decay.

2 Experimental method

JYFLTRAP [9] is a double cylindrical Penning trap mass spectrometer that is situated at the Ion Guide Isotope Separator On-Line (IGISOL) facility [10, 11] in the Accelerator Laboratory of the University of Jyväskylä, Finland. Combination of these two setups, IGISOL and JYFLTRAP together, gives perfect means to perform Q value measurements that are crucial for the studies of Majorana nature of neutrino.

With an electric discharge ion source placed inside the IGISOL target chamber one can produce simultaneously the stable or quasi stable decay parent and daughter ions as continuous ion beam. This beam is then cooled and bunched in a radio frequency quadrupole trap (RFQ) [12] and injected into the purification Penning trap that applies buffer-gas cooling method [13]. Very often this standard purification technique is not enough to remove the other unwanted ion species away and therefore an additional purification step called Ramsey cleaning is performed in the precision trap where the contaminant ions are excited in larger orbits and subsequently injected back to the purification trap through a narrow diaphragm allowing for only the ion of interest to survive [14]. After an additional cooling the ions are retransfered to the precision trap for the cyclotron frequency determination ($\nu_c = \frac{1}{2\pi}\frac{q}{m}B$) by applying time-of-flight ion-cyclotron resonance technique [15]. Nowadays this is done by using a Ramsey-type ion motion excitation pattern [16, 17].

3 Experimental results

3.1 Neutrinoless double-β decay

3.1.1 Q value of ^{150}Nd

We have performed double-β decay Q value measurement of ^{150}Nd since it is considered to be one of the most prominent cases where to look for the $0\nu\beta^-\beta^-$ decay due to its relatively high decay energy $Q_{\beta\beta}$ =3367.7(22) keV [18] (Atomic Mass Evaluation 2003 (AME2003)). Three large scale experiments DCBA [19],

Table 1 Collection of data of the measured candidates for resonant 0νECEC decay

Atom	Q value /keV	$Q - E_H - E_{H'}$/keV	AME Q value/keV	E_{Daughter}
^{112}Sn	1919.82(16)	1866.42(16)	1919(4)	1871.00(19)
^{74}Se	1209.169(49)	1206.704(49)	1210(2)	1204.205(7)
^{136}Ce	2378.53(27)	2303.65(27)	2419(13)	2315.32(7)

Shown are the atom, the measured decay Q value, the decay Q value after subtracting the energies of the captured electrons, the AME2003 Q value and the energy of the level that was considered to be in resonance

SuperNEMO [20], and SNO+ [21] are using ^{150}Nd in their setups. A recent search for the $0\nu\beta^-\beta^-$ decay in ^{150}Nd was done by using NEMO-3 detector in [22]. Therefore, a new more accurate determination of the decay Q value $^{150}\text{Nd} \rightarrow {}^{150}\text{Sm}$ was important. The new JYFLTRAP value was determined to be 3371.38(20) keV [23] which deviates 3.7 keV from the old adopted value and is a factor of 10 more precise.

3.1.2 Q value of ^{136}Xe

We have confirmed the ^{136}Xe double-β decay Q value to be 2457.86(48) keV [24]. This result agrees with the value 2457.83(37) keV measured at the Florida State University [25]. The search for the $0\nu\beta^-\beta^-$ decay of ^{136}Xe is boosted by starting new large-scale underground experiments the Enriched Xenon Observatory (EXO) [26] and the Neutrino Experiment with a Xenon TPC (NEXT) [27]. A review of the related experiments is given in [28].

3.2 Resonant neutrinoless double-electron-capture decay

We have studied and eliminated three cases of prominent candidates of a resonant 0νECEC decay ^{112}Sn [29], ^{74}Se [30] and ^{136}Ce [24]. In Table 1 it is show the Q values measured at JYFLTRAP, Q values after subtracting the energies of the captured electrons, the old adopted Q values (AME2003), and the energy of the excited state of the decay daughter that was considered to be in resonance.

3.2.1 Decay of ^{112}Sn

The ^{112}Sn double-electron-capture Q value was determined to be 1919.82(16) keV [29] which is 25 times more precise than the previous value 1919(4) keV [18]. The $Q - E_H - E_{H'}$ value depends on the atomic shells where the two electrons are captured from, i.e. if the process is a KK, KL or LL capture. The closest match with the 1871.00(19) keV 0^+ state of ^{112}Cd is obtained in KL capture. Using 26.7 keV and 4.02 keV for the K- and L-shell electrons, respectively, we obtain $Q - E_H - E_{H'} = 1866.42(16)$ keV. $Q - E_H - E_{H'} - E_{\text{Daughter}} = -4.58$ keV would imply that the resonant decay is forbidden or at least very unlikely to occur. Similar conclusion were drawn in [31].

3.2.2 Decay of ^{74}Se

The measured JYLTRAP Q value for the ^{74}Se double-electron-capture decay is 1209.169(49) keV [30]. The old adopted value was 1210(2) [18]. Using L_2L_3 electron capture assuming that spin and parity are conserved one gets by using values $E_H = 1.2478$ for L_2 and $E_{H'} = 1.2167$ for L_3 (http://ie.lbl.gov/atomic/bind.pdf) that $Q - E_H - E_{H'} = 1206.704(49)$ keV, which practically excludes the energy degeneracy with the second 2^+ state of ^{74}Ge at 1204.205(7) keV [32].

3.2.3 Decay of ^{136}Ce

The determined double-electron-capture Q value of ^{136}Ce is 2378.53(27) keV. This new Q value differs from the old adopted value of 2419(13) [18] and is 50 times more precise. Using KK electron capture and $E_H = E_{H'} = 37.4406$ keV (http://ie.lbl.gov/atomic/bind.pdf) one obtains $Q - E_H - E_{H'} = 2303.53(27)$ keV. This differs by $\Delta = -11.67$ keV from the excitation energy of the fourth 0^+ state of ^{136}Ba at 2315.32(7) keV [33]. Therefore this resonant double-electron-capture of ^{136}Ce cannot be detected.

4 Conclusions

We have performed double-β decay Q value measurements of ^{150}Nd and ^{136}Xe and eliminated three candidates for the resonant 0νECEC decay: ^{112}Sn, ^{74}Se and ^{136}Ce. This work has been now paused due to the upgrade of the IGISOL experiment and will continue after commissioning of the new IGISOL IV setup. Other groups, like SHIPTRAP, have carried out this search for the resonant double-electron capture decay [34].

Acknowledgements This work has been supported by the Academy of Finland under the Finnish Center of Excellence Programme 2006-2011 (Nuclear and Accelerator Based Physics Programme at JYFL).

References

1. Maltoni, M., et al.: New J. Phys. **6**, 122 (2004)
2. Schwetz, T., Tórtola, M., Valle, J.W.F.: New J. Phys. **10**, 113011 (2008)
3. Nunokawa, H., Parke, S., Valle, J.W.F.: Prog. Part. Nucl. Phys. **60**, 338 (2008)
4. Suhonen, J., Civitarese, O.: Phys. Rep. **300**, 123 (1998)
5. Faessler, A., Šimkovic, F.: J. Phys. G: Nucl. Part. Phys. **24**, 2139 (1998)
6. Avignone III, F.T., Elliott, S.R., Engel, J.: Rev. Mod. Phys. **80**, 481 (2008)
7. Bernabeu, J., De Rujula, A., Jarlskog, C.: Nucl. Phys. B **223**, 15 (1983)
8. Sujkowski, Z., Wycech, S.: Phys. Rev. C **70**, 052501(R) (2004)
9. Kolhinen, V.S., et al.: Nucl. Instrum. Methods Phys. Res., Sect. A **528**, 776 (2004)
10. Äystö, J.: Nucl. Phys. A **693**, 477 (2001)
11. Penttilä, H., et al.: Eur. Phys J. A **25**(Suppl. 1), 745 (2005)
12. Nieminen, A., et al.: Nucl. Instrum. Methods Phys. Res., Sect. A **469**, 244 (2001)
13. Savard, G., et al.: Phys. Lett. A **158**, 247 (1991)
14. Eronen, T., et al.: Nucl. Instrum. Methods Phys. Res., Sect. B **266**, 4527 (2008)
15. Gräff, G., Kalinowsky, H., Traut, J.: Z. Phys. A **297**, 35 (1980)
16. George, S., et al.: Int. J. Mass. Spectrom. **264**, 110 (2007)
17. Kretzschmar, M.: Int. J. Mass. Spectrom. **264**, 122 (2007)
18. Audi, G., Wapstra, A.H., Thibault, C.: Nucl. Phys. **A729**, 337 (2003)
19. Ishihara, N., et al.: J. Phys.: Conf. Series **203**, 012071 (2010)
20. Piquemal, F.: Phys. Atom Nucl. **69**, 2096 (2006)
21. Zuber, K.: AIP Conf. Proc. **942**, 101 (2007)

22. Argyriades, J., et al.: Phys. Rev. C **80**, 032501(R) (2009)
23. Kolhinen, V.S., et al.: Phys. Rev. C **82**, 022501(R) (2010)
24. Kolhinen, V.S., et al.: Phys. Lett. B **697**, 116 (2011)
25. Redshaw, M., Wingfield, E., McDaniel, J., Myers, E.G.: Phys. Rev. Lett. **98**, 053003 (2007)
26. Kaufman, L.J.: J. Phys.: Conf. Series **203**, 012067 (2010)
27. Novella, P., et al.: J. Phys.: Conf. Series **203**, 012068 (2010)
28. Cremonesi, O.: Hyperfine Interact. **193**, 261 (2009)
29. Rahaman, S., et al.: Phys. Rev. Lett. **103**, 042501 (2009)
30. Kolhinen, V.S., et al.: Phys. Lett. B **684**, 17 (2010)
31. Green, K.L., et al.: Phys. Rev. C **80**, 032502(R) (2009)
32. Singh, B., Farhan, A.R.: Nucl. Data Sheets **107**, 1923 (2006)
33. Sonzogni, A.A.: Nucl. Data Sheets **95**, 837 (2002)
34. Eliseev, S., et al.: Phys. Rev. C **83**, 038501 (2011)

DOI 10.1007/978-94-007-5555-0_19
Reprinted from *Hyperfine Interactions* Volume 223, Issues 1-3, 2014
and *The European Physical Journal A* Volume 48, Issue 4, 2012

THE EUROPEAN PHYSICAL JOURNAL A

Regular Article – Theoretical Physics

Nuclear matrix elements for the resonant neutrinoless double electron capture

J. Suhonen[a]

Department of Physics, Post Office Box 35(YFL), FIN-40014 University of Jyväskylä, Finland

Received: 6 February 2012 / Revised: 22 February 2012
Published online: 19 April 2012 –
Communicated by J. Äystö

Abstract. The rate of the neutrinoless double electron capture (0νECEC) decay with a resonance condition depends sensitively on the mass difference between the initial and final nuclei of decay. This is where the JYFLTRAP Penning-trap measurements at the JYFL become invaluable in estimation of the half-lives of these decays. In this work the resonant 0νECEC decay is discussed from the point of view of its theoretical aspects, in particular regarding the resonance condition and the involved nuclear matrix elements (NME). The associated decay amplitudes are derived and the calculations of the NMEs by the microscopic many-body approach of the multiple-commutator model are outlined. The resonant 0νECEC decays of ^{74}Ge and ^{136}Ce are discussed as applications of the theory framework.

1 Introduction

At present the neutrinoless double-beta ($0\nu\beta\beta$) decay is considered to be the most easily accessible means of extracting information on the possible Majorana mass of the neutrino. The reason for this is that the $0\nu\beta\beta$ decay occurs only if the neutrino is a massive Majorana particle. The search for the $0\nu\beta\beta$ decay is mostly concentrated on the $0\nu\beta^-\beta^-$ decays due to their favorable Q values [1]. Assuming that the neutrino carries Majorana mass also the positron emitting modes of $0\nu\beta\beta$ decay, $0\nu\beta^+\beta^+$, $0\nu\beta^+$EC and 0νECEC, are possible but they are hard to detect owing to their small decay Q values [2]. The neutrinoless double electron capture, 0νECEC, can only be realized as a resonant decay [3] or a radiative process with or without a resonance condition [4]. The resonant 0νECEC decay has attracted a lot of experimental attention recently [5–9].

The resonance condition —close degeneracy of the initial and final (excited) atomic states— can enhance the decay rate by a factor as large as 10^6 [3]. Possible candidates for the resonant decays are many [3,4]. Verification of the fulfillment of the resonance condition is of utmost importance before expensive experiments are conducted in the aim to detect the resonant 0νECEC. The half-life of the resonant 0νECEC depends sensitively on the mass difference between the initial and final nuclei of decay. The verification of the resonance enhancement thus calls for accurate measurements of these mass differences. At the JYFL the way to achieve the necessary accuracy —in the 100 eV range— is to engage the Penning ion trap, JYFLTRAP, for these measurements. In fact there is a campaign at the JYFL to run the Penning trap to sort out potential candidates for measurements of resonant 0νECEC decays by dedicated underground experiments.

In this work two examples —the decay of ^{74}Se to the second 2^+ state in ^{74}Ge and the decay of ^{136}Ce to the fourth 0^+ state in ^{136}Ba— are discussed from the point of view of the nuclear-structure aspects of resonant 0νECEC decays.

[a] e-mail: jouni.suhonen@phys.jyu.fi

2 Theory framework of the resonant neutrinoless double electron capture

2.1 Decay half-life

In this work we study a particular type of neutrinoless double electron capture (0νECEC) decay, namely the resonant 0νECEC process of the form

$$e^- + e^- + (A,Z) \rightarrow (A,Z-2)^* \rightarrow (A,Z-2) + \gamma + 2X, \quad (1)$$

where the capture of two atomic electrons leaves the final nucleus in an excited state that decays by one or more gamma-rays and the atomic vacancies are filled by outer electrons with emission of X-rays. The daughter state $(A, Z-2)^*$ is a virtual state with energy

$$E = E^* + E_H + E_{H'}, \quad (2)$$

including the nuclear excitation energy and the binding energies of the two captured electrons. For the half-life of the parent atom the resonance condition can be written [3, 10] in the Breit-Wigner form

$$\frac{\ln 2}{T_{1/2}} = \frac{\mathcal{M}^2}{(Q-E)^2 + \Gamma^2/4}\Gamma, \quad (3)$$

where $\mathcal{M}$ contains the leptonic phase space and the nuclear matrix element, Γ denotes the combined nuclear and atomic radiative widths (few tens of electron volts [11]) and $|Q-E|$ is the so-called degeneracy parameter containing the energy (2) of the virtual final state and the difference between the initial and final atomic masses (the Q value of resonant 0νECEC). The Q value needs to be measured very accurately in order to judge whether the resonant 0νECEC is detectable or not. For this, the JYFLTRAP Penning trap is suited in a perfect manner.

To be able to estimate the half-life we have to evaluate the quantity $\mathcal{M}$ in (3). In the mass mode of the resonant 0νECEC, it can be written as

$$\mathcal{M} = G_{0\nu}^{\text{ECEC}} M_{0\nu}^{\text{ECEC}} \langle m_\nu \rangle, \quad (4)$$

where $\langle m_\nu \rangle$ is the effective neutrino mass of the neutrinoless double-beta decay [2] and $G_{0\nu}^{\text{ECEC}}$ is the leptonic phase-space factor

$$G_{0\nu}^{\text{ECEC}} = \left(\frac{G_{\text{F}}\cos\theta_{\text{C}}}{\sqrt{2}}\right)^2 \frac{g_{\text{A}}^2}{2\pi}\frac{(Z\alpha m_{\text{e}})^3}{\pi R_A}\eta_x, \quad R_A = 1.2A^{1/3}, \quad (5)$$

with G_{F} the Fermi coupling constant, θ_{C} the Cabibbo angle, $g_{\text{A}} = 1.25$ the axial-vector weak coupling constant, Z the charge of the mother nucleus, α the fine-structure constant, m_{e} the electron rest mass and R_A the radius of a nucleus with mass number A. The quantity η_x in (5) is a suppression factor depending on the atomic orbitals where the two electrons are captured from. Solution of the Schrödinger equation for a point-charge nucleus produces the values

$$\eta_{\text{KK}} = 1, \quad \eta_{\text{KL}_1} = \frac{1}{\sqrt{8}}, \quad \eta_{\text{L}_1\text{L}_1} = \frac{1}{8},$$
$$\eta_{\text{L}_2\text{L}_2} = \eta_{\text{L}_2\text{L}_3} = \eta_{\text{L}_3\text{L}_3} = \frac{1}{96}. \quad (6)$$

This approximation is reasonable and compares well with the Dirac solution for a homogeneously charged spherical nucleus [12]. Expressions for the involved nuclear matrix element $M_{0\nu}^{\text{ECEC}}$ are given in sect. 2.2 and 2.3.

2.2 Decays to 0^+ final states

The nuclear matrix element (NME) involved in (4) for decays to the 0^+ final states is defined as

$$M_{0\nu}^{\text{ECEC}} = \frac{1}{R_A}\left[M_{\text{GT}}^{\text{ECEC}} - \left(\frac{g_{\text{V}}}{g_{\text{A}}}\right)^2 M_{\text{F}}^{\text{ECEC}}\right], \quad (7)$$

where $R_A = 1.2A^{1/3}\,\text{fm}$ is the nuclear radius, g_{V} (g_{A}) is the vector (axial-vector) weak coupling constant, M_{GT} is the Gamow-Teller and M_{F} is the Fermi NME. The two matrix elements are given by

$$M_{\text{F}}^{\text{ECEC}} = \sum_a \left(0_{\text{f}}^+ || \sum_{m,n} t_m^+ t_n^+ h_{\text{F}}(r_{mn}, E_a) || 0_{\text{i}}^+\right), \quad (8)$$

$$M_{\text{GT}}^{\text{ECEC}} = \sum_a \left(0_{\text{f}}^+ || \sum_{m,n} t_m^+ t_n^+ h_{\text{GT}}(r_{mn}, E_a) \boldsymbol{\sigma}_m \cdot \boldsymbol{\sigma}_n || 0_{\text{i}}^+\right). \quad (9)$$

Here the summation runs over all the states J^π of the intermediate nucleus and $r_{mn} = |\mathbf{r}_m - \mathbf{r}_n|$ is the relative coordinate between the nucleons m and n. The isospin raising operators τ_n^+ convert the n-th proton to a neutron and $\boldsymbol{\sigma}$ are the usual Pauli spin matrices. Here 0_{i}^+ (0_{f}^+) is the ground state of the even-even mother (daughter) nucleus and the neutrino potential $h_K(r_{mn}, E_a)$, $K = \text{F}, \text{GT}$, is defined as

$$h_K(r_{mn}, E_a) = \frac{2}{\pi}R_{\text{A}} \int \text{d}q \frac{q h_K(q^2)}{q + E_a - (E_{\text{i}} + E_{\text{f}})/2} j_0(q r_{mn}), \quad (10)$$

where j_0 is the spherical Bessel function. The term $h_K(q^2)$ in (10) includes the contributions arising from the short-range correlations, nucleon form factors and higher-order terms of the nucleonic weak current [13–15].

Next, we write the nuclear matrix elements explicitly. They are given by

$$\begin{aligned} M_K^{\text{ECEC}} = & \sum_{J^\pi, k_1, k_2, J'} \sum_{pp'nn'} (-1)^{j_p + j_{n'} + J + J'} \sqrt{2J'+1} \\ & \times \begin{Bmatrix} j_p & j_n & J \\ j_{n'} & j_{p'} & J' \end{Bmatrix} (nn' : J' || \mathcal{O}_K || pp' : J') \\ & \times (0_{\text{f}}^+ || [c_{n'}^\dagger \tilde{c}_{p'}]_J || J_{k_1}^\pi) \\ & \times \langle J_{k_1}^\pi | J_{k_2}^\pi \rangle (J_{k_2}^\pi || [c_n^\dagger \tilde{c}_p]_J || 0_{\text{i}}^+), \; K = \text{F}, \text{GT}, \end{aligned} \quad (11)$$

where k_1 and k_2 label the different nuclear-model solutions for a given multipole J^π, the set k_1 stemming from the calculation based on the final nucleus and the set k_2 stemming from the calculation based on the initial nucleus. The operators $\mathcal{O}_K$ inside the two-particle matrix element derive from (8) and (9) and they can be written as

$$\begin{aligned} \mathcal{O}_{\text{F}} &= h_{\text{F}}(r_{12}, E_k), \\ \mathcal{O}_{\text{GT}} &= h_{\text{GT}}(r_{12}, E_k) \boldsymbol{\sigma}_1 \cdot \boldsymbol{\sigma}_2, \\ r_{12} &= |\mathbf{r}_1 - \mathbf{r}_2|, \end{aligned} \quad (12)$$

where E_k is the average of the k-th eigenvalues of the nuclear-model calculations based on the initial and final nuclei of the decay. Here the one-body transition densities are $(0_{\text{f}}^+ || [c_{n'}^\dagger \tilde{c}_{p'}]_J || J_{k_1}^\pi)$ and $(J_{k_2}^\pi || [c_n^\dagger \tilde{c}_p]_J || 0_{\text{i}}^+)$, and they are given separately for the different nuclear-structure formalisms in sect. 3. The quantity $\langle J_{k_1}^\pi | J_{k_2}^\pi \rangle$ is the overlap factor and its purpose is to match the the two sets of J^π states that stem from the two different nuclear-model calculations —one starting from the mother nucleus (the k_2

states) and the other starting from the daughter nucleus (the k_1 states). Its explicit form for the pnQRPA nuclear model is given in sect. 3.1.

The two-particle matrix element of (11) can be written as

$$(nn':J'||\mathcal{O}_K||pp':J') = \widehat{J'}\widehat{j_p}\widehat{j_{p'}}\widehat{j_n}\widehat{j_{n'}}\sum_{\lambda S}(2\lambda+1)(2S+1)F_K \times \begin{Bmatrix} l_p & l_{p'} & \lambda \\ \frac{1}{2} & \frac{1}{2} & S \\ j_p & j_{p'} & J' \end{Bmatrix} \begin{Bmatrix} l_n & l_{n'} & \lambda \\ \frac{1}{2} & \frac{1}{2} & S \\ j_n & j_{n'} & J' \end{Bmatrix} \times \sum_{n_1n_2lNL} M_\lambda(n_1lNL;n_nl_nn_{n'}l_{n'})M_\lambda(n_2lNL;n_pl_pn_{p'}l_{p'}) \times \int \mathrm{d}^3r\phi_{n_1l}(\mathbf{r})h_K(r_{12},E_a)\phi_{n_2l}(\mathbf{r}), \tag{13}$$

where $\widehat{j} = \sqrt{2j+1}$, the neutrino potentials are those of (10) and we have defined

$$F_\mathrm{F} = 1, \quad F_\mathrm{GT} = 6(-1)^{S+1}\begin{Bmatrix} \frac{1}{2} & \frac{1}{2} & S \\ \frac{1}{2} & \frac{1}{2} & 1 \end{Bmatrix}. \tag{14}$$

The quantities M_λ are the Moshinsky brackets that mediate the transformation from the laboratory coordinates $\mathbf{r}_1$ and $\mathbf{r}_2$ to the center-of-mass coordinate $\mathbf{R} = \frac{1}{\sqrt{2}}(\mathbf{r}_1+\mathbf{r}_2)$ and the relative coordinate $\mathbf{r} = \frac{1}{\sqrt{2}}(\mathbf{r}_1-\mathbf{r}_2)$. In this way, the short-range correlations of the two decaying protons are easily incorporated in the theory. The wave functions $\phi_{nl}(\mathbf{r})$ are taken to be the eigenfunctions of the isotropic harmonic oscillator.

2.3 Decays to 2^+ final states

The 0νECEC transition to the 2^+_f final states is mediated by a spherical thensor of rank two. Denoting this tensor by $\mathbf{T}_2$ one can write the transition matrix element as

$$\begin{aligned} M_{0\nu}^\mathrm{ECEC} &= \sum_a \left(2^+_\mathrm{f}||\sum_{m,n} t^+_m t^+_n \mathbf{T}_2(mn;a)||0^+_\mathrm{i}\right) \\ &= \frac{1}{\sqrt{5}}\sum_{\substack{\mathcal{J}^\pi,k_1,k_2\\J,J',J_1}}\sum_{nn'pp'}\widehat{J}\widehat{J'}\widehat{J_1}\begin{Bmatrix} j_n & j_p & \mathcal{J} \\ j_{n'} & j_{p'} & J_1 \\ J & J' & 2 \end{Bmatrix} \\ &\quad\times(nn':J||\mathcal{O}_2||pp':J') \\ &\quad\times\left(2^+_\mathrm{f}||\left[c^\dagger_{n'}\tilde{c}_{p'}\right]_{J_1}||\mathcal{J}^\pi_{k_1}\right) \\ &\quad\times\langle\mathcal{J}^\pi_{k_1}|\mathcal{J}^\pi_{k_2}\rangle\left(\mathcal{J}^\pi_{k_2}||\left[c^\dagger_n\tilde{c}_p\right]_\mathcal{J}||0^+_\mathrm{i}\right). \end{aligned} \tag{15}$$

The most straightforward way to build $\mathbf{T}_2$ is to use a double Gamow-Teller type of operator

$$\mathbf{T}_2(mn;a) = h_\mathrm{GT}(r_{mn},E_a)[\boldsymbol{\sigma}_m\boldsymbol{\sigma}_n]_2, \tag{16}$$

where the neutrino potential is given in (10). Then the two-nucleon operator $\mathcal{O}_2$ in (15) becomes

$$\mathcal{O}_2 = h_\mathrm{GT}(r_{12},E_k)[\boldsymbol{\sigma}_1\boldsymbol{\sigma}_2]_2, \quad r_{12} = |\mathbf{r}_1-\mathbf{r}_2|, \tag{17}$$

where E_k is again the average of the k-th eigenvalues of the two nuclear-model calculations. The resulting two-particle matrix element is given by

$$(nn':J||\mathcal{O}_2||pp':J') = 6\sqrt{5}\widehat{J}\widehat{J'}\widehat{j_p}\widehat{j_{p'}}\widehat{j_n}\widehat{j_{n'}}\sum_\lambda\widehat{\lambda}^2(-1)^{J+\lambda+1} \times\begin{Bmatrix} \lambda & 1 & J \\ 2 & J' & 1 \end{Bmatrix}\begin{Bmatrix} l_p & l_{p'} & \lambda \\ \frac{1}{2} & \frac{1}{2} & 1 \\ j_p & j_{p'} & J' \end{Bmatrix}\begin{Bmatrix} l_n & l_{n'} & \lambda \\ \frac{1}{2} & \frac{1}{2} & 1 \\ j_n & j_{n'} & J' \end{Bmatrix} \times\sum_{n_1n_2lNL} M_\lambda(n_1lNL;n_nl_nn_{n'}l_{n'})M_\lambda(n_2lNL;n_pl_pn_{p'}l_{p'}) \times\int \mathrm{d}^3r\phi_{n_1l}(\mathbf{r})h_\mathrm{GT}(r_{12},E_a)\phi_{n_2l}(\mathbf{r}), \tag{18}$$

where the neutrino potential is the usual one (10) and the other quantities have been defined in sect. 2.2.

3 Nuclear-structure models

The starting point in the present calculations is the theoretical framework of the quasiparticle random-phase approximation (QRPA). This framework is based on the quasiparticles that are created in a BCS calculation via the Bogoliubov-Valatin transformation [16]. We use two kinds of QRPA approaches. The proton-neutron QRPA (pnQRPA) is used to produce the J^π states of the intermediate odd-odd nucleus and the charge-conserving QRPA (ccQRPA) is used to produce the excited states of the even-even daughter nucleus. The states of the pnQRPA and the ccQRPA are connected by decay amplitudes that are calculated in the higher-QRPA framework called the multiple-commutator model (MCM). All these approaches are discussed below.

3.1 Proton-neutron quasiparticle random-phase approximation

We use the pnQRPA to form the wave functions that represent states of an odd-odd nucleus. The pnQRPA state is obtained by acting by a pnQRPA creation operator on the QRPA ground state $|\mathrm{QRPA}\rangle$. This action can be expressed in the form

$$\begin{aligned} |J^\pi_k\,M\rangle &= Q^\dagger(J^\pi_k,M)|\mathrm{QRPA}\rangle \\ &= \sum_{pn}\left(X^{J^\pi_k}_{pn}\left[a^\dagger_pa^\dagger_n\right]_{JM} - Y^{J^\pi_k}_{pn}\left[a^\dagger_pa^\dagger_n\right]^\dagger_{JM}\right). \end{aligned} \tag{19}$$

Here J^π is the multipolarity of the nuclear state and and k enumerates the phonons. The operator $a^\dagger_p$ ($a^\dagger_n$) creates a proton (neutron) quasiparticle in the orbital p (n). The sum runs over all proton-neutron configurations in the chosen valence space.

Making use of the ansatz (19) one can derive (see, *e.g.*, [16]) the pnQRPA equations of motion in the quasiboson approximation. The equations can be cast in the matrix form

$$\begin{pmatrix} \mathbf{A}(J) & \mathbf{B}(J) \\ \mathbf{B}(J) & \mathbf{A}(J) \end{pmatrix} \begin{pmatrix} \mathbf{X}^J \\ \mathbf{Y}^J \end{pmatrix} = E_J \begin{pmatrix} \mathbf{X}^J \\ -\mathbf{Y}^J \end{pmatrix}, \tag{20}$$

where J is the angular momentum and the matrix elements of the $\mathbf{A}(J)$ and $\mathbf{B}(J)$ matrices read

$$\begin{aligned} A_{pn,p'n'}(J) = {} & \delta_{pp'}\delta_{nn'}(E_p + E_n) \\ & + g_{\rm pp}(u_p u_n u_{p'} u_{n'} + v_p v_n v_{p'} v_{n'}) \\ & \times \langle p\,n\,;\,J|V|p'\,n'\,;\,J\rangle \\ & + g_{\rm ph}(u_p v_n u_{p'} v_{n'} + v_p u_n v_{p'} u_{n'}) \\ & \times \langle p\,n^{-1}\,;\,J|V_{\rm RES}|p'\,n'^{-1}\,;\,J\rangle \end{aligned} \tag{21}$$

and

$$\begin{aligned} B_{pn,p'n'}(J) = {} & -g_{\rm pp}(u_p u_n v_{p'} v_{n'} + v_p v_n u_{p'} u_{n'}) \\ & \times \langle p\,n\,;\,J|V|p'\,n'\,;\,J\rangle \\ & + g_{\rm ph}(u_p v_n v_{p'} u_{n'} + v_p u_n u_{p'} v_{n'}) \\ & \times \langle p\,n^{-1}\,;\,J|V_{\rm RES}|p'\,n'^{-1}\,;\,J\rangle, \end{aligned} \tag{22}$$

with v and u corresponding to the occupation and unoccupation amplitudes stemming from the BCS calculation.

Above $\langle V\rangle$ is the normalized, J-coupled and antisymmetrized two-body interaction matrix element (see eq. (8.16) of [16]) and the particle-hole matrix element $\langle V_{\rm RES}\rangle$ of the residual interaction is obtained from $\langle V\rangle$ by the Pandya transformation

$$\langle p\,n^{-1}\,;\,J|V_{\rm RES}|p'\,n'^{-1}\,;\,J\rangle = -\sum_{J'} \hat{J'}^2 \begin{Bmatrix} j_p & j_n & J \\ j_{p'} & j_{n'} & J' \end{Bmatrix} \langle p\,n'\,;\,J'|V|p'\,n\,;\,J'\rangle. \tag{23}$$

The quantities E_p and E_n in (21) are the quasiparticle energies for the proton orbital p and neutron orbital n, respectively. The $j_{p,n}$ in (23) are the total angular momenta of the proton or neutron single-particle orbitals.

In (21) and (22) the particle-hole and particle-particle parts of the pnQRPA matrices are separately scaled by the particle-hole parameter $g_{\rm ph}$ and particle-particle parameter $g_{\rm pp}$ [17–19]. The particle-hole parameter affects the position of the Gamow-Teller giant resonance and its value was fixed by the available systematics [16] on the location of the giant state. The parameter $g_{\rm pp}$ affects only weakly the neutrinoless double electron capture rates and it value was set to the default $g_{\rm pp} = 1.0$.

The pnQRPA transition densities are given by

$$\begin{aligned} (0_f^+||[c_{n'}^\dagger \tilde{c}_{p'}]_J||J_{k_1}^\pi) = {} & \hat{J}(-1)^{j_{p'}+j_{n'}+J+1} \\ & \times \left[\bar{u}_{p'}\bar{v}_{n'}\bar{X}_{p'n'}^{J_{k_1}^\pi} + \bar{v}_{p'}\bar{u}_{n'}\bar{Y}_{p'n'}^{J_{k_1}^\pi}\right], \end{aligned} \tag{24}$$

$$\begin{aligned} (J_{k_2}^\pi||[c_n^\dagger \tilde{c}_p]_J||0_i^+) = {} & \hat{J}(-1)^{j_p+j_n+J+1} \\ & \times \left[v_p u_n X_{pn}^{J_{k_2}^\pi} + u_p v_n Y_{pn}^{J_{k_2}^\pi}\right], \end{aligned} \tag{25}$$

where v ($\bar{v}$) and u ($\bar{u}$) correspond to the BCS occupation and unoccupation amplitudes of the initial (final) even-even nucleus. The amplitudes X and Y ($\bar{X}$ and $\bar{Y}$) come from the pnQRPA calculation starting from the initial (final) nucleus of the 0νECEC decay. The overlap factor in (11) and (15) is given by

$$\langle J_{k_1}^\pi | J_{k_2}^\pi \rangle = \sum_{pn} \left[X_{pn}^{J_{k_2}^\pi} \bar{X}_{pn}^{J_{k_1}^\pi} - Y_{pn}^{J_{k_2}^\pi} \bar{Y}_{pn}^{J_{k_1}^\pi} \right] \tag{26}$$

and it takes care of the matching of the corresponding states in the two sets of states based on the initial and final even-even reference nuclei.

3.2 The charge-conserving QRPA and the multiple-commutator model

The multiple-commutator model (MCM) [19,20] is designed to connect excited states of an even-even reference nucleus to states of the neighboring odd-odd nucleus. Earlier the MCM has been used extensively in the calculations of double–beta-decay rates [21,22]. The states of the odd-odd nucleus are given by the pnQRPA in the form (19). The excited states of the even-even nucleus are generated by the (charge conserving) quasiparticle random-phase approximation (ccQRPA) described in detail in [16]. Here the symmetrized form of the phonon amplitudes is adopted contrary to ref. [16] so that the k-th J^π state can be written as a QRPA phonon in the form

$$\begin{aligned} |J_k^\pi\, M\rangle &= Q^\dagger(J_k^\pi, M)|{\rm QRPA}\rangle \\ &= \sum_{ab} \left(Z_{ab}^{J_k^\pi} \left[a_a^\dagger a_b^\dagger\right]_{JM} - W_{ab}^{J_k^\pi} \left[a_a^\dagger a_b^\dagger\right]_{JM}^\dagger \right) |{\rm QRPA}\rangle. \end{aligned} \tag{27}$$

The symmetrized amplitudes Z and W are obtained from the usual ccQRPA amplitudes X and Y [16] through the following transformation:

$$Z_{ab}^{J_k^\pi} = \begin{cases} X_{ab}^{J_k^\pi}, & \text{if } a = b \\ \frac{1}{2}X_{ab}^{J_k^\pi}, & \text{if } a < b\ , \\ \frac{1}{2}X_{ba}^{J_k^\pi}, & \text{if } a > b \end{cases} \tag{28}$$

and similarly for W in terms of Y. The amplitudes X and Y are obtained by solving the QRPA equations of motion that formally look like the matrix equation (20). In the case of the ccQRPA the matrix elements of the $\mathbf{A}(J)$ and $\mathbf{B}(J)$ matrices read

$$\begin{aligned} A_{ab,cd}(J) = {} & \delta_{ac}\delta_{bd}(E_a + E_b) \\ & + g_{\rm pp}(u_a u_b u_c u_d + v_a v_b v_c v_d) \\ & \times \langle a\,b\,;\,J|V|c\,d\,;\,J\rangle \\ & + g_{\rm ph}\mathcal{N}_{ab}(J)\mathcal{N}_{cd}(J)\big[(u_a v_b u_c v_d + v_a u_b v_c u_d) \\ & \times \langle a\,b^{-1}\,;\,J|V_{\rm RES}|c\,d^{-1}\,;\,J\rangle \\ & - (-1)^{j_c+j_d+J}(u_a v_b v_c u_d + v_a u_b u_c v_d) \\ & \times \langle a\,b^{-1}\,;\,J|V_{\rm RES}|d\,c^{-1}\,;\,J\rangle\big] \end{aligned} \tag{29}$$

and

$$B_{ab,cd}(J) = -g_{\mathrm{pp}}(u_a u_b v_c v_d + v_a v_b u_c u_d)\langle a\,b;\,J|V|c\,d;\,J\rangle \\ +g_{\mathrm{ph}}\mathcal{N}_{ab}(J)\mathcal{N}_{cd}(J)\big[(u_a v_b v_c u_d + v_a u_b u_c v_d) \\ \times\langle a\,b^{-1};\,J|V_{\mathrm{RES}}|c\,d^{-1};\,J\rangle \\ -(-1)^{j_c+j_d+J}(u_a v_b u_c v_d + v_a u_b v_c u_d) \\ \times\langle a\,b^{-1};\,J|V_{\mathrm{RES}}|c\,d^{-1};\,J\rangle\big], \tag{30}$$

where $\mathcal{N}_{ab}(J)$ is the quasiparticle-pair normalization constant

$$\mathcal{N}_{ab}(J) = \frac{\sqrt{1+\delta_{ab}(-1)^J}}{1+\delta_{ab}}, \tag{31}$$

and the amplitudes v and u correspond to the occupation and unoccupation amplitudes of the BCS. Definitions of the two-body interaction matrix elements appearing in (29) and (30) were given in sect. 3.1.

The ccQRPA phonon defines a state in the final nucleus of the resonant 0νECEC decay. In particular, if this final state is the k-th J^+ state the related transition density, to be inserted in (11) or (15), becomes

$$(J_k^+||[c_{n'}^\dagger\tilde{c}_{p'}]_L||\mathcal{J}_{k_1}^\pi) = 2\widehat{\mathcal{J}}\widehat{L}\widehat{\mathcal{J}}(-1)^{L+J+\mathcal{J}} \\ \times\Bigg(\sum_{p_1}\Big[\bar{v}_{p'}\bar{v}_{n'}\bar{X}_{p_1n'}^{\mathcal{J}_{k_1}^\pi}\bar{Z}_{p'p_1}^{J_k^+} - \bar{u}_{p'}\bar{u}_{n'}\bar{Y}_{p_1n'}^{\mathcal{J}_{k_1}^\pi}\bar{W}_{p'p_1}^{J_k^+}\Big]\begin{Bmatrix}\mathcal{J} & J & L\\ j_{p'} & j_{n'} & j_{p_1}\end{Bmatrix} \\ +\sum_{n_1}(-1)^{J+j_{n'}+j_{n_1}}\Big[\bar{u}_{p'}\bar{u}_{n'}\bar{X}_{p'n_1}^{\mathcal{J}_{k_1}^\pi}\bar{Z}_{n'n_1}^{J_k^+} - \bar{v}_{p'}\bar{v}_{n'}\bar{Y}_{p'n_1}^{\mathcal{J}_{k_1}^\pi}\bar{W}_{n'n_1}^{J_k^+}\Big] \\ \times\begin{Bmatrix}\mathcal{J} & J & L\\ j_{n'} & j_{p'} & j_{n_1}\end{Bmatrix}\Bigg) \tag{32}$$

instead of the expression (24) for the ground-state transition. Again v ($\bar{v}$) and u ($\bar{u}$) correspond to the BCS occupation and unoccupation amplitudes of the initial (final) even-even nucleus. The amplitudes X and Y ($\bar{X}$ and $\bar{Y}$) come from the pnQRPA calculation starting from the initial (final) nucleus of the 0νECEC decay. The amplitudes $\bar{Z}$ and $\bar{W}$ are the amplitudes of the k-th J^+ state in the 0νECEC daughter nucleus.

One can take a $J_k^\pi = 2_1^+$ phonon of (27) and build an ideal two-phonon J^+ state of the form

$$\left|J_{2\text{-ph}}^+\right\rangle = \frac{1}{\sqrt{2}}\left[Q^\dagger(2_1^+)Q^\dagger(2_1^+)\right]_J|\mathrm{QRPA}\rangle. \tag{33}$$

An ideal two-phonon state consists of partner states $J^\pi = 0^+, 2^+, 4^+$ that are degenerate in energy, and exactly at an energy twice the excitation energy of the 2_1^+ state. In practice, this degeneracy is always lifted by the residual interaction between the one- and two-phonon states [23]. The related transition density of the MCM, that can be inserted in (11) or (15), attains the form

$$\left(J_{2\text{-ph}}^+||[c_{n'}^\dagger\tilde{c}_{p'}]_L||\mathcal{J}_{k_1}^\pi\right) = \frac{40}{\sqrt{2}}\widehat{J}\widehat{L}\widehat{\mathcal{J}}(-1)^{J+\mathcal{J}+j_{p'}+j_{n'}} \\ \times\sum_{p_1n_1}\Big[\bar{v}_{p'}\bar{u}_{n'}\bar{X}_{p_1n_1}^{\mathcal{J}_{k_1}^\pi}\bar{Z}_{p'p_1}^{2_1^+}\bar{Z}_{n'n_1}^{2_1^+} + \bar{u}_{p'}\bar{v}_{n'}\bar{Y}_{p_1n_1}^{\mathcal{J}_{k_1}^\pi}\bar{W}_{p'p_1}^{2_1^+}\bar{W}_{n'n_1}^{2_1^+}\Big] \\ \times\begin{Bmatrix} j_{p'} & j_{p_1} & 2\\ j_{n'} & j_{n_1} & 2\\ L & \mathcal{J} & J\end{Bmatrix}, \tag{34}$$

where, as usual, the barred quantities denote amplitudes obtained for the 0νECEC daughter nucleus.

4 Nuclear-structure calculations

In this section we discuss as examples the nuclear-structure calculations related to the 0νECEC decays of two nuclei, ^{74}Se and ^{136}Ce, that are located in quite different mass regions. Since 0νECEC decays involve charge-changing decay transitions it is advantageous to perform also an auxiliary analysis by engaging theoretical description of the lateral beta-decay feeding of the low-lying states of the final nuclei of the resonant 0νECEC decays.

4.1 Determination of the model parameters

In the case of the decay of ^{74}Se the nuclear-structure calculation was performed in the 1p-0f-2s-1d-0g (9 orbitals) single-particle valence space for both protons and neutrons. For the decay of ^{136}Ce the proton single-particle space was chosen as 1p-0f-2s-1d-0g-0h$_{11/2}$ (10 orbitals) and that of the neutrons consisted of 2s-1d-0g-2p-1f-0h (11 orbitals). The single-particle energies of all these orbitals were generated by the use of a spherical Coulomb-corrected Woods-Saxon (WS) potential with a standard parametrization [24], optimized for nuclei near the line of beta stability.

As the two-body interaction we have used the Bonn-A G-matrix and we have renormalized it in the standard way [17–19]: The quasiparticles are treated in the BCS formalism and the pairing matrix elements are scaled by a common factor, separately for protons and neutrons. In practice these factors are fitted such that the lowest quasiparticle energies obtained from the BCS match the experimental pairing gaps for protons and neutrons, respectively. The average behavior of both the proton and neutron pairing gaps is given by the three-point formulae (see, *e.g.*, [16])

$$\Delta_{\mathrm{p}} = \frac{1}{4}(-1)^{Z+1}\left[S_{\mathrm{p}}(A+1,Z+1)\right. \\ \left.-2S_{\mathrm{p}}(A,Z) + S_{\mathrm{p}}(A-1,Z-1)\right], \tag{35}$$

$$\Delta_{\mathrm{n}} = \frac{1}{4}(-1)^{N+1}\left[S_{\mathrm{n}}(A+1,Z)\right. \\ \left.-2S_{\mathrm{n}}(A,Z) + S_{\mathrm{n}}(A-1,Z)\right], \tag{36}$$

where S_p and S_n are the experimental separation energies obtained from the measured nuclear masses (see, *e.g.*, [25]) for protons and neutrons, respectively, and Z and N are the proton and neutron numbers.

The intermediate J^π states were generated both from the mother and daughter ground states by the use of the pnQRPA, discussed in sect. 3.1. The two obtained sets of J^π states were matched by using the overlap (26). The particle-hole strength parameter g_{ph} of the pnQRPA, shown in (21) and (22), determines the energy of the Gamow-Teller giant resonance (GTGR) and its value was fixed by fitting the empirical location of the GTGR [16]. The particle-particle parameter g_{pp} has little effect on the β^+/EC type of transitions and we have used the default value $g_{pp} = 1.0$.

The value of the NME in (4) is affected by both the value of the axial-vector coupling constant g_A and the nucleon-nucleon short-range correlations (SRC). In this work the quenched value $g_A = 1.0$ is assumed and we use the unitary correlation operator method (UCOM) to account for the SRC. The UCOM has recently been applied successfully in calculations of the neutrinoless double-beta decay [13–15,26]. The effect of the UCOM SRC is to slightly reduce the magnitude of the NME by gently shifting the wave function of the relative motion away from the touching point of the two decaying nucleons. Also the dipole form factors of the nucleons and higher-order nucleonic currents have been included in the present calculations in the way described in [13–15].

The final states of the presently discussed decays are the second 2^+ state, 2^+_2, in ^{74}Ge and the fourth 0^+ state, 0^+_4, in ^{136}Ba (the 0^+ ground state is counted as the first 0^+ state). The 2^+_2 state in ^{74}Ge is assumed to be a member of a two-phonon triplet with a wave function (33). The experimental energy of the involved first 2^+ state was reproduced in the ccQRPA calculation by varying the value of the particle-hole strength g_{ph} that scales the particle-hole matrix elements in the way presented in eqs. (29) and (30). The value of the particle-particle strength can safely be taken to be $g_{pp} = 1.0$ since the properties of the ccQRPA states depend only very weakly on the value of this parameter. The 0^+ states are a special case in a ccQRPA calculation: The first ccQRPA root is spurious and has to be removed by setting its value to zero. It has to be noted that in the ccQRPA the other 0^+ states are not contaminated by this spuriosity [27]. To bring the first ccQRPA root to zero one has to change the values of both g_{ph} and g_{pp}. The values of these parameters can be fixed uniquely by reproducing at the same time the low-energy spectrum of the 0^+ states. The low-energy spectra of ^{74}Ge and ^{136}Ba are discussed in sect. 4.2.

Finally, it should be stressed that the present calculations assume spherical shape of the involved nuclei. In fact many of the nuclei under study here possess a non-zero static oblate or prolate deformation in their ground state. The corresponding deformation parameter is, however, not very large, of the order of $|\beta| \leq 0.25$. Thus one should expect to encounter effects arising from the nuclear deformation itself or the possible deformation differences between different nuclei participating the weak decay processes. These effects should show up already at the level of single beta decays and that is why it pays to study the lateral beta-decay feeding of the nuclei involved in the double-beta transitions. This aspect, among others, is addressed in sect. 4.2.

4.2 Auxiliary analysis through lateral beta decays

The available data on beta-decay rates allow for studies of the lateral beta-decay feeding of the low-energy states in the 0νECEC daughter nuclei ^{74}Ge and ^{136}Ba. The transitions are first-forbidden for $^{74}\mathrm{As}(2^-_1) \to {}^{74}\mathrm{Ge}(0^+_f, 2^+_f, 4^+_f)$ and allowed Gamow-Teller for $^{136}\mathrm{La}(1^+_1) \to {}^{136}\mathrm{Ba}(0^+_f, 2^+_f)$.

The allowed Gamow-Teller beta-decay transitions of interest in this work are of the type $1^+ \to 0^+, 2^+$. For them the $\log ft$ value is defined as [16]

$$\log ft = \log(f_0 t_{1/2}) = \log\left[\frac{6147}{B_{\mathrm{GT}}}\right],$$
$$B_{\mathrm{GT}} = \frac{g_A^2}{2J_i+1}(J^+\|\boldsymbol{\sigma}\|1^+)^2, \tag{37}$$

for the initial 1^+ and final $J^+ = 0^+, 2^+$ states. Here $f_0 = f_0^+ + f_0^{\mathrm{EC}}$ is the the leptonic phase-space factor for the allowed β^+/EC decays defined in [16]. The first-forbidden decay transitions can be divided into two categories: those with an angular-momentum change of two units (first-forbidden unique, FFU) and those with an angular-momentum change of at most one unit (first-forbidden non-unique, FFNU). For the first-forbidden unique transitions $2^- \to 0^+, 4^+$ we can define [16]

$$\log ft = \log(f_{1u} t_{1/2}) = \log\left[\frac{6147}{\frac{1}{12}B_{1u}}\right],$$
$$B_{1u} = \frac{g_A^2}{2J_i+1}\mathcal{M}_{1u}^2, \tag{38}$$

where

$$\mathcal{M}_{1u} = \frac{m_e c^2}{\sqrt{4\pi}}(J^+\|[\boldsymbol{\sigma}\mathbf{r}]_2\|2^-), \tag{39}$$

for the initial 2^- and final $J^+ = 0^+, 4^+$ states. Here $f_{1u} = f_{1u}^+ + f_{1u}^{\mathrm{EC}}$ is the the leptonic phase-space factor for the FFU β^+/EC decays as given in [16]. For the first-forbidden non-unique transitions $2^- \to 2^+$ one defines [16,28,29]

$$\log ft = \log(f_0 t_{1/2}) = \log\left[f_0\frac{6147}{\mathcal{S}_1}\right],$$
$$\mathcal{S}_1 = \mathcal{S}_1^+ + \mathcal{S}_1^{\mathrm{EC}}, \tag{40}$$

where $\mathcal{S}_1$ is the integrated shape factor that is a complex combination of the various nuclear matrix elements and phase-space factors, see ref. [29].

Results of the beta-decay calculations for ^{74}Ge are shown in table 1. There the experimental and computed energies of the 2^+_1 state and the two-phonon triplet in

Table 1. Experimental and theoretical level energies in ^{74}Ge and $\log ft$ values for the feeding of these levels through the first-forbidden β^+/EC decay of the 2^- ground state of ^{74}As. The experimental data is taken from [30].

State	Experiment		Theory	
	E(MeV)	$\log ft$	E(MeV)	$\log ft$
0_1^+	0.000	9.7	0.000	8.95
2_1^+	0.596	6.96	0.596	6.77
2_2^+	1.204	8.25	1.195	7.43
4_1^+	1.464	11.28	1.195	10.45
0_2^+	1.483	10.34	1.195	9.03

Table 2. Experimental and theoretical level energies in ^{136}Ba and $\log ft$ values for the feeding of these levels through the allowed β^+/EC decay of the 1^+ ground state of ^{136}La. The experimental data is taken from [31].

State	Experiment		Theory	
	E(MeV)	$\log ft$	E(MeV)	$\log ft$
0_1^+	0.000	4.56	0.000	4.02
2_1^+	0.818	5.90	0.818	4.93
2_2^+	1.551	7.39	1.636	8.12
0_2^+	1.579	6.18	1.636	7.54
2_3^+	2.080	6.51	1.971	4.29
2_4^+	2.129	5.97	2.256	4.41
0_3^+	2.141	5.72	1.946	4.62
0_4^+	2.315	6.25	2.358	5.11
2_5^+	2.486	6.28	2.458	5.44

^{74}Ge are shown in columns two and four. In this table also the experimental and computed $\log ft$ values corresponding to the β^+/EC-decay feeding of these states by the 2^- ground state of ^{74}As are shown. Decays to the 0^+ and 4^+ states are first-forbidden unique and the decays to the 2^+ states are non-unique. The correspondence of the computed $\log ft$ values with the experimental ones is reasonable, *i.e.* what could be expected from a simple MCM description of β^+/EC-decay transitions. In particular, the theoretical decay rate to the two-phonon 2^+ state, 2_2^+, is too high pointing to slight overestimation of the magnitude of the corresponding 0νECEC NME.

The 0_4^+ state in ^{136}Ba is presumed to be a ccQRPA phonon of the form (27). The experimental and computed energies of this state and a number of other low-energy 0^+ and 2^+ states in ^{136}Ba are shown in table 1 in columns two and four. As can be seen the computed energies agree very well with the experimental ones. In the same table we show also the $\log ft$ values of the β^+/EC-decay feeding of these states via allowed Gamow-Teller transitions from the 1^+ ground state of ^{136}La. As can be seen the computed $\log ft$ values are usually a bit too small indicating that the associated transitions are predicted to be too fast by the MCM formalism. In particular, the theoretical decay rate to the 0_4^+ state is too high pointing to a possible overestimation of the magnitude of the corresponding 0νECEC NME.

One possible source of the noticed differences between the computed and measured $\log ft$ values could be the omission of the deformation degree of freedom in the present calculations. In particular the deformation differences between the mother and daughter nuclei could drive the suppression of beta-decay transition rates. Such tendencies are visible in tables 1 and 2, especially for the decay of ^{136}La in table 2. However, it seems that the deformation effects are not so strong as to ruin the overall compatibility of the trends visible in the computed and measured $\log ft$ values.

4.3 Results for the NMEs of the resonant 0νECEC decays

We can use relations (3) and (4) to write the half-life of the resonant 0νECEC in the form

$$T_{1/2} = \frac{C^{\mathrm{ECEC}}}{(\langle m_\nu \rangle [\mathrm{eV}])^2}\ \mathrm{years}, \tag{41}$$

where the effective neutrino mass has to be inserted in units of eV. The 0νECEC-decay half-life of ^{74}Se can now be obtained directly by the use of eqs. (41), (4), (15), (25) and (34), where the second leg of the decay is evaluated by the use of the MCM-calculated transition density (34). The corresponding value for the matrix element in (15) is $M_{0\nu}^{\mathrm{ECEC}}(0^+ \to 2_2^+) = 0.624\,\mathrm{MeV}$. However, additional considerations, discussed extensively in [32], reduce this value by some two orders of magnitude. Such a small value of $M_{0\nu}^{\mathrm{ECEC}}$ results in an extremely long 0νECEC half-life [32] irrespective of the fulfillment of the resonance condition (3). Also the uncertainty in the value of the final NME should be accounted for. In this case, the uncertainty is considerable due to the recoil structure of the NME. The decay rate is further suppressed by the JYFLTRAP-measured Q value that yields a rather large value $|Q-E|_{\mathrm{min}} = 2.4\,\mathrm{keV}$ for the degeneracy parameter. An additional suppression is imposed by the necessary capture from the L_2 ($2\mathrm{p}_{1/2}$) and L_3 ($2\mathrm{p}_{3/2}$) atomic orbitals (see (6)). Summing up all these effects produces a half-life value (41) with $C^{\mathrm{ECEC}} \approx 2\times10^{42} - 1\times10^{45}$ being indicative of the large uncertainty in the recoil structure of the NME, not so much of the uncertainties in the nuclear wave functions.

The 0νECEC-decay half-life of ^{136}Ce can be obtained by the use of (41), (4), (7), (11), (25) and (32), where the second leg of the decay uses the MCM-calculated transition density (32). The corresponding value for the matrix element (7) is computed to be $M_{0\nu}^{\mathrm{ECEC}}(0^+ \to 0_4^+) = 8.0 - 22\,\mathrm{MeV}$ depending on which of the ccQRPA roots would correspond to the high-lying experimental 0_4^+ state. The JYFLTRAP-measured Q value is $|Q-E|_{\mathrm{min}} = 11.7\,\mathrm{keV}$ [33] for the atomic KK capture. In this case then $C^{\mathrm{ECEC}} \approx (3-23)\times10^{32}$ in (41).

5 Conclusions

A general theoretical framework of the resonant neutrinoless double electron capture is presented from the point of view of the involved nuclear matrix elements. Use of this theory framework together with the Q value measurements by the JYFLTRAP produces information about the nuclei whose 0νECEC decays could be detected in dedicated underground experiments. This, in turn, is important in pinning down the value of the possible Majorana mass of the neutrino.

In this work explicit expressions for the amplitudes of the resonant 0νECEC decays have been derived. The nuclear matrix elements have been evaluated within the higher-QRPA framework of the multiple commutator model. The associated transition densities have been derived and presented in the article. An auxiliary analysis has been performed by the use of the lateral beta-decay feeding of the final nuclei of the double electron capture decays. The related allowed and first-forbidden transitions have been defined and computed by the use of the multiple-commutator model. As an example, the developed formalism is applied to the decays of ^{74}Se and ^{136}Ce. The values of the computed matrix elements and the Q values measured by the JYFLTRAP suggest that the resonant double electron captures in ^{74}Se and ^{136}Ce are impossible to detect in foreseeable future.

This work was supported by the Academy of Finland under the Finnish Center of Excellence Program 2006-2011 (Nuclear and Accelerator Based Program at JYFL).

References

1. F.T. Avignone III, S.R. Elliott, J. Engel, Rev. Mod. Phys. **80**, 481 (2008).
2. J. Suhonen, O. Civitarese, Phys. Rep. **300**, 123 (1998).
3. J. Bernabeu, A. De Rujula, C. Jarlskog, Nucl. Phys. B **223**, 15 (1983).
4. Z. Sujkowski, S. Wycech, Phys. Rev. C **70**, 052501(R) (2004).
5. A.S. Barabash, Ph. Hubert, A. Nachab, V. Umatov, Nucl. Phys. A **785**, 371 (2007).
6. A.S. Barabash *et al.*, Nucl. Phys. A **807**, 269 (2008).
7. J. Dawson *et al.*, Phys. Rev. C **78**, 035503 (2008).
8. M.F. Kidd, J.H. Esterline, W. Tornow, Phys. Rev. C **78**, 035504 (2008).
9. D. Frekers, hep-ex/0506002v2.
10. R.G. Winter, Phys. Rev. **100**, 142 (1955).
11. B. Crasemann, *Atomic Inner-Shell Processes* (Academic Press, New York, 1975).
12. M. Doi, T. Kotani, Prog. Theor. Phys. **87**, 1207 (1992).
13. M. Kortelainen, O. Civitarese, J. Suhonen, J. Toivanen, Phys. Lett. B **647**, 128 (2007).
14. M. Kortelainen, J. Suhonen, Phys. Rev. C **75**, 051303(R) (2007).
15. M. Kortelainen, J. Suhonen, Phys. Rev. C **76**, 024315 (2007).
16. J. Suhonen, *From Nucleons to Nucleus: Concepts of Microscopic Nuclear Theory* (Springer, Berlin, 2007).
17. J. Suhonen, A. Faessler, T. Taigel, T. Tomoda, Phys. Lett. B **202**, 174 (1988).
18. J. Suhonen, T. Taigel, A. Faessler, Nucl. Phys. A **486**, 91 (1988).
19. J. Suhonen, Nucl. Phys. A **563**, 205 (1993).
20. O. Civitarese, J. Suhonen, Nucl. Phys. A **575**, 251 (1994).
21. M. Aunola, J. Suhonen, Nucl. Phys. A **602**, 133 (1996).
22. M. Aunola, J. Suhonen, Nucl. Phys. A **643**, 207 (1998).
23. D.S. Delion, J. Suhonen, Phys. Rev. C **67**, 034301 (2003).
24. A. Bohr, B.R. Mottelson, *Nuclear Structure*, Vol. **I** (Benjamin, New York, 1969).
25. G. Audi, A.H. Wapstra, C. Thibault, Nucl. Phys. A **729**, 337 (2003).
26. J. Suhonen, M. Kortelainen, Int. J. Mod. Phys. E **17**, 1 (2008).
27. M. Baranger, Phys. Rev. **120**, 957 (1960).
28. M.T. Mustonen, M. Aunola, J. Suhonen, Phys. Rev. C **73**, 054301 (2006).
29. E. Ydrefors, M.T. Mustonen, J. Suhonen, Nucl. Phys. A **842**, 33 (2010).
30. B. Singh, A.R. Farhan, Nucl. Data Sheets **107**, 1923 (2006).
31. A.A. Sonzogni, Nucl. Data Sheets **95**, 837 (2002).
32. V.S. Kolhinen, V.-V. Elomaa, T. Eronen, J. Hakala, A. Jokinen, M. Kortelainen, J. Suhonen, J. Äystö, Phys. Lett. B **684**, 17 (2010).
33. V.S. Kolhinen, T. Eronen, D. Gorelov, J. Hakala, A. Jokinen, A. Kankainen, J. Rissanen, J. Suhonen, J. Äystö, Phys. Lett. B **697**, 116 (2011).

DOI 10.1007/978-94-007-5555-0_20
Reprinted from *Hyperfine Interactions* Volume 223, Issues 1-3, 2014
and *The European Physical Journal A* Volume 48, Issue 4, 2012

Decay studies for neutrino physics

Electron capture decays of ^{100}Tc and ^{116}In

**A. García · S. Sjue · E. Swanson · C. Wrede ·
D. Melconian · A. Algora · I. Ahmad**

Published online: 30 March 2012

Abstract Both ^{100}Mo and ^{116}Cd are zero-neutrino double-β-decay candidates with $J^{\pi} = 0^{+}$ and the ground states of the respective intermediate nuclei, ^{100}Tc and ^{116}In, have $J^{\pi} = 1^{+}$. This makes it possible to measure these matrix elements, which turn out to provide a valuable benchmark for testing models. The electron-capture (EC) decay of ^{100}Tc can also be used to directly determine the efficiency of a potential neutrino detector. However, the experimental determinations of the EC decay branches of interest are challenging because they are only small fractions of the main decays and co-produced radioactivities contribute significant background counts in the regions of interest. This article summarizes our work at Jyväskylä, where we have taken advantage of the high-purity beams produced at IGISOL in combination with JYFLTRAP to achieve significant improvements in the determinations of the EC branches of ^{100}Tc and ^{116}In.

Keywords Neutrinos · Double-beta decay · Electron-capture decay · Nuclear structure

A. García (✉) · S. Sjue · E. Swanson · C. Wrede
CENPA, University of Washington, Seattle, WA 98195, USA
e-mail: agarcia3@uw.edu

C. Wrede
Department of Physics and Astronomy and National Superconducting Cyclotron Laboratory, Michigan State University, East Lansing, MI 48824, USA

D. Melconian
Texas A&M University, College Station, TX 77843, USA

A. Algora
Instituto de Fisica Corpuscular, University of Valencia, Valencia, Spain

I. Ahmad
Argonne National Laboratory, Argonne, IL 60439, USA

1 Introduction

The observation of zero-neutrino double β decay would establish the Majorana nature of neutrinos. Zero-neutrino double β decays could also help determine the effective neutrino mass if the corresponding matrix elements were known. The nuclear matrix elements are calculated using theoretical methods such as the shell model (SM) [1, 2], the quasi-particle random phase approximation (QRPA) [3, 4], and the interacting boson approximation (IBA) [5]. Since these calculations depend on the relevant nuclear structure it is important to provide experimental benchmarks. For example, the measured two-neutrino double β decay rates are commonly used to constrain the SM and QRPA calculations. In the cases of ^{100}Mo and ^{116}Cd (both $J^\pi = 0^+$) an additional benchmark is available in the single β-decay rates of the ground states of the respective intermediate nuclides ^{100}Tc and ^{116}In (both $J^\pi = 1^+$). Although the double β decay can, in principle, proceed by virtual transitions through all $J^\pi = 1^+$ states in the intermediate nucleus, the rough magnitudes of the strengths are such that the contributions from only the ground states almost reproduce the two-neutrino double β-decay halflives for both cases. This single-state dominance is particularly pronounced for the two-neutrino double β decays to excited final states in the $A = 100$ and $A = 116$ systems [6].

The cases of ^{100}Mo and ^{116}Cd present an interesting opportunity to determine the effects of nuclear deformations on the matrix elements by comparing the calculations to measurements in two sequences that show different deformation properties: ^{100}Ru is known to be in a region where deformation plays a significant role, while both ^{116}Cd and ^{116}Sn have been shown to be dominated by vibrations [7, 8].

Additional motivation to determine the electron capture (EC) decay branching ratio in ^{100}Tc decay comes from a proposal to build a solar neutrino detector from natural molybdenum based on neutrino absorption by ^{100}Mo [9]. Calculations of the neutrino absorption cross section of ^{100}Mo are based on the strength of the EC transition.

The intermediate nuclei ^{100}Tc and ^{116}Cd have large β^- decay Q values but only small Q values for their allowed electron capture (EC) decays. Therefore, the branching ratios for the latter are expected to be less than 10^{-4} and it is challenging to determine them experimentally. The first experiments to measure these EC branching ratios were performed at Lawrence Berkeley National Laboratory [10] and the University of Notre Dame [11] using He-jet systems to transport the radioactivity. However, these experiments were limited by background from co-produced radioactivities. The high selectivity of IGISOL overcomes these limitations by increasing the signal-to-noise ratio. In the following, we present a brief description of the setup at IGISOL (Section 2), outline the analysis for the $A = 100$ case (Section 3), give results on the EC branching ratio of ^{100}Tc, and comment on the implications of our results for both cases (Section 4).

2 Apparatus

The ions of interest, ^{100}Tc and ^{116}In, were produced at IGISOL via the ^{100}Mo(p, n) and ^{115}In(d, p) reactions, respectively, using beams provided by the K130 cyclotron. The reaction products were extracted using the ion guide technique, mass separated

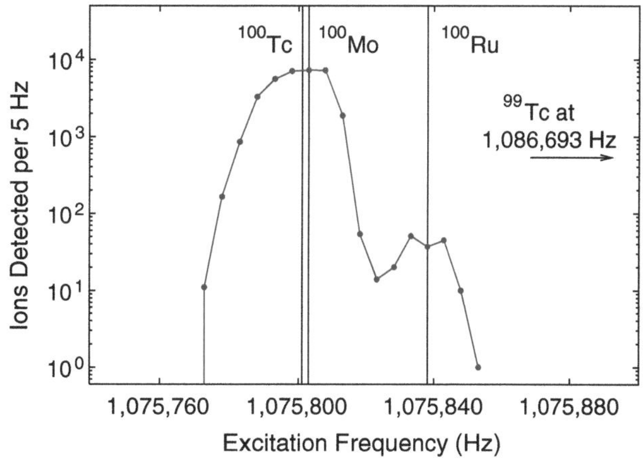

Fig. 1 From ref. [12]: mass scan showing the mass resolving power of ≈25,000 for A = 100 obtained with JYFLTRAP. The most harmful contaminant, ^{99}Tc, was filtered out efficiently

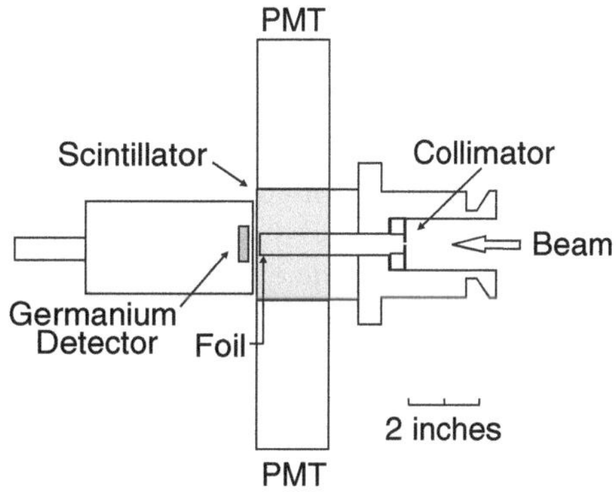

Fig. 2 Setup to determine small EC branching ratios at IGISOL. The PMTs are photomultipliers used to read out the anti-coincidence scintillator and allow to improve the signal-to-noise ratio

with a dipole magnet and injected to JYFLTRAP for final purification. A mass resolving power of 25,000 was reached. In this way the isotopes of interest were efficiently selected, while the co-produced radioactivities were essentially removed from the beam. As an example, in the case of ^{100}Tc an impurity of concern was the 6-h halflife isomer ^{99}Tc. Figure 1 shows the high degree of separation achieved via the JYFLTRAP setup [12] for this case.

The ions of interest were extracted from JYFLTRAP and implanted into a thin Al foil located inside of a scintillator that provided a geometrical efficiency >99%. Only 3 mm of scintillator separated the foil from a planar Ge x-ray detector that abutted the scintillator, as shown in Fig. 2. The Ge detector had a geometrical efficiency of 17% and an energy resolution of 420 eV-FWHM at $E_\gamma = 17$ keV.

X-ray signals were processed using a high-gain amplification of the Ge-detector signal and γ-ray signals were simultaneously processed using a low-gain amplification. The scintillator was used as a veto to further enhance the signal-to-noise ratio because the β^- decays generate background from (1) direct deposition of their energy into the Ge detector, (2) bremsstrahlung, and (3) conversion and Compton scattering of the accompanying γ rays.

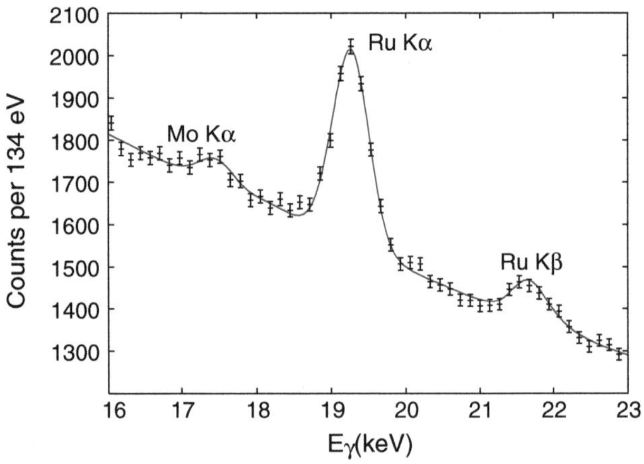

Fig. 3 From ref. [12]: x-ray spectrum for five runs for the decay of ^{100}Tc (*black data points*) and corresponding fit (*red line*). The intensity of the Mo x-rays was used to determine the EC branch

3 Analysis

The EC branching ratio is the ratio of the number of EC decays, N_{EC}, to the total number of decays, N_{tot}. We used ^{100}Mo K-shell x-rays (Fig. 3) as a proxy for ^{100}Tc EC events to determine N_{EC} and the intensities of known γ-ray transitions to determine N_{tot}. Specifically, we used the equation,

$$B(\text{EC}) = \frac{A(\text{Mo-K})}{A(590.8\text{ keV})} \frac{\eta(590.8\text{ keV})}{\eta(\text{Mo-K})} \frac{(1-c)I_\gamma(590.8\text{ keV})}{f_K \omega_K}, \tag{1}$$

where $A(\text{Mo-K})$ and $A(590.8\text{ keV})$ are the photopeak areas for the Mo-K and 590.8-keV transitions; $\eta(590.8\text{ keV})/\eta(\text{Mo-K})$ is the relative efficiency between the 590.8-keV and Mo-K transitions; c is the fraction of 590.8-keV γ rays lost because of summing from coincident β-particles and 539.5-keV γ rays; $I_\gamma(590.8\text{ keV})$ is the absolute intensity of the 590.8-keV γ ray; $f_K = 0.88$ is the fraction of EC decays that produce a vacancy in the K shell; and $\omega_K = 0.765$ is the total K fluorescence yield, i.e., the probability of emission of a K-shell Mo x ray per K-shell vacancy.

The ratio of areas was determined by integrating the experimental spectra. The relative efficiency of the Ge detector between x rays and γ rays, and the value of c, were determined using calibration sources (^{92}Tc and ^{134}Cs) and a Monte Carlo simulation based on the experimental geometry. The value for $I_\gamma(590.8\text{ keV})$ was measured to be $5.5 \pm 0.3\%$, in excellent agrement with the literature value [13, 14]. The values for f_K and ω_K were adopted from the literature [15].

4 Results

We determined the ^{100}Tc EC branching ratio to be $B(\text{EC}) = (2.60 \pm 0.34 \pm 0.20) \times 10^{-5}$, where the first uncertainty is statistical and the second is from the calibration of the Ge detector. We combined the two uncertainties to produce a final result of $B(\text{EC}) = (2.6 \pm 0.4) \times 10^{-5}$ that was used for subsequent calculations.

From our ^{100}Tc EC branching ratio, we deduced the ground-state to ground-state Gamow-Teller strength to be $B(\mathrm{GT};^{100}\mathrm{Mo} \rightarrow {}^{100}\mathrm{Tc}) = 0.95 \pm 0.16$ [12]. The data from the ^{116}In EC decay experiment at the IGISOL are still under analysis [16].

4.1 Efficiency of a ^{100}Mo neutrino detector

The value for the $^{100}\mathrm{Mo} + \nu \rightarrow {}^{100}\mathrm{Tc} + e^-$ transition strength determined using the IGISOL data is approximately 80% larger than the indirectly determined value $B(\mathrm{GT};^{100}\mathrm{Mo} \rightarrow {}^{100}\mathrm{Tc}) = 0.52 \pm 0.06$ that was estimated from measurements of the (^{3}He,t) charge-exchange reaction [17]. Revising the estimate of ref. [9], a solar neutrino detector would require $\sim 1.6 \times 10^3$ kg of ^{100}Mo (17×10^3 kg of natural Mo) based on our measurement. This is about a factor of 2 smaller and less expensive than the estimate based on the results of the charge-exchange reaction.

4.2 Nuclear structure of double β decays

Our IGISOL value for the $^{100}\mathrm{Tc} + e^- \rightarrow {}^{100}\mathrm{Mo} + \nu$ transition strength is significantly lower than the QRPA predictions [18, 19], which are in the range $4 \leq B(\mathrm{GT};^{100}\mathrm{Mo} \rightarrow {}^{100}\mathrm{Tc}) \leq 6$. Prior to our work, Faessler et al. [20] have shown that all of the observables in both the $A = 100$ and $A = 116$ cases could be reproduced using QRPA calculations by allowing the 'effective' axial vector coupling constant to be smaller than unity. Alternatively, Kotila et al. [21] have found good agreement with our ^{100}Tc measurements by using the anharmonic vibrator approach called MAVA. The improved precision for the ^{100}Tc and ^{116}In EC branches from our measurements at the IGISOL will contribute to more stringent tests for these calculations, and will hopefully contribute to improving calculations for zero-neutrino double β decays.

In summary, the EC decay branching ratios of ^{100}Tc and ^{116}In can be used as benchmarks for nuclear structure calculations of double β decays. IGISOL and JYFLTRAP have contributed to producing much cleaner results.

Acknowledgements We gratefully acknowledge the contributions of the IGISOL group and funding from the U.S. Department of Energy, Office of Nuclear Physics, under Contract Nos. DE-FG02-97ER41020 (CENPA) and DE-AC02-06CH11357 (ANL), to the National Science Foundation Grant No. PHY-06-06007, and the Spanish grants FPA 2005-03993 and FPA2008-06419-C02-01.

References

1. Haxton, W.C., Stephenson Jr., G.J.: Prog. Part. Nucl. Phys. **12**, 409 (1989)
2. Caurier, E., Nowacki, F., Poves, A., Retamosa, J.: Phys. Rev. Lett. **77**, 1954 (1996)
3. Rodin, V.A., Faessler, A., Simkovic, F., Vogel, P.: Phys. Rev. C **68**, 044302 (2003); see also arXiv:nucl-th/0503063
4. Suhonen, J., Civitarese, O.: Phys. Rep. **300**, 123 (1998)
5. Barea, J., Iachello, F.: Phys. Rev. C **79**, 044301 (2009)
6. Civitarese, O., Suhonen, J.: Phys. Rev. C **58**, 1535 (1998)
7. Kumpulainen, J., Julin, R., Kantele, J., Passoja, A., Trzaska, W.H., Verho, E., Väärmäki, J., Cutoiu, D., Ivascu, M.: Phys. Rev. C **45**, 640 (1992)
8. Raman, S., Walkiewicz, T.A., Kahane, S., Jurney, E.T., Sa, J., Gácsi, Z., Weil, J.L., Allaart, K., Bonsignori, G., Shriner Jr., J.F.: Phys. Rev. C **43**, 521 (1991)
9. Ejiri, H., Engel, J., Hazama, R., Krastev, P., Kudomi, N., Robertson, R.G.H.: Phys. Rev. Lett. **85**, 2917 (2000)

10. García, A., et al.: Phys. Rev. C **47**, 2910 (1993)
11. Bhattacharya, M., García, A., Hindi, M.M., Norman, E.B., Ortiz, C.E., Kaloskamis, N.I., Davis, C.N., Civitarese, O., Suhonen, J.: Phys. Rev. C **47**, 2910 (1998)
12. Sjue, S.K.L., et al.: Phys. Rev. C **78**, 064317 (2008)
13. Furutaka, K., Nakamura, S., Harada, H., Katoh, T., Fujii, T., Yamana, H.: J. Nucl. Sci. Technol., **38**, 1035 (2001)
14. Furutaka, K., Harada, H., Nakamura, S., Katoh, T., Fujii, T., Yamana, H., Ramanc, S., et al.: J. Nucl. Radiochem. Sci., **6**, 283 (2005)
15. Firestone, R.B., Shirley, V.S. (eds.): Table of the Isotopes. Wiley, New York (1996)
16. Wrede, C., et al.: in preparation
17. Akimune, H., Ejiri, H., Fujiwara, M., Daito, I., Inomata, T., Hazama, R., Tamii, A., Toyokawa, H., Yosoi, M.: Phys. Lett. B **394**, 23 (1997)
18. Griffiths, A., Vogel, P.: Phys. Rev. C, **46**, 181 (1992)
19. Suhonen, J., Civitarese, O.: Phys. Rev. C **49**, 3055 (1994)
20. Faessler, A., Fogli, G.L., Lisi, E., Rodin, V., Rotunno, A.M., Šimkovic, F.: J. Phys. G. **35**, 075104 (2008)
21. Kotila, J., Suhonen, J., Delion, D.S.: J. Phys. G **37**, 015101-12 (2010)

DOI 10.1007/978-94-007-5555-0_21
Reprinted from *Hyperfine Interactions* Volume 223, Issues 1-3, 2014
and *The European Physical Journal A* Volume 48, Issue 4, 2012

Physics highlights from laser spectroscopy at the IGISOL

D. H. Forest · B. Cheal

Published online: 30 March 2012

Abstract Laser spectroscopy provides model-independent access to a variety of radioactive nuclear ground state and isomeric state properties. These include the nuclear moments, changes in mean-square charge radii, and direct measurements of the nuclear spin. At the IGISOL laboratory, the collinear laser spectroscopy programme is able to access cases, such as refractory elements and short-lived states, not available at conventional facilities. A summary of physics highlights is presented here.

Keywords Collinear · Laser · Spectroscopy · IGISOL · Cooler

1 Introduction

Collinear laser spectroscopy [1] has long been applied to radioactive ion beams from the IGISOL [2] at the JYFL Accelerator Laboratory, Finland. By overlapping the laser and fast (30 keV) ion beams in a parallel geometry, high resolution optical spectra can be obtained from ion fluxes as low as a few hundred ions per second. From the hyperfine structure, model-independent measurements of the magnetic dipole moment, μ, and electric quadrupole moment, Q_s, provide an insight into both single particle and collective phenomena. Shell model calculations can interpret such data to provide a sensitive probe of the nuclear wave function, level migrations and

D. H. Forest (✉)
School of Physics and Astronomy, University of Birmingham, B15 2TT, Birmingham, UK
e-mail: D.H.Forest@bham.ac.uk

B. Cheal
School of Physics and Astronomy, University of Manchester, M13 9PL, Manchester, UK
e-mail: bradley.cheal@manchester.ac.uk

shell occupancies. In other forms of spectroscopy experiment, spin assignments may be inferred from, for example, nuclear decay rates and gamma ray multi-polarities. Laser spectroscopy provides an unambiguous measurement of the nuclear ground state or isomeric state spin directly from the hyperfine structure and irrespective of the other quantities extracted. Such measurements then underpin the assignments of the excited levels explored by other techniques.

A fourth property obtained is the change in nuclear mean-square charge radius, $\delta\langle r^2\rangle^{A,A'}$, between a pair of isotopes A and A'. Although obtained in a model-independent manner, it can be expressed as:

$$\langle r^2\rangle = \langle r^2\rangle_s\left(1+\frac{5}{4\pi}\sum_i\langle\beta_i^2\rangle\right)+3\sigma^2 \tag{1}$$

where $\langle r^2\rangle_s$ is a spherical, sharp-surfaced, nucleus of the same volume, calculated sufficiently accurately by the droplet model [3, 4]. This equation reveals a dependence of $\langle r^2\rangle$ not only upon the nuclear size, but also the deformation, $\langle\beta_i^2\rangle$, and surface diffuseness, σ. This latter parameter is usually considered constant along an isotope chain and thus the diffuseness contribution cancels for changes in $\langle r^2\rangle$ between isotopes. As can be seen from (1), $\langle r^2\rangle$ depends on all types of deformation such as quadrupole and octupole. Generally the summation is dominated by the first (quadrupole) term.

From the spectroscopic quadrupole moment, Q_s, extracted from the hyperfine structure, the intrinsic quadrupole moment, Q_0, can be determined via:

$$Q_0=\frac{(I+1)(2I+3)}{I(2I-1)}Q_s. \tag{2}$$

Whereas the mean-square charge radius depends on the quadrupole deformation parameter in mean-square form, $\langle\beta_2^2\rangle$, the intrinsic quadrupole moment can be expressed as:

$$Q_0=\frac{5Z\langle r^2\rangle_{sph}}{\sqrt{5\pi}}\langle\beta_2\rangle(1+0.36\langle\beta_2\rangle) \tag{3}$$

where $\langle r^2\rangle_{sph}$ is for a spherical nucleus of the same volume and calculated using the droplet model [3, 4]. Two indicators of quadrupole deformation, $\langle\beta_2\rangle$ and $\langle\beta_2^2\rangle$, therefore arise from the optical spectra. By making the following definitions:

$$\beta^2=\langle\beta_2\rangle^2+\left(\langle\beta_2^2\rangle-\langle\beta_2\rangle^2\right)=\beta^2_{static}+\beta^2_{dynamic} \tag{4}$$

a comparison of these complementary measures can be used to indicate dynamic effects in the deformation or "β-softness".

Collinear laser spectroscopy is applicable to nuclides far from stability down to half-lives which are limited only by the need to produce a beam of sufficient flux. At the IGISOL laboratory, universal access is available to elements irrespective of chemical or physical properties [2, 5, 6]. These are produced with sub-millisecond extraction times as a result of the thin target foils and gas stopper arrangement. An overview of the production and laser spectroscopic techniques employed at JYFL are given in a separate article in this volume.

Figure 1 shows the nuclear chart with nuclides indicated where optical methods have been used. Measurements performed at JYFL, which have focussed on refrac-

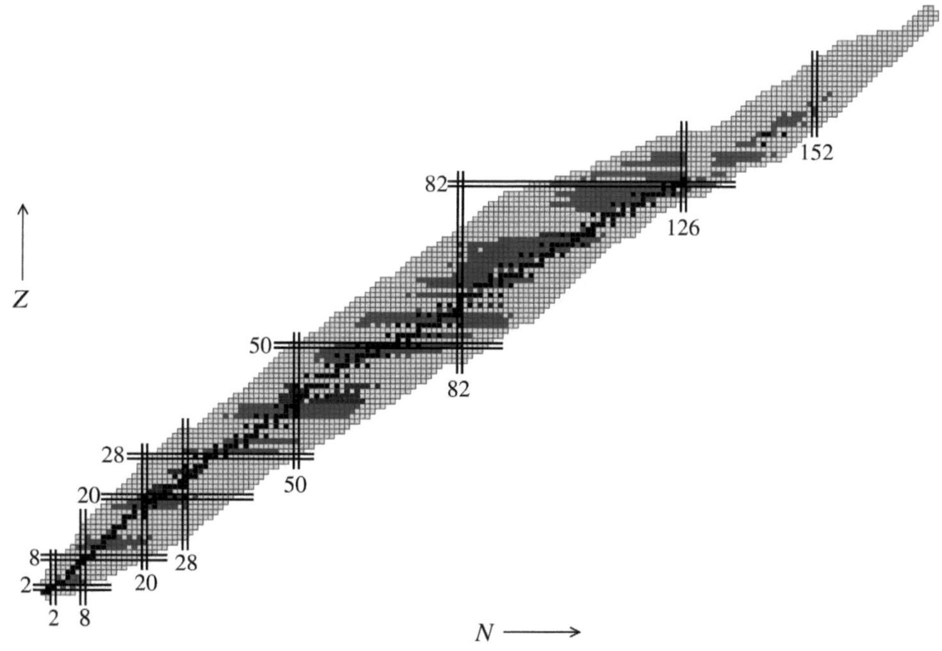

Fig. 1 Chart of the nuclides indicating all isotopes measured using optical techniques (*red*) and those specifically at JYFL (*blue*)

tory and other elements unavailable elsewhere, are listed in Table 1. The principal areas of study are outlined below.

2 The sudden onset of deformation at $N = 60$

In the mass $A \approx 100$ region, a sudden onset of deformation has long been established at $N \approx 60$. Spectroscopic data appear to give a conflicting picture of the nature of the transition. For example, the 2^+ state lifetimes suggested an abrupt transition in the nuclear shape [25] whereas studies using the EUROGAM2 array pointed to a more gradual onset of deformation from $N = 56$ [26]. Later, experiments using the GAMMASPHERE detector indicated sharp variations in the yrast structure despite a gradual change in quadrupole deformation [27]. In the optical data, the changing behaviour with neutron number manifests itself not only in the quadrupole moments but also as an increase in mean-square charge radius. A sudden increase was first observed in the rubidium isotopes ($Z = 37$) [28] and later in strontium ($Z = 38$) [29–31], but its absence is noted in Kr ($Z = 36$) [32].

Production of elements above strontium were hindered by their refractory nature, but the development of the IGISOL and ion beam cooler [12, 23] enabled such optical spectroscopy to be performed for the first time. Neutron-rich isotopes of yttrium ($Z = 39$), zirconium ($Z = 40$), niobium ($Z = 41$) and molybdenum ($Z = 42$) were produced from proton- and deuteron-induced fission, and neutron-deficient

Table 1 All isotopes and isomers measured by laser spectroscopy at JYFL

Element	Z	Measured isotopes	Reference
Sc	21	42–46	[7]
		$44^{m}, 45^{m}$	
Ti	22	44–50	[8]
Mn	25	50–56	[9]
		$50^{m}, 52^{m}$	
Y	39	86–90, 92–99, 101, 102	[10, 11]
		$87^{m}, 88^{m}, 89^{m}, 90^{m}, 93^{m}, 96^{m}, 97^{m}, 97^{m2}, 98^{m}, 100^{m}$	
Zr	40	87–92, 94, 96–102	[12–14]
		$87^{m}, 89^{m}$	
Nb	41	90–93, 99, 101, 103	[15]
		$90^{m}, 91^{m}$	
Mo	42	90–92, 94–98, 100, 102–106, 108	[16]
Ba	56	140, 142, 144	[17, 18]
		130^{m}	
Ce	58	136–148 (evens only)	[19]
Yb	70	176^{m}	[20]
Hf	72	170–180	[20–23]
		$171^{m}, 178^{m}$	
Ta	73	181	[24]
		180^{m}	

isotopes using a variety of light-ion induced fusion evaporation reactions. These covered both the $N = 50$ shell closure and spanned the $N = 60$ onset of deformation.

2.1 Zirconium

One of the initial goals of the laser/IGISOL collaboration was to measure the ground and long-lived isomeric state properties of unstable isotopes of zirconium ($Z = 40$). The isotopes from $A = 87$ to $A = 102$ (except for $A = 93$ and $A = 95$) were measured, along with long-lived isomers in ^{87}Zr and ^{89}Zr [12–14]. The mean-square charge radii were observed to increase at a steady rate both before and after a clear discontinuity at $N = 60$. Relativisitic mean-field calculations [33] underestimated the sharpness of the transition at $N = 60$, which was more accurately reproduced by global macroscopic–microscopic calculations [34] but with poorer overall performance. Calculations of the deformation, $\langle\beta_2^2\rangle$, can be made, both from the charge radii and from $B(E2)$ values [35]. Beyond $N = 60$, the approximations used in each case, namely that the charge radii are dominated by quadrupole deformation and secondly, that the $B(E2)$ strength to all excited 2^+ states is dominated by the first such state, are valid and a close correspondence is seen. For $N < 60$ the lack of such a correspondence was judged to be due to the "missing" $B(E2)$ strength [12], since calculations of octupole deformation [25] failed to reproduce the observed trend.

2.2 Yttrium

While the neutron-rich zirconium isotopes after $N = 60$ were shown clearly to possess a rigid, strongly prolate deformation, it was not possible to determine the evolution of the static deformation below the shape change. This was because the

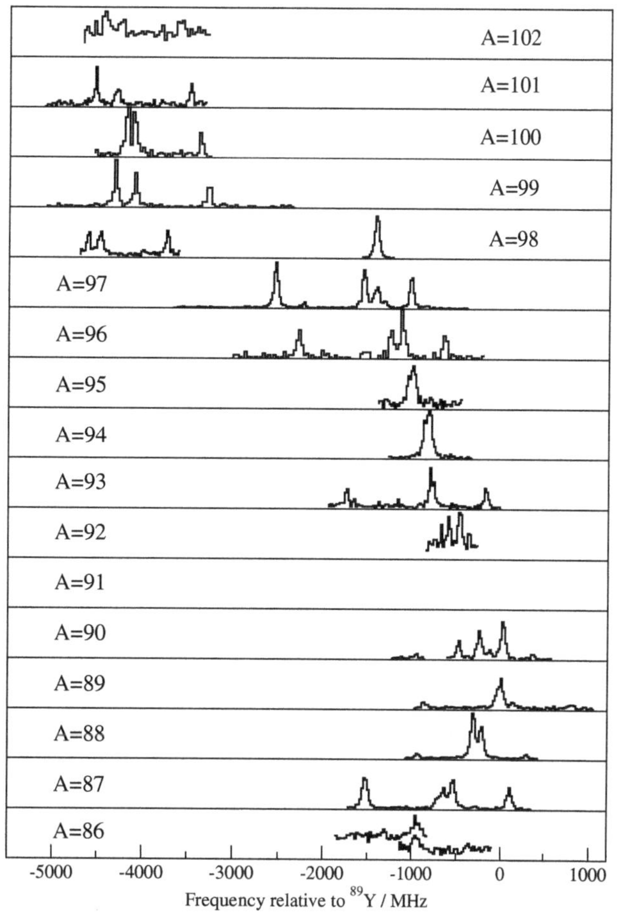

Fig. 2 Sample resonance fluorescence spectra for the studied yttrium isotopes and isomers

nuclear spins were either $I = 0$ (for ^{96}Zr and ^{98}Zr) or $I = 1/2$ (for ^{97}Zr and ^{99}Zr), which precluded gaining any quadrupole moment information from the hyperfine structure. In order to address this lack of knowledge, it was necessary to examine a neighbouring system, that should exhibit similar systematic trends, but allowed the static deformation to be determined. The yttrium ($Z = 39$) chain of isotopes provided just such a system.

While many of the odd-A yttrium isotopes have an $I^{\pi} = 1/2^{-}$ ground state, due to the unpaired proton occupying the $p_{1/2}$ orbital, only ^{95}Y lacks a low-lying, $I^{\pi} = 9/2^{+}$, isomeric state coming from the proton occupying the $g_{9/2}$ orbital. A long chain of isotopes from $A = 86$ to $A = 102$ (with the exception of ^{91}Y) were measured, and the data are shown in Fig. 2 [10].

As yttrium has just the one stable isotope, ^{89}Y, it was not possible to calibrate the atomic factors of the isotope shift by using the standard King Plot technique [1]. However, it was found that modified King Plots [10], that used $\delta\langle r^2\rangle$ values from neighbouring isotones, were highly linear. This indicated that the trends in (near-stability) charge radii systematics were very similar across this region, and allowed an approximate calibration to be performed, and $\delta\langle r^2\rangle$ values for yttrium to be extracted.

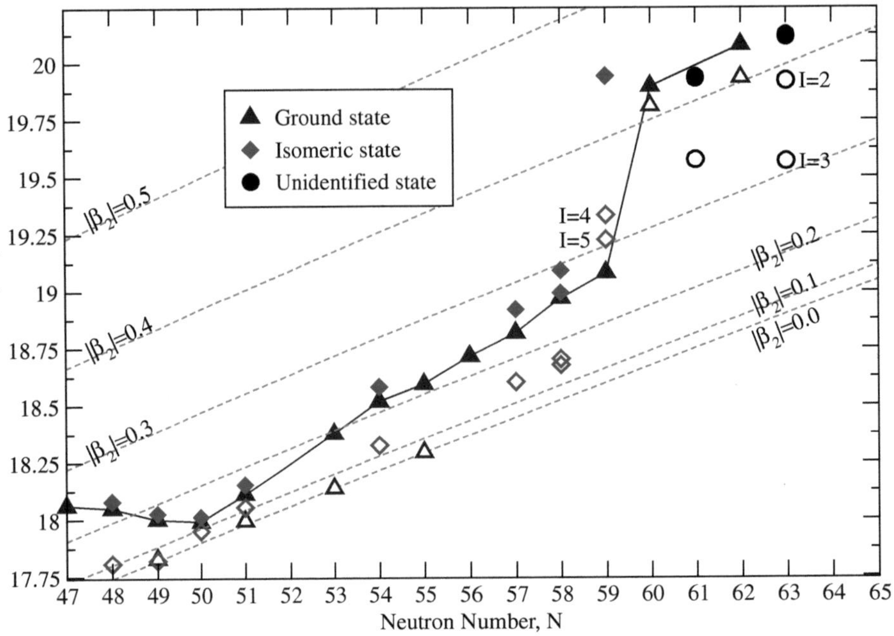

Fig. 3 The yttrium mean-square charge radii for the ground and isomeric states. Experimentally determined mean-square charge radii are indicated by filled symbols. Open symbols indicate mean-square charge radii calculated from the droplet model [3, 4] using only the contribution from the static deformation. Isodeformation contours are included as *dashed lines*

The experimental radii are shown in Fig. 3 together with estimates of the radii assuming a static deformation. These were calculated using (1), but substituting $\langle\beta_2^2\rangle$ with $\langle\beta_2\rangle^2$, calculated from the quadrupole moments. A discrepancy between these and experimental $\delta\langle r^2\rangle$ values is therefore an indication of β-softness. From the $N = 50$ shell closure, a weak but increasing oblate deformation is seen accompanied by a proportionate increase in softness. Following the shape change at $N = 60$, ^{99,101}Y adopt strongly deformed prolate shapes, with the $\langle\beta_2^2\rangle \approx \langle\beta_2\rangle^2$ correspondence indicating a rigid deformation for these isotopes.

Unfortunately it was not possible to unambiguously assign spins to all observed nuclear states, as the transition used—5s^2 $^1S_0 \rightarrow$ 4d5p $^1P_1$—only gives a maximum of three hyperfine components per nuclear state. As there are four unknowns—μ, Q_s, $\delta\langle r^2\rangle$ and I—it was not possible to simultaneously determine all of them from this data. As this restriction applies to all transitions from the ground state, a transition from a higher-lying metastable state is required in order to address the spin assignments [11].

The technique of optical pumping in the cooler/buncher [15] was therefore used to populate a low-lying metastable state, and spectroscopy was then performed using a transition from this level. This allowed additional data on the $A = 100$ structure to be obtained, which, in conjunction with the previous results, showed unambiguously that the spin of the observed state was $I = 4$ [11].

This result immediately indicated that the observed nuclear state was not the ground state, but the $\tau_{1/2} = 0.94$ s isomeric state at 145 keV [36], and that it

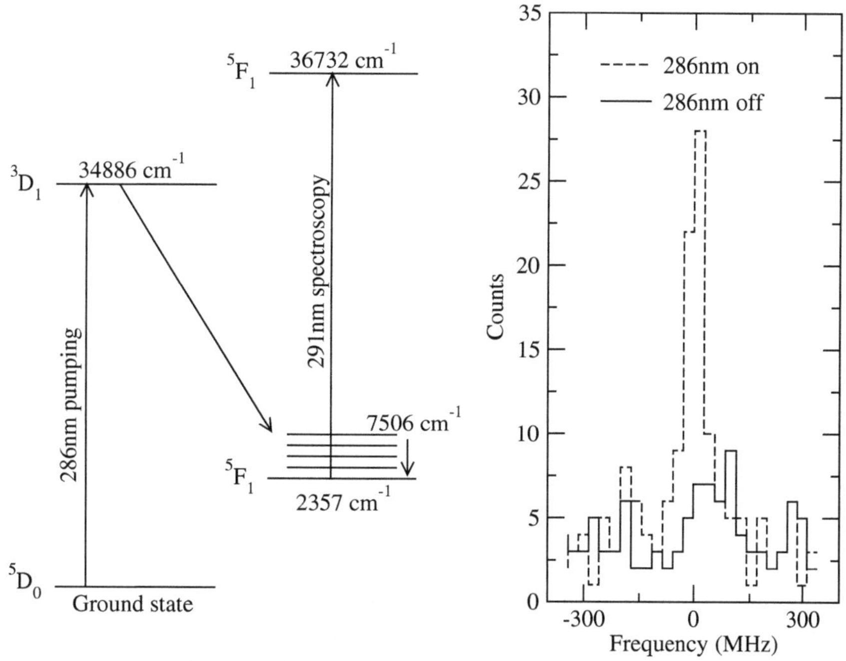

Fig. 4 Optical pumping scheme used for Nb. The effect of the 286.6 nm pumping light on the 290.9 nm fluoresence signal from the $F = 9 \rightarrow F = 9$ component of ^{90g}Nb, is shown

had an unexpectedly soft nature to its deformation, demonstrated by there being a significant difference between the static deformation—from the hyperfine structure—and the static + dynamic deformation—from the mean-square charge radius. Analysis of the magnetic moment lead to the two-quasiparticle configuration $\{\pi[422]5/2 \otimes \nu[411]3/2\}4^+$ being proposed for this isomer.

2.3 Niobium

There were two main problems to overcome in order to extend the study of the $Z \sim 40$, $A \sim 100$ region to the isotopes of niobium ($Z = 41$). Firstly, singly-charged niobium ions have an electronic ground state spin of $J = 0$, which restricts the nuclear information that can be measured. Secondly, while there is a transition from the ground state that can be accessed by the high resolution dye laser system, 4d^4 ^{5}D$_0$ (0 cm^{-1}) → 4d^{3}5p ^{3}D$_1$ (34,886 cm^{-1}) at 286.6 nm, it proved to be too inefficient for spectroscopic study.

However, while this ground state transition was not suitable for spectroscopy, it could be used by the optical pumping technique to efficiently enhance the population of a metastable state at 2,357 cm^{-1}, allowing spectroscopy to be performed on the 5s ^{5}F$_1$ (2,357 cm^{-1}) → 5p ^{5}F$_1$ (36732 cm^{-1}) transition at 290.9 nm, as shown in Fig. 4.

This selective enhancement of the spectroscopic efficiency allowed spectra to be measured for the neutron-deficient isotopes and isomers, ^{90,90m,91,91m,92}Nb—spanning the $N = 50$ neutron shell closure—and the neutron-rich isotopes, ^{99,101,103}Nb, which cover the $N = 60$ shape change region. The nuclei undergo a similar transition as in yttrium, but the magnitude of the effect is lower and a significant degree of β-softness is observed both before and after the shape change.

2.4 Molybdenum

The sudden onset of deformation at $N = 60$ for isotope chains with $37 \leq Z \leq 41$ has been well established by a variety of techniques, including measurements of the lifetimes of the first excited 2^+ states [25] and optical measurements of the charge radii. However, no such sharp change at $N = 60$ is seen in either of these observables for ruthenium ($Z = 44$) [37].

To investigate where this alteration in the behaviour at $N = 60$ occurs, molybdenum isotopes from $A = 90$ to $A = 108$ were studied on the $4d^4 5s\ ^6D_{1/2}$ (11,783 cm^{-1}) $\rightarrow 4d^4 5p\ ^6F_{1/2}$ (45,853 cm^{-1}) transition at 293.4 nm. In this case, a transition from a metastable state is required as all transitions from the ground state lie in the deep UV ($\sim$ 210 nm), which is unreachable with the cw laser system.

It was discovered during tests to develop a suitable optical pumping scheme to populate excited states that there was a naturally large metastable population that survived for many hundreds of milliseconds in the cooler/buncher, independent of any optical pumping or production method. While metastable states have been observed to be populated by passage through the SPIG [5], the persistence of population in such a high-lying state through the cooler/buncher is unique to Mo^+ out of all the elements studied so far at the laser-IGISOL facility. Usually these states relax to the ground state via collisions with the buffer gas in the cooler/buncher, but relaxation from the excited states to the ground state seems to be hindered greatly in Mo^+.

The surprising longevity of the metastable population in the cooler/buncher allowed the neutron-deficient isotopes $^{90,91}Mo$ and the neutron-rich, $^{102-106,108}Mo$, isotopes to be studied [16]. These, along with the stable isotopes, again covered the $N = 50$ shell closure and, more importantly, the $N = 60$ region. The molybdenum ($Z = 42$) isotope chain displays a more gradual evolution of its charge radius, with the size increasing steadily from $N = 50$ across $N = 60$ with no sudden jump. There is also an indication that the deformation has saturated in this chain, as the deformation for ^{108}Mo ($N = 66$) appears to be lower than that for ^{106}Mo ($N = 64$). Unfortunately there is currently no information on the static deformation for the radioisotopes of molybdenum, as the transition used, $4d^4 5s\ ^6D_{1/2} \rightarrow 4d^4 5p\ ^6F_{1/2}$, has $J = 1/2$ for both lower and upper levels and hence no quadrupole contribution to the hyperfine structure.

2.5 Regional systematics

The charge radii variations across the $Z \sim 40$ region are shown in Fig. 5a. As the neutron number is increased from $N = 50$ to $N = 60$, the isotopes become softer, with an increasing dynamic contribution to the quadrupole deformation. For elements between $Z = 37$ (Rb) and $Z = 41$ (Nb), a sharp transition to strongly prolate, predominantly rigid, deformed nuclei is observed at $N = 60$. This transition is centred symmetrically around $Z = 39$ (Y), disappearing at $Z = 36$ (Kr) and $Z = 42$ (Mo). An overview is given in reference [38].

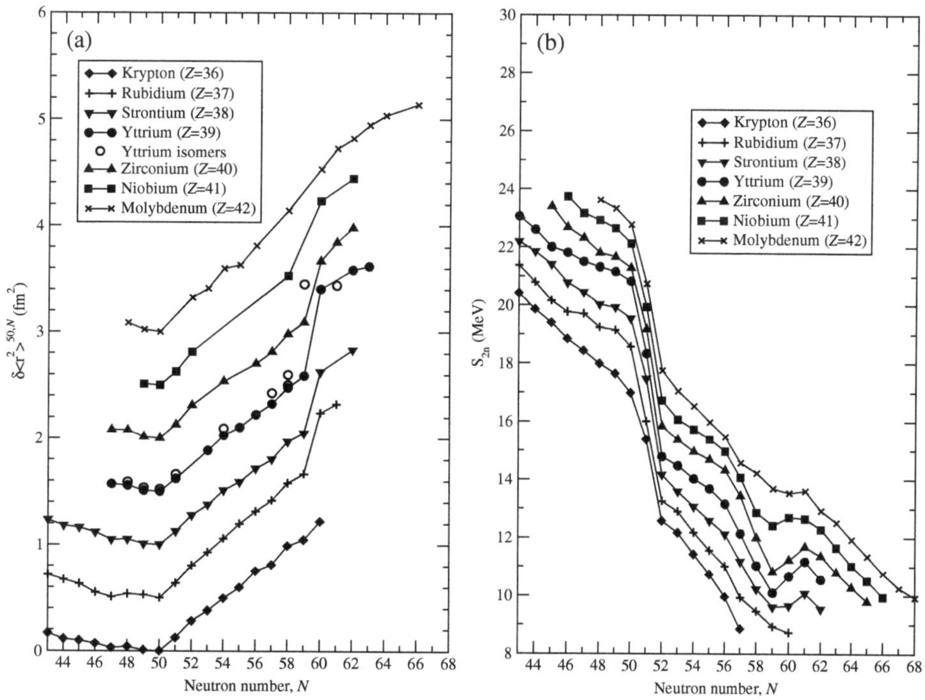

Fig. 5 **a** Difference in mean-square charge radii in the $N = 60$ region given relative to $N = 50$. Each isotone chain is set apart by 0.5 fm^2 for clarity. **b** Two neutron separation energies. Relative errors are smaller than the symbol sizes. See text and reference [16] for sources of data

Figure 5b shows the two-neutron separation energies for the same region, determined from high precision Penning trap mass measurements (e.g. [39, 40]). The correspondence in behaviour with that of the charge radii is striking. The shape change at $N = 60$ is clearly seen, being most prominent for yttrium. The effect decreases symmetrically to either side, becoming insignificant at krypton and molybdenum.

There are several different theoretical explanations for the origin of the sudden shape change. Shell model descriptions attribute the onset of deformation to the isoscalar component of the $n - p$ interaction between nucleons in the spin-orbit partner orbitals $(1g_{9/2})_\pi$ and $(1g_{7/2})_\nu$ [41]. This lowers the energy of the $g_{7/2}$ state, thereby increasing the valence space which drives the deformation. Alternatively, mean-field models explain the deformation as being due to the occupation of strongly downsloping low-K components of the $(1h_{11/2})_\nu$ orbital [42]. More recent projected shell model calculations suggest that both of these mechanisms play a role in the shape change at $N = 60$ [43]. Meanwhile, a study of both mean-square charge radii and two-neutron separation energies using the D1S-Gogny interaction attributes the smooth behaviour in molybdenum to emerging triaxiality [44].

2.6 Cerium ($Z = 58$, $N = 90$)

A similar change in the charge radii, though much less pronounced than the effect at $N = 60$, is also observed around $N = 90$. Previous data had shown that Xe ($Z = 54$)

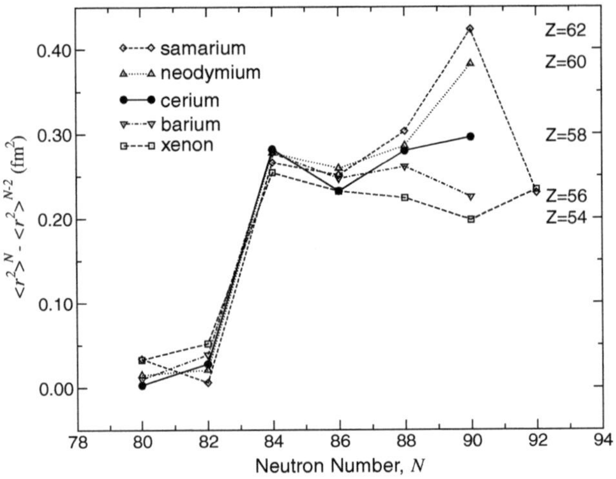

Fig. 6 Differences in mean-square charge radii, $\langle r^2\rangle^N - \langle r^2\rangle^{N-2}$, in cerium ($Z = 58$). The neighbouring chains of samarium ($Z = 62$), neodymium ($Z = 60$), barium ($Z = 56$) and xenon ($Z = 54$) are plotted for comparison. See publication [19] for data references

and Ba ($Z = 56$) charge radii exhibited very similar behaviour, and that Nd ($Z = 60$) charge radii varied in much the same way as those for Sm ($Z = 62$). However, where a large increase in charge radius is seen at $N = 90$ for $Z > 58$, no such increase is seen for $Z < 58$. In order to investigate this phenomenon, the neutron rich even-even isotopes of cerium ($Z = 58$), from $A = 144 - 148$, were studied [19].

The changes in charge radii for the region are shown in Fig. 6, where the cerium data reveals that the shape change develops smoothly from $Z = 56$ to $Z = 60$. This behaviour is interpreted as arising from a weakening of the $Z = 64$ sub-shell gap as $N = 90$ is reached. At $N = 90$, the neutrons have started to occupy the $h_{9/2}$ orbital, which interacts with the $\pi h_{11/2}$ orbital via the monopole part of the $n - p$ tensor interaction [45], lowering its effective energy. This removes the sub-shell gap at $Z = 64$, increasing the available valence space for the nucleons. This allows for greater quadrupole deformations, and hence charge radii, to be reached for isotopes with $N \sim 90$ than for those with $N < 90$, with the deformation dependent on the number of protons outside the $Z = 50$ shell closure [45]. This explains both the lack of effect below $N = 90$, and the smooth Z-dependence seen at $N = 90$. Beyond the shape change, a more moderate and uniform rate of increase in the charge radius is seen for the different isotopic chains.

3 Multi-quasi particle isomers

Laser spectroscopy of multi-quasi particle isomers began with the measurement of the 31 year 16^+ isomer in ^{178}Hf [46]. Despite the isomer possessing higher quadrupole deformation, the mean-square charge radius was significantly smaller than that of the ground state. Since this measurement, seven other cases have been studied and all show the same surprising effect. Of these eight isomers, half have been measured at the JYFL IGISOL: ^{130m}Ba [18], ^{176m}Yb [20], ^{97m2}Y [10, 20] and ^{178m1}Hf [20].

3.1 ^{130m}Ba

Initial work on multi-quasi particle isomers at the IGISOL began with the study of ^{130m}Ba, looking at the $I^\pi = K^\pi = 8^-$ isomer. The half-life of this state was determined in this experiment to be 9.54(14) ms. As this short half-life would limit the bunch accumulation time of the bunching technique [12, 23], the photon-ion coincidence method [47, 48] was used to suppress the background. The measured nuclear moments demonstrated that the isomer has a predominantly 2-quasi neutron character, resolving an ambiguity in the gamma-ray systematics of the isotonic chain [18].

3.2 ^{176m}Yb

The ^{176m}Yb $I^{\pi} = 8^{-}$ isomer has a pure 2-quasi neutron configuration. This is evidenced by the magnetic moment being what would be expected from adding the single neutron moments of ^{175}Yb and ^{177}Yb. As with ^{130m}Ba, the deformation of the $I = 0$ ground state was extracted from B(E2) measurements, while that of the isomer was calculated from the quadrupole moment using (3). In both of these cases, the quadrupole moment-derived static deformation (cf. (4)) of the isomer appears to exceed the B(E2)-derived total (static and dynamic) deformation of the ground-state. However, this conclusion would depend on the confidence with which $\langle\beta_2^2\rangle$ and $\langle\beta_2\rangle$ can be extracted from B(E2) and $\delta\langle r^2\rangle$, respectively.

3.3 ^{97m2}Y

This $I = (27/2^{-})$ isomer can be considered to be a 2-quasiparticle excitation of the first isomer, ^{97m1}Y ($I = 9/2^{+}$). While the spin could not be unambiguously determined, the magnetic moments are consistent with the previously assigned configuration and the conclusions remain unchanged. It is in going from the first to the second isomer that a reduction in $\langle r^2\rangle$ is seen, despite an increase in static deformation. Both isomers are assigned aligned (stretched) configurations whose nuclear spins define the symmetry axes.

3.4 ^{178m1}Hf

Much interest in the behaviour of a lower-lying isomer, $I^{\pi} = 8^{-}$, $t_{1/2} = 4$ s followed the measurement of the 31 y, $I^{\pi} = 16^{+}$ isomer. Both were produced using a ^{176}Yb(α,2n)^{178}Hf reaction. The measured magnetic moment reveals a mix of 40% 2-quasi proton and 60% 2-quasi neutron structures, agreeing with band mixing calculations from gamma-ray measurements. The 2-quasi neutron configuration is identical to that in ^{176m}Yb, allowing the contributions from the 2-quasi neutron and 2-quasi proton configurations to be disentangled. A second $I^{\pi} = 8^{-}$ state in ^{178}Hf has the opposite mixture (60% 2-quasi proton and 40% 2-quasi neutron). The $I^{\pi} = 16^{+}$ isomer is a superposition of these two $I^{\pi} = 8^{-}$ states, has an excitation energy almost equal to the sum and is a pure 4-quasi particle state [49]. The 2-quasi particle $I^{\pi} = 8^{-}$ isomer showed a decrease in mean-square charge radius (relative to the ground state) that was approximately half of that observed for the $I^{\pi} = 16^{+}$ state.

3.5 Trends and conclusions

A summary of all measurements to date are contained in reference [20]. From these it may be concluded that the magnitude of the $\langle r^2\rangle$ reduction appears to scale with

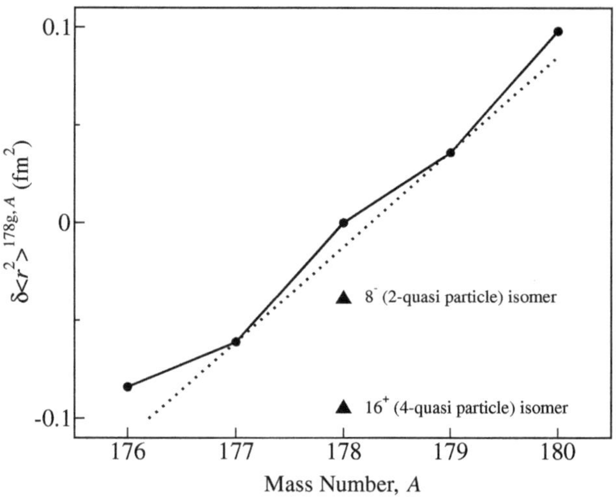

Fig. 7 The mean-square charge radii, $\delta\langle r^2\rangle^{178,A}$, in hafnium [20]. The *dotted line* is shown to guide the eye and note the odd-even staggering effect

multi-quasiparticle number. This is most clearly illustrated in Fig. 7, which shows the mean-square charge radii of the two ^{178}Hf isomers relative to the ground state. Also shown in the figure is the usual odd-N/even-N staggering of the ground state radii, which could correspond to a "1-quasi particle effect" of the same phenomenon.

Both the yttrium and hafnium measurements performed at Jyväskylä indicate that the charge radius decreases despite a definite increase in static deformation. From (1), it would appear that the effect is due to an increase in nuclear rigidity or a decrease in diffuseness for the multi-quasi particle state. Although theoretical studies of the ^{178g,m2}Hf system (incorporating pairing effects) predict a decrease in mean-square charge radius [50], the scale has so far not been reproduced.

4 The $Z = 20$–28 region

The region framed by the magic numbers $20 \leq Z \leq 28$ and $20 \leq N \leq 28$ forms an ideal testing ground for the study of microscopic structural effects. Nuclear properties in the vicinity of magic numbers are likely to be single-particle in nature and provide a stringent test of the ability of effective inter-nucleon interactions to reproduce the experimental observables. Studies of Sc, Ti and Mn have commenced at JYFL as outlined below.

4.1 Charge radii behaviour in Sc and Ti

Exploration of the $f_{7/2}$ shell between $N = 20$ and $N = 28$ has been of long standing interest due to the unusual behaviour of the mean-square charge radii in the calcium chain [51]. A symmetrical parabolic shape is observed whereby the two doubly-magic nuclei, ^{40}Ca and ^{48}Ca have nearly identical mean-square charge radii. In Fig. 8 it can be seen that this is different from lower-Z chains where the mean-square charge radius increases with neutron number across the $\nu f_{7/2}$ shell.

A scandium target irradiated by protons or deuterons produced beams of ^{44}Ti and ^{45}Ti [8], and, in a later experiment, ^{42,43,44,45,46}Sc and ^{44m,45m}Sc [7]. Nuclear moments

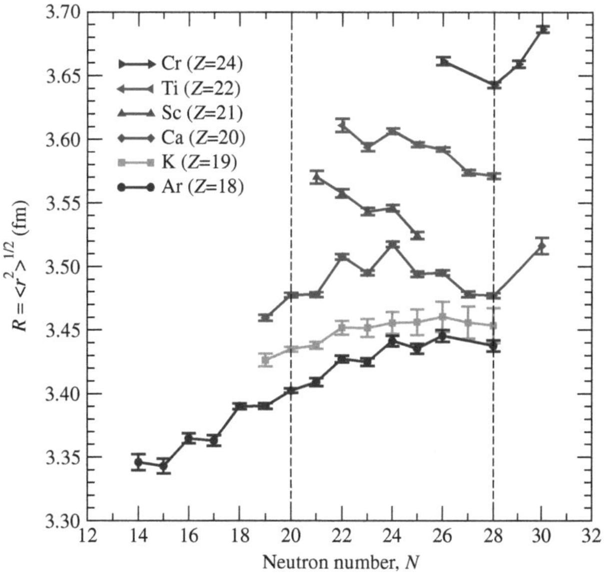

Fig. 8 Changes in nuclear charge radii in the $Z \sim 20$ region. For data, see reference [7] and references therein

and mean-square charge radii were extracted in each case. To calibrate the two atomic factors (required to extract changes in mean-square charge radii from the isotope shift measurements) existing radii measurements may be used, if available from other techniques for three isotopes. As with yttrium and niobium, scandium has only one stable isotope and such information is therefore not available. It was not possible to make an approximation using neighbouring chains since the systematics vary widely. New *ab initio* calculations were therefore performed to determine the atomic factors [7].

Both chains show an increasing trend in charge radius as isotopes approach $N = 20$. This is in contrast to the trend seen for $Z < 20$, where the charge radii increase towards $N = 28$. Calcium ($Z = 20$) can then be seen as partaking of both trends, leading to the parabolic shape. In the case of titanium, the increase in charge radius at ^{44}Ti is closely reproduced by the relativistic mean field model, which suggests that a proton skin is being formed [33]. For scandium, shell model calculations were performed and compared with the nuclear moments and mean-square charge radii. Although the calculations reproduced the charge radii trends in calcium, the correspondence for scandium was notably worse. This suggests that the effective interaction used does not lower the gap at $Z = 20$ properly for scandium and so underestimates the amount of excitation across the *sd* and *pf* shells.

4.2 Spectroscopy of manganese

From the work around $N = 60$, a very close correspondence between the two-neutron separation energy (S_{2n}) trends and the charge radii systematics was noted. Examination of the region around the $N = 28$ magic number revealed that manganese ($Z = 25$) showed no shell effects in its S_{2n}, thus providing the motivation to study this element. However, all transitions from the ground state of the ion were at too short a wavelength (< 280 nm) to be excited by the cw dye laser system. By

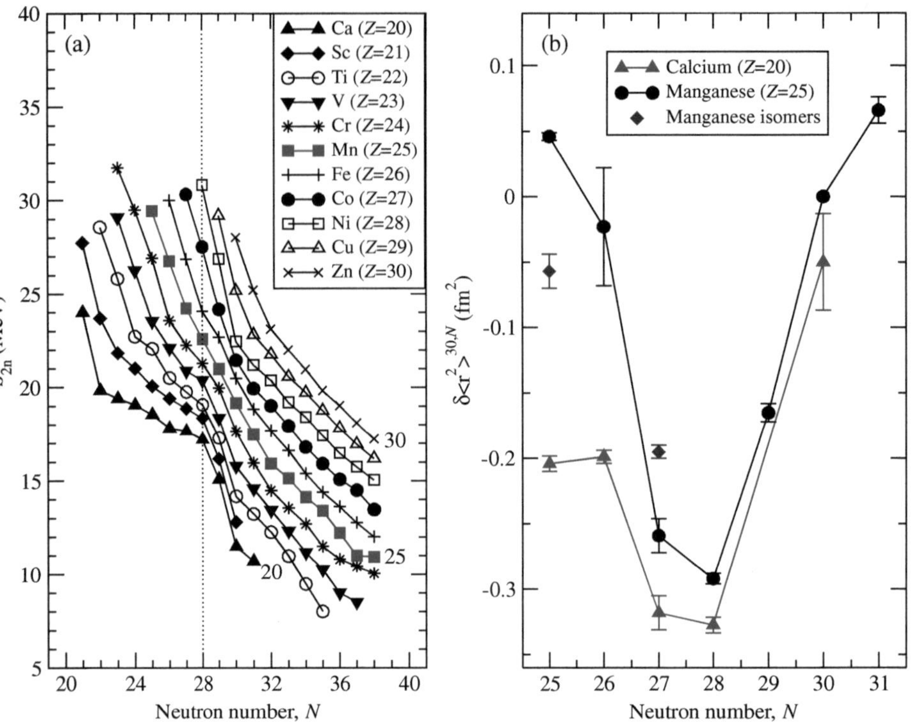

Fig. 9 **a** Two-neutron separation energies across the $N = 28$ shell closure shown for isotopic chains from $Z = 20 - 30$. The relative errors are smaller than the symbol sizes. **b** Differences in the manganese and calcium mean-square charge radii as a function of neutron number. For clarity the isotone chains are set apart by an arbitrary 0.05 fm^2. Isomeric states in $^{50,52}Mn$ are denoted by the *diamonds*. See reference [9] for data

frequency tripling the output of a Ti:Sa pulsed laser, it was possible to optically pump the population of the ion from its ground state to the metastable state at 9,472.97 cm^{-1}. This then gave access to the $3d^54s\ ^5S_3 \rightarrow 3d^54p\ ^5P_1$ (43,557 cm^{-1}) transition at 293.3 nm [9]. Furthermore, the large hyperfine splitting of the ground state transition, relative to the resolution of the optical pumping, allowed the hyperfine components to be individually excited while in the cooler. Populating particular hyperfine F states (of the 9,472.97 cm^{-1} level) could therefore be biased to suit the hyperfine transition under study [9].

Proton and deuteron induced reactions with Cr, Mn and Fe targets were used to produce and measure the $^{50,51,52,53,54,55,56}Mn$ ground states and $^{50m,52m}Mn$. Figure 9 shows the extracted mean-square charge radii from this work together with those of the calcium ($Z = 20$) chain for comparison. Also shown in the figure are the values of S_{2n} for the $Z = 20$–30 region. Although the two-neutron separation energies show a smooth behaviour through $N = 28$, the mean-square charge radii by contrast show a distinct minimum and thus a clear shell closure effect. Mean-square charge radii measurements will now be used to investigate subshell candidates in the neutron-rich region. In the $A = 100$ region, and for other higher-Z chains, a close correspondence is seen between the radius and separation energy and can be understood in terms of nuclear deformation effects perturbing both sets of macroscopic parameters in a similar fashion. In this region, by contrast, such a correspondence is not seen, and a microscopic explanation appears to be required.

References

1. Cheal, B., Flanagan, K.T.: J. Phys. G: Nucl. Part. Phys. **37**(11), 113101 (2010)
2. Äystö, J.: Nucl. Phys. A **693**(1–2), 477 (2001)
3. Myers, W.D., Schmidt, K.H.: Nucl. Phys. A **410**(1), 61 (1983)
4. Berdichevsky, D., Tondeur, F.: Z. Phys. A: Hadrons Nucl. **322**(1), 141 (1985)
5. Karvonen, P., Moore, I.D., Sonoda, T., Kessler, T., Penttilä, H., Peräjärvi, K., Ronkanen, P., Äystö, J.: Nucl. Instrum. Methods Phys. Res., Sect. B **266**(21), 4794 (2008)
6. Karvonen, P., et al.: Nucl. Instrum. Methods Phys. Res., Sect. B **266**(19–20), 4454 (2008)
7. Avgoulea, M., et al.: J. Phys. G: Nucl. Part. Phys. **38**(2), 025104 (2011)
8. Gangrsky, Y.P., et al.: J. Phys. G: Nucl. Part. Phys. **30**(9), 1089 (2004)
9. Charlwood, F.C., et al.: Phys. Lett. B **690**(4), 346 (2010)
10. Cheal, B., et al.: Phys. Lett. B **645**(2–3), 133 (2007)
11. Baczynska, K., et al.: J. Phys. G: Nucl. Part. Phys. **37**(10), 105103 (2010)
12. Campbell, P., et al.: Phys. Rev. Lett. **89**(8), 082501 (2002)
13. Forest, D.H., et al.: J. Phys. G: Nucl. Part. Phys. **28**(12), L63 (2002)
14. Thayer, H.L., et al.: J. Phys. G: Nucl. Part. Phys. **29**(9), 2247 (2003)
15. Cheal, B., et al.: Phys. Rev. Lett. **102**(22), 222501 (2009)
16. Charlwood, F.C., et al.: Phys. Lett. B **674**(1), 23 (2009)
17. Cooke, J.L., et al.: J. Phys. G: Nucl. Part. Phys. **23**(11), L97 (1997)
18. Moore, R., et al.: Phys. Lett. B **547**(3–4), 200 (2002)
19. Cheal, B., et al.: J. Phys. G: Nucl. Part. Phys. **29**(11), 2479 (2003)
20. Bissell, M.L., et al.: Phys. Lett. B **645**(4), 330 (2007)
21. Levins, J.M.G., et al.: Phys. Rev. Lett. **82**(12), 2476 (1999)
22. Yeandle, G., et al.: J. Phys. G: Nucl. Part. Phys. **26**(6), 839 (2000)
23. Nieminen, A., et al.: Phys. Rev. Lett. **88**(9), 094801 (2002)
24. Bissell, M.L., et al.: Phys. Rev. C **74**(4), 047301 (2006)
25. Mach, H., Wohn, F.K., Molnr, G., Sistemich, K., Hill, J.C., Moszynski, M., Gill, R.L., Krips, W., Brenner, D.S.: Nucl. Phys. A **523**(2), 197 (1991)
26. Urban, W., et al.: Nucl. Phys. A **689**(3–4), 605 (2001)
27. Wu, C.Y., Hua, H., Cline, D., Hayes, A.B., Teng, R., Clark, R.M., Fallon, P., Goergen, A., Macchiavelli, A.O., Vetter, K.: Phys. Rev. C **70**(6), 064312 (2004)
28. Thibault, C., et al.: Phys. Rev. C **23**(6), 2720 (1981)
29. Buchinger, F., et al.: Phys. Rev. C **41**(6), 2883 (1990)
30. Buchinger, F., et al.: Phys. Rev. C **42**(6), 2754 (1990)
31. Lievens, P., Silverans, R.E., Vermeeren, L., Borchers, W., Neu, W., Neugart, R., Wendt, K., Buchinger, F., Arnold, E.: Phys. Lett. B **256**(2), 141 (1991)
32. Keim, M., Arnold, E., Borchers, W., Georg, U., Klein, A., Neugart, R., Vermeeren, L., Silverans, R.E., Lievens, P.: Nucl. Phys. A **586**(2), 219 (1995)
33. Lalazissis, G.A., Raman, S., Ring, P.: At. Data Nucl. Data Tables **71**(1), 1 (1999)
34. Möller, P., Nix, J.R., Myers, W.D., Swiatecki, W.J.: At. Data Nucl. Data Tables **59**(2), 185 (1995)
35. Raman, S., Nestor Jr., C.W., Tikkanen, P.: At. Data Nucl. Data Tables **78**(1), 1 (2001)
36. Singh, B.: Nuclear Data Sheets **109**(2), 297 (2008)
37. King, W.H.: Proc. R. Soc. **280**, 430 (1964)
38. Charlwood, F.C., et al.: Hyperfine Interact. **196**, 143 (2010)
39. Hager, U., et al.: Phys. Rev. Lett. **96**(4), 042504 (2006)
40. Hager, U., et al.: Nucl. Phys. A **793**(1–4), 20 (2007)
41. Federman, P., Pittel, S.: Phys. Rev. C **20**(2), 820 (1979)
42. Werner, T.R., Dobaczewski, J., Guidry, M.W., Nazarewicz, W., Sheikh, J.A.: Nucl. Phys. A **578**(1–2), 1 (1994)
43. Verma, S., Dar, P.A., Devi, R.: Phys. Rev. C **77**(2), 024308 (2008)
44. R. Rodríguez-Guzmán, Sarriguren, P., Robledo, L., Perez-Martin, S.: Phys. Lett. B **691**(4), 202 (2010)

45. Wolf, A., Casten, R.F.: Phys. Rev. C **36**(2), 851 (1987)
46. Boos, N., et al.: Phys. Rev. Lett. **72**(17), 2689 (1994)
47. Eastham, D.A., Walker, P.M., Smith, J.R.H., Griffith, J.A.R., Evans, D.E., Wells, S.A., Fawcett, M.J., Grant, I.S.: Opt. Commun. **60**(5), 293 (1986)
48. Eastham, D.A., Gilda, A., Warner, D., Evans, D., Griffiths, J., Billowes, J., Dancey, M., Grant, I.: Opt. Commun. **82**(1–2), 23 (1991)
49. Soloviev, V.G., Sushkov, A.V.: J. Phys. G: Nucl. Part. Phys. **16**(3), L57 (1990)
50. Pillet, N., Quentin, P., Libert, J.: Nucl. Phys. A **697**(1–2), 141 (2002)
51. Palmer, C.W.P., Baird, P.E.G., Blundell, S.A., Brandenberger, J.R., Foot, C.J., Stacey, D.N., Woodgate, G.K.: J. Phys. B: At. Mol. Opt. Phys. **17**(11), 2197 (1984)

DOI 10.1007/978-94-007-5555-0_22
Reprinted from *Hyperfine Interactions* Volume 223, Issues 1-3, 2014
and *The European Physical Journal A* Volume 48, Issue 4, 2012

Collinear laser spectroscopy at the new IGISOL 4 facility

B. Cheal · D. H. Forest

Published online: 28 March 2012

Abstract In 2011 the collinear laser spectroscopy programme at the University of Jyväskylä Accelerator Laboratory (JYFL), Finland, will move to the new IGISOL 4 facility. With its own dedicated cyclotron, this new laboratory will offer unparalleled access to beam time for both technique development and exploitation. Production of sub-millisecond states is available, including elements of a refractory nature.

Keywords Collinear · Laser · Spectroscopy · IGISOL · Cooler

1 Introduction

Laser spectroscopy [1] provides model-independent access to both macroscopic and microscopic nuclear properties, including the magnetic dipole moment, spectroscopic electric quadrupole moment and changes in mean–square charge radius. In addition, laser spectroscopy enables measurements of ground state and isomeric state spins, underpinning assignments for excited states made using nuclear decay spectroscopy. The nuclear moments are a sensitive probe of the wave function, while the mean–square charge radius is a measure not only of nuclear size but of deformation (including dynamical effects) and surface diffuseness. Together with mass measurements [2], a comprehensive picture of the nuclear ground state properties is formed.

B. Cheal (✉)
School of Physics and Astronomy,
University of Manchester, M13 9PL, Manchester, UK
e-mail: Bradley.Cheal@manchester.ac.uk

D. H. Forest
School of Physics and Astronomy,
University of Birmingham, B15 2TT Birmingham, UK

At the IGISOL facility [3–5], radioactive ion beams are available from production techniques insensitive to chemical or physical properties. Reaction products recoil from thin foil targets into a buffer gas where they are transported out of the ion guide in a fast–flowing gas jet. This process can take place in less than a milli-second, permitting studies of isotopes which are too short–lived to diffuse out of conventional thick target ion sources with sufficient flux. Even if the in–target production could be high enough to compensate, such ion beams would then be swamped by longer–lived isobars.

Collinear laser spectroscopy [1] has been applied by the Birmingham–Manchester–Jyväskylä collaboration to IGISOL beams since the late 1990s [6]. During this time, much technique development has taken place, most notably the pioneering use of an ion beam cooler–buncher for laser spectroscopic studies [7–9]. Cooled beams enable a better overlap of the laser and ion beams (which interact in an anti–parallel geometry) and reduced longitudinal energy spread. With the latter within a few eV, the use of fast–ion beams compresses the velocity spread in the forward direction, and the Doppler broadening of the resonances to within the natural line-width. A photo-multiplier tube (PMT) is mounted at 90° to both beams to detect the fluorescence photons as a frequency scan is made by applying a Doppler tuning potential to the region of interaction. Continuos (non-resonant) scattering of photons from the laser beam into the PMT dominates the photon background and is critically suppressed by accumulating and releasing the ions from the cooler in bunches. The PMT signal is then gated to accept the resonant signal from the ion bunch but with a background count rate proportional to the gate width (15 μs) rather than the sample accumulation time (eg. 150 ms) [7, 8].

In a more recent development, optical manipulation of ions inside the cooler has been used to extend the reach of collinear laser spectroscopy to elements previously inaccessible for efficient study [9, 10]. The extended laser–ion interaction time allows ground–state transitions which are spectroscopically weak to be efficiently excited. This has the result of populating metastable states, making collinear laser spectroscopy possible using transitions from these rather than from the ground state. Transitions for high–resolution (collinear) study can then be chosen on the basis of spectroscopic efficiency and hyperfine structure, without being constrained by lower–state population. An added flexibility is that although continuous-wave lasers with a narrow line-width are used for spectroscopy (with a collinear geometry) pulsed lasers, which can readily access a greater range of wavelengths (via production of higher harmonics), can be used for optical pumping in the cooler.

Recent experiments which have utilised the technique include yttrium [11, 12] (where the hyperfine structure of the $J = 0 \rightarrow J = 1$ transitions from the ground–state did not permit measurements of nuclear spin), niobium [9] (where in addition the ground state transition was spectroscopically weak) and manganese [13] (where all transitions from the ground state had wavelengths which were too short for high-resolution continuous-wave production—also true for molybdenum [14]).

Overviews of techniques employed at JYFL and physics cases addressed in recent years are given in separate articles in this volume. In 2011, the IGISOL facility (including the collinear laser spectroscopy apparatus) will move to a new laboratory. Further details of this and future plans are given below.

2 The new IGISOL laboratory

A new laboratory has recently been constructed adjacent to the existing facility which will house, and be virtually dedicated to, IGISOL experiments. The 51 × 13 m area contains an MCC 30/15 cyclotron, which will provide 18–30 MeV proton beams and 9–15 MeV deuteron beams with currents up to 100 μA and 50 μA, respectively. Recent tests however, demonstrated currents of 200 μA for protons and 62 μA for deuterons—well beyond guaranteed specification. For heavier ions, eg. alpha particles, beam delivery from the K130 cyclotron of the existing facility will be possible when necessary. For laser spectroscopy, this development will provide enhanced yields and unparalleled beam time.

Beam time is a vital ingredient for the study of the most exotic cases, many chosen on the basis of a known physics case. In addition, laser spectroscopy can explore systematic effects over long isotopic chains and map the nuclear chart with model–independent data. Unexpected phenomena often then provide the basis for future, targeted studies. Beam time is also essential for technique development. Smaller scale laboratories, such as JYFL, provide a dynamic and flexible environment for speculative projects. These can then be exploited at JYFL and exported to other facilities. At ISOLDE, the cooler–buncher ISCOOL has been installed and already used for the laser spectroscopic study of level migrations in copper and gallium isotopes [15, 16]. With free access to the cooler, in-cooler optical pumping could be tested at JYFL with ease. This technique is now used for most JYFL collinear laser spectroscopy experiments and is in the process of being developed for use at ISOLDE.

The proposed layout of the main experimental area is shown in Fig. 1. Primary beams from the MCC 30/15 (or K130) are transported to the IGISOL chamber, where a variety of reaction mechanisms can be used. Radioactive products are extracted in the form of a 30 keV ion beam and passed through a dipole magnet for mass selection. To ensure continuous running of the facility during IGISOL cooling and target changes, an offline source of stable beams will also be available. For laser spectroscopy (and mass measurements) the switchyard will deliver beams to the ion beam cooler. The cooler sits on a high voltage platform at a potential just below that of the ion source. On extraction from the cooler, the ions will pass through a low energy (800 eV) transfer section, through two quadrupole bends, each deflecting the beam by 90°, before re-acceleration to 30 keV. This arrangement allows optical access to the central axis of the cooler by shining laser light through the quadrupole bend. Previously, the Penning trap was mounted in-line with the cooler and had to be disconnected each time for optical pumping. Such an arrangement will also allow optical pumping in the low–energy transfer section, rather than the cooler's gas volume, to be explored.

Although both the isotope yields and beam time will be much improved at the new facility, the sensitivity of the technique will continue to be improved. Efforts will be made to extend the effective interaction time, ensuring each ion undergoes excitation at least once. To reduce the background from laser scatter, shorter time gates can be placed around the resonant counts. In the bunching technique, this will require a shorter effective bunch length. Background is also caused by the ions themselves, which undergo collisions with residual gas atoms in the beam line. This becomes

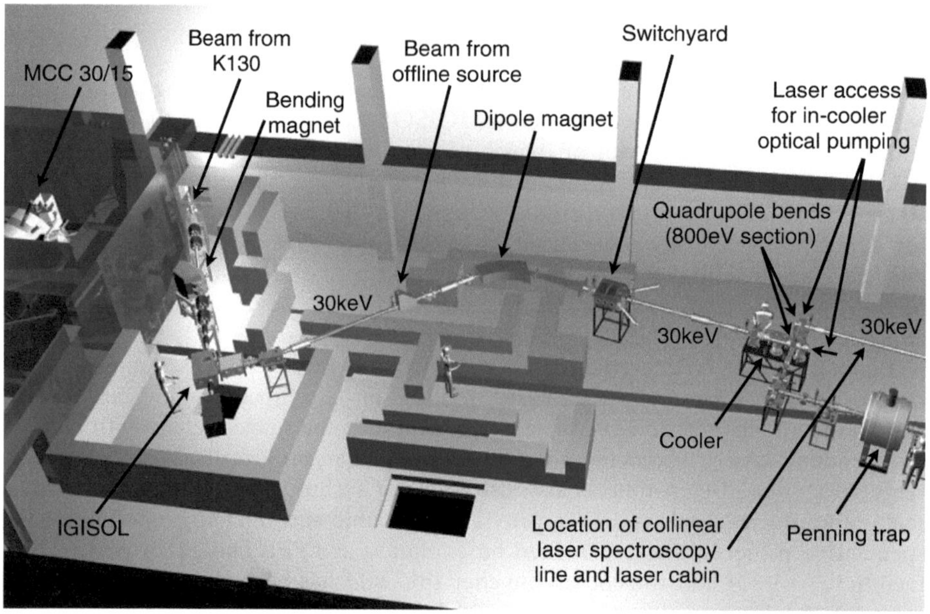

Fig. 1 Proposed layout of the new IGISOL laboratory

increasingly evident with high levels of isobaric contamination in the beam and exotic species may form less than a part per million of the total flux. An alternative method, used for all experiments at JYFL prior to the installation of the cooler [6], is to tag single photons in coincidence with the detection of the corresponding ion [17, 18]. With the use of segmented photomultiplier tubes and pencil–lead–like cooled beams, measurements are theoretically possible on fluxes down to an ion per second if sufficient beam purity can be achieved.

Much effort will be targeted towards the production of pure beams of a single isotope or even isomer. A new technique is proposed which will utilize the efficient optical pumping in the cooler to achieve this. When ions are released from the cooler in bunches, their time–of–flight depends on the mass–to–charge ratio, m/q, which has been selected upstream by the dipole magnet. Using a multi-step excitation process, lasers will be used to resonantly ionise a selected element from a singly–charged to a doubly–charged state while inside the cooler. These ions alone will then have half the mass–to–charge ratio of the others and leave the cooler more quickly. The ion bunch that they form will become completely separated from the bunch of contaminant ions and the latter electrostatically deflected from the beam path. An ultra–pure beam of single (N, Z) will then remain and be available for laser spectroscopy or any other post–cooler experiment.

An alternative, but complementary, technique will use the Penning trap to purify the sample before re-ejecting the bunch back into the low–energy section and to the collinear laser spectroscopy apparatus. Switching units will be employed to execute this cycle. Laser spectroscopy is used to measure and link the nuclear spin, moments and charge radius, and in some cases the half–life (by changing the bunch accumulation time in the cooler). If, however, a ground–state and isomeric–state are both observed without sufficient spin information to be correctly labelled, and the half–lives are similar or both long, then the Penning trap can be used to identify which state is which. Even if the states lie close in energy and one cannot be entirely filtered, the Penning trap can be used to bias the relative amounts and each hyperfine structure specifically identified with each state.

3 Physics priorities

The new facility and the available beam time will allow systematic studies of model-independent data across nuclear chart, in addition to many targeted physics cases. To date, several regions of the nuclear chart remain unexplored, even relatively close to stability. As highlighted in a recent review [19], most of these lie in the refractory regions around nickel ($Z = 28$), ruthenium ($Z = 44$) and rhenium ($Z = 75$). Only at an IGISOL facility are all of these available, with fast extraction times, for laser spectroscopy. These cases will therefore be targeted as priorities. The extended access to fission fragments will allow studies of neutron-rich nuclei to be pushed further from stability. This will enable studies of nuclear structure with more extreme proton–neutron ratios, including r-process nuclei. A new ion guide and a converter will be designed for neutron-induced fission. For some cases this will enhance the yield while reducing isobaric contamination (and also the plasma density). Light-ion (p, n, d, α) induced fusion evaporation reactions provide access to elements too light or heavy to be produced effectively from fission, in addition to neutron-deficient isotopes. Although a specialised ion guide for reactions induced with heavier ions has been in use at JYFL for some time, typical yields have been too small for collinear laser spectroscopy. With planned sensitivity enhancements, this will no longer be the case and more exotic neutron-deficient isotopes will be accessible.

Elements between calcium ($Z = 20$) and nickel ($Z = 28$) have been of much interest due to the region being rich in nuclear structure effects and providing stringent tests of shell model interactions. Studies of titanium [20] and scandium [21] revealed a trend whereby the mean-square charge radius increased with decreasing neutron number towards $N = 20$. These studies will be extended to cover lighter isotopes such as ^{42}Ti, and neutron-rich scandium isotopes (using (d,2p) reactions from titanium targets). This work was initially motivated by the observation that the two doubly-magic nuclei, ^{40}Ca and ^{48}Ca have near-identical mean-square charge radii [22]. Preparatory work for the study of exotic calcium isotopes is planned.

In the middle of the $20 < Z < 28$ region, neutron-deficient isotopes of manganese were studied to investigate the $N = 28$ shell closure [13]. Unlike the two-neutron separation energies, which are smooth through $N = 28$, the mean-square charge radii show a distinct minimum and therefore a clear signature of a shell closure effect. An efficient optical pumping scheme was developed at Jyväskylä which will now be exported to ISOLDE, CERN. At Jyväskylä, optical pumping techniques will now be used to study new sub-shell closure candidates and an apparent onset of collectivity towards $N = 40$ in iron [23] and chromium [24].

In recent years, a mapping of the $N = 60$ sudden onset of deformation has commenced at JYFL. Laser spectroscopic measurements of yttrium [11, 12], zirconium [8, 25, 26], niobium [9] and molybdenum [14]—elements not available at conventional facilities—revealed that the magnitude of the effect is centred around the yttrium chain [10]. In the case of yttrium, a weak but increasing oblate deformation is seen from $N = 50$ to $N = 59$, whereas for ^{99,101}Y ($N = 60, 62$) strongly prolate, rigidly deformed shapes are observed [11]. For $N \geq 60$ the strontium and zirconium isotones show similar static deformation, and the isotopes ^{100,102}Y were assumed to follow this trend. Only in rubidium and niobium were comparatively soft shapes observed for $N \geq 60$, before the effect becomes insignificant for krypton and molybdenum.

Due to the $J = 0$ ground-state spin of the yttrium ion (limiting all optical transitions to $J = 0 \rightarrow J = 1$) it was not possible to determine the nuclear spin. This was only later possible [12] using the optical pumping technique [9], allowing the nuclear moments (and charge radius) of ^{100}Y to be unambiguously determined. A sizeable dynamic component to the deformation is indicated, unlike the rigidity observed in the neighbouring even-N isotopes. A similar softness is also observed in the ^{98m}Y isomer, which lies at a critical point in the transition, showing the largest dynamical component in absolute terms and yet a prolate deformation. Moreover, the state measured in ^{100}Y could be unambiguously assigned $I = 4$, suggesting it is an isomeric state and not the ground state [27]. IGISOL production of the two states is reported to be similar [28] and yet only one state was observed. A search for the ground state will now be undertaken, which if weakly deformed (like the ^{98}Y ground state) would lie in a much different scanning region due to the isotope shift. The optical pumping technique will now allow the spin of the ^{100g}Y ground state, and ^{98m}Y and ^{102}Y to be determined. As described in the previous section, the Penning trap [28] may be used to identify and choose between ground and isomeric states prior to laser spectroscopy, where ambiguities exist.

Laser spectroscopy of molybdenum was performed from a high-lying metastable state at 11783 cm^{-1}, using a $J = 1/2 \rightarrow J = 1/2$ optical transition [14]. The charge radii indicate that ^{108}Mo is past the peak of deformation. Optical pumping will allow this population to be enhanced, permitting measurements still further from stability. In addition, a $J = 1/2 \rightarrow J = 3/2$ transition (from the same state) will be used to yield the quadrupole moments and nuclear spin measurements. Shapes above molybdenum, in neutron-rich Tc, Ru, Rh and Pd isotopes will later be investigated.

The transitional region of $Z = 73 - 76$ (Ta, W, Re, Os) lies between the well-deformed rare earth isotopes and the spherical lead isotopes. In the region from platinum to lead, the transition to deformed shapes is different for each isotope chain measured, including a reversal of the normal odd-N/even-N staggering of the mean-square charge radii [1, 29]. Neutron-deficient isotopes near stability with $Z = 73 - 76$ are readily produced in light ion reactions at Jyväskylä and in-cooler optical pumping [9] will provide access to these elements for the first time.

In tungsten, the shortest-lived isomers produced at the IGISOL have quadrupole moments previously measured by non-optical techniques but with the crystal field gradient not accurately known. In the non-optical work, such isomers were the longest-lived that could be studied (the work spanned isomer lifetimes from nanoseconds to milliseconds). Measurement of the 5.5 ms ^{180m}W isomer using optical and non-optical methods, together with a calculation of the atomic electric field gradient, will allow accurate calibration of all quadrupole moments in the tungsten chain. Besides the measurements obtained by laser spectroscopy, this will aid other studies in the region (see for example, reference [30]).

A wealth of produceable multi-quasi particle isomers occur in the $A \approx 180$ region, including seven in the near-stability $^{178-182}$Ta isotopes. In isomers of this type, which have been studied at the IGISOL, the mean-square charge radius is observed to be smaller than in the corresponding ground state radius, despite an increase in static deformation [31]. The effect appears to scale with quasi-particle number but it is unclear whether the phenomenon is related to a decrease in the dynamic component of the deformation or nuclear surface diffuseness. Either of these can reduce the mean-square charge radius, but quantitative estimates have so far been unsuccessful in reproducing the magnitude of the effects. Whatever the mechanism, it is likely to influence the normal odd-even staggering which, although observed throughout the nuclear chart in deformed and spherical regions, is still poorly understood in a quantitative way.

Spectroscopy of $^{178-182}$Ta alone would more than double the number of multi-quasi particle isomers where such effects have been studied. Although optical transitions from the ground state are available, the upper states have been found to be perturbed by second order effects [32] and the ground state possesses a large hyperfine anomaly. Strong second order mixing would require many additional (weak) hyperfine peaks to be measured. An optical pumping scheme has therefore been developed [9] to avoid such states and efficiently measure the nuclear moments and charge radii. However, with these initial measurement greatly simplifying the problem, selected states could then be remeasured using transitions from the ground state. This would provide a direct measurement of the hyperfine anomaly and therefore the distribution of magnetisation over the nuclear volume.

Bismuth contains three multi-quasi particle isomers which are available for study at the IGISOL [33]. The close proximity to doubly magic ^{208}Pb should ensure that the ground states of these isotopes are near-spherical, with low total (static + dynamic) deformation. If large reductions in mean-square charge radii are still observed for the isomeric states, then increases in nuclear rigidity can be ruled out as the origin of the effect.

Isotopes of thorium are available from the IGISOL using a variety of reactions and mechanisms. Of particular interest is the elusive ultra low-lying isomeric state in ^{229m}Th. With a reported excitation energy of only a few eV above the ground state [34], it would be the lowest-lying nuclear isomeric state in nature. Much evidence has been accumulated, but only of an indirect nature. A direct measurement of the hyperfine structure will prove the existence of such a state beyond question. The high resolution of the collinear technique will allow the resonances of the ground and isomeric states to be individually resolved.

4 Summary

A new era of collinear laser spectroscopy developments and measurements will begin with the construction of the IGISOL 4 facility. The principal advantage of the new MCC 15/30 cyclotron will be a many-fold increase in the beam time available both for development and exploitation. While the previous laboratory evolved to incorporate advances in methodology, the layout of the new facility has been optimised with these in mind.

References

1. Cheal, B., Flanagan, K.T.: J. Phys. G: Nucl. Part. Phys. **37**(11), 113101 (2010)
2. Jokinen, A., Eronen, T., Hager, U., Moore, I., Penttilä, H., Rinta-Antila, S., Äystö, J.: Int. J. Mass Spectrom. **251**(2–3), 204 (2006)
3. Äystö, J.: Nucl. Phys. A **693**(1–2), 477 (2001)
4. Karvonen, P., et al.: Nucl. Instrum. Methods Phys. Res., Sect. B **266**(19–20), 4454 (2008)
5. Karvonen, P., Moore, I.D., Sonoda, T., Kessler, T., Penttilä, H., Peräjärvi, K., Ronkanen, P., Äystö, J.: Nucl. Instrum. Methods Phys. Res., Sect. B **266**(21), 4794 (2008)
6. Levins, J.M.G., et al.: Phys. Rev. Lett. **82**(12), 2476 (1999)
7. Nieminen, A., et al.: Phys. Rev. Lett. **88**(9), 094801 (2002)
8. Campbell, P., et al.: Phys. Rev. Lett. **89**(8), 082501 (2002)
9. Cheal, B., et al.: Phys. Rev. Lett. **102**(22), 222501 (2009)
10. Charlwood, F.C., et al.: Hyperfine Interact. **196**, 143 (2010)
11. Cheal, B., et al.: Phys. Lett. B **645**(2–3), 133 (2007)
12. Baczynska, K., et al.: J. Phys. G: Nucl. Part. Phys. **37**(10), 105103 (2010)
13. Charlwood, F.C., et al.: Phys. Lett. B **690**(4), 346 (2010)
14. Charlwood, F.C., et al.: Phys. Lett. B **674**(1), 23 (2009)
15. Flanagan, K.T., et al.: Phys. Rev. Lett. **103**(14), 142501 (2009)
16. Cheal, B., et al.: Phys. Rev. Lett. **104**(25), 252502 (2010)
17. Eastham, D.A., Walker, P.M., Smith, J.R.H., Griffith, J.A.R., Evans, D.E., Wells, S.A., Fawcett, M.J., Grant, I.S.: Opt. Commun. **60**(5), 293 (1986)
18. Eastham, D.A., Gilda, A., Warner, D., Evans, D., Griffiths, J., Billowes, J., Dancey, M., Grant, I.: Opt. Commun. **82**(1–2), 23 (1991)
19. Klüge, H.J., et al.: Hyperfine Interact. **196**, 295 (2010)
20. Gangrsky, Y.P., et al.: J. Phys. G: Nucl. Part. Phys. **30**(9), 1089 (2004)
21. Avgoulea, M., et al.: J. Phys. G: Nucl. Part. Phys. **38**(2), 025104 (2011)
22. Palmer, C.W.P., Baird, P.E.G., Blundell, S.A., Brandenberger, J.R., Foot, C.J., Stacey, D.N., Woodgate, G.K.: J. Phys. B: At. Mol. Opt. Phys. **17**(11), 2197 (1984)
23. Ljungvall, J., et al.: Phys. Rev. C **81**(6), 061301 (R) (2010)
24. Gade, A., et al.: Phys. Rev. C **81**(5), 051304 (R) (2010)
25. Forest, D.H., et al.: J. Phys. G: Nucl. Part. Phys. **28**(12), L63 (2002)
26. Thayer, H.L., et al.: J. Phys. G: Nucl. Part. Phys. **29**(9), 2247 (2003)
27. Singh, B.: Nucl. Data Sheets **109**(2), 297 (2008)
28. Hager, U., et al.: Nucl. Phys. A **793**(1–4), 20 (2007)
29. Billowes, J.: Nucl. Phys. A **682**(1–4), 206 (2001)
30. Balabanski, D.L., et al.: Phys. Rev. Lett. **86**(4), 604 (2001)
31. Bissell, M.L., et al.: Phys. Lett. B **645**(4), 330 (2007)
32. Bissell, M.L., et al.: Phys. Rev. C **74**(4), 047301 (2006)
33. Moore, I.D., Kessler, T., Äystö, J., Billowes, J., Campbell, P., Cheal, B., Tordoff, B., Bissel, M.L., Tungate, G.: Hyperfine Interact. **171**(1–3), 135 (2006)
34. Reich, C.W., Helmer, R.G.: Phys. Rev. Lett. **64**(3), 271 (1990)

DOI 10.1007/978-94-007-5555-0_23
Reprinted from *Hyperfine Interactions* Volume 223, Issues 1-3, 2014
and *The European Physical Journal A* Volume 48, Issue 4, 2012

THE EUROPEAN PHYSICAL JOURNAL A

Special Article – Tools for Experiment and Theory

Laser developments and resonance ionization spectroscopy at IGISOL

M. Reponen[1,a], I.D. Moore[1,b], T. Kessler[2], I. Pohjalainen[1], S. Rothe[3], and V. Sonnenschein[1]

[1] Department of Physics, University of Jyväskylä, P.O. Box 35 (YFL), FI-40014 Jyväskylä, Finland
[2] Physikalisch-technische Bundesanstalt, Bundesalle 100, D-38116 Braunschweig, Germany
[3] CERN, CH-1211 Genève 23, Switzerland

Received: 31 January 2012
Published online: 18 April 2012 –
Communicated by J. Äystö

Abstract. We present an overview of recent laser ion source developments at the IGISOL facility, Jyväskylä. Technological advances in the lasers have led to a considerable increase in second-harmonic laser power with the use of intra-cavity second-harmonic generation, as well as to narrow linewidth capability by applying an injection-locking technique to a Ti:sapphire laser. The use of a diffraction grating for frequency selection in a new laser resonator has dramatically improved the wide-range tunability of the laser system, resulting in an ideal tool for the development of new ionization schemes. The role of different laser bandwidths, laser intensity and environmental broadening mechanisms on the experimental width of the measured spectral line have been studied using bismuth, silver and nickel, in the gas cell and expanding gas jet. Applications of novel ion guide nozzle design has led to remarkably collimated gas jets which overcome the current limitations in the gas cell-based laser ion source trap (LIST) method. Detailed planning is under way to optimize the new laser laboratory and laser transport path in order to fully exploit the unique opportunities afforded by the new IGISOL-4 facility.

1 Introduction

During the last few years a modern laser ion source facility has been constructed at the IGISOL (Ion Guide Isotope Separator On-Line) facility, Jyväskylä, Finland, motivated by the need to improve the selectivity and, in some cases, the efficiency of the ion guide technique. Throughout this period the laser ion source has been continually developed, often in close collaboration with colleagues from the University of Mainz, Germany, and several milestones have been achieved *en route* towards the expected installation in the extended JYFL Accelerator Laboratory in 2011. Much work has been performed for the development of new resonance ionization schemes of elements of interest for future study, and in using the laser ion source as a sensitive tool to probe the complex chemical and plasma-related processes occurring in the ion guide during the extraction of chemically active elements in the presence of the primary cyclotron beam [1,2].

The Fast Universal Resonant laser IOn Source (FURIOS) consists of a solid-state laser system which can provide an almost universal coverage of ionization schemes throughout the periodic table [3]. The laser system is currently under expansion in order to improve the ability to perform sensitive in-source resonance ionization spectroscopy, a key goal towards bringing an additional spectroscopic tool to bear on the wealth of exotic nuclei available at IGISOL. FURIOS consists of four broadband Ti:sapphire lasers pumped by a single 100 W Nd:YAG laser operating at a repetition rate of 10 kHz. Three of the lasers form the backbone of the laser ion source for on-line operation, all wavelength tunable with etalon and birefringent filters. The fourth laser is based on a grating concept and is primarily used for developing new ionization schemes. By utilizing a higher-harmonic generation, a wavelength coverage from 205 nm to 990 nm can be accessed. The FURIOS set-up also includes an atomic beam unit (ABU) which allows the off-line development of ionization schemes and acts as a reference station during on-line runs. Most recently, an ion guide quadrupole mass spectrometer has been built which will increase the flexibility of performing resonance ionization spectroscopy without the need for utilizing the IGISOL target chamber. It is envisaged that the study of the low-lying isomeric state of ^{229}Th, populated in the alpha decay of ^{233}U, will be a first candidate for the new spectrometer system.

In general, two laser ion source development programs are underway, both of which aim to couple the high repetition rate laser system to the ion guide for the selective and

[a] e-mail: mikael.h.t.reponen@jyu.fi
[b] e-mail: iain.d.moore@jyu.fi

efficient production of radioactive isotopes. One method produces photoions within the ion guide volume while the second, the so-called LIST (Laser Ion Source Trap) method, within the expanding supersonic gas jet emitted from the the gas cell. Previous studies at JYFL have shown that on-line conditions prevent efficient laser ionization in the ion guide volume due to the high ion-electron pair density of the buffer gas caused by the passage of the primary cyclotron beam [2]. Recently, this problem was addressed by the LISOL (Laser Ion Source On-Line) group of the University of Leuven with a novel ion guide design [4] which prevents the recombination of laser ions by spatially separating the laser ionization zone from the target region in what has been termed a dual-chamber cell or "shadow gas cell". Such an ion guide has been adapted and built in Jyväskylä to be used in the IGISOL target chamber. The LIST approach, in which neutral radioactive atoms are selectively ionized upon exit from the ion guide within the expanding gas jet, is a relatively new, novel concept, which aims for ultra-high purification of low-energy radioactive ion beams (via the repulsion of any non-neutral fraction from the ion guide) [5]. Recent work in Jyväskylä has concentrated on addressing the current limitation of this technique, namely the poor geometrical overlap between the expanding gas jet and the counter-propagating laser beams [6]. In this article, we briefly summarize the successful results obtained with a Laval nozzle, and discuss the possibility of using the gas jet environment (reduced gas pressure and temperature) for laser spectroscopy of short-lived exotic nuclei.

Two laser-related developments are described which have direct implications for laser ion source activities: improvements in laser power and a reduction in the broadband laser linewidth. In the former, the available laser power for second-harmonic (SHG) laser light has been greatly increased by switching from external generation via a single pass configuration to intra-cavity generation which uses multiple passes. This modification in higher-harmonic generation should enable the lasers to operate well beyond the saturation for any given transition. The second major development, performed in collaboration with the LARISSA group of the University of Mainz, Germany, is an injection-locked Ti:sapphire laser with a reduction in laser linewidth from approximately 5 GHz to a few tens of MHz. The use of this laser for the study of the ground-state hyperfine structure of ^{229}Th at the University of Mainz is detailed elsewhere in this Hyperfine Portrait. Recent funding has been obtained from the Academy of Finland to construct a similar laser at the University of Jyväskylä for future efficient high-resolution in-source spectroscopy applications at IGISOL.

2 Laser developments

2.1 Intra-cavity second harmonic generation of a ns pulsed Ti:sapphire laser

Intra-cavity second-harmonic generation (SHG) is a well-established method used in both continuous wave (CW) Ti:sapphire and dye laser cavities [7,8]. The FURIOS pulsed Ti:sapphire lasers have enabled second- and higher-harmonic generation in a single pass configuration outside the main resonator for efficient off-line use. In an on-line situation, however, the main goal is to achieve the highest possible ionization efficiency, which translates into the requirement of complete saturation of all resonant transitions in the scheme of interest with a reasonable overhead. This overhead must compensate for the losses stemming from laser transport to the ion source. The current laser transport path (~ 15 m) from the laser laboratory to the IGISOL target chamber is non-optimal due to the late installation of the laser ion source with respect to the IGISOL facility. Overall distance and complexity of the path lead to high losses in laser beam intensity that in the worst case have reached 70%. This is particularly evident for the first resonant transitions which are often in the ultra-violet (UV) region. In general it is not possible to transport all needed wavelengths (UV to IR) on single, ultra-broadband dielectric mirrors (99.5% reflectivity from 300 nm to 1100 nm) and the current transport path involves a rather convoluted process with a loss of about 5% to 10% per mirror when operating deep in the UV region. The future Accelerator facility has involved detailed planning for the laser laboratory in the initial stage, thus minimizing the number of mirrors, operating the mirrors at right angles and reducing the transport path to ~ 11 m. For a thorough discussion on issues related to laser beam transport we refer to the article by Lassen and collaborators [9].

An efficient conversion of the fundamental infrared Ti: sapphire laser output to shorter wavelengths by second-, third- and fourth-harmonic generation (SHG, THG, FHG) is needed in cases of weak atomic transitions and/or large interaction volumes with atom-loaded buffer gas cells. In the external single pass SHG with non-linear crystals such as beta-barium borate (BBO) used at IGISOL, conversion efficiencies of typically 15% to 25% have been achieved. Efficiencies for the subsequent THG/FHG are even lower and typical output powers in the VUV to UV (210 nm to 320 nm) range from 20 mW to 100 mW for a fundamental Ti:sapphire laser output of 3 W. Limitations of this method stem from the damage threshold of the crystal surface, angular acceptance of the SHG process and birefringent walk-off in the crystal. Most of these problems can be alleviated by the use of a multi-pass setup. One way to do this is to place the SHG crystal directly into the Ti:sapphire laser resonator, benefiting from the cavity-enhanced laser pulse intensity.

Intra-cavity SHG has been adapted for the existing Ti:sapphire laser system with minimal changes to the current cavity design as seen in fig. 1 (for the standard design see fig. 3.2 in [10]). The intra-cavity intensity of the fundamental wave is enhanced by replacing the output coupler (M1) (reflectivity $R = 70\%$ to 80%) with a highly reflective mirror, $R = 98\%$. The higher-quality factor (Q-factor) of the laser resonator increases the fundamental wave intensity and cavity lifetime supporting the SHG process. As the second harmonic is supposed to leave the cavity without incurring further losses, the end mirror (M4) is

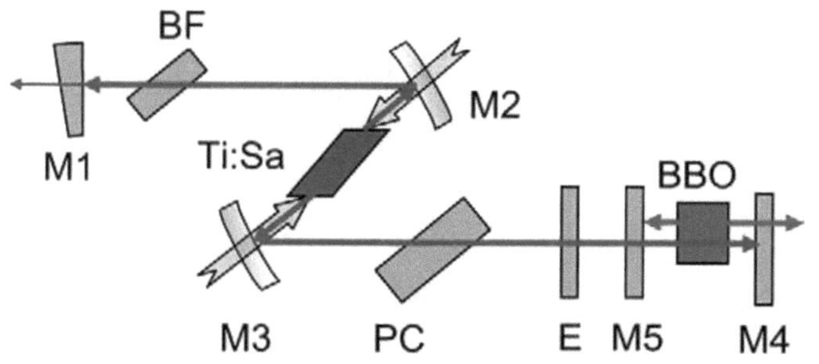

Fig. 1. (Colour on-line) Schematic drawing of the intra-cavity SHG set-up. M1 to M4 are cavity mirrors in the standard Z-shaped resonator configuration. PC, BF and E refer to the Pockels cell, birefringent filter and etalon, respectively. BBO (beta-barium borate doubling crystal) and M5 (dichroic mirror) are the elements needed for the intra-cavity SHG process.

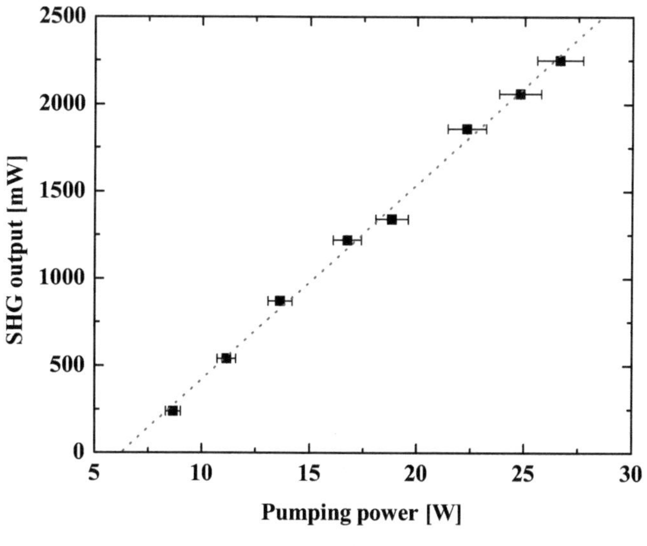

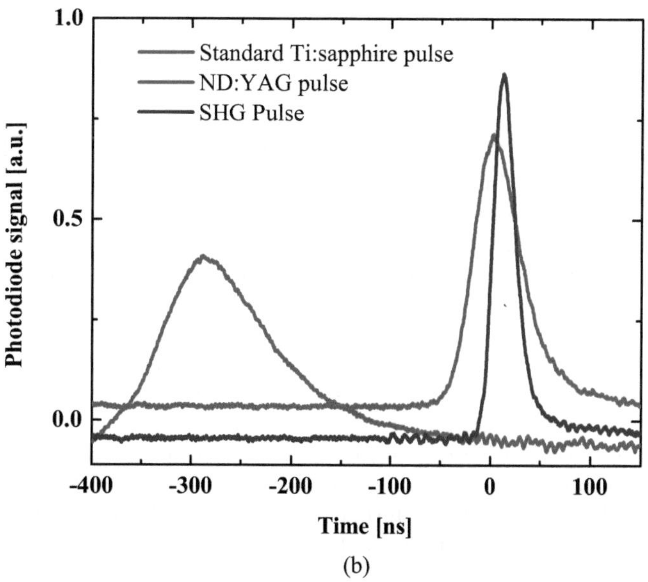

Fig. 2. (Colour on-line) (a) Intra-cavity SHG output power as a function of Nd:YAG pumping power. The error bars of the SHG power are within the symbols. (b) Temporal profiles of the 532 nm Nd:YAG pump laser pulse, the intra-cavity SHG Ti:sapphire pulse and the fundamental infrared pulse from a standard Ti:sapphire.

replaced by a dichroic mirror with high reflectivity for the fundamental wave (> 99.9% for 750 nm to 870 nm) and an antireflection coating for the second harmonic (< 0.25% for 380 nm to 430 nm). This makes the cavity almost closed with respect to the fundamental light with only a small leakage remaining through M1 allowing for cavity tuning and wavelength monitoring. Due to the Z-shaped laser resonator design, the second harmonic propagates in both directions from the BBO crystal. Therefore, for an optimal use of the SHG laser light a second dichroic mirror (M5, reflectivity <0.25% for 680 nm to 950 nm, > 99.9% for 360 nm to 435 nm) inside the cavity is installed.

The losses in the cavity resulting from the SHG crystal and the additional mirror are compensated by the higher reflectivity of the output coupler which leads to a higher total Q-factor of the cavity. The dependence of the SHG output on the Nd:YAG pumping power is shown in fig. 2a. The threshold for lasing action is close to 6 W, which is almost identical to the normal Ti:sapphire set-up without intra-cavity SHG. A conversion efficiency of ∼ 10% has been determined from the slope of the figure. With this efficiency, a dramatic improvement in the second-harmonic output power of up to 2.45 W has been achieved with the available pumping power (for safety reasons this was not tested further). This efficiency could be improved even further by introducing additional focusing into the BBO crystal however space limitations would require a redesign of the basic resonator geometry resulting in costly changes to the current laser system.

The intra-cavity SHG can be useful only if the SHG pulse can be time synchronized with the pulses from the other lasers using Pockels cells. If the pulses are too far apart in time, the power gained from intra-cavity doubling is lost as the Pockels cells introduce increasing losses with increasing time delay. Pulse timing was recorded with a set of fast photodiodes (Roithner LaserTechnik SSO-PD-Q-0.25-5-SMD). Although the photodiodes were rather insensitive to the frequency doubled laser light, a sufficient signal was visible at high powers. When using a free running cavity the pulse of the SHG output arrived quite soon after the Nd:YAG pump pulse in comparison to the normal Ti:sapphire system. This is expected due to the faster pulse build-up in a resonator with a higher Q-factor. Figure 2b shows the individual pulse timings of the laser systems following synchronization. The shorter Full Width at Half Maximum (FWHM) of the SHG output pulse can be explained by the square-law intensity dependence of the SHG process.

Intra-cavity SHG has been applied in the study of two different elements, samarium and silver. In the case of samarium, above 1 W of SHG power was extracted from the cavity and used for the first resonant step. In the study of silver, both the fundamental infrared leaking through M1 and the frequency doubled light were extracted from the laser. These beams were subsequently combined in a third-harmonic generation process. With an IR laser power of ~ 1.6 W and a blue power of ~ 320 mW it was possible to generate approximately 24 mW of 328 nm UV light. It should be noted that in this case the output coupler had a 90% reflectivity as high power was needed for both the fundamental and the SHG beam. Such examples have proven the intra-cavity SHG to be a more efficient and compact tool than the external single pass SHG, requiring only simple mirror changes to the existing resonator configuration. The method is also inherently safer for the doubling crystals as the need for hard focusing of fundamental laser light is no longer required. Furthermore, in addition to a greatly improved conversion efficiency, the intra-cavity SHG benefits third- and fourth-harmonic generation due to a much improved beam shape.

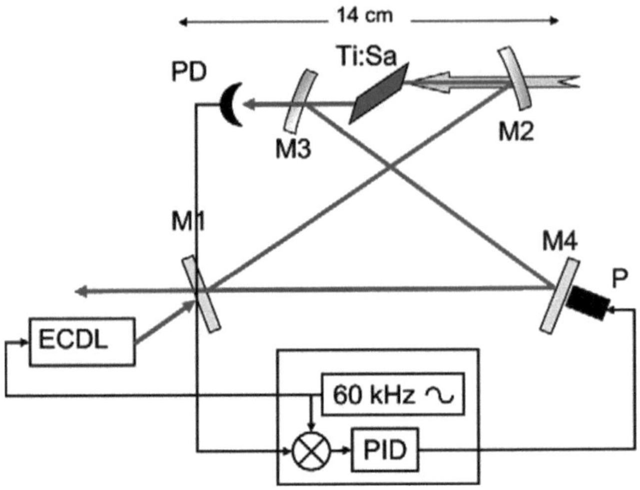

Fig. 3. (Colour on-line) Schematic diagram of the injection-locked Ti:sapphire laser, see text for details.

2.2 Injection-locked Ti:sapphire laser spectroscopy

The FURIOS solid-state laser system provides sufficient laser power and flexibility to efficiently ionize a variety of atomic species. The Ti:sapphire resonator, originally designed by the University of Mainz, has been duplicated for laser resonant ionization applications at other ISOL facilities worldwide. The use of lasers in nuclear physics is however not only limited to the application of a laser ion source. With high-resolution laser spectroscopy, fundamental information on the ground-state properties of nuclei, including nuclear spin, electromagnetic moments and changes in the mean-square charge radii along long chains of isotopes can be extracted in a model-independent manner [11]. For this kind of application, continuous wave laser systems with a linewidth of a few MHz are used, overlapping in a collinear or anticollinear geometry [12] with a fast (tens of keV) atomic or ionic beam. A fluorescence signal from a strong ground-state or metastable-state transition is typically used for detection. The low signal-to-noise ratio of such a detection method limits the performance of the spectroscopy and therefore alternative techniques have been developed including photon-ion/atom coincidence methods and bunched beam spectroscopy utilizing radio-frequency quadrupole cooler-buncher devices [13].

An alternative to photon counting is ion counting following photoionization, as performed regularly with the FURIOS laser system. However, in order to resolve possible hyperfine components of an atomic transition a reduced laser linewidth would be mandatory in most cases (the typical laser linewidth of the FURIOS laser system is ~ 5 GHz). Such a laser system would not only allow for atomic spectroscopy in a collinear beam line, as for example illustrated in [14], but for any other source of atoms whereby the various broadening effects on the atomic transition are appropriately reduced. In the environment of the IGISOL gas cell the atomic lines are typically broadened to a few GHz via a combination of collisional and Doppler broadening processes. To perform high-resolution resonance ionization spectroscopy in the gas cell infers a need to reduce the gas pressure to a few mbar. Such a scenario is however impractical for on-line conditions in which several hundred mbar of gas pressure is required to stop and thermalize the nuclear reaction products. Alternate methods have recently been proposed which include spectroscopy in the environment of the expanding supersonic gas jet immediately downstream from the gas cell or to perform the spectroscopy in a trap.

In order to be able to perform high-resolution spectroscopy with a pulsed solid-state laser system, a novel laser has been developed which is capable of bridging the gap between high-resolution spectroscopy and highly efficient photoionization, in collaboration with the University of Mainz. The laser design is based on a pulsed Ti:sapphire resonator injection-locked to a narrow-linewidth extended cavity diode laser (ECDL). A detailed description of the cavity and its performance can be found in [15]. The general idea is to amplify the narrow-linewidth light of the ECDL laser (Master laser) in a pulsed resonant optical amplifier (Slave laser). To achieve the maximum amplification, the resonator of the slave has to be tuned to the frequency of the master in a so-called injection-locking technique. A schematic diagram of the setup is shown in fig. 3. The pulsed cavity is formed by an output coupler with a transmission of typically 20% (M1), two curved mirrors (M2 and M3) and a high reflector (M4) mounted on a piezo actuator (P). The Ti:sapphire crystal is placed in the focus between mirrors M2 and M3. The cavity is locked to the diode laser frequency via a standard dither lock system. The ECDL diode current is modulated by a frequency generator operating at 60 kHz. The same signal is fed into a phase-sensitive detector together with a photodiode signal (PD) of the laser light leakage behind mirror M3. The resulting feedback signal is fed through a proportional-integral-derivative (PID) controller into the piezo actuator, which regulates the cavity length to maximize the transmission of the seeding light onto the photodiode and thus to keep the slave in resonance with the master frequency.

In order to demonstrate the suitability of the injection-locked Ti:sapphire laser for high-resolution spectroscopic

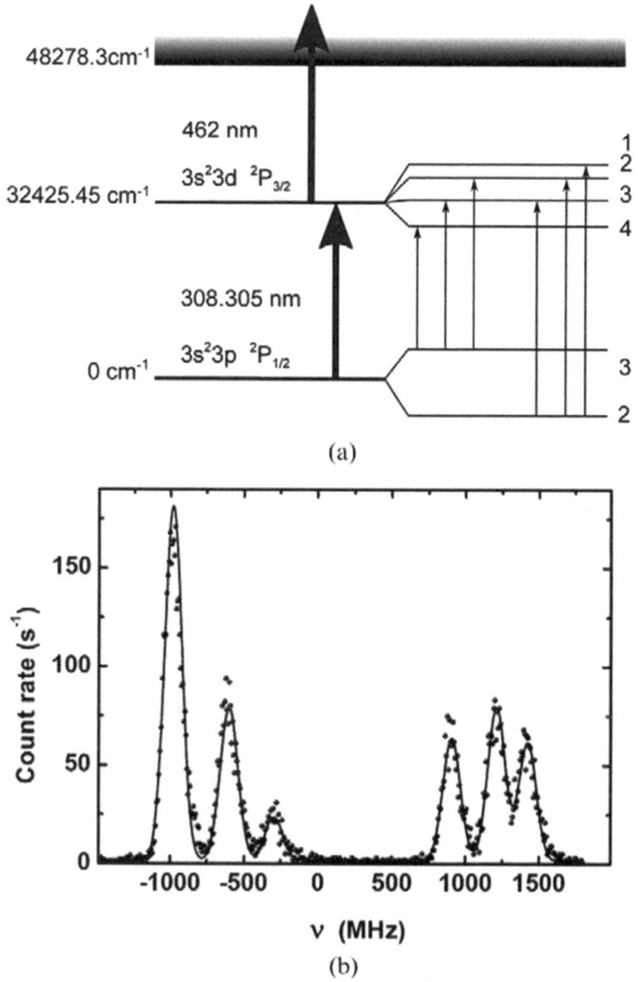

Fig. 4. The resonant ionization scheme for aluminium (a) and the measured hyperfine structure for ^{27}Al (b). The spin of the nuclear ground state $I = 5/2$.

applications, the well-known hyperfine structure of the stable isotope of aluminum, ^{27}Al, was investigated. A sample of stable aluminum was evaporated from a furnace and probed in a crossed-beams geometry with a Quadrupole Mass Spectrometer (QMS) used as a mass-selective filter. The strong D2 transition from the $3s^23p$ $^2P_{1/2}$ atomic ground state to the $3s^23d$ $^2D_{3/2}$ excited level at 32 435 cm^{-1} was chosen as a benchmark test case, exhibiting a relatively small splitting between the hyperfine components of about 300 MHz. The excitation and ionization scheme for the RIS experiment is shown in fig. 4a. The required laser radiation of 308 nm was achieved by sequential frequency doubling and tripling of the fundamental light (924 nm) from the injection-locked Ti:Sapphire laser using standard non-linear optics. The remaining non-converted second-harmonic laser radiation was available for efficient non-resonant post-ionization as indicated in fig. 4a. The Ti:sapphire laser was locked and scanned over the hyperfine structure with reference to a frequency-stabilized He-Ne laser, and the photoions were detected as a function of the laser wavelength. The resulting laser scan over the hyperfine structure of ^{27}Al is shown in fig. 4b. The experimentally observed linewidth of $\sim$ 145 MHz is dominated by Doppler broadening, nevertheless the individual hyperfine components are still well resolved. This width can be compared to a natural linewidth of $\sim$ 12 MHz for the transition used and a laser bandwidth of only 20 MHz.

We note that an average UV power of $\sim$ 0.3 mW was sufficient to obtain the hyperfine components shown in fig. 4b. While this experiment has demonstrated the capabilities of the injection-locked Ti:sapphire laser under off-line conditions, the possibilities for application of the system at IGISOL have yet to be exploited. In the immediate future, new infrastructure funding from the Academy of Finland will be used to construct a next-generation narrow-linewidth Ti:sapphire system for permanent use in the new FURIOS laser ion source facility. The major improvement to the Mainz laser system will be the replacement of the external cavity diode laser (with a rather narrow wavelength tuning range) with a new CW Ti:Sapphire laser, which will match the wavelength tuning range of the current pulsed lasers.

2.3 A grating-based Ti:sapphire laser and its application to the spectroscopy of samarium

On-line, the production of the most exotic isotopes of interest is often overwhelmed by less exotic isobaric species (in the case of a fission reaction) or by stronger fusion-evaporation channels when producing very neutron-deficient nuclei. It is important to have a high overall efficiency of the ion source (including subsequent transportation) however the lack of selectivity in many cases still hinders current experiments. Although a resonant ionization laser ion source (RILIS) provides Z selectivity, high laser power does not necessarily lead to high ionization efficiencies if the ionization scheme is not optimal. It is therefore imperative that all facilities utilizing a RILIS ion source have access to off-line ionization scheme development.

The search for ionization schemes in the heavier elements is hindered by the lack of information about high-lying intermediate states, autoionizing (AI) and Rydberg states. In particular, information regarding the relative ionization strengths of AI states is of high importance for the establishment of a highly efficient scheme. Approximately 35 elements are available from Ti:sapphire-based laser ion sources [16], and databases exist providing information on energy levels which can be used as a first step in planning new ionization schemes [17,18]. Still, for some elements, the information about Ti:sapphire accessible levels is scarce.

The conventional Ti:sapphire systems in use at IGISOL and other on-line facilities provide only a limited mode-hop free wavelength scanning range due to the mode selection with an etalon and birefringent filter combination, hence making the search for new transitions tedious. An automated long-range scan would require precise synchronized motor control of these two devices. Replacing the end mirror of the Ti:sapphire cavity with a grating in Littrow configuration enables a continuous wave-

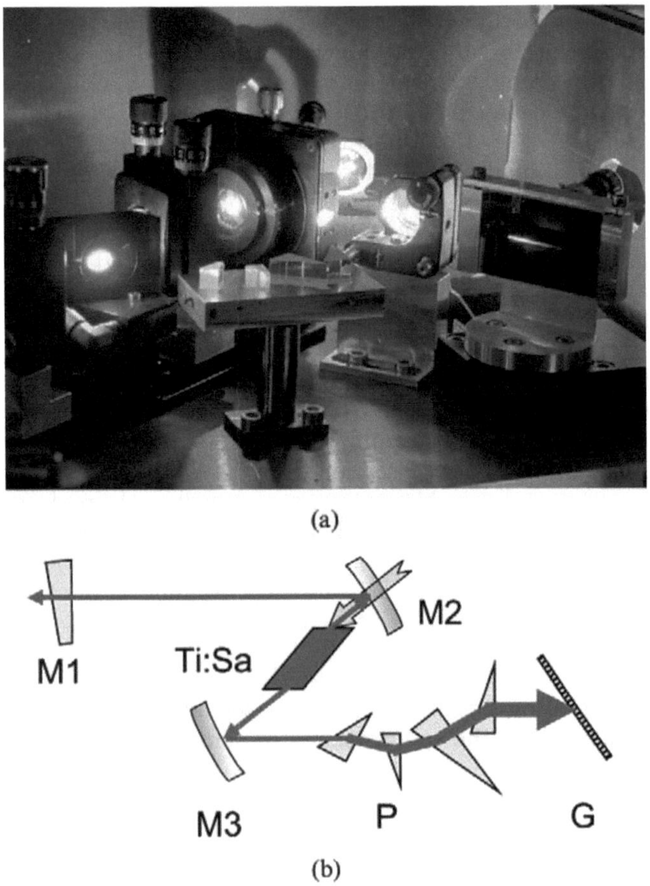

Fig. 5. (a) A photograph of the grating-based Ti:sapphire laser and (b) a schematic drawing of the resonator cavity. P represents the anamorphic prism pairs used for beam magnification and G is the motorised grating.

length selection over nearly the whole gain spectrum of a Ti:sapphire crystal. Wavelength selection is achieved just by tuning the angle of the grating, as expressed through the relation:

$$2d \sin \theta_m = m\lambda, \quad (1)$$

where θ_m is the grating angle, d is the fringe spacing and m is an integer.

A grating-based Ti:sapphire laser has been built in-house using a Horiba gold-coated grating with 480 lines per mm, optimized for first-order reflection at a wavelength of 800 nm. Figure 5a shows a photograph of the laser and a schematic diagram of the resonator can be seen in fig. 5b. The effective wavelength resolution of a grating depends upon the number of lines overlapping with the incident laser beam. Therefore, two sets of anamorphic prism pairs are inserted into the cavity in order to expand the beam waist by a factor of ten such that the laser beam covers a greater surface area of the grating (~ 1.5 cm). Each of the prisms has one uncoated and one anti-reflection coated surface, with the uncoated surface facing the laser beam at Brewster's angle to minimize losses. A precision rotation stage (DT-65 N from miCos GmbH) equipped with a stepping motor with a typical angle resolution of 0.002° is used to tune the grating angle.

A proof-of-principle study using the grating-based Ti:sapphire laser has been made on the rare earth metal samarium ($Z = 62$). Samarium has several stable and extremely long-lived isotopes, and its nucleus undergoes a significant shape change from spherical ^{144}Sm (due to the magic neutron number $N = 82$), to a strongly deformed prolate shape at ^{152}Sm. The isotope shifts, as direct indicators of a change in the mean-square charge radii, have already been measured with high precision using continuous wave lasers [19]. Therefore, a repeat of the measurement would be a benchmark case to test the spectroscopic capabilities of the FURIOS laser system without the need for valuable on-line cyclotron beam time.

The search for an appropriate ionization scheme for samarium started from known transitions from the atomic ground state to energy levels at 22 914 cm^{-1} and 23380 cm^{-1}. The first transitions were driven by a Ti:sapphire laser using the newly developed intra-cavity SHG technique discussed in sect. 2.1. The grating laser was scanned from 34 000 cm^{-1} to 39 000 cm^{-1} (740 nm to 880 nm) in order to find intermediate states by recording the ion count rate as a function of the wavelength of the laser. Some of the scans were performed with and without a non-resonant ionization step performed with a Copper Vapour Laser (CVL). Figure 6 shows excerpts from two grating-laser scans from the same first excited state, where it is noticeable that while the majority of the peaks are strongly enhanced by the CVL, one peak at 34648 cm^{-1} does not appear to be affected. This could be a sign of a near-resonant transition from the intermediate state into an AI, outweighing the gain from the non-resonant ionization step driven by the CVL.

Laser scans searching for AI states were performed from particularly strong intermediate states which at the same time showed an improvement in ion rate when adding the CVL, as these were likely to exhibit pronounced AI structures. An example of an AI scan is shown in fig. 7. In the near future, the most efficient ionization schemes will be used in studies of the isotope shifts of Sm in order to benchmark the capability of FURIOS to perform in-source resonance ionization spectroscopy.

3 Towards in-source and in-jet laser spectroscopy

For the majority of analytical studies involving high-resolution laser spectroscopy, not only is elemental selectivity required but also high isotopic selectivity. For this purpose narrow-bandwidth ($\sim$ 1 MHz) continuous-wave (CW) laser systems are used, delivering only low power yet permitting the resolving of isotope shifts and hyperfine structure throughout long chains of isotopes. For ensuring the highest sensitivity in trace detection, or alternatively high efficiency of an on-line resonance ionization laser ion source, powerful pulsed lasers often operating in a high repetition rate mode ($\sim$ kHz regime) with small duty cycle

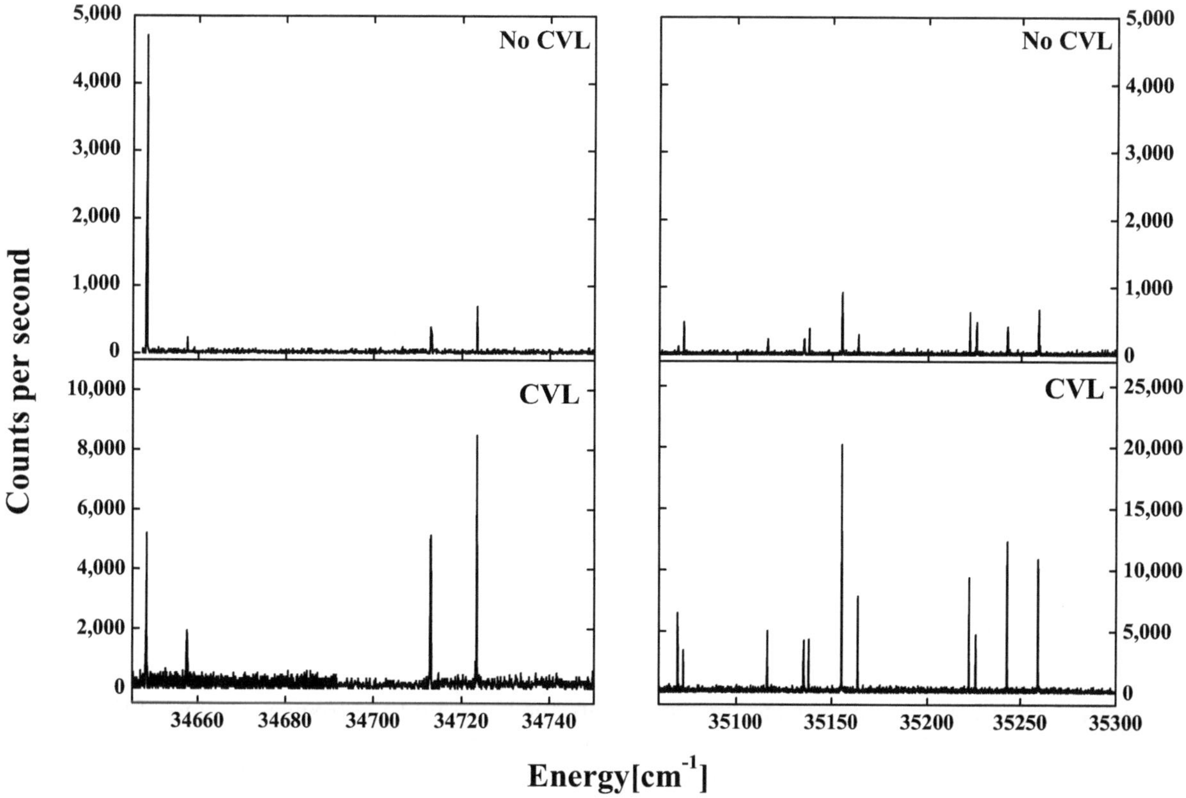

Fig. 6. Two regions scanned by the grating-based Ti:sapphire laser showing a number of intermediate states. A comparison is made illustrating the benefit to the ion count rate of adding an additional non-resonant ionization step performed with a CVL laser. On the left panels, a peak at 34 648 cm^{-1} is not affected by the CVL suggesting the intermediate transition is in near-resonance with an AI state.

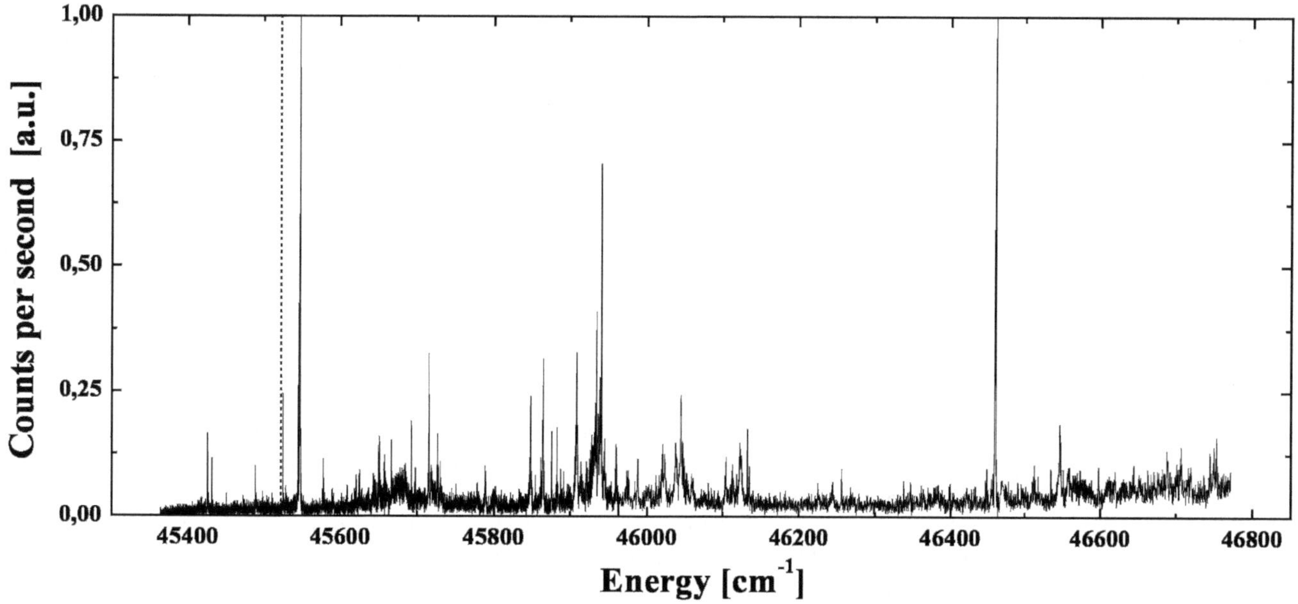

Fig. 7. A grating-laser scan for AI states in samarium showing two particularly strong peaks and a relatively complex structure in between. The dotted line indicates the ionization potential of samarium at 45 519 cm^{-1}.

losses, are most suitable to minimize ionization efficiency losses. Such laser systems need to deliver sufficiently high laser power in order to saturate each individual excitation step. Standard broadband pulsed laser systems have spectral linewidths of a few GHz, therefore isotopic selection is usually provided via a subsequent mass-selective system, *e.g.*, the magnetic sector field of a high-transmission mass separator or a quadrupole mass spectrometer.

In-source laser ionization spectroscopy in a buffer gas cell coupled to an on-line isotope separator allows the study of short-lived isotopes, and has been demonstrated recently at the Leuven Isotope Separator On-Line (LISOL) facility, Louvain-la-Neuve, in a successful measurement of the magnetic dipole moment of ^{57}Cu [20]. In contrast to spectroscopy in high-temperature ISOL target-ion source systems, the buffer gas cell provides higher sensitivity and accuracy due to a smaller total laser linewidth. Furthermore, it allows laser spectroscopy measurements of short-lived radioactive isotopes of refractory elements not available from high-temperature ion sources. The experimental width of the measured spectral line is a convolution of several effects which add to the intrinsic laser linewidth, namely the pressure broadening from collisions with the buffer gas, the power broadening from the lasers and the Doppler broadening due to the atomic velocity distribution. The linewidth of the FURIOS Ti:sapphire laser system depends to some extent on the resonator mirrors and Nd:YAG pump laser power [21], and has been estimated to be 3 GHz to 6 GHz via the study of atomic spectral lines of several elements. In sect. 2.2 we discussed the development of a novel Ti:sapphire laser resonator with the attractive features necessary to satisfy both high power and high resolution. Pressure and Doppler broadening effects are discussed in detail in [10,22]. In the following subsections we present results on the effect of laser power broadening, followed by developments towards in-source resonance ionization spectroscopy in the IGISOL gas cell and gas jet. Laser ionization spectroscopy in the expanding gas jet after the gas cell is an extension to our work on the Laser Ion Source Trap (LIST) method which is discussed in sect. 4.

3.1 Power broadening

In isotope separation applications, both maximum ionization efficiency and highest selectivity are desired. However, this results in a quandary as a higher laser intensity improves the ionization but at the cost of selectivity and vice versa, a good selectivity requires low intensity in order to avoid power broadening which implies a sacrifice of ionization. For a practical discussion on the factors affecting selectivity in resonant multi-photon ionization we refer the reader to [23].

In practice, in-source spectroscopy is performed with the laser power adjusted to as low a value as possible while still allowing sufficient ionization to perform the measurement. One of the candidate elements for in-source spectroscopy at the IGISOL facility is silver, with particular interest in the production and study of the $N = Z$ isotope ^{94}Ag [24]. As part of the laser development programme, the effect of power broadening has been investigated in the atomic beam unit via frequency scans of the first resonant step (5s $^2S_{1/2} \rightarrow$ 5p $^2P_{3/2}$) in atomic silver using a UV-UV-field ionization scheme. The laser beam was focused to a waist of about $\sim 3\,\mathrm{mm}^2$ in order to achieve as high intensity as possible with the available laser power of 22 mW. The intensity was changed by rotating a quarter wave plate on the third-harmonic generation unit which allows the directional stability of the laser to be preserved. At the lowest laser intensities the oven temperature was increased accordingly in order to compensate for the rapidly decreasing ion count rate.

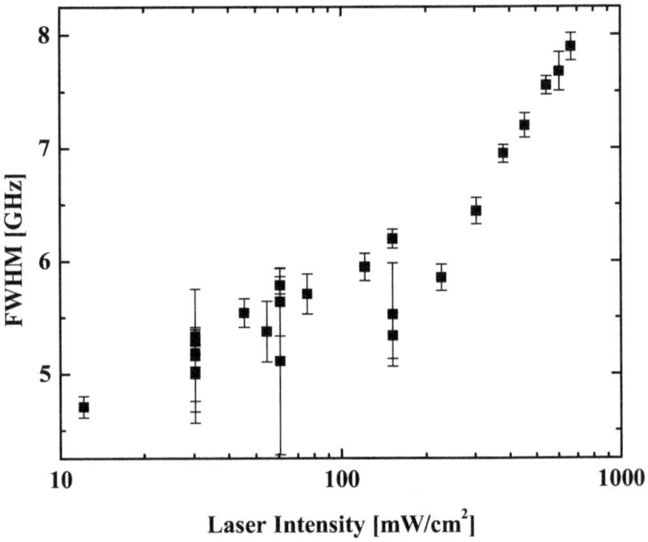

Fig. 8. The FWHM of the resonant 5s $^2S_{1/2} \rightarrow$ 5p $^2P_{3/2}$ transition lineshape in silver as a function of laser intensity.

The overall resonance line is described by a Voigt profile in which the Gaussian and Lorentzian contributions arise from the intrinsic laser bandwidth and the power broadening, respectively. We note that the pressure broadening (Lorentzian contribution) and Doppler broadening (Gaussian contribution) are negligible as the atomic beam unit operates in high vacuum (10^{-6} mbar) and ionization is performed in a crossed-beams geometry. The total FWHM of the Voigt profile can be approximated numerically [25]:

$$w_{tot} \approx 0.5346 \cdot w_L + \sqrt{0.2166 \cdot w_L^2 + w_G^2}\,, \qquad (2)$$

where w_L and w_G are the Lorentzian and the Gaussian components, respectively. Figure 8 illustrates the overall atomic resonance linewidth as a function of the laser intensity, using the deduced Gaussian and Lorentzian widths from the Voigt fitting routine and inputting these into eq. (2). With an excited-state lifetime of ~ 7 ns and a resonant transition wavelength of 328.162 nm (in vacuum), the saturation intensity of the transition used is calculated to be $\sim 80\,\mathrm{mW/cm}^2$.

At intensities below $\sim 100\,\mathrm{mW/cm}^2$, the total transition linewidth appears to converge albeit the individual measurements fluctuate. At these intensities the

Lorentzian component of eq. (2) becomes negligible and an average linewidth value of 5.3 ± 0.2 GHz can be deduced, dominated by the Gaussian laser linewidth. From a $\sqrt{3}$ dependence of the third-harmonic linewidth, the fundamental laser linewidth is estimated to be ~ 3 GHz, in agreement with typical Ti:sapphire linewidths stated elsewhere [21]. At a laser intensity above $\sim 200\,\mathrm{mW/cm^2}$, the total atomic transition linewidth starts to increase, reaching ~ 8 GHz at an intensity of $\sim 700\,\mathrm{mW/cm^2}$. At high intensities, the linewidth values tend towards saturation (seen more clearly when the data is shown on a linear scale). A similar trend has been seen in a lineshape study by Cavalieri and Eramo [26], in which the effect of the laser bandwidth on the power broadening was investigated. The bending of the experimental curve was reproduced theoretically by introducing a cutoff parameter in the laser spectrum.

3.2 In-source spectroscopy

A programme of resonance ionization spectroscopy at the IGISOL facility both in the buffer gas volume of the ion guide and in the radiofrequency (rf) sextupole commenced in 2006 with studies of the bismuth atomic system. The general aim of this work is to measure optical isotope/isomer shifts in particularly short-lived nuclei, which are directly proportional to variations in the nuclear mean-square charge radius. The physics motivation behind the bismuth atomic system is described fully in [27]. The development work included a linewidth resolution study of the hyperfine structure of stable ^{209}Bi utilizing two different ionization schemes, both using a single mode frequency doubled CW dye laser for the first excitation step with either a non-resonant ionization step to the continuum using 355 nm laser light from a frequency tripled 50 Hz Nd:YAG laser, or excitation to an autoionizing level with a 10 kHz frequency doubled Ti:sapphire laser [28].

Atoms of ^{209}Bi were introduced into the ion guide from a resistively heated Ta filament ($\sim 7\,\mathrm{mg/cm^2}$ thickness), with bismuth sputtered onto the surface. Laser light entered the ion guide longitudinally through a sapphire window, ionizing neutral atoms along the axis of the cell. A helium gas pressure of 50 mbar was maintained throughout the tests. Following extraction of the photoions, guidance through the rf sextupole and injection into the mass separator, the ions were finally detected on a set of micro channel plates downstream from the separator focal plane. The hyperfine spectrum of ^{209}Bi is composed of six components, separated into two groups of three due to the large splitting of the $J = 1/2$ upper atomic level of 24.6 GHz. Results from both ionization schemes are shown in fig. 9, illustrating scans over the upper three hyperfine components which have a total splitting of ~ 5 GHz. It is difficult to make direct comparisons between the two spectra as the repetition rates and hence pulse energies of the ionization lasers are significantly different. The increase in the count rate in the scan on the right however is most certainly due to the increase in the ionization probability when using an autoionizing level compared with a non-resonant transition.

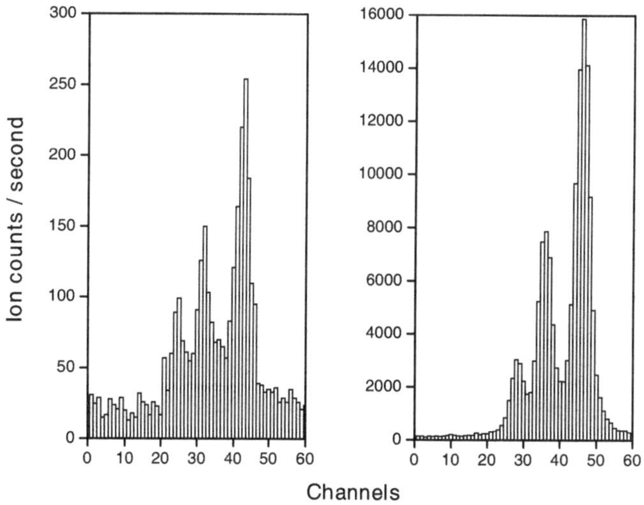

Fig. 9. Hyperfine structure of ^{209}Bi following resonance ionization spectroscopy in the ion guide at 50 mbar He pressure. A CW dye laser was scanned over the upper hyperfine triplet. The left panel shows results obtained using a non-resonant 50 Hz ionization step, while the Ti:Sapphire operating at 10 kHz was used to reach an autoionizing state resulting in the figure on the right.

Although the CW laser was successfully used to obtain the spectra shown in fig. 9, the power available is not sufficient to probe the atomic yields expected for the online production of exotic nuclei. In later work, the 50 Hz Nd:YAG laser was used to pump a pulsed dye amplifier (Spectra Physics) seeded by a single-mode CW dye laser in order to achieve the first resonant transition. However, the low repetition rate (and thus high energy per pulse) had the disadvantage of hugely power broadening the hyperfine structure. Pumping the dye amplifier cells with the high repetition rate copper vapour laser (CVL) had an immediate benefit in reducing the resultant linewidth of the atomic structure. With this arrangement it was possible to search for pressure shifts in the hyperfine centroids and pressure-broadening effects. The former was negligible for a pressure change in helium from 40 mbar to 200 mbar, the latter resulting in a broadening of approximately 10 MHz/mbar [29]. In a similar manner, the pressure broadening and pressure shift were studied at the LISOL facility, Louvain-La-Neuve, for resonant transitions in copper and nickel in argon gas [22]. The resulting differences between nickel and copper highlights the sensitivity to the electronic transition studied. We also note that the broadening and shift in the case of bound electrons in states of high principal quantum number exceed the effects on low-lying bound states by an order of magnitude. This has been studied in Rydberg spectroscopy of gallium in the IGISOL gas cell for both argon and helium buffer gas [10].

More recently, laser spectroscopy has been performed in and around a gas catcher at the LISOL facility, using the low repetition rate excimer-based laser system to probe the atomic system of nickel [22]. In that work, even though the linewidth of the transition is dominated by

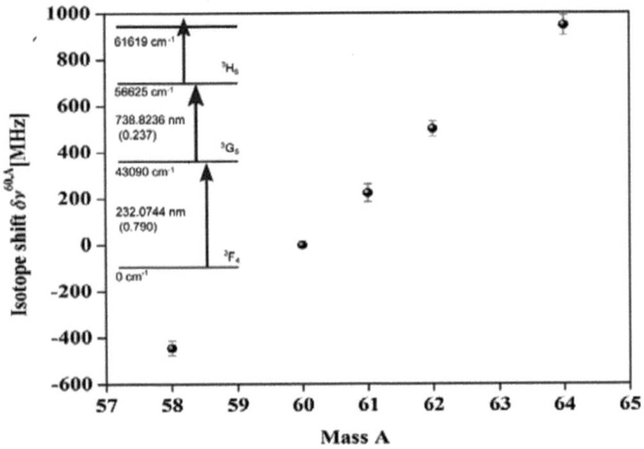

Fig. 10. Isotope shift of the stable nickel isotopes. The inset figure shows the relevant partial atomic level diagram. Below the transition wavelengths the log-ft (transition strength) is listed [17].

the laser bandwidth and power broadening, the isotope shift of the even-A stable isotopes of nickel was observed by performing spectroscopy in the expanding gas jet (this environment will be discussed in the following subsection). It was not possible to extract nuclear-structure information from the data due to the limited contribution to the isotope shift of the nuclear effects, combined with the low resolution of the pulsed laser system. Elements with larger hyperfine parameters (copper and bismuth among them) are better candidates for in-source spectroscopy. Nevertheless, in order to provide a consistency check of the LISOL data and to benchmark the progress of the IGISOL laser system, nickel is also a candidate test element for FURIOS.

A resistively heated nickel filament ($11\,\mathrm{mg/cm^2}$) mounted within the gas volume of an ion guide acted as an atomic source. In a similar laser ionization geometry to the previously discussed bismuth work, nickel atoms were resonantly excited and ionized by a three step scheme with two resonant and a final non-resonant ionization step. The first step transition from the $^3F_4$ ground-state to the $^3G_5$ intermediate level at $43\,090\,\mathrm{cm^{-1}}$ ($232.0744\,\mathrm{nm}$) is the same as the transition used at the LISOL facility. This radiation was realized with fourth-harmonic generation, while the second transition to the $^3H_6$ state at $56\,625\,\mathrm{cm^{-1}}$ was achieved using the fundamental Ti:sapphire radiation at $738.8236\,\mathrm{nm}$. Finally, a non-resonant ionization step was provided by the CVL. Wavelength scans over the first step resonant transition were performed for all stable nickel isotopes ($A = 58, 60, 61, 62, 64$) in 50 mbar argon gas, with mass selection provided by the IGISOL dipole magnet. The resulting resonance lines were fitted with Voigt profiles and the isotope shifts with respect to ^{60}Ni were determined from the centroids, as shown in fig. 10. In order to extract changes in the mean-square charge radius, the electronic F-factor in the Field Shift and the Specific Mass Shift have to be known. A relative measurement of these parameters may be made via the so-called King Plot method, plotting the isotope shift values in one transition versus the shifts in a second transition. However, the isotope shifts in this mass range are dominated by the mass shift component and the limited contribution from the nuclear volume effect is small compared to the experimental resolution. Therefore, the extraction of the changes in the mean-square charge radius in this mass range is impossible with the current experimental setup.

3.3 In-jet spectroscopy

The prospect of utilizing the expanding gas jet environment immediately after the exit hole of a gas cell is highly attractive for performing laser spectroscopy on short-lived exotic nuclei. As discussed in preceding sections, for precise laser-spectroscopic studies the width of the spectral line should be as close to the natural linewidth as possible. In addition to the intrinsic laser linewidth and effects of power broadening, in-source spectroscopy using a buffer gas cell suffers from the effects of pressure broadening (in general $\sim 10\,\mathrm{MHz/mbar}$ for a low-lying bound-state resonance) and in a hot-cavity ion source, Doppler broadening (typical cavity temperatures may reach $2500\,\mathrm{K}$). Although the resolution for laser spectroscopy in a gas cell is better than with a hot cavity, the environment still prohibits the natural linewidth of an atomic transition to be reached. The gas jet environment however limits the impact of pressure and Doppler broadening effects. The supersonic adiabatic expansion into vacuum substantially reduces the Doppler broadening and the gas density is too low to contribute in a meaningful way. We note however that the velocity distribution in the gas jet must be taken into account and will convolute with the laser lineshape to broaden the optical resonance. This effect may be reduced by a dedicated optimization of the gas jet, which has recently been undertaken at JYFL.

The LISOL group of the University of Leuven have performed studies on the impact of the gas jet on the resonant linewidth of nickel and have compared this with the effect of pressure conditions inside the gas cell and in a reference cell under vacuum [22]. The studies have shown that the gas jet environment is rather comparable to that of vacuum conditions, whereas the main parameter for determining the resonance linewidth inside the gas cell is the gas pressure. In addition, using laser Doppler-shift velocimetry in a longitudinal ionization geometry, the group were able to determine the flow velocity of moving atoms within the jet in comparison to atoms in the reference cell. A nickel resonance peak displacement of $\sim 7\,\mathrm{GHz}$ in helium and $\sim 2.5\,\mathrm{GHz}$ in argon resulted in jet velocities of $1663\,\mathrm{m/s}$ and $550\,\mathrm{m/s}$ for helium and argon, respectively.

In a similar manner, and using nickel as a source of atoms, the first proof-of-principle experiments have recently taken place at the IGISOL facility. Figure 11 shows the resonance of the first step transition of nickel under three different conditions: 1) inside the atomic beam unit (vacuum, 10^{-6} mbar); 2) inside the gas cell (helium at 50 mbar); 3) inside the gas jet. A slightly different excitation and ionization scheme was used in this work

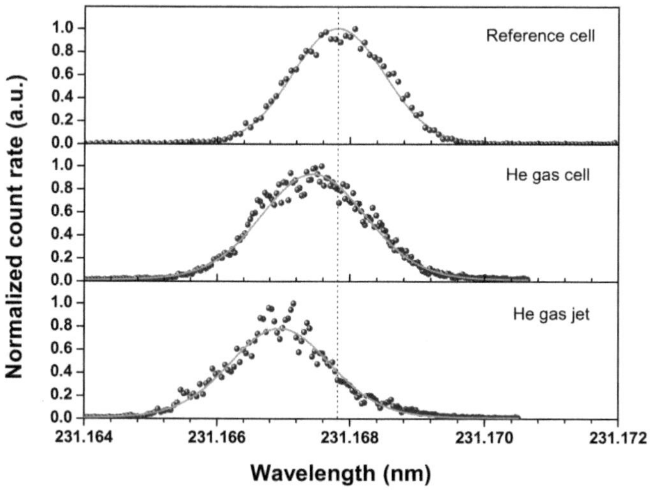

Fig. 11. Scan of the first step transition of Ni at 231.1669 nm in the atomic beam unit (top), in the gas cell (middle) and in the gas jet (bottom). A helium buffer gas pressure of 50 mbar was used. The solid line represents the best fit of a Voigt profile.

compared to the in-source spectroscopy of sect. 3.2. From the $^3F_4$ ground state, an excited state at 43259 cm^{-1} ($^3F_4$, 231.1669 nm, log-ft 0.42) was reached with a frequency quadrupled Ti:sapphire laser, followed by a second resonant step at 738.3963 nm (log-ft −0.06) to the $^3F_4$ excited state at 56802 cm^{-1}. The ionization step was realized non-resonantly with a mixture of the laser radiation from the first two steps. In the gas jet mode, the lasers are introduced through the backside of the ion guide and pass through the exit hole (1.3 mm in diameter) into the SPIG. In order to suppress unwanted photoions created inside the ion guide a positive DC potential of +160 V was applied to the repeller electrode of the rf sextupole. This geometry limits the ionization region where the laser beams and jet atoms overlap. Although the mass separator window provides a line of sight for lasers to be introduced in a counter-propagating mode, the laser radiation currently experiences severe losses at the window prohibiting efficient ionization. This is caused by the difficulty of accessing the window for cleaning. In the future IGISOL 4 facility, access for the laser light will be through a 15 degree electrostatic deflector located before the mass separator dipole magnet.

The displacement of the resonance centroid acquired in the jet compared to the gas cell or atomic beam unit reflects the Doppler shift of the moving atoms while they are ionized inside the SPIG. A measured displacement of 4.9 GHz with respect to the reference cell translates into a jet velocity of ∼ 1128 m/s. This differs from the velocity of ∼ 1663 m/s determined by the LISOL group, however the size and geometry of the rf sextupole guides are different at both facilities. In JYFL, an extra electrode acts as a repeller for unwanted ions, whereas the LISOL SPIG rods themselves are biased with respect to the gas cell. The ionization zone in the JYFL SPIG is further downstream and thus the jet velocity is different. A shift is also apparent in the resonance centroid of the in source data with respect to the atomic beam unit, which results in a frequency displacement of ∼ 2 GHz. We are not sensitive to pressure shifts in the helium pressure range tested (50 mbar to 200 mbar) and therefore this shift is currently not understood. However, the data on the low wavelength side of the gas cell nickel centroid exhibits a shoulder which is not accounted for in the Voigt fit. This is most likely due to photoions contributing from in-jet ionization. Finally, we note that the Lorentzian width from the Voigt-fitting routine is negligible in the atomic beam unit and the total width, dominated by the Gaussian laser linewidth, is 9.1(3) GHz. The fundamental laser linewidth in this setup is therefore ∼ 4.5 GHz, which can be compared to the LISOL laser linewidth of ∼ 1.6 GHz [22]. Although the resonance width of both the gas cell and gas jet profiles are larger, 10.5(3) GHz and 10.1(3) GHz, respectively, the laser linewidth is dominant and the contribution due to changes in environmental pressure is masked.

4 Status of the Laser Ion Source Trap (LIST) project

The Laser Ion Source Trap (LIST) project is a variant of the Ion Guide Laser Ion Source concept, IGLIS. Originally, the LIST was proposed to improve the beam quality from a hot-cavity ion source [30] by decoupling the hot production and evaporation regions from the ionization volume, thus addressing the problem of an often poor selectivity due to the surplus of unwanted surface ions. A recent update on the hot cavity LIST progress can be found in [31,32]. Later, the idea of coupling this concept to a gas cell was suggested [5] and provided the motivation for the development of the rf sextupole at JYFL [33]. In this approach, the reaction products neutralize within the gas cell and upon exit they are selectively re-ionized with counter-propagating lasers within the adiabatically expanding supersonic gas jet. The photoions are captured by the rf field of the SPIG and are transported to the high-vacuum region of the mass separator by a voltage gradient. A positive DC voltage applied to the first electrode of the SPIG acts to repel any non-neutral fraction thus ensuring the highest possible beam purity is maintained for subsequent experiments.

The first proof of principle of the LIST method was demonstrated at the IGISOL facility in 2006 during the in-source spectroscopy studies of the bismuth atomic system discussed in sect. 3.2 [34]. Due to the low laser power available from the frequency doubled CW dye laser coupled with the poor transmission of the dipole separator window, a 50 Hz Nd:YAG laser was used to pump a pulsed dye amplifier which was frequency doubled to produce the required 306.7 nm light for the resonant excitation step. This laser was transported into the gas cell through the the rear sapphire window and was focused into the exit hole region. The final step, realized using non-resonant 355 nm laser light from the same pump laser, was introduced through the mass separator window albeit with

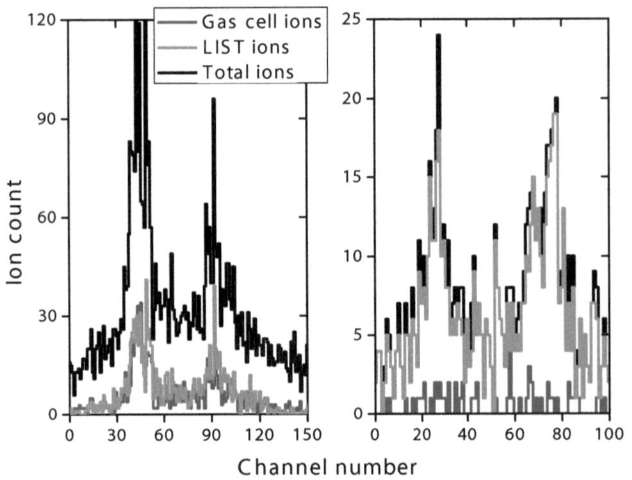

Fig. 12. Laser wavelength scan over the hyperfine components of ^{209}Bi. In the left-hand panel the repeller electrode was set to -10 V, in the right-hand panel it was set to $+100$ V. The different spectra represent ions created in the gas cell or in the SPIG. Additionally, the total number of ions is given.

losses due to the poor window transmission. Figure 12 illustrates the ion rate observed on channel plates downstream from the mass separator as a function of the resonant step wavelength (depicted as channel number). Both upper and lower groups of hyperfine components are visible. The resolution of the upper group can be compared with the hyperfine structure in the left-hand panel of fig. 9, the difference being due to the power-broadening effect caused by the low repetition rate of the pump laser.

Timing gates were set on the different arrival periods of the laser ions, triggered by the pump laser, which are represented by the different spectra in fig. 12. In the left-hand panel, the repeller electrode was set to -10 V and the SPIG was allowed to operate in a conventional "ion transport mode". The spectra show that the number of ions created in the SPIG is almost equivalent to the number of laser ions created inside the gas cell. The summation of both spectra has also been plotted. In the right-hand panel, the repeller electrode has been switched to $+100$ V. The ions created in the gas cell are repelled from entering the SPIG and only those ions created in the SPIG (in the "LIST mode") are counted, confirmed by the summation signal. We note that when the repeller voltage was simply reversed from -10 V to $+10$ V the ion signal remains unaffected, suppression of the ions needed a larger potential barrier. Secondly, LIST ions were only seen when the usual 5 mm gap between the gas cell and repeller electrode was closed, suggesting gas/atom transport to the SPIG is rather poor.

Later, the proof of principle of the LIST method was demonstrated at the LISOL facility, Louvain-la-Neuve, in which systematic studies of resonant laser ionization of atoms within the cold jet were performed in a longitudinal as well as transverse geometry with respect to the atom flow [22]. The enhancement of beam purity was investigated in both off-line and on-line conditions, the latter with ^{94}Rh isotopes produced in fusion-evaporation reactions. Due the LISOL laser system (low repetition rate) and LIST geometry, the resulting statistics was limited however the work illustrated that high selectivity is indeed achievable. A common limiting factor to both JYFL and LISOL which currently hinders the applicability of the LIST technique is the geometrical overlap efficiency between the expanding gas jet and the laser photons. If the gas jet can be sufficiently collimated, for example, over a distance of 10 cm without expansion, a 10 kHz laser repetition rate is sufficient for one atom-photon encounter for a jet velocity of 1000 m/s.

A detailed gas jet study has been undertaken in JYFL in order to obtain low divergence, uniform jets over a distance of several cm. A number of exit nozzle types have been tested under varying gas cell-to-background pressures, information being obtained by imaging the light emitted from excited helium or argon gas atoms, from the detection of radioactive ^{219}Rn recoils and directly via Pitot probe measurements [6]. Initially it was observed that a gas jet emitted from a simple exit hole could be collimated when the background pressure in the expansion vessel reached approximately 10% of the ion guide pressure. Although this approach achieved the goal, the pressure conditions required are not ideal for on-line IGISOL operation due to an increase in the probability of discharge between electrodes of the ion guide system. In order to address this concern a proof-of-concept Laval nozzle was subsequently manufactured, based on existing designs in the literature (see fig. 13b). Such a nozzle is formed with a carefully shaped contour and has a specific ratio between the area of the nozzle throat and exit. These features isentropically transform the jet from a high-pressure low-velocity region, into a low-pressure high-velocity region. In an ideal case all the radial velocity in the jet is transformed into axial velocity hence removing the primary cause for the jet expansion. On the other hand, because of the need for a precisely calculated and manufactured design, the nozzle works ideally at a fixed pressure ratio. Figure 13a shows one example of the resultant jet structure visualized as a contour map. The ion guide was operated with argon at a pressure of 250 mbar resulting in a background pressure 0.33 mbar, very close to the nozzle design requirements. A collimated supersonic jet is seen with a remarkably consistent width even as far as 14 cm downstream from the nozzle. Details related to the structure visible in the jet can be found in [6].

The advantage of the LIST method compared to the standard laser ionization within the gas cell (or, for that matter, within a hot cavity) lies in the ability to successfully suppress all ions produced inside the cell. The suppression voltage required in a hot-cavity laser ion source is rather low, typically a few volts [35]. However, as mentioned in the in-jet studies of nickel and the first proof of principle of the LIST technique using bismuth, suppression voltages applied to the repeller electrode in the JYFL LIST experiments have been of the order of $+100$ V to $+150$ V. The LIST coupled to the LISOL gas catcher used suppression voltages of about 20 V in He and 50 V in Ar, respectively [22]. One possible consequence of working

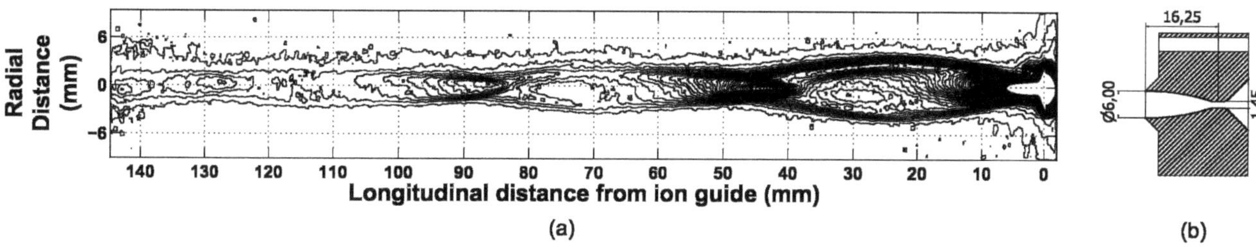

Fig. 13. Schematic of the prototype Laval nozzle (right) followed by a contour map of the resulting argon gas jet (left). A series of expansion and compression waves are visible, however, the jet remains collimated over the 14 cm imaging region.

with electric voltages in poor vacuum conditions is that of discharging. The operational conditions of the JYFL SPIG when working in the LIST mode have occasionally resulted in discharging when argon gas has been used and therefore further studies of the repelling conditions have been performed.

The JYFL SPIG is described in [33] and therefore will not be discussed here. However, with a relatively large repeller electrode aperture (6 mm) and a short distance to the gas cell (a few mm), the longitudinal electric field generated by the repeller is not very efficient at suppressing the ion beam. Moreover, since the axial electric field inside the SPIG is rather low, the DC potential difference between the repeller electrode and the rods of the SPIG is relatively high, a few 100 V. Simulations of the repelling effect have been performed to study the repelling behaviour as a function of suppression voltage, carrier gas type and distance between the ion guide and repeller electrode [36], using Simion 8.0 simulation software [37] and two additional gas models. The two carrier gases He and Ar were simulated at an ion guide pressure of 60 mbar, and the ion of interest was nickel. Additionally, two gap sizes were studied, 1 mm and 1 cm. In parallel, experiments were performed using resonant laser ionization of nickel, in source, using the same parameters as detailed in the simulations.

Figure 14 shows the results of both simulations and experiments for He and Ar at both gap sizes, indicated by the symbols and lines in the figure legend. The ion yield is normalized to the number of ions obtained with a suppression potential of 0 V. In each case the voltages needed for a complete suppression of laser-ionized nickel are considerably higher than 1 V which would represent the kinetic energy of the Ni^+ ions without buffer gas collisions. As expected, the argon buffer gas can more effectively push the Ni^+ ions over the potential barrier than helium, even though the flow velocity of argon is a factor of two lower than helium. This is due to both the heavier mass of an argon atom, and to the larger van der Waals radius resulting in a collision cross section approximately 30% higher. The most significant discrepancy between simulation and experiment is with the 1 mm gap result, illustrated in the right-hand panel. This can be understood as arising from a combination of an underestimation of the pressure downstream of the gas jet as well as an incomplete understanding of the gas flow velocity vectors [36]. Further simulations suggest that the ion suppression may be more effective at lower repelling voltages if it is done perpendicular

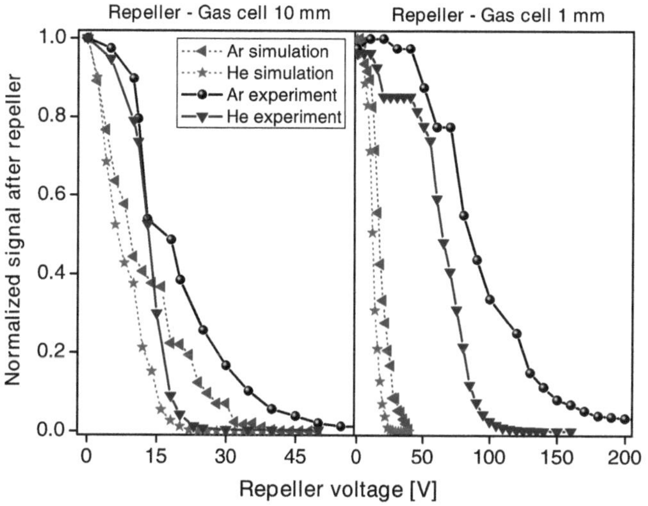

Fig. 14. Experimental and simulated laser-ionized nickel ion count rate as a function of repelling voltage for argon and helium buffer gas with a 1 cm and 1 mm gap between the ion guide and repeller electrode.

to the optical axis. Such a transverse ion collector system has been successfully used at the LISOL facility in the dual chamber gas cell, which separates the stopping- and laser-ionization volumes. This allows an electrical field to be applied to the ion collector, located behind the exit hole of the gas cell in the laser-ionization volume, in the presence of the primary beam passing through the stopping chamber [4]. This concept has been also been developed for the FURIOS laser ion source and first results from the dual chamber gas cell will be discussed elsewhere in this Portrait.

5 Summary and outlook

In summary, we have presented an overview of recent laser ion source activities immediately prior to the current IGISOL shutdown which commenced in 2010. The laser hardware developments described in this article have improved the capabilities of the FURIOS laser system for the use not only in the laser ion source, but also as a sensitive and versatile spectroscopic tool. Intra-cavity second-harmonic generation has led to a greatly increased higher-harmonic laser power which will directly benefit the saturation of atomic transitions in atom-loaded buffer gas following transportation of the laser radiation over almost 15 m. The commissioning of a new Ti:sapphire laser based upon the use of a diffraction grating for frequency selection has dramatically improved the wide-range tunability of the laser system resulting in an ideal tool for spectroscopy and the development of new excitation schemes. In order to bridge the gap between high-resolution spectroscopy and highly efficient photoionization, a novel injection-locked Ti:sapphire laser resonator has been developed in collaboration with the University of Mainz. In the current design the laser radiation from a narrow-linewidth extended cavity diode laser is amplified in a pulsed resonant optical amplifier and has resulted in a measured laser bandwidth of ~ 20 MHz. A benchmark experiment to resolve the hyperfine components of ^{27}Al has been successfully performed. In the near future a new injection-locked laser will be built at the University of Jyväskylä, improving the existing design by replacing the diode laser with a CW Ti:sapphire laser to better match the wavelength tuning range of the current pulsed laser system.

The application of resonance ionization spectroscopy within the buffer gas cell has and will continue to be an important programme of research in the future. With minor changes to the operation of a laser ion source, sensitive spectroscopy can be performed in source on low yields of short-lived radioactive isotopes and isomers. A study of a number of elements including silver, nickel and bismuth has resulted in a more complete understanding as to the role of different laser bandwidths, environmental broadening mechanisms (Doppler, pressure) and laser power broadening which all convolute to the experimental width of the measured spectral line. The prospect of utilizing the expanding gas jet environment following the ion guide is highly attractive, Doppler and pressure broadening effects being substantially reduced. First proof-of-principle experiments have taken place using nickel as a source of atoms and a measurement of the Doppler shift of the moving atoms ionized in the jet has been made and compared to atoms in the gas cell or reference atomic beam unit operating in vacuum. Although the velocity distribution in the gas jet must be taken into account this effect may be reduced by a dedicated optimization of the jet.

Indeed, as part of the Laser Ion Source Trap (LIST) programme at the IGISOL facility, such jet optimization is underway. The LIST method offers a novel means with which to attain ultra-high selectivity in radioactive ion beam production. The current limitation in comparison to a more standard ion guide-based laser ion source is a poor geometrical overlap efficiency between the expanding gas jet (and thus atoms of interest) and the laser photons. This limitation has recently been overcome with the application of a Laval nozzle to the gas cell. Complications are associated with the fabrication of such nozzles and they are designed to work best at only one gas cell-to-background pressure ratio. Despite these challenges, the first design applied at IGISOL has resulted in a gas jet retaining a remarkably collimated structure as far as 14 cm downstream from the nozzle.

Finally, a new era of laser ion source developments and in-source/in-jet spectroscopic measurements will begin with the completion of the new IGISOL-4 facility. Detailed planning for the laser laboratory and laser beam transport has been undertaken in the initial stage, overcoming the shortfalls in the old laboratory which resulted in complications associated with a long, complex transport path to the gas cell. A full exploitation of the gas jet for spectroscopy as well as the LIST technique will commence with the addition of a new 15 degree electrostatic bender, allowing direct access of laser radiation towards the ion guide nozzle without the current prohibitive losses experienced via the mass separator window. Ongoing developments in gas cell technology, including the commissioning of the dual chamber gas cell, and the use of a novel radiofrequency inductively heated hot-cavity graphite catcher, will exploit the unique opportunities afforded by the new facility. The availability of primary beams from both the $K = 130$ cyclotron and the new MCC 15/30 cyclotron will result in a manyfold increase in the available beam time for development and exploitation, and with the expecting increase in intensities the laser ion source will be a vital tool to provide a complimentary increase in selectivity.

This work has been supported by the Academy of Finland under the Finnish Center of Excellence Program 2006-2011 (Nuclear and Accelerator Based Physics Program at JYFL).

References

1. T. Kessler, I.D. Moore, Y. Kudryavtsev *et al.*, Nucl. Instrum. Methods Phys. Res. B **266**, 681 (2008).
2. I.D. Moore, T. Kessler, T. Sonoda *et al.*, Nucl. Instrum. Methods Phys. Res. B **268**, 657 (2010).
3. I.D. Moore, A. Nieminen, J. Billowes *et al.*, J. Phys. G: Nucl. Part. Phys. **31**, S1499 (2005).
4. Y. Kudryavtsev, T. Cocolios, J. Gentens *et al.*, Nucl. Instrum. Methods Phys. Res. B **267**, 2908 (2009).
5. I.D. Moore, J. Billowes, P. Campbell *et al.*, AIP Conf. Proc. **831**, 511 (2006).
6. M. Reponen, I.D. Moore, I. Pohjalainen *et al.*, Nucl. Instrum. Methods Phys. Res. A **635**, 24 (2011).
7. W.-L. Zhou, Y. Mori, T. Sasaki *et al.*, Appl. Phys. Lett. **66**, 2463 (1995).
8. A. Ferguson, M. Dunn, A. Maitland, Opt. Commun. **19**, 10 (1976).
9. J. Lassen, P. Bricault, M. Dombsky *et al.*, AIP Conf. Proc. **1099**, 769 (2009).
10. T. Kessler, *Development and application of laser technologies at radioactive ion beam facilities*, Ph.D. thesis, University of Jyväskylä (2008).
11. E. Otten, D. Bromley, *Nuclear Radii and Moments of Unstable Nuclei*, Vol. **8** (Plenum Press, New York, 1988).
12. J. Billowes, P. Campbell, J. Phys. G: Nucl. Part. Phys. **21**, 707 (1995).
13. B. Cheal, K.T. Flanagan, J. Phys. G: Nucl. Part. Phys. **37**, 113101 (2010).
14. R. Neugart, W. Klempt, K. Wendt, Nucl. Instrum. Methods Phys. Res. B **17**, 4, 354 (1986).
15. T. Kessler, H. Tomita, C. Mattolat *et al.*, Laser Phys. **18**, 842 (2008).

16. T. Gottwald, C. Havener, J. Lassen *et al.*, Rev. Sci. Instrum. **81**, 02A514 (2010).
17. P. Smith, C. Heise, J. Esmond *et al.*, *Kurucz database*, `http://www.pmp.uni-hannover.de/cgi-bin/ssi/test/kurucz/sekur.html`.
18. Yu. Ralchenko, A.E. Kramida, J. Reader, *NIST Database*, `http://www.nist.gov/pml/data/asd.cfm`.
19. S. Beiersdorf, W. Heddrich, K. Kesper *et al.*, J. Phys. G: Nucl. Part. Phys. **21**, 215 (1995).
20. T.E. Cocolios, A.N. Andreyev, B. Bastin *et al.*, Phys. Rev. Lett. **103**, 102501 (2009).
21. R. Horn, *Aufbau eines Systems gepulster, abstimmbarer Festörpelaser zum Einzats in der Resonanzionisations-Massenspektrometrie*, Master's thesis, Johannes Gutenberg University Mainz (2003).
22. T. Sonoda, T. Cocolios, J. Gentens *et al.*, Nucl. Instrum. Methods Phys. Res. B **267**, 2918 (2009).
23. B. Dai, P. Lambropoulos, Phys. Rev. A **34**, 3954 (1986).
24. M. Reponen, T. Kessler, I.D. Moore *et al.*, Eur. Phys. J. A **42**, 509 (2009).
25. J. Olivero, R. Longbothum, J. Quant. Spectrosc. Radiat. Transfer **17**, 233 (1977).
26. S. Cavalieri, R. Eramo, Opt. Commun. **118**, 245 (1995).
27. M. Bissell, K. Flanagan, M. Gardner *et al.*, Phys. Lett. B **645**, 330 (2007).
28. I.D. Moore, T. Kessler, J. Äystö *et al.*, Hyperfine Interact. **171**, 135 (2006).
29. B. Tordoff, *Development of Resonance Ionization Techniques at the Jyväskylä IGISOL*, Ph.D. thesis, University of Manchester (2007).
30. K. Blaum, C. Geppert, H.J. Kluge *et al.*, Nucl. Instrum. Methods Phys. Res. B **204**, 331 (2003) (14th International Conference on Electromagnetic Isotope Separators and Techniques Related to their Applications).
31. K. Wendt, T. Gottwald, D. Hanstorp *et al.*, Hyperfine Interact. **196**, 151 (2010).
32. F. Schwellnus, K. Blaum, R. Catherall *et al.*, Rev. Sci. Instrum. **81**, 02A515 (2010).
33. P. Karvonen, I.D. Moore, T. Sonoda *et al.*, Nucl. Instrum. Methods Phys. Res. B **266**, 4794 (2008).
34. B. Marsh, *In-Source Laser Resonance Ionization at ISOL Facilities*, Ph.D. thesis, University of Manchester (2007).
35. K. Wendt, K. Blaum, K. Brück *et al.*, Nucl. Phys. A **746**, 47 (2004) (Proceedings of the Sixth International Conference on Radioactive Nuclear Beams (RNB6)).
36. I. Pohjalainen, *Experimental gas jet studies for the IGISOL LIST method and simulation modeling*, Master's thesis, University of Jyväskylä (2010).
37. S.I.S Inc., Simion 8.0, 1027 Old York Rd., Ringoes, NJ, USA.

DOI 10.1007/978-94-007-5555-0_24
Reprinted from *Hyperfine Interactions* Volume 223, Issues 1-3, 2014
and *The European Physical Journal A* Volume 48, Issue 4, 2012

THE EUROPEAN PHYSICAL JOURNAL A

Regular Article – Experimental Physics

The search for the existence of ^{229m}Th at IGISOL

V. Sonnenschein[1,a], I.D. Moore[1,b], S. Raeder[2], A. Hakimi[2], A. Popov[3], and K. Wendt[2]

[1] Department of Physics, University of Jyväskylä, P.O. Box 35 (YFL), FI-40014 Jyväskylä, Finland
[2] Institute of Physics, University of Mainz, Staudingerweg 7, 50099 Mainz, Germany
[3] Petersburg Nuclear Physics Institute, Gatchina, St-Petersburg 188350, Russia

Received: 30 March 2012
Published online: 23 April 2012 –
Communicated by J. Äystö

Abstract. An overview of preparatory work aiming at the identification of the low-lying 7.6 eV isomer in ^{229}Th through a measurement of its hyperfine structure is presented. A ^{233}U recoil gas cell has been developed and has undergone several iterations in order to improve the efficiency of extracting a low-energy beam of ^{229}Th$^+$ ions. Spectroscopic studies on stable ^{232}Th have been carried out to establish an efficient laser ionization scheme. The latter will be applied in connection with the gas cell in order to improve the extraction efficiency by accessing the neutral atomic fraction. In addition to that of the ^{229}Th ground state, the hyperfine structure of the isomer can then be determined either by collinear laser spectroscopy on the extracted and accelerated ions or by direct in-source/in-jet high-resolution Resonance Ionization Spectroscopy (RIS). The first experiments using high-resolution RIS have been performed and isotope shifts of ^{228}Th, ^{229}Th and ^{230}Th relative to ^{232}Th were measured on an atomic $7s^2 \to 7s7p$ ground-state transition at 380.42 nm. A template of the ground-state hyperfine structure of ^{229}Th has been established for a 261.24 nm UV transition. This is an important step towards identification of the isomeric state.

1 Introduction

The existence of a low-lying isomeric state at an energy of 7.6 ± 0.5 eV in ^{229}Th, $I = \frac{3}{2}^+$, was inferred from high-resolution γ-ray spectroscopy experiments using a differencing technique of the resulting decay paths [1]. This is by far the lowest excitation energy of any nuclear state known, lying in the range of typical atomic excitation and ionization energies. Direct observation of the decay of the isomer, including its M1 transition to the nuclear ground state, is regarded as an experimental challenge of the highest importance but unfortunately has thus far been unsuccessful. The isomeric half-life is theoretically estimated to be between 1 to 5 hours by extrapolation from a comparable transition of the ^{233m}U isomer connecting the same Nilsson states but with an energy difference of 312 keV [1], as well as from evaluations of the nuclear matrix element based on another M1 transition in the ^{229}Th nucleus [2]. Most recently, Tkalya has revised his theoretical half-life estimate to be $\sim$ 25 min following the new estimate for the energy of the isomeric state [3].

Aside from the fundamental nuclear structure interest which includes a measurement of the mean-square charge radius and electromagnetic moments, the low-lying isomer has attracted much attention in recent years due to the exciting possibility that a relatively simple table-top laser system might be able to access the required transition wavelength of about 160 nm in order to directly probe the isomeric transition. This would provide a unique opportunity for fundamental research on laser-optical atom-to-nucleus coupling, leading to a number of interesting applications. For example, proposals detail the potential use of the ^{229m}Th isomer as a nuclear time standard of highest precision [4]. This so-called nuclear clock would exhibit a number of advantages compared with current state-of-the-art atomic clocks. Line shifts induced by any external fields would be dramatically reduced by the electron screening of the nucleus. The currently estimated half-life of the isomeric state would result in a linewidth of less than 1 mHz, leading to a theoretical precision of $\Delta\lambda/\lambda \approx 10^{-19}$, about two orders of magnitude more precise than current high-precision clocks. In addition, the strong dependence of the transition on the balance of individual fundamental forces within the nucleus could yield an amplification of possible frequency shifts induced by a variation of the electromagnetic coupling constant α with time [5], or of the constituent quark masses [6], by several orders of magnitude. In this manner, long-term comparisons of a hypothetical ^{229m}Th clock to a conventional atomic clock could stimulate the ongoing highest priority research on the cosmological evolution of the fundamental force coupling constants. Furthermore, other proposals

[a] e-mail: volker.t.sonnenschein@jyu.fi
[b] e-mail: iain.d.moore@jyu.fi

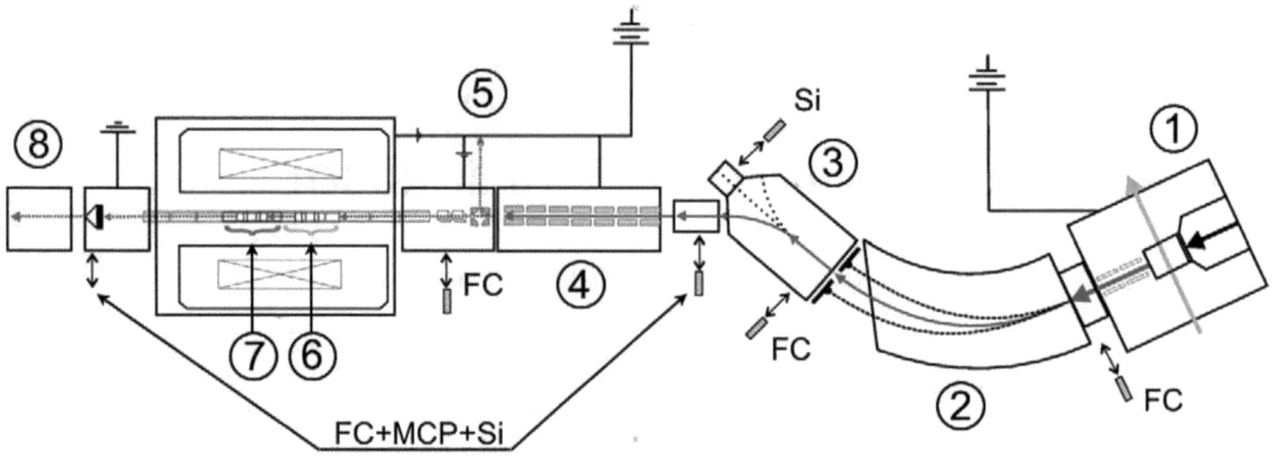

Fig. 1. (Colour on-line) Schematic representation of the IGISOL-3 facility as utilized in the preparation of the low-energy ion beam of ^{229}Th. The labels are as follows: ① Target chamber (the cyclotron beam is represented by the blue arrow, however it was not used in the work discussed here). ② 55° dipole magnet. ③ Electrostatic switchyard. ④ RF cooler/buncher. ⑤ Beam line to collinear laser spectroscopy station. ⑥ Preparation trap. ⑦ Precision trap. ⑧ Post-trap setups. Beam diagnostic elements are labeled as FC (Faraday cup), MCP (micro-channel plate) and Si (Si detector).

suggest the use of the isomer as a qubit for quantum computing [4], as stepping stone towards a γ-ray laser [3], as a system in which tests of the effect of the chemical environment on nuclear decay rates can be studied [7], or for investigations on NEET (Nuclear Excitation by Electron Transition) [8], a process which could help in pumping the isomeric state from the ground state.

An alternative approach to a direct measurement of the decay of the low-lying isomeric state was proposed in Jyväskylä a number of years ago, involving a measurement of its unique hyperfine structure with respect to the known structure of the nuclear ground state [9]. In this article we describe the evolution of a gas cell-based concept for the production of a low-energy (several tens of keV) ion beam of ^{229}Th at the Ion Guide Isotope Separator On-Line (IGISOL) facility of the University of Jyväskylä, to be used in a future laser spectroscopic investigation of the isomer. The advantage of the IGISOL method lies in the manipulation of the ions in a gas media immediately following the moment of production. This may alleviate problems in searching for the isomer in atomic form as the excitation energy of the isomeric state is currently higher than the first ionization potential of neutral thorium and thus the lifetime of the state may be considerably reduced due to internal conversion.

In addition, we present the parallel activity at the University of Mainz, Germany, to develop a suitable laser excitation and ionization scheme which could be used to enhance the selectivity and efficiency of the gas cell for the production of a radioactive ion beam of thorium. Finally, we provide an overview of recent resonance ionization spectroscopy on the ground-state structure of ^{229}Th, performed in Mainz. This is an important step towards the search for the isomeric state using either in-source or in-jet resonance ionization spectroscopy at the new IGISOL-4 facility in the future. An unambiguous identification of the existence of the isomer via laser spectroscopy would be a milestone towards any possible future applications.

2 Development of a low-energy ion beam of ^{229m}Th

The IGISOL method, developed in the Accelerator Laboratory of the University of Jyväskylä in the early 1980s, is described in detail in [10,11]. A thin (typically a few mg/cm^2) target is bombarded by an energetic primary beam. The reaction products recoil out from the target and are thermalized in a noble gas (usually He) with a typical pressure of a few hundred mbar. Due to the high ionization potential of helium, the majority of the reaction products end up in a charge state of 1^+ and are subsequently extracted from the stopping volume by gas flow. The ions are guided into a differentially pumped electrode structure either via a skimmer electrode (used in all experiments discussed in this article) or via a sextupole RF ion guide (SPIG) [12] before being accelerated to nominally 30 kV potential towards the mass separator. Following acceleration, the ions are separated by their mass-to-charge ratio with a 55° dipole magnet with a typical mass resolving power (MRP) $\frac{M}{\Delta M} = 500$, whereby the isobar is delivered to an electrostatic switchyard for analysis or is prepared for experiments with further ion manipulation devices. In general, the yield of activity produced from the IGISOL system is measured at the focal plane of the mass separator using a Si-detector setup (with 33% absolute efficiency) or multi-channel plates. For activities with particularly long half lives, for example, ^{229}Th ($\tau_{1/2} =$ 7880 years), particle identification may be carried out by isobaric mass separation using the JYFLTRAP Penning trap [13], situated after the rf-quadrupole cooler [14] as illustrated in fig. 1.

The ground and isomeric state of ^{229}Th may be produced at IGISOL either in a direct nuclear reaction, for example, via a light-ion induced fusion-evaporation reaction such as ^{232}Th(p,p3n)^{229g,m}Th, or via the α-decay of ^{233}U, whereby approximately 2% of the decay is expected to populate the isomer [15]. In the former method the population of the ground and isomeric states is expected to be rather similar, however the drawback could be the severe plasma effect caused by the competing fission channel as well as by the presence of primary beam in the gas cell. Plans to pursue this production mode exist at JYFL in connection with a dual chamber gas cell based on a design from Louvain-la-Neuve [16]. In the following, we summarize the latter approach of producing a low-energy ion beam of ^{229g,m}Th, initiated in a gas cell development programme in 2005.

2.1 The ^{229}Th source and a summary of the first production tests

A source of ^{233}U was purchased from Isotope Products Laboratory, Burbank, USA by McGill University, produced through the neutron capture of naturally abundant ^{232}Th via the reaction $^{232}\mathrm{Th}(\mathrm{n},\gamma)^{233}\mathrm{Th} \rightarrow {}^{233}\mathrm{Pa} \rightarrow {}^{233}\mathrm{U}$, in which the decays from ^{233}Th proceed via beta emission to ^{233}U. At McGill, the source was dissolved in isopropanol and via electrolysis was electroplated onto 12 stainless-steel strips which were shipped to the University of Jyväskylä. The activity of the source, stated by the supplier, was 10^6 Bq. Through a subsequent measurement of the α-spectrum the number of ^{229}Th recoils was estimated to be $\sim 4 \times 10^5\,\mathrm{s}^{-1}$ [17].

The initial ion guide design and testing has been discussed thoroughly in [18] and will therefore only be summarized briefly. The geometry of the gas cell was chosen in such a way as to accomodate the electroplated strips along the inside surface and to maximize the stopping and subsequent extraction of thorium recoils. Due to the large volume in which the recoils are stopped (the gas cell volume is $\sim 1200\,\mathrm{cm}^3$), gas flow alone is an inadequate means of transporting the recoils to the exit hole and thus the extraction end of the gas cell was designed to be adaptable to different electric field arrangements. Different extraction methods were tested via measurements of the gas cell efficiency using an α-recoil source of ^{223}Ra. Following the alpha decay of ^{223}Ra, recoils of ^{219}Rn are ejected from the source, extracted from the gas cell, accelerated and mass separated. Alpha decays from the daughter of ^{219}Rn, ^{215}Po, are detected after implantation into a foil in front of a silicon detector mounted in the focal plane of the 55° mass separator, illustrated in fig. 1. The efficiency is determined by comparing the detected activity with the total number of α-recoils emanating from the source.

These measurements led to the choice of an electron emitter as a means of providing electric-field guidance and thus efficient ion extraction. The principle of the emitter is rather simple: a circular focussing electrode followed by a loop of tungsten wire doped with 1% of thorium (W(Th)) is mounted close to the exit nozzle. By passing an electric current through the wire it can be Joule heated to a temperature of $\sim 1800\,^\circ$C, resulting in the emission of electrons. The focussing electrode guides the electrons away from the exit nozzle region and into the gas cell, creating a potential difference inside the chamber. Ions are then drawn through the exit hole due to the potential gradient and are extracted to form a beam.

Following this, a study of mass-separated alpha spectra from decays in the ^{233}U chain was performed. Alpha peaks attributed to ^{221}Fr ($\tau_{1/2} = 4.9$ min) and ^{217}At ($\tau_{1/2} = 32.3$ ms) were observed and through calibration of the source strength resulted in a gas cell efficiency of $6 \pm 0.5\%$ for both nuclei. With a ^{229}Th half life of 7880 years, particle identification was carried out with the JYFLTRAP Penning trap mass spectrometer [13]. Only the purification trap was used and a mass resolving power ($\mathrm{m}/\Delta\mathrm{m}$) of 23,500 was obtained for $^{229}\mathrm{Th}^+$. Accounting for transmission losses between the ion guide and Penning trap, the gas cell efficiency for $^{229}\mathrm{Th}^+$ was estimated to be 0.06% of the source strength, two orders of magnitude lower than for $^{221}\mathrm{Fr}^+$ and $^{217}\mathrm{At}^+$ [18]. In that work, the discrepancy between the gas cell efficiency of $^{229}\mathrm{Th}^+$ and that for the daughter activities of $^{221}\mathrm{Fr}^+$ and $^{217}\mathrm{At}^+$ was attributed to a substantial fraction of thorium extracted from the gas cell in molecular form (namely $^{229}\mathrm{Th(H_2O)}_2^+$ and $^{229}\mathrm{ThO}^+$ in a relative abundance to $^{229}\mathrm{Th}^+$ of 55% and 27%, respectively) and to an unknown neutral fraction.

In summary, the first gas cell design resulted in the successful production of a low-energy beam of $^{229}\mathrm{Th}^+$ however impurity molecules limited the fraction available in atomic form, favouring the formation of water and oxide compounds. Delivery of the ion beam to experimental setups, for example, to the collinear laser spectroscopy station for a high-resolution measurement of the hyperfine structure, is also reduced by the transmission efficiency of the IGISOL system. For example, in ref. [18], the electrostatic switchyard to Penning-trap transmission efficiency of $^{229}\mathrm{Th}^+$ was determined to be $\sim 30\%$, dependent on the formation of molecules inside the rf-quadrupole cooler and thus sensitive to the bunching time. The main molecule formed in the cooler is $^{229}\mathrm{ThO}^+$, primarily arising from oxygen present in the system after the breakup of water molecules in the rf field. By combining the gas cell efficiency of 0.06% and the typical rf cooler transmission efficiency, a rate of $50\,\mathrm{s}^{-1}$ $^{229}\mathrm{Th}^+$ ions is available for study. Assuming a 2% branching ratio to the isomeric state, this ion rate is prohibitively low for a collinear laser spectroscopic measurement. To address this, new gas cell designs were proposed in order to improve the efficiency.

2.2 Towards a higher gas cell efficiency for the production of $^{229}\mathrm{Th}^+$

In early 2007 a new gas cell design was developed, one which was optimized in order to better suppress impurities as well as to maintain an efficient extraction electric field (as previously, the extraction side of the gas cell is adaptable to a variety of devices). In the first design (ref. [18]),

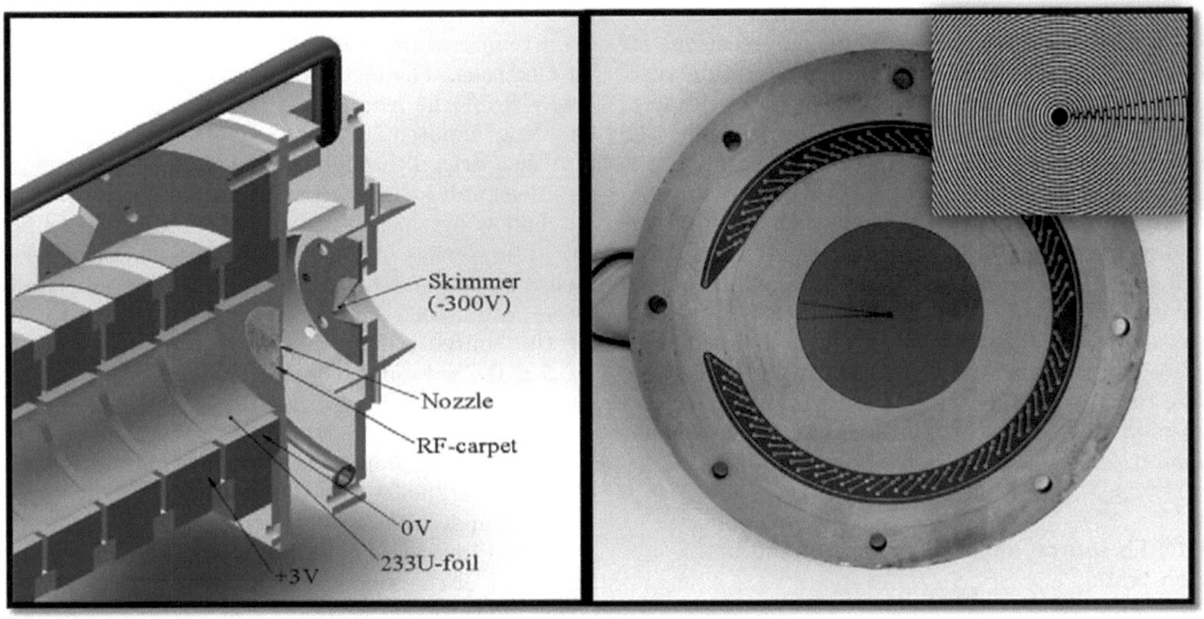

Fig. 2. (Colour on-line) Schematic illustration of the new gas cell coupled to an rf carpet. Typical operating potentials for the skimmer electrode and the final axial segments are given, those for the rf carpet are discussed in the text. The grey tube is used for circulating liquid nitrogen. The right panel shows a photograph of the front face of the rf carpet with the inset showing a closer view of the electrode structure.

the main body of the gas cell was constructed from a single piece of machined aluminium. In order to improve the gas cell efficiency, longitudinal dc field guidance was required along the axis of the cell and thus the body was replaced with a number of segmented stainless-steel rings, separated by ceramic insulators. A resistor chain network linking the segmented rings was used to provide the axial dc field. All pieces were carefully cleaned in an ultrasonic bath before being assembled. Additional in-situ baking of the gas cell can be performed via a set of heater lamps and for further suppression of impurities, it is possible to cool the gas cell at the front and rear with circulating liquid nitrogen. It was also realized that due to the high temperatures involved in the use of the electron emitter, molecular impurities are likely to be emitted from both the tungsten wire and surrounding chamber walls. By avoiding the use of the emitter this problem can be solved.

2.2.1 Gas cell coupled with a radiofrequency (rf) carpet

It is well known that a simple static electric field guidance in a gas cell would likely result in losses of the ions as they follow the line of the electric force until termination on the gas cell exit wall. If the gas flow is fast enough to overcome the electric force at the nozzle then this problem may be alleviated. A novel method to reduce such losses utilizes a fine electrode structure (carpet or funnel) in which a radiofrequency field applied to a series of rotationally symmetric ring electrodes produces an effective electric field which drives ions away from the electrodes. A superimposed dc field guides the ions towards the exit nozzle whereby gas flow extraction takes over. This so-called rf ion-guide system was originally pioneered by the group of Wada and colleagues at the Atomic Physics Laboratory in RIKEN, Japan [19]. By coupling this technology to a large gas catcher, Wada showed that it was possible to convert an energetic radioactive ion beam from a projectile fragment separator into a low-energy beam required for experiments such as collinear laser spectroscopy or ion trapping.

An rf carpet was therefore designed and built for the thorium gas cell, replacing the electron emitter in order to reduce possible impurities caused by the high temperature of the tungsten filament. Figure 2 illustrates the coupling of the carpet to the extraction side of the gas cell. The segmented structure of the body of the cell is clearly visible and the position of the electroplated uranium foils is shown. A skimmer electrode forms an intrinsic part of the extraction system. The carpet consists of 70 ring electrodes with a 0.3 mm pitch and an exit hole 1.2 mm in diameter. The rf electrodes consist of a layer of planar printed circuit board (teflon base and gold plating), with the rf divider circuits soldered to the back surface. The right hand panel of fig. 2 shows a photograph of the rf carpet surface which faces towards the gas cell. Typical rf operating parameters for the extraction of $^{217}At^+$ ions were 2.1 MHz and ~ 60 V_{pp}. A voltage of ~ 20 V dc was applied across the carpet.

The first tests with the rf carpet coupled to the gas cell concentrated on collecting mass spectra from different mass regions in order to probe the chemistry of thorium and the final charge state distribution. An example of the mass spectra obtained via ion counting using a set of micro-channel plates mounted in front of the rf

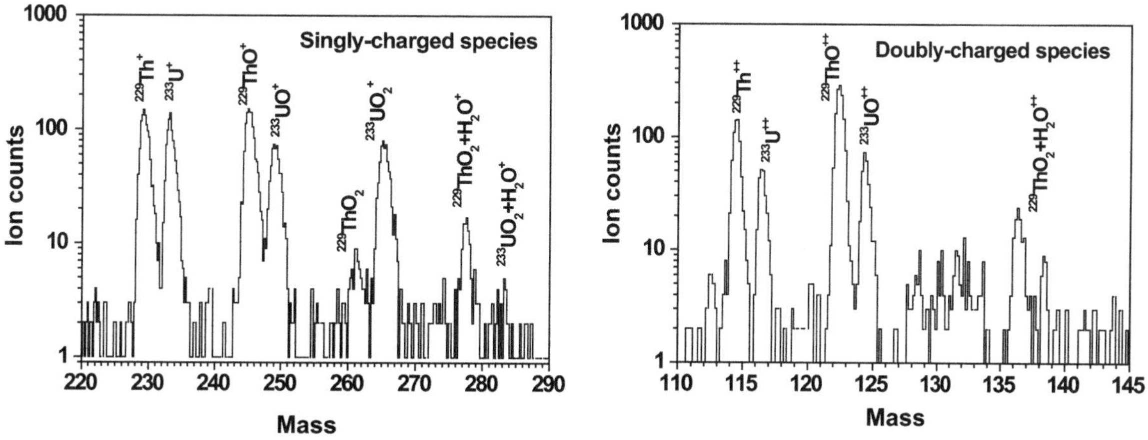

Fig. 3. Mass spectra of the singly and doubly charged regions of ^{229}Th using the gas cell coupled to the rf carpet. Uranium is produced directly from the electroplated source strips. The similar ion chemistry of thorium and uranium is apparent in the molecular distribution.

cooler-buncher (fig. 1) is shown in fig. 3. These results were collected following cooling of the gas cell to approximately $-70\,°$C and illustrate comparable rates of singly- and doubly charged thorium and uranium (the latter recoils directly from the source). The formation of oxygen and water-based molecular adducts visible in the mass spectra is likely to arise from impurities in the helium buffer gas. We note that in almost all experiments at IGISOL, grade 4.6 (99.996%) purity helium gas is used in combination with a cold trap made of activated carbon, cooled with liquid nitrogen, to remove impurities. Thorium and uranium are both particularly reactive elements and exhibit a very similar ion chemistry as reflected by the molecular distribution.

The gas cell was also baked *in situ* with the heater lamps to $\sim 100\,°$C (the teflon carpet restricted the use of higher temperatures). Interestingly, the doubly charged atomic species showed count rates over an order of magnitude greater than the singly charged atomic species. This reflects the improvement in the purity of the gas cell system. For most elements (including thorium) the second ionization potential is lower than the first ionization potential of helium (24.6 eV). Therefore, in a pure atomic helium gas, ions can only reach a 2^+ charge state in collisions with helium atoms. However, impurities such as O_2, N_2 and H_2O often present in the system lead to charge transfer reactions with doubly charged ions.

Finally, the gas cell efficiency was compared to that of the original electron emitter design discussed in sect. 2.1. Following analysis of the alpha peak attributed to $^{217}At^+$, an efficiency of $\sim 0.6\%$ was obtained. This is one order of magnitude lower than the earlier design which resulted in an efficiency of $6 \pm 0.5\%$ for both $^{221}Fr^+$ and $^{217}At^+$. On the other hand, the efficiency of the gas cell for $^{229}Th^+$ was increased to 0.36%, a factor of six improvement over the original design. This result reflects a better control over the level of impurities. The distribution of thorium extracted in singly and doubly charged form is an issue which has yet to be addressed.

2.2.2 Gas cell coupled with an electron emitter

Due to the rather low gas cell efficiency for the extraction of $^{217}At^+$ (even though an improvement in the extraction of $^{229}Th^+$ was observed), a third gas cell iteration was tested in early 2008, illustrated in a schematic drawing in fig. 4. This combined the original electric field extraction provided by the electron emitter with the new, segmented stainless-steel body. Water cooling was used to remove the heat generated by the hot W(Th) filament, resulting in a measured temperature of 40–60 °C on the front electrode of the cell body during experiments. An additional heating of up to 150 °C was available from the heat lamps in order to bake the cell, the temperature restriction having been relaxed following removal of the rf carpet. An α-recoil source of ^{223}Ra was mounted between two of the gas cell axial electrodes towards the helium inlet side of the chamber.

In a similar manner to the studies using the rf carpet gas cell, mass spectra were taken and analyzed in order to understand the effects of ion chemistry and charge state distribution. Unlike the earlier studies of Tordoff and collaborators [18], water adducts were not observed. A dedicated effort was subsequently made to study the thorium charge state dependence as a function of the electron emitter current, summarized in fig. 5. The results are compiled from two sets of data taken two months apart. In addition, results obtained from the rf carpet gas cell have been included to provide a reference measurement at zero emitter current. All data were obtained at a gas cell pressure of 25 mbar unless otherwise stated, however we noted that the ratio of doubly charged to singly charged ions of ^{229}Th was not influenced by the gas pressure. The solid line is a fit of the data using a simple decay function and has been drawn to guide the eye. The results in fig. 5 suggest that the emitter current may be used to tune the charge state distribution however it is likely that the gas cell efficiency will to some extent always be distributed between the singly and doubly charged states of ^{229}Th.

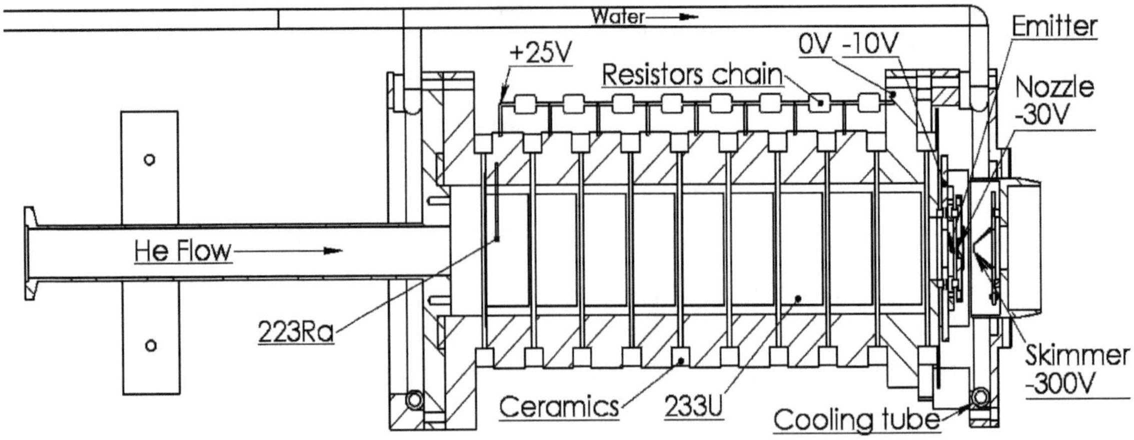

Fig. 4. A schematic drawing of the third iteration of gas cell design. An axial dc electric field is applied to the segmented body of the thorium gas cell, coupled to an electron emitter-type field extraction. Relevant dc potentials are shown. The ions are extracted through the nozzle on the right and are accelerated through the skimmer electrode into the IGISOL mass separator.

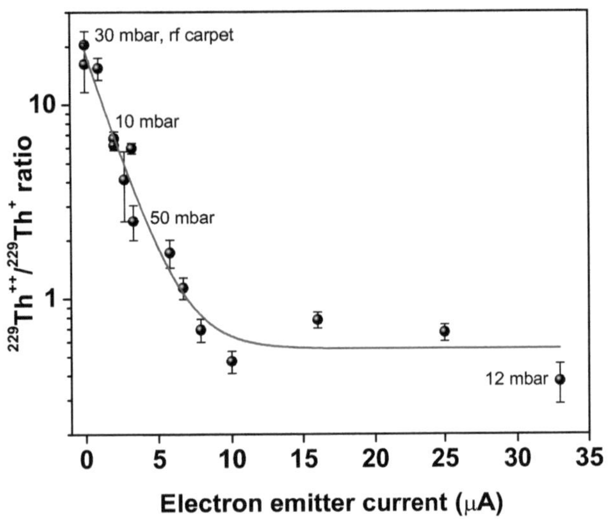

Fig. 5. Charge state ratios of $^{229}Th^{2+}/^{229}Th^{+}$ as a function of electron emitter current. All results were obtained with 25 mbar He pressure unless otherwise stated. Data plotted from the rf carpet gas cell represent a reference measurement at zero emitter current.

During the same experiment the ratio of ThO^+/Th^+ was measured as a function of emitter current, and was found to be relatively stable over a range from 2 to 25 μA. Therefore, the steady reduction in the charge state ratio as a function of filament current cannot be caused by any impurity variation arising from additional heating and is likely to be due to an increase in the electron density. Additionally, the elemental dependence of the charge state distribution was briefly checked. At a current of 2.0 μA, the ratio of $^{229}Th^{2+}/^{229}Th^{+} \sim 6.2$ whereas that of $^{233}U^{2+}/^{233}U^{+} \sim 1.1$. At a current of 0.9 μA the ratio of $^{229}Th^{2+}/^{229}Th^{+} \sim 15.5$ and that of $^{233}U^{2+}/^{233}U^{+} \sim 1.7$. In the future, further systematic testing will be performed in order to expand the study to other available activities including ^{233}U, ^{221}Fr, ^{219}Rn and ^{217}At.

With an electron emitter current of 16 μA, a measurement of the yields of $^{221}Fr^+$ and $^{217}At^+$ at the focal plane of the IGISOL mass separator resulted in a gas cell efficiency of $\sim$ 16% for both nuclei (this can be compared to $\sim$ 6% obtained with the previously published gas cell design [18]). Using the Penning trap and accounting for transmission losses, the gas cell efficiency for $^{229}Th^+$ was estimated to be 1.6% of the source strength. This can be directly compared with the original gas cell efficiency of 0.06% and with the efficiency of 0.36% obtained with the current gas cell design coupled to an rf carpet.

2.3 An overview of the three gas cell geometries and current status

Table 1 provides a summary of the results obtained from the three gas cell iterations discussed in the current article, detailing the main design features and the method of ion extraction. As can be seen, the final iteration of gas cell design (gas cell C) provides the highest efficiency for extracting a beam of $^{229}Th^+$. This efficiency translates into a usable $^{229}Th^+$ ion yield of $\sim 2400\,s^{-1}$ and assuming a 2% branching into the isomeric state, a $^{229m}Th^+$ isomeric yield of $\sim 50\,s^{-1}$. In comparison to the extraction of the daughter activities of $^{221}Fr^+$ and $^{217}At^+$, the $^{229}Th^+$ gas cell efficiency is currently a factor of 10 lower. It is known that this is partly due to the mixture of charge state distribution and to molecular formation with residual oxygen in the gas.

With the recent construction of the new IGISOL-4 facility, the gas handling system is undergoing a complete reconstruction. All gas lines are being built out of polished stainless steel and VCR fittings will replace the previous swagelock connectors. A new getter purifier is planned to be used in combination with zeolite cold traps and it is expected that this will reduce the impurity level of reactive

Table 1. Gas cell efficiences for the measured decay products within the ^{233}U decay chain. A summary of the different extraction methods and gas cell designs is given.

	Gas cell A			Gas cell B			Gas cell C		
Technical design	original cell aluminium body no baking, no cooling			new cell stainless steel bakeable, liquid N_2 cooling			new cell stainless steel bakeable, water cooling		
Extraction technique	electron emitter			RF carpet, longitudinal DC			electron emitter, longitudinal DC		
Efficiency	$^{229}Th^+$	$^{221}Fr^+$	$^{217}At^+$	$^{229}Th^+$	$^{221}Fr^+$	$^{217}At^+$	$^{229}Th^+$	$^{221}Fr^+$	$^{217}At^+$
	0.06%	6%	6%	0.36%	0.6%	0.6%	1.6%	16%	16%

gases such as oxygen. Additionally, losses due to molecular formation in the rf cooler will hopefully be minimized as the gas line system is also to be upgraded in the new facility. Finally, and most importantly, it is suspected that the main cause of the difference in efficiency between $^{229}Th^+$ its α-decay daughters is due to a considerable neutral fraction. The methods developed with which to access this neutral fraction are described in the following section.

3 Establishing an efficient laser ionization scheme for enhanced production of a thorium ion beam

A high-resolution collinear laser spectroscopy experiment aimed at unfolding the hyperfine structure of the isomeric state from that associated with the ground state will be extremely challenging if the isomeric yield is limited to $\sim 50\,s^{-1}$ as currently estimated (see the discussion in the previous section). In addition, the population of the isomeric state will be distributed over several hyperfine components differing in their total angular momentum. Combined with a typical fluorescence detection efficiency of 0.02% and the background associated with laser scattered light in a collinear setup, it is highly unlikely to expect such an experiment to succeed in its current form.

There are two lines of research being pursued by our collaboration to improve the sensitivity towards a successful detection of the isomer. The first, discussed in the following, aims to use resonant laser ionization to increase the $^{229}Th^+$ ion yield from the recoil gas cell by accessing the remaining neutral fraction. Multi-step resonant excitation and ionization has been increasingly employed as a tool for selective and efficient radioactive ion beam production at on-line facilities such as ISOLDE at CERN, Switzerland, ISAC at TRIUMF, Canada and elsewhere [20]. The second project involves high-resolution resonance ionization spectroscopy (RIS) either inside the gas cell or in the expanding gas jet using an injection-locked Ti:sapphire laser system ($\sim$ 20 MHz bandwidth). In order to identify the hyperfine structure of the isomer and to distinguish it from that of the corresponding ground state, a hyperfine template of a pure ^{229g}Th source is required. The first steps in the preparation of such laser-based experiments have recently been carried out in the LARISSA laboratory at the University of Mainz, Germany. Ti:sapphire laser RIS on stable ^{232}Th has been performed in order to search for a suitable resonant laser excitation and ionization scheme [21]. In addition, a ^{229g}Th source has been acquired from the Nuclear Chemistry department in Mainz and several transitions from the atomic ground state were investigated for a possible hyperfine structure.

The technique of resonance ionization spectroscopy (RIS) has been previously applied to thorium using pulsed as well as continuous-wave dye lasers within a simple two-step excitation/ionization scheme [22,23]. Our collaboration has recently developed a laser ionization scheme involving three resonant steps, which promises comparable or even higher efficiency than the earlier work. The higher selectivity afforded by the increased number of laser excitations provides a high suppression of background from abundant uranium, which could be produced either by non-resonant ionization or via fragmentation of UO induced by any blue to UV laser radiation.

Information about atomic level and transition data, especially close to the ionization potential (IP) of thorium at $50867\,cm^{-1}$ [24] is still fragmentary and was insufficient to initially build a three-step scheme. Direct searches for unknown intermediate as well as autoionizing (AI) states were performed using a set of three tunable Ti:sapphire lasers pumped by a frequency doubled Nd:YAG laser with a repetition rate of 10 kHz. Two of the lasers are equipped with a combination of birefringent filter and etalon for frequency selection, resulting in a laser line narrowing to a spectral bandwidth of $\sim$ 5 GHz. The third laser had a Littrow-configured grating mounted on a motorized rotation stage which replaced the high reflecting resonator mirror, allowing for a broad continuous wavelength tunability from 710 nm to 950 nm. This laser is ideally suited for long-range scans searching for unknown atomic transitions. Output powers for the standard laser resonators were in the 3 W range, and 1 W for the grating-assisted laser. Single-pass second-harmonic generation using nonlinear BBO crystals enhances the wavelength capability of the Ti:sapphire lasers, yielding up to 500 mW output power in the blue to UV spectral range. Synchronization

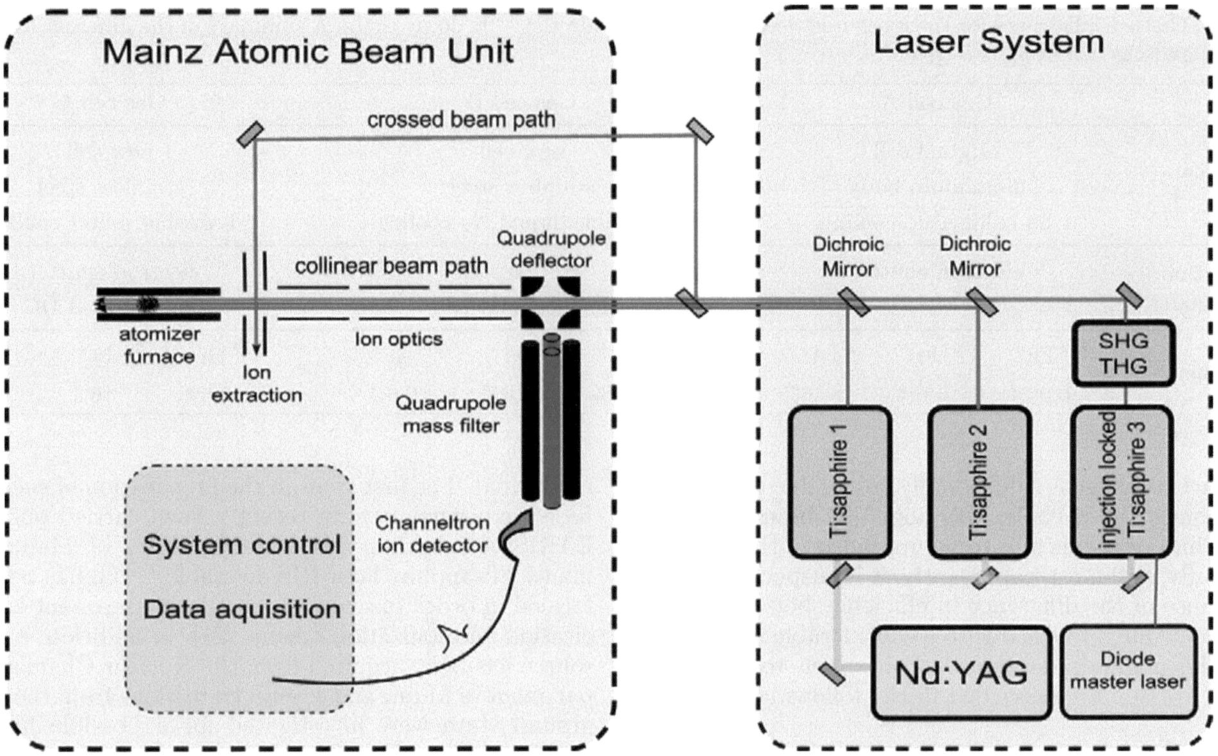

Fig. 6. (Colour on-line) Schematic of the experimental setup including three tunable Ti:sapphire lasers and the Mainz Atomic Beam Unit (MABU). For studies of the hyperfine structure of ^{229g}Th, an injection-locked Ti:sapphire laser replaced the grating-assisted laser and an optional crossed-beam path and refocussing mirror were added to the standard setup.

of the approximately 40 ns long laser pulses is performed using fast Q-switching techniques with Pockels cells and the resulting time-synchronized lasers are monitored with fast photodiodes. The laser beams are spatially overlapped and focused to a beam waist of $\sim 100\,\mu$m within the centre of the Mainz Atomic Beam Unit (MABU) graphite oven (inner diameter 2.2 mm, 50 mm length) in the so-called longitudinal (in-source) beam path geometry, illustrated in fig. 6. For high-resolution spectroscopy the first step laser may also be introduced in a crossed-beam geometry to minimize Doppler broadening of the transition.

In our analysis, samples containing approximately 10^{15} atoms of ^{232}Th in nitric acid solution were dried on a zirconium foil which was folded and inserted into the graphite oven. Zirconium was chosen in order to improve the reduction of thorium oxide to neutral thorium. A second solution containing about $2\cdot 10^{12}$ atoms of ^{229g}Th, supplied by the Mainz Nuclear Chemistry department, was divided into smaller subsamples of $\sim 10^{11}$ atoms. A long-term stable atom source was achieved by resistively heating the oven to 1600 K. The evaporated thorium atoms are then resonantly excited and subsequently ionized by laser radiation inside the oven. The resulting photo ions, as well as thermally ionized surface ions, were extracted and guided by ion optics to a static quadrupole deflector. Deflecting the ion beam by 90° separated the ions from the neutral particles and allowed for a convenient anti-collinear overlap geometry of laser radiation and atomic beam. A subsequent quadrupole mass filter was used to suppress surface ions of different masses produced in the hot cavity. It was operated in a high transmission mode that resulted in a somewhat reduced mass resolution of $m/\Delta m \approx 200$ during the spectroscopic experiments. Ions of interest were finally detected by an off-axis channeltron detector operating in single ion counting mode. Data acquisition was performed with a LabView program and included a recording of the ion count rate, monitoring of the laser wavelengths and control of the grating laser and the quadrupole mass filter.

Three strong transitions from the atomic ground state were chosen from literature as first resonant excitation steps [25,26]. From each first excited level, long-range wavelength scans were performed with the grating-assisted laser resulting in 24 newly found even parity excited states in an energy region from 37500 cm^{-1} to 40500 cm^{-1}. A third laser operating at a wavelength around 750 nm provided non-resonant photo-ionization into the continuum from each level populated by the scanning laser. Both the laser for the first excitation step as well as the grating-assisted scanning laser were kept at low-to-medium power levels (~ 10 mW for the first step, ~ 50 mW for the second step) to reduce single colour multi-step excitation processes, while the third Ti:sapphire laser was operated at maximum power in order to enhance the non-resonant ionization. Time limitations prevented a full systematic study of the saturation behaviour of each newly found intermediate transition and therefore absolute transition strengths could not be obtained. Lacking further information, a few

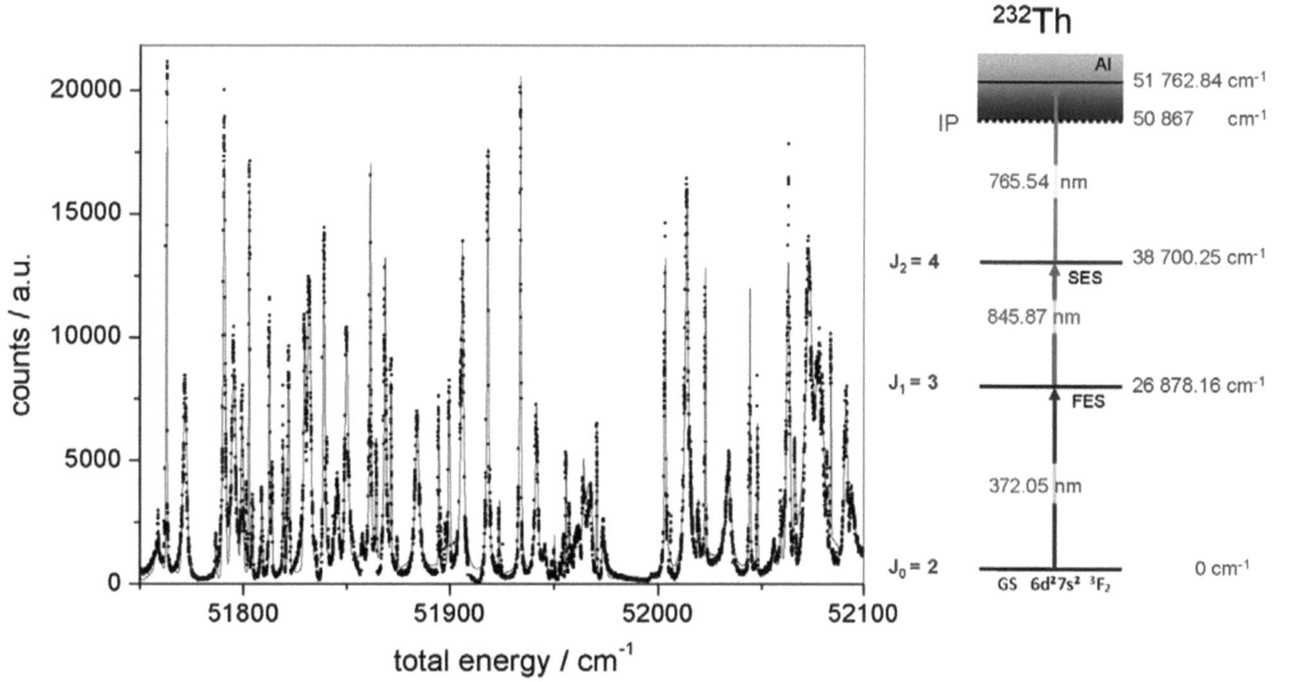

Fig. 7. (Colour on-line) A laser scan from an intermediate state at $38700\,cm^{-1}$ resulting in a multitude of autoionizing states. The ionization potential (IP) of thorium is at $50867\,cm^{-1}$. One of the strongest transitions in this spectrum was chosen for the final excitation and ionization scheme shown on the right.

of the most promising intermediate transitions were therefore selected as starting points for further scans in the search for Rydberg and autoionizing states.

Figure 7 illustrates an example of a scan from a level at $38700\,cm^{-1}$ showing a complex series of strong autoionizing resonances in a region of high-level density, with peaks of differing linewidths. One of the two strongest transitions was chosen for the final ionization scheme, illustrated in the figure. Although several other transitions provided similar ionization strengths, the relatively narrow autoionizing transition chosen for the final scheme is expected to provide further enhancement of selectivity in the presence of contamination.

Saturation measurements for the final scheme have been carried out in the limited volume of the graphite oven, suggesting a full saturation of all three resonant transitions with moderate laser powers of 2.2 mW, 68 mW and 125 mW for the first, second and third step, respectively. It is expected that it will be possible to expand the interaction area of the laser beams while maintaining saturation in the larger volume of the recoil ion guide. Additionally, with the relatively slow extraction time from the gas cell compared to the laser repetition rate it will be possible to interrogate the thorium atoms with several laser pulses. A paper providing further details of these spectroscopic investigations including a thorough documentation of all newly found intermediate and autoionizing states, along with isotope shift and total efficiency measurements for the final ionization scheme, has recently been published [21].

4 Towards high-resolution RIS of ^{229}Th

With the assumption that the new laser ionization scheme discussed in sect. 3 can improve the ion guide efficiency for the production of a low-energy beam of $^{229}Th^+$ to a level similar to that of $^{221}Fr^+$ and $^{217}At^+$, we can expect a $^{229}Th^+$ ion rate of $2.4 \times 10^4\,s^{-1}$ at the JYFL collinear laser spectroscopy station.

Tordoff and collaborators used collinear laser spectroscopy at JYFL to investigate ionic transition efficiencies in Th^+ ions using a discharge source of ^{232}Th [9]. Following the now standard technique of cooling and bunching in an rf-cooler device, spectroscopy was performed from the ionic ground state and a chosen metastable state. The latter state was populated via the technique of optical manipulation in the cooler-buncher which has been shown to efficiently redistribute the ground-state population to a preferred excited state chosen according to transition strength, accessibility for continuous wave narrow linewidth lasers or to angular momentum considerations [27]. The off-line tests resulted in a collinear technique efficiency of 1 photon detected per 5000 ions illuminated on the 322.871 nm transition from the ionic ground state to the $30972.162\,cm^{-1}$ level. This line was compared to a 297.309 nm transition from the $1521.896\,cm^{-1}$ metastable state, poorly populated in the discharge source. Without optical pumping, only 1 photon in 64000 ions illuminated was observed. Following illumination of the rf-cooler axis with 374.225 nm laser light from a pulsed Ti:sapphire laser and subsequent relaxation from the pumped $26721.891\,cm^{-1}$ state, the population of

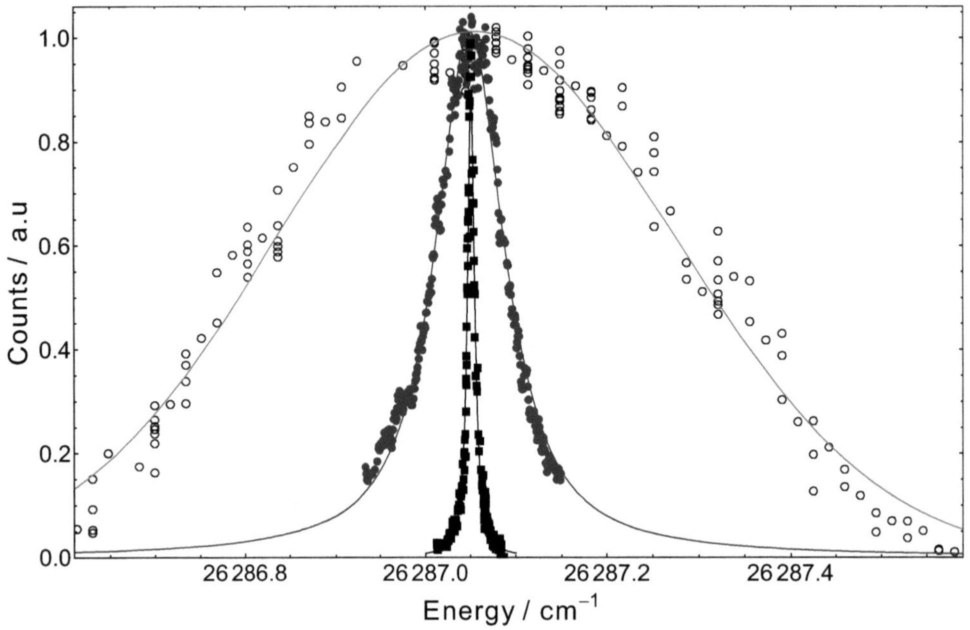

Fig. 8. Laser wavelength scans using the 380.416 nm transition from the thorium atomic ground state to an excited state at 26287.049 cm^{-1}. The figure shows a comparison of line profiles using the broadband Ti:sapphire laser (light-gray open circles), the injection-locked laser in an anti-collinear geometry (dark-grey filled circles) and in a crossed-beam geometry (black filled squares).

the metastable state increased, leading to an improved efficiency of 1 photon detected per 13000 ions illuminated. In this particular instance, the tested ground-state transition is a factor of three more efficient.

An alternative method to photon counting is that of ion counting following photo-ionization. Although the detection technique can be orders of magnitude more efficient, the drawback for resonance ionization spectroscopy (RIS) is the often prohibitive laser linewidth (the typical laser linewidth of the FURIOS laser system is ~ 5 GHz). Nevertheless, recent laser developments have resulted in a dramatic reduction of the linewidth and thus direct in-source/in-jet high-resolution RIS of the atomic hyperfine structure has become an attractive alternative to collinear laser spectroscopy.

4.1 Improvement of spectral resolution

The typical hyperfine splitting of high-Z elements such as thorium is in the multi-GHz range, however the bandwidth of pulsed laser systems is still too large for high-resolution spectroscopy. In order to address this problem, injection-locking of the pulsed Ti:sapphire laser by a cw diode laser acting as a frequency defining master laser can provide a narrow bandwidth of ~ 20 MHz. A detailed description of this arrangement including the laser resonator and its performance can be found in [28]. The dependence of the injection-locked laser system on available laser diodes is a current drawback and in the future will be addressed at JYFL with the use of a cw Ti:sapphire master laser. Unfortunately, in the current work the laser diode did not reach the fundamental wavelength of $\sim$ 744 nm required for the first resonant excitation step of the efficient ionization scheme established in sect. 3 (the diode tuning range was $\sim$ 755–790 nm).

A search was therefore made for accessible transitions from the atomic ground state using the frequency doubled wavelength range of the laser diode. Following this, a particularly strong transition with a wavelength of $\sim$ 380.42 nm to a state at 26287.049 cm^{-1} ($6d^27s7p\,^3D_1$) was selected for further study. The complete scheme included a second step at $\sim$ 808.28 nm to a higher-lying excited state at 38659 cm^{-1}, and a final transition at $\sim$ 784.99 nm to an AI state at 51398 cm^{-1}. A comparison of the spectral line shape of the first resonant transition using either a broadband Ti:sapphire laser or the injection-locked laser in an anti-collinear or crossed-beam geometry (illustrated in fig. 6) is shown in fig. 8. The choice of laser-atom interaction geometry has no visible effect when using the broadband Ti:sapphire laser and the resultant linewidth of 15 GHz (frequency doubled) is dominated by the Gaussian laser bandwidth. Using the injection-locked laser in an anti-collinear geometry, the spectral linewidth fitted with a Voigt profile reduces to 2.7 GHz, dominated by the Lorentzian contribution arising from power broadening. We note that the signal does not go to zero due to the limited mode-hop free tuning range of the laser diode. Finally, in a crossed-beam geometry, the spectral distribution is best reproduced by a Voigt profile with a total linewidth of 260 MHz, again dominated by the Lorentzian component with only a small Gaussian contribution arising from the imperfectly collimated atomic beam and the laser bandwidth of ~ 20 MHz.

An important source of spectral line broadening in high resolution measurements is that of power broadening.

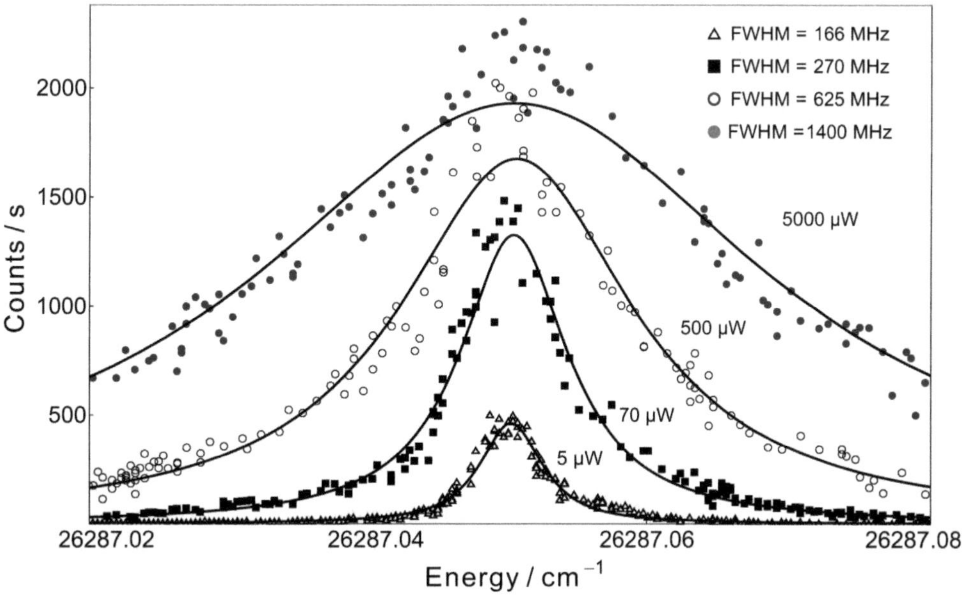

Fig. 9. Laser wavelength scans illustrating the power broadening of the transition to the 26287.049 cm^{-1} atomic state. The laser power and resulting spectral linewidth are indicated.

Typically the laser intensity is adjusted to be as low as possible while still allowing for sufficient ionization to perform the measurement. The effect of power broadening is illustrated in fig. 9 as part of the study of the 380.42 nm transition. The laser scans were performed in the crossed-beam geometry in order to be most sensitive to the Lorentzian component. We note that even at the lowest laser power of 5 μW the Lorentzian wings are considerable, reflecting the large natural linewidth of 135 MHz of the transition to 26287 cm^{-1}. At a laser power of 5 mW the linewidth increased to $\sim$ 1.4 GHz.

Additional broadening mechanisms must be accounted for when performing in-source laser ionization spectroscopy in a buffer gas cell. In addition to the intrinsic laser linewidth and power broadening, pressure broadening from collisions with the buffer gas and Doppler broadening due to the atomic velocity distribution are effects which convolute and add to the experimental width of the measured spectral line. Therefore, the prospect of utilizing the expanding gas jet environment following the gas cell is highly attractive as Doppler and pressure broadening effects are substantially reduced.

The ^{229g}Th solution supplied by the Nuclear Chemistry department in Mainz contained a contaminating fraction of the isotopes ^{228}Th ($\tau_{1/2} = 1.9$ a) and ^{230}Th ($\tau_{1/2} = 7.5 \times 10^4$ a) which allowed for a measurement of their isotope shifts relative to stable ^{232}Th on the transition discussed above. To reach a sufficient signal strength for these less abundant isotopes the injection-locked laser was used in an anti-collinear geometry as the crossed-beam geometry results in a reduction in yield of 10^3 due to losses from beam collimation and reduced laser-atom interaction time. Wavelength scans of each isotope are shown in fig. 10 along with the Gaussian fit curves used for determination of the centroids. The resulting centroids for each

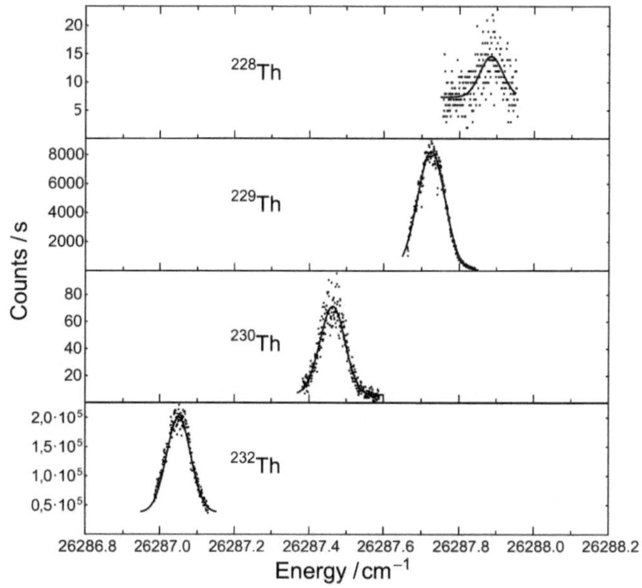

Fig. 10. Laser wavelength scans of the 380.416 nm transition from the thorium atomic ground state to the excited level at 26287.049 cm^{-1} for the isotopes $^{228-232}$Th.

isotope as well as level isotope shifts relative to ^{232}Th are detailed in table 2. In the literature, only the isotope shift for the $^{232-230}$Th pair is given, $-0.4158(19)$ cm^{-1} [29], in agreement with our results.

In order to provide a consistency check of the measurements made in the thorium atomic system the King Plot method may be used [30]. Figure 11 shows a plot of the isotope shifts in the 380.42 nm atomic line (this work) compared with those measured by Kälber and colleagues in the 583.9 nm ($6d^27s \rightarrow 5f(2F)6d^2(1D)$) ionic line [31]. Both sets of shifts have been modified to account

Table 2. Energy level centroids and isotope shifts of the first excited level at 26287.049 cm^{-1}(Lit) for the isotopes $^{228-232}$Th. The errors for the isotope shifts are statistical only, while the errors for the absolute transition energies include an accuracy error of about 0.002 cm^{-1} of the used wavemeter.

A	E (^{A}Th) / cm^{-1}	T($^{232-A}$Th) / cm^{-1}
232	26287.0504(26)	0
230	26287.4629(28)	−0.413(1)
229	26287.7245(26)	−0.674(1)
228	26287.8851(42)	−0.835(2)

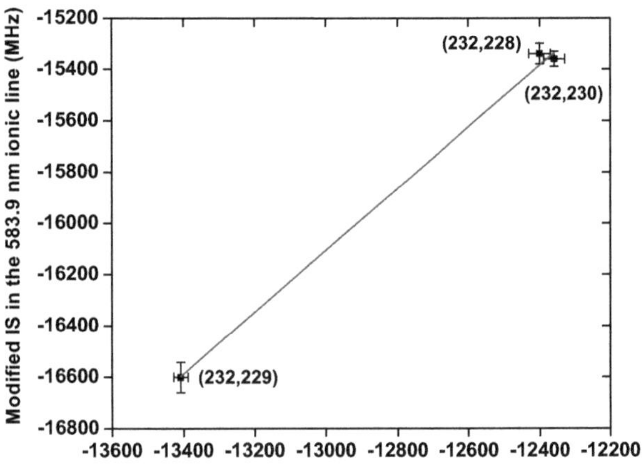

Fig. 11. King Plot of isotope shifts in the Th 380.42 nm atomic line and 583.9 nm ionic line.

for the different masses involved using the 232,230 mass pair as a reference. Plotting the modified isotope shift in one transition for all isotope pairs A, A' against those of another transition should yield a straight line with a gradient F_1/F_2 and intercept $M_1-(F_1/F_2)M_2$, where F and M are the field shift (FS) and mass shift (MS) respectively for transitions 1 and 2. Considering that the mass shift in the heavy elements nearly vanishes in comparison to the field shift, one would expect that the linear dependence should go through the zero intercept within errors. Indeed, this is the case as the resultant fit yields an intercept value of −349(876) MHz and a gradient of +1.212(69). Neglecting the mass effect, the ionic isotope shift values from Kälber can be combined with the change in mean-square charge radii to provide a value for the field shift of the ionic transition. The weighted average from the isotopes measured in that work yields a value of −74.9(6) GHz/fm^2. Combining this with the gradient from the King Plot in fig. 11, we extract a field shift value for the atomic transition of −62(4) GHz/fm^2. The negative value of the FS indicates a change in s-electron density to a reduced density in the excited state, as expected from the atomic configuration.

Finally, as seen in fig. 10, we note that the ground-state transition to the 26287.049 cm^{-1} level has no sensitivity to the hyperfine structure of ^{229}Th and therefore further transitions were studied, discussed in the following section.

4.2 First results of the hyperfine structure of ^{229}Th

If the atomic spin, J, and nuclear spin, I, are both greater than zero, then the fine structure level is split into hyperfine levels. The levels are identified by the quantum number F ($F = I + J$), which arises from the coupling of I and J. The hyperfine splitting of each level is given by [32]

$$W_{\mathrm{HFS}} = A \cdot \frac{K}{2} + B \cdot \frac{\frac{3}{4} \cdot K(K+1) - I(I+1)J(J+1)}{2I(2I-1)J(2J-1)}, \quad (1)$$

with A the magnetic dipole coupling constant, B the electric quadrupole coupling constant and K the so-called Casimir Factor,

$$K = F(F+1) - J(J+1) - I(I+1). \quad (2)$$

The intensities $S_{FF'}$ of transitions between individual hyperfine levels of two different J states can be estimated with the help of the 6J symbol,

$$S(F \rightarrow F') = \frac{(2F+1)(2F'+1)}{2I+1} \cdot \begin{Bmatrix} J' & F' & I \\ F & J & 1 \end{Bmatrix}^2, \quad (3)$$

wherein the prime notation indicates the quantum numbers of the second atomic state involved in the transition. The actual observed intensities can slightly deviate from theory due to saturation effects as well as to transition rates to higher-lying excited states used to ionize the atom, which depend on the spectral profiles of the subsequent excitation and ionization lasers and the unknown hyperfine level splitting of the higher states.

In the case of ^{229}Th, the nuclear ground state has a nuclear spin $I = 5/2$ and the atomic ground state, with a configuration 6d^{2}7s^2 $^3F_2$, has a total angular momentum spin $J = 2$, leading to a hyperfine splitting of the ground state into 5 levels with

$$|I-J| < F < |I+J| \longrightarrow F = 1/2, 3/2, 5/2, 7/2, 9/2. \quad (4)$$

For an excited state with $J = 1$ (chosen as an example based on the results discussed later) and the selection rule $\Delta F = \pm 1, 0$ this leads to nine allowed hyperfine transition lines as shown in fig. 12. The relative transition strength of each line as deduced by eq. (3) is indicated. The hyperfine levels were ordered arbitrarily, the actual ordering and energy splitting is dependent on the A and B coupling constants.

In preparation for detailed ^{229}Th scans, several first step transitions in the frequency doubled range of the diode laser (listed in table 3) were studied with a broadband Ti:sapphire laser in the anti-collinear geometry in a search for indications of broadening due to an underlying hyperfine structure. However, of all transitions exhibiting a strong ion signal, not one showed clear signs of such an effect and only few very weak transitions had indications of a slight asymmetry or broadening in comparison to the resonance of ^{232}Th. A repeat of the scanning over several resonances was performed using the injection-locked laser in the same anti-collinear geometry. Even with the reduction in linewidth from 10 GHz to ∼ 2 GHz, no sigificant

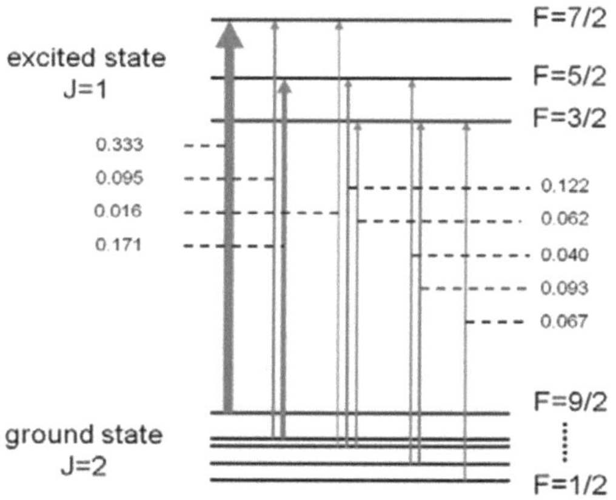

Fig. 12. Hyperfine splitting and transitions for ^{229g}Th with a nuclear spin $I = 5/2^+$. The atomic spins of the relevant J states are given. Estimated relative transition strengths of the individual hyperfine transitions are shown and highlighted with the arrow thicknesses.

Table 3. Energy levels investigated with the injection-locked laser for HFS in the blue (left) and in the UV regime (right).

E (^{232}Th) / cm^{-1}	J
25306.763	2
25321.922	3
25526.314	1
26096.986	3
26113.270	2
26287.049	1
26508.031	3
26651.831	2

E (^{232}Th) / cm^{-1}	J
38088.999	2
38278.875	1
38355.339	2
38675.348	3
38814.244	2
38840.553	1

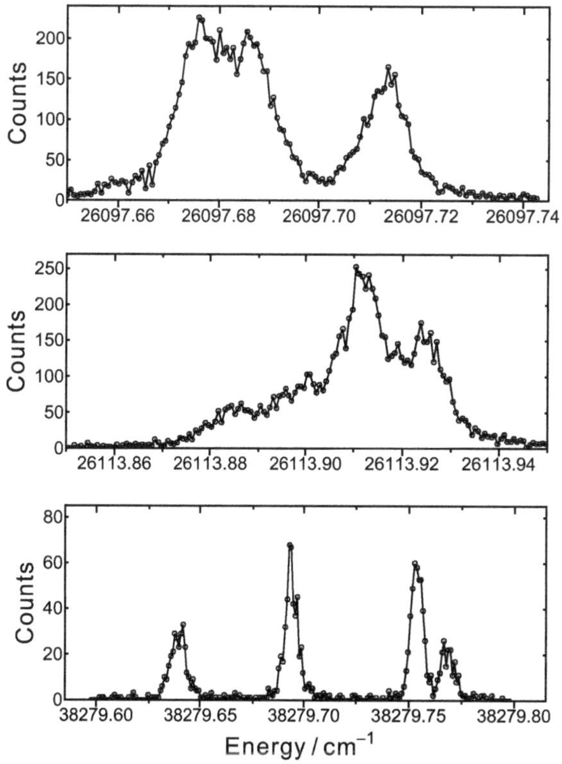

Fig. 13. Partially resolved hyperfine structures of transitions from the atomic ground state to excited levels at 26096 cm^{-1} (top) and 26113 cm^{-1} (middle). The structure of both states spans a width of $\sim$ 1500 MHz. The bottom panel shows a much improved sensitivity to the HFS using a ground-state transition to a $J = 1$ state at 38279 cm^{-1}. The total width of the HFS is approximately 4 GHz.

asymmetry or hyperfine structure could be resolved. Finally, a crossed-beam geometry was tested following the insertion of a collimating slit approximately 2 cm from the exit of the oven. Although the signal in this geometry is reduced by a factor of 10^3, two transitions to states at 26096 cm^{-1} ($J = 3$) and 26113 cm^{-1} ($J = 2$) with sufficient signal strength resulted in partially resolved hyperfine structures, as shown in the top and middle panels of fig. 13.

With multiple hyperfine components expected for both transitions, the current spectral resolution is still insufficient to probe the seemingly unresolved structures. In order to probe configurations with a possible larger hyperfine splitting, the injection-locked Ti:sapphire laser was frequency tripled to the ultraviolet (UV) spectral region (255–265 nm) in order to access states around 38000 cm^{-1}. Only a limited UV laser power of 1–2 mW could be produced with the current setup, as the pumping power for the laser was rather low, approximately 3 W.

Six different levels listed in the literature were accessible with the frequency tripled laser, summarized in table 3. Only for the transition leading to a state at 38279 cm^{-1}, $J = 1$, is the transition strength known ($3 \times 10^7\,\text{s}^{-1}$). In addition, the available electronic configuration assignments for all states was incomplete. Rather than using a three-step excitation and ionization scheme, the higher energy of the first step allowed for a simple two-step process, with the second transition directly populating an AI state. Due to the different parity however, the earlier AI scans were not suitable and those of correct parity listed in literature did not provide a good ion signal. Therefore, laser scans were made to search for identification of so far unknown states. For all states found, even in the best cases the overall ion signal was a factor of 5–10 lower than using the earlier blue-IR-IR three-step schemes, most likely due to the lack of available UV power. The finally chosen AI state was located at 51427(2) cm^{-1}.

All six first step transitions were scanned with the injection-locked laser in an anti-collinear geometry and of these, only the transition to the state at 38279 cm^{-1} showed a significant increase in the peak width of ^{229}Th in comparison to ^{232}Th, as well as a strongly asymmetric peak shape. By moving to a crossed-beams geometry and increasing the oven temperature, a much improved sensitivity to the hyperfine structure (HFS) was obtained. The summation result of seven laser scans is shown in the bottom panel of fig. 13.

Even though this latter transition clearly illustrates a well-resolved signal exhibiting several peaks, a full separation of the expected nine hyperfine components was not achieved. Nevertheless, using the magnetic dipole moment [33] and electrical quadrupole moment of the ground state [34] and the hyperfine factors extracted from the analysis of the spectra shown in fig. 13, together with the estimates of the nuclear moments of the isomeric state as given in ref. [3], the hyperfine structure of the isomeric state is expected to collapse to a total width of less than 800 MHz. With such a collapsed structure, the primary aim of a laser spectroscopic experiment will be to identify the isomeric state via the isomeric shift with respect to that of the ground state, rather than initially attempting a determination of the nuclear moments of the isomeric state. A full analysis of the hyperfine structure of the ^{229}Th atomic ground state as well as the isotopic shift will be published elsewhere.

5 Outlook

This article has summarized the extensive developments of the gas cell technology aimed at improving the extraction efficiency and selectivity of producing a low-energy beam of ^{229}Th at the IGISOL facility, Jyväskylä. In parallel, a dedicated programme of Resonance Ionization Spectroscopy (RIS) has been initiated in close collaboration with the LARISSA group at the University of Mainz. A multitude of intermediate states as well as high-lying autoionizing states were discovered and recently published. With the expected completion of the new IGISOL-4 facility in 2012, a collinear laser spectroscopy experiment on both the ground state and isomeric state of ^{229}Th will be attempted. This will utilize the most efficienct RIS scheme to improve the thorium ion guide efficiency.

An alternative method to search for the isomeric state using either in-source or in-jet spectroscopy is expected to commence in late 2012. An important milestone in this activity has recently been attained: a template for the hyperfine structure of the ground state of atomic ^{229}Th. This high-resolution measurement was realized utilizing an injection-locked narrow bandwidth Ti:sa laser and will be published elsewhere. Currently, a similar laser system is planned for use at IGISOL-4, improving the existing Mainz design by replacing the diode laser (used in this work) with a cw Ti:sapphire laser system. This will increase the flexibility in targeting other atomic lines of interest which may show enhanced sensitivity to hyperfine structure. Nevertheless, by combining the estimated nuclear moments of the isomeric state with the known ground-state moments and the newly determined ground-state hyperfine factors, an expected isomeric hyperfine structure spanning only $\sim$ 800 MHz will pose a serious challenge to the in-source method. Once a narrow bandwidth laser system is in operation at IGISOL-4 then the prospect of spectroscopy in the cold gas jet offers the most attractive environment in which to continue these studies.

Interestingly, in the application of the 7.6 eV electromagnetic transition to sensitive tests of the variation of the fine structure constant α, it has been proposed that direct laboratory measurements of the change in nuclear mean-square charge radius between the isomer and ground-state nucleus can provide a reliable empirical enhancement factor with little or no dependence on the assumed nuclear model [35]. The isotope shift of ^{229}Th on the two transitions (blue and UV) determined in this work is an important step to providing such needed information.

This work has been supported by the Academy of Finland under the Finnish Center of Excellence Program 2006-2011 (Nuclear and Accelerator Based Physics Program at JYFL) and by the GRASPANP Graduate School in Particle and Nuclear Physics of Finland. We also acknowledge support from the "Bundesministerium für Bildung und Forschung" under contract number BMBF-06Mz228.

References

1. B.R. Beck, J.A. Becker, P. Beiersdorfer, G.V. Brown, K.J. Moody, J.B. Wilhelmy, F.S. Porter, C.A. Kilbourne, R.L. Kelley, Phys. Rev. Lett. **98**, 142501 (2007).
2. A.M. Dykhne, E.V. Tkalya, JETP Lett. **67**, 549 (1998).
3. E.V. Tkalya, Phys. Rev. Lett. **106**, 162501 (2011).
4. E. Peik, Chr. Tamm, Europhys. Lett. **61**, 181 (2003).
5. E. Litvinova, H. Feldmeier, J. Dobaczewski, V. Flambaum, Phys. Rev. C **79**, 064303 (2009).
6. V. Flambaum, R. Wiringa, Phys. Rev. C **79**, 034302 (2009).
7. E.V. Tkalya, Phys. Usp. **46**, 315 (2003).
8. I. Izosimov, J. Nucl. Sci. Technol. Suppl. **6**, 1 (2008).
9. B. Tordoff, J. Billowes, P. Campbell, B. Cheal, D.H. Forest, T. Kessler, J. Lee, I.D. Moore, A. Popov, G. Tungate, Hyperfine Interact. **171**, 197 (2007).
10. P. Dendooven, Nucl. Instrum. Methods B **126**, 182 (1997).
11. J. Äystö, Nucl. Phys. A **693**, 477 (2001).
12. P. Karvonen, I.D. Moore, T. Sonoda, T. Kessler, H. Penttilä, K. Peräjärvi, P. Ronkanen, J. Äystö, Nucl. Instrum. Methods Phys. Res. B **266**, 4794 (2008).
13. V.S. Kolhinen, S. Kopecky, T. Eronen, U. Hager, J. Hakala, J. Huikari, A. Jokinen, A. Nieminen, S. Rinta-Antila, J. Szerypo, J. Äystö, Nucl. Instrum. Methods A **528**, 776 (2004).
14. A. Nieminen, J. Huikari, A. Jokinen, J. Äystö, P. Campbell, E.C.A. Cochrane, Nucl. Instrum. Methods Phys. Res. A **469**, 244 (2001).
15. Y.A. Akovali, Nucl. Data Sheets **58**, 555 (1989).
16. Yu. Kudryavtsev, T.E. Cocolios, J. Gentens, M. Huyse, O. Ivanov, D. Pauwels, T. Sonoda, P. Van den Bergh, P. Van Duppen, Nucl. Instrum. Methods Phys. Res. B **267**, 2908 (2009).
17. B. Tordoff, *Development of Resonance Ionization Techniques at the Jyväskylä IGISOL*, Ph.D. thesis, University of Manchester (2007).
18. B. Tordoff, T. Eronen, V.V. Elomaa, S. Gulick, U. Hager, P. Karvonen, T. Kessler, J. Lee, I. Moore, A. Popov, S. Rahaman, T. Sonoda, Nucl. Instrum. Methods B **252**, 347 (2006).
19. M. Wada *et al.*, Nucl. Instrum. Methods B **204**, 570 (2003).
20. C. Geppert, Nucl. Instrum. Methods B **266**, 4354 (2008).
21. S. Raeder, V. Sonnenschein, T. Gottwald, I. Moore, M. Reponen, S. Rothe, N. Trautmann, K. Wendt, J. Phys. B: At. Mol. Opt. Phys. **44**, 165005 (2011).
22. T. Billen, K. Schneider, T. Kirsten, A. Mangini, A. Eisenhauer, Appl. Phys. B **57**, 109 (1993).

23. S.G. Johnson, B.L. Fearey, Spectrochim. Acta B **48**, 1065 (1993).
24. S. Köhler, R. Deißenberger, K. Eberhardt, N. Erdmann, G. Herrmann, G. Huber, J.V. Kratz, M. Nunnemann, G. Passler, P.M. Rao, J. Riegel, N. Trautmann, K. Wendt, Spectrochim. Acta B **52**, 717 (1997).
25. J. Blaise, J.-F. Wyart, *Tables Internationales de Constantes*, in *Selected Constants, Energy Levels and Atomic Spectra* Vol. **20** (Universitè P. et M. Curie, Paris, 1992).
26. R.L. Kurucz, *Atomic spectral line database CD-ROM 23* (1995) available at `http://www.pmp.uni-hannover.de/cgi-bin/ssi/test/kurucz/sekur.html`.
27. B. Cheal, K. Baczynska, J. Billowes, P. Campbell, F.C. Charlwood, T. Eronen, D.H. Forest, A. Jokinen, T. Kessler, I.D. Moore, M. Reponen, S. Rothe, M. Rüffer, A. Saastamoinen, G. Tungate, J. Äystö, Phys. Rev. Lett. **102**, 222501 (2009).
28. T. Kessler, H. Tomita, C. Mattolat, S. Raeder, K. Wendt, Laser Phys. **18**, 1 (2008).
29. R. Engleman, B.A. Palmer, J. Opt. Soc. Am. **78**, 694 (1983).
30. W.H. King, *Isotope Shifts in Atomic Spectra* (Plenum Publishing Co., N.Y., 1984).
31. W. Kaelber, J. Rink, K. Beck, W. Faubel, S. Goring, G. Meisel, H. Rebel, R.C. Thompson, Z. Phys. A **334**, 103 (1989).
32. M.B.G. Casimir, *On the Interaction between Atomic Nuclei and Electrons* (W.H. Freeman, 1963).
33. S. Gerstenkorn, P. Luc, J. Verges, D.W. Englekemeir, J.E. Gindler, F.S. Tomkins, J. Phys. France **35**, 483 (1974).
34. C.J. Campbell, A.G. Radnaev, A. Kuzmich, Phys. Rev. Lett. **106**, 223001 (2011).
35. J.C. Berengut, V.A. Dzuba, V.V. Flambaum, S.G. Porsev, Phys. Rev. Lett. **102**, 210801 (2009).

DOI 10.1007/978-94-007-5555-0_25
Reprinted from *Hyperfine Interactions* Volume 223, Issues 1-3, 2014
and *The European Physical Journal A* Volume 48, Issue 4, 2012

THE EUROPEAN
PHYSICAL JOURNAL A

Regular Article – Experimental Physics

Penning-trap mass measurements on $^{92,94-98,100}$Mo with JYFLTRAP

A. Kankainen[a], V.S. Kolhinen, V.-V. Elomaa[b], T. Eronen[c], J. Hakala, A. Jokinen, A. Saastamoinen, and J. Äystö

Department of Physics, P. O. Box 35(YFL), FI-40014 University of Jyväskylä, Finland

Received: 31 January 2012
Published online: 18 April 2012 – © Società Italiana di Fisica / Springer-Verlag 2012
Communicated by E. De Sanctis

Abstract. Penning-trap measurements on stable $^{92,94-98,100}$Mo isotopes have been performed with relative accuracy of $1 \cdot 10^{-8}$ with the JYFLTRAP Penning-trap mass spectrometer by using ^{85}Rb as a reference. The Mo isotopes have been found to be about 3 keV more bound than given in the Atomic Mass Evaluation 2003 (AME03). The results confirm that the discrepancy between the ISOLTRAP and JYFLTRAP data for $^{101-105}$Cd isotopes was due to an erroneous value in the AME03 for ^{96}Mo used as a reference at JYFLTRAP. The measured frequency ratios of Mo isotopes have been used to update mass-excess values of 30 neutron-deficient nuclides measured at JYFLTRAP.

1 Introduction

Recently, a discrepancy was found between cadmium mass measurements performed at JYFLTRAP, where ^{96}Mo$^+$ was used as a reference [1], and SHIPTRAP [2] and ISOLTRAP results, where ^{85}Rb$^+$ was used as a reference ion [3]. Earlier JYFLTRAP has shown to be capable of performing very accurate mass measurements. Therefore, it was assumed that the ^{96}Mo mass-excess value would be about 3 keV off in the Atomic Mass Evaluation 2003 (AME03) [4]. The mass evaluation done in ref. [3] showed that there is a -3.2 keV shift between the AME03 value and the evaluated value for ^{96}Mo. In this work, we wanted to confirm this evaluation result by a direct mass measurement of ^{96}Mo. If the mass of ^{96}Mo is off by 3 keV, also the neighbouring isotopes connected by (n, γ) reactions in the AME03 are likely to be off. Thus, we decided to check the mass-excess values of all stable molybdenum isotopes and investigate where the possible 3 keV offset could come from. These measurements have a direct effect on previous JYFLTRAP results since stable molybdenum isotopes (^{94,96,97,98}Mo) have been used as references for 30 neutron-deficient nuclides at JYFLTRAP.

2 Mass measurements

JYFLTRAP [5] is a double cylindrical Penning-trap mass spectrometer for accurate mass measurements at the Ion-Guide Isotope Separator On-Line (IGISOL) facility [6,7] in Jyväskylä, Finland. The setup consists of a Radio-Frequency Quadrupole (RFQ) cooler and buncher [8] and a double Penning-trap [9] spectrometer (see fig. 1). In this experiment, we used an off-line electric discharge ion source at IGISOL to create singly-charged ^{85}Rb$^+$ and $^{92,94-98,100}$Mo$^+$ ions.

The first trap of JYFLTRAP (purification trap) is used for the isobaric purification of the injected ion bunches by using the buffer-gas cooling technique [10]. The mass measurement is carried out in the second Penning trap (precision trap). The measurement is based on the determination of the sideband frequency of the ions of interest $\nu_+ + \nu_-$, where ν_+ and ν_- are the reduced cyclotron frequency and the magnetron frequency, respectively. In an ideal Penning trap, this sideband frequency matches with the true cyclotron frequency $\nu_c = \frac{1}{2\pi}\frac{q}{m}B$ of ions with charge state q and mass m in the magnetic field B [11]. The frequency determination was done by using the time-of-flight ion-cyclotron-resonance (TOF-ICR) technique [12]. In this method, the ion's radial energy is increased in the trap by using an azimuthal quadrupole radio-frequency (RF) field with the cyclotron frequency of the ions. Since the radial energy of the ions is converted to axial energy in the gradient of the magnetic field when extracted from the trap, the increased energy leads to a shorter flight time to the micro-channel plate (MCP) detector. In this experiment, a Ramsey-type ion motion excitation was used [13,14] with two 25 ms long fringes separated by a 750 ms long waiting time. Figure 2 shows two examples of Ramsey TOF resonances for ^{85}Rb$^+$ and ^{96}Mo$^+$. The cyclotron frequency of an ion and its uncertainty are obtained from the experimental TOF data by fitting the theoretical fit function.

[a] e-mail: anu.k.kankainen@jyu.fi

[b] *Present address*: Turku PET Centre, Accelerator Laboratory, Åbo Akademi University, FI-20500 Turku, Finland.

[c] *Present address*: Max-Planck-Institut für Kernphysik, Saupfercheckweg 1, D-69117 Heidelberg, Germany.

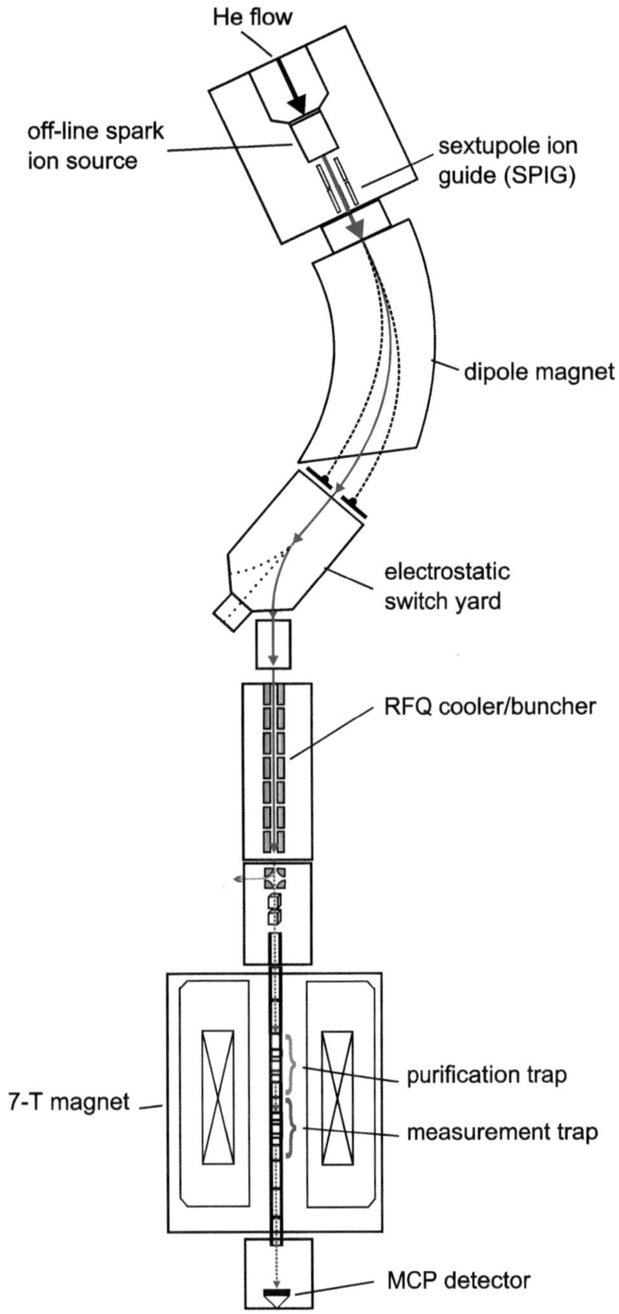

Fig. 1. Schematic illustration of the JYFLTRAP beam line for off-line experiments.

The mass measurement in a Penning trap is based on the measurement of the frequency ratio r between a reference ion with a well-known mass and an ion of interest:

$$r = \frac{\nu_{c,ref}}{\nu_c}. \quad (1)$$

This frequency ratio and its uncertainty are used to deduce the mass of the ion of interest by using the equation

$$m_{meas} = r(m_{ref} - m_e) + m_e, \quad (2)$$

where m_{meas} is the mass of the atom of interest, m_{ref} is the mass of the reference atom and m_e is the mass of the electron. Thus, the uncertainty in the mass of the reference atom contributes to the final mass value, and it can be re-evaluated by using the most accurate knowledge for the mass of the reference atom.

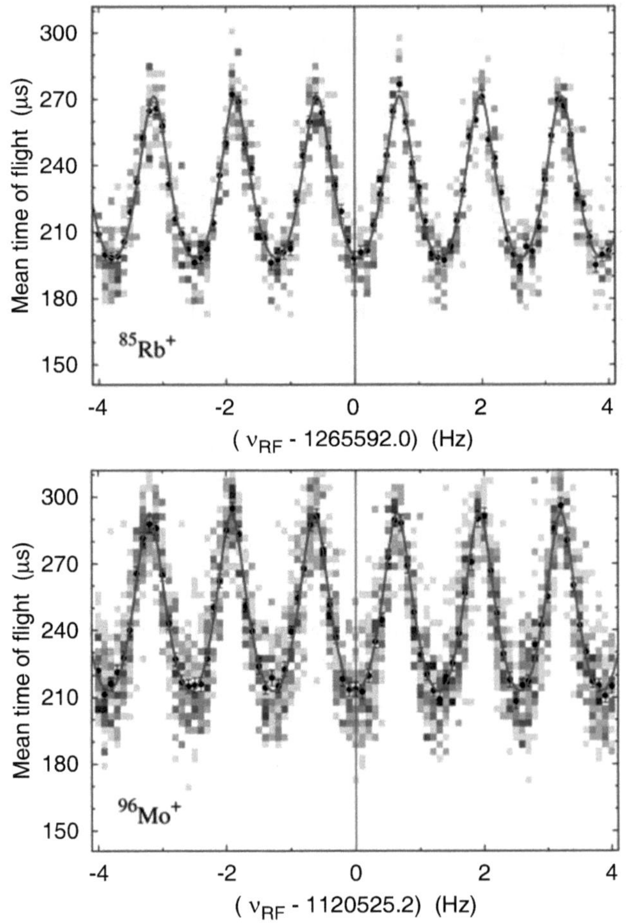

Fig. 2. Ramsey TOF resonances of ^{85}Rb and ^{96}Mo with 25-750-25 ms timing containing 15 scan cycles ($\approx$ 22 min). Shadowed boxes show the density of detected ions, black dots are the average time of flights with the uncertainty limits and solid line is the theoretical fitted line shape.

3 Data analysis and results

To minimize frequency shifts coming from the drifting magnetic field, the measurements were performed by running 2 scan cycles of the ion of interest (Mo^+) and then 2 scan cycles of the reference ion ($^{85}Rb^+$) and repeating this pattern. The measured data were divided into 15-cycle-long runs. A count-rate class analysis [15], where the data were divided into classes according to the number of detected ions, was carried out. The frequency was extrapolated to 0.6 ions in the trap due to the 60% detection efficiency.

The average frequency ratios were calculated by using the weighted means. The internal errors (σ_{int}, statistical uncertainties of the weighted means) were compared to external errors (σ_{ext}, weighted standard deviations) and so-called Birge ratios $R = \sigma_{ext}/\sigma_{int}$ were determined [16].

Table 1. Isotope, number of measurements, frequency ratio r, mass excess and the literature value [4] for the mass excess. Each measurement contains 15 scan cycles. The reference ion was $^{85}Rb^+$.

Isotope	#	r	ME(keV)	AME03(keV)	JYFL-AME03(keV)
^{92}Mo	16	1.082380355(10)	−86807.8(8)	−86805(4)	−2.8(39)
^{94}Mo	21	1.105914072(12)	−88412.7(9)	−88410(2)	−3.0(21)
^{95}Mo	17	1.117699950(12)	−87710.8(1.0)	−87707(2)	−3.3(21)
^{96}Mo	26	1.129463266(13)	−88793.4(1.0)	−88790(2)	−2.9(22)
^{97}Mo	23	1.141256082(14)	−87542.8(1.1)	−87540(2)	−2.3(22)
^{98}Mo	16	1.153025857(15)	−88114.5(1.2)	−88112(2)	−2.7(23)
^{100}Mo	11	1.176604185(16)	−86190.9(1.3)	−86184(6)	−6.6(60)

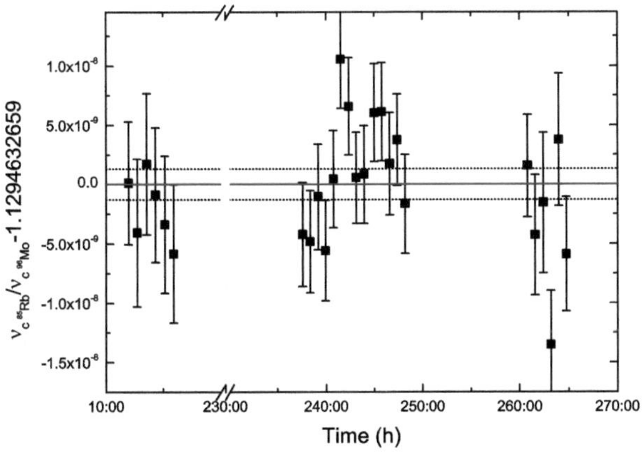

Fig. 3. Measured frequency ratios between $^{85}Rb^+$ and $^{96}Mo^+$ obtained by using 25-750-25 ms Ramsey excitation. Each data point contains 15 scan cycles ($\approx$ 22 min) of both masses. Solid line is the weighted average of these points and the dotted lines show one standard deviation limits.

In the ideal case, the ratio should be close to 1. For the uncertainty of r we took always the larger one of the internal and external uncertainties. For this value σ_r, a mass-dependent uncertainty $\sigma_m(r) = (7.5\cdot 10^{-10}/u)\cdot \Delta m\cdot r$ [17] and a residual uncertainty $\sigma_{res}(r) = (7.9 \cdot 10^{-9}) \cdot r$ [17] were added quadratically.

Measured frequency ratios and their uncertainties are shown in table 1. Birge ratios were close to 1 in each mass measurement set. This means that the deviations in the data are statistical. The used atomic mass unit is u = 931 494.009 0(71) keV [18], the electron mass m_e = 510.998910(13) keV [19], and the value for ^{85}Rb mass excess ME = −82167.331(11) keV [4]. An example of the measured frequency ratios is shown in fig. 3.

4 Comparison to previous measurements

In the Atomic Mass Evaluation 2003 (AME03) the molybdenum masses had three main sources. Masses 92-93, 94-95, 95-96, 96-97, 97-98 were linked together with (n,γ) reaction measurements [20,21]. Masses 92, 94, 95, 96, 98 were measured with a mass spectrometer by using different CH molecules as references [22]. Masses 92-94 and 98-100 were measured by comparing mass differences of molybdenum oxide chlorides [23]. Actually, also other molybdenum pairs were measured in ref. [23] but only two links were left in the AME03 sheets. Moreover, there have been several β-decay measurements from both sides [24–31]. Different reaction studies, such as (p,n) [32–34], (p,d) [35], $(d,{}^3\text{He})$ [36], $({}^3\text{He},p)$ [37], $({}^3\text{He},d)$ [38,39], $({}^3\text{He},{}^6\text{He})$ [40], (t,p) [41], (t,α) [42], and (n,α) [43], have also yielded information on molybdenum isotopes.

In fig. 4 all the links influencing the Mo mass-excess values in the AME03 are shown. Since the JYFLTRAP values disagreed with the AME03 values (except for ^{92}Mo), a thorough comparison to earlier measurements was carried out and all possible links from and to the Mo isotopes in the AME03 were checked out in this work. The results are given below nuclide by nuclide.

The main result is that the values from Bishop *et al.* [23] disagree with the JYFLTRAP values and explain most of the difference between the JYFLTRAP results and the AME03 values. Deviations were also found to C_7H_{10}–^{94}Mo [22], $^{95}\text{Nb}(\beta^-)^{95}\text{Mo}$ [24], $^{98}\text{Mo}(n,\gamma)^{99}\text{Mo}$ [20,21], and $^{100}\text{Mo}({}^3\text{He},p)^{102}\text{Tc}$ [37]. The (n,γ) results between ^{94}Mo and ^{98}Mo agree nicely with the JYFLTRAP values. JYFLTRAP mass-excess values for Mo isotopes suggest that these Mo isotopes are systematically too weakly bound in the AME03. This will also have an effect on the nuclides which have main influences coming from these isotopes, such as for neighbouring Nb and Tc nuclides or ^{101}Mo.

^{92}Mo

The JYFLTRAP mass value for ^{92}Mo agrees with the values from C_7H_8–^{92}Mo [22] and $^{92}\text{Mo}(n,\gamma)^{93}\text{Mo}$ [20] experiments, which have altogether a 78.3% influence on the ^{92}Mo value in the AME03. The JYFLTRAP value for ^{92}Mo agrees with the AME03 value and other experimental data except with the data from ref. [23]. The JYFLTRAP mass value for ^{92}Mo is 6.3(22) keV higher than the value obtained from the mass difference of $^{94}\text{Mo}^{35}\text{Cl}^{16}\text{O}$–$^{92}\text{Mo}^{37}\text{Cl}^{16}\text{O}$ [23] employing the JYFLTRAP mass value of ^{94}Mo (see fig. 5). For ^{92}Mo, some (p,n), $({}^3\text{He},t)$, (p,α) and $(\alpha,{}^8\text{He})$ experiments having uncertainties bigger than 20 keV have been omitted from

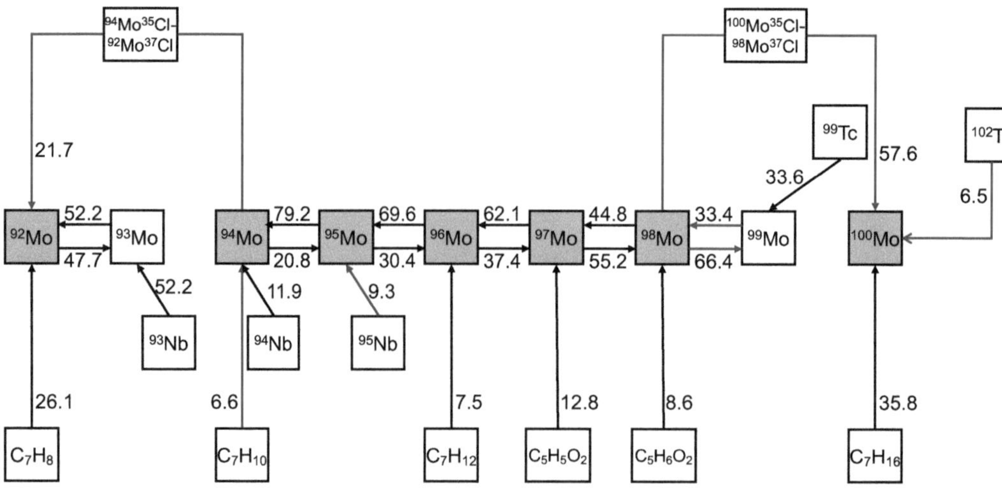

Fig. 4. (Colour on-line) Influences (%) of different reactions on the molybdenum mass-excess values in the AME03 [4]. The red arrows show the reactions which disagree with the results of this work. The massses of highlighted Mo isotopes were measured in this work.

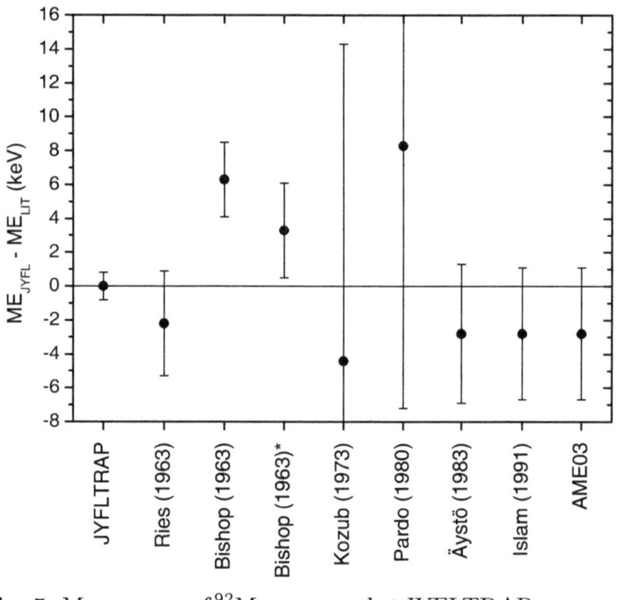

Fig. 5. Mass excess of ^{92}Mo measured at JYFLTRAP compared to earlier experiments of C_7H_8–^{92}Mo (Ries *et al.* [22]), ^{94}Mo ^{35}Cl^{16}O–^{92}Mo^{37}Cl^{16}O (Bishop *et al.* [23]), ^{92}Mo$(p,d)^{91}$Mo (Kozub and Youngblood [35]), ^{92}Mo(^{3}He, ^{6}He)^{89}Mo (Pardo *et al.* [40]), ^{93}Tc S_p value (Äystö *et al.* [25]), ^{92}Mo$(n,\gamma)^{93}$Mo (Islam *et al.* [20]) and AME03 [4]. Bishop (1963)* employs the AME03 mass value [4] for ^{94}Mo instead of the value from this work.

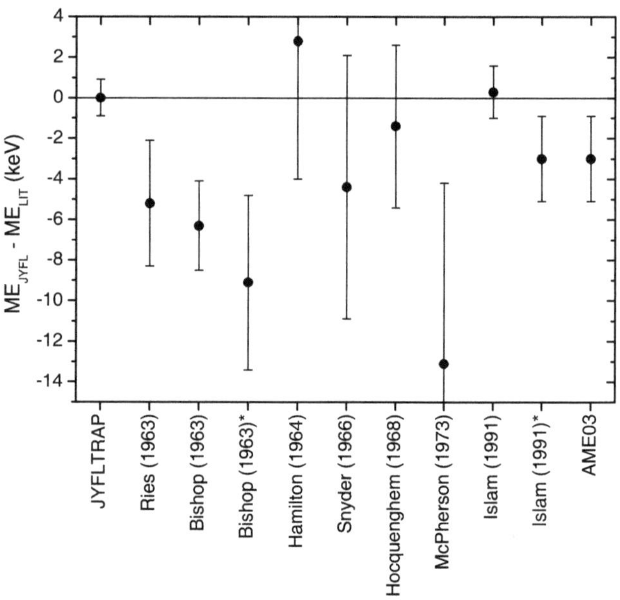

Fig. 6. Mass excess of ^{94}Mo measured at JYFLTRAP compared to earlier experiments of C_7H_{10}–^{94}Mo (Ries *et al.* [22]), ^{94}Mo^{35}Cl^{16}O–^{92}Mo^{37}Cl^{16}O (Bishop *et al.* [23]), ^{94}Tc$(\beta^+)^{94}$Mo (Hamilton *et al.* [26]), ^{94}Nb$(\beta^-)^{94}$Mo (Snyder *et al.* [27], Hocquenghem *et al.* [28]), ^{94}Mo$(p,n)^{94}$Tc (McPherson and Gabbard [32]), ^{94}Mo$(n,\gamma)^{95}$Mo (Islam *et al.* [20]) and AME03 [4]. The values marked with * are based on the AME03 [4] mass values of ^{92}Mo and ^{95}Mo.

fig. 5, where the deviation from different experiments (ME_{LIT}) to the JYFLTRAP mass-excess value (ME_{JYFL}) is shown.

^{94}Mo

The mass value measured for ^{94}Mo at JYFLTRAP disagrees with the AME03 value by −3.0(21) keV. Similarly to ^{92}Mo, there is a 6.3(22) keV difference between the JYFLTRAP value and the value obtained from the mass difference of ^{94}Mo^{35}Cl^{16}O–^{92}Mo^{37}Cl^{16}O [23] employing the JYFLTRAP value for ^{92}Mo (see fig. 6). Actually, even bigger disgareement is found when the AME03 value for ^{92}Mo is used. In addition, the value from C_7H_{10}–^{94}Mo [22] gives a 5.3(31) keV higher mass-excess value for ^{94}Mo than measured at JYFLTRAP. The beta-decay experiments [26–28] agree with the JYFLTRAP value but the value from (p,n) reactions [32] disagrees with it. The

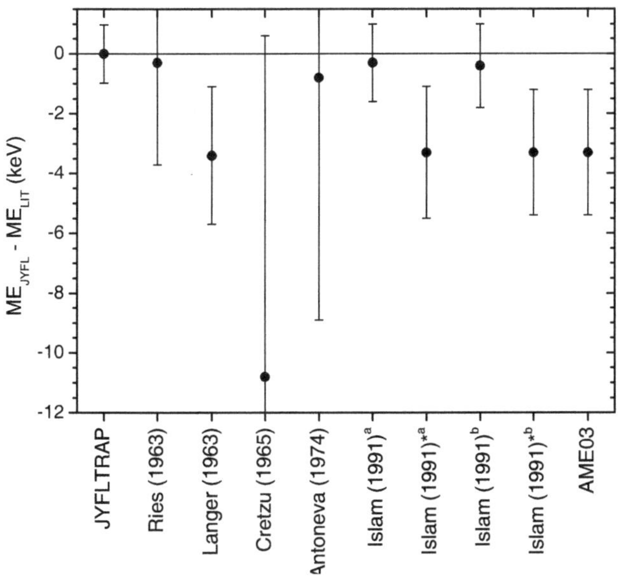

Fig. 7. Mass excess of ^{95}Mo measured at JYFLTRAP compared to earlier experiments of C_7H_{11}–^{95}Mo (Ries *et al.* [22]), ^{95}Nb$(\beta^-)^{95}$Mo (Langer and Wortman [24]), ^{95}Tc$(\beta^+)^{95}$Mo (Cretzu *et al.* [29], Antoneva *et al.* [30]), ^{94}Mo$(n,\gamma)^{95}$Mo (Islam *et al.* [20]), ^{95}Mo$(n,\gamma)^{96}$Mo (Islam *et al.* [20]), and AME03 [4]. The values marked with * are based on the AME03 [4] mass values of ^{94}Mo and ^{96}Mo. Islam (1991)^a refers to ^{94}Mo$(n,\gamma)^{95}$Mo value and ^{b} to the ^{95}Mo$(n,\gamma)^{96}$Mo value.

^{94}Mo$(n,\gamma)^{95}$Mo [20] agrees nicely with the JYFLTRAP results for ^{94}Mo and ^{95}Mo. A shift of about 3 keV is found when the AME03 value for ^{95}Mo is used instead of the JYFLTRAP value.

^{95}Mo

The JYFLTRAP mass value for ^{95}Mo is 3.3(21) keV lower than the AME03 value. The AME03 value is mainly based on the ^{94}Mo$(n,\gamma)^{95}$Mo [20] and ^{95}Mo$(n,\gamma)^{96}$Mo [20] values, which agree well with the JYFLTRAP results for ^{94}Mo, ^{95}Mo, and ^{96}Mo. A shift is observed when the AME03 value is applied for ^{94}Mo and ^{96}Mo, again indicating that these Mo isotopes are systematically less bound in the AME03 (see fig. 7). Also the value based on the beta decay ^{95}Nb$(\beta^-)^{95}$Mo [24] disagreeing with the JYFLTRAP value has an influence on the ^{95}Mo value in the AME03.

^{96}Mo

The JYFLTRAP mass-excess value for ^{96}Mo is 2.9(22) keV lower than the AME03 value. All reaction links in the AME03 agree with the JYFLTRAP value when JYFLTRAP mass-excess values for ^{95}Mo and ^{97}Mo are used (see fig. 8). The disagreement between the AME03 value and the JYFLTRAP value comes from the erroneous mass values of ^{95}Mo and ^{97}Mo in the AME03.

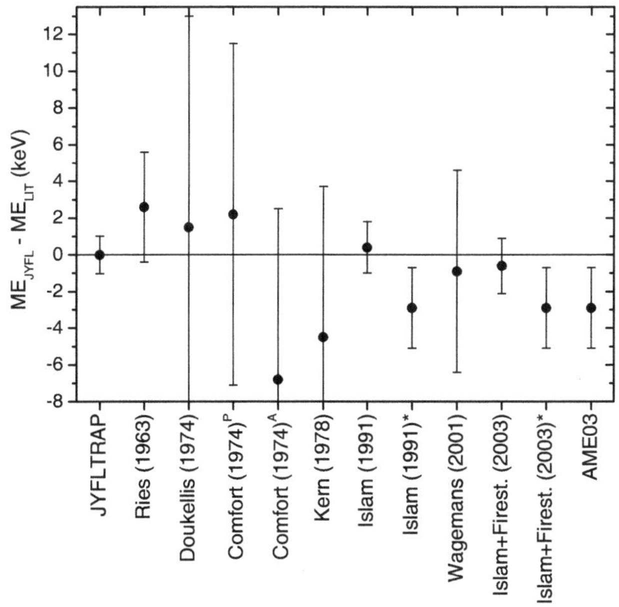

Fig. 8. Mass excess of ^{96}Mo measured at JYFLTRAP compared to earlier experiments of C_7H_{12}–^{96}Mo (Ries *et al.* [22]), ^{96}Mo$(p,n)^{96}$Tc (Doukellis *et al.* [33], Kern *et al.* [34]), ^{96}Mo$(^3$He$,d)^{97}$Tc (Comfort *et al.* [38]) measured at Pittsburgh (P) and at Argonne (A), ^{95}Mo$(n,\gamma)^{96}$Mo (Islam *et al.* [20]), ^{99}Ru$(n,\alpha)^{96}$Mo (Wagemans *et al.* [43]), ^{96}Mo$(n,\gamma)^{97}$Mo (Islam *et al.* [20], Firestone *et al.* [21]), and AME03 [4]. The values marked with * are based on the AME03 [4] mass values of ^{95}Mo and ^{97}Mo. The value from ^{96}Nb$(\beta^-)^{96}$Mo having an uncertainty of 20 keV has been left out.

^{97}Mo

The JYFLTRAP mass-excess value for ^{97}Mo is 2.3(22) keV lower than the AME03 value. The JYFLTRAP value agrees with all the reaction links in the AME03 except with the values from ^{97}Nb$(\beta^-)^{97}$Mo [31], ^{96}Mo$(n,\gamma)^{97}$Mo and ^{97}Mo$(n,\gamma)^{98}$Mo when AME03 values for ^{96}Mo and ^{98}Mo are applied (see fig. 9). The ^{97}Nb beta decay does not have an influence on the ^{97}Mo mass value in the AME03.

^{98}Mo

The JYFLTRAP value for the mass excess of ^{98}Mo disagrees with the AME03 value by 2.7(22) keV. The JYFLTRAP mass-excess result is in agreement with the results from $C_5H_6O_2$–^{98}Mo [22] and ^{97}Mo$(n,\gamma)^{98}$Mo [20], when JYFLTRAP value for ^{97}Mo is applied (see fig. 10). Here, again a disagreement is found when the AME03 values of ^{97}Mo and ^{99}Mo are used for the ^{97}Mo$(n,\gamma)^{98}$Mo and ^{98}Mo$(n,\gamma)^{99}$Mo reactions, indicating that the Mo isotopes have generally too high mass-excess values in the AME03.

^{100}Mo

The JYFLTRAP mass-excess value slightly disagrees with the AME03 value for ^{100}Mo. An almost perfect agreement is found with the value based on C_7H_{16}–^{100}Mo [22],

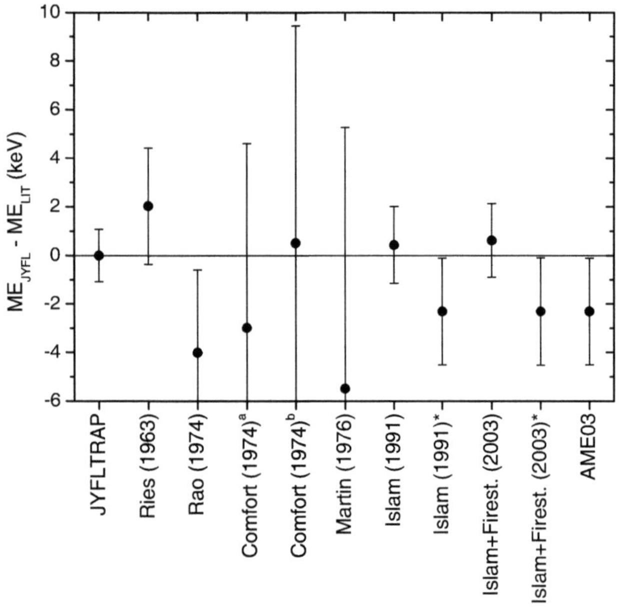

Fig. 9. Mass excess of ^{97}Mo measured at JYFLTRAP compared to earlier experiments of $C_5H_5O_2$–^{97}Mo (Ries *et al.* [22]), $^{97}Nb(\beta^-)^{97}Mo$ (Rao *et al.* [31]), $^{97}Mo(p,n)^{97}Tc$ (Comfort *et al.* [38]), $^{97}Mo(^3He,d)^{98}Tc$ (Comfort *et al.* [38], Martin and Macphail *et al.* [39]), $^{97}Mo(n,\gamma)^{98}Mo$ (Islam *et al.* [20]), $^{96}Mo(n,\gamma)^{97}Mo$ (Islam *et al.* [20], Firestone *et al.* [21]), and AME03 [4]. Comfort (1974)^a refers to $^{97}Mo(p,n)^{97}Tc$ and ^{b} to $^{97}Mo(^3He,d)^{98}Tc$. The values marked with * are based on the AME03 [4] mass values of ^{96}Mo and ^{98}Mo.

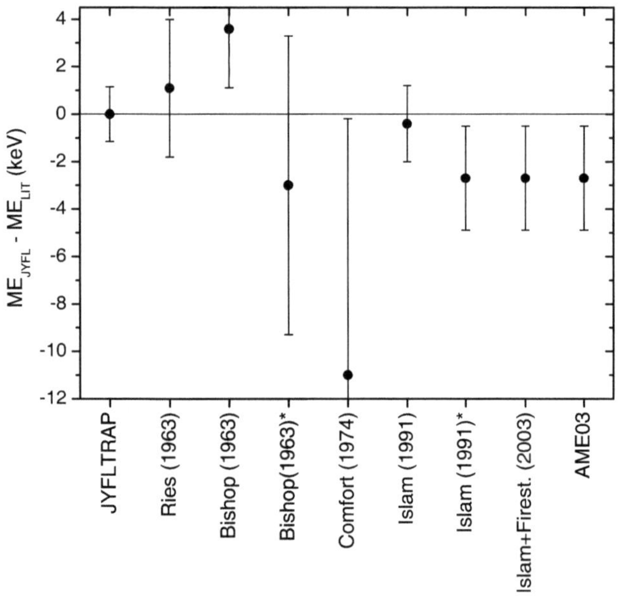

Fig. 10. Mass excess of ^{98}Mo measured at JYFLTRAP compared to earlier experiments of $C_5H_6O_2$–^{98}Mo (Ries *et al.* [22]), $^{100}Mo^{35}Cl^{16}O$–$^{98}Mo^{37}Cl^{16}O$ (Bishop *et al.* [23]), $^{98}Mo(p,n)^{98}Tc$ (Comfort *et al.* [38]), $^{97}Mo(n,\gamma)^{98}Mo$ (Islam *et al.* [20]), $^{98}Mo(n,\gamma)^{99}Mo$ (Islam *et al.* [20], Firestone *et al.* [21]), and AME03 [4]. The values marked with * are based on AME03 [4] mass values of ^{100}Mo and ^{97}Mo.

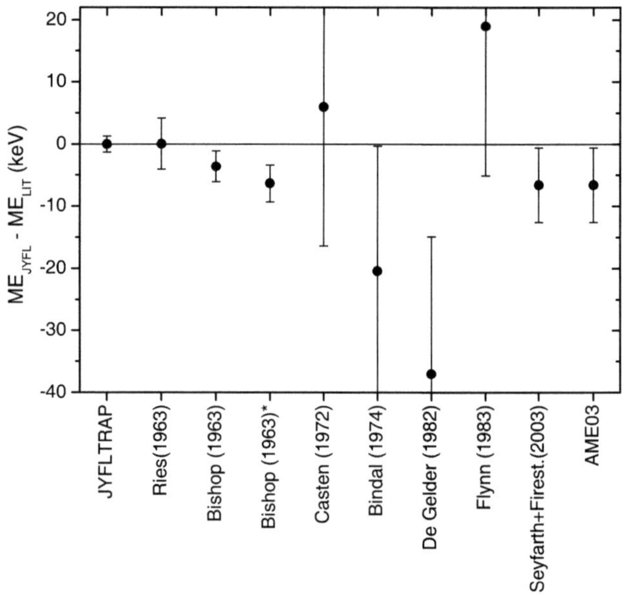

Fig. 11. Mass excess of ^{100}Mo measured at JYFLTRAP compared to earlier experiments of C_7H_{16}–^{100}Mo (Ries *et al.* [22]), $^{100}Mo^{35}Cl^{16}O$–$^{98}Mo^{37}Cl^{16}O$ (Bishop *et al.* [23]), $^{100}Mo(t,p)^{102}Mo$ (Casten *et al.* [41]), $^{100}Mo(d,^3He)^{99}Nb$ (Bindal *et al.* [36]), $^{100}Mo(^3He,p)^{102}Tc$ (De Gelder *et al.* [37]), $^{100}Mo(t,\alpha)^{99}Nb$ (Flynn *et al.* [42]), $^{100}Mo(n,\gamma)^{101}Mo$ (Seyfarth *et al.* [44], Firestone *et al.* [21]), and AME03 [4]. The value marked with * is based on the AME03 [4] mass value of ^{98}Mo. The values from $(t,^3He)$ and β^- experiments have been left out due to large uncertainties.

but the values from $^{100}Mo^{35}Cl^{16}O$–$^{98}Mo^{37}Cl^{16}O$ [23] and $^{100}Mo(^3He,p)^{102}Tc$ [37] influencing the AME03 value of ^{100}Mo, deviate from our Penning-trap mass measurement (see fig. 11). In addition, the value derived from the $^{100}Mo(n,\gamma)$ ^{101}Mo [44] reaction gives a similar deviation as the AME03 value. This suggests that the AME03 value for ^{101}Mo should be about 6.6 keV lower in order to agree with the (n,γ) data.

5 Updated mass values of the nuclides measured at JYFLTRAP using Mo references

Up to date, 30 neutron-deficient nuclides have been measured with respect to molybdenum reference ions at JYFLTRAP (see refs. [1,45,46]). Thus, the results of this paper have an effect on these mass-excess values. The values can be easily updated by multiplying the old frequency ratio measured against a molybdenum isotope (r_{old}) by the frequency ratio of the corresponding molybdenum ion to $^{85}Rb^+$ measured in this work (r_{Mo-Rb}):

$$m_{new} = r_{old} r_{Mo-Rb} \cdot \left[m(^{85}Rb^+) - m_e\right] + m_e. \quad (3)$$

Table 2. Old mass-excess values for several neutron-deficient nuclides measured at JYFLTRAP employing Mo isotopes as references and the new values updated with the frequency ratios measured in this work. The difference between the old and new values is also tabulated. The isomer contribution has been taken into account for ^{83,84}Y and ^{85,87,88}Nb.

Isotope	Reference	ME_{old} (keV)	ME_{new} (keV)	$ME_{new} - ME_{old}$ (keV)
^{80}Y	^{96}Mo	−61144(7) [45]	−61146.8(6.1)	−2.8
^{81}Y	^{97}Mo	−65709(6) [45]	−65711.4(5.5)	−2.4
^{82}Y	^{98}Mo	−68060(6) [45]	−68062.8(5.5)	−2.8
^{83}Y	^{98}Mo	−72201(19)[1] [45]	−72204(19)[1]	−3.2
^{84}Y	^{97}Mo	−73922(20)[2] [46]	−73925(20)[2]	−2.4
^{83}Zr	^{98}Mo	−65908(7) [45]	−65910.6(6.1)	−2.6
^{84}Zr	^{98}Mo	−71418(6) [45]	−71420.3(5.6)	−2.3
^{85}Zr	^{98}Mo	−73170(6) [45]	−73173.3(5.8)	−3.3
^{86}Zr	^{98}Mo	−77958(7) [45]	−77961.0(5.9)	−3.0
^{87}Zr	^{97}Mo	−79341.4(5.3) [46]	−79343.9(5.0)	−2.5
^{88}Zr	^{98}Mo	−83624(7) [45]	−83627.2(6.2)	−3.2
^{85}Nb	^{98}Mo	−66308(21)[3] [45]	−66310(21)[3]	−2.4
^{86}Nb	^{98}Mo	−69129(6)[4] [45]	−69132.2(5.6)[4]	−3.2
^{87}Nb	^{98}Mo	−73870(7)[5] [45]	−73873.0(6.5)[5]	−3.1
^{88}Nb	^{98}Mo	−76169(14)[6] [45]	−76172(13)[6]	−2.5
^{96}Mo	^{97}Mo	−88789(6) [45]	−88791.3(5.5)	−2.3
^{98}Mo	^{97}Mo	−88111(6) [45]	−88113.0(5.6)	−2.0
^{91}Tc	^{94}Mo/^{85}Rb	−75984.8(3.3)[7] [46]	−75986.4(3.2)[7]	−1.6
^{91}Ru	^{94}Mo/^{85}Rb	−68237.1(2.9)[8] [46]	−68238.2(2.8)[8]	−1.1
^{94}Pd	^{94}Mo	−66097.9(4.7) [46]	−66100.9(4.4)	−3.0
^{95}Pd	^{94}Mo	−69961.6(4.8) [46]	−69964.6(4.5)	−3.0
^{96}Pd	^{94}Mo	−76179.0(4.7) [46]	−76182.0(4.3)	−3.0
^{99}Pd	^{96}Mo	−82178.9(5.1) [1]	−82181.9(4.8)	−3.0
^{101}Pd	^{96}Mo	−85427.1(5.2) [1]	−85430.2(4.9)	−3.1
^{101}Cd	^{96}Mo	−75827.8(5.6) [1]	−75830.9(5.2)	−3.1
^{102}Cd	^{96}Mo	−79655.6(5.3) [1]	−79658.7(4.9)	−3.1
^{103}Cd	^{96}Mo	−80648.5(5.3) [1]	−80651.6(5.0)	−3.1
^{104}Cd	^{96}Mo	−83962.9(5.6) [1]	−83966.1(5.2)	−3.2
^{105}Cd	^{96}Mo	−84330.1(5.5) [1]	−84333.4(5.1)	−3.3
^{102}In	^{96}Mo	−70690.4(5.4) [1]	−70693.5(5.1)	−3.1

[1] The original values of −72170(6) keV (old) and −72173.2(5.6) keV (new) were modified for an unknown mixture of isomeric states (^{83}Y^m at 61.98(11) keV [47]).

[2] The original values of −73888.8(5.2) keV (old) and −73891.2(4.9) keV (new) were modified for an unknown mixture of isomeric states (^{84}Y^m at 67 keV [48]).

[3] The original values of −66273(7) keV (old) and −66275.4(6.0) keV (new) were modified for an unknown mixture of isomeric states (^{85}Nb^m at $E_x \geq 69$ keV [49]).

[4] Possible contribution from an isomer at 250(160)# keV [47] has not been taken into account.

[5] The original values of −73868(7) keV (old) and −73871.1(6.4) keV (new) were modified for an unknown mixture of isomeric states (^{87}Nb^m at 3.84(14) keV [47]).

[6] The original values of −76149(7) keV (old) and −76151.5(6.1) keV (new) were modified for an unknown mixture of isomeric states (^{88}Nb^m at 40(140) keV [47]).

[7] A weighted mean of JYFLTRAP and SHIPTRAP values (see ref. [46]). The JYFLTRAP values with ^{94}Mo reference are −75983.4(45) keV (old) and −75986.3(4.2) keV (new).

[8] A weighted mean of JYFLTRAP and SHIPTRAP values (see ref. [46]). The JYFLTRAP values with ^{94}Mo reference are −68235.3(48) keV (old) and −68238.2(4.5) keV (new).

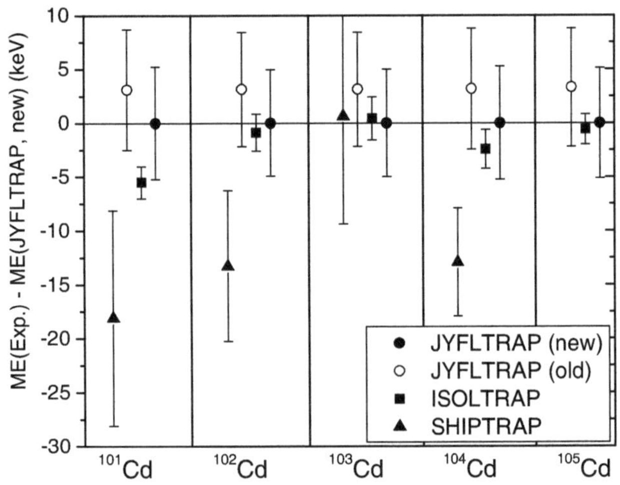

Fig. 12. Comparison of different Penning-trap measurements for $^{101-105}$Cd with respect to the new JYFLTRAP value. After adjusting the old mass-excess values measured against ^{96}Mo at JYFLTRAP with the ^{96}Mo value measured in this work, a new JYFLTRAP value is obtained. The new value agrees with the values from ISOLTRAP [3]. The SHIPTRAP values for ^{101,102,104}Cd still disagree with JYFLTRAP but the value for ^{103}Cd agrees well with JYFLTRAP and ISOLTRAP.

The updated values are collected in table 2. The Y, Zr, and Nb isotopes were measured as oxides. The uncertainties due to the isomers ^{83}Y^m ($E_x = 61.98(11)$ keV [47]), ^{84}Y^m ($E_x = 67$ keV [48]), ^{85}Nb^m ($E_x \geq 69$ keV [49]), ^{87}Nb^m ($E_x = 3.84(14)$ keV [47]), and ^{88}Nb^m ($E_x = 40(140)$ keV [47]) have been taken into account according to eq. (14) of ref. [50] and added quadratically to the experimental uncertainties. No correction due to the isomer in ^{86}Nb ($E_x = 250(160)\#$ keV [47]) has been done since this isomer is considered as uncertain. It should also be noted that for ^{85}Nb, the energy of the isomer is only a lower limit [49]. The previous values for ^{91}Tc and ^{91}Ru were published in ref. [46], which was a joint publication of JYFLTRAP and SHIPTRAP. Here, the JYFLTRAP values measured against ^{94}Mo have been updated and new weighted means of JYFLTRAP and SHIPTRAP values have been calculated for ^{91}Tc and ^{91}Ru. As can be seen from table 2, the updated values are on the average about 2.8 keV lower than the old values. This is well within the error bars.

Although the 3 keV shift in the mass excesses of Mo isotopes is less than 1σ, it is important to take it into account. For example, the Cd isotopes have been measured at SHIPTRAP [2] and ISOLTRAP [3] by using ^{85}Rb as a reference. The mass-excess values determined with JYFLTRAP for Cd isotopes employing ^{96}Mo as a reference [1] disagreed in some cases with ISOLTRAP and SHIPTRAP (see fig. 12). The shift from the AME03 value in the mass of ^{96}Mo was already observed in the mass evaluation performed in ref. [3]. In this work, we have experimentally determined this mass value. The updated Cd values (see table 2) agree within one standard deviation with the ISOLTRAP data. However, the SHIPTRAP values for ^{101,102,104}Cd still deviate from the ISOLTRAP and JYFLTRAP data.

6 Conclusions

In this paper, we have reported frequency ratios between $^{92,94-98,100}$Mo and ^{85}Rb measured with the JYFLTRAP setup. The mass-excess values of the Mo isotopes have been determined with about 1 keV precision, which is at least by a factor of 2 more precise than in the AME03. In addition, all measured stable Mo isotopes have been found to be more bound than given in the AME03. This will also have an effect on the nuclides which have main influences coming from these Mo isotopes in the AME03, such as for neighbouring Nb, Mo and Tc nuclides.

^{94,96,97,98}Mo have been used as references for 30 neutron-deficient nuclides measured at JYFLTRAP, and thus, these values have been updated with the new molybdenum values. Although the difference to the previous values is less than 1σ, it is worthwhile to take it into account for example when comparing to results from other facilities. In addition, proton-capture rates relevant for astrophysical *rp* [51,52] and νp [53,54] processes depend exponentially on proton separation energies, and already a small change will have an effect on the rate. In any case, the stable molybdenum isotopes are now more accurate references for future mass measurements of neutron-deficient nuclides.

This measurement was motivated by the discrepancy in the cadmium mass-excess values between JYFLTRAP, ISOLTRAP and SHIPTRAP. An inaccurate mass-excess value of ^{96}Mo in the literature has now been confirmed to be the reason for the deviation. This gives a perfect example why the main result from a Penning-trap measurement should be rather the frequency ratio between the reference ion and the ion of interest and its uncertainty rather than the mass-excess value itself. This way one can always use the most accurate value for the mass of the reference ion and recalculate the mass values of ions of interest.

This work has been supported by the EU 6th Framework programme "Integrating Infrastructure Initiative - Transnational Access", Contract Number: 506065 (EURONS) and by the Academy of Finland under the Finnish Centre of Excellence Programme 2006-2011 (Nuclear and Accelerator Based Physics Programme at JYFL). AK acknowledges the support from the Accademy of Finland under the project 127301.

References

1. V. Elomaa *et al.*, Eur. Phys. J. A **40**, 1 (2009).
2. A. Martín *et al.*, Eur. Phys. J. A **34**, 341 (2007).
3. M. Breitenfeldt *et al.*, Phys. Rev. C **80**, 035805 (2009).
4. G. Audi, A.H. Wapstra, C. Thibault, Nucl. Phys. A **729**, 337 (2003).
5. A. Jokinen *et al.*, Int. J. Mass Spectrom. **251**, 204 (2006).
6. J. Äystö, Nucl. Phys. A **693**, 477 (2001).
7. H. Penttilä *et al.*, Eur. Phys. J. A **25**, 745 (2005).
8. A. Nieminen *et al.*, Nucl. Instrum. Methods Phys. Res. A **469**, 244 (2001).
9. V.S. Kolhinen *et al.*, Nucl. Instrum. Methods Phys. Res. A **528**, 776 (2004).
10. G. Savard *et al.*, Phys. Lett. A **158**, 247 (1991).
11. L.S. Brown, G. Gabrielse, Rev. Mod. Phys. **58**, 233 (1986).

12. M. König *et al.*, Int. J. Mass Spectrom. Ion Process. **142**, 95 (1995).
13. S. George *et al.*, Int. J. Mass Spectrom. **264**, 110 (2007).
14. M. Kretzschmar, Int. J. Mass Spectrom. **264**, 122 (2007).
15. A. Kellerbauer *et al.*, Eur. Phys. J. D **22**, 53 (2003).
16. R.T. Birge, Phys. Rev. **40**, 207 (1932).
17. V.-V. Elomaa *et al.*, Nucl. Instrum. Methods Phys. Res. A **612**, 97 (2009).
18. G. Audi, Hyperfine Interact. **132**, 7 (2001).
19. Particle Data Group, Phys. Lett. B **667**, 103 (2008).
20. M.A. Islam, T.J. Kennett, W.V. Prestwich, Can. J. Phys. **69**, 658 (1991).
21. R. Firestone *et al.*, *Database of prompt gamma rays from slow neutron capture for elemental analysis* (Lawrence Berkeley National Laboratory, 2004) lBNL Paper LBNL-55199.
22. R.R. Ries, R.A. Damerow, W.H. Johnson, Phys. Rev. **132**, 1662 (1963).
23. R. Bishop *et al.*, Can. J. Phys. **41**, 1532 (1963).
24. L.M. Langer, D.E. Wortman, Phys. Rev. **132**, 324 (1963).
25. J. Äystö *et al.*, Nucl. Phys. A **404**, 1 (1983).
26. J. Hamilton, K. Löbner, A. Sattler, R.V. Lieshout, Physica **30**, 1802 (1964).
27. R.E. Snyder, G.B. Beard, Phys. Rev. **147**, 867 (1966).
28. J.C. Hocquenghem, S. André, P. Liaud, J. Phys. France **29**, 138 (1968).
29. T. Cretzu, K. Hohmuth, J. Schintlmeister, Nucl. Phys. **70**, 129 (1965).
30. N. Antoneva *et al.*, Izv. Akad. Nauk SSSR, Ser. Fiz. **38**, 48 (1974).
31. C.N. Rao, B.M. Rao, P.M. Rao, K.V. Reddy, *Proceedings of the 17 Nuclear Physics and Solid State Physics Symposium, Bombay* Vol. **17** (Department of Atomic Energy, Government of India, 1974) p. 10.
32. M.R. McPherson, F. Gabbard, Phys. Rev. C **7**, 2097 (1973).
33. G. Doukellis *et al.*, Nucl. Phys. A **229**, 47 (1974).
34. B.D. Kern *et al.*, Phys. Rev. C **18**, 1938 (1978).
35. R.L. Kozub, D.H. Youngblood, Phys. Rev. C **7**, 404 (1973).
36. P.K. Bindal, D.H. Youngblood, R.L. Kozub, Phys. Rev. C **10**, 729 (1974).
37. P. De Gelder *et al.*, Phys. Rev. C **25**, 146 (1982).
38. J.R. Comfort, R.W. Finlay, C.M. McKenna, P.T. Debevec, Phys. Rev. C **10**, 1236 (1974).
39. D.J. Martin, M.R. Macphail, Phys. Rev. C **13**, 1117 (1976).
40. R.C. Pardo *et al.*, Phys. Rev. C **21**, 462 (1980).
41. R.F. Casten, E.R. Flynn, O. Hansen, T.J. Mulligan, Nucl. Phys. A **184**, 357 (1972).
42. E.R. Flynn, R.E. Brown, F. Ajzenberg-Selove, J.A. Cizewski, Phys. Rev. C **28**, 575 (1983).
43. C. Wagemans, J. Wagemans, G. Goeminne, Hyperfine Interact. **132**, 323 (2001).
44. H. Seyfarth *et al.*, Fizika (Croatia) **22**, 183 (1990).
45. A. Kankainen *et al.*, Eur. Phys. J. A **29**, 271 (2006).
46. C. Weber *et al.*, Phys. Rev. C **78**, 054310 (2008).
47. G. Audi, O. Bersillon, J. Blachot, A.H. Wapstra, Nucl. Phys. A **729**, 3 (2003).
48. J. Döring, A. Aprahamian, M. Wiescher, J. Res. Natl. Inst. Stand. Technol. **105**, 43 (2000).
49. A. Kankainen *et al.*, Eur. Phys. J. A **25**, 355 (2005).
50. A.H. Wapstra, G. Audi, C. Thibault, Nucl. Phys. A **729**, 129 (2003).
51. H. Schatz *et al.*, Phys. Rep. **294**, 167 (1998).
52. H. Schatz, Int. J. Mass Spectrom. **251**, 293 (2006).
53. C. Fröhlich *et al.*, Phys. Rev. Lett. **96**, 142502 (2006).
54. J. Pruet *et al.*, Astrophys. J. **644**, 1028 (2006).

DOI 10.1007/978-94-007-5555-0_26
Reprinted from *Hyperfine Interactions* Volume 223, Issues 1-3, 2014
and *The European Physical Journal A* Volume 48, Issue 4, 2012

THE EUROPEAN
PHYSICAL JOURNAL A

Regular Article – Experimental Physics

High-precision Q_{EC}-value measurements for superallowed decays

T. Eronen[1,a] and J.C. Hardy[2,b]

[1] University of Jyväskylä, Department of Physics, P.O. Box 35 (YFL), FI-40014, Finland
[2] Cyclotron Institute, Texas A&M University, College Station, Texas 77843, USA

Received: 30 January 2012 / Revised: 23 February 2012
Published online: 19 April 2012 –
Communicated by J. Äystö

Abstract. The superallowed β-decay $Q_{\rm EC}$-value measurement program at JYFLTRAP has been very fruitful with 14 $Q_{\rm EC}$ values of outstanding precision measured between 2005 and 2010, when the IGISOL and JYFLTRAP facilities were shut down for relocation.

1 Introduction

There are few topics in nuclear physics that have been studied intensively for nearly 60 years. Yet the first measurement of superallowed $0^+ \to 0^+$ β transitions between $T = 1$ nuclear analog states was made on the decays of ^{10}C and ^{14}O in 1953 by Sherr and Gerhart [1], while the most recent superallowed-decay $Q_{\rm EC}$-value measurement, coincidentally also involving ^{10}C, was published in 2011 [2]. In 1953, the objective was to see if the ft values for ^{10}C and ^{14}O were equal, as predicted by the Fermi theory of β decay. Today, with many more superallowed transitions available, the consistency of their ft values is still of interest, but it is the opportunity they offer to test the universality of the weak interaction and the validity of the Standard Model that has become a driving force for the field.

As it was in 1953, the experimental focus of the field is still on precision. If the weak vector current is conserved, how precisely is it conserved? If the Standard Model depends on the absence of scalar currents and on the unitarity of the Cabibbo-Kobayashi-Maskawa (CKM) matrix, how definitively can these requirements be tested experimentally? This quest for higher and higher precision has pushed experimenters and theorists to greater and greater heights. In 1953, the ft values for ^{10}C and ^{14}O were determined to $\pm 41\%$ and $\pm 23\%$, respectively, a *tour de force* for the time. Today, there are ten different $0^+ \to 0^+$ superallowed transitions known to $\sim 0.1\%$ precision and three more known to between 0.1% and 0.3%. In recent years IGISOL and JYFLTRAP have played an important role in helping to achieve the current level of precision.

2 Superallowed $0^+ \longrightarrow 0^+\beta$ decay

What makes superallowed $0^+ \to 0^+\beta$ transitions between $T = 1$ nuclear analog states so attractive is that their measured ft values are nearly independent of nuclear-structure ambiguities, and they depend uniquely on the vector (and scalar, if it exists) part of the weak interaction. Neglecting for now the possibility of any scalar current, the ft value for such a transition can be related directly to the vector coupling constant, $G_{\rm V}$, by the following equation [3]:

$$\mathcal{F}t \equiv ft(1+\delta_R')(1+\delta_{NS}-\delta_C) = \frac{K}{2G_V^2(1+\Delta_V^R)}, \quad (1)$$

where $\mathcal{F}t$ is defined to be the "corrected" ft value and $K/(\hbar c)^6 = 2\pi^3\hbar\ln 2/(m_e c^2)^5 = 8120.2787(11) \times 10^{-10}\,\text{GeV}^{-4}$s. There are four small theoretical correction terms: δ_C is the isospin-symmetry-breaking correction; Δ_V^R is the part of the radiative correction that is transition independent; and the terms δ_R' and δ_{NS} comprise the transition-dependent part of the radiative correction, the former being a function only of the maximum positron energy and the atomic number, Z, of the daughter nucleus, while the latter, like δ_C, depends in its evaluation on the details of nuclear structure. All four correction terms are of the order of one percent, with their uncertainties thought to be about one tenth of their calculated values.

Superficially, based on eq. (1), the precisely measured ft value for a single $0^+ \to 0^+$ superallowed β transition would seem to be enough to obtain a precise value for $G_{\rm V}$.

[a] *Present address*: Max-Planck-Institut für Kernphysik, Saupfercheckweg 1, 69117 Heidelberg, Germany;
e-mail: `tommi.o.eronen@jyu.fi`
[b] e-mail: `hardy@comp.tamu.edu`

However, there is a problem with such a simple approach: the result for G_V depends on four calculated correction terms, two of which are nuclear-structure dependent. Can one really be sure that the result from a single transition is reliable at a precision of 0.1% or below without any independent confirmation of the calculated correction terms? To respond to this question and give credibility to the nuclear result for G_V, many superallowed transitions have been measured precisely, all of which have so far turned out to yield statistically identical results for $\mathcal{F}t$, and hence for G_V. Since the uncorrected ft values actually scatter over a relatively wide range, it is the structure-dependent corrections that are principally responsible for bringing the $\mathcal{F}t$ values into agreement with one another.

Obviously, this is a powerful experimental validation of the calculated corrections and fully justifies the extraction of G_V from the average value of $\mathcal{F}t$ via eq. (1). However, it also means that if the precision on G_V is to be improved, it is not enough to improve one ft value; there must be an improvement of the input data across the board. This is a challenging task but the ultimate payoff is considerable. With G_V reliably determined from the average of a set of consistent $\mathcal{F}t$ values, a value for the up-down quark-mixing element of the CKM matrix, V_{ud}, can be obtained via the relationship $V_{ud} = G_V/G_F$, where G_F is the weak-interaction constant for the purely leptonic muon decay. From there, one can test the unitarity of the CKM matrix by evaluating the sum of squares of its top-row elements: *viz.* $V_{ud}{}^2 + V_{us}{}^2 + V_{ub}{}^2$. In the most recent published review [4] of the subject, this sum was evaluated to equal 0.99990(60), showing CKM unitarity to be satisfied to a precision of 0.06%. The same review also describes how this result and its uncertainty constrain the possibilities for new physics beyond the Standard Model. An even more precise value would further constrain new physics or could possibly identify its presence. With nuclear superallowed β decay playing a key role in the unitarity test result, there is much motivation to strive for improvements in the experimental data that contribute to the determination of V_{ud}.

The experimental input to eq. (1) is entirely contained within the value of ft. Three measured quantities are required: the total transition energy, Q_{EC}; the half-life, $t_{1/2}$, of the parent state; and the branching ratio, R, for the particular transition of interest. The Q_{EC} value is required to determine the statistical rate function, f, while the half-life and branching ratio combine to yield the partial half-life, t. It is important to recognize, though, that f varies approximately with the fifth power of Q_{EC}: If the relative uncertainty in the measured Q_{EC} value is 1×10^{-4}, the corresponding relative uncertainty in f is $\sim5\times10^{-4}$. Thus, the precision required for Q_{EC}-value measurements is substantially higher than that required for half-lives and branching ratios.

Until the advent of on-line Penning traps, the Q_{EC} values for superallowed transitions were measured with nuclear reactions, most frequently by (p,n) or (^{3}He, t) reactions on the β-decay–daughter nuclei, which are stable for all the cases studied until recently. Figure 1 shows the relative precision, $\Delta Q/Q$, obtained for various measurements and techniques plotted as a function of publication date, starting in 1960. It can be seen that the relative precision obtained with such reaction measurements improved slowly but steadily over the years, from $\sim5\times10^{-4}$ in the early 1960s to $\sim3\times10^{-5}$ by 1990 but, at that point, a limit seemed to have been reached: for the next fifteen years there were no further improvements in Q_{EC}-value precision.

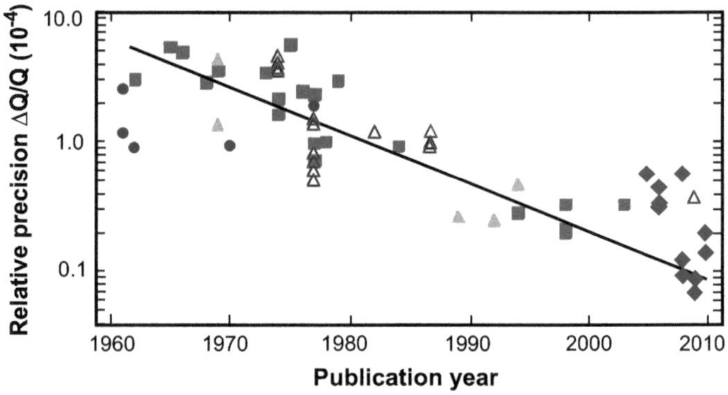

Fig. 1. (Color online) The relative precision, $\Delta Q/Q$, for Q_{EC}-value measurements of superallowed transitions is plotted against their publication date, where Q is the measured Q_{EC} value and ΔQ is its quoted uncertainty. The data encompass the superallowed transitions from ^{10}C, ^{14}O, ^{26}Al^m, ^{34}Cl, ^{38}K^m, ^{42}Sc, ^{46}V, ^{50}Mn and ^{54}Co, and are taken from a series of survey articles [3,5–7]. Each point is identified by the experimental method used in the corresponding measurement: solid green squares denote (p,n) reactions; open triangles, (^{3}He, t) reactions; solid blue circles, two-nucleon transfer reactions (p, t) or (^{3}He, n); solid yellow triangles, combined (p, γ) and (n, γ) reactions; and solid red diamonds, Penning-trap measurements. The line simply illustrates the decreasing trend.

This situation did not change even with the first appearance in 2005 of Q_{EC}-value results from on-line Penning traps: initially, their relative precision was no lower than the value at which the reaction measurements had been stuck for fifteen years. However, as can be seen in fig. 1, in the space of only five years since their first contribution, Penning traps have improved the relative precision for measured Q_{EC} values by a factor of 5, to as low as $\sim7\times10^{-6}$. A large fraction of these measurements — and, in fact, all the highest-precision ones— were made with JYFLTRAP at the Jyväskylä Cyclotron Laboratory, where superallowed Q_{EC}-value measurements were begun almost immediately after the JYFLTRAP Penning-trap setup had been first commissioned. Since then, Q_{EC} values of 14 different superallowed β emitters have been measured and published [2,8–14]. The chart of nuclides in fig. 2 shows the parent nuclei of a number of superallowed β transitions and indicates for each the Penning trap(s) that have measured its Q_{EC} value.

Since JYFLTRAP began operation, the setup has undergone several upgrades —for example, implementation of ion-motion excitation with Ramsey's method of time-separated oscillatory fields [27,28]— which have improved the precision of the measurements considerably. The methods used and their improvements will be described in the following section.

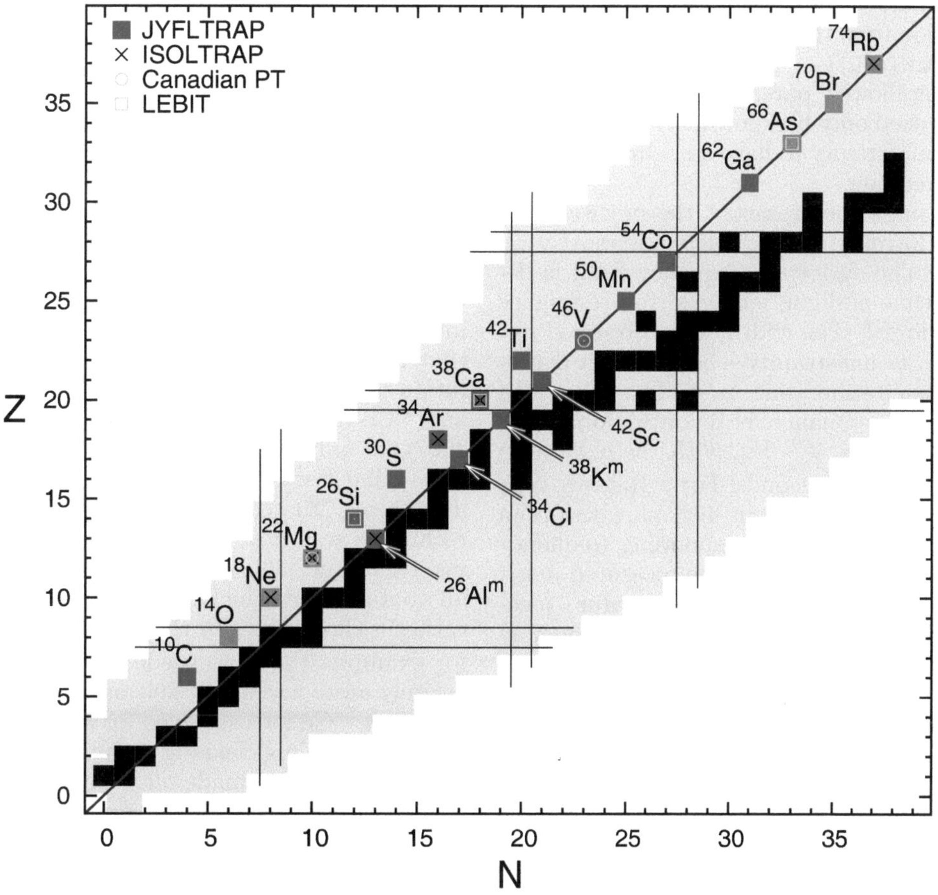

Fig. 2. (Color online) Partial nuclear chart in which the most relevant superallowed β emitters are shown. Those whose Q_{EC} values have been measured with a Penning trap are indicated. So far JYFLTRAP (solid red symbols), ISOLTRAP (crosses) [15–21], Canadian PT (circles) [22,23] and LEBIT (open squares) [24–26] have contributed to these studies.

3 Experimental methods

The IGISOL and JYFLTRAP setups are thoroughly described elsewhere in this special issue of the European Physical Journal A. What is most important for the Q_{EC}-value measurements of interest here is that the IGISOL technique makes possible the simultaneous availability of ion beams corresponding to both the parent and the daughter of a superallowed transition. This enables a direct measurement of their cyclotron frequency ratio, which yields their mass difference —the Q_{EC} value— without recourse to any independent calibration standard. With one exception (the Canadian Penning-trap measurement of the ^{46}V Q_{EC} value [23]), all other Penning-trap measurements have determined the mass of the superallowed parent relative to a different calibration mass (see caption of fig. 2 for references). Although in another case [21], the daughter's mass was also measured, it was in a separate beam-time period and with a separate calibration. If a Q_{EC} value is determined in this way from two independent mass measurements, its precision is degraded both by systematic mass-dependent frequency shifts encountered when a calibration mass having a different A-value is used, and by the uncertainty of the calibration mass itself. Of course, if the daughter mass is not remeasured at all, then the uncertainty in its value adopted from the literature contributes as well.

With the A/q doublet technique used at JYFLTRAP, the frequency ratio is taken between ions with the same A value so mass-dependent effects become negligible. Furthermore, the result is essentially unaffected by any reference masses. A measured cyclotron frequency relates to the corresponding mass by the relation

$$\nu_c = \frac{1}{2\pi}\frac{q}{m}B, \tag{2}$$

where q/m is the charge-to-mass ratio of the ions and B is the magnetic field. The Q_{EC} value for singly charged ions can then be written as

$$\begin{aligned} Q_{\mathrm{EC}} &= M_p - M_d \\ &= \left(\frac{\nu_{c,\mathrm{d}}}{\nu_{c,\mathrm{p}}} - 1\right)(M_d - m_e) + \Delta_{p,d}, \end{aligned} \tag{3}$$

where $\nu_{c,\mathrm{p}}$ and $\nu_{c,\mathrm{d}}$ are the measured cyclotron frequencies of the parent and daugher ions, respectively; m_e is the electron rest mass, and the term $\Delta_{p,d}$ arises from atomic-electron binding-energy differences between the parent and daughter known to sub-eV accuracy. At JYFLTRAP, so far, only singly charged ions have been used and the maximum contribution from the $\Delta_{p,d}$ term has been about 20 eV. Since the term $(\nu_{c,d}/\nu_{c,p} - 1)$ in eq. (4) is always $\ll 10^{-3}$ for the superallowed parent-daughter pairs, the daughter mass, M_d, need only be known to few keV precision in order for its uncertainty to have a negligible impact on the Q_{EC}-value precision.

In Penning-trap mass spectrometry, the precision obtainable for the cyclotron frequency depends mostly on the excitation time. The cyclotron frequency itself is determined with the time-of-flight ion-cyclotron-resonance (TOF-ICR) technique [29]. The width of the observed resonance —and hence its uncertainty— is established predominantly by the excitation time used. The full width at half maximum of the resonance for a conventional type of excitation is $\Delta\nu_{FWHM} \approx \frac{0.9}{T_{RF}}$ Hz, with the excitation time T_{RF} being expressed in seconds. For a Ramsey-type excitation [27,28], the width is about 40% narrower than that. Also, since the width and corresponding frequency uncertainty is constant irrespective of the measured mass, the uncertainty achievable in mass (or Q_{EC}-value) measurements is smaller for lighter nuclei. The natural way to report performance of the spectrometer is to provide the Q_{EC}-value uncertainty relative to the mass of the ion: *viz.* $\Delta Q/M$. Note, however, that the ratio $\Delta Q/Q$ is better suited to any assessment of the contribution of a measured Q_{EC} value to the ft value of the associated β-transition (see fig. 1).

3.1 The early JYFLTRAP Q_{EC}-value measurements

Soon after the JYFLTRAP setup had been commisioned and the very first TOF-ICR observed in 2003, the experimental program to measure superallowed Q_{EC} values was initiated. The first Q_{EC}-value result was for ^{62}Ga, which was measured in 2005 and published in 2006 [8]. At that time, the data-acquisition system did not allow single-bunch data to be recorded, but instead data from several bunches were collected and only the total read out. Unfortunately, this prohibited any so-called countrate class analysis [30], which exposes cyclotron frequency shifts that may occur as a function of the number of ions simultaneously stored in the trap. The frequency is expected to shift if impurity ions are present. Being unable to extract this information from the ^{62}Ga measurement itself, we subsequently studied the behavior of ^{62}Cu ions as a function of the number of trapped ions and used that information to apply a systematic uncertainty to the ^{62}Ga result. The final uncertainty on the latter was ±540 eV, which corresponds to $\Delta Q/M \approx 9 \times 10^{-9}$.

The next cases measured were the Q_{EC} values of ^{26}Al^m, ^{42}Sc and ^{46}V, all three measured in 2006 and published together in the same year [9]. The ^{46}V result confirmed a measurement made with the Canadian Penning trap and published a year earlier [23], which had deviated considerably from a reaction-based Q_{EC}-value result that had been accepted for three decades. In the same year, the ^{26}Si and ^{42}Ti Q_{EC} values were also measured but not published until later [11,13]. All five measurements used a control system that allowed single-bunch recording [31] and countrate class analysis. This reduced considerably our uncertainties compared to the earlier ^{62}Ga results, down to $\Delta Q/M \approx 5 \times 10^{-9}$, which corresponds to about ±250 eV for ^{42}Sc.

3.2 Later measurements

In many cases the $T_z = 0$ nucleus in a parent-daughter pair has a high-spin isomeric state near the $T = 1$ state of interest. Sometimes these two states are so close in energy that the sideband cooling technique has inadequate resolution [32] to separate them. This was the case for ^{50}Mn and ^{54}Co [10], which were measured in late 2006 and early 2007: not only was high resolution required to separate the close-lying states, but the half-lives of both nuclei were under 300 ms. To deal with this situation, a completely new technique was developed —the so-called Ramsey cleaning technique [33]— which allowed complete removal of all contaminants, including the nearby isomeric state, in rather a short time. In the case of ^{54}Co ($T_{1/2} \approx 200$ ms), for example, the time needed to clear the isomer ^{54}Co^m entirely away was about 200 ms, a short enough time that sufficient ^{54}Co ions survived for a TOF-ICR measurement to be made. This cleaning technique was also essential to the measurements made on ^{34}Cl and ^{38}K^m in 2009 [12].

Undoubtedly the most significant boost to our precision came from our use of ion-motion excitation with Ramsey's method of time-separated oscillatory fields [34]. This became possible as soon as the ideal resonance lineshape became available as an analytical function in 2007 [27]. A refined version of the function, which incorporates ion-motion damping due to friction-like ion collisions with the buffer gas, appeared in 2010 [35]. As already described, most of our later measurements, from 2007 on, have been done with Ramsey's method, thus reducing our statistical uncertainty by a factor of three compared with the conventional excitation used previously. Certainly, precision can also be improved by an increase in the excitation time, but the short half-lives of the superallowed β emitters preclude that option for them. The Ramsey method offers improved precision with no increase in the excitation time. The best precision, $\Delta Q/M$, we have achieved so far is $\approx 1 \times 10^{-9}$.

3.3 Other precision improvements

Although the major boosts to precision have already been described, other small improvements have been implemented as well over the years.

3.3.1 Recooling and recentering

We performed our earliest measurements in a rather straightforward manner by first accumulating ions in the RFQ trap [36], then purifying them in the purification trap [37], and finally measuring their ν_c in the precision

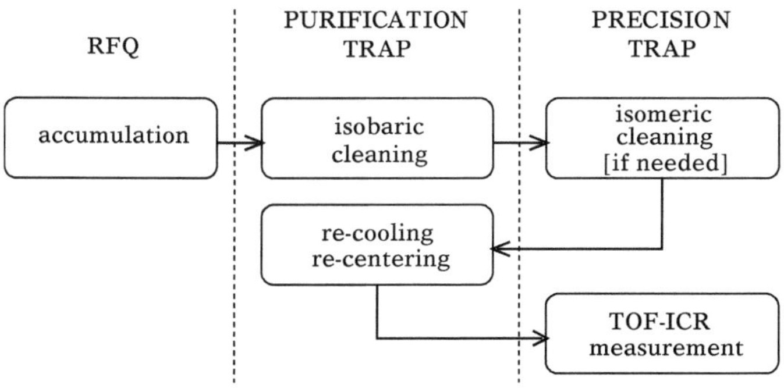

Fig. 3. The measurement cycle.

trap. As we gained experience, though, we realized that the physical size of an ion bunch and its energy spread are affected by how many ions are present in the purification trap during the purification process. If 5 ions are purified from a bunch containing 10^3 ions, the resulting 5 ion bunch is much bigger than if the 5 ions had been purified from a bunch containing only 10 ions. A ratio of approximately one in 10^3 is typical for superallowed parent and daughter yields at IGISOL.

If the bunch sizes are different, the ions within them are exposed to different parts of the magnetic field in the trap. Since some inhomogeneity in the field is unavoidable, the resonance curves obtained from two sets of ions prepared under even slightly different conditions can exhibit a relative frequency shift between them. To remove this systematic effect and obtain the highest precision in measuring the frequency ratio for a mass doublet, one must ensure that the initial conditions in the measurement trap for both are as nearly identical as possible.

In our later measurements, after introduction of the Ramsey cleaning technique [33], we reset any effects caused by excessive loading of the purification trap by adding an additional cooling step in the purification trap. In fact, whether we use Ramsey cleaning or not, we now resend each purified ion bunch from the precision trap back to the purification trap. At that stage, the only ions present are the few that will ultimately be used for the TOF-ICR measurement, so the centering quadrupole excitation in the purification trap efficiently centers the ions and makes all bunches equally small. After they are sent again to the precision trap the subsequent quadrupole excitation there results in a high-quality TOF-ICR resonance curve obtained under well-controlled conditions, which are the same for both parent and daughter nuclei. A typical measurement cycle —taken from our most recent measurement [2]— appears in fig. 3.

3.3.2 Resonance quality factor

We generated our first TOF-ICR resonance curves using only the average time of flight (TOF) from many frequency scans plotted against the applied frequency; however, we have since discovered that a visualization of the

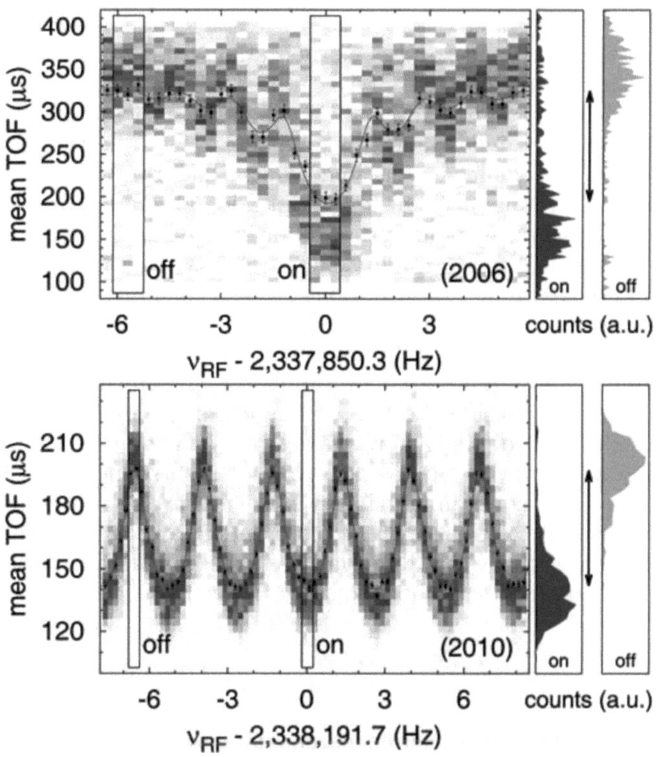

Fig. 4. (Color online) Top: TOF-ICR curve obtained for ^{46}V in 2006 with conventional excitation techniques [38]. An excitation time of 700 ms was used. Bottom: TOF-ICR curve obtained for ^{46}V in 2010 with the Ramsey technique [27]. In this case the excitation time pattern was 25-350-25 ms (on-off-on). In both panels, the blue pixels depict the numbers of detected ions: the darker the pixel the more ions it represents. The two small panels located on the right of each large panel show the distribution of on- and off-resonance ions corresponding to the data inside the rectangles in the main plot. The arrow between the panels shows the difference of the average time of flights. The calculated quality factor F in the 2006 data is about 1; the quality factor obtained for its daughter ^{46}Ti was somewhat better (~ 2). However the quality factor for both ^{46}V and ^{46}Ti obtained in 2010 was about 3. In the latter case no contaminant ions were present.

quality of the data can also be built in by having the plot include all the individual TOF points from each frequency scan. Figure 4 uses this method to illustrate the extent to which the quality of our resonance curves for ^{46}V has improved between 2006 and 2010. This method of plotting the data has proven to be very informative since it shows not only the resonant ions but also the non-resonant contaminant ions as well.

A more quantitative measure of data quality is provided by the so-called quality factor F of a TOF-ICR curve, which is defined as

$$F = \frac{|TOF_{\mathrm{ON}} - TOF_{\mathrm{OFF}}|}{\sqrt{\sigma^2_{\mathrm{ON}} + \sigma^2_{\mathrm{OFF}}}}, \qquad (4)$$

where the subscript "ON" refers to ions close to the resonance frequency ν_c, and "OFF" to ions far from resonance. The value of TOF_x is the average time of flight of the ions of type x, and σ_x is their standard deviation. For example, in fig. 4 the "ON" and "OFF" resonance ions used in the right-hand panels are those enclosed by the labeled rectangles in the main panels. The selection for "OFF" resonance ions is somewhat arbitrary: Any region with slow time-of-flight ions would do. The F factor of a resonance measurement thus quantifies the quality of that resonance. With a quality factor of less than 1, the on- and off-resonance distributions are less than one standard deviation apart. The higher the factor becomes, the better is the separation between the on- and off-resonance ions. This is illustrated by the side panels in fig. 4. For those, data inside the two rectangles in the main plot is used. The vertical arrow between the two panels shows the difference of the average time of flights. In the 2006 data some contaminant ions were evidently present since all ions are not distributed close to the average time-of-flight points. Naturally, a larger F also means improved precision in fitting the resonance frequency itself.

3.3.3 Fast switching between parent and daughter ions

Until 2009, we measured parent-daughter pairs by scanning the frequency with parent ions for as long as one hour to obtain sufficient statistics for resonance-curve fitting, and then switching to daughter-ion frequency scans for a similar length of time. There could be as much as two hours between the beginning of the parent scans and the end of the daughter scans. Because the measured magnetic field drift is $3.22 \times 10^{-11}\,\text{min}^{-1}$ [39], it could reach the 10^{-9} level over that length of time, and would thus cause frequency shifts that needed to be accounted for in determining the Q_{EC} value and its uncertainty.

Beginning with our measurements on ^{34}Cl and ^{38}K^m in 2009 we have effectively eliminated any dependence on frequency shifts caused by a drifting magnetic field. We have done that by switching back and forth between parent and daughter ions after every complete frequency scan, typically every minute or so. One scan, however, does not contain enough statistics to perform reliable data fitting. Thus the results from a number of scans need to be combined for both the parent and the daughter. Even though the data from a time span of one hour (in the worst case) may need to be included, the data from both the parent and daughter ions have nevertheless been collected under essentially identical field conditions and no corrections are required. It should be noted, however, that magnetic-field drift still contributes to the statistical uncertainty, but its effect can ultimately be reduced simply by the collection of more data.

3.4 Summary of experimental improvements

Up to now, the Q_{EC} values of 14 different superallowed β transitions, with parent nuclei ranging from ^{10}C to ^{62}Ga, have been measured with JYFLTRAP. The error bars, which began at ± 540 eV have dropped to as low as ± 40 eV.

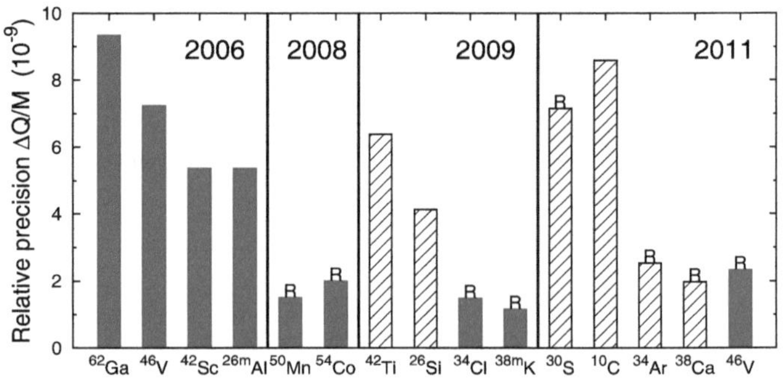

Fig. 5. (Color online) The $\Delta Q/M$ ratio obtained for each of the the 15 Q_{EC}-value measurements of 14 superallowed β transitions made at JYFLTRAP. The emitter nuclei are ordered according to their publication year, and those Q_{EC} values measured with the Ramsey method are indicated by "R". The isospin $T_z = -1$ nuclei are plotted with hatched bars and $T_z = 0$ emitters with solid red bars.

In terms of $\Delta Q/M$, the best measure for Penning-trap performance, the relative precision has been improved from $\approx 9 \times 10^{-9}$ to $\approx 1 \times 10^{-9}$ in 5 years. These results are summarized in fig. 5 where they appear chronologically. The available production rate of the ions of interest and the technique used to measure them, whether by conventional or Ramsey excitation, both play an important role in determining the precision achievable. There may also be specific problems associated with a particular case. Thus, although the best results have tended to improve from year to year, not all cases in any one year are necessarily better that those in the previous year.

4 Impact on superallowed β decay

The advent of Penning-trap Q_{EC}-value measurements has had two major impacts on superallowed β decay. First, these measurements, though at first not significantly more precise than the previous reaction-based results, were certainly subject to very different systematic uncertainties. If the results from both methods agreed, then their average resulted in a Q_{EC} value in which much confidence could be placed. Conversely, when the reaction-based results for several transitions, all published in one long-accepted reference [40], were found to disagree consistently with recent Penning-trap results, that reference was rejected with cause from the accepted data base. (See ref. [3] for a detailed discussion of this issue.)

Second, as the precision of Penning-trap measurements surpassed that of the reaction-based ones, these new results served to decrease the uncertainties on the ft values for the corresponding β transitions, or, in cases for which the measured half-life or branching ratio results dominated the overall uncertainty, they cleared the way for future improvements in those measurements to lead to immediate reductions in the ft-value uncertainty. We will deal with these two impacts of Penning-trap measurements separately.

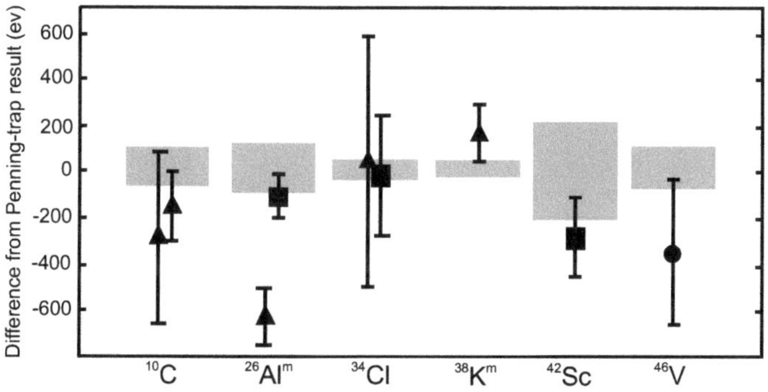

Fig. 6. Differences between precise reaction-based and Penning-trap Q_{EC}-value measurements for ^{10}C, ^{26}Al^m, ^{34}Cl, ^{38}K^m, ^{42}Sc and ^{46}V. Only measurements with uncertainties ≤ 540 eV are included. The (p,n)-threshold measurements are shown as triangles and are from refs. [41–45] (in order from $A = 10$ to $A = 38$). The $(p,\gamma)+(n,\gamma)$ measurements appear as squares and are from groups of references, which are given in ref. [3]. The (^{3}He, t) measurement for ^{46}V appears as a circle and is from ref. [46]. The grey bands about the zero line represent the uncertainty of the Penning-trap measurements, which are taken from the following references: ^{10}C [2]; ^{26}Al^m [9,17]; ^{34}Cl [12]; ^{38}K^m [12]; ^{42}Sc [9]; and ^{46}V [23,9,2].

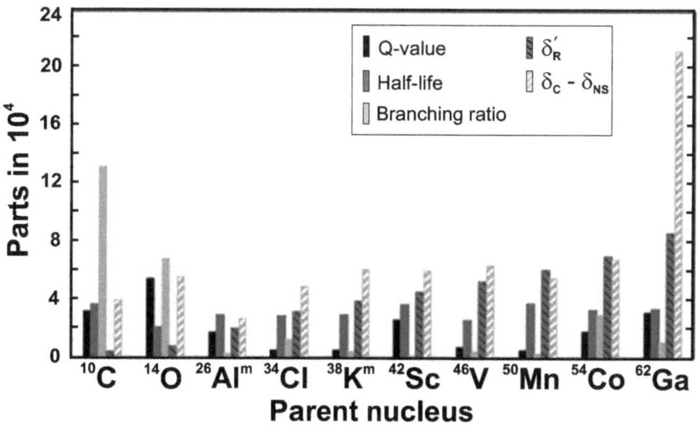

Fig. 7. Summary histogram of the fractional uncertainties attributable to each experimental and theoretical input factor that contributes to the final $\mathcal{F}t$ values for the ten superallowed transitions whose ft values are currently known with $\sim 0.1\%$ precision or better.

4.1 Comparison of reaction and Penning-trap results

There are six superallowed $0^+ \to 0^+$ transitions for which both Penning-trap and comparably precise reaction-based measurements exist. The results from both types of measurement are plotted for comparison in fig. 6. There it can be seen that the agreement between the reaction-based values (points with error bars) and Penning-trap ones (grey bands) is very good, with one exception: the (p,n)-threshold measurement for the ^{26}Al^m transition [43]. That case is a particularly difficult one to measure by the (p,n) reaction since there is a pronounced resonance in yield within a few keV of threshold, and a target impurity added further complications to the published measurement. Although great care was obviously taken in the experiment, the difficulties were apparently not fully overcome in this particular case. Clearly, though, it is not indicative of any large systematic discrepancy between results from the two experimental techniques.

The other four (p,n) results, the three $(p,\gamma)+(n,\gamma)$ ones and the one obtained from the (^{3}He, t) reaction all agree with the Penning-trap results within their uncertainties. Their weighted average difference (reaction result minus Penning-trap result) is $-64(69)$ eV. Any systematic difference between reaction and trap measurements, if it exists at all, must be less than ~ 130 eV, which is below —and usually well below— the uncertainties quoted on the reaction-based measurements themselves. At the level of precision currently required in dealing with world data for the evaluation of weak-interaction parameters, it appears that one can safely combine the results of both types of measurement without including any additional systematic uncertainties. This is an important conclusion.

4.2 Error budget for superallowed ft values

The best way to assess the impact of Penning-trap Q_{EC}-value measurements on the $\mathcal{F}t$ values for superallowed β decay is to examine the current status of the error budgets for the ten best-known superallowed transitions. These are presented in fig. 7, where the reader may also be reminded that Q_{EC} contributes to f approximately with its fifth power (see sect. 2). Thus, to make the same contribution to the error budget the precision required for a Q_{EC} value is five times more demanding than that required for a half-life or branching ratio.

Of the ten cases shown in fig. 7, only ^{14}O has yet to have its Q_{EC} value measured by a Penning trap. The fractional uncertainties for the Q_{EC} values of the remaining nine can be seen to range from 5×10^{-5} to 3×10^{-4}. In the 2005 survey of superallowed β decay [7], completed before Penning-trap measurements had appeared for any of these nuclei, the comparable range of fractional uncertainties was from 1×10^{-4} to 6×10^{-3}. The improvement has been significant in all cases, and dramatic in a few.

Looking in more detail at fig. 7, we find that for six cases the fractional uncertainties contributed by the transition Q_{EC} value are less than 2×10^{-4}, an amount below —and in the cases of ^{34}Cl, ^{38}K^m, ^{46}V and ^{50}Mn, well below— the contributions from other experimental and theoretical factors. Only for ^{26}Al^m in this group of six would a more precise measurement have an immediate effect in reducing the overall $\mathcal{F}t$-value uncertainty. The Q_{EC} values for three of the remaining four cases —^{10}C, ^{42}Sc and ^{62}Ga— all contribute $\sim 3 \times 10^{-4}$ to their respective budgets. From this group of three, only ^{42}Sc could have its overall uncertainty impacted by an improved Q_{EC} value; for ^{10}C and ^{62}Ga other factors, experimental in the former case and theoretical in the latter, dominate their error budgets. Clearly, the remaining case presented in the figure, ^{14}O, which has not yet been measured by Penning trap at all, could benefit significantly from a measurement with that technique.

5 Conclusions and outlook

The study of superallowed β decay has benefited enormously in the past five years from Penning-trap $Q_{\rm EC}$-value measurements. While a number of on-line Penning-trap facilities around the world have contributed to this new wave of measurements, the leading edge of precision has resided with the combined IGISOL and JYFLTRAP facility, which has unique capabilities for the measurement of precise $Q_{\rm EC}$ values. We have described the properties of this facility and the improvements that have been implemented since 2005, when the first $Q_{\rm EC}$ value was measured there. We have also illustrated how much these new measurements have affected the $\mathcal{F}t$ values used in testing the Standard Model.

Out of the most precisely known superallowed β emitters shown in fig. 7, ^{14}O is the only one for which the $Q_{\rm EC}$ value has not been measured with a Penning trap. Indeed ^{14}O is the only exception even if one includes all the thirteen superallowed β emitters that currently contribute to the world average $\mathcal{F}t$ [3]. In looking to the future, a Penning-trap measurement of the $Q_{\rm EC}$ value for the ^{14}O decay must obviously take the highest priority. However, given the proper yield many of the other cases could still be significantly improved with the doublet measurement technique. Not all the improved results would immediately impact the corresponding $\mathcal{F}t$ value since in many cases the half-life and/or branching-ratio uncertainties predominate. Nevertheless, these measurements should be made: ultimately, high precision will never be wasted.

This work was supported by the European Union (RTD project EXOTRAPS, Contract No. ERBFMGECT980099; RTD project NIPNET, Contract No. HPRI-CT-2001-50034; Fifth Framework Programme "Improving Human Potential-Access to Research Infrastructure", Contract No. HPRI-CT-1999-00044; 6th Framework programme "Integrating Infrastructure Initiative- Transnational Access" Contract No. 506065 (EURONS)) and by the academy of Finland under the Finnish Centre of Excellence Programmes 2000-2005 and 2006-2011. JCH was supported by the U.S. Department of Energy under Grant DE-FG03-93ER40773 and by the Robert A. Welch Foundation under Grant no. A-1397.

References

1. R. Sherr, J.B. Gerhart, Phys. Rev. **91**, 909 (1953).
2. T. Eronen *et al.*, Phys. Rev. C **83**, 055501 (2011).
3. J.C. Hardy, I.S. Towner, Phys. Rev. C **79**, 055502 (2009).
4. I.S. Towner, J.C. Hardy, Rep. Prog. Phys. **73**, 046301 (2010).
5. J.C. Hardy, I.S. Towner, Nucl. Phys. A **254**, 221 (1975).
6. J.C. Hardy *et al.*, Nucl. Phys. A **509**, 429 (1990).
7. J.C. Hardy, I.S. Towner, Phys. Rev. C **71**, 055501 (2005).
8. T. Eronen *et al.*, Phys. Lett. B **636**, 191 (2006).
9. T. Eronen *et al.*, Phys. Rev. Lett. **97**, 232501 (2006).
10. T. Eronen *et al.*, Phys. Rev. Lett. **100**, 132502 (2008).
11. T. Eronen *et al.*, Phys. Rev. C **79**, 032802 (2009).
12. T. Eronen *et al.*, Phys. Rev. Lett. **103**, 252501 (2009).
13. T. Kurtukian Nieto *et al.*, Phys. Rev. C **80**, 035502 (2009).
14. J. Souin *et al.*, Eur. Phys. J. A **47**, 1 (2011).
15. K. Blaum *et al.*, Nucl. Phys. A **746**, 305 (2004).
16. M. Mukherjee *et al.*, Phys. Rev. Lett. **93**, 150801 (2004).
17. S. George *et al.*, EPL **82**, 50005 (2008).
18. F. Herfurth *et al.*, Nucl. Instrum. Methods Phys. Res. A **469**, 254 (2001).
19. F. Herfurth *et al.*, Eur. Phys. J. A **15**, 17 (2002).
20. S. George *et al.*, Phys. Rev. Lett. **98**, 162501 (2007).
21. A. Kellerbauer *et al.*, Phys. Rev. Lett. **93**, 072502 (2004).
22. G. Savard *et al.*, Phys. Rev. C **70**, 042501 (2004).
23. G. Savard *et al.*, Phys. Rev. Lett. **95**, 102501 (2005).
24. A. A. Kwiatkowski *et al.*, Phys. Rev. C **81**, 058501 (2010).
25. R. Ringle *et al.*, Phys. Rev. C **75**, 055503 (2007).
26. P. Schury *et al.*, Phys. Rev. C **75**, 055801 (2007).
27. M. Kretzschmar, Int. J. Mass Spectrom. **264**, 122 (2007).
28. S. George *et al.*, Int. J. Mass Spectrom. **264**, 110 (2007).
29. G. Gräff, H. Kalinowsky, J. Traut, Z. Phys. A **297**, 35 (1980).
30. A. Kellerbauer *et al.*, Eur. Phys. J. D **22**, 53 (2003).
31. J. Hakala *et al.*, Nucl. Instrum. Methods Phys. Res. B **266**, 4628 (2008).
32. G. Savard *et al.*, Phys. Lett. A **158**, 247 (1991).
33. T. Eronen *et al.*, Nucl. Instrum. Methods Phys. Res. B **266**, 4527 (2008).
34. N.F. Ramsey, Rev. Mod. Phys. **62**, 541 (1990).
35. S. George *et al.*, Int. J. Mass Spectrom. **299**, 102 (2011).
36. A. Nieminen *et al.*, Nucl. Instrum. Methods Phys. Res. A **469**, 244 (2001).
37. V.S. Kolhinen *et al.*, Nucl. Instrum. Methods Phys. Res. A **528**, 776 (2004).
38. M. König *et al.*, Int. J. Mass Spectrom. **142**, 95 (1995).
39. S. Rahaman *et al.*, Eur. Phys. J. A **34**, 5 (2007).
40. H. Vonach *et al.*, Nucl. Phys. A **278**, 189 (1977).
41. P.H. Barker, R.E. White, Phys. Rev. C **29**, 1530 (1984).
42. P.H. Barker, P.A. Amundsen, Phys. Rev. C **58**, 2571 (1998).
43. S.A. Brindhaban, P.H. Barker, Phys. Rev. C **49**, 2401 (1994).
44. P.H. Barker, R.E. White, H. Naylor, N.S. Wyatt, Nucl. Phys. A **279**, 199 (1977).
45. P.D. Harty, N.S. Bowden, P.H. Barker, P.A. Amundsen, Phys. Rev. C **58**, 821 (1998).
46. T. Faestermann *et al.*, Eur. Phys. J. A **42**, 339 (2009).

DOI 10.1007/978-94-007-5555-0_27
Reprinted from *Hyperfine Interactions* Volume 223, Issues 1-3, 2014
and *The European Physical Journal A* Volume 48, Issue 4, 2012

THE EUROPEAN
PHYSICAL JOURNAL A

Regular Article – Experimental Physics

Mass measurements of neutron-deficient nuclei and their implications for astrophysics

A. Kankainen[1,a], Yu.N. Novikov[2,3], H. Schatz[4], and C. Weber[5]

[1] Department of Physics, University of Jyväskylä, P.O. Box 35, FI-40014 University of Jyväskylä, Jyväskylä, Finland
[2] Petersburg Nuclear Physics Institute, 188300 Gatchina, Russia
[3] GSI Helmholtzzentrum für Schwerionenforschung GmbH, Planckstrasse 1, D-64291 Darmstadt, Germany
[4] Department of Physics and Astronomy, National Superconducting Cyclotron Laboratory, and Joint Institute for Nuclear Astrophysics, Michigan State University, East Lansing, Michigan 48824, USA
[5] Faculty of Physics, Ludwig-Maximilians-Universität München, Am Coulombwall 1, D-85748 Garching, Germany

Received: 31 January 2012
Published online: 19 April 2012 –
Communicated by J. Äystö

Abstract. During the years 2005–2010 the double–Penning-trap mass spectrometer JYFLTRAP has been used to measure the masses of 90 ground and 8 isomeric states of neutron-deficient nuclides with a typical precision of better than 10 keV. The masses of 14 nuclides —^{84}Zr, ^{88,89}Tc, $^{90-92}$Ru, $^{92-94}$Rh, ^{94,95}Pd, ^{106,108,110}Sb— have been experimentally determined for the first time. This article gives an overview on these measurements and their impact on the modeling of the astrophysical *rp*-process.

1 Introduction

Very neutron-deficient nuclei are synthesized in astrophysical environments, where extreme temperatures and densities as well as an abundance of hydrogen can lead to a rapid proton capture process (*rp*-process) [1,2]. As proton capture rates become faster than β^+ decays, unstable neutron-deficient nuclei are built up. The most common occurrences of such a scenario are type-I X-ray bursts, which occur on the surface of neutron stars that accrete hydrogen-rich matter from a companion star in a stellar binary system. The hydrogen-rich fuel layer on the neutron star surface builds up for hours to days before it explodes giving rise to a bright X-ray burst typically lasting for 10–100 seconds. Such bursts are frequently observed with modern X-ray observatories and are directly powered by the nuclear energy generated in the *rp*-process reaching nuclei up to tellurium [3–5].

Recently, a new astrophysical process with rapid proton captures on neutron-deficient nuclei has been proposed, the νp-process [6–8]. It takes place in supernovae and possibly in gamma-ray bursts where proton-rich outflows are created by strong neutrino fluxes [6]. In the νp-process, the flow towards heavier elements is accelerated via fast (n,p) reactions on long-lived nuclides along the path of the *rp*-process. The neutrons that induce these reactions are created via antineutrino absorptions on protons. It has been shown that the reaction sequence can reach nuclei up to $A = 108$ or $A = 152$ depending on the electron fraction [9].

Nova explosions are a third astrophysical environment where a mild form of the *rp*-process takes place. They occur when an accreted fuel layer on the surface of a white dwarf star explodes. The *rp*-process in novae typically proceeds near stability and is predicted to end in the $A = 40$ region [10–13], though recent model calculations have identified parameters that can lead to much more violent explosions with a stronger *rp*-process reaching the iron region [14]. Accurate nuclear physics is needed to predict the contributions of novae and the νp-process to Galactic nucleosynthesis, to interpret X-ray burst light curves in terms of neutron star properties and to predict the composition of the ashes of X-ray bursts needed to model neutron star crusts. With accurate nuclear physics one can also predict new observables that observational programs can search for.

Nuclear masses play a central role in rapid proton capture processes [3,15–21]. When proton capture rates are high, it is photodisintegration that limits further proton captures and forces the reaction sequence to proceed via a β^+ decay or a (n,p) reaction. Such photodisintegration rates depend exponentially on the energy required to remove a proton from the nucleus, and therefore on the nuclear masses. In addition, many of the important proton capture rates in the *rp*-process are governed by resonances. Rates depend exponentially on resonance energies, which, in most cases, are determined by measuring excitation energies and nuclear masses. Typically, a precision of around

[a] e-mail: anu.k.kankainen@jyu.fi

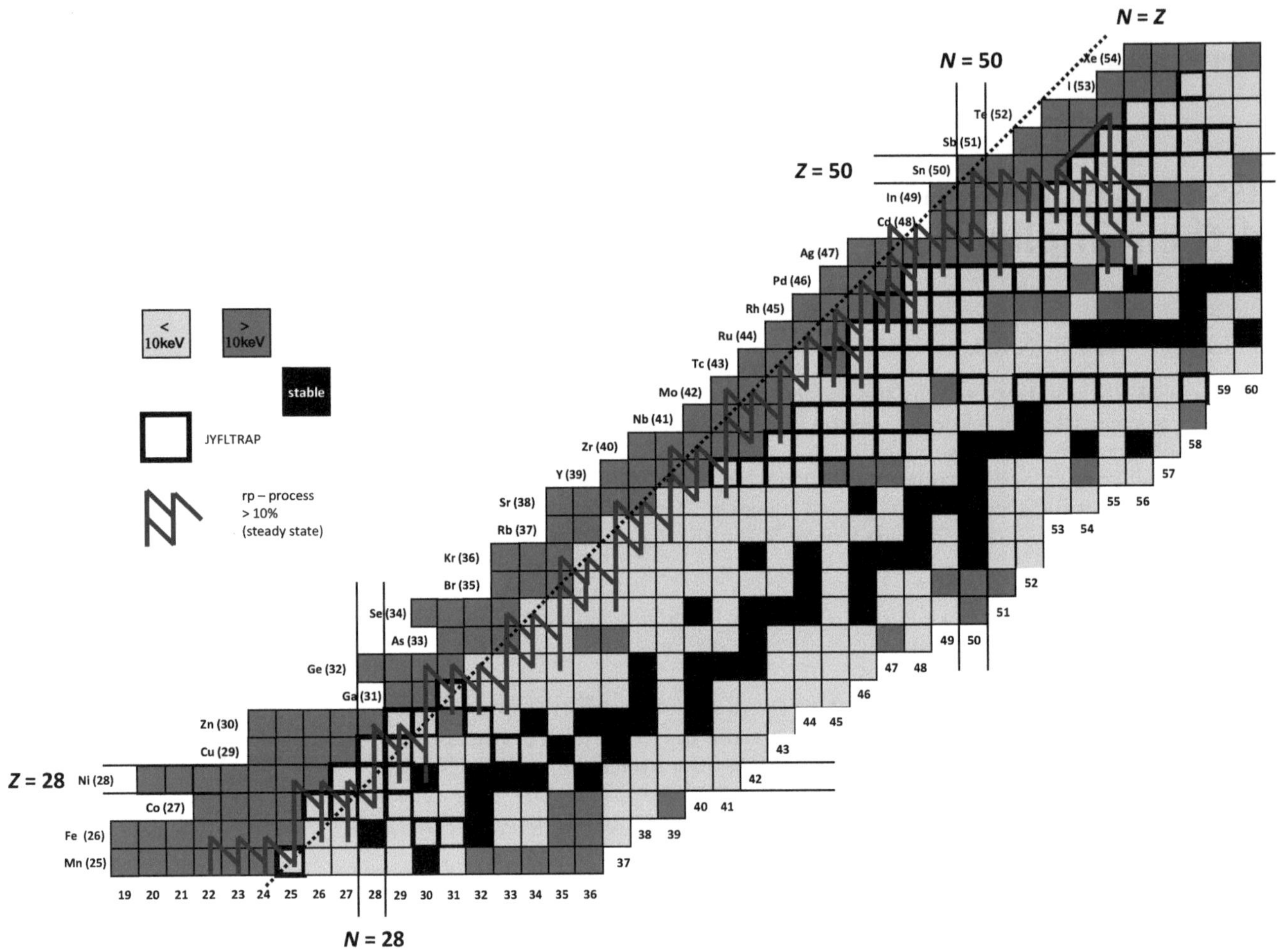

Fig. 1. (Color online) Chart of the nuclides showing the neutron-deficient nuclides measured at JYFLTRAP (bold black squares) and the nuclides whose mass excesses are known better than 10 keV (in yellow) and worse than 10 keV (in red, including the nuclides not yet experimentally determined). The main *rp*-process path (in blue) is plotted for conditions where the thermonuclear burning proceeds in steady state.

10 keV is required, for some resonant reaction rates even much better than 10 keV [16]. Penning-trap mass measurements have had a major impact on nuclear astrophysics, as they can easily reach or surpass such precision even for the most unstable nuclei within reach. This is illustrated in fig. 1, where Penning-trap mass measurements since the Atomic Mass Evaluation 2003 (AME03) are highlighted. Clearly, the Penning-trap approach has addressed the long-standing problem of unreliable and uncertain experimental masses of unstable nuclei and has been applied to a large number of neutron-deficient nuclei close to or in the *rp*-process. In some cases, the effort to push to the most exotic nuclei has led to the paradox situation that now some of the masses of exotic isotopes are better known than their more stable counterparts. The program to measure masses of unstable nuclei with JYFLTRAP has been a major contributor to this success together with CPT [22] at ANL, ISOLTRAP [23] at CERN-ISOLDE, LEBIT [24] at NSCL, and SHIPTRAP [25] at GSI.

In addition to nuclear astrophysics, neutron-deficient nuclei offer a wealth of other interesting physics phenomena to be studied. For example, Q_{EC}-value measurements at JYFLTRAP have been useful for testing the isobaric multiplet mass equation [26,27] and the Conserved Vector Current (CVC) hypothesis [28]. Masses of neutron-deficient nuclei provide also important data for developing and testing different mass models. Energy systematics, such as two neutron or proton separation energies, yield information, *e.g.*, on the possible onset of deformation and the evolution of shell-gap energies as well as on neutron-proton pairing and the Wigner energy [29].

2 Experimental methods

The ions of interest have been produced with beams from the K-130 cyclotron of the JYFL Accelerator Laboratory impinging on a thin (few mg/cm^2) target situated at the Ion-Guide Isotope Separator On-Line (IGISOL) facility [30]. Which ion guide is used depends on the applied beam: a heavy primary beam would cause plasma effects inside the gas cell and thus it is stopped before entering the gas cell whereas light beams, such as p or ^{3}He, can pass through the gas cell. In the light-ion ion guide, the

Table 1. The production methods of the neutron-deficient nuclides measured at JYFLTRAP. LI stands for the light-ion guide and HI for the heavy-ion ion guide technique.

E_{beam}	Beam	Target	Ion guide	Measured nuclides	Ref.
40 MeV	p	^{nat}Mg	LI	^{23}Al, ^{23}Mg	[26]
35 MeV	p	^{27}Al	LI	^{26}Si, ^{26}Al, $^{26}Al^{m}$	[33]
40 MeV	p	ZnS	LI	^{31}S	[27]
15 MeV	p	ZnS	LI	^{34}Cl, $^{34}Cl^{m}$	[34]
20 MeV	p	KCl	LI	^{38}K, $^{38}K^{m}$	[34]
20 MeV	^{3}He	^{nat}Ca	LI	^{42}Sc, $^{42}Sc^{m}$	[35]
17 MeV	^{3}He	^{nat}Ca	LI	^{42}Ti	[36]
20 MeV	p	^{46}Ti	LI	^{46}V	[35]
13–15 MeV	p	^{50}Cr	LI	^{50}Mn, $^{50}Mn^{m}$	[37]
40 MeV	p	^{54}Fe	LI	^{53}Co, $^{53}Co^{m}$	[38]
13–15 MeV	p	^{54}Fe	LI	^{54}Co, $^{54}Co^{m}$	[37]
105 MeV	^{20}Ne	^{nat}Ca	HI	^{56}Co	[38]
25 MeV	^{3}He	^{54}Fe	LI	$^{55,56}Ni$	[38]
75 MeV	^{20}Ne	^{nat}Ca	HI	^{57}Ni	[38]
40 MeV	p	^{58}Ni	LI	^{56}Ni, $^{57,58}Cu$	[38]
25 MeV	^{3}He	^{58}Ni	LI	$^{59,60}Zn$	[38]
48 MeV	p	^{64}Zn	LI	^{62}Cu, ^{62}Zn, ^{62}Ga	[39]
150–170 MeV	^{32}S	^{54}Fe	HI	$^{80-82}Y$	[40]
150–170 MeV	^{32}S	^{nat}Ni	HI	^{83}Y, $^{83-86,88}Zr$, $^{85-88}Nb$	[40]
222 MeV	^{36}Ar	^{nat}Ni	HI	^{84}Y, ^{87}Zr, ^{89}Mo, ^{88}Tc	[19]
189 MeV	^{40}Ca	^{nat}Ni	HI	^{88}Mo, ^{92}Tc, $^{91,94}Ru$, $^{94-95}Rh$	[19]
205 MeV	^{40}Ca	^{nat}Ni	HI	^{91}Tc, ^{91}Ru	[19]
220 MeV	^{40}Ca	^{nat}Ni	HI	$^{89-90}Tc$, $^{90,92-93}Ru$, $^{92-94}Rh$	[19]
170 MeV	^{40}Ca	^{nat}Ni	HI	$^{94-96}Pd$, $^{95}Pd^{m}$	[19]
70 MeV	^{3}He	^{nat}Ru	LI	$^{97-98}Pd$	[41]
62 MeV	p	^{106}Cd	LI	$^{99,101}Pd$, $^{101-105}Cd$, ^{102}In	[41]
70 MeV	^{3}He	^{106}Cd	LI	^{104}In	[41]
100 MeV	^{3}He	^{106}Cd	LI	^{100}Ag	[41]
295–330 MeV	^{58}Ni	^{nat}Ni	HI	$^{104-108}Sn$, $^{106-110}Sb$, $^{108-109}Te$, ^{111}I	[3]

target is inside the IGISOL gas cell, where the product recoils are stopped, and a good fraction of the ions end up with charge state 1^{+}. In the heavy-ion ion guide, also known as HIGISOL [31,32], the target is in front of the gas cell and the products have to recoil at small angles in order to pass through a Havar window around the beam stopper before thermalizing in the gas cell. The employed production methods are summarized in table 1.

After thermalization, the ions are extracted from the ion guide with the help of differential pumping and an electric field. The ions are accelerated to 30 keV and mass-separated by a dipole magnet before entering the radio-frequency quadrupole (RFQ) [42], which is used for cooling and bunching of the ions. After the RFQ, the ions are injected into the JYFLTRAP [43] cylindrical double Penning trap situated inside a 7 T superconducting solenoid. The first trap is called the purification trap since it is used for isobaric purification via mass-selective buffer gas cooling [44]. The precision mass measurements are carried out in the second trap, known as the precision trap, via the time-of-flight ion cyclotron resonance (TOF-ICR) method [45,46].

In the TOF-ICR method, the cyclotron frequency of the ion of interest with mass m and charge q in a magnetic field B, $\nu_c = qB/(2\pi m)$, is compared to the cyclotron frequency of a reference ion ($\nu_{\rm ref}$) with a well-known atomic mass $m_{\rm ref}$. Since the mass-separated ions have typically the same charge state of 1^{+} at IGISOL, the mass of the ion of interest is obtained as: $m = (\nu_{\rm ref}/\nu_c)(m_{\rm ref} - m_e) + m_e$. Each measurement of the ion of interest is sandwiched between two reference measurements in order to determine the value of the magnetic field B at the time of the actual measurement. Since the IGISOL method provides a large variety of possible reference ions to be measured on-line, the ion with a superior precision closest to the ion of interest is typically chosen for reference. In some cases, the production rate of oxides can be higher or comparable to the rate of the ions. For example, the masses of neutron-deficient yttrium, niobium and zirconium isotopes have been measured as oxides.

The data have been fitted and analysed mainly with the programs LAKRITSI and COMA [28]. A count-rate-class analysis [47] has been performed whenever possible. The uncertainty contribution of possible magnetic field fluctuations has been added, and mass-dependent and residual uncertainties have been included in the final atomic-mass values. A detailed description of the data analysis can be found in ref. [19]. The latest results for the mass-dependent and residual uncertainties applicable to the measurements performed after the June 2007 magnet quench are reported in the JYFLTRAP paper on carbon-cluster cross-reference measurements [48].

3 Mass measurements of neutron-deficient nuclides at JYFLTRAP

Measurements performed at JYFLTRAP in 2005–2010 have yielded 90 published ground-state and 8 published isomeric-state masses of neutron-deficient nuclides (see fig. 1). These measurements have been motivated, for example, by the astrophysical *rp*- and νp-processes [3,19, 38,40,41] or nucleosynthesis in novae [26,27,33] or isospin symmetry studies. The results of mass measurements related to the superallowed $0^+ \to 0^+$ β decays [33–37,39] are not discussed in this review. In the following, the measured masses are discussed from lighter to heavier nuclides and related energy systematics (S_p-, Q_α-, and S_2p-values) are reviewed. The astrophysical implications are discussed separately in sect. 4.

3.1 Mass measurements at A = 23 and A = 31

The mass measurements at $A = 23$ and $A = 31$ have been motivated by nova nucleosynthesis [12] and the Isobaric Multiplet Mass Equation (IMME) (see, *e.g.*, ref. [49,50]). According to the IMME, the masses of members with an isospin projection $T_Z = (N-Z)/2$ of an isobaric multiplet T should lie along a parabola $M(T,T_Z) = a+bT_Z+cT_Z^2$. In general, the IMME has worked well and it has been used to predict masses of more exotic members of the multiplet. A breakdown of the IMME could be explained, for example, by higher-order perturbations or by the importance of including three-body terms. In addition, a significant component of isospin mixing could also result in deviations from the quadratic behaviour [51].

At JYFLTRAP, the masses of ^{23}Al and ^{23}Mg have been determined precisely against the reference nuclide ^{23}Na. The new mass value for ^{23}Al is 22(19) keV lower and 55 times more precise than the AME03 [52] value. The JYFLTRAP mass value for ^{23}Mg agrees with the AME03 value [52] and almost doubles the precision. The most recent ground-state masses for ^{23}Al [26], ^{23}Mg [26, 52], ^{23}Na [52] and ^{23}Ne [52], and excitation energies for the $T = 3/2$ isobaric analog states [53,54] were adopted and a quadratic IMME fit ($\chi^2/n = 0.28$) was performed for the $T = 3/2$ quartet at $A = 23$ in ref. [26]. The cubic fit yielded a cubic term $d = 0.22(42)$ keV consistent with zero. Thus, the IMME was found to work well in the $A = 23$ quartet.

Another interesting $T = 3/2$ quartet lies at $A = 31$: ^{31}Si, ^{31}P, ^{31}S, and ^{31}Cl. For this quartet, the mass of ^{31}S has been precisely measured at JYLFTRAP and was found to deviate from the AME03 value [52]. Unfortunately, an attempt to measure the mass of the much more uncertain ^{31}Cl nucleus failed due to overwhelming nitrogen oxide background at $A = 31$ caused by poor vacuum conditions in the IGISOL facility during that run. In the future, a mass measurement of ^{31}Cl would be very interesting at the new IGISOL-4 facility.

Currently, the only isobaric quintet that has been measured precisely enough to test the IMME lies at $A = 32$. Although there have been no mass measurements at $A = 32$ at JYFLTRAP, the IMME at $A = 32$ has been studied in one of our publications [27]. A recent ^{32}S(^{3}He, t)^{32}Cl measurement [55] showed a discrepancy to the ^{32}Cl mass obtained from the proton separation energy [56] and the ^{31}S mass from AME03. The mass of ^{31}S measured at JYFLTRAP was found to deviate from AME03 [52] by 2.1(15) keV. When our new mass value is combined with the proton separation energy of ref. [56], a mass excess value, which is more precise than the previous measurements and agrees with the adopted AME03 value, is obtained for ^{32}Cl. In ref. [27], the quadratic IMME fit was performed with six different data sets corresponding to two different values of ^{32}P [52,57] and three different values of ^{32}Si [52,57–59]. The overall result of the quadratic fits is that the IMME fails significantly ($\chi^2/n > 6.5$) in all data sets. The cubic term d was found to depend strongly on the data set: the values varied from $d = 0.51(15)$ keV to $d = 1.00(13)$ keV. In the future, further mass measurements of ^{32}Ar, ^{32}Cl, ^{32}P, and ^{32}Si, should solve these discrepancies and really validate the breakdown of the IMME.

3.2 Mass measurements around ^{56}Ni

Mass measurements around the doubly magic $N = Z$ nucleus ^{56}Ni provide essential data for the studies of, *e.g.*, isospin symmetry, mirror nuclei, Coulomb displacement energies, neutron-proton pairing, and shell-gap energies. At JYFLTRAP, Q_EC-values of the mirror nuclei ^{53}Co, ^{55}Ni, ^{57}Cu, and ^{59}Zn, as well as of the $N = Z$ nuclei ^{54}Co, ^{56}Ni, ^{58}Cu, ^{60}Zn, and ^{62}Ga have been precisely measured (see refs. [37–39]). In addition to Q_EC-values, a few proton separation energies (S_p) have been directly measured by using the $(A-1, Z-1)$ nuclide as a reference for the (A, Z) nuclide [38]. In order to obtain more accurate mass measurements, a network calculation of 17 frequency ratio measurements between 13 nuclides has been performed around ^{56}Ni [38]. In this way, also the masses of reference nuclides, such as ^{55}Co and ^{58}Ni, and ^{59}Cu, could be evaluated. The obtained masses for ^{55}Ni, ^{56}Ni, and ^{57}Cu agree well with the AME03 values [52], but are 15, 26, and 31 times more precise, respectively.

The most significant deviations (around 2σ) to the AME03 values in this mass region have been found for copper isotopes ^{58}Cu ($-3.6(17)$ keV) and ^{62}Cu (10.6(42) keV). The measured JYFLTRAP mass of ^{62}Cu is higher than the adopted AME03 value, which was based on measurements using the ^{62}Ni$(p,n)^{62}$Cu reaction [60, 61] and the β-decay energies of ^{62}Cu [62–64]. The first two of the three β-decay experiments [62,63] have underestimated the β-decay energy, which results in a too low adopted AME03 value.

The mass of ^{58}Cu has already been measured with the JYFLTRAP purification trap in 2004 [43]. The new JYFLTRAP mass value for ^{58}Cu agrees with the old purification trap measurement [43] but differs from the precise (p,n) threshold energy measurements [65–67]. However, when the (p,n) measurements are revised with the updated mass values [68] an agreement with JYFLTRAP is obtained.

Absolute deviations to AME03 are rather small in this mass region. The biggest deviations (in keV) belong to ^{59}Zn (47(40) keV) and ^{60}Zn (15(11) keV). For those cases the Q values derived from ^{58}Ni$(p,\pi)^{59}$Zn [69] and ^{58}Ni$(^{3}$He$,n)^{60}$Zn [70] lead to slightly underestimated masses for ^{59}Zn and ^{60}Zn. In addition to these nuclides, a difference was found for the proton-decaying high-spin isomer in ^{53}Co: the new JYFLTRAP value is 36(22) keV lower than in AME03 [52].

Q_{EC} values of several $T = 1/2$ mirror nuclei have been measured at JYFLTRAP. These experiments have been motivated by recent studies [71,72] where corrected ft values have been calculated for $T = 1/2$ mirror transitions and the Conserved Vector Current Hypothesis has been tested. The Q_{EC} values of the $T = 1/2$ nuclei ^{31}S, ^{53}Co, ^{55}Ni, ^{57}Cu, and ^{59}Zn have been measured with a precision of better than 0.7 keV at JYFLTRAP [27,38]. In addition, the Q_{EC} value of ^{53}Co^{m} [38] has been measured with a precision of 1.2 keV. The largest differences to the AME03 values have been found at ^{53}Co^{m} and ^{59}Zn. The Q_{EC} values of ^{45}V and ^{49}Mn have been measured in 2010 but the results have not yet been published. In the future, these measurements will be continued with ^{47}Cr and ^{51}Fe.

In addition to the $T = 1/2$ mirror nuclei, the Q_{EC} values of the $T = 1$ nuclei ^{56}Ni, ^{58}Cu, and ^{60}Zn have been measured at JYFLTRAP. Interestingly, the JYFLTRAP Q_{EC} value of ^{58}Cu was found to be 4.6(15) keV lower than in AME03. This Q_{EC} value has to be taken into account in studies of isospin symmetry among $A = 58$ nuclei [73].

There have been no mass measurements for the rp-process between ^{62}Ga and ^{80}Y at JYFLTRAP. This region has been quite well covered by other Penning traps. At ISOLTRAP, Ga, Se, Br, Kr, Rb, and Sr isotopes have been extensively studied [74–79]. Ga, Ge, As, Se, and Br isotopes relevant for the rp-process have been determined with LEBIT [18,80], and Se, As, and Ge isotopes at $A = 68$ with CPT [81].

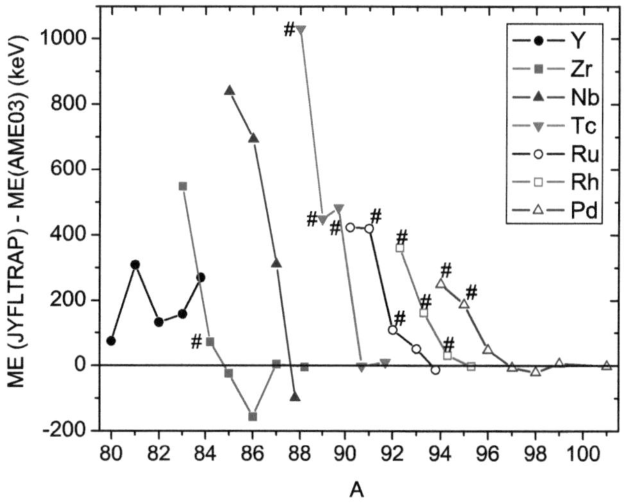

Fig. 2. (Color online) JYFLTRAP mass excess values compared to the AME03 values for the measured Y, Zr, Nb, Tc, Ru, Rh and Pd isotopes. "#" denotes that the AME03 value has been based on extrapolations. No error bars have been included in order to show more clearly the trend of the differences.

3.3 Mass measurements around A = 80–100

Neutron-deficient nuclei filling up neutron and proton $g_{9/2}$ shells in the mass region $A = 80$–100 provide an abundant region for nuclear structure studies. Strong proton-neutron interaction and competition between single-particle and collective motions result in the presence of many isomers and collective features. Spin-gap isomers, such as in ^{94}Ag [82–84] or ^{95}Pd [85], or shape isomers like in ^{72}Kr [86] have gained a lot of attention. Shell-model calculations are possible for these nuclides: in many cases it is sufficient to include only the $1p_{1/2}$ and $0g_{9/2}$ shells. Experimental data are needed to test and develop shell-model calculations and more sophisticated model spaces, such as $(1p_{3/2}, 0f_{5/2}, 1p_{1/2}, 0g_{9/2})$ [84] or $(0g_{9/2}, 1d_{5/2}, 0g_{7/2}, 1d_{3/2}, 2s_{1/2})$ [83].

The mass measurements conducted at JYFLTRAP in this mass region have revealed large deviations to the AME03 values. Many of the measured mass excess values, for example, ^{83,85}Zr and $^{85-88}$Nb, have earlier been based on β-decay endpoint energies, which have a tendency to underestimate the masses due to unobserved feeding to higher-lying states in the daughter nuclides (pandemonium effect [87]). This problem is likely to play a stronger role in nuclides further from stability. Thus, the more exotic the nuclide, the more it typically deviates from the JYFLTRAP value (see fig. 2). The yttrium isotopes do not follow this trend since the AME03 value for ^{80}Y is based on a direct time-of-flight mass measurement at the ISN SARA cyclotron [88]. The mass of ^{86}Zr in AME03

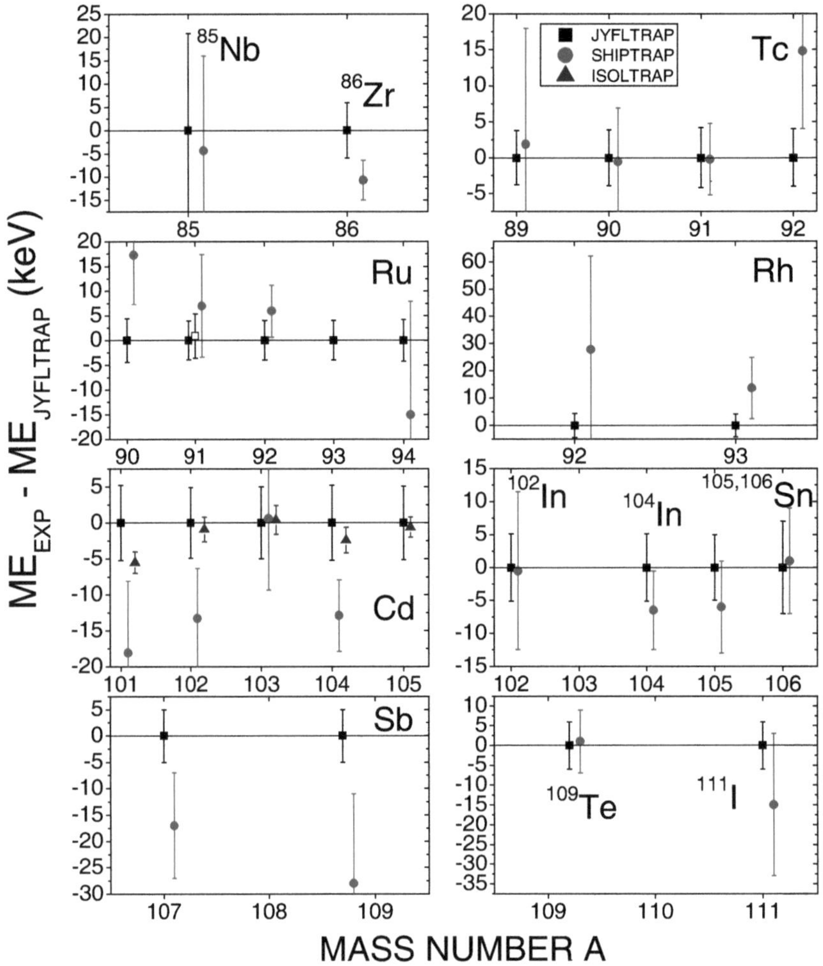

Fig. 3. (Color online) Differences between the JYFLTRAP mass excess values (black squares) and the values obtained at SHIPTRAP (red circles) [19,90] and ISOLTRAP (blue triangles) [17]. The error bars represent the original experimental uncertainties and not the uncertainties of the differences. The isomer or unknown level scheme corrections have not been applied for ^{92}Rh and ^{104}In in this figure. ^{91}Ru has been measured against ^{85}Rb (solid square) and ^{94}Mo (open square) at JYFLTRAP. The masses measured against Mo isotopes at JYFLTRAP have been corrected with the new molybdenum mass values from JYFLTRAP (for details, see the previous article of this special issue of the European Physical Journal A).

is based on ^{90}Zr(α, ^{8}He)^{86}Zr reactions [89] and not on β-decay energies. Many nuclides —^{84}Zr, ^{88,89}Tc, $^{90-92}$Ru, $^{92-94}$Rh, and $^{94-95}$Pd— have been measured for the first time at JYFLTRAP. As can be seen from fig. 2, the extrapolated mass excess values of AME03 are typically too small. In other words, JYFLTRAP has found these nuclei to be less bound.

In the mass region below $Z = 50$, the mass excesses of $^{89-92}$Tc, $^{90-92,94}$Ru, $^{92-93}$Rh as well as ^{99,101}Ag, $^{101-104}$Cd, $^{102-105}$In have been measured with the SHIPTRAP Penning trap at GSI [19,90]. In general, SHIPTRAP and JYFLTRAP results agree well with each other (see fig. 3). Small differences are found for ^{92}Tc ($\Delta_{S-J} =$ 15(12) keV) and ^{90}Ru ($\Delta_{S-J} =$ 17(11) keV). Differences are also found for ^{101,102,104}Cd, where ISOLTRAP data [17] are in agreement with JYFLTRAP, not with SHIPTRAP. Many of these masses have also been measured with CPT but the preliminary graphical data [91] for ^{90,91}Mo, $^{90-93}$Tc, $^{93-94}$Ru, ^{94,95}Rh, $^{104-107}$In, $^{104-108}$Sn, $^{107-108}$Sb could not be included in this comparison[1]. The JYFLTRAP mass excess values used here have been updated with the new molybdenum masses measured at JYFLTRAP discussed in the previous article of this special issue of the European Physical Journal A.

Many open questions concerning the neutron-deficient nuclides around $A = 80$–100 remain. For example, whether an observed state is a pure ground state, an isomeric state, or a mixture of these should be investigated and verified for some of the measured nuclides. Corrections due to a possible mixture of ground and isomeric states or due to unknown level scheme have been applied according to eq. (14) of ref. [93] and have been added quadratically to the experimental uncertainties for the nuclei listed in table 2. However, in some cases isomers might not be known,

[1] The CPT results on Nb, Mo, Tc, Ru, and Rh isotopes [92] were published at proof stage, and thus could not be included here.

Table 2. Isomers taken into account in the determination of the ground-state masses in the $A \approx 80$–100 region.

Isomer	E_x (keV)	Ref.	Isomer	E_x (keV)	Ref.
^{83}Y^m	61.98(11)	[94]	^{88}Tc^m	300#	[94]
^{84}Y^m	67	[95]	^{92}Rh^m	50#	[96,97]
^{85}Nb^m	≥ 69	[98]	^{100}Ag^m	15.52(16)	[94]
^{87}Nb^m	3.84(14)	[94]	^{104}In^m	93.48(10)	[94]
^{88}Nb^m	40(140)	[94]			

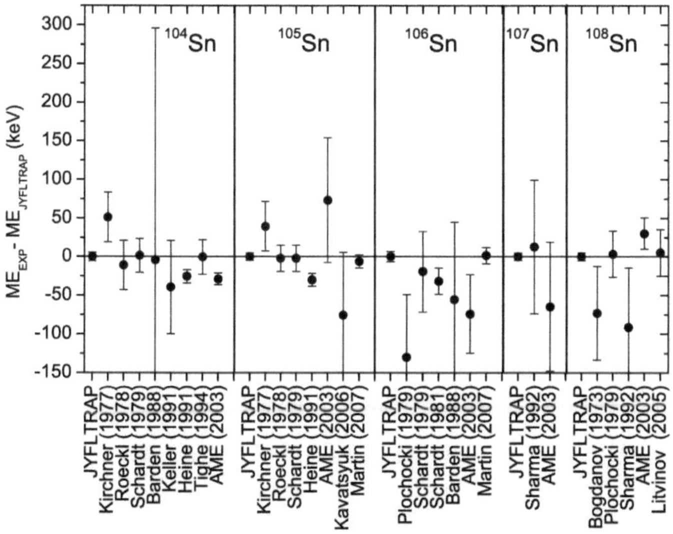

Fig. 4. JYFLTRAP mass excess values for Sn isotopes compared to other experiments. Beta [99–102], alpha [103–107], proton [108] and β-delayed proton decay [109] energies have been used with the latest mass values obtained at JYFLTRAP whenever possible. Direct mass measurements performed at SHIPTRAP [90] for ^{105,106}Sn, and at ESR for ^{108}Sn [110] as well as the values based on the mass ratios of ^{107}Sn/^{106}Sn and ^{107}Sn/^{108}Sn [111] are also shown in the figure. The AME03 values [52] deviate from the JYFLTRAP values of ^{104,106,108}Sn.

or information is uncertain. For example, no correction has been applied to ^{86}Nb ($E_x = 250(160)\#$ keV [94]) since this isomer is considered as uncertain. Clearly more experiments are needed to clarify the low-lying level structure of nuclides in this mass region, and revisions of some mass values might become necessary as a consequence.

3.4 Mass measurements above ^{100}Sn

Masses of $^{104-108}$Sn, $^{106-110}$Sb, ^{108,109}Te, and ^{111}I in the endpoint region of the *rp*-process have been measured with JYFLTRAP [3]. The masses of ^{106,108,110}Sb have been experimentally determined for the first time, and found to agree with the AME03 extrapolations. Overall agreement of the JYFLTRAP mass values with the AME03 values in this region is fairly good: the only deviations occur at ^{104,106,108}Sn, and ^{109}Te. In the following, JYFLTRAP results are compared to previous experiments nuclide by nuclide (see also figs. 4, 5, and 6).

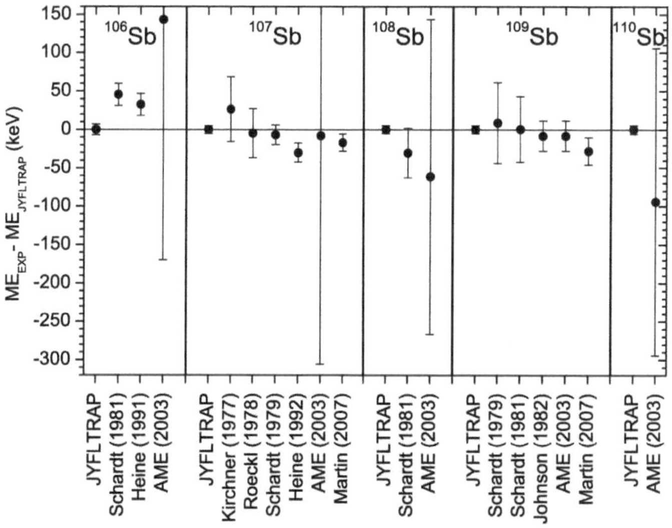

Fig. 5. JYFLTRAP mass excess values for Sb isotopes compared to other experiments. Beta [112] and alpha-decay [103–105,107,113] energies have been combined with the latest JYFLTRAP values in the calculations. SHIPTRAP mass values for ^{107,109}Sb [90] deviate from the new JYFLTRAP values. The AME03 [52] mass values of $^{106-108,110}$Sb are based only on extrapolations. The JYFLTRAP values agree with these AME03 values.

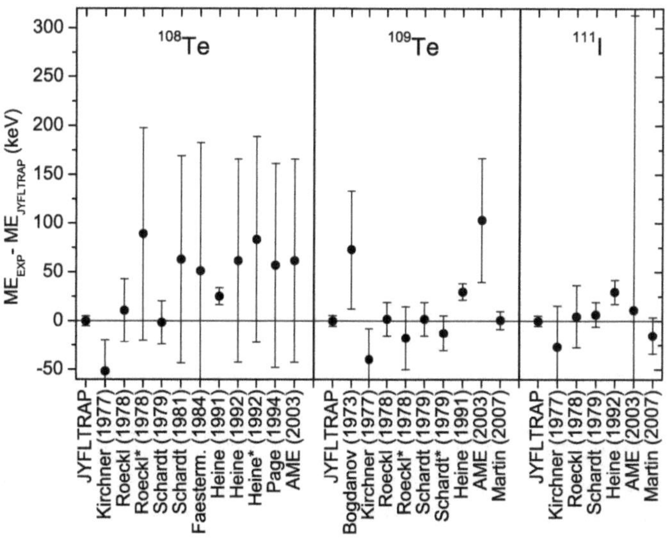

Fig. 6. JYFLTRAP mass excess values for Te and I isotopes compared to other experiments. Alpha [103–107,113,114], proton [113,115] and β-delayed proton decay [109] energies have been used with the latest JYFLTRAP values in the calculations of the mass values. The SHIPTRAP measurements of ^{109}Te and ^{111}I [90] agree well with the JYFLTRAP data. The AME03 value [52] for ^{109}Te deviates from the JYFLTRAP data.

^{104}Sn: The mass of ^{104}Sn has been previously measured via β-decay endpoint energies [99,100]. It can also be determined from the α-decay energy of ^{108}Te [103–106]. The β-decay results and the Q_α-values from refs. [104,105] agree with the JYFLTRAP result but the α-decay energies from refs. [103,106] differ from it. The AME03 value, which is mainly based on the α-decay energies, is significantly lower than the JYFLTRAP value.

^{105}Sn: The mass of ^{105}Sn can be determined from the Q_α-values of ^{109}Te [103–106]. The Q_α-values from refs. [104,105] agree with the JYFLTRAP result but as in the case of ^{104}Sn the values from refs. [103,106] differ: Q_α from ref. [103] is around 40 keV smaller and the value from ref. [106] is around 30 keV higher than the Q_α-value obtained with JYFLTRAP. The JYFLTRAP result agrees with the β-endpoint measurement of ^{105}Sn [101] and the SHIPTRAP value [90].

^{106}Sn: The $Q_{\rm EC}$ value of ^{106}Sn from ref. [99], the Q_α-value of ^{110}Te from ref. [105] and the SHIPTRAP value [90] agree well with the result obtained at JYFLTRAP. However, the values based on the $Q_{\rm EC}$-value from ref. [102] and the Q_α-value of ^{110}Te from ref. [107] differ slightly from the JYFLTRAP mass excess value for ^{106}Sn. As a result, the AME03 value differs from the JYFLTRAP value.

^{107}Sn: The mass of ^{107}Sn has earlier been based on the mass ratio measurement of ^{107}Sn to ^{106}Sn [111]. The mass value determined at JYFLTRAP agrees well with it.

^{108}Sn: The β-delayed proton decay of ^{109}Te [109], the mass ratio of ^{108}Sn to ^{107}Sn [111], and the AME03 value [52] disagree slightly from the JYFLTRAP value. The result from the β decay of ^{108}Sn [102] and the direct mass measurement performed at ESR [110] agree very well with JYFLTRAP. The differences between the JYFLTRAP and the AME03 mass excess values for ^{109}Te and ^{107}Sn explain why the mass excess values for ^{108}Sn derived from refs. [109,111] differ from the AME03 value (see fig. 4).

^{106}Sb: The mass of ^{106}Sb has been measured for the first time with JYFLTRAP and the result agrees with the extrapolation based on systematic trends in AME03 [52].

^{107}Sb: The mass of ^{107}Sb has been directly measured at SHIPTRAP [90]. The value differs from the JYFLTRAP value by 17(12) keV. The mass can also be determined from the α-decay energies of ^{111}I [103–105,113] and the mass of ^{111}I [3]. All α-decay results except the one from ref. [113] agree with JYFLTRAP.

^{108}Sb: The mass of ^{108}Sb has been experimentally determined at JYFLTRAP for the first time. The obtained value agrees with the extrapolated value [52]. The mass can also be calculated from the Q_α of ^{112}I [107] and the mass of ^{112}I measured at SHIPTRAP [90]. The value agrees with the JYFLTRAP result.

^{109}Sb: The mass of ^{109}Sb has been measured at SHIPTRAP. The JYFLTRAP value differs from it by 28(18) keV. The β-endpoint energy of ^{109}Sb [112] and the α-decay energies of ^{113}I [105,107] together with the mass of ^{113}I measured at SHIPTRAP [90] agree with JYFLTRAP

^{110}Sb: The mass of ^{110}Sb has been measured for the first time at JYFLTRAP. The mass excess value agrees with the extrapolated value of AME03 [52].

^{108}Te: The mass of ^{108}Te can be determined in several ways. Firstly, the experimental Q_α-values of ^{108}Te [103–106] and the mass excess of ^{104}Sn yield similar agreements and disagreements as for ^{104}Sn for which the mass excess values were derived from these same Q_α-values and the mass excess of ^{108}Te. Secondly, the Q_α-values of ^{112}Xe [104,107,113,114] and the AME03 value for ^{112}Xe can be used to estimate the mass of ^{108}Te. These values agree with the JYFLTRAP result. Thirdly, the proton decay data of ^{109}I [113,115] combined with the AME03 value for ^{109}I agree with the JYFLTRAP measurement.

^{109}Te: The mass of ^{109}Te has previously been measured at SHIPTRAP [90] and the values agree with the JYFLTRAP results. It can also be derived from the Q_α-values of ^{109}Te [103–106] with the ^{105}Sn mass excess value, Q_α-values of ^{113}Xe [104,105] with the ^{113}Xe mass measured at SHIPTRAP, and from the β-delayed proton data of ^{109}Te [109] and the mass of ^{108}Sn. Of these results, only refs. [104,105] agree with JYFLTRAP. The AME03 value [52] based mainly on ref. [109] is significantly higher than the JYFLTRAP mass excess value.

^{111}I: The mass of ^{111}I has also been measured at SHIPTRAP [90] and it agrees with JYFLTRAP. It can also be determined from the Q_α-values of ^{111}I [103–105,113] with the mass of ^{107}Sb. The measurements agree with the present data except for the result from ref. [113].

3.5 Energy systematics: S_p-, Q_α-, and S_{2p}-values

Proton separation energies ($S_{\rm p}$) are important for the calculations of proton capture rates and for the modeling of the astrophysical *rp*- or νp-processes, since the proton capture rates and obtained abundances depend exponentially on the $S_{\rm p}$-values. At JYFLTRAP, the mass excesses for both the (A,Z) and $(A-1,Z-1)$ nuclides, and thus the $S_{\rm p}$-values, have been determined for 41 nuclides. The measured masses have an impact on altogether 138 $S_{\rm p}$-values. For ^{56}Ni, ^{57}Cu, ^{59}Zn, and ^{60}Zn, a sub-keV precision in the proton separation energy has been achieved by using the $(A-1,Z-1)$ nuclide as a reference in the mass measurement. Figure 7 shows a comparison of the $S_{\rm p}$-values calculated entirely from JYFLTRAP masses with the values of AME03.

The biggest deviations in the proton separation energies compared to the AME03 values are found for Nb, Tc, Zr and Mo isotopes (see table 3). In addition, the $S_{\rm p}$-values of ^{58}Cu, ^{59}Zn, ^{60}Zn, ^{88}Nb, ^{104}In, and ^{109}Sb differ slightly from AME03 (see fig. 7). The deviations to the AME03 values are $-3.5(26)$, 50(40), 15(11), $-100(100)$, $-110(90)$, and 40(30) keV, respectively. For the proton separation energies, for which either the mass of the (A,Z) or $(A-1,Z-1)$ nuclide has been determined at JYFLTRAP, the values for ^{82}Zr, ^{84}Nb, ^{89}Ru, ^{91}Rh, ^{93}Pd, and ^{95}Ag are partly based on extrapolated AME03 values, and thus,

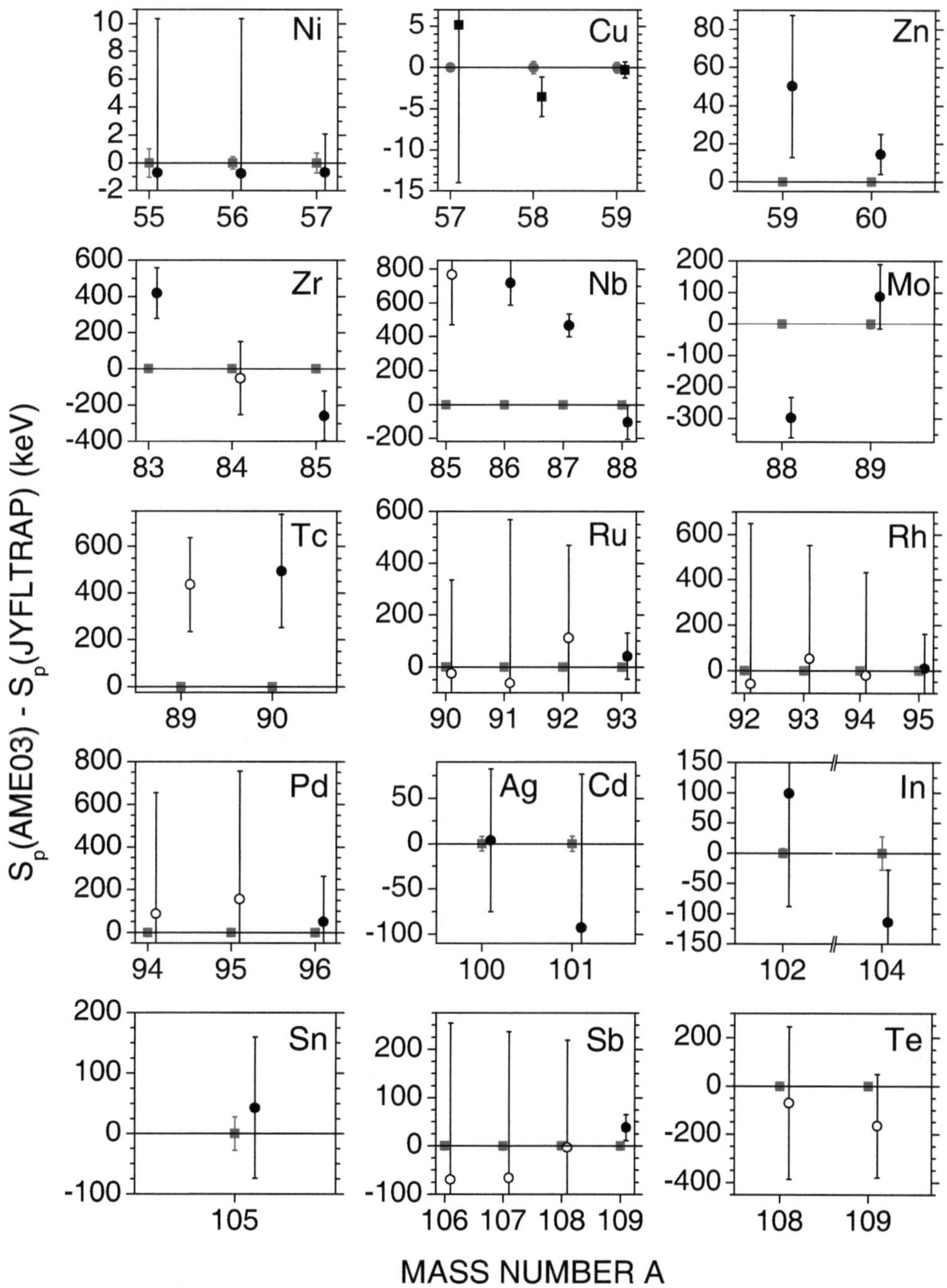

Fig. 7. (Color online) A comparison of proton separation energies measured at JYFLTRAP (masses both for (A, Z) and $(A-1, Z-1)$ measured) to the values from AME03 [52]. The JYFLTRAP values are shown as red squares, the AME03 values as black circles and the extrapolated values in AME03 by open black circles. The uncertainty of the JYFLTRAP S_p-value is in many cases much smaller than the size of the red square.

new experimental data may change those values dramatically. In addition, the S_p-values for ^{62}Cu, ^{63}Zn, $^{82-83}$Y, ^{86}Zr, and ^{106,108}Sn disagree with the AME03 values. In the lighter mass region, the S_p-values for ^{23,26}Al, $^{26-27}$Si, ^{31}S, ^{34}Cl, ^{42}Ti, and ^{46}V differ from the AME03 values.

In the SnSbTe region, α separation energies become sufficiently small for α decay to become energetically possible, and for (p, α) and (γ, α) reactions to play an important role in astrophysical environments. Q_α-values are therefore needed. The Q_α-values of ^{108,109}Te and ^{111}I have been precisely determined at JYFLTRAP by measuring the masses of the corresponding mother and daughter nuclides (see table 4). The JYFLTRAP Q_α-value for ^{108}Te is significantly lower than in AME03. By combining the results of JYFLTRAP [3,41] with SHIPTRAP [90], also the Q_α-values for $^{105-108}$Sn and $^{106-109}$Sb, $^{110-112}$Te, $^{112-113}$I, and ^{113}Xe have been obtained. For these Q_α-values, deviations to AME03 are found for ^{108}Sn, ^{109}Sb, and $^{110-112}$Te.

Two-proton separation energies S_{2p} plotted against the proton number Z often show continuous and smooth

Table 3. The largest observed differences (in keV) to the AME03 proton separation energies at JYFLTRAP. For the nuclides marked with ^{a}, the $Z-1$ nuclide has not been measured at JYFLTRAP and AME03 has been adopted. The mass excesses for the nuclides marked with ^{b} have been adopted from AME03 but the corresponding $Z-1$ nuclides have been measured with JYFLTRAP. "#"; denotes that the used AME03 value is based on extrapolations. The big deviations for ^{86,87}Mo and ^{88}Tc vanish when the new mass values from ref. [116] are adopted.

Nuclide	S_p (keV)	$S_{p,lit.}$ [52] (keV)	$S_p - S_{p,lit.}$ (keV)
$^{81}Y^a$	2692(9)	3000(60)	−310(60)
$^{84}Y^a$	4418(23)	4650(90)	−230(100)
$^{82}Zr^b$	5770(230)#	5460(230)#	310(320)#
^{83}Zr	5137(9)	5560(140)	−420(140)
^{85}Zr	6538(21)	6280(140)	260(140)
$^{84}Nb^b$	3260(300)#	2710(310)#	550(430)#
^{85}Nb	2178(22)	2950(300)#	−770(300)#
^{86}Nb	3248(8)	3970(130)	−720(130)
^{87}Nb	3201(9)	3670(70)	−470(70)
$^{86}Mo^b$	5540(440)	4700(490)	840(660)
$^{87}Mo^b$	5850(220)	5160(240)	690(330)
^{88}Mo	6102(8)	5810(60)	300(60)
$^{88}Tc^a$	1270(240)	2300(300)#	−1030(380)#
^{89}Tc	1997(6)	2430(200)#	−440(200)#
^{90}Tc	2998(6)	3490(240)	−500(240)
$^{89}Ru^b$	5120(510)#	4090(540)#	1030(740)#
$^{91}Rh^b$	1510(400)#	1090(500)#	420(640)#
$^{93}Pd^b$	3990(400)#	3630(570)#	360(700)#
$^{95}Ag^b$	1290(400)#	1040(570)#	250(700)#

Table 4. Q_α-values from JYFLTRAP mass measurements [3, 41] combined with the SHIPTRAP results [90], and comparison to the literature values [52]. Only alpha decays for which both the mother and daughter have been measured at JYFLTRAP and/or SHIPTRAP and at least either of them has been measured at JYFLTRAP have been taken into account.

Nuclide	Q_α (keV)	$Q_{\alpha,lit.}$ (keV)	$Q_\alpha - Q_{\alpha,lit.}$ (keV)
^{105}Sn	74(5)	60(170)	14(170)
^{106}Sn	−116(6)	−170(60)	54(60)
^{107}Sn	−286(6)	−350(80)	62(90)
^{108}Sn	−527(6)	−491(22)	−36(23)
^{106}Sb	1795(9)	1950(330)#	−155(340)
^{107}Sb	1558(10)	1520(300)#	38(300)
^{108}Sb	1309(7)	1170(220)#	139(220)
^{109}Sb	963(11)	797(26)	166(28)
^{108}Te	3416(8)	3445(4)	−29(9)
^{109}Te	3198(6)	3230(50)	−32(50)
^{110}Te	2696(9)	2723(16)	−27(19)
^{111}Te	2500(8)	2670(110)	−170(110)
^{112}Te	2079(10)	2310(170)	−231(170)
^{111}I	3265(7)	3280(50)	−15(50)
^{112}I	2956(10)	2990(50)	−34(50)
^{113}I	2708(10)	2710(50)	−2(50)
^{113}Xe	3085(9)	3090(50)	−5(50)

behavior over several isotopes. Deviations from the trend reflect possible subshell or shell closures or an onset of deformation. The shell closure at $Z = 50$ is seen in fig. 8, where two-proton separation energies obtained at JYFLTRAP (combined with the Penning trap data from refs. [17,90]) are plotted against proton number Z: the energies fall down steeply at $Z = 50$. The trend is very similar to the AME03 values for the isotonic chains above $N = 47$. Below $N = 47$, the isotonic chains lie at lower energies than in AME03, since the mass excess values of Nb, Mo and Tc isotopes have been too low in AME03. In addition, these lines are steeper when proceeding towards more proton-rich nuclides. The new Penning-trap measurements have shown that these nuclides are less proton bound than in AME03.

After the JYFLTRAP masses for ^{83}Zr and ^{85}Nb had been determined, the S_{2p} for ^{85}Mo ($Z = 42$, $N = 43$, blue symbol) and ^{87}Tc ($Z = 43$, $N = 44$, red symbol) obtained with the AME03 masses for ^{85}Mo and ^{87}Tc show an extreme deviation from the systematic trend in their respective isotonic chains (blue line for $N = 43$ and red line for $N = 44$). The masses of ^{85}Mo and ^{87}Tc have then been measured at SHIPTRAP and big deviations of 1590(280) keV and 1430(300) keV from AME03 have been found [116]. When the SHIPTRAP values are adopted, the S_{2p} values follow largely the systematic trend (see fig. 8).

The strong deviations between AME03 and new mass measurements in this region, and in particular the observed dramatic change in systematic trends and the disappearance of irregularities when using new Penning-trap masses casts doubt on the remaining AME03 masses and the resulting AME03 S_{2p} trends in this region. More mass measurements towards more exotic isotones are urgently needed to verify or correct the AME03 data. Such measurements would also be important to show whether the linear trend in S_{2p}-values continues or whether there are indications of deformation. For example, in-beam gamma spectroscopy experiments have shown that almost all nuclei from krypton ($Z = 36$) to niobium ($Z = 41$) have permanent deformation when $N < 44$ [117].

Mass measurements around ^{100}Sn offer a possibility to determine the $Z = 50$ proton shell-gap energy $E_{gap,Z=50}$ from two-proton binding energies: $E_{gap,Z=50} = S_{2p}(Z = 50) - S_{2p}(Z = 52)$ (see fig. 9). Penning-trap mass measurements performed at JYFLTRAP [3,41], SHIPTRAP [90], and ISOLTRAP [17] improve the precisions of the shell-gap energies a lot and reveal deviations from the AME03

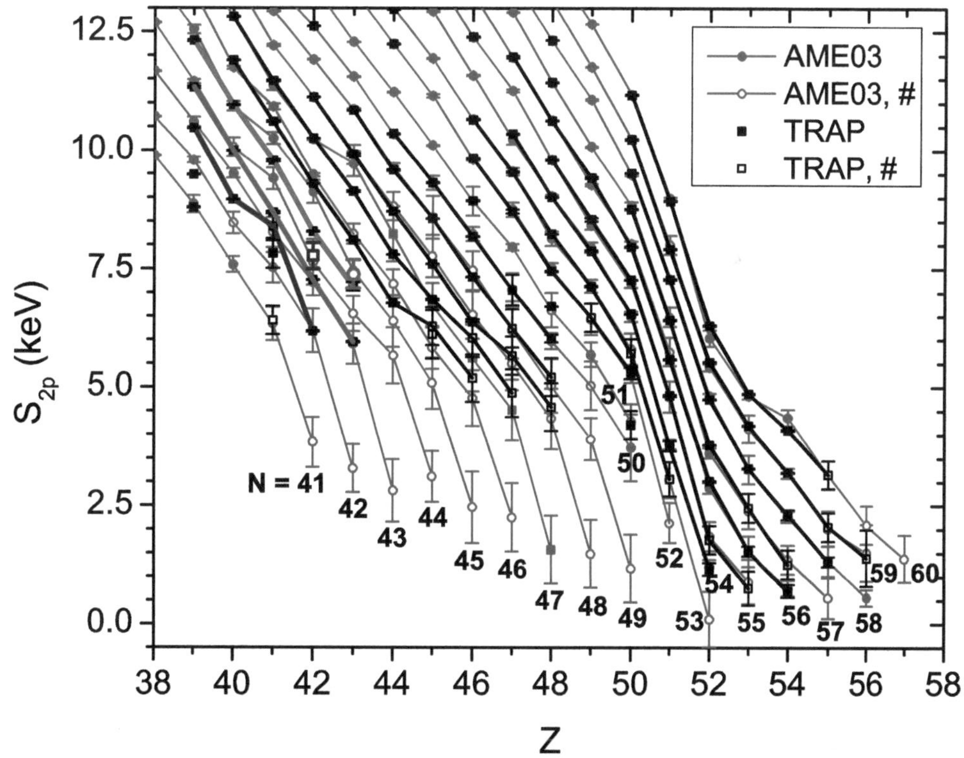

Fig. 8. (Color online) Two-proton separation energies determined at JYFLTRAP and comparison to the AME03 values [52]. The data from SHIPTRAP [90,116] and ISOLTRAP [17] have been taken into account in the Penning-trap data whenever possible. The "TRAP" data cover the S_{2p}-values where at least one data set comes from a Penning-trap measurement and the missing data have been adopted from AME03. The "AME03" values are based only on AME03. "#" marks that the used AME03 value is based on extrapolations. The isotonic chains of $N = 43$, $N = 44$, and $N = 45$ are shown in blue, red, and magenta, respectively. The open blue and red squares show the S_{2p}-values for ^{85}Mo and ^{87}Tc based on the JYFLTRAP and (erroneous) AME03 values.

values at $N = 56$–60. With the new trap data a minimum in the shell-gap energies is achieved at $N = 59$, and it is broader than with the AME03 data where a deeper minimum at $N = 58$ is observed. Both data show an increasing shell-gap energy when proceeding towards the magic neutron number $N = 50$.

4 Astrophysical aspects

4.1 Nucleosynthesis in novae

Novae occur in binary systems consisting of a white dwarf accreting hydrogen-rich matter from a main-sequence companion. Thermonuclear explosions of the accreted envelope drive the nova phenomenon. Novae on particularly massive oxygen-neon white dwarfs (ONe novae) are thought to reach peak temperatures of up to 4×10^8 K. Under these conditions the explosive hydrogen burning occurs as a sequence of proton captures and β^+ decays (and sometimes (p, α) reactions) proceeding up to $A = 40$ due to the presence of NeNa-MgAl seed nuclei (see refs. [10–13,118]). Of special interest in ONe novae is the production of ^{22}Na and ^{26}Al, which are sufficiently long-lived radioactive isotopes for their decay γ-radiation to be potentially observable. ^{22}Na ($T_{1/2} = 2.6019(4)$ y [119]) decays into a short-lived excited state of ^{22}Ne which de-excites to its ground state by emitting a 1.275 MeV γ-ray. Although several attempts to observe these γ-rays from nearby novae have been made, only an upper limit of the ejected ^{22}Na has been obtained [118]. The ^{26}Al ground state ($T_{1/2} = 7.17(24) \times 10^5$ y [119]) decays to an excited state of ^{26}Mg at 1.809 MeV. The γ-rays following the de-excitation of this state have been observed with γ-ray telescopes but the half-life is too long to associate them with a particular astrophysical event. The general distribution of Galactic 1.809 MeV activity indicates an origin mainly associated with massive stars, but it is important to determine a possible nova contribution.

^{22}Na is produced in a so-called NeNa cycle where ^{20}Ne$(p,\gamma)^{21}$Na is followed either by proton capture ^{21}Na $(p,\gamma)^{22}$Mg$(\beta^+)^{22}$Na or β decay ^{21}Na(β^+) ^{21}Ne(p,γ) ^{22}Na$(\beta^+)^{22}$Ne$(p,\gamma)^{23}$Na$(p,\alpha)^{20}$Ne. In order to model the production of ^{22}Na, the destruction channels, such as ^{22}Mg$(p,\gamma)^{23}$Al and ^{22}Na$(p,\gamma)^{23}$Mg have to be known precisely. At JYFLTRAP, ^{23}Al was found to be 22(19) keV more proton-bound [26] than in AME03 [52]. Thus, the resonant contribution for the rate of ^{22}Mg$(p,\gamma)^{23}$Al is little higher, and ^{23}Al is more resilient to destruction through photodissociation. Also the mass of ^{23}Mg has been measured at JYFLTRAP [26] but its impact on the

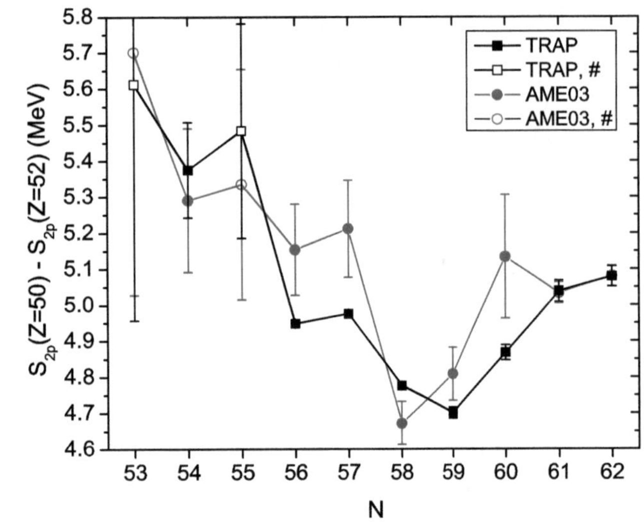

Fig. 9. Proton shell-gap energies at $Z = 50$ determined from two-proton separation energies. Recent Penning-trap data from JYFLTRAP [3,41], SHIPTRAP [90], and ISOLTRAP [17] have been adopted whenever possible. The "TRAP" data cover the values where at least one data set comes from a Penning-trap measurement and the missing data have been adopted from AME03. The "AME03" data points are based only on AME03. "#" marks that the used AME03 value is based on extrapolations.

calculated resonant rate for the ^{22}Na$(p,\gamma)^{23}$Mg reaction has not been investigated.

^{26}Al is produced in a so-called MgAl cycle: ^{24}Mg(p,γ) ^{25}Al$(\beta^+)^{25}$Mg$(p,\gamma)^{26}$Al$_{g.s.}(\beta^+)^{26}$Mg$(p,\gamma)^{27}$Al$(p,\alpha)^{24}$Mg. The production of ^{26}Al$_{g.s.}$ can be bypassed via ^{25}Al(p,γ) ^{26}Si$(\beta^+)^{26}$Al$^m(\beta^+)^{26}$Mg. Therefore, the rate for the proton capture reaction ^{25}Al$(p,\gamma)^{26}$Si is extremely important to constrain the model [118]. The JYFLTRAP mass value of ^{26}Si results in a 3.7(31) keV lower proton separation energy for ^{26}Si than in AME03 [52]. This changes the calculated stellar reaction rates of ^{25}Al$(p,\gamma)^{26}$Si by about 10% [33] compared to the rates calculated with the values from ref. [120].

The reaction ^{30}P$(p,\gamma)^{31}$S plays a crucial role governing the flow towards ^{32}S and heavier species in novae [11,118]. At ^{30}P, the reaction flow has to proceed either via ^{30}P(p,γ) ^{31}S(p,γ) ^{32}Cl(β^+) ^{32}S or via ^{30}P(p,γ) ^{31}S$(\beta^+)^{31}$P$(p,\gamma)^{32}$S. The ^{30}P$(p,\gamma)^{31}$S rate also has an effect on the ^{30}Si abundance [11]. The lower the proton capture rate, the more favorable is the β^+ decay of ^{30}P and more ^{30}Si is produced. A more accurate reaction rate and ^{30}Si abundance (or ^{30}Si/^{28}Si abundance ratio) helps in the identification of presolar grains with a possible nova origin [121]. The proton separation energy obtained at JYFLTRAP [27] is 2.1(16) keV lower than the adopted value [52]. Although the difference is quite small, it has an effect on the calculated reaction rate of ^{30}P$(p,\gamma)^{31}$S which has been studied, for example, in refs. [122–125].

Although in both cases the impact on the reaction rates is small, and there are much larger uncertainties from poorly known excited states, the measurements are a step towards more reliable rates that will be of particular importance once resonance parameters have been measured in future experiments. More importantly however, precise masses are essential for carrying out planned direct reaction rate measurements at radioactive beam facilities. Such experiments inherently suffer from limited beam intensities and resonance energies must be known precisely to make experiments feasible.

4.2 νp-process

The astrophysical νp-process has been suggested to occur in supernovae and possibly in gamma-ray bursts where proton-rich ejecta are created by strong neutrino fluxes [6, 7]. In principle, it proceeds similarly to the rp-process as a sequence of proton captures and β^+ decays but the flow towards heavier elements is accelerated via fast (n,p) reactions bridging the slow β^+ decays along the path of the rp-process. The neutrons needed for the (n,p) reactions are created by antineutrino absorptions on protons. Nuclei with mass numbers $A > 64$ are produced in the νp-process, which has been proposed to be a candidate for the origin of solar abundances of light p nuclei ^{92,94}Mo and ^{96,98}Ru [6].

The νp-process has been modeled with the JYFLTRAP mass values [19,40] in ref. [19]. The JYFLTRAP mass excess value for ^{88}Tc was found to be 1031(218) keV higher than in AME03 [52]. The heavier ^{88}Tc means a lower proton separation energy for ^{88}Tc, which increases the reaction rate of ^{88}Tc$(\gamma,p)^{87}$Mo and suppresses the flow through ^{87}Mo$(p,\gamma)^{88}$Tc. The main reaction flow to nuclei with $A > 88$ proceeds then through ^{87}Mo$(n,p)^{87}$Nb $(p,\gamma)^{88}$Mo$(p,\gamma)^{89}$Tc. The stronger reaction flow through ^{87}Mo$(n,p)^{87}$Nb and a higher abundance of ^{87}Nb enables also a flow through ^{87}Nb$(n,p)^{87}$Zr which is seen in the abundance pattern as an increased abundance of ^{87}Sr [19].

A recent mass measurement of ^{87}Mo [116] revealed a deviation of 810(220) keV to the AME03 value. With the new mass values for ^{87}Mo [116] and ^{88}Tc [19], the proton separation energy of ^{88}Tc is in agreement with the AME03 value: the difference is 220(320) keV. This obviously changes the modeling results and further demonstrates the importance of accurate mass measurements.

Another example for the impact of the new mass values is ^{90}Tc. The average mass measured at JYFLTRAP and SHIPTRAP [19] is 486(240) keV higher than the AME03 value based on the β-decay energies. A lower proton separation energy increases the reaction rate of ^{90}Tc$(\gamma,p)^{89}$Mo slightly. This, in turn, shifts some abundance from the $A = 90$ chain into the $A = 89$ chain and shows up in the abundance pattern as a decrease for ^{90}Zr [19].

The νp-process has been proposed to be a candidate for the origin of the solar abundances of the light p nuclei ^{92,94}Mo and ^{96,98}Ru [6]. Of special interest is the relative production of ^{92}Mo and ^{94}Mo governed by the proton separation energy of ^{93}Rh. The more proton-bound ^{93}Rh, the more ^{94}Pd, and in the end, more ^{94}Mo will be produced. In order to obtain the observed solar abundance ratio of ^{92}Mo and ^{94}Mo in current νp-process models it has been suggested that the proton separation energy of ^{93}Rh should be very close to 1.65 MeV [126]. The JYFLTRAP and SHIPTRAP mass measurements [19] yield a proton separation energy $S_{\mathrm{p}} = 2001(5)$ keV in agreement with the result from the Canadian Penning Trap $S_{\mathrm{p}} = 2007(9)$ keV [21]. However, the experimental results disagree with the value needed for solar ^{92}Mo/^{94}Mo ratio in supernova outflows. This suggests that supernova outflows do not exclusively produce both molybdenum isotopes or that the winds are qualitatively different from the current supernova models [126].

The impact of mass variations on the abundances and production path of the νp-process has been investigated in ref. [20]. There, ^{93}Pd, ^{100}Cd, and ^{101}In were found to have highest influence on the final abundances and path. In addition, varying the mass excess of ^{80}Zr by 2σ has a huge impact on the νp-process modeling due to large experimental mass uncertainty of ^{80}Zr (1490 keV) [20]. These nuclides will be searched for at IGISOL-4.

4.3 rp-process

Explosive hydrogen burning at temperatures in excess of 10^8 K via the rp-process was first introduced in ref. [1]. Astrophysical sites where such burning occurs are X-ray bursts and, to a limited extent, novae. In this rp-process [1, 2], the rapid proton captures on seed nuclei or on the products of helium burning lead to a production of heavier elements. Proton captures will proceed until they are inhibited by a low or negative Q-value, which leads to substantial (γ, p) photodisintegration. At those points, the rp-process has to wait for a much slower β^+ decay to happen. If the half-life of the β-decaying nucleus is particularly long the nucleus is called a waiting point.

Nuclear masses are relevant for the modeling of the rp-process due to the exponential dependence of effective waiting-point lifetimes on binding energy differences. The isotonic abundance ratios are exponentially dependent on proton separation energies as shown by the Saha equation (see, *e.g.*, ref. [16]). The impact of mass uncertainties on the rp-process has been studied, for example, in refs. [15,20]. In ref. [15], a list of nuclides whose masses should be measured in order to more reliably model the rp-process in X-ray bursts is given. From that list, the masses of ^{106}Sb and ^{107}Sb have already been determined at JYFLTRAP [3], and interesting (but also experimentally challenging) candidates for future mass measurements at JYFLTRAP are ^{31}Cl, ^{56}Cu, ^{61}Ga, ^{83,84}Nb, ^{86}Tc, ^{89}Ru, ^{90}Rh, ^{99}In, ^{96}Ag, ^{97}Cd, and ^{103}Sn.

The impact of mass variations on the light curve, the final abundances (ashes) and path of the rp-process in X-ray burst was modeled in ref. [20]. ^{94}Ag, ^{93}Pd, and ^{91}Rh were found to have highest influence on the nucleosynthesis modeling. In particular, the reaction $^{93}\mathrm{Pd}(p,\gamma)^{94}\mathrm{Ag}$ has the strongest effect on the simulation of X-ray burst lightcurve. The sensitivity of the flow ratio between ^{93}Pd and ^{94}Ag is an indication of forward and backward flows of similar order of magnitude, and thus, suggests that ^{93}Pd is a waiting point [20]. JYFLTRAP aims to measure the masses of ^{93}Pd and ^{94}Ag in future.

4.3.1 Doubly magic waiting point ^{56}Ni

A typical duration of an X-ray burst is 10–100 s. ^{56}Ni decays under terrestrial conditions mainly by electron capture. Electron densities during the rp-process are much smaller than in atoms, leading to a very long half-life of the order of at least 10 h making ^{56}Ni essentially stable for typical rp-process timescales. Because of its doubly magic character the proton capture Q-value of ^{56}Ni is relatively low, making it a potentially unique rp-process waiting point that cannot be overcome by β^+ decay. Indeed, historically ^{56}Ni was considered as an endpoint of the rp-process [1]. However, with modern nuclear masses the proton capture Q-value is sufficienly high that ^{56}Ni only becomes a waiting point for temperatures in excess of around 2 GK [2,127]. Nevertheless, such temperatures are reached in some X-ray burst models, leading to a temporary stalling of the rp-process. The onset of processing beyond ^{56}Ni during the cooling of the burst depends sensitively on the exact Q-value for the ^{56}Ni proton capture. This Q-value has been directly measured at JYFLTRAP by using ^{56}Ni as a reference for ^{57}Cu in the mass measurement. In this way, the accuracy of the proton-capture Q-value was improved from 695(19) keV [52] to 689.69(51) keV [38].

Since the calculated resonant reaction rate to a state at an energy E_{x} is exponentially dependent on the resonance energies $E_{\mathrm{r}} = E_{\mathrm{x}} - Q_{p,\gamma}$, the uncertainty of the proton capture Q value has a large effect on the uncertainty of the $^{56}\mathrm{Ni}(p,\gamma)$ reaction rate. With the new Q-value, a factor of four in the uncertainty of the reaction rate at temperatures around 1 GK shown in ref. [127] is removed and the new rate is a little higher than the one calculated with the old Q value. The new Q-value supports the conclusions of ref. [127] that the lifetime of ^{56}Ni against proton capture is much shorter than in the previous works. This reduces the minimum temperature required for the rp-process to proceed beyond ^{56}Ni. This temperature threshold coincides with the temperature for the break out of the hot CNO cycles with the rates of ref. [127]. Therefore, the rp-process can always proceed beyond ^{56}Ni provided a sufficient amount of hydrogen is present.

4.3.2 Quenching of the SnSbTe cycle

The JYFLTRAP value for the proton separation energy of ^{106}Sb, $S_{\mathrm{p}} = 424(8)$ keV, disagrees considerably with the value $S_{\mathrm{p}} = 930(210)$ keV [128] based on the alpha-decay energies and the mass of ^{114}Cs determined from its β-delayed proton decay. The value of ref. [128] has been considered erroneous already in AME03 [52], where an extrapolated value of $S_{\mathrm{p}} = 360(320)\#$ keV agreeing with JYFLTRAP is given for ^{106}Sb. The consequences of the new proton separation energies of ^{106}Sb and other nuclides in the SnSbTe region at JYFLTRAP, have been investigated in ref. [3] with the same one-zone model as in ref. [129]. In ref. [129], ^{105}Sb was found proton unbound with $S_p(^{105}\mathrm{Sb}) = -356(22)$ keV based on α-decay energies and the dependence of the branching into the SnSbTe cycle was plotted against the proton decay Q-value. Due to proton-unbound ^{104,105}Sb the *rp*-process has to proceed along the tin isotopes to ^{105}Sn. There, the proton capture probability depends on its $Q_{(p,\gamma)}$-value, in other words, on the proton separation energy of ^{106}Sb. With the new JYFLTRAP value for $S_p(^{106}\mathrm{Sb})$, only 3% of the reaction flow branches into the SnSbTe cycle at ^{105}Sn, thus considerably attenuating the cycling via the chain ^{105}Sn–^{106}Sb–^{107}Te–^{103}Sn–^{103}In–^{104}Sn–^{104}In–^{105}Sn.

The proton-separation energy of ^{107}Sb is also quite low, 588(7) keV, and only 13% branching to the SnSbTe cycle is found at ^{106}Sn. Since the β-decay half-life of ^{106}Sn is long (2.1 min), it inhibits further processing towards ^{107}Sn and ^{108}Sb, although the proton separation energy of ^{108}Sb (1222(8) keV) would be high enough for branching into the SnSbTe cycle.

Previous *rp*-process calculations had assumed there is a strong SnSbTe cycle where proton captures on ^{105}Sn and ^{106}Sb lead to ^{107}Te, which then α decays back to ^{103}Sn creating helium and cycling the matter between Sn, Sb and Te isotopes. This cycle resulted in a large accumulation of the longest-lived isotope in the SnSbTe cycle, ^{104}Sn. This was also seen in the final composition of the burst ashes where ^{104}Pd was the most abundant element.

With the new JYFLTRAP mass values the SnSbTe cycle develops closer to stability and requires therefore longer processing to be reached. Model calculations indicate that even under the most favorable conditions it is unlikely that a substantial SnSbTe cycle can develop in an X-ray burst. As a result, the final composition of the ashes is characterized by a much broader distribution of ^{68}Zn, ^{72}Ge, ^{104}Pd, ^{105}Pd, and residual helium with comparable abundances. The absence of a strong SnSbTe cycle also reduces late-time ^{4}He production, which reduces the late-time boost of hydrogen consumption and the associated rise in energy production. This leads to a slightly longer, less luminous tail. However, the effect is small since the Sn isotopes are reached at a very late stage in the burst. The quenching of the SnSbTe cycle leads also to a reduction of residual ^{4}He and ^{12}C.

Although the masses needed to constrain the reaction flow in the SnSbTe cycle are now mostly well known, uncertainties remain in the proton capture rates on the antimony isotopes, in particular ^{105}Sb and ^{106}Sb. It has been estimated that the proton-capture rate uncertainties vary up to a factor of 3 close to stability and the uncertainties far from stability might be larger [130]. For example, the $^{106}\mathrm{Sb}(p,\gamma)$ rate calculated with the code from ref. [131] is about a factor of 4 larger than the NON-SMOKER [132] rate used in ref. [3].

4.3.3 *rp*-process modeling with updated masses

To explore the impact of recent Penning-trap mass measurements, including the JYFLTRAP measurements, on the *rp*-process in X-ray bursts, and to explore the impact of remaining uncertainties, we carried out model calculations with the one-zone X-ray burst model from [20,133]. Models employing the one-zone approximation reproduce the composition of the burst ashes and some general features of the burst light curve quite well, and at the same time allow to explore the impact of variations in the nuclear physics input in a computationally efficient way. The impact of mass uncertainties on X-ray burst models has been studied before using the post-processing approximation [15], which neglects the impact of modified nuclear physics on energy generation. Other work has been based on the same model employed here, but masses were varied within quoted 1σ errors in a random way [20].

Here we choose a different approach with the goal to illustrate the maximum possible impact that mass uncertainties can have on burst model predictions. To that end, we carry out two calculations for each set of masses, one where all proton capture Q-values are simultaneously increased by 3σ, and one where they are all simultaneously decreased by 3σ. 3σ might seem like a large variation. However, for the Penning-trap mass measurements, error bars are so small that even a 3σ variation has no significant effect on the model. On the other hand, for theoretically predicted masses, for example from Coulomb shifts, or for masses determined with other experimental methods, such as β-endpoint measurements, systematic errors make such large deviations more likely than they might appear based on a Gaussian probability distribution. In addition, the effects of masses on the *rp*-process can be highly non-linear. Changes in the observables for 1σ mass variations can therefore not simply be scaled to estimate the impact of large variations due to systematic errors. Previous sensitivity studies therefore likely underestimated the impact of mass uncertainties and have probably not identified all critical masses. Our study is therefore complementary to previous approaches —while it is not intended to provide statistically correct error bars for observables, it will identify all possible mass uncertainties and provide an envelope for observables that takes into account non-linearities and the possibility of systematic, non-Gaussian errors.

We ran calculations with three sets of masses: AME03, AME10, and NOJYFL. AME03 includes experimental and extrapolated masses from AME03, and uses Coulomb Shifts [134] to calculate the masses of more exotic nuclei with $Z > N$. AME10 uses in addition all Penning-trap mass measurements in the *rp*-process region that have been published through 2010 (see fig. 1). Recently published results [79,116,135] were not included in the calculations. NOJYFL is similar to AME10 but we removed all masses that have been measured by JYFLTRAP to illustrate the impact of the JYFLTRAP program.

The *rp*-process reaction paths for the AME10 upper and lower proton capture Q values are shown in fig. 10. Clearly, current mass uncertainties still allow for very large changes in the reaction paths. For the low Q-values the path below ruthenium is shifted by one mass unit, above ruthenium by 2 mass units closer to stability. With the low Q-values the ^{80}Zr waiting point is completely bypassed as the reaction flow proceeds towards stability in

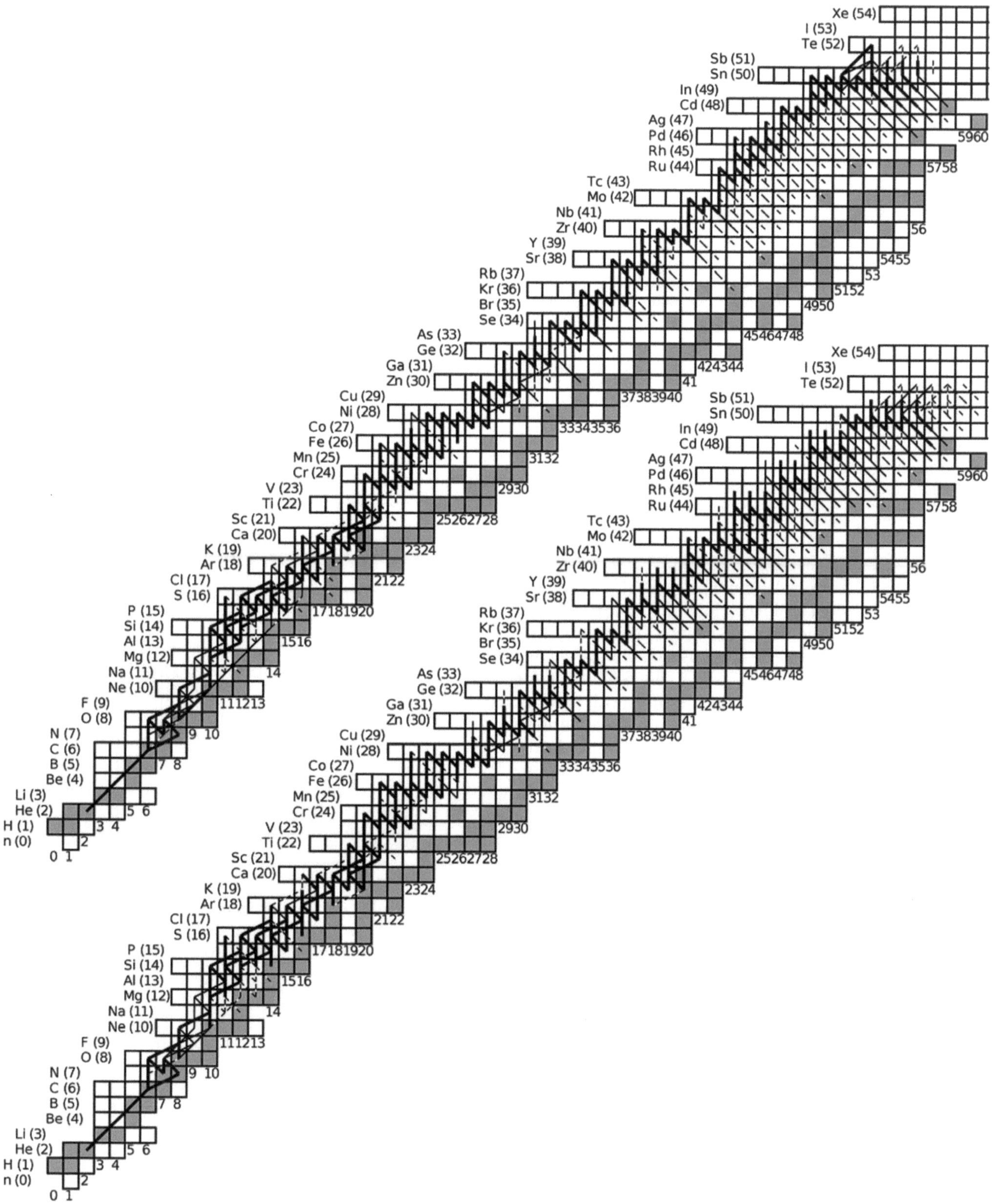

Fig. 10. The *rp*-process reaction flows corresponding the AME10 upper ($+3\sigma$) (top) and lower (-3σ) proton capture Q-values (bottom). Plotted are net flows of $> 10\%$ (thick solid line), 1%–10% (thin solid line), and 0.1%–1% (dashed line) of the flow through the 3α reaction. Note that flows for (p,γ) reactions are numerically not accurate for cases of (p,γ)-(γ,p) equilibrium near the proton drip line.

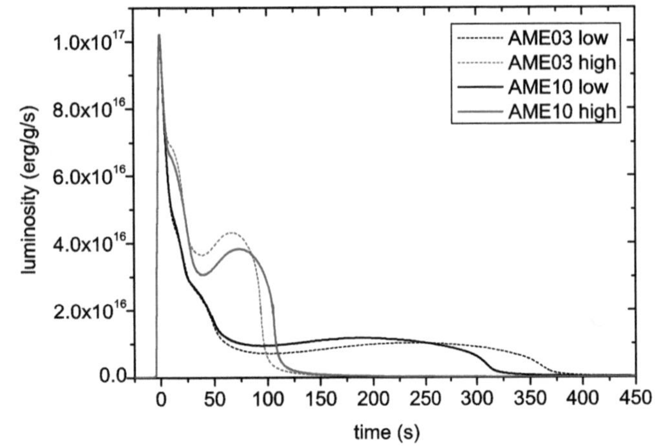

Fig. 11. (Color online) Calculated emitted luminosity as a function of time during an X-ray burst. The burst light curve is based on proton-capture Q-values from i) AME03 values varied 3σ up (AME03 high, red dashed line) or down (AME03 low, black dashed line) and ii) AME10 values (including all new Penning-trap measurements before 2011) varied 3σ up (AME10 high, red solid line) or down (AME10 low, black solid line).

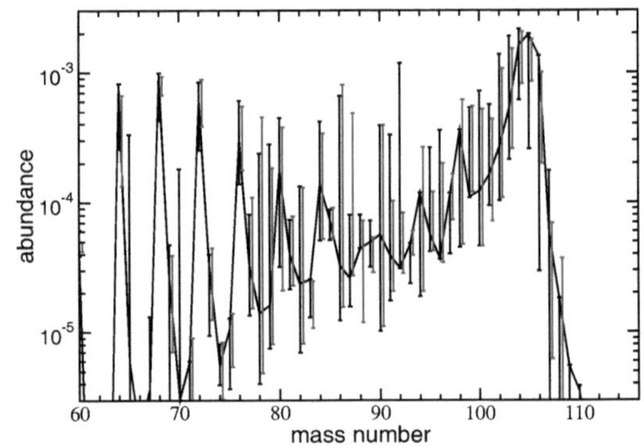

Fig. 12. (Color online) Final abundance (mass fraction divided by mass number) distribution as a function of mass number for an X-ray burst for data sets AME03 and AME10. The vertical bars represent the variation of the abundances obtained with upper ($+3\sigma$) and lower (-3σ) proton-capture Q-values. The left part of the data pairs belongs to AME03 (black) and the right part to AME10 (red).

the yttrium isotopic chain. While JYFLTRAP mass measurements have made the formation of a SnSbTe cycle much more unlikely, the high Q-value calculation does develop such a cycle at ^{103}Sn because of the large uncertainties (300# keV and 360# keV) that still exist for the ^{103}Sn and ^{104}Sb masses. While a significantly proton bound ^{104}Sb, which would be required to form a SnSbTe cycle, seems unlikely given the trend of the measured proton separation energies of less neutron-deficient Sb isotopes, tentative experimental results indeed hint at ^{104}Sb not being a fast proton emitter [129].

Figure 11 demonstrates that the burst light curve can be influenced dramatically by nuclear masses. At the extreme of very high Q-values, the major waiting points can be bypassed efficiently and processing towards heavy nuclei during the burst cooling phase when the luminosity is decreasing is greatly accelerated leading to a significant increase in energy generation, and fast exhaustion of fuel. The result is the development of a shoulder in the light curve during the early cooling phase, and a second peak in the burst profile after about 70 s. On the other hand, very low Q-values increase photodisintegration, which hampers proton capture and results in a slower rp-process closer to stability. As a consequence, energy is generated at a reduced rate, but for a longer time leading to a very long burst tail lasting about 5 minutes past the initial burst peak. Clearly Penning-trap mass measurements since 2003 have not yet reached the majority of rp-process masses and do reduce the uncertainty only somewhat. It should be noted that the systematic AME03 mass extrapolations play an important role in the model calculations. These extrapolations were based on the AME03 experimental data sets. Mass measurements since 2003 would likely lead to improved extrapolations and more reliable burst calculations. This effect is not included here, and its exploration

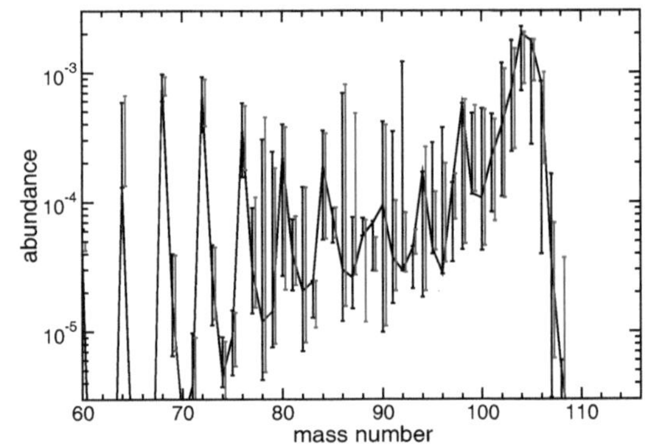

Fig. 13. (Color online) Same as fig. 12 but for X-ray burst calculations with AME10 and NOJYFL. The left part of the data pairs belongs to NOJYFL (black) and the right part to AME10 (red).

needs to wait for the publication of the next Atomic Mass Evaluation.

Figures 12 and 13 illustrate corresponding variations in the final composition of the ashes. The composition is given as a function of mass number, as the ashes decay along a mass chain until the first stable isotope is reached. Clearly significant variations up to an order of magnitude are possible, but not more. For many mass numbers, the maximum variations are much smaller. Among the important most abundant isotopes in the ashes, Penning-trap mass measurements since 2003 had the largest impact on the $A = 68$, 91, 92, 105, 106 isobars where uncertainties in the final abundance were drastically reduced (see fig. 12). The JYFLTRAP mass measurements had a major impact on all of these, with the exception of $A = 68$ (see fig. 13).

The reduction of the $A = 92$ abundance uncertainty is particularly important in light of the interest in a possible production of ^{92}Mo in the *rp*-process.

5 Outlook

Mass measurements on neutron-deficient nuclides of astrophysical interest will continue with JYFLTRAP at the new IGISOL-4 facility. There will be more beam time available since the two cyclotrons, MCC-30 and K-130, can work in parallel. The new facility should also offer better and cleaner conditions for mass measurements. A permanent yield station and a post-trap spectroscopy setup will help in monitoring the experiments. Besides IGISOL-4 there have been major developments in the JYFLTRAP mass measurements since the first measurements of *rp*-process nuclei [40]. A new fast cleaning procedure to produce isomerically pure ion samples has been developed [37]. The Ramsey method of time-separated oscillatory fields [136,137] has been successfully applied to short-lived ions of astrophysical interest. Precise data on mass-dependent and residual uncertainties of JYFLTRAP have been obtained via carbon cluster measurements [48]. Interleavedly performed measurements [34] reduce the uncertainty due to temporal B-field fluctuations, which will result in a promising improvement on future mass measurements for neutron-deficient nuclei.

With the light-ion guide at IGISOL-4, nuclides important for modeling the explosive hydrogen burning in ONe novae as well as for testing the IMME, such as ^{27}P, ^{31}Cl, and ^{32}Cl, will be produced. Also the Q_{EC}-value measurements of mirror nuclei will be continued. In the heavier mass region, proton or ^{3}He beams on a ^{92}Mo target could help in populating the low-spin isomers in ^{91}Tc and ^{93}Ru. More exotic species, such as the $N = 50$ isotones ^{93}Tc, ^{97}Ag, ^{98}Cd, ^{99}In, and ^{100}Sn could be searched for with the heavy-ion ion guide. In addition, a mass measurement of the waiting-point nucleus ^{80}Zr would be very interesting. The masses of ^{94}Ag and ^{93}Pd have been shown to have a high impact on *rp*-process models, and thus should be measured. ^{93}Pd could be produced at HIGISOL whereas for ^{94}Ag, a special hot cavity laser ion source has been developed [138]. Since the half-life of the ^{94}Ag ground state is only around 30 ms, it is too short-lived for measurements at JYFLTRAP. However, the much longer-lived isomeric states (7^+, 0.55(6) s [139]) and (21^+, 0.40(4) s [139]) could be measured. Of these, the 21^+ spin-gap isomer has gained much interest recently due to its claimed two-proton decay [140].

Taking into account all the developments carried out at JYFLTRAP and IGISOL, new mass measurements of Y, Nb, Mo, and Tc isotopes at the heavy-ion ion guide at IGISOL-4 would be relevant. These measurements would confirm the nature of the measured state (ground state or isomer) in previous experiments [19,40,116]. In addition, the few nuclides (^{92}Tc, ^{90}Ru, ^{107}Sb, and ^{109}Sb) for which deviations to the SHIPTRAP values have been found, could be reinvestigated.

More accurate mass values would be obtained via networks of mass measurements, where a single reference atom would not play such a big role. This would also give better information on the mass surface and be useful for mass predictions. In addition, direct frequency ratio measurements between proton capture mother and daughter would result in more precise proton capture Q-values. Identification of the ground and isomeric states and/or production ratios for possible isomer corrections should be investigated via post-trap spectroscopy. In many cases, simple half-life measurements based on β particles would enlighten the situation a lot. In summary, there is a wealth of fascinating experiments on neutron-deficient nuclei of astrophysical interest to be performed with JYFLTRAP in future.

This work has been supported by the Academy of Finland under the Finnish Centre of Excellence Programme 2006-2011 (Nuclear and Accelerator Based Physics Programme at JYFL). AK acknowledges the support from the Academy of Finland under the project 127301.

References

1. R.K. Wallace, S.E. Woosley, Astrophys. J. Suppl. Ser. **45**, 389 (1981).
2. H. Schatz *et al.*, Phys. Rep. **294**, 167 (1998).
3. V.-V. Elomaa *et al.*, Phys. Rev. Lett. **102**, 252501 (2009).
4. S.E. Woosley *et al.*, Astrophys. J. Suppl. S. **151**, 75 (2004).
5. J. José, F. Moreno, A. Parikh, C. Iliadis, Astrophys. J. Suppl. S. **189**, 204 (2010).
6. C. Fröhlich *et al.*, Phys. Rev. Lett. **96**, 142502 (2006).
7. J. Pruet *et al.*, Astrophys. J. **644**, 1028 (2006).
8. S. Wanajo, Astrophys. J. **647**, 1323 (2006).
9. S. Wanajo, H.-T. Janka, S. Kubono, Astrophys. J. **729**, 46 (2011).
10. J. José, A. Coc, M. Hernanz, Astrophys. J. **520**, 347 (1999).
11. J. José, A. Coc, M. Hernanz, Astrophys. J. **560**, 897 (2001).
12. J. José, M. Hernanz, J. Phys. G **34**, R431 (2007).
13. S. Starrfield *et al.*, Astrophys. J. **692**, 1532 (2009).
14. S.A. Glasner, J.W. Truran, Astrophys. J. Lett. **692**, L58 (2009).
15. A. Parikh *et al.*, Phys. Rev. C **79**, 045802 (2009).
16. H. Schatz, Int. J. Mass Spectrom. **251**, 293 (2006).
17. M. Breitenfeldt *et al.*, Phys. Rev. C **80**, 035805 (2009).
18. J. Savory *et al.*, Phys. Rev. Lett. **102**, 132501 (2009).
19. C. Weber *et al.*, Phys. Rev. C **78**, 054310 (2008).
20. T. Fleckenstein, Diploma thesis, Justus-Liebig-Universität Gießen, 2008.
21. J. Fallis *et al.*, Phys. Rev. C **78**, 022801 (2008).
22. G. Savard *et al.*, Hyperfine Interact. **132**, 221 (2001).
23. M. Mukherjee *et al.*, Eur. Phys. J. A **35**, 1 (2008).
24. R. Ringle *et al.*, Nucl. Instrum. Methods Phys. Res. A **604**, 536 (2009).
25. M. Block *et al.*, Eur. Phys. J. D **45**, 39 (2007).
26. A. Saastamoinen *et al.*, Phys. Rev. C **80**, 044330 (2009).
27. A. Kankainen *et al.*, Phys. Rev. C **82**, 052501 (2010).
28. T. Eronen, Ph.D. thesis, Department of Physics, University of Jyväskylä, 2008.

29. E. Wigner, Phys. Rev. **51**, 947 (1937).
30. J. Äystö, Nucl. Phys. A **693**, 477 (2001).
31. M. Oinonen *et al.*, Nucl. Instrum. Methods Phys. Res. A **416**, 485 (1998).
32. J. Huikari *et al.*, Nucl. Instrum. Methods Phys. Res. B **222**, 632 (2004).
33. T. Eronen *et al.*, Phys. Rev. C **79**, 032802 (2009).
34. T. Eronen *et al.*, Phys. Rev. Lett. **103**, 252501 (2009).
35. T. Eronen *et al.*, Phys. Rev. Lett. **97**, 232501 (2006).
36. T. Kurtukian Nieto *et al.*, Phys. Rev. C **80**, 035502 (2009).
37. T. Eronen *et al.*, Phys. Rev. Lett. **100**, 132502 (2008).
38. A. Kankainen *et al.*, Phys. Rev. C **82**, 034311 (2010).
39. T. Eronen *et al.*, Phys. Lett. B **636**, 191 (2006).
40. A. Kankainen *et al.*, Eur. Phys. J. A **29**, 271 (2006).
41. V. Elomaa *et al.*, Eur. Phys. J. A **40**, 1 (2009).
42. A. Nieminen *et al.*, Nucl. Instrum. Methods Phys. Res. A **469**, 244 (2001).
43. V.S. Kolhinen *et al.*, Nucl. Instrum. Methods Phys. Res. A **528**, 776 (2004).
44. G. Savard *et al.*, Phys. Lett. A **158**, 247 (1991).
45. G. Gräff, H. Kalinowsky, J. Traut, Z. Phys. A **297**, 35 (1980).
46. M. König *et al.*, Int. J. Mass Spectrom. Ion Processes **142**, 95 (1995).
47. A. Kellerbauer *et al.*, Eur. Phys. J. D **22**, 53 (2003).
48. V.-V. Elomaa *et al.*, Nucl. Instrum. Methods Phys. Res. A **612**, 97 (2009).
49. W. Benenson, E. Kashy, Rev. Mod. Phys. **51**, 527 (1979).
50. M. Bentley, S. Lenzi, Prog. Part. Nucl. Phys. **59**, 497 (2007).
51. N. Auerbach, Phys. Rep. **98**, 273 (1983).
52. G. Audi, A.H. Wapstra, C. Thibault, Nucl. Phys. A **729**, 337 (2003).
53. R. Firestone, Nucl. Data Sheets **108**, 1 (2007).
54. V.E. Iacob *et al.*, Phys. Rev. C **74**, 045810 (2006).
55. C. Wrede *et al.*, Phys. Rev. C **81**, 055503 (2010).
56. M. Bhattacharya *et al.*, Phys. Rev. C **77**, 065503 (2008).
57. M. Redshaw, J. McDaniel, E.G. Myers, Phys. Rev. Lett. **100**, 093002 (2008).
58. A.A. Kwiatkowski *et al.*, Phys. Rev. C **80**, 051302 (2009).
59. A. Paul, S. Röttger, A. Zimbal, U. Keyser, Hyperfine Interact. **132**, 189 (2001).
60. P. Rikmenspoel, D.V. Patter, Nucl. Phys. **24**, 494 (1961).
61. G. Rickards, B.E. Bonner, G.C. Phillips, Nucl. Phys. **86**, 167 (1966).
62. R. Nussbaum *et al.*, Physica **20**, 571 (1954).
63. K. Sato, Mass Spectrom. (Japan) **5**, 54 (1964).
64. S. Antman, H. Pettersson, A. Suarez, Nucl. Phys. A **94**, 289 (1967).
65. J.M. Freeman *et al.*, Nucl. Phys. **65**, 113 (1965).
66. B.E. Bonner, G. Rickards, D.L. Bernard, G.C. Phillips, Nucl. Phys. **86**, 187 (1966).
67. J. Overley, P. Parker, D. Bromley, Nucl. Instrum. Methods **68**, 61 (1969).
68. L. Erikson *et al.*, Phys. Rev. C **81**, 045808 (2010).
69. B. Sherrill *et al.*, Phys. Rev. C **28**, 1712 (1983).
70. M.B. Greenfield, C.R. Bingham, E. Newman, M.J. Saltmarsh, Phys. Rev. C **6**, 1756 (1972).
71. N. Severijns, M. Tandecki, T. Phalet, I.S. Towner, Phys. Rev. C **78**, 055501 (2008).
72. O. Naviliat-Cuncic, N. Severijns, Phys. Rev. Lett. **102**, 142302 (2009).
73. H. Fujita *et al.*, Phys. Rev. C **75**, 034310 (2007).
74. C. Guénaut *et al.*, Phys. Rev. C **75**, 044303 (2007).
75. D. Rodríguez *et al.*, Nucl. Phys. A **769**, 1 (2006).
76. F. Herfurth *et al.*, Eur. Phys. J. A **15**, 17 (2002).
77. F. Herfurth *et al.*, Nucl. Phys. A **746**, 487 (2004).
78. G. Sikler *et al.*, Nucl. Phys. A **763**, 45 (2005).
79. F. Herfurth *et al.*, Eur. Phys. J. A **47**, 1 (2011).
80. P. Schury *et al.*, Phys. Rev. C **75**, 055801 (2007).
81. J.A. Clark *et al.*, Phys. Rev. Lett. **92**, 192501 (2004).
82. M.L. Commara *et al.*, Nucl. Phys. A **708**, 167 (2002).
83. C. Plettner *et al.*, Nucl. Phys. A **733**, 20 (2004).
84. K. Kaneko, Y. Sun, M. Hasegawa, T. Mizusaki, Phys. Rev. C **77**, 064304 (2008).
85. E. Nolte, H. Hick, Z. Phys. A **305**, 289 (1982).
86. E. Bouchez *et al.*, Phys. Rev. Lett. **90**, 082502 (2003).
87. J.C. Hardy, L.C. Carraz, B. Jonson, P.G. Hansen, Phys. Lett. B **71**, 307 (1977).
88. S. Issmer *et al.*, Eur. Phys. J. A **2**, 173 (1998).
89. S. Kato *et al.*, Phys. Rev. C **41**, 1276 (1990).
90. A. Martín *et al.*, Eur. Phys. J. A **34**, 341 (2007).
91. J.A. Clark *et al.*, Eur. Phys. J. A **25, s01**, 629 (2005).
92. J. Fallis *et al.*, Phys. Rev. C **84**, 045807 (2011).
93. A.H. Wapstra, G. Audi, C. Thibault, Nucl. Phys. A **729**, 129 (2003).
94. G. Audi, O. Bersillon, J. Blachot, A.H. Wapstra, Nucl. Phys. A **729**, 3 (2003).
95. J. Döring, A. Aprahamian, M. Wiescher, J. Res. Natl. Inst. Stand. Technol. **105**, 43 (2000).
96. S. Dean *et al.*, Eur. Phys. J. A **21**, 243 (2004).
97. D. Kast *et al.*, Z. Phys. A **356**, 363 (1996).
98. A. Kankainen *et al.*, Eur. Phys. J. A **25**, 355 (2005).
99. R. Barden *et al.*, Z. Phys. A **329**, 319 (1988).
100. H. Keller *et al.*, Z. Phys. A **340**, 363 (1991).
101. M. Kavatsyuk, L. Batist, M. Karny, E. Roeckl, Int. J. Mass Spectrom. **251**, 138 (2006).
102. A. Płochocki *et al.*, Nucl. Phys. A **332**, 29 (1979).
103. R. Kirchner *et al.*, Phys. Lett. B **70**, 150 (1977).
104. E. Roeckl *et al.*, Phys. Lett. B **78**, 393 (1978).
105. D. Schardt *et al.*, Nucl. Phys. A **326**, 65 (1979).
106. F. Heine *et al.*, Z. Phys. A **340**, 225 (1991).
107. D. Schardt *et al.*, Nucl. Phys. A **368**, 153 (1981).
108. R.J. Tighe *et al.*, Phys. Rev. C **49**, R2871 (1994).
109. D.D. Bogdanov, V.A. Karnaukhova, L.A. Petrov, Yad. Fizika **17**, 457 (1973).
110. Y. Litvinov *et al.*, Nucl. Phys. A **756**, 3 (2005).
111. K. Sharma *et al.*, in *Proceedings of 9th International Conference on Atomic Masses and Fundamental constants AMCO-9*, and *6th International Conference on Nuclei far from Stability NUFAST-6*, edited by R. Neugart, A. Wöhr (IOP Publishing, Bristol, Philadelphia, 1992) p. 31.
112. M.G. Johnston *et al.*, J. Phys. G **8**, 1405 (1982).
113. F. Heine *et al.*, in *Proceedings of 9th International Conference on Atomic Masses and Fundamental constants AMCO-9*, and *6th International Conference on Nuclei far from Stability NUFAST-6*, edited by R. Neugart, A. Wöhr (IOP Publishing, Bristol, Philadelphia, 1992) p. 331.
114. R.D. Page *et al.*, Phys. Rev. C **49**, 3312 (1994).
115. T. Faestermann *et al.*, Phys. Lett. B **137**, 23 (1984).
116. E. Haettner *et al.*, Phys. Rev. Lett. **106**, 122501 (2011).
117. S.M. Fischer *et al.*, Phys. Rev. C **75**, 064310 (2007).
118. J. José, M. Hernanz, C. Iliadis, Nucl. Phys. A **777**, 550 (2006).
119. P.M. Endt, Nucl. Phys. A **633**, 1 (1998).

120. A. Parikh *et al.*, Phys. Rev. C **71**, 055804 (2005).
121. S. Amari *et al.*, Astrophys. J. **551**, 1065 (2001).
122. D.G. Jenkins *et al.*, Phys. Rev. C **73**, 065802 (2006).
123. Z. Ma *et al.*, Phys. Rev. C **76**, 015803 (2007).
124. C. Wrede *et al.*, Phys. Rev. C **76**, 052802 (2007).
125. C. Wrede *et al.*, Phys. Rev. C **79**, 045803 (2009).
126. J.L. Fisker, R.D. Hoffman, J. Pruet, Astrophys. J. Lett. **690**, L135 (2009).
127. O. Forstner *et al.*, Phys. Rev. C **64**, 045801 (2001).
128. A. Płochocki *et al.*, Phys. Lett. B **106**, 285 (1981).
129. C. Mazzocchi *et al.*, Phys. Rev. Lett. **98**, 212501 (2007).
130. W. Rapp *et al.*, Astrophys. J. **653**, 474 (2006).
131. M. Aikawa *et al.*, Astron. Astrophys. **441**, 1195 (2005).
132. T. Rauscher, F.-K. Thielemann, At. Data Nucl. Data Tables **79**, 47 (2001).
133. H. Schatz *et al.*, Phys. Rev. Lett. **86**, 3471 (2001).
134. B.A. Brown *et al.*, Phys. Rev. C **65**, 045802 (2002).
135. X.L. Tu *et al.*, Phys. Rev. Lett. **106**, 112501 (2011).
136. S. George *et al.*, Int. J. Mass Spectrom. **264**, 110 (2007).
137. M. Kretzschmar, Int. J. Mass Spectrom. **264**, 122 (2007).
138. M. Reponen *et al.*, Eur. Phys. J. A **42**, 509 (2009).
139. D. Abriola, A. Sonzogni, Nucl. Data Sheets **107**, 2423 (2006).
140. I. Mukha *et al.*, Nature **439**, 298 (2006).

DOI 10.1007/978-94-007-5555-0_28
Reprinted from *Hyperfine Interactions* Volume 223, Issues 1-3, 2014
and *The European Physical Journal A* Volume 48, Issue 4, 2012

Diffusion studies with radioactive ions

J. Räisänen · H. J. Whitlow

Published online: 30 March 2012

Abstract An overview of the modified radiotracer based diffusion studies carried out at IGISOL is provided. The experimental procedures are briefly described followed by examples involving IGISOL as the key facility. In this respect the studies related to silicon-germanium ($Si_{1-x}Ge_x$) alloys and on the related diffusion systematics are summarized. Another group of examples is related to mobility determination of lead isotopes in glass for verifying retrospective radon measurements. Finally an outlook to future possibilities related to employing radiotracers in solid state research is provided.

Keywords Diffusion · Radiotracers · SiGe · Rn dosimetry · Glass

1 Introduction

Utilization of radioactive isotopes in solid state physics research has gained a stable position as an established research tool and several nuclear characterization methods are nowadays available. For applications in solid state since, ion implantation is the most convenient way of introducing the radioactive probe atoms into the material to be studied since it allows, e.g., precise control of the impurity concentration through a mathematically well defined distribution below the surface.

In the development of new materials for a wide range of technological applications, the role of understanding material diffusion properties has played a key role. The procedure of the modified radiotracer technique incorporates nuances enabled by

J. Räisänen (✉)
Department of Physics, University of Helsinki, P.O.Box 43, 00014, Helsinki, Finland
e-mail: raisanen@mappi.helsinki.fi

H. J. Whitlow
Department of Physics, University of Jyväskylä, P.O.Box 35 (YFL), 40014, Jyväskylä, Finland
e-mail: harry.j.whitlow@jyu.fi

development of modern experimental techniques; radioactive ion beams followed by ion implantation and ion beam sputtering [1]. Tracers with a short half-life enable diffusion experiments with peak tracer concentrations in the range of 1×10^{10}–1×10^{12} atoms/cm^3. The superb sensitivity of the radio tracer technique makes it easier to maintain the intrinsic conditions. High sensitivity is also useful in cases where the element under study has low solid solubility. Furthermore, the high sensitivity means that the separation of the implanted atoms is large (370–8000 atomic spacing for the peak concentrations above) which on the other hand implies that the measurements involve isolated atoms diffusing in a pure substrate without the influence of cluster formation. This is fundamentally a more clean measurement than at higher concentrations where cluster formation may influence the diffusion properties. Use of the modified radiotracer technique for studying diffusion is presently not very widespread and only few laboratories have the potential for performing self-sustaining experiments.

The fact that the radioactive atoms are introduced by ion implantation and that suitable isotopes can be made using the IGISOL technique for almost every element heavier than chlorine, as well as many important lighter elements, implies the method can be applied to a wide range of diffusion and other problems such as corrosion and mechanical wear erosion. New constellations of materials for nanoelectronics [2] with strained interfaces imply that diffusion may not behave as in bulk materials and trapping may be concentration dependant by e.g. saturation of interface sites. To elucidate the diffusion behaviour of realistic low-concentrations of impurities the modified radioactive tracer method is well suited. Other types of measurement where the extreme sensitivity and short half-lives are beneficial include in-vitro corrosion and wear studies of surgical proteases and functional surfaces.

The success of nuclear solid state physics research is strongly dependent on the availability of a broad range of appropriate radioactive isotopes. In this article an overview of the modified radiotracer based diffusion studies carried out at IGISOL is provided. The survey starts with brief description of the experimental procedures followed by examples involving IGISOL as the key facility. Finally, an outlook to future possibilities related to employing radiotracers in solid state research will be provided.

2 Experimental procedures

The basic principles of the modified radiotracer technique are presented in Fig. 1. The technique includes following steps: (i) Production and implantation of the radioactive isotopes of the diffusing element, the tracers. (ii) Sample annealing. (iii) Serial sectioning by ion beam sputtering followed by measurements of the activity of eroded material and the depth profile construction, and (iv) Quantitative analysis of the experimental depth profile.

2.1 Implantation of radiotracers

The tracer with convenient half-life and decay mode is selected first. In practice the isotopes with the shortest half-life employed in diffusion experiments in general are ^{8}Li ($T_{1/2} = 0.84$ s) and ^{11}C ($T_{1/2} = 20.3$ min). The implantation energy is chosen so

Fig. 1 Experimental steps of the modified radiotracer technique: 1) implantation of the radioactive tracer, 2) diffusion annealing, 3) sample serial sectioning and 4) activity measurements

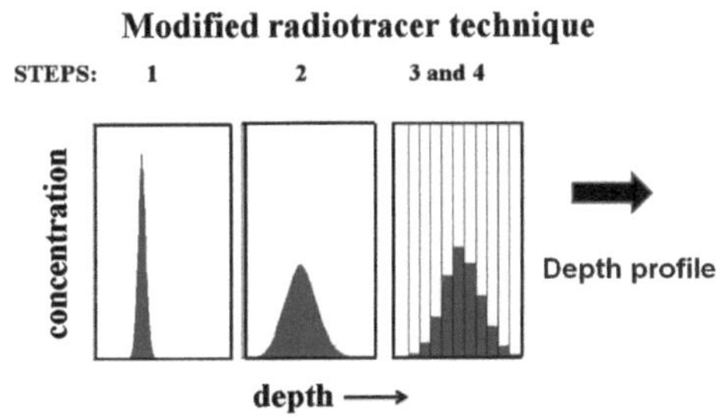

that the migration of the tracer atoms is optimized. Typical values are a few tens of keV, which are also sufficient for tracer implantation well below the surface (10–50 nm). This ensures avoidance of unwanted processes such as surface oxidation. Proper implantation energy is also important for setting the well-defined diffusion conditions in annealing. Too high energies would lead to flatter implantation profiles and therefore to less sensitive diffusion coefficient extraction. It should be noted that ion implantation has also an unwanted side effect of inducing defects in the studied material. Fortunately, the implantation fluence necessary for the diffusion studies is rather low.

2.2 Sample annealing

Generally, the diffusion process is too slow at room temperature. To ensure exploitable diffusion data, it should be determined within as wide temperature range as feasible. Therefore the sample temperature must be elevated by using a high temperature resistance furnace. During annealing, it should be taken care of that all possible chemical reactions taking place at the sample surface are avoided or controlled. To achieve this, annealing is usually carried out under inert gas atmosphere (argon) or vacuum (10^{-8}–10^{-9} mbar). The sample is also commonly encapsulated between protective caps (often of same material as the sample) to minimize the harmful surface reactions and contamination. During annealing, the temperature is monitored using a thermocouple attached to the sample holder. For short annealing times, heating and cooling corrections must be incorporated.

2.3 Depth profiling

The samples are serial sectioned by sputtering in order to deduce the concentration versus depth profiles. Using ion beam sputtering, sectioning of even monolayers is feasible providing possibilities for clearly improved depth resolution. This enables diffusion studies at lower temperatures than with conventional radiotracer techniques. The principle of serial sectioning by sputtering is widely applied e.g. in Secondary Ion Mass Spectrometry (SIMS).

The eroded material must be collected for further analysis and this is realized by collecting the sputtered material on a mylar foil, placed as close as possible to the sputtered sample. The collection efficiency and depth resolution are optimized by keeping the energy of the primary beam as low as 1–2 keV and the angle between the incoming beam and the normal of the sample surface at 70 degrees to minimize ion beam mixing. With intense primary beams, typical eroding speeds are few tens of

Fig. 2 Set-up for sputtering-based sample depth profiling

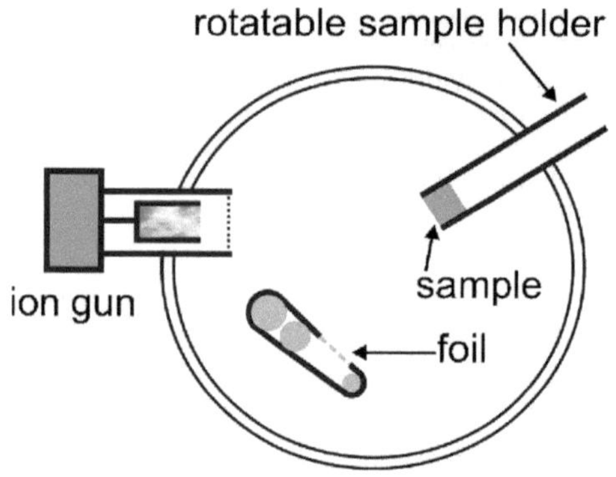

nm/min; sufficient that the profiling can be extended to few microns in depth. When one layer has been eroded, another collection foil segment is moved in the collection position and the previous foil segment is wrapped on a roll like a film in a camera. This way each segment corresponds to certain depth slice of the sample. During the erosion process, the beam current is monitored for normalization purposes. The activity collected on each frame of the film is subsequently measured in a detector system (described below). The depth scale is fixed by measuring the eroded crater depth by a stylus profiler and utilizing the total integrated charge. The eroded area must be larger than the implantation area, so that the distortion of the determined profile due to the edges of the formed crater can be avoided. The half-life of the isotope to be profiled plays also a role. Typically, depth profiling at the resolution of about 0.4 nm is achievable for an isotope with 1–4 hour half-life when implantation fluences of the order of 10^9 cm^{-2} are employed. The set-up utilized in the studies carried out employing the IGISOL facility is shown schematically in Fig. 2.

2.4 Activity measurement

In the activity measurements special attention should be paid to efficient radiation detection as this decreases the required amount of implanted ions and/or reduces the measurement time. Often the IGISOL-employed radiotracers decay via β-decay, so that the activity of the mylar foils can be measured using large area silicon detectors (active volume thickness typically 0.5 mm). To maximize the efficiency, two of these detectors have been placed face to face with a separation distance of only 5 mm. Further, to minimize the noise level, the detectors are cooled to −5°C. Another alternative being employed is the use of two large PIN-diode arrays for β activity measurement [3]. Background radiation is minimized using lead shielding and unwanted counts due to muons are rejected using anticoincidence logic.

3 Diffusion coefficient determination

After the depth distributions of the tracer atoms (before and after annealing) have been determined via serial sectioning, quantitative analysis is needed for determining the diffusion coefficients D at given temperatures.

If D is constant with respect to depth x and concentration C, Fick's second law reduces to:

$$\frac{\partial C}{\partial t} = D\left(\frac{\partial^2 C}{\partial x^2}\right) \tag{1}$$

The diffusion coefficients can be determined by observing the time evolution of the experimental diffusion profiles. If the as-implanted profile can be approximated by a Gaussian function, $C_0(x, t = 0)$, the analytical solution for the concentration independent diffusion equation is:

$$C(x,t) = \frac{C_0}{2\sqrt{1+\frac{2Dt}{w^2}}} \left[\begin{array}{l} erfc\left(-\dfrac{\frac{x_c}{2w^2}+\frac{x}{4Dt}}{\sqrt{\frac{1}{2w^2}+\frac{1}{4Dt}}}\right)\exp\left(\dfrac{(x-x_c)^2}{2w^2+4Dt}\right)+ \\ k\cdot erfc\left(-\dfrac{\frac{x_c}{2w^2}-\frac{x}{4Dt}}{\sqrt{\frac{1}{2w^2}+\frac{1}{4Dt}}}\right)\exp\left(\dfrac{(x+x_c)^2}{2w^2+4Dt}\right) \end{array} \right]. \tag{2}$$

The parameter $k\,(-1 < k < 1)$ represents different boundary conditions at the sample surface ($x = 0$). If the sample surface acts as a perfect sink, $k = -1$, whereas $k = +1$ corresponds to the case of an ideally reflecting surface. The parameters $C_{0,}\, w$ and x_c can be determined by fitting the Gaussian function to an as-implanted profile (when $t = 0$ and $k = 0$, (2) reduces to a Gaussian function (3)).

$$C(x, t = 0) = C_0 \exp\left[-\frac{(x-x_c)^2}{2w^2}\right]. \tag{3}$$

In reality, the shape of the experimental as-implanted profile is not exactly Gaussian. Due to this fact, the diffusion (1) is often solved numerically.

In most cases, the temperature dependence of the diffusion coefficient follows the Arrhenius law:

$$D = D_0 \exp\left(-\frac{H}{kT}\right), \tag{4}$$

where T is temperature and k is the Boltzmann constant. The pre-exponential factor D_0 and the activation enthalpy H, are found by fitting (4) to the experimentally determined D values (e.g. Fig. 4 below).

4 Examples of radiotracer diffusion studies at IGISOL

4.1 Diffusion in SiGe alloys

In diffusion studies during the recent years one strong emphasis has been on silicon-germanium ($Si_{1-x}Ge_x$) alloys and studies on the related diffusion systematics. In addition to its technological importance [2], $Si_{1-x}Ge_x$ alloy is an interesting material also on the point of view of basic diffusion research. The diffusion properties of impurity atom and self-diffusion in Si and Ge are different. In this light it is expected that the diffusion properties of $Si_{1-x}Ge_x$ alloys will experience changes as a function

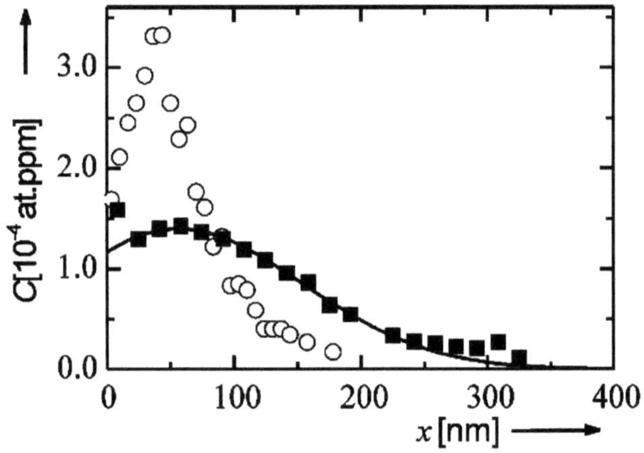

Fig. 3 The as-implanted (○)^{31}Si profile in $Si_{0.20}Ge_{0.80}$ and the profile obtained after 1.5 hour annealing at 850°C (■). The implantation energy is 37 keV. The diffusion coefficient at the specified temperature is obtained by fitting the appropriate solution of the diffusion equation to the curve shown by the solid line. (From Ref. [11])

of composition. At which compositions and how these changes happen have been of major interest. There are clear differences in dopant diffusion between Si and Ge and by measuring the diffusivity of the different elements as a function of concentration it has been possible to gain information on the origin of these differences [4, 5]. This information is particularly useful in developing the existing models of point-defect mediated diffusion.

As an example of the potential of the modified radiotracer technique for studying diffusion, typical depth profiles for silicon self-diffusion in $Si_{1-x}Ge_x$ alloys are presented in Fig. 3. The employed method allows experiments under intrinsic conditions due to its superior sensitivity; furthermore, self-diffusion studies are not feasible by most experimental techniques. The merits of utilizing the IGISOL facility for radiotracer diffusion studies are summarized in Ref. [6].

4.2 Mobility determination of lead isotopes in glass for retrospective radon measurements

In many countries exposure to radon in dwelling places is significant health issue. A radio-epidemiological technique to retrospectively measure the radon exposure is based on determining the content of ^{210}Pb in the surface of soda-lime glass [7, 8]. The ^{210}Pb is incorporated in the surface layer of glass by recoil implantation during single or double α-decay in the ^{222}Rn decay series. A critical assumption in these measurements is that the recoil implanted ^{210}Pb is immobile in the outer 100 nm or so of the glass surface [9, 10]. This has been verified by measuring the diffusion of the short-lived ^{209}Pb ($t_{1/2} = 3.253$ h) as a model for the long-lived ^{210}Pb ($t_{1/2} = 22$ a) isotope in the surface of glass and testing different cleaning techniques. The high sensitivity of this method was essential to measure realistic low concentrations of implanted Pb because lead is a network-forming element in glass.

Figure 4 shows the Arrhenius fit to the diffusion coefficients. Extrapolating to room temperature gives a diffusion length of 10^{-22}–10^{-23} m which confirms that the recoil implanted ^{210}Pb is indeed immobile in the 100 nm thick surface layer of glass. Furthermore, normal domestic glass cleaning did not result in any detectable loss of Pb [10].

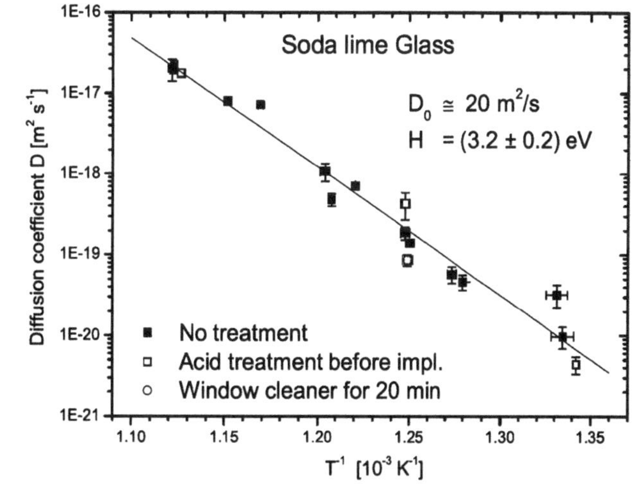

Fig. 4 Arrhenius fit to the diffusion data for ^{209}Pb in soda-lime glass. (From Ref. [10])

Table 1 Radiotracer-based diffusion studies carried out at IGISOL

Tracer	Half-life	Studied material	Reference
^{31}Si	2.6 h	SiGe, Ge, FeSi, $MoSi_2$	[11–14]
^{72}As	26.0 h	SiGe, Si, Ge	[15, 16]
^{66}Ga	9.49 h	SiGe, Si, Ge	[17, 18]
^{209}Pb	3.25 h	Soda-lime glass	[9]

The diffusion studies carried out at IGISOL are summarized in Table 1.

5 Future prospects

Nuclear solid state techniques which rely on specific properties of excited nuclei may be used to study the interaction of nuclear moments with local fields in solids. After doping of the material with radioactive probe atoms via ion implantation (as the most versatile procedure) these techniques exploit the hyperfine interactions or the emission of radiation following the decay process. This way information on defect-probe atom interactions, internal electric and magnetic fields in crystals and thin films as well as on probe atom lattice sites is obtained. The application of such techniques is restricted by the limited number of adequate probe isotopes and as IGISOL is capable of producing dedicated isotopes with sufficient beam currents, it can be utilized by new ways in this respect.

An example of such use of the IGISOL facility is the recently initiated project aiming at studying ferromagnetism in GaN, ZnO and AlN by perturbed angular correlation (PAC) [19]. Utilization of the IGISOL facility is expected to enable production of a novel PAC probe which is perfectly suitable for the planned experiments. The project will be realized in collaboration with the PAC-group of Bonn University who has long-term and profound experience in the method.

Acknowledgement The work of HJW was carried out under the auspices of the Academy of Finland, Centre of Excellence in Nuclear and Accelerator Based Physics, Ref. 213503 and Grant 129999.

References

1. Räisänen, J.: Nanometre science with radioactive ion beams. In: Hellborg, R., Whitlow, H.J., Zhang, Y. (eds.) Ion Beams in Nanoscience and Technology, pp. 219–235. Springer, Heidelberg (2009)
2. Östling, M., Malm, B.G.: High speed electronics. In: Hellborg, R., Whitlow, H.J., Zhang, Y. (eds.) Ion Beams in Nanoscience and Technology, pp. 21–35. Springer, Heidelberg (2009)
3. Laitinen, P., Tiourine, G., Touboltsev, V., Räisänen, J.: Detection system for depth profiling of radiotracers. Nucl. Instrum. Meth. B **190**, 183–185 (2002)
4. Riihimäki, I.: Point-defect mediated diffusion in intrinsic SiGe-alloys. PhD Thesis, JYFL Research Report 2 (2010)
5. Laitinen, P.: Self- and impurity diffusion in intrinsic relaxed silicon-germanium. PhD Thesis, JYFL Research Report 1 (2004)
6. Laitinen, P., Riihimäki, I., Huikari, J., Räisänen, J.: Versatile use of ion beams for diffusion studies by the modified radiotracer technique. Nucl. Instrum. Meth. B **219–220**, 530–533 (2004)
7. Samuelsson, C.: Retrospective determination of radon in houses. Nature **334**, 338–340 (1988)
8. Roos, B., Whitlow, H.J.: Computer simulation and experimental studies of implanted ^{210}Po in glass resulting from radon exposure. Health Phys. **84**, 72–81 (2003)
9. Ekman, J., Helgesson, J., Karlsson, L., Mohsenpour, M., Riihimäki, I., Tuboltsev, V., Jalkanen, P., Virtanen, A., Kettunen, H., Huikari, J., Nieminen, A., Moore, I., Penttilä, H., Arstila, K., Äystö, J., Räisänen, J., Whitlow, H.J.: Retention of Pb isotopes in glass surfaces for retrospective assessment of radon exposure. Nucl. Instrum. Meth. B **249**, 544–547 (2006)
10. Laitinen, M., Riihimäki, I., Ekmann, J., Anada Sagari, A.R., Karlsson, L.B., Sangyuenyongpipat, S., Gorelick, S., Kettunen, H., Penttilä, H., Hellborg, R., Sajavaara, T., Helgesson, J., Whitlow, H.J.: Mobility determination of lead isotopes in glass for retrospective radon measurements. Radiat. Prot. Dosim. **131**, 212–216 (2008)
11. Laitinen, P., Strohm, A., Huikari, J., Nieminen, A., Voss, T., Grodon, C., Riihimäki, I., Kummer, M., Äystö, J., Dendooven, P., Räisänen, J., Frank, W.: ISOLDE collaboration: self-diffusion of ^{31}Si and ^{71}Ge in relaxed $Si_{0.20}Ge_{0.80}$ layers. Phys. Rev. Lett. **89**, 085902 (2002)
12. Strohm, A., Voss, T., Frank, W., Laitinen, P., Räisänen, J.: Self-diffusion of ^{71}Ge and ^{31}Si in Si-Ge alloys. Z. Metallkde **93**, 737–744 (2002)
13. Riihimäki, I., Virtanen, A., Pusa, P., Räisänen, J., Salamon, M., Mehrer, H.: Si self-diffusion in cubic B20 -structured FeSi. Europhys. Lett. **82**, 66005 (2008)
14. Salamon, M., Strohm, A., Voss, T., Laitinen, P., Riihimäki, I., Divinski, S., Frank, W., Räisänen, J., Mehrer, H.: Self-diffusion of silicon in molybdenum disilicide. Philos. Mag. A **84**, 737–756 (2004)
15. Laitinen, P., Räisänen, J., Riihimäki, I., Likonen, J., Vainonen-Ahlgren, E.: Fluence effect on ion implanted As diffusion in relaxed SiGe. Europhys. Lett. **72**, 416–422 (2005)
16. Laitinen, P., Riihimäki, I., Räisänen, J.: ISOLDE collaboration: arsenic diffusion in relaxed $Si_{1-x}Ge_x$. Phys. Rev. B **68**, 155209 (2003)
17. Riihimäki, I., Virtanen, A., Kettunen, H., Pusa, P., Räisänen, J.: Diffusion properties of Ga in $Si_{1-x}Ge_x$ alloys. J. Appl. Phys. **104**, 123510 (2008)
18. Riihimäki, I., Virtanen, A., Rinta-Anttila, S., Pusa, P., Räisänen, J., ISOLDE Collaboration: Vacancy-impurity complexes and diffusion of Ga and Sn in intrinsic and p-doped germanium. Appl. Phys. Lett. **91**, 91922 (2007)
19. Vianden, R., Räisänen, J., Riihimäki, I., Virtanen, A.: Ferromagnetism in GaN, ZnO and AlN studied by perturbed angular correlation. JYFL Research Proposal (2010)

DOI 10.1007/978-94-007-5555-0_29
Reprinted from *Hyperfine Interactions* Volume 223, Issues 1-3, 2014
and *The European Physical Journal A* Volume 48, Issue 4, 2012

Production of pure ^{133m}Xe for CTBTO

K. Peräjärvi · T. Eronen · D. Gorelov · J. Hakala · A. Jokinen · H. Kettunen · V. Kolhinen · M. Laitinen · I. D. Moore · H. Penttilä · J. Rissanen · A. Saastamoinen · H. Toivonen · J. Turunen · J. Äystö

Published online: 6 April 2012

Abstract Underground nuclear weapon detonations release gaseous species into the atmosphere. The most interesting isotopes/isomers from the detection point of view are ^{131m}Xe, ^{133m}Xe, ^{133}Xe and ^{135}Xe. We have developed a method that employs high-precision Penning trap mass spectrometry at the JYFLTRAP facility, the University of Jyväskylä, to produce pure calibration samples of these isotopes/isomers. Among developments this work required a new mass resolution record of a few parts-per-million. Here the status and future plans of the project are reviewed.

Keywords Penning trap · ^{133m}Xe · CTBTO

1 Introduction

The Comprehensive Nuclear-Test-Ban Treaty (CTBT) is a convention that bans all nuclear weapon detonations (www.ctbto.org). The Treaty plays an important role in non-proliferation of nuclear weapons. It was opened for signatures in New York in 1996 and it will enter into force 180 days after it has been ratified by all the States listed in its Annex 2. So far, 35 countries out of 44 have ratified the Treaty. In the meantime, the Preparatory Commission for the Comprehensive Nuclear-Test-Ban

K. Peräjärvi (✉) · H. Toivonen · J. Turunen
STUK-Radiation and Nuclear Safety Authority, P.O. Box 14, 00881 Helsinki, Finland
e-mail: kari.perajarvi@stuk.fi

T. Eronen · D. Gorelov · J. Hakala · A. Jokinen · H. Kettunen · V. Kolhinen · M. Laitinen · I. D. Moore · H. Penttilä · J. Rissanen · A. Saastamoinen · J. Äystö
Department of Physics, University of Jyväskylä, P.O. Box 35 (YFL), 40014 Jyväskylä, Finland

Present Address:
T. Eronen
Max-Planck-Institut für Kernphysik, Saupfercheckweg 1, 69117 Heidelberg, Germany

Treaty Organization (CTBTO) was established to prepare for the Treaty's entry into force.

The CTBTO has a strong technical focus. One of its main tasks is to develop an International Monitoring System (IMS) that is capable of detecting nuclear explosions occurring at any place or environment around the world at any time [1]. Monitoring techniques include infrasound, hydroacoustics, seismology and radionuclide methods. Upon completion, the network will contain 321 permanent monitoring stations. 80 of them will be monitoring radioactive substances in air. Particulate samples collected by these stations are first analyzed at the collection site and then are sent for second counting and analysis to one of the 16 radionuclide laboratories. Data provided by these stations and laboratories are transmitted via the International Data Centre located in Vienna, Austria, to all States Signatories.

Half of the radionuclide monitoring stations will also have the capability of detecting radioactive noble gases. This is important for detecting the underground nuclear weapons tests, since formed refractory elements tend to stay underground, i.e., they cannot pass through rock and soil into the atmosphere [2, 3]. Noble gas surveillance on the other hand relies on the detection of ^{131m}Xe ($T_{1/2} = 11.84$ d), ^{133m}Xe ($T_{1/2} = 2.19$ d), ^{133}Xe ($T_{1/2} = 5.243$ d) and ^{135}Xe ($T_{1/2} = 9.14$ h) because of their significant production cross sections in fission and optimal half-lives (not too long or short) [4]. Also their relative amounts within a sample are important. Namely, activity ratios can be used to disentangle releases from nuclear power plants and weapons testing. The ^{133m}Xe/^{133}Xe ratio is a key indicator [5].

Analysis of collected xenon samples is typically based on either 4π electron-gamma coincidence counting using scintillators [6] or high resolution gamma-ray singles counting with HPGe detectors [7]. In the case of the electron-gamma coincidence approach, for example, the determination of ^{131m}Xe and ^{133m}Xe activities are based on coincidences between K X-rays and the corresponding conversion electrons. In particular, since the 198.7 keV conversion electrons of ^{133m}Xe may generate background in the domain of ^{131m}Xe and because ^{133}Xe and ^{135}Xe also contribute to K X-ray–conversion electron coincidences it is obvious why monoisotopic/isomeric xenon samples are important and desired for the calibration and testing of these instruments.

Two types of xenon release may be associated with underground nuclear weapon detonations: fast release in which fission gases are rapidly vented to the atmosphere or slow diffusion-based seeping. For the analysis of the first (second) scenario independent (cumulative) fission yields are important input parameters. Seismic data together with atmospheric transport modelling are applied to tell to which of the scenarios, if any, the detected radioactive xenon may be attributed.

In addition to nuclear spin, parity and lifetime, the only difference between ^{133m}Xe and ^{133}Xe is their 2 parts-per-million or 233 keV/c^2 mass difference. The beta decay of ^{133}I mainly (97.1%) feeds the ground state of ^{133}Xe. Partly because of this, pure samples of ^{133m}Xe have not been previously available for the calibration and development of noble gas sampler stations. Due to the non-existence of these centrally important ^{133m}Xe samples and the recent breakthroughs in mass purification at the IGISOL/JYFLTRAP facility [8] a project to study the feasibility of ^{133m}Xe sample fabrication at Jyväskylä was initiated in 2009. Current status and the future plans of the project employing an upgraded IGISOL/JYFLTRAP facility are discussed below. We note that this is not the first application of the IGISOL technique to the

CTBT field. Namely, the decay properties of ^{133m}Xe have earlier been studied using IGISOL at Jyväskylä [9].

2 Experimental techniques and results

This section is divided into two sub-sections: the first section summarizes the present experimental status of the project and the second part discusses remaining challenges related to the possible industrialization of the xenon production at the upgraded IGISOL/JYFLTRAP facility.

2.1 Current experimental status

Xenon atoms of interest are produced in fission reactions induced with 25 MeV protons. Both uranium and thorium targets have successfully been used. Fission products are thermalized in flowing helium gas that transports them out from the stopping chamber. The majority of products are evacuated as singly charged ions that can be separated from the neutral helium buffer gas using electric fields. They are mass separated with a dipole magnet, accumulated and bunched with the RFQ cooler [10] and then guided to a double Penning trap system, where the final mass purification prior to the implantation takes place [11]. The final cleaning step reaching part-per-million resolution uses ion motion excitation with time-separated rf electric fields (the so-called Ramsey method) which utilizes both the purification and the precision trap [8]. The typical time required for the purification, i.e. time from fission to implantation in the collection foil, is about 0.5 s. A more detailed description of the xenon sample production can be found from [12].

An achieved mass resolving power ($m/\Delta m$) of $\sim 10^6$ is enough for the separation of the isomeric and ground state of ^{133}Xe [12]. The typical implantation rate of ^{133m}Xe is about 60 ions/s, i.e., a sample of 10^5 atoms can be prepared in less than 30 min. In Ref. [12] we also show that 30 keV implantation energy is sufficient such that xenon atoms are not released easily from the aluminium matrix. To efficiently extract xenon from the Al matrix, the foil needs to be melted.

The capability to produce ^{131m}Xe and ^{135}Xe samples has recently been studied at Jyväskylä along with the xenon release investigation from graphite foils. Since there are also other successful techniques for the production of pure samples of ^{131m}Xe and ^{135}Xe [13], the motivation of our effort on these nuclei was to confirm that we have that capability available, if requested, using the technique discussed above. In the mass purification sense the production of ^{131m}Xe and ^{135}Xe samples is easier compared to ^{133m}Xe. While preparing ^{131m}Xe samples, it is crucially important not to let any traces of stable xenon gas enter to the system. Namely, ^{131}Xe is a stable isotope that enters the system not only from the fission reaction but also as an impurity in the used helium buffer gas and thus cannot be avoided in the trap while purifying nuclides in the mass chain A = 131. Large amounts of stable ^{131}Xe will saturate the Penning trap and render ^{131m}Xe production impossible.

A study around graphite foils was made in order to search for material that releases xenon easier than aluminum. Complete results of these more recent investigations will be published elsewhere.

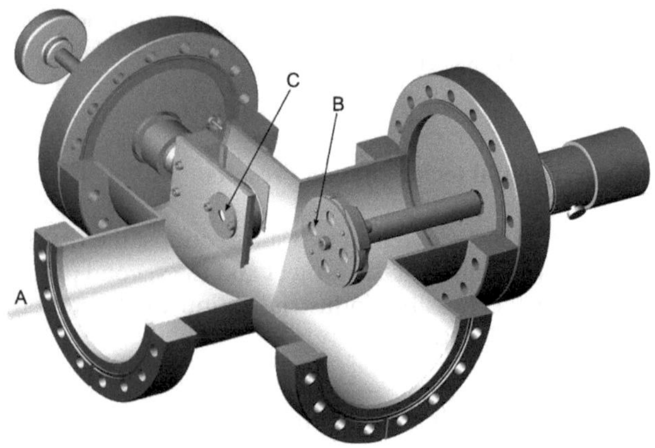

Fig. 1 Illustration of implantation setup used in the initial tests showing (*A*) the beam hitting (*B*) the collection foil through the collimator and (*C*) the collimated micro-channel-plate detector for the beam tuning

2.2 Future plans

The timing of this project could not be better due to the upgrade of the Accelerator Laboratory. The new MCC30/15 cyclotron will primarily serve the upgraded IGISOL/JYFLTRAP facility making the scheduling of xenon production runs easier and more flexible in the future. The new location of the IGISOL/JYFLTRAP facility will also be more spacious. Because of that, there will be room after the Penning trap for a permanent or semi-permanent xenon implantation setup. Such a setup has significant advantages for industrial xenon production.

Figure 1 presents the configuration of existing implantation setup. As shown, it does not allow for on-line monitoring of sample production. On-line ion counting will in the future be realized by detecting secondary electrons sputtered away from the foil while ions are being implanted using a channeltron or micro-channel-plate detector. Such a setup will enable the generation of implantation time profiles and also a calculation of the number of implanted ions. The capability of producing gaseous xenon samples still requires some development work even though it is known how the implanted xenon can be efficiently released to the gas phase.

3 Discussion

The achieved control on xenon sample purity allows the characterization of the response of a noble-gas sampler station for well known xenon compositions. Nearly mono-isotopic or -isomeric samples can be provided if the sample can be delivered to the station fast enough. Obviously, another possible mode of operation is to bring a noble gas sampler station to Jyväskylä for tests.

An enhanced version of the implantation setup together with gaseous xenon sample production capability are currently been worked on. After finishing these developments, together with the upgrade of the Accelerator Laboratory, the commercial production of xenon samples will be possible. All these developments should be finished before the end of 2012.

It should be noted that the described technology can also be used to study in detail the independent fast neutron induced fission yields of nuclei such as $^{235,238}U$ and

^{239}Pu. As discussed earlier, such yield information would also be significant while interpreting the data provided by the IMS. So far, these trap-assisted independent fission yield measurements have only been made for the proton-induced fission at Jyväskylä [14]. A break-up of a deuterium beam would be used as a source of fast neutrons. The IGISOL technique has two features that need to be mentioned in this context. Firstly, it is chemically insensitive, meaning that the refractory and gaseous species have a similar behaviour and therefore there are no restrictions in extracting nuclei of all elements, unlike more standard ISOL facilities such as ISOLDE at CERN. This also means that pure samples of almost any isotope or isomer can be prepared using the technique. Secondly, the system is very fast; about 10 ms delay times are feasible for extraction from the gas filled stopping volume. These two features simplify the data analysis significantly since very seldom one needs to worry about the chemistry or the radioactive decay corrections. One should also emphasize the fact that since the identification is based on atomic mass one does not need to measure the decay to be able to do the identification. This is of importance since some of the nuclei that are produced in fission are very long lived or even stable.

References

1. Clery, D.: Science **325**(5939), 382 (2009)
2. Carrigan, C.R., et al.: Nature **382**, 528 (1996)
3. De Geer, L.-E.: Nature **382**, 491 (1996)
4. Bowyer, T.W., et al.: J. Environ. Radioact. **59**, 139 (2002)
5. Reeder, P.L., Bowyer, T.W.: Nucl. Instrum. Methods Phys. Res. A **408**, 573 (1998)
6. Ringbom, A., et al.: Nucl. Instrum. Methods Phys. Res. A **508**, 542 (2003)
7. Stocki, T.J., et al.: Appl. Radiat. Isotopes **61**, 231 (2004)
8. Eronen, T., et al.: Nucl. Instrum. Methods Phys. Res. B **266**, 4527 (2008)
9. Peräjärvi, K., et al.: Appl. Radiat. Isotopes **66**, 530 (2008)
10. Nieminen, A., et al.: Nucl. Instrum. Methods Phys. Res. A **469**, 244 (2001)
11. Kolhinen, V.S., et al.: Nucl. Instrum. Methods Phys. Res. A **528**, 776 (2004)
12. Peräjärvi, K., et al.: Appl. Radiat. Isotopes **68**, 450 (2010)
13. Haas, D.A., et al.: J. Radioanal. Nucl. Chem. **282**, 677 (2009)
14. Penttilä, H., et al.: Eur. Phys. J. A **44**, 147 (2010)

DOI 10.1007/978-94-007-5555-0_30
Reprinted from *Hyperfine Interactions* Volume 223, Issues 1-3, 2014
and *The European Physical Journal A* Volume 48, Issue 4, 2012

Decay heat studies for nuclear energy

A. Algora · D. Jordan · J. L. Taín · B. Rubio · J. Agramunt · L. Caballero · E. Nácher · A. B. Perez-Cerdan · F. Molina · E. Estevez · E. Valencia · A. Krasznahorkay · M. D. Hunyadi · J. Gulyás · A. Vitéz · M. Csatlós · L. Csige · T. Eronen · J. Rissanen · A. Saastamoinen · I. D. Moore · H. Penttilä · V. S. Kolhinen · K. Burkard · W. Hüller · L. Batist · W. Gelletly · A. L. Nichols · T. Yoshida · A. A. Sonzogni · K. Peräjärvi

Published online: 27 April 2012

Abstract The energy associated with the decay of fission products plays an important role in the estimation of the amount of heat released by nuclear fuel in reactors. In this article we present results of the study of the beta decay of some refractory isotopes that were considered important contributors to the decay heat in reactors. The measurements were performed at the IGISOL facility of the University of Jyväskylä, Finland. In these studies we have combined for the first time a Penning trap (JYFLTRAP), which was used as a high resolution isobaric separator, with

A. Algora (✉) · D. Jordan · J. L. Taín · B. Rubio · J. Agramunt · L. Caballero · E. Nácher · A. B. Perez-Cerdan · F. Molina · E. Estevez · E. Valencia
IFIC (CSIC-Univ. Valencia), Valencia, Spain
e-mail: algora@ific.uv.es

A. Algora · A. Krasznahorkay · M. D. Hunyadi · J. Gulyás · A. Vitéz · M. Csatlós · L. Csige
Institute of Nuclear Research, Debrecen, Hungary

T. Eronen · J. Rissanen · A. Saastamoinen · I. D. Moore · H. Penttilä · V. S. Kolhinen
University of Jyväskylä, Jyväskylä, Finland

K. Burkard · W. Hüller
GSI, Darmstadt, Germany

L. Batist
PNPI, Gatchina, Russia

W. Gelletly · A. L. Nichols
University of Surrey, Guildford, UK

T. Yoshida
Tokyo City University, Tokyo, Japan

A. A. Sonzogni
NNDC, Brookhaven National Laboratory, Upton, NY, USA

K. Peräjärvi
STUK, Helsinki, Finland

a total absorption spectrometer. The results of the measurements as well as their consequences for decay heat summation calculations are discussed.

Keywords Decay heat · Total absorption · Trap-assisted spectroscopy

1 Introduction

Many stages of the nuclear fuel cycle require a sound estimate of the energy (or heat) associated with the radioactive decay of fission products. Independent of which reactor system is under consideration, this knowledge is fundamental for the design and safe operation of the reactor, for the evaluation of shielding requirements on fuel discharge and transport routes and for the safe management of radioactive waste products extracted from spent fuel during reprocessing. This energy, commonly called decay heat, can be measured directly or can be calculated theoretically. The theoretical estimates are in general preferred in view of their flexibility. For example theoretical estimates of the decay heat can play an important role in the study of possible new nuclear fuel mixtures for functioning reactors and those of next generation.

There are two approaches for calculating the decay heat of fission products, namely, the statistical method and the summation method. From the historical point of view it is interesting to note that one of the first ways to estimate the decay heat arose from the work of Way and Wigner [1]. This method considered fission products as a statistical ensemble and used empirical relations for both radioactive half-lives and atomic masses to obtain the gamma and gamma plus beta component of the decay heat following an instantaneous fission burst. The statistical method of Way and Wigner was the only method available in the early years and was extensively used for decay times in the range of a few seconds to one hundred days. Because of its nature it is less accurate for longer cooling times.

The alternative theoretical approach to the statistical method is the so-called summation calculation. In the application of this method the first step is the determination of the inventory of radioactive isotopes produced during the irradiation period of the fuel and at any time following shutdown. The inventory is obtained from the solution of a linear system of coupled first order differential equations which describe the buildup and decay of fission products. Once the inventory of fission products is determined, the decay heat can be simply derived by summing the nuclide activities weighted with the mean gamma, beta and alpha energies released per decay as follows:

$$\mathbf{f(t)} = \sum_i \left(\overline{\mathbf{E}}_{\beta,\mathbf{i}} + \overline{\mathbf{E}}_{\gamma,\mathbf{i}} + \overline{\mathbf{E}}_{\alpha,\mathbf{i}}\right) \lambda_\mathbf{i} \mathbf{N_i(t)} \tag{1}$$

where $f(t)$ is the power function, $\overline{E_i}$ is the mean decay energy of the ith nuclide (β, γ and α components), λ_i is the decay constant of the ith nuclide, and $N_i(t)$ is the number of nuclide i at cooling time t.

The summation method is more general and flexible but it requires large databases that contain all the information needed: half-lives, neutron-capture cross sections, mean energies released in the decay, etc. Naturally, the lack of information in the early days favoured the use of the statistical method. However continuously since the 1950s the availability of decay data has increased and there have been national and international efforts to create databases for nuclear applications including the decay heat, see for example the European JEFF, the Japanese JENDL and the American initiative ENDF (http://www.oecd-nea.org/dbdata/jeff/; http://www.ndc.jaea.go.jp/jendl/jendl.html; http://www.nndc.bnl.gov/endf/). Thanks to these efforts summation calculations now constitute the most commonly used method for decay heat calculations.

The decay data available from databases is commonly the result of the evaluation of high resolution measurements, which are based on the use of Ge detectors. In these experiments the beta probability to a certain level in the daughter nucleus is determined from the intensity balance of the gamma-feeding and de-excitation of the level. If there is a large fragmentation in the gamma intensity due to the high level density at high excitation energy or if in the decay there are gamma rays of high energy involved, many gamma rays can remain undetected using conventional Ge detectors. As a consequence, feeding at high excitation is not observed and incorrectly assigned to low-lying levels. This systematic uncertainty is called the pandemonium effect [2]. The problem is related to the relatively low efficiency of Ge detectors and to the resulting difficulty in detecting all the individual gamma rays that follow the beta decay. The pandemonium effect obviously has major consequences for decay heat calculations. Due to the effects of missing levels at high excitation energy there is an underestimate of the total γ energy and an overestimate of the total β energy released in the decay, if these values are obtained from high resolution data.

There is a solution to the pandemonium effect: the application of the total absorption technique. This technique is based on the detection of the gamma cascades that follow the beta decay instead of detecting the individual gamma rays as in conventional high resolution spectroscopy. With the use of a total absorption spectrometer, in essence a calorimeter placed around the source, an almost 100% efficiency for detecting gamma cascades can be achieved and then the pandemonium effect can be avoided. This technique has been applied by our groups in experiments at GSI and ISOLDE [3–7], and new methods of analysis have been developed [8–11]. The decay of neutron rich nuclei of relevance for the decay heat problem have also been measured by Greenwood et al. [12] with the total absorption technique, but using a different analysis method.

Our interest in this subject was triggered by the work of Yoshida et al. [13]. This work shows that independently of the improvements achieved in the World's major data libraries for summation calculations, there remains a substantial discrepancy between the calculations and the experiments in the γ-ray component of the decay heat in ^{239}Pu in the cooling range from 300 to 3,000 s (after an instantaneous fission event). The authors called this effect *the γ-ray discrepancy* [13]. The γ-ray discrepancy occurs not just for the decay heat calculations for ^{239}Pu, but for ^{233,235,238}U as well [13]. This is the reason why the identification of the nuclei responsible for the discrepancy was considered a very important task by the authors. A careful study showed that the best candidates are ^{102,104,105}Tc. According to Yoshida these nuclei were suspected to suffer from the pandemonium effect.

Shortly after the recognition of the need for a new study of the beta decay of ^{102,104,105}Tc using the total absorption technique, specialists in the decay heat field were contacted who identified a set of additional nuclides of interest. The result was a high priority list that contains nuclei that are large contributors to the decay heat and also to the inconsistencies in the different decay data libraries used for summation calculations [14]. From the list the cases with highest priority were selected (^{102,104,105,106,107}Tc, ^{105}Mo, and ^{101}Nb). The resulting measurement, because of its connection with the IGISOL facility, will be discussed in more detail in the next section.

2 First trap-assisted TAS experiment at IGISOL

Some of the selected isotopes of interest are refractory elements which are difficult to extract from conventional ion sources. For that reason our experiment was performed at the Ion Guide Isotope Separator On-Line (IGISOL) facility of the University of Jyväskylä [15]. The ion guide method is chemically insensitive and allows the extraction of also the refractory elements.

In our experiment we have combined for the first time a total absorption spectrometer with the Penning trap available at the IGISOL facility (JYFLTRAP) [16] and measured the beta decay of ^{101}Nb, ^{102,104,105,106,107}Tc and ^{101,105}Mo. The Penning trap is an excellent tool for this kind of experiment where the purity of the sources is of great importance.

In our measurements proton beams of 30 and 50 MeV were used to produce fission in a natural U target of 15 mg/cm^2 thickness. Typical primary beam currents were about 4 μA. The nuclei of interest were first separated using the IGISOL facility and then injected into the JYFLTRAP. After the second separation with the Penning trap the activity was implanted on a tape outside the Total Absorption Gamma Spectrometer (TAGS), which was then used to transport it to the internal measurement position. The TAGS used in the measurements was designed at the Nuclear Institute of St. Petersburg. It consists of two NaI(Tl) cylindrical crystals. The larger crystal has dimensions of $\oslash = 200$ mm $\times\ l = 200$ mm, and it has a longitudinal hole of $\oslash = 43$ mm. The other crystal completes the 4π geometry and has dimensions of $\oslash = 200$ mm $\times\ l = 100$ mm. The total gamma efficiency of this setup was estimated to be 70% at 5 MeV using Monte Carlo (MC) simulations. The TAGS detector was used in combination with a Si detector to measure coincidences with the electrons from the β^- decay. The efficiency of the Si detector in this particular setup was estimated to be $\sim 25\%$. In addition, a Ge detector was placed at the collection point to continuously monitor the composition of the sources.

The first step in the analysis is to evaluate and remove possible distortions and contaminations of the measured spectra. The use of the Penning trap, which guarantees sources of very high isobaric purity, only solves part of the problem, since during the measurement daughter activity can also contaminate our spectrum. The contribution of the daughter activity can be reduced by proper selection of collection/measuring cycles but in some cases cannot be avoided completely. In such cases, where a contribution from the daughter was expected, separate measurements of the daughter activity were performed. During our measurements we also measured the room background for one hour every two hours. Another possible distortion of the spectra comes from pulse pileup. We have calculated this effect as described in [11].

3 Analysis and conclusions

The analysis of the total absorption spectra requires the solution of the so-called "TAS inverse problem" $\mathbf{d} = \mathbf{R}(B)\mathbf{f}$ where $\mathbf{d}$ represents the measured data, $\mathbf{R}(B)$ is the response matrix of the detector, and $\mathbf{f}$ is the sought after feeding distribution. The response function $\mathbf{R}(B)$ depends on the detector and on the branching ratios of the levels in the daughter nucleus (B). Since radioactive sources are not available for measuring all the necessary responses, they can only be calculated using MC techniques. For all the analyzed cases the response function was determined using the code GEANT4 [17]. The quality of the MC simulations was tested by comparing the results of the simulations and measured spectra of radioactive sources of ^{22,24}Na, ^{60}Co, and ^{137}Cs.

Once the response matrix is defined and calculated, the inverse problem has to be solved. In the analysis of the decays studied we have used the Expectation Maximisation Algorithm. This algorithm has been adapted to the TAS inverse problem in [8, 9]. As mentioned earlier the response matrix of the detector depends on the level scheme of the daughter nucleus we analyse. As part of the analysis several assumptions can be made for the level schemes populated in the decay depending on the knowledge available from high resolution measurements. This information is used for the construction of the low-lying part of the branching ratio matrix (B). For the upper (or unknown) part of the branching ratio matrix, as well as for its connection with the low-lying part a statistical model is used, which is based on level densities and E1, M1, E2 -decay strength functions with parameters taken from (http://www-nds.iaea.org/ripl2/).

As an example of the general procedure we will discuss in more detail the ^{105}Tc case. In the ^{105}Tc decay analysis, we have used the known adopted level scheme of ^{105}Ru up to the excitation 1325 keV [18] and the statistical nuclear model was used to generate the average branching ratio matrix from the excitation of 1360 keV up to the Q value of 3746(6) keV [19]. Compared to the ^{104}Tc case [20] the ^{105}Tc case is more uncertain since the spin-parity assignment to the ground state of the parent is $(3/2^-)$ and most of the low-lying levels of ^{105}Ru fed in the decay according to [18] have positive parity. For that reason we have allowed feeding to all low-lying levels in the analysis excluding only levels at excitation 229.5 and 301.7 keV that have $7/2^+$ spin-parity assignment. Another uncertainty of this analysis is related to the fact that in our measurement we were not able to determine the ground state feeding and in the latest compilation [18] an upper limit to this feeding is given (<9%). For that reason the results of two analyses as limiting cases were studied in which the feeding to the ground state of ^{105}Ru is set to 9%, or to 0%. The results of our final analysis for ^{105}Tc is shown in Fig. 1 for the first assumption of fixing the ground state feeding to 9%.

In the upper panel of Fig. 1 the analysed spectrum (in gray) is compared with the spectrum generated from the feeding pattern determined in our analysis (black) for the decay of ^{105}Tc. The curves are almost indistinguishable. Also in the upper panel the contribution of the contaminants to the analysed spectrum is presented. In the lower panel the results for the deduced beta feeding are compared with the feeding previously known from high resolution measurements. It is clear from Fig. 1 that a large amount of beta feeding is observed in our experiment at high excitation in the daughter from the ^{105}Tc decay which was not previously seen in

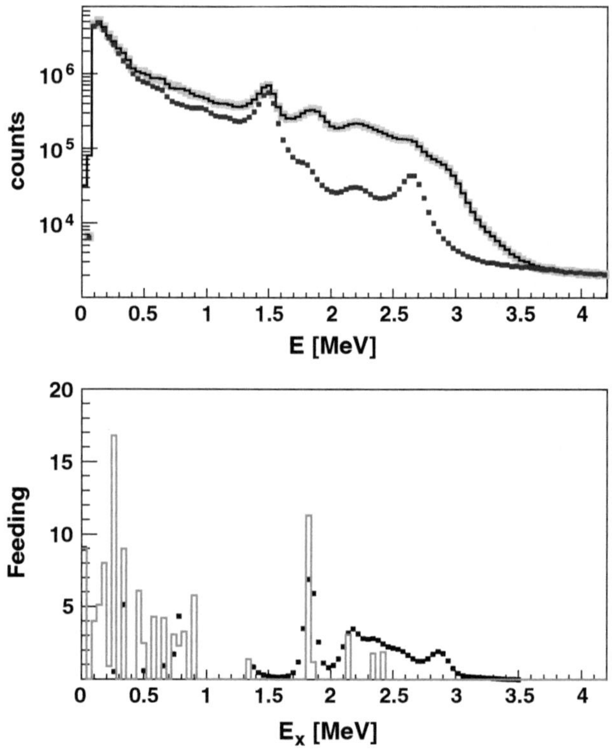

Fig. 1 TAS results obtained for ^{105}Tc (see details in the text)

high resolution experiments [18]. This indicates that the level scheme suffers from the Pandemonium effect. Our results show a similar pattern for the decays of ^{104,106,107}Tc and ^{105}Mo.

4 Conclusions and future

The results of our analysis were used to calculate the mean gamma and beta energies released in the decay, which have been recently published in [20]. They show that there is a large increase in the mean gamma energy released in the decay of $^{104-107}$Tc and ^{105}Mo compared with the high resolution results. In the case of ^{102}Tc and ^{101}Nb the mean energies calculated are similar to those obtained in high resolution experiments, which means that in these specific cases the decay data do not suffer from the pandemonium effect.

When our mean gamma energies for all the nuclei studied (^{101}Nb, ^{102}Tc, ^{104}Tc, ^{105}Tc, ^{106}Tc, ^{107}Tc and ^{105}Mo) are incorporated in the ENDF/B-VII decay data library (Sonzogni A., private communication), a large part of the discrepancy in the gamma component of the decay heat for ^{239}Pu is resolved (Fig. 2). Not only the discrepancy in the 300–3,000 s cooling time interval is reduced, but the discrepancy that existed with the ENDF/B-VII database in the range of 4–200 s is essentially removed. The effect of our data is less dramatic for ^{235}U because of the difference of the fission

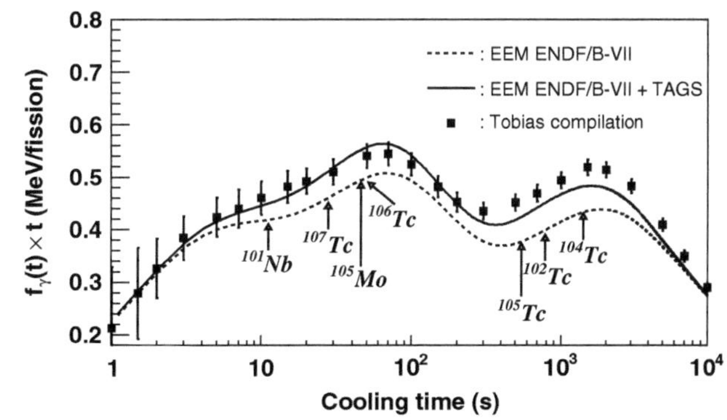

Fig. 2 Comparison of the calculated electromagnetic decay heat component for ^{239}Pu before and after the inclusion of our measurements with the data of Tobias et al. [21]. The time of the peak contribution of each of the studied isotopes is marked with *arrows*

yields and means that other nuclei will have to be identified and studied for solving discrepancies for this fuel. Similar conclusions were obtained recently for the JEFF database (Mills R.W., private communication). Concerning the ^{235}U case we are in contact with specialists in the field (Yoshida T., private communication) and new measurements are planned. The decay data needs for ^{232}Th fuel have also been recently identified [22] with emphasis on the need for TAS and beta delayed neutron measurements. Our results [20] and recent studies, which demonstrate the need for additional measurements, (Yoshida T., private communication) [22] show the potential for the application of the TAS technique in combination with IGISOL and JYFLTRAP for decay heat studies. These measurements could also have an impact in the prediction of the neutrino spectra from reactors, which may be relevant for neutrino oscillation experiments and non-proliferation applications [23].

Acknowledgements This work was supported by the following projects: Spanish FPA 2005-03993 and FPA2008-06419-C02-01; OTKA K72566; the EC contract MERG-CT-2004-506849, the EU 6th Fram. Program (contract No: 506065 (EURONS) and the Spanish-Hungarian collaboration program.

References

1. Way, K., Wigner, E.P.: Phys. Rev. **73**, 1318 (1948)
2. Hardy, J.C., Carraz, L.C., Jonson, B., Hansen, P.G.: Phys. Lett. **71B**, 307 (1977)
3. Hu, Z., et al.: Phys. Rev. C **60**, 024315 (1999)
4. Hu, Z., et al.: Phys. Rev. C **62**, 064315 (2000)
5. Algora, A., et al.: Nucl. Phys. A **654**, 727c (1999)
6. Nacher, E., et al.: Phys. Rev. Lett. **92**, 232501-1 (2004)
7. Poirier, E., et al.: Phys. Rev. C **69**, 034307 (2004)
8. Cano-Ott, D.: Ph.D. Thesis, University of Valencia (2000)
9. Tain, J.L., Cano-Ott, D.: NIM A **571**, 719 and 728 (2007)
10. Cano-Ott, D., et al.: Nucl. Instrum. Methods A **430**, 333 (1999)
11. Cano-Ott, D., et al.: Nucl. Instrum. Methods A **430**, 488 (1999)
12. Greenwood, R.C., et al.: Nucl. Instrum. Methods A **314**, 514 (1992)
13. Yoshida, T., et al.: Nucl. Sci. Technol. **36**, 135 (1999)
14. Nuclear Science NEA, Report No. 6284 (2007)
15. Äystö, J.: Nucl. Phys. A **693**, 477 (2001)
16. Kolhinen, V., et al.: Nucl. Instrum. Methods Phys. Res., Sect. A **528**, 776 (2004)
17. Agostinelli, S., et al.: Nucl. Instrum. Methods A **506**, 250 (2003)
18. de Frenne, D., Jacobs, E.: Nucl. Data Sheets **105**, 775 (2005)

19. Audi, G., et al.: Nucl. Phys. A **729**, 337 (2003)
20. Algora, A., et al.: Phys. Rev. Lett. **105**, 202501 (2010)
21. Tobias, A.: CEGB report RD/B/6210/R89 (1989)
22. Gupta, M., et al.: INDC(NDS)-0577
23. Fallot, M., et al.: In: Proc. of the Int. Conf. on Nuclear Data for Science and Technology, 2007, p. 1273. EDP Sciences, Nice (2008)

DOI 10.1007/978-94-007-5555-0_31
Reprinted from *Hyperfine Interactions* Volume 223, Issues 1-3, 2014
and *The European Physical Journal A* Volume 48, Issue 4, 2012

Measurements of nuclide yields in neutron-induced fission of natural uranium for SPIRAL2

G. Lhersonneau · T. Malkiewicz · W. H. Trzaska

Published online: 11 July 2012

Abstract Cross-sections for nuclide production in fast-neutron induced fission of natural uranium are part of the input for predictions of yields of neutron-rich nuclides obtainable at Radioactive Ion Beam facilities. We first describe the neutron spectra produced according to the scheme once envisaged for SPES (protons on an enriched ^{13}C target) and the one adopted for SPIRAL2 (deuterons on natural carbon), which both have been measured at JYFL. We then present the measurements of Z-splits in isobaric chains performed at IGISOL. When coupled with the fission cross-section and A-splits for the relevant neutron spectrum, they allow estimates of nuclide cross-sections. It looks that calculations, even those based on modern libraries, are too optimistic by about a factor of two.

Keywords Deuteron-induced reactions · Neutron-induced fission · Isotope separation and enrichment

G. Lhersonneau (✉)
GANIL, Caen, France
e-mail: lhersonneau@ganil.fr

T. Malkiewicz
LPSC, Grenoble, France
e-mail: malkiewicz@lpsc.in2p3.fr

W. H. Trzaska
JYFL, Jyväskylä, Finland
e-mail: wladyslaw.h.trzaska@jyu.fi

Present Address:
T. Malkiewicz
CSC - IT Center for Science, Helsinki, Finland
e-mail: tmalkiew@csc.fi

1 Introduction

The largest number of nuclei assumed to be particle bound are neutron-rich. However, the curvature of the valley of stability prevents accelerator-based experiments on these nuclei with stable beams and targets. Since the 90's physicists have been thinking of ways to produce sufficiently large amounts of unstable neutron-rich nuclei so that it could be worth to accelerate them as secondary beams. Such experiments have been possible for some time, e.g. at ISOLDE where Coulomb excitation experiments are being performed. Still, higher intensities and post-accelerator energies are needed. A very large scale project in Europe is EURISOL of which a concept has been worked out [1]. Yet, its realisation certainly is far in the future. Intermediate European projects are SPIRAL2 at GANIL [2], HIE-ISOLDE at CERN [3] and SPES [4] at Legnaro. The foundation stone of the SPIRAL2 building has been laid on October 17th 2011. Like foreseen with EURISOL, neutron-rich nuclei will be produced by neutron-induced fission in a massive target of natural material, e.g. uranium in chemical form of a carbide. The fast neutrons are generated by stopping a deuteron beam delivered by a powerful accelerator (several mA, 40 MeV deuterons) in a thick graphite target, the so-called converter.

Benefits of neutron-induced fission are both technical and physical. Targets are not overheated by electronic stopping power and can take very high fluxes. These targets can be very heavy, e.g. up to 2 kg of natural uranium for SPIRAL2 in the so-called high-density UC_x. Their temperature is controlled by external heating which makes it more homogeneous, avoiding trapping of nuclei in colder spots, and safer. Neutron-induced fission produces a more neutron-rich distribution than charged particles at similar excitation energy. However, it remains a rather open question how these huge targets would be able to efficiently release the short-lived fission products. It is here worth to mention the encouraging results at PNPI-Gatchina obtained with targets up to 700 g weight with a new type of uranium carbide of density 11 g/cm^3 that allows to pack 4 times more target material in the same volume than the standard UCx. Fast release as well as the long-term stability of the performance are proved for alkaline elements Rb, Cs and Fr [5]. Yet, the properties for other elements are awaiting to be thoroughly tested.

The in-target production rate for a nuclide (Z,A) involves the product of neutron flux times the cross-section. Neutron generation at intermediate projectile energies turns out to be a rather inefficient process. Measurements carried out with various beams and targets, mostly for the purpose of neutron physics and applications, but also more specifically for SPES and SPIRAL2 show that the conversion factor at intermediate projectile energy hardly can exceed few percents. It is also to be noted that cross-section data for nuclide production in fission induced by the energetic neutrons generated in converters are scarce. Libraries are based for a large part on models extrapolated from neutron reactions at d + D and d + T neutron generators.

In order to quantify both neutron yields and nuclidic cross-sections in fast-neutron-induced fission for their planned facilities, the SPIRAL2 and SPES groups asked for dedicated experiments. The choice of JYFL was natural regarding the available beams and the infrastructure. The ion-production mechanism at IGISOL makes a new type of cross-section measurement possible, including those of very short-lived isotopes of refractory elements. It is to be noted that a cross-section discussed here has to be taken in the context of the converter method. One measures a weighted cross-section that depends on the composition of the neutron spectrum. It is therefore relevant to consider neutron production and cross-section together. The neutron measurements will be presented, yet briefly. They cannot be ignored because of their impact in the evolution of the designs of SPES and SPIRAL2. The cross-section measurements made at IGISOL will be discussed in some detail.

2 Neutron measurements

Neutron measurements have been carried out at JYFL for SPIRAL2 using d + C, d + Be at 50 MeV [6] and for SPES using p + ^{13}C at 30 MeV [7]. These measurements were using the time-of-flight (TOF) method and the HENDES setup. The Legnaro group later performed a p + ^{13}C TOF and activation run at 90 MeV at the KVI-Groningen [8]. This energy was chosen because it was near the 100 MeV proton energy of the driver accelerator proposed in the SPES design. The neutron yield turned out to be comparable with the neutron yield at 30 MeV reported in [7]. The absence of an increase suggested a mistake and prompted a new campaign at JYFL. The new measurements confirmed the angular and energy distributions of [7] but showed that the neutron conversion factor had been overestimated by a factor of about 7 [9, 10]. The actual conversion factor of 0.1% for 30 MeV protons turned out to be too low for reaching a level competitive with SPIRAL2. Consequently, SPES now follows the direct production option of uranium fission by 40 MeV protons delivered by a recently bought cyclotron [11].

Other neutron measurements for 40 MeV deuterons were carried out for SPIRAL2, with the aim to see if a heavy-water converter would be worth to be studied [10, 12]. The measured conversion factor, 1.5 times higher than for carbon, was not as high as hoped based on Monte-Carlo simulations. The gain in number of fissions had been even lower since the moderating effect of deuterium moves the neutron spectrum to lower energies where the fission cross-section of natural uranium (99.3% of it being ^{238}U) is low. In addition, the measurement showed that neutron angular distributions were more forward-peaked than the calculated ones. It appeared not to be rewarding to have a target of such a large radius (4 cm) as initially planned. As a consequence, at least to start with, the SPIRAL2 target should have a weight of about 700 g (radius 2.5 cm) if filled with high-density uranium carbide. This is indeed the weight of the targets successfully tested at Gatchina. Another surprise was that Monte-Carlo simulations had predicted an overall flux 2.5 times higher than the measured one.

Figure 1 shows neutron spectra seen by a disk of radius (r) centered on the beam axis and placed at a distance (d) from the converter such that the maximum angle of neutrons is $\theta = \mathrm{atan}(r/d) = 30°$. They are calculated from experimental energy and angular distributions, all measured at JYFL. The curve for 50 MeV deuterons on carbon is obtained by interpolating and integrating the distributions shown by Radivojevič et al. [6]. The other curves were measured by activation [10, 12]. The points mark the pairs (E, $\partial^2\phi/\partial E\partial\Omega$), ϕ being the neutron flux per deuteron, used as parameters for a numerical description of the spectra, see in the original papers for more. The flux of neutrons above 1 MeV per projectile and within a maximum angle of 30° to the beam axis is 0.20%, 0.86% and 1.68%, respectively.

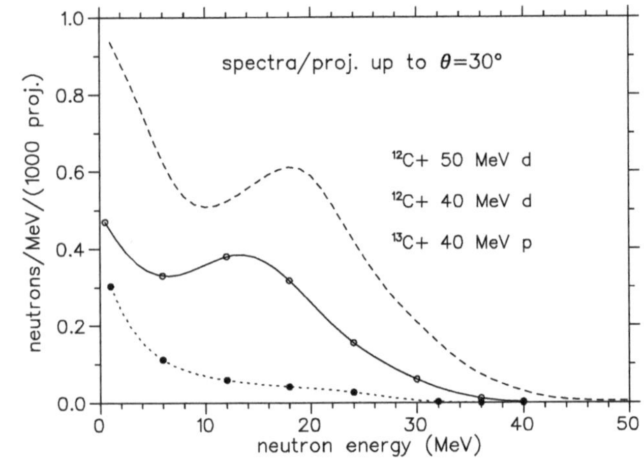

Fig. 1 Neutron spectra impinging on a disk (see text) for selected production schemes; *lower curve*: 40 MeV protons on ^{13}C (SPES-like), *middle curve*: 40 MeV deuterons (SPIRAL2) and *upper curve*: 50 MeV deuterons (used for nuclide yield measurement at IGISOL). See text for more

Finally, we mention a neutron measurement for 55 MeV deuterons on carbon and heavy water converters performed in 2011 [13]. It suggests an overall gain of 2.3 in neutron flux with respect to the design value of 40 MeV. The number of fissions increases by a factor of 2.7, owing to better overlap of the spectrum with the fission cross-section.

3 Cross-section measurements at IGISOL

3.1 Principle of the experiment

The crucial and specific to the ion-guide feature of IGISOL used for these measurements is that all elements are mass separated and that it can be assumed that the efficiency of separating neighouring elements is the same. Other conclusions have been reported by the Leuven group, though with Ar instead of He as stopping gas [14]. That report led to criticism about the validity of JYFL experiments. As a matter of fact, during the analysis of two experiments at IGISOL, within the accuracy achievable, nothing pointed out that one should give up this assumption. Discontinuities have been occasionally noticed, but only when crossing closed atomic shells.

The IGISOL efficiency is in fact not known to the accuracy needed for an absolute cross-section measurement. What is measured is a distribution of production rates of isobars. It is subsequently converted into a distribution of cross-sections owing to the proportionality. Several steps for normalisation are necessary. We assumed the distribution of isobars is Gaussian with a constant Z-width, not depending on the mass. This property is established from the analysis of several data at comparable excitation energies, e.g. 14 MeV neutrons [15] or proton-induced fission at 24 MeV [16]. Thus, if, ideally, at least 3 cross-sections are measurable per mass chain (A), one can fit a Gaussian and scale the integral of the distribution to be equal to the cross-section for that mass σ(A). The latter depends on the neutron spectrum n(E_n) impinging on the uranium target.

There are several data sets [17] giving $\sigma(A,E_n)$ as function of neutron energy (E_n). An experiment at JYFL by the HENDES group [18] contributed to these data.

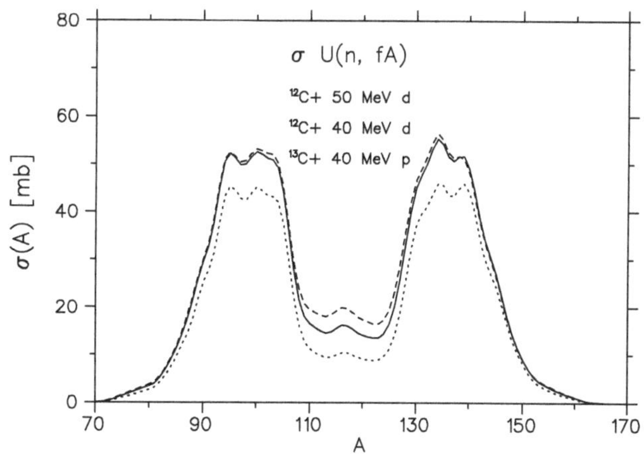

Fig. 2 Cross-sections for mass chains in neutron-induced fission of uranium after weighting by the neutron spectra for the selected production schemes shown in Fig. 1. Half of the sum on the masses correspond to the effective σ(n,f), giving 0.90, 1.09 and 1.14 b, respectively

A deuteron beam was sent onto a thin uranium target. The fission fragments were detected in position-sensitive avalanche counters. The TOF of fragments with respect to a start detector for the δ-electrons defined the mass splits. The emerging proton was detected in a CsI scintillator. It allowed to reconstruct the energy of the neutron. This experiment therefore measured fragment mass splits rather than cross-section versus neutron energy. Yet, there is a wealth of data for the total fission cross-section versus neutron energy [17] to allow for proper normalisation.

For a neutron spectrum we must then calculate

$$\sigma(A) = \frac{\int \sigma(A, E_n)\, n(E_n)\, dE_n}{\int n(E_n)\, dE_n},$$

where the neutron spectrum is the one on the uranium target, calculated from the independently measured distributions. Figure 2 shows σ(A) for the selected neutron spectra taken from Fig. 1. The σ_A values for p + ^{13}C are approximately 20% lower than those by d + C. This results from the lower average neutron energy of the spectrum. Indeed, the neutron-induced fission cross-section increases rather quickly (up to 1.4 b at 20 MeV) before tending to saturate (1.6 b at 40 MeV). It is interesting to note that there is hardly a difference in σ(A) when the neutrons are generated by deuterons of 40 or 50 MeV, except in the symmetric-fission region. However, the neutron yield, and consequently the in-target production, is twice higher at 50 MeV, for the same deuteron beam intensity.

3.2 Experimental methods and analysis

At IGISOL the independent cross-section σ(Z,A) for a nuclide is thus deduced from the rate the nucleus impinges on the spot where it is implanted. The nuclei were collected on a tape that was stationary during collection, and only moved for measuring a new mass chain. A germanium detector recorded the γ-rays emitted by the isobars versus collection time. A typical measurement for a mass chain was lasting one hour, during which one can reasonably assume the separated beam to be stable. This duration is long enough to reach an equilibrium regime of activities of the nuclides on the neutron-rich side of the σ(Z) curve and a slightly beyond the top.

Corrections for radioactive filiation did play a role for the less neutron-rich isobars, whose half-life becomes comparable with the collection time. These isobars could not be used to extract precise independent cross-sections. They still were used as an extra check for consistency based on the cumulative yield.

Detection of γ-rays for yield measurements is implying problems. Many γ-intensities per decay listed in literature appeared to be inaccurate for our purpose, especially when there is a large β-branch to the daughter ground state. It has been a difficult evaluator task to accept, reject or rescale using local systematics some of the nuclear data. This had not been possible without knowing the a priori Gaussian shaped Z-distribution. Conversely, a posteriori, if one is confident in the cross-sections, one has a very good way to measure γ-branchings.

After unnormalised cross-sections have been obtained for a large number of nuclei spread over enough masses, a global set of parameters has been searched. The analysis was repeated with several Z-widths (σ_Z) common to all masses to obtain the best one by χ^2 method. With this best value kept fixed, new fits defined the position of maxima for each mass. In order to be able to include data of mass chains with less than three evaluated isobars, a global description of the position of maxima was searched. The location of the maxima $Z_m(A)$ in the (Z,A) plane turned out to be well described by a surprisingly simple linear function, $Z_m = a\,A + b$. Finally, with the adopted 3 parameters (a, b, σ_z) kept fixed, the calculated Gaussians were properly scaled to the expected $\sigma(A)$, i.e.,

$$\sigma(Z, A) = \sigma(A)\ \frac{1}{\sqrt{2\pi}\,\sigma_Z}\ \exp\left(-\frac{1}{2}\left(\frac{Z - a\,A - b}{\sigma_Z}\right)^2\right)$$

3.3 Results

Experiments have been carried out at IGISOL first in 1999 on request of GANIL-IPN Orsay for neutrons of 50 MeV d + C [19] and then in 2001 on request of Legnaro with neutrons of 55 MeV protons on a carbon target enriched in ^{13}C [20]. We note that, since the scale of $n(E_n)$ does not affect the calculated $\sigma(A)$, the nuclide cross-sections measured for neutrons produced by p + ^{13}C [20] remain correct in spite of the overestimated neutron flux of [7]. The uranium target did not need to be tilted unlike during irradiation with protons, which is more favourable for the helium flow in the IGISOL chamber. Cross-sections in fission induced by 24 MeV protons were also measured in [19] to provide a realistic comparison under same experimental conditions. The distributions obtained in both neutron-induced fission experiments are similar, even if the neutron spectra are different. However, they both are clearly shifted towards the neutron-rich side with respect to those by p-induced fission. In other words, for a given mass the shift ΔZ is about 0.7, which, by accident, is also the σ-width parameter of the $\sigma(Z)$ Gaussian. It may be worth to comment on the existence of shell effects. They are not included in the global fit (a, b) parameters, but they seem to be visible in the 14 MeV data of [15]. The deviations Z_m(14 MeV)-Z_m(IGISOL) versus A exhibit a S-shape of 0.5 Z-units amplitude. The Z_m positions fitted from the 14 MeV data favour neutron-rich isotopes near the double shell closure in ^{132}Sn, while it is opposite in the region of strong deformations near ^{98}Sr.

The three parameters (a, b, σ_Z) from the IGISOL mass-splits and a normalisation to the mass cross-section calculated independently (with a knowledge of the

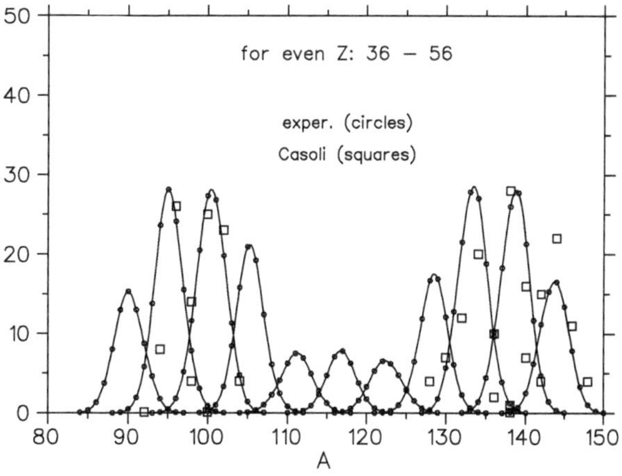

Fig. 3 Experimental cross-sections for selected even-even nuclides in neutron-induced fission of uranium. The *solid lines* connect the values calculated from the IGISOL Z-splits parameters and $\sigma(A)$ with neutron spectrum for 40 MeV deuterons (*small circles*). *Open squares* are deduced from the direct measurements using yrast cascades by Casoli et al. at Los Alamos

neutron spectrum on target) allow to simply estimate cross-sections for many nuclei in the range A = 85–155 in fast-neutron induced fission. The only comparable measurement of cross-sections with a neutron spectrum in the same energy range we are aware of has been conducted at Los Alamos [21, 22]. The technique was very different since based on the yield of the yrast cascades of excited fission fragments. The authors reported data for some even-even nuclei for several neutron energy bins. After mixing these data according to the neutron spectrum composition at IGISOL there is surprisingly good agreement. It gives confidence that the IGISOL data are correct. Figure 3 shows $\sigma(Z,A)$ for even-even nuclei obtained from these two experiments.

Calculations have been carried out at GANIL to simulate the IGISOL experiment and test the cross-section libraries. The position of maxima and width are well reproduced when using the most recent JEFF-3.1 cross-section library that distinguishes between fission by thermal, 0.2–5 MeV and > 5 MeV neutrons. In contrast, the amplitudes are overestimated by factors of 2–3 in the peaks of the mass distributions, while the agreement in the symmetric fission region is rather good. It may indicate a deficiency in the libraries at the highest neutron energies where experimental data are lacking and cross-sections have to be calculated by models.

In early 2011 a new kind of experiment devoted to cross-sections has been carried out at JYFL [23]. Motivated by concerns about the element-independent assumption of the efficiency at IGISOL and the assumptions about feeding patterns in prompt fission, a decay method to count the fission products remaining in the target has been used. This experiment is under evaluation. So far, one can state that it is able to detect a cascade of γ-rays with a 10% probability per decay in isotopes of all elements for which the decay half-life is longer than 1 s and the cross-section larger than 1 mb. A very large number of cross-sections is therefore potentially within reach.

4 Outlook

The design of challenging Radioactive Ion Beam facilities like SPES or SPIRAL2 necessarily involves complex simulations of the alternative solutions in order to define the best option. The neutron-distribution measurements we carried out at JYFL have shown that the neutron flux had been overestimated by the Monte-Carlo code during the design of the converter+target unit of SPIRAL2. The universality of the mass-separated beams obtainable at IGISOL have demonstrated to be a unique asset for dedicated measurements of nuclide yields. As a result, it seems that the codes for nuclide production in neutron-induced fission in the energy range adopted for SPIRAL2 are also too optimistic. Experiments are therefore recommended whenever possible with reasonable effort in order to validate the calculations.

Acknowledgements The work presented here results from contributions of many people listed in original publications to whom those who wrote this report are much indebted. Furthermore, the authors wish to thank the IGISOL, K-130 cyclotron and workshop teams for the excellent working conditions. These works were supported by EURISOL and NUSTAR European programs.

References

1. EURISOL Design Study: See www.eurisol.org/
2. SPIRAL2 at GANIL: http://pro.ganil-spiral2.eu/spiral2
3. HIE: High Intensity and Energy ISOLDE. http://hie-isolde.web.cern.ch/hie-isolde/
4. SPES: Selective Production of Exotic Species. http://www.lnl.infn.it/~spes/
5. Barzakh, A., et al.: Eur. Phys. J. A **47**, 70 (2011)
6. Radivojevič, Z., et al.: Nucl. Instr. Methods Phys. Res. B **183**, 212 (2001)
7. Radivojevič, Z., et al.: Nucl. Instr. Methods Phys. Res. B **194**, 251 (2002)
8. Alyakrinskiy, O., et al.: Nucl. Instr. Methods Phys. Res. A **547**, 616 (2005)
9. Lhersonneau, G., et al.: Nucl. Instr. Methods Phys. Res. A **576**, 371 (2007)
10. Malkiewicz, T.: Ph.D. thesis, Jyväskylä, Finland. Research report No.10/2009 (2009)
11. Prete, G.: LNL Annual Report 2009. INFN-LNL Report 230, ISSN 1828-8545, p. 199. Legnaro (2010)
12. Lhersonneau, G., et al.: Nucl. Instr. Methods Phys. Res. A **603**, 228 (2009)
13. Lhersonneau, G., et al.: Eur. Phys. J. A (2012, submitted)
14. Kudryavtsev, Yu., et al.: Nucl. Instr. Methods Phys. Res. B **266**, 4368 (2008)
15. England, T.R., Rider, B.F.: Los Alamos National Laboratory. LA-UR-94-3106; ENDF-349 (1993). Available at http://ie.lbl.gov/fission.html
16. Kudo, H., et al.: Phys. Rev. C **57**, 178 (1998)
17. ENDF and EXFOR data bases of NNDC: http://www.nndc.bnl.gov/
18. Trzaska, W.H., et al.: Seminar on fission Pont d'Oye IV. Habaye-La-Neuve, Belgium, 6–8 Oct 1999, p. 257. World Scientific (2000)
19. Lhersonneau, G., et al.: Eur. Phys. J. A **9**, 385 (2000)
20. Stroe, L., et al.: Eur. Phys. J. A **17**, 57 (2003)
21. Casoli, P.: Ph.D. thesis, Bordeaux University, France, No 2710 (2003)
22. Ethvignot, T., et al.: In: Shibata, K. (ed.) Proc. of Int. Conf. of Nuclear Data for Science and Technology (ND2001), Tsukuba, Japan, 7–12 Oct 2001. Atomic Energy Society of Japan, vol. 1, p. 254 (2002)
23. Lhersonneau, G., et al.: NIM A (2012, accepted)

MIX
Papier aus verantwortungsvollen Quellen
Paper from responsible sources
FSC® C105338

If you have any concerns about our products,
you can contact us on
ProductSafety@springernature.com

In case Publisher is established outside the EU,
the EU authorized representative is:
Springer Nature Customer Service Center GmbH
Europaplatz 3, 69115 Heidelberg, Germany

Printed by Libri Plureos GmbH
in Hamburg, Germany